W9-ADQ-779

Applied Geophysics

Applied Geophysics

W. M. TELFORD
Professor of Applied Geophysics, McGill University

L. P. GELDART
Project Director, Canadian International Development Agency, Salvador, Brazil

R. E. SHERIFF
Senior Vice President, Seiscom Delta Inc.; Lecturer, University of Houston

D. A. KEYS
formerly, Vice President (Scientific) National Research Council of Canada

CAMBRIDGE UNIVERSITY PRESS
CAMBRIDGE
LONDON   NEW YORK   NEW ROCHELLE
MELBOURNE   SYDNEY

Published by the Press Syndicate of the University of Cambridge
The Pitt Building, Trumpington Street, Cambridge CB2 1RP
32 East 57th Street, New York, NY 10022, USA
296 Beaconsfield Parade, Middle Park, Melbourne 3206, Australia

Library of Congress catalogue card number: 74–16992

ISBN 0 521 20670 7 hard covers
ISBN 0 521 29146 1 paperback

First published 1976
Reprinted 1977, 1978 (twice), 1980, 1981, 1982

Printed in the United States of America
Typeset in Great Britain
Printed and bound by Vail-Ballou Press, Inc., Binghamton, New York

Contents

7. Electromagnetic methods

Preface

This book began as a revision of the classic text, Eve and Keys – *Applied Geophysics in the Search for Minerals*. However, it soon became obvious that the great advances in exploration geophysics during the last two decades have so altered not only the field equipment and practise but also the interpretation techniques that revision was impractical and that a completely new textbook was required.

Although a number of years ago several excellent textbooks on exploration geophysics existed in English, very few have appeared during the past fifteen years. As a result none exist today (1974) which provide a thorough account including the major advances of the past decade or so, especially those in seismic and electromagnetic prospecting. Thus, there is need for a book in English incorporating the latest state of the art.

Readers of textbooks in applied geophysics will often have a background which is strong either in physics or geology but not in both. This book has been written with this in mind so that the physicist may find to his annoyance a detailed explanation of simple physical concepts (for example, energy density) and step-by-step mathematical derivations; on the other hand the geologist may be amused by the over-simplified geological examples and the detailed descriptions of elementary concepts.

The textbook by Eve and Keys was unique in that it furnished a selection of problems for classroom use. This feature has been retained in the present book.

Since the labour of writing the book was not divided equally among the four authors, it is only fair to point out that the senior author (WMT) is primarily responsible for chapters 2, 5–9, chapter 4 is the responsibility of LPG and RES, while chapters 3 and 10 are due to WMT and DAK jointly, and chapter 11 is due to WMT and RES. The entire book was revised, edited and collated by LPG.

The authors wish to take this opportunity to thank all those who have helped in the preparation of this book. In particular we wish to express our gratitude to the various companies and organizations and the individuals within these entities who contributed information and illustrations during the course of the preparation of the book. We wish to make special mention of the following who were especially helpful: Waldemar Assis, Alex Becker, A. Brown, Amalendu Roy, A. E. Singleton, Chevron Oil Company, Geophysical Division, Petrobras, Barringer Geophysics, Crone Geophysics, Geonics, McPhar Geophysics, SOQUEM.

December 1974

W. M. TELFORD   R. E. SHERIFF
L. P. GELDART   D. A. KEYS

1. Introduction

Geophysics, as its name indicates, has to do with the physics of the earth and its surrounding atmosphere: Gilbert's discovery that the earth behaves as a great and rather irregular magnet and Newton's theory of gravitation may be said to be the beginning of geophysics. Mining and the search for metals date from the earliest times but the scientific record began with the publication in 1556 of the famous treatise *De re metallica* by Georgius Agricola (1965), which for many years was the authoritative work on mining. The initial step in the application of geophysics to the search for minerals probably was taken in 1843 when Von Wrede pointed out that the magnetic theodolite, used by Lamont to measure variations in the earth's magnetic field, might also be employed to discover magnetic ore bodies. However, this idea was not acted upon until the publication of Professor Robert Thalén's book in 1879, entitled *On the Examination of Iron Ore Deposits by Magnetic Methods*. The Thalén-Tiberg magnetometer manufactured in Sweden, and later the Thomson-Thalén instrument, furnished the means of locating the strike, dip and depth below surface of magnetic dikes.

The continued expansion in the demand for metals of all kinds and the enormous increase in the use of oil and natural gas during the past fifty years have led to the development of many geophysical techniques of ever-increasing sensitivity for the detection and mapping of unseen deposits and structures. Advances have been especially rapid during the past decade or so because of the development of new electronic devices for field equipment and the widespread application of the digital computer in the interpretation of geophysical data.

Since the great majority of mineral deposits are beneath the surface, their detection depends upon those characteristics which differentiate them from the surrounding media. Methods based upon variations in the elastic properties of the rocks have been developed for determining structures associated with oil and gas, such as faults, anticlines and synclines, even though these are often thousands of feet below the surface. The variation in electrical conductivity and in natural currents in the earth, the rates of decay of artificial potential differences introduced into the ground, local changes in gravity, magnetism and radioactivity – all provide information to the geophysicist about the nature of the structures below the surface, thus permitting him to determine the most favourable places for locating the mineral deposits he seeks.

Several of the devices now used by geophysicists were developed from methods used for locating guns, submarines and aircraft during the two World Wars. Guns were located in France during the First World War by measuring the arrival

1

times of the elastic waves generated in the earth by the recoil of the guns; this led directly to the refraction method of seismic prospecting. Submarines were located by transmitting sound pulses underwater and measuring the interval between the emission and return of the pulses; knowing the velocity of sound in the water, the distance to the reflecting object could be determined. Radar, which was developed during the Second World War, utilizes radio pulses in a similar manner; a modified form of radar called Shoran has been widely used for navigation purposes in marine and airborne geophysical surveys. Ships, submarines and mines were also detected in both wars by their magnetic properties. In geophysical exploration we have one great advantage, namely that the structure or deposit to be located remains fixed, whereas in wartime the submarine, ship or aircraft is in motion.

It should be pointed out that geophysics techniques can only detect a discontinuity, that is, where one region differs sufficiently from another in some property. This, however, is a universal limitation, for man cannot perceive that which is homogeneous in nature, he can only discern that which has some variation in time and space.

Geophysics has to do with all aspects of the physics of the earth, atmosphere and – one might add today – space. The latter application has come about now that man is able to land on the moon and to investigate the atmospheres and surfaces of other planets by means of data obtained by unmanned spacecraft. One might argue that geophysics includes much of what is known as geology, glaciology and astronomy, but the argument would be futile since these venerable sciences have established themselves long ago as separate fields of study. A partial list of the various divisions of geophysics is given below; many of these fields have been investigated for many years for purely scientific interest, to increase our knowledge of the world in which we live.

Seismology
Thermal properties of the earth
Terrestrial magnetism
Geodesy and gravitation
Radioactivity of the earth, sea and atmosphere, cosmic rays
Atmospheric electricity
Meteorology

Many of these divisions are artificial and they overlap considerably. This holds for investigations in applied geophysics as well, in that several different approaches may be made to determine more accurately the exact location of a structure or deposit. The purely scientific investigation of such subjects as the rate of evaporation of water from lakes, the chemical constitution of different rocks and waters from streams and ponds, the measurement of natural earth currents, potential variations and impurities in the atmosphere – all have a definite influence on methods of locating the deposits which the applied geophysicist seeks. For example, the concentration of radon in the air or streams may be associated with deposits of uranium. Electromagnetic waves caused by distant thunderstorms have been used to locate conducting ores at great depths below the surface.

Applied geophysics in the search for minerals, oil and gas may be divided into the following general methods of exploration:

Gravitational
Magnetic
Electrical
Electromagnetic
Seismic
Radioactivity
Well Logging
Miscellaneous chemical, thermal and other methods.

Clearly certain geological conditions are generally associated with metallic ores, others with gas and oil. Ore deposits are usually found in areas where extensive igneous activity occurred, after which the rocks may or may not have undergone metamorphosis; ultimately the area was sufficiently eroded to bring the deposits close enough to the surface to be discovered and exploited. On the other hand coal is the result of rapid burial of luxurious vegetation which existed near a sea or large lake while gas and oil are usually due to the deposition and subsequent burial of marine organisms.

Thus it is natural for mineral exploration to take place in areas suitable for the occurrence of the mineral being sought. The search for metallic ores is generally concentrated in known areas of igneous and metamorphic rocks such as the Canadian Shield and parts of many of the great mountain chains, for example, the Rocky Mountains, the Andes, the Alps and the Urals; however, important exceptions occur since: (1) minerals can be transported away from the place of original formation, perhaps by mechanical transport as in the case of alluvial gold, perhaps in solution, (2) some minerals such as salt and gypsum are deposited originally from aequeous solution and hence occur in sedimentary areas. The search for coal, oil and gas is confined to sedimentary basins except for the rare instance where oil or gas can migrate into fractured igneous or metamorphic rocks.

The choice of technique or techniques to locate a certain mineral depends upon the nature of the mineral and of the surrounding rocks. Sometimes a method may give a direct indication of the presence of the mineral being sought, for example, the magnetic method when used to find magnetic ores of iron or nickel; at other times the method may only indicate whether or not the conditions are favourable to the occurrence of the mineral sought, for example, the magnetic method is often used in petroleum exploration as a reconnaissance tool to find the depth to the igneous basement rocks and so determine where the sediments are thick enough to warrant exploration for petroleum.

Surveys using aircraft carrying magnetic, electromagnetic and other devices are the most rapid method of finding geophysical anomalies. Such areal surveys are also the most inexpensive methods of covering large areas and hence are frequently used for reconnaissance surveys; any anomalies of interest are later investigated using more detailed aerial surveys and/or ground surveys. Seismic exploration is another technique which has been used to explore large areas, both on land and offshore, however at considerably greater cost, both in time and money.

Table 1.1 shows the expenditures for the acquisition of geophysical data during the year 1972. The total expenditure of slightly less than $900,000,000 is divided

Table 1.1. *Costs of acquisition and processing of geophysical data. Data are for the year 1972 and are in millions of dollars*

	US $ (million)	Percentage of total
Petroleum exploration		
Seismic—land	603	67·4
Seismic—marine	199	22·3
Surface gravity/magnetic	17	1·9
Airborne magnetic	6	0·7
Sonic/velocity logging	3	0·3
Total, petroleum	828	92·6
Mineral exploration		
Airborne	19	2·1
Ground methods	13	1·5
Total, minerals	32	3·6
Miscellaneous		
Engineering, construction and ground water	20	2·2
Oceanography	14	1·6
Total, miscellaneous	34	3·8
Overall total	894	100·0

Based on data in Fig. 1, p. 97, *Geophysics*, **39** (Feb. 1974).

among petroleum exploration (93%), mining (3·5%) and engineering, etc. (3·5%). The expenditures for petroleum exploration are largely for seismic surveys (97%) while expenditures in mining are mainly for non-seismic methods. The fact that almost one billion dollars are expended annually to acquire geophysical data attests to the confidence placed in geophysical techniques. Another indication of the value of geophysical exploration is the fact that oil companies have paid as much as $36,800 per acre for oil concessions (Florida offshore, 1973) on the basis primarily of the results of geophysical exploration.

The science of applied geophysics is relatively new and the design of instruments, field techniques and interpretation of the data are undergoing rapid development. The following chapters will provide the reader with a survey of the different methods currently employed to acquire and interpret geophysical data as an aid in the exploration for minerals and petroleum and in the planning of large construction projects.

References

The following brief list contains books which are mainly of historical interest in applied geophysics and earth physics. It illustrates the fact that this is a comparatively young discipline and that it is growing rapidly.

Agricola, Georgius (1965). *De re metallica*. New York, Dover.
Ambronn, R. (1928). *Elements of geophysics* (tr. by M. C. Cobb). New York, McGraw-Hill.
Broughton Edge, A. B. and Laby, T. H. (1931). *Principles of geophysical prospecting*. London, Cambridge.
Dix, C. H. (1952). *Seismic prospecting for oil*. New York, Harper.
Eve, A. S. and Keys, D. A. (1956). *Applied geophysics*. London, Cambridge.
Gutenberg, B. (1929). *Lehrbuch der Geophysik*. Berlin, Gebruder Borntraeger.
Gutenberg, B. (1951). *Physics of the earth*. New York, Dover
Heiland, C. A. (1940). *Geophysical exploration*. New York, Prentice-Hall.
Jakosky, J. J. (1950). *Exploration geophysics*. Los Angeles, Trija.
Jeffreys, H. (1970). *The earth*. London, Cambridge.
Landsberg, H. E. (1952). *Advances in geophysics*. New York, Academic.
Nettleton, L. L. (1940). *Geophysical prospecting for oil*. New York, McGraw-Hill.
Rayleigh, J. W. S. (1945). *The theory of sound*, vols. I and II. New York, Dover.
Rothe, E. (1930). *Les methodes de prospection du sous-sol*: Paris, Gauthier-Villars.
Shaw, H. (1931). *Applied geophysics*: London, H.M. Stationery Office.
Wien, W. and Harms, F. (eds.) (1930). *Handbucher Experimental-physic*, **25**, Geophysik: Leipzig, Akademische Verlagsgesellschaft.

Many texts have been used, in many cases quite freely, for general references throughout this book. In addition to several of the older texts listed above, such as Heiland, Nettleton and Eve and Keys, many of the more recent texts have been used; these are listed below. Where these texts are particularly useful in the study of certain chapters *of this book*, the latter are noted at the end of the reference.

Dobrin, M. (1976). *Geophysical prospecting*. New York, McGraw-Hill. (Chapters 2, 3.)
Grant, F. S. and West, G. F. (1965). *Interpretation theory in applied geophysics*. New York, McGraw-Hill. (Chapters 2, 3, 4, 7, 8.)
Keller, G. V. and Frischknecht, F. C. (1966). *Electrical methods in geophysical prospecting*. London, Pergamon. (Chapters 5, 7, 8, 9.)
Parasnis, D. S. (1962). *Principles of applied geophysics*. London, Methuen.
Parasnis, D. S. (1966). *Mining geophysics*. Amsterdam, Elsevier. (Chapters 2, 3, 5, 7, 8, 9.)

Russell, W. L. (1951). *Principles of petroleum geology.* New York, McGraw-Hill.

Seguin, Maurice K. (1971). *La geophysique et les proprietes physiques des roches.* Quebec, Les Presses de l'Universite Laval.

Society of Exploration Geophysicists (1966). *Mining geophysics,* vols. I and II. Tulsa, Soc. of Explor. Geophys. (Chapters 3, 5, 7, 9.)

Society of Exploration Geophysicists (1967). *Seismic refraction prospecting.* Tulsa, Soc. of Explor. Geophys. (Chapter 4.)

2. Gravity methods

2.1 Introduction

Gravity prospecting involves the measurement of variations in the gravitational field of the earth. The observations are normally made within a few feet of the earth's surface or on a ship; airborne tests and underground surveys have also been carried out occasionally. Like magnetics, radioactivity and a few of the minor electrical techniques, this is a natural source method in which local variations in density of rocks near the surface cause minute changes in the main gravity field.

Gravity and magnetic methods are similar in several ways. Both attempt to measure small differences in a force field which is relatively huge. The main fields in both cases vary with position and to a lesser extent with time. In both methods it is possible to determine the absolute fields.

However, there are several distinct basic differences between gravity and magnetic prospecting. Because density variations are relatively small and uniform compared to changes in magnetic susceptibility, gravity anomalies are smaller and much smoother than magnetic. The instruments used in gravity must be considerably more sensitive than in magnetics (one part in 10^8 compared to 1 part in 10^4). The time variation of the magnetic field is much more complex and more rapid than that of the gravity field. The corrections to observed readings are much more complicated in gravity work than in magnetic or, in fact, than in any other geophysical method. The accuracy of measurement of the anomalous field is much greater in magnetics and the same instrument can be employed to measure the main field and the small variations; this is not the case in gravity work. Furthermore, gravity instruments and field operations are more expensive than magnetic and the field work is slower and requires more skilled personnel.

Gravity prospecting is used as a reconnaissance tool in oil exploration; although expensive, it is still considerably cheaper than the seismic method. In mineral exploration it has usually been employed as a secondary method, although recently it has become more popular for detailed follow-up of magnetic and electromagnetic anomalies during integrated base-metal surveys.

2.2 Principles and elementary theory

2.2.1 *Newton's law of gravitation*

The expression for the force of gravitation is given by Newton's law which is the basis for gravity work. This law states that the force between two particles of mass m_1 and m_2 is directly proportional to the product of the masses and inversely

7

proportional to the square of the distance between the centres of mass. This force
is given by the equation

$$\mathbf{F} = -\gamma \frac{m_1 m_2}{r^2} \mathbf{r}_1, \tag{2.1}$$

where $\mathbf{F}$ is the force on m_2, $\mathbf{r}_1$ is a unit vector directed from m_1 towards m_2, r is
the distance between m_1 and m_2, and γ is the *universal gravitational constant*.
The minus sign arises because the force is always attractive (see discussion of
eq. (3.1), §3.2.1*a*). In cgs units $\mathbf{F}$ is in dynes, m_1 and m_2 in grams, r in centimetres.
The value of γ is then 6.67×10^{-8} dyne-cm^2/g^2 which is equal to the force in
dynes between two small uniform spheres, each of mass 1 g, placed so that the
centres are 1 cm apart.

Obviously the gravitational force is one of the so-called weak forces existing in
nature. Recently it has been suggested that the quantity γ is not constant, but is
decreasing slowly with time. There are many possible consequences of such a
variation, one of which would be an increase in the earth's radius with time. This
in turn would have a profound effect on the geophysical history of the earth.
However, the postulated rate of change of γ, if it exists at all, is so small (about 1%
in the earth's lifetime of several billion years) that it has no significance whatever
in gravity prospecting.

2.2.2 *Acceleration of gravity*

Returning to eq. (2.1), the acceleration of m_2 due to the presence of m_1 can be
found by dividing $\mathbf{F}$ by m_2. In particular, if m_1 is the mass of the earth, M_e, the
acceleration of the mass m_2 at the surface of the earth is

$$\mathbf{g} = \frac{\mathbf{F}}{m_2} = -\gamma \frac{M_e}{R_e^2} \mathbf{r}_1, \tag{2.2}$$

R_e being the radius of the earth and $\mathbf{r}_1$ extending outward from the centre of the
earth along the radius. This acceleration, which we call the *acceleration of gravity*,
was first measured by Galileo in his famous experiment at Pisa, where he dropped
objects from the top of the leaning tower. The numerical value at the earth's
surface is about 980 cm/sec^2. In honour of Galileo, the unit of acceleration of
gravity, 1 cm/sec^2, is called the *gal*.

Modern gravity meters, which measure extremely small variations in this
acceleration, have a sensitivity of about 10^{-5} gals, or 0.01 milligals (mgal). As
a result, they are capable of distinguishing changes in the absolute value of g
with a precision of one part in 10^8.

2.2.3 *Gravitational potential*

(a) *General.* Gravitational fields are conservative, that is to say, the work done
moving a mass in a gravitational field is independent of the path traversed and
depends only on the end points. In fact if the mass is eventually returned to its
original position the net energy expenditure is zero, regardless of the path followed.

Another way of expressing this is to say that the sum of kinetic (motion) and potential (position) energy is constant within a closed system.

The gravitational force is a vector whose direction is along the line joining the centres of the two masses. The force giving rise to a conservative field may be derived from a scalar potential function

$$\nabla U(\mathbf{r}) = \mathbf{F}(\mathbf{r})/m_2 = \mathbf{g}(\mathbf{r}). \qquad (2.3a)$$

Alternatively, we can solve this equation for the gravity potential in the form

$$U(\mathbf{r}) = \int_\infty^R \mathbf{g} \cdot d\mathbf{r} = -\gamma M \int_\infty^R \frac{dr}{r^2} = \frac{\gamma M}{R}, \qquad (2.3b)$$

which is a statement of the work done in moving unit mass from a very distant point (mathematically from infinity) by any path at all to a point distant R from the centre of gravity of M.

It is often simpler to solve gravity problems by calculating the scalar potential U, rather than the vector $\mathbf{g}$. It is then relatively easy to obtain $\mathbf{g}$ from eq. (2.3a).

(b) *Newtonian or three-dimensional potential.* Considering a three-dimensional mass of arbitrary shape as in fig. 2.1, the potential – and the acceleration of gravity – at a point some distance away can be calculated by dividing the mass

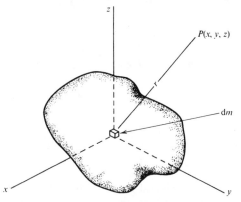

Fig. 2.1 Potential of three-dimensional mass.

into small elements and integrating to get the total effect. Clearly it is easier to use the potential, if the problem can be solved at all. The potential due to an element of mass dm at a distance r from P is, from (2.3b),

$$dU = \gamma dm/r = \gamma\sigma \, dx \, dy \, dz/r,$$

where σ is the density and $r^2 = x^2 + y^2 + z^2$. Then the potential of the total mass m will be

$$U = \gamma\sigma \int_x \int_y \int_z \frac{1}{r} \, dx \, dy \, dz. \qquad (2.4a)$$

Sometimes it is more convenient to use cylindrical coordinates. Since $dx\,dy\,dz = r\,dr\,d\phi\,dz$, the above expression becomes

$$U = \gamma\sigma \int_r \int_\phi \int_z dr\,d\phi\,dz. \tag{2.4b}$$

In spherical coordinates, $dx\,dy\,dz = r^2 \sin\theta\,dr\,d\phi\,d\theta$; hence we have

$$U = \gamma\sigma \int_r \int_\phi \int_\theta r \sin\theta\,dr\,d\phi\,d\theta. \tag{2.4c}$$

The acceleration in the direction of the z-axis (that is, the vertical which is the only direction in which g can be measured directly) is given by

$$g_z = \frac{\partial U}{\partial z} = -\gamma\sigma \int_x \int_y \int_z \frac{z}{r^3} dx\,dy\,dz, \tag{2.5a}$$

or, in the other coordinate systems,

$$g_z = -\gamma\sigma \int_r \int_\phi \int_z \frac{z}{r^2} dr\,d\phi\,dz, \quad \text{(cylindrical)} \tag{2.5b}$$

$$g_z = -\gamma\sigma \int_r \int_\phi \int_\theta \frac{z}{r} \sin\theta\,dr\,d\phi\,d\theta = -\gamma\sigma \int_r \int_\phi \int_\theta \sin\theta \cos\theta\,dr\,d\phi\,d\theta.$$

$$\text{(spherical)} \quad (2.5c)$$

(c) *Logarithmic or two-dimensional potential.* If the mass is very long in the y-direction, and has a uniform cross-section of arbitrary shape in the xz-plane, the gravity attraction derives from a logarithmic, rather than Newtonian, potential. Then eq. (2.4a) becomes

$$U = \gamma\sigma \int_x \int_z dx\,dz \int_{-\infty}^{+\infty} \frac{dy}{r}.$$

In order to keep the last integral finite, we replace the limits of $\pm\infty$ by $\pm L$, where L is finite. Later we shall let L approach infinity. Then writing U_L for the value of this integral, we have

$$U_L = \int_{-L}^{L} \frac{dy}{r} = \int_{-L}^{L} \frac{dy}{\sqrt{(x^2 + z^2 + y^2)}} = \int_{-L}^{L} \frac{dy}{\sqrt{(a^2 + y^2)}},$$

where $a^2 = x^2 + z^2$. Thus

$$U_L = \log\left\{\frac{L + \sqrt{(L^2 + a^2)}}{-L + \sqrt{(L^2 + a^2)}}\right\}.$$

Now we change the potential of the two-dimensional cross-section by subtracting a constant, say the potential when $a^2 = 1$. This is necessary to maintain U_L finite. The expression is then modified to

$$U_L = \log \left\{ \frac{L + \sqrt{(L^2 + a^2)}}{-L + \sqrt{(L^2 + a^2)}} \right\} - \log \left\{ \frac{L + \sqrt{(L^2 + 1)}}{-L + \sqrt{(L^2 + 1)}} \right\}$$

$$= \log \left[\left\{ \frac{L + \sqrt{(L^2 + a^2)}}{-L + \sqrt{(L^2 + a^2)}} \right\} \left\{ \frac{-L + \sqrt{(L^2 + 1)}}{L + \sqrt{(L^2 + 1)}} \right\} \right]$$

$$\approx \log \left[\left\{ \frac{1 + (1 + a^2/2L^2)}{a^2/2L^2} \right\} \left\{ \frac{1/2L^2}{2 + 1/2L^2} \right\} \right]$$

(since $L \gg a$).

Now we allow L to go to infinity and obtain

$$U_L = \log \frac{2}{a^2} \times \frac{1}{2} = - \log (x^2 + z^2) = -2 \log r,$$

where now $r^2 = x^2 + z^2$.

The resulting logarithmic potential expression becomes:

$$U = 2\gamma\sigma \int_x \int_z \log \left(\frac{1}{r} \right) dx\, dz. \tag{2.6}$$

The gravity effect for the two-dimensional body is

$$g_z = \frac{\partial U}{\partial z} = -2\gamma\sigma \int_x \int_z \frac{z}{r^2} dx\, dz. \tag{2.7}$$

In eqs (2.4) to (2.7), we have assumed the density to be constant throughout the volume. This is not generally the situation in the field. If σ is a function of the coordinates as well, the potential can be calculated only for a few simple shapes.

2.2.4 *Potential field equations*

The divergence theorem (Gauss' theorem) states that the integral of the divergence of a vector field over a region of space is equivalent to the integral of the outward normal component of the field over the surface enclosing the region. Mathematically, we have

$$\int_v \nabla \cdot \mathbf{g} \, dv = \int_s g_n \, dS. \tag{2.8a}$$

If there is no attracting matter contained within this volume, the integrals are zero, and

$$\nabla \cdot \mathbf{g} = 0.$$

But from eq. (2.3a) the gravitational force is the gradient of scalar potential, so that

$$\nabla \cdot \mathbf{g} = \nabla \cdot \nabla U = \nabla^2 U = 0, \tag{2.9a}$$

i.e., the potential satisfies Laplace's equation in free space. In the respective co-ordinate systems (Cartesian, cylindrical, spherical) Laplace's equation is

$$\nabla^2 U = \partial^2 U/\partial x^2 + \partial^2 U/\partial y^2 + \partial^2 U/\partial z^2 = 0, \tag{2.9b}$$

$$= \frac{1}{r}\frac{\partial}{\partial r}\left(\frac{r\partial U}{\partial r}\right) + \frac{1}{r^2}\frac{\partial^2 U}{\partial \phi^2} + \frac{\partial^2 U}{\partial z^2} = 0, \tag{2.9c}$$

$$= \frac{1}{r^2}\frac{\partial}{\partial r}\left(r^2\frac{\partial U}{\partial r}\right) + \frac{1}{r^2 \sin\theta}\frac{\partial}{\partial\theta}\left(\sin\theta\frac{\partial U}{\partial\theta}\right) + \frac{1}{r^2 \sin^2\theta}\left(\frac{\partial^2 U}{\partial\phi^2}\right) = 0. \tag{2.9d}$$

If, on the other hand, there is a particle of mass m within the volume and, in particular, if we consider it to be at the centre of a spherical surface of radius r, the right-hand side of (2.8a) is

$$\int_s g_n \, \mathrm{d}S = -(\gamma m/r^2)(4\pi r^2) = -4\pi\gamma m.$$

It can be shown that this result holds regardless of the shape of the surface and the position of the particle m within the surface. If the surface encloses several particles of total mass M, we can write

$$\int_v \nabla \cdot \mathbf{g} \, \mathrm{d}v = \int_s g_n \, \mathrm{d}S = -4\pi\gamma M. \tag{2.8b}$$

If this volume v is made very small, enclosing a point in the region, we may re-move the integral sign to give

$$\nabla \cdot \mathbf{g} = -4\pi\gamma\sigma,$$

where σ is the density at the point. Then from eq. (2.3a),

$$\nabla^2 U = -4\pi\gamma\sigma, \tag{2.10}$$

which is Poisson's equation.

From eqs. (2.9) and (2.10), we see that the gravity potential satisfies Laplace's equation in free space and Poisson's equation in a region containing attracting material.

2.2.5 *Derivatives of the potential*

(a) *Vertical gravity component.* As mentioned in §2.2.3a, it is often-simpler to obtain gravity effects by employing the potential first and then taking the derivative. Other quantities useful in gravity analyses may also be obtained by differentiating the potential in a variety of ways.

As noted in eqs. (2.5) and (2.7), vertical gravity g_z is the derivative of potential in the direction of the vertical axis. This is the quantity measured by gravity meters.

(b) *Vertical gravity gradient and second derivative.* If we calculate the first vertical derivative of g_z from eq. (2.5a) we find

$$\frac{\partial g_z}{\partial z} = \frac{\partial^2 U}{\partial z^2} = -\gamma\sigma \iiint \left(\frac{1}{r^3} - \frac{3z^2}{r^5}\right) dx\, dy\, dz, \qquad (2.11a)$$

while from eq. (2.7), for the two-dimensional case,

$$\frac{\partial g_z}{\partial z} = 2\gamma\sigma \iint \left(\frac{2z^2 - r^2}{r^4}\right) dx\, dz, \qquad (2.11b)$$

with similar expressions from (2.5b) and (2.5c). The second derivatives for the same equations are, for the three-dimensional body,

$$\frac{\partial^2 g_z}{\partial z^2} = \frac{\partial^3 U}{\partial z^3} = -3\gamma\sigma \iiint \left(\frac{5z^3}{r^7} - \frac{3z}{r^5}\right) dx\, dy\, dz, \qquad (2.12a)$$

while in two dimensions,

$$\frac{\partial^2 g_z}{\partial z^2} = 4\gamma\sigma \iint \left(\frac{3z}{r^4} - \frac{4z^3}{r^6}\right) dx\, dz, \qquad (2.12b)$$

again with similar results from (2.5b) and (2.5c).

Although the first derivative is not much used at present, attempts have been made to measure vertical gradient in one type of airborne instrument, as well as much earlier with a modified torsion balance. However, since the torsion balance (the original field instrument for gravity surveys, now replaced by the gravity meter) measures the horizontal components of the second derivative of potential, it is theoretically possible to calculate $\partial g_z/\partial z$ from Laplace's equation.

The second vertical derivative of g_z has been employed considerably in gravity interpretation work for upward and downward continuation and for the enhancement of small anomalies at the expense of large-scale effects. Obviously both derivatives tend to magnify near-surface features and discriminate between them, by increasing the power of the linear dimension in the denominator. That is to say, for three-dimensional bodies, since the gravity effect varies inversely as the distance squared, the first and second derivatives vary as the inverse third and fourth power respectively; for two-dimensional bodies this exponent is reduced by one in each case.

(c) *Horizontal gravity gradient (HGG).* By taking the derivative of g_z in eq. (2.5a) along the x- or y-axis, we obtain the components of the *horizontal gradient of gravity*. Using subscripts to denote differentiation, we get for the x-component

$$U_{xz} = \frac{\partial^2 U}{\partial x\, \partial z} = 3\gamma\sigma \iint \frac{xz}{r^5} dx\, dy\, dz, \qquad (2.13a)$$

for the three-dimensional body, while for two dimensions (see eq. (2.7)),

$$U_{xz} = 4\gamma\sigma \int\int \frac{xz}{r^4} \, dx \, dz. \tag{2.13b}$$

The horizontal gradient can be obtained by torsion balance measurement. Otherwise it can be determined from gravity profiles or contours, as the slope or rate of change of g_z with horizontal displacement. This is a very significant parameter in gravity interpretation, since the sharpness of a gravity profile is an indication of the depth of the anomalous mass.

(d) *Horizontal directive tendency (HDT) – differential curvature.* Another second derivative of potential, this quantity is the horizontal gradient of the horizontal components of gravity, e.g., from eg. (2.4a),

$$U_{xx} = \frac{\partial g_x}{\partial x} = \frac{\partial^2 U}{\partial x^2} = \gamma\sigma \int\int\int \left(\frac{3x^2}{r^5} - \frac{1}{r^3}\right) dx \, dy \, dz. \tag{2.14}$$

Other components are $U_{yy} = \partial^2 U/\partial y^2$ and $U_{xy} = \partial^2 U/\partial x \, \partial y$. The general expression is (see also §2.4.3c)

$$\text{HDT} = \{(U_{yy} - U_{xx})^2 + (2U_{xy})^2\}^{\frac{1}{2}}. \tag{2.14}$$

Since g_x and g_y cannot be measured directly, the differential curvature may only be obtained by using the torsion balance. It is a measure of the warped shape of the potential surface in the vicinity of a gravity anomaly.

We will make use of some of these derivatives later in gravity interpretation.

2.3 Gravity of the earth

2.3.1 *Figure of the earth*

(a) *General.* Gravity prospecting evolved from the study of the earth's gravitational field, a subject of interest to geodesists for the past 250 years in determining the shape of the earth. For instance, when the French scientist Pierre Bouguer found by surveying that the length of a degree of latitude near the equator was shorter than at Paris, it was immediately obvious that the earth was not a perfect sphere and hence that the gravitational acceleration would not be constant over the earth's surface. Other examples of the development in knowledge of the earth's gravity field as a result of geodetic work could be cited.

It is now known that the magnitude of gravity on the earth's surface depends on five factors: latitude, elevation, topography of the surrounding terrain, earth tides and variations in density in the subsurface. The last factor is the only one of significance in gravity exploration and its effect is generally very much smaller than that of the other four combined. For example, the change in gravity from equatorial to polar regions amounts to about 5 gals, or 0·5% of the average value of g (980 gals), while the effect of elevation in some cases might be as large as 0·1 gal or 0·01% of g. A large gravity anomaly in oil exploration, on the other hand, would be 10 mgals (0·001% of *g*), while in mineral areas the value would perhaps be one tenth of this.

Thus it is clear that variations in *g* which are significant in prospecting are not only minute in comparison with the value of *g* itself, but also in comparison with the effects of large changes in latitude and elevation. Fortunately, it is generally possible to remove these large variations with good accuracy.

The shape of the earth, established as a result of geodetic measurements and more recently by satellite tracking, is practically spheroidal, bulging at the equator and flattened at the poles, such that the difference between equatorial and polar radii, divided by the former, is 1/298·25. This ratio is known as the *polar flattening*. Theoretically it is possible to calculate this shape by assuming the earth to be a fluid mass, rotating about its polar axis and having a density which increases with depth ($\approx$ 3 g/cm³ at surface to about 12 g/cm³ at the centre, although the variation is not uniform). The surface of this theoretical shape is an equipotential of the gravity field plus the centripetal acceleration.

(b) *The reference spheroid.* The surface of the earth is defined as a mathematical figure in terms of the gravity values at all surface points. This mathematical figure is known as the *reference spheroid*. It is related to the mean sea-level surface with excess land masses removed and ocean deeps filled. Thus it is an equipotential surface, that is, the force of gravity is everywhere normal to this surface, or the plumb line is vertical at all points. The formula, adopted by the International Association of Geodesy in 1967, gives the value of *g* at any point on this spheroid as

$$g = g_0(1 + \alpha \sin{}^2\phi + \beta \sin{}^2 2\phi), \qquad (2.15)$$

where g_0 = equatorial gravity = 978·0318 gals, ϕ = latitude, and the constants α and β are equal to 0·0053024 and −0·0000058 respectively.

[Note that the above value of g_0 does not have the minus sign required by eq. (2.2); since this minus sign should always be present, that is, the force of gravity is always attractive, it is customary to avoid it by using the absolute value of *g*. This practise will be followed in the balance of this chapter; however, in §3.2.2*c* where eqs. (3.14*a*), (3.14*b*) and (3.14*c*) relate the gravitational potential and field to their magnetic counterparts, the minus sign must be taken into account.]

Although eq. (2.15) is still the standard for gravity latitude reductions, considerable data acquired since 1930 from gravity measurements at sea and particularly from observations of exact paths of satellites tend to modify the constants. It appears that the earth is slightly pear-shaped and that there may be minute distortions in the equatorial sphericity. The latter could introduce a longitude term into the formula. However, all these modifications have dimensions in the order of meters, compared to the polar flattening of about 20 km. Thus eq. (2.15) is quite accurate enough to be used for calculating the variation of gravity with latitude.

(c) *The geoid*. Even in its most refined state, the gravity relation in eq. (2.15) is a very crude approximation. It assumes that there are no undulations in the earth's surface, whereas in fact we have mean continental elevations of about 500 metres and maximum land elevations and ocean depressions of the order ±9000 metres, all referred to sea level. Obviously the true sea level is influenced by these variations and the surveyor must take them into account when measuring elevations.

The geodesist defines a practical mean sea level (equipotential) surface for making these measurements. It is known as the *geoid* and is defined as average sea level over the oceans and over the surface of sea water which would lie in canals, say, cut through the land masses.

Clearly the geoid and reference spheroid surfaces do not – in fact, never could – coincide at all points, since the geoid is warped upward under the continental masses due to attracting material above and downward over the ocean basins. However, the deviations between the two surfaces are not more than 50 metres. Traces of the geoid and spheroid surfaces, greatly exaggerated as to irregularities, are illustrated in fig. 2.2.

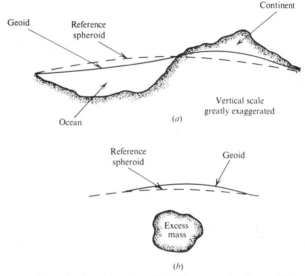

Fig. 2.2 Comparison of spheroid and geoid. (*a*) Large-scale effects; (*b*) local effects.

This simplified figure of the earth, described by either the mathematical or geodetic surface, although allowing for increasing density with depth, assumes no lateral density variations. These, of course, when they occur near surface, are the object of search in gravity exploration. The simple picture, however, is sufficient to make the standard gravity reductions, necessary for analysis of the gravity survey.

2.3.2 *Gravity reductions*

(a) *General*. Field-work is carried out on land by taking gravity readings at grid stations covering an area of interest. These readings will generally be influenced by topography and other factors, hence must be corrected for variations in latitude, elevation and topography to reduce them to the values they would have on some datum equipotential surface, such as the geoid, or a surface everywhere parallel to it. Two additional corrections may occasionally be required: these are due to earth tides and the effect of isostasy.

(b) *Latitude correction.* Both the rotation of the earth and its slight equatorial bulge produce an increase of gravity with latitude. The centrifugal acceleration due to the spinning earth – maximum at the equator and zero at the poles – opposes the gravitational acceleration, while the polar flattening also increases gravity at the poles. The latter effect is partly counteracted by the increased attracting mass at the equator. Thus it is necessary to apply a *latitude correction* where there are any appreciable north-south excursions of the grid stations. It is obtained by differentiating eq. (2.15):

$$\frac{dg_L}{ds} = \frac{1}{R_e} \frac{dg_L}{d\phi} \approx \frac{1}{R_{eq}} \frac{dg_L}{d\phi} \approx 1 \cdot 307 \sin 2\phi \text{ mgal/mile}, \tag{2.16}$$

where ds = N–S horizontal distance, R_e = radius of the earth at latitude ϕ, R_{eq} = equatorial radius.

The maximum value occurs at latitude $45°$ where the correction amounts to $0 \cdot 01$ mgal/40 ft. It is obviously zero at the equator and poles. The correction is linear over north–south distances of about one mile. For larger surveys it is necessary to take account of the changes in ϕ. Since gravity increases with latitude (either north or south) the correction is added as we move towards the equator.

(c) *Free-air correction.* Since gravity varies inversely with the square of distance, it is necessary to correct for changes in elevation between stations so that all field readings are reduced to a datum surface. This is known as the *free-air correction,* since it takes no account of the material between the station and the datum plane. It is obtained by differentiating the scalar equation equivalent to eq. (2.2); the result is

$$\frac{dg_{FA}}{dR_e} = -\frac{2\gamma M_e}{R_e^3} \approx -\frac{2g}{R_{eq}} \approx -0 \cdot 09406 \text{ mgal/ft} = -0 \cdot 3085 \text{ mgal/m.} \tag{2.17}$$

As the negative sign implies, the free-air correction is added to the field reading when the station is above the datum plane and subtracted when below it.

It is clear from a consideration of the latitude and free-air corrections that the station position must be precisely known if one is to make proper use of the gravity data; it must be located within 40 ft in horizontal N–S distance and its elevation should be measured to an accuracy of at least 2 inches in order that the survey may be accurate to within $0 \cdot 01$ mgal – which is the sensitivity of present instruments.

(d) *Bouguer correction.* The *Bouguer correction* accounts for attraction of material between the station and the datum plane, which was ignored in the free-air calculation. If the station were centrally located on a plateau of large horizontal extent and uniform thickness and density (see fig. 2.3*a*) the gravity readings would

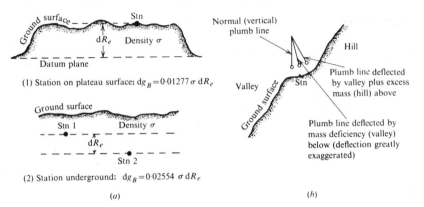

Fig. 2.3 Gravity corrections. (*a*) Bouguer correction; (*b*) terrain correction.

be increased by the effect of this slab between the station and the datum. The *Bouguer correction*, derived by assuming the slab to be of infinite horizontal extent, is given by

$$\frac{dg_B}{dR_e} \approx \frac{dg_B}{dR_{eq}} = 2\pi\gamma\sigma \text{ mgal/ft} = 0.01277 \, \sigma \text{ mgal/ft} = 0.04188 \, \sigma \text{ mgal/m} \quad (2.18a)$$

where σ = slab density (see eq. (2.43*c*)).

If we assume an average density for crustal rocks of 2·67 g/cm³, the numerical value is

$$dg_B/dR_{eq} = 0.0341 \text{ mgal/ft.} = 0.112 \text{ mgal/m.} \quad (2.18b)$$

The Bouguer correction is applied in the opposite sense to free-air, that is, it is subtracted when the station is above the datum plane and vice-versa.

When gravity measurements are made at underground stations, as in fig. 2.3*a*, the slab between stations 1 and 2 exerts an attraction downward on station 1, upward on 2. Thus the difference in gravity between them is $4\pi\gamma\sigma \, dR_e$, i.e., the Bouguer correction is doubled.

The Bouguer and free-air corrections may be combined. From eqs. (2.17) and (2.18*b*) the resultant is −0·060 mgal/ft, a convenient, though not necessarily trustworthy, figure.

Two assumptions are made in deriving the Bouguer correction: first that the slab is of uniform density, second that it is of infinite horizontal extent. Neither is really valid. In order to modify the first, one would need to have considerable knowledge of the local geology as to rock type and actual densities. The second is taken care of in the next reduction.

(e) *Terrain correction.* This correction allows for surface irregularities in the vicinity of the station, that is, hills rising above the gravity station and valleys (or lack of material) below it. From fig. 2.3*b* it is obvious that both of these topographic undulations affect the gravity measurement in the same sense, reducing the readings because of upward attraction (hills) or lack of downward attraction (valleys). Hence the *terrain correction* is always added to the station reading.

There are several graphical methods for calculating terrain corrections. All of them require a good topographical map of the area (50 ft contours or smaller) extending if possible considerably beyond the survey grid (within which, of course, the elevations are known more precisely). The usual procedure is to divide the area into compartments and compare the average elevation within each compartment with the station elevation. This is best done by outlining the compartments on a transparent sheet which overlies the topographic map.

The most commonly used template is a set of concentric circles and radial lines, making sectors whose areas increase with distance from the centre. The gravity effect of a single sector can be calculated from the following formula (see eq. (2.43*d*)):

$$\mathrm{d}g_T = \gamma\sigma\theta\{(r_0 - r_i) + \sqrt{(r_i^2 + z^2)} - \sqrt{(r_0^2 + z^2)}\}, \qquad (2.19a)$$

where θ = sector angle (radians), $z = |e_s - e_a|$, e_s = station elevation, e_a = average elevation in sector, r_0, r_i = outer and inner sector radii. When $\sigma = 2\cdot67$ g/cm³ and linear dimensions are in feet, this becomes (in milligals)

$$\mathrm{d}g_T = 5\cdot35 \times 10^{-3}\theta\{(r_0 - r_i) + \sqrt{(r_i^2 + z^2)} - \sqrt{(r_0^2 + z^2)}\}. \qquad (2.19b)$$

The use of a terrain chart of this type is illustrated in fig. 2.4. The transparent template is placed over the topographic map with the centre of the circles at the

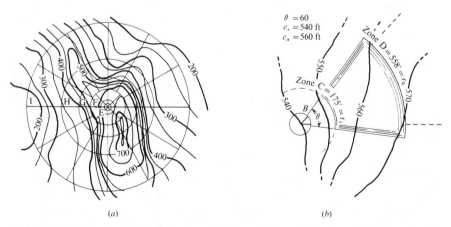

(a)   (b)

Fig. 2.4 Use of terrain chart with topographic map. (*a*) Terrain chart overlying topographic map; (*b*) enlarged view of a single zone.

gravity station. The average elevation within a single compartment is estimated from the contours within it and subtracted from the known station elevation. The difference is z in eq. (2.19), from which the contribution to dg_T can be calculated for the compartment. Tables of terrain corrections for zone charts of particular dimensions, developed by Hammer (1939), facilitate this operation considerably.

Other methods for segmenting the topography map have been applied occasionally; for instance, when the contours are practically linear there is no advantage in using circular sectors. An alternative scheme, due to Jung, uses elementary areas so proportioned that the gravity effect of each at the station is the same, regardless of distance or azimuth.

In rugged terrain it is often necessary to make topographic corrections for ground considerably beyond the boundary of the survey area. In addition, two or more map scales may be required, as the zone dimensions in Table 2.1 clearly indicate. Some reduction in the labour of computing the effect of outer zones, however, is usually possible, since these dimensions, particularly in mineral exploration work, are large compared to the survey grid spacing. Thus there is very little change in the average elevations of the outer compartments as one shifts the template from station to station. By using a few detailed calculations to plot an empirical curve of dg_T versus z, the total effect of the outer rings on each station can be found with reasonable accuracy.

Terrain corrections may also be made on a computer by using an appropriate set of grid station elevations taken off the map enclosing the survey area. This method is clearly best suited to a completed survey; during the field-work, when the area of operation may be expanding or changing from day to day, the computations must be done by hand.

Regardless of the approach, the topography reduction is a slow and very boring task. Furthermore in areas of steep and erratic slopes it is not necessarily accurate, particularly in the vicinity of the station itself, the zone which normally contributes most to the correction. Thus at the edge of a steep cliff or gorge the terrain correction obtained from a zone chart is almost inevitably in error. Sometimes dot charts are employed (see §2.6.4*a*) in addition to, or instead of, the zone charts. However, a better solution would be to move the gravity station, if this is possible.

Table 2.1 contains a set of terrain corrections for a Hammer chart, using an average rock density of 2·67. It can be seen from this table that if $z/r \leqslant 1/20$, r being the average distance from the compartment to the station, the correction is insignificant.

(f) *Earth-tide correction.* Instruments for measuring gravity are quite sensitive enough to record the changes in g caused by movement of the sun and moon. These variations have amplitudes as large as 0·3 mgal. Like sea tides, they depend on latitude and time, although the amplitude of the earth motion is only a fraction of that of the ocean tides. Figure 2.5 shows the earth tidal variations measured with a gravity meter in Montreal in April 1969.

Table 2.1. Terrain corrections

From Hammer's Tables, but based on average density $\sigma = 2.67$ g/cm³

$$dg_T = \theta\gamma\sigma \{r_o - r_i + \sqrt{(r_i^2 + z^2)} - \sqrt{(r_o^2 + z^2)}\} \qquad r_i, r_o = \text{inner, outer sector radii}$$

$\gamma = 6.67 \times 10^{-8}$, dg_T in milligals, z = average sector elevation

z, r_i, r_o in feet;

Zone B 4 sectors 6·56'–54·6'		Zone C 6 sectors 54·6'–175'		Zone D 6 sectors 175'–558'		Zone E 8 sectors 558'–1280'		Zone F 8 sectors 1280'–2936'		Zone G 12 sectors 2936'–5018'		Zone H 12 sectors 5018'–8578'		Zone I 12 sectors 8578'–14612'	
±z	dg_T	±z	dg_T	±z	dg_T	±z	dg_T	±z	dg_T	±z	dg_T	±z	dg_T	±z	dg_T
0·0-1·1	0·00000	0·0-4·3	0·00000	0·0-7·7	0·00000	0-18	0·00000	0-27	0·00000	0-58	0·00000	0-75	0·00000	0-99	0·00000
1·1-1·9	0·00133	4·3-7·5	0·00133	7·7-13·4	0·00133	18-30	0·00133	27-46	0·00133	58-100	0·00133	75-131	0·00133	99-171	0·00133
1·9-2·5	0·00267	7·5-9·7	0·00267	13·4-17·3	0·00267	30-39	0·00267	46-60	0·00267	100-129	0·00267	131-169	0·00267	171-220	0·00267
2·5-2·9	0·0040	9·7-11·5	0·0040	17·3-20·5	0·0040	39-47	0·0040	60-71	0·0040	129-153	0·0040	169-200	0·0040	220-261	0·0040
2·9-3·4	0·0053	11·5-13·1	0·0053	20·5-23·2	0·0053	47-53	0·0053	71-80	0·0053	153-173	0·0053	200-226	0·0053	261-296	0·0053
3·4-3·7	0·0067	13·1-14·5	0·0067	23·2-25·7	0·0067	53-58	0·0067	80-88	0·0067	173-191	0·0067	226-250	0·0067	296-327	0·0067
3·7-7	0·0133	14·5-24	0·0133	25·7-43	0·0133	58-97	0·0133	88-146	0·0133	191-317	0·0133	250-414	0·0133	327-540	0·0133
7-9	0·0267	24-32	0·0267	43-56	0·0267	97-126	0·0267	146-189	0·0267	317-410	0·0267	414-535	0·0267	540-698	0·0267
9-12	0·040	32-39	0·040	56-66	0·040	126-148	0·040	189-224	0·040	410-486	0·040	535-633	0·040	698-827	0·040
12-14	0·053	39-45	0·053	66-76	0·053	148-170	0·053	224-255	0·053	486-552	0·053	633-719	0·053	827-938	0·053
14-16	0·067	45-51	0·067	76-84	0·067	170-189	0·067	255-282	0·067	552-611	0·067	719-796	0·067	938-1038	0·067
16-19	0·080	51-57	0·080	84-92	0·080	189-206	0·080	282-308	0·080	611-666	0·080	796-866	0·080	1038-1129	0·080
19-21	0·0935	57-63	0·0935	92-100	0·0935	206-222	0·0935	308-331	0·0935	666-716	0·0935	866-931	0·0935		
21-24	0·107	63-68	0·107	100-107	0·107	222-238	0·107	331-353	0·107	716-764	0·107	931-992	0·107		
24-27	0·120	68-74	0·120	107-114	0·120	238-252	0·120	353-374	0·120	764-809	0·120	992-1050	0·120		
27-30	0·133	74-80	0·133	114-120	0·133	252-266	0·133	374-394	0·133	809-852	0·133	1050-1105	0·133		
		80-86	0·147	120-127	0·147	266-280	0·147	394-413	0·147	852-894	0·147				
		86-91	0·160	127-133	0·160	280-293	0·160	413-431	0·160	894-933	0·160				
		91-97	0·174	133-140	0·174	293-306	0·174	431-449	0·174	933-972	0·174				
		97-104	0·187	140-146	0·187	306-318	0·187	449-466	0·187	972-1009	0·187				
		104-110	0·200	146-152	0·200	318-331	0·200	466-483	0·200	1009-1046	0·200				

In spite of the fact that this change in *g* is quite significant and can be calculated theoretically for any time and place, it is not general practice to obtain the correction directly. This is because the variation is smooth and relatively slow; as a result it is easily taken out in the instrument drift correction, which is described in §2.5.3*b*.

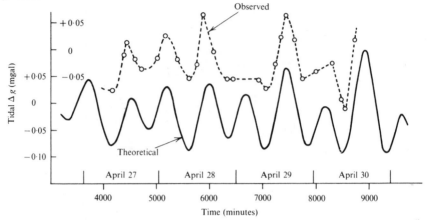

Fig. 2.5 Earth-tide variations, Montreal, April 1969. (Gravity readings have been corrected for drift.)

(g) *Isostasy and the isostatic correction.* If the earth crust had no lateral variations in density we would find that a set of gravity readings, after suitable corrections for the effects described above, would be identical. Any differences in the corrected values constitute a gravity anomaly, known as the *Bouguer anomaly*, the result of lateral variations in density.

From world-wide gravity measurements, it has been found that the average Bouguer anomaly on land near sea level is approximately zero. In oceanic areas it is generally positive, while in regions of large elevation it is mainly negative. These large-scale effects are due to density variations in the crust and indicate that the material beneath the oceans is more dense than normal, while in regions of elevated land masses it is less.

Two hypotheses were put forward about one hundred years ago to account for this large-scale systematic density variation; one was proposed by the scientist, G. B. Airy, the other by an English archdeacon, J. H. Pratt. During a British Army geodetic survey of India, a rather large discrepancy between astronomic and geodetic results south of the Himalayas could not be resolved by assuming that the mountains had the same density as the adjacent flat lands. Airy proposed a rigid crust of uniform density floating on a liquid substratum of higher density. Four years later Pratt, who had been involved in the original calculations, suggested a crust, also floating on a uniform liquid, the density of the crust varying with the topography, being lower in mountain regions and higher where the crust was thin.

These two hypotheses are illustrated in fig. 2.6. There is clearly a fundamental difference between them, since in Airy's scheme the crustal thickness varies more, deep roots being present under the mountains. Even today it is not definitely established that either hypothesis is correct, although Airy's version is preferred; probably both are true to some extent. In general, however, Pratt's assumption of the uniform crustal bottom makes the *isostatic correction* (which is in reality a large-scale terrain correction) easier to apply. Heiskanen has presented a modified form of the hypothesis, in which he combines a lateral variation of crustal density with a variable depth, plus a gradual increase of density with depth.

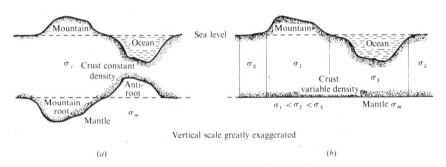

Fig. 2.6 Isostasy hypotheses. (*a*) Airy's hypothesis; (*b*) Pratt's hypothesis.

The term *isostasy* was not used in connection with these gravity variations until 1889, when it was first introduced by the geologist C. E. Dutton. Isostasy may be defined in two ways: (i) a condition of the earth such that continual adjustments are made to approach gravitational equilibrium; (ii) a variation in density of the earth's crust systematically related to surface elevations.

The isostatic correction is of secondary importance in gravity prospecting, since few survey areas would be large enough to require its application.

(h) *Bouguer gravity anomaly.* When all of the preceding corrections have been applied to the observed gravity reading, we obtain the value of *Bouguer gravity* for the station:

$$g_B = g_{obs} \pm dg_L + dg_{FA} - dg_B + dg_T, \qquad (2.20a)$$

where g_{obs} = station reading, dg_L = latitude correction, dg_{FA} = free-air correction, dg_B = Bouguer correction, dg_T = terrain correction.

Putting in numerical values, eq. (2.20a) becomes

$$g_B = g_{obs} \pm dg_L + 0.094h - (0.01277h - T)\sigma, \qquad (2.20b)$$

where dg_T is replaced by $T\sigma$ (obviously the density is the same for both corrections). The *Bouguer anomaly*, δg_B, is

$$\delta g_B = g_B - g_r = g_{obs} \pm dg_L + dg_{FA} - dg_B + dg_T - g_r, \qquad (2.21a)$$

or, in numerical form,

$$\delta g_B = g_{obs} \pm dg_L + 0.094h - (0.01277h - T)\sigma - g_r, \qquad (2.21b)$$

where g_r is generally some particular station value in the survey area rather than a theoretical result from eq. (2.15). Note that the signs of dg_{FA} and dg_B change when the station is below the datum plane.

Another quantity which is sometimes seen in the literature is the *free-air anomaly*; this is the value of δg_B when dg_B and dg_T in eq. (2.21a) are omitted.

2.3.3 *Densities of rocks and minerals*

(a) *General.* As mentioned above, the significant parameter in gravity exploration – the anomaly source – is a local variation in density. Two problems arise in connection with this parameter.

(i) The maximum density variation between different rocks and between rocks and minerals is approximately two. This is a very small change compared to the range of magnetic susceptibility ($\sim 10^5$), electrical conductivity ($\sim 10^{10}$), radio-activity (~ 100) and even elastic properties or acoustic wave velocity (10–20).

(ii) Generally it has not been possible to measure density *in situ*. A density borehole logger (see §11.3.3) has been used to a limited extent in oil exploration (well logging) for the last ten years. Bulk density may also be estimated where the seismic velocity in a formation is known. In general P-wave velocity increases with formation density; very roughly, in the velocity range 6000–20,000 ft/sec, the density increases from 1·9 to 2·8 g/cm^3. These methods of density determination, however, are neither applicable to mineral exploration, nor generally available for gravity reconnaissance in oil work.

Thus it is necessary to make density measurements in the laboratory on small samples of outcrop or drill core. However, the laboratory results do not necessarily give the true bulk density of the formation, since the samples may be weathered, fragmented or dehydrated.

Consequently the density, which is, of course, all important (see eqs. (2.5), (2.7), (2.18), (2.19)), is usually not very well known in a particular field situation. For this reason it is practical to collect and tabulate the densities of rocks and minerals in some detail.

(b) *Sedimentary rocks.* These have, on the average, lower densities than igneous and metamorphic rocks. Among the sediments the average density varies with the composition, being lowest for conglomerate and sandstone, followed by shale, limestone and dolomite in that order, However, actual values differ widely from the average, hence there is considerable overlap. Dolomites and shales are the most uniform in density, but even here the variation may be as large as 0·6 g/cm^3; in limestones and sandstones it is about 1·0 g/cm^3.

This wide range in density is due primarily to variations in porosity of the sedimentary rocks. Furthermore pore fluids may affect the density by as much as 10%. Thus, as mentioned previously, the laboratory density measurement should be made, if possible, with the sample in somewhat the same conditions as those prevailing in the formation from which it was removed. This requires information about the regional climate and location of the water table, if the formation is near the surface.

The density of sedimentary rocks is also influenced by their age, previous history and depth below surface. Obviously a porous rock buried under a heavy load will be compacted and consolidated to a degree which depends on the size and duration of the load. The density thus increases with depth and time. This effect is more pronounced in clays and shales than in sandstones and limestones. Sample measurements on shale drill core have shown an increase in density from 2·0 to 2·6 g/cm³ as depth of burial increased by 5000 ft.

Table 2.2 lists densities for various sediments and sedimentary rocks and approximate density range for each type. It is worth noting that the normal

Table 2.2. *Densities of sediments and sedimentary rocks*

Rock type	Range	Average (wet)	Range	Average (dry)
	(g/cm³)		(g/cm³)	
Alluvium	1·96–2·0	1·98	1·5–1·6	1·54
Clays	1·63–2·6	2·21	1·3–2·4	1·70
Glacial drift	—	1·80	—	—
Gravels	1·7–2·4	2·0	1·4–2·2	1·95
Loess	1·4–1·93	1·64	0·75–1·6	1·20
Sand	1·7–2·3	2·0	1·4–1·8	1·60
Sands and clays	1·7–2·5	2·1	—	—
Silt	1·8–2·2	1·93	1·2–1·8	1·43
Soils	1·2–2·4	1·92	1·0–2·0	1·46
Sandstones	1·61–2·76	2·35	1·6–2·68	2·24
Shales	1·77–3·2	2·40	1·56–3·2	2·10
Limestones	1·93–2·90	2·55	1·74–2·76	2·11
Dolomite	2·28–2·90	2·70	2·04–2·54	2·30

density contrast between adjacent sedimentary formations in the field is seldom more than 0·25 g/cm³, much less than the maximum possible indicated by the range in the table. Clearly this is connected with the geological history of the region.

(c) *Igneous rocks.* Although on average, igneous rocks are more dense than sedimentary, there is considerable overlap. Among the volcanics lavas have rather low, intrusive rocks somewhat higher, densities.

Generally basic igneous rocks have larger densities than acidic forms. Porosity, which affects the density of sediments so greatly, is of minor significance in igneous and metamorphic rocks, unless they are highly fractured. Table 2.3 lists a number of igneous rocks in order of increasing density.

(d) *Metamorphic rocks.* Density usually increases with the degree of metamorphism since the process tends to fill pore spaces and recrystallize the rock in a denser form. Thus metamorphosed sediments, such as marble, slate and quartzite, are generally denser than the original limestone, shale and sandstone. The same

Table 2.3. *Densities of igneous rocks*

Rock type	Range (g/cm³)	Average	Rock type	Range (g/cm³)	Average
Rhyolite glass	2·20–2·28	2·24	Quartz diorite	2·62–2·96	2·79
Obsidian	2·2–2·4	2·30	Diorite	2·72–2·99	2·85
Vitrophyre	2·36–2·53	2·44	Lavas	2·80–3·0	2·90
Rhyolite	2·35–2·70	2·52	Diabase	2·50–3·20	2·91
Dacite	2·35–2·8	2·58	Essexite	2·69–3·14	2·91
Phonolite	2·45–2·71	2·59	Norite	2·70–3·24	2·92
Trachyte	2·42–2·8	2·60	Basalt	2·70–3·30	2·99
Andesite	2·4–2·8	2·61	Gabbro	2·70–3·50	3·03
Nephelite-Syenite	2·53–2·70	2·61	Hornblende-Gabbro	2·98–3·18	3·08
Granite	2·50–2·81	2·64	Peridotite	2·78–3·37	3·15
Granodiorite	2·67–2·79	2·73	Pyroxenite	2·93–3·34	3·17
Porphyry	2·60–2·89	2·74	Acid igneous (av.)	2·30–3·11	2·61
Syenite	2·60–2·95	2·77	Basic igneous (av.)	2·09–3·17	2·79
Anorthosite	2·64–2·94	2·78			

may be said for the metamorphic forms of igneous rocks – gneiss versus granite, amphibolite versus basalt, etc., although the density differences are not so significant.

As in igneous rocks, the density of metamorphic types usually increases as the acidity decreases. However, density variations are more erratic in metamorphic rocks than in sedimentary and igneous varieties because of their complicated histories. Thus a definite classification on the basis of density is difficult. Examples are given in Table 2.4.

Table 2.4. *Densities of metamorphic rocks*

Rock type	Range (g/cm³)	Average	Rock type	Range (g/cm³)	Average
Quartzite	2·5–2·70	2·60	Serpentine	2·4–3·10	2·78
Schists	2·39–2·9	2·64	Slate	2·7–2·9	2·79
Graywacke	2·6–2·7	2·65	Gneiss	2·59–3·0	2·80
Granulite	2·52–2·73	2·65	Chloritic slate	2·75–2·98	2·87
Phyllite	2·68–2·80	2·74	Amphibolite	2·90–3·04	2·96
Marble	2·6–2·9	2·75	Eclogite	3·2–3·54	3·37
Quartzitic slate	2·63–2·91	2·77	Metamorphic-Av.	2·4–3·1	2·74

(e) *Densities of minerals, miscellaneous materials.* With a few exceptions the nonmetallic minerals are of lower density than the average for rocks (2·67). Metallic minerals, on the other hand, are mainly heavier than this average, but since they rarely occur pure in large volumes, their positive density contrasts are normally not as large as the theoretical maximum. Tables 2.5 and 2.6 list densities for various materials and minerals.

Table 2.5. *Densities of nonmetallic minerals and miscellaneous materials*

Type	Range	Average	Type	Range	Average
	(g/cm³)			(g/cm³)	
Snow	—	0·125	Gypsum	2·2–2·6	2·35
Petroleum	0·6–0·9	—	Bauxite	2·3–2·55	2·45
Ice	0·88–0·92	—	Kaolinite	2·2–2·63	2·53
Sea Water	1·01–1·05	—	Orthoclase	2·5–2·6	—
Peat	—	1·05	Quartz	2·5–2·7	2·65
Asphalt	1·1–1·2	—	Calcite	2·6–2·7	—
Lignite	1·1–1·25	1·19	Talc	2·7–2·8	2·71
Soft coal	1·2–1·5	1·32	Anhydrite	2·9–3·0	2·93
Anthracite	1·34–1·8	1·50	Biotite	2·7–3·2	2·92
Brick	—	1·50	Magnesite	2·9–3·12	3·03
Carnallite	1·6–1·7	—	Fluorite	3·01–3·25	3·14
Sulphur	1·9–2·1	—	Epidote	3·25–3·5	—
Chalk	1·53–2·6	2·01	Diamond	—	3·52
Graphite	1·9–2·3	2·15	Corundum	3·9–4·1	4·0
Rock salt	2·1–2·6	2·22	Barite	4·3–4·7	4·47
			Zircon	4·0–4·9	4·57

As mentioned previously, seismic wave velocities usually increase with formation density. It might seem attractive to establish bulk densities of massive sulphide deposits by the seismic method. Although test results to date are inconclusive, this probably will not prove to be a very fruitful approach. In spite of the large possible density contrast (1–2 g/cm³) with the host rock, the sulphides are in most deposits inhomogeneous and full of fractures. Thus the resultant positive velocity contrast is likely to be smaller than one would expect and may even be negative. In any case, seismic velocities vary with the elastic constants, as well as the densities, of solids.

2.3.4 *Density estimates from field results*

(a) *General.* As mentioned at the beginning of §2.3.3a, it is not generally possible to measure density directly and tables 2.2 to 2.6 were introduced to provide a guide in situations where some value must be assumed. Obviously the wide range for most rocks and minerals makes this estimate rather risky. Several procedures have been employed for estimating density from gravity measurements in the field.

Table 2.6. *Densities of minerals*

Mineral	Range (g/cm³)	Average	Mineral	Range (g/cm³)	Average
Copper	—	8·7	Sulphides, arsenides		
Silver	—	10·5	Sphalerite	3·5–4·0	3·75
Gold	15·6–19·4	—	Covellite	—	3·8
Oxides, carbonates			Malachite	3·9–4·03	4·0
Limonite	3·5–4·0	3·78	Chalcopyrite	4·1–4·3	4·2
Siderite	3·7–3·9	3·83	Stannite	4·3–4·52	4·4
Rutile	4·18–4·3	4·25	Stibnite	4·5–4·6	4·6
Manganite	4·2–4·4	4·32	Pyrrhotite	4·5–4·8	4·65
Chromite	4·3–4·6	4·36	Molybdenite	4·4–4·8	4·7
Ilmenite	4·3–5·0.	4·67	Marcasite	4·7–4·9	4·85
Pyrolusite	4·7–5·0	4·82	Pyrite	4·9–5·2	5·0
Magnetite	4·9–5·2	5·12	Bornite	4·9–5·4	5·1
Franklinite	5·0–5·22	5·12	Millerite	5·3–5·65	5·4
Hematite	4·9–5·3	5·18	Chalcocite	5·5–5·8	5·65
Cuprite	5·7–6·15	5·92	Cobaltite	5·8–6·3	6·1
Cassiterite	6·8–7·1	6·92	Arsenopyrite	5·9–6·2	6·1
Wolframite	7·1–7·5	7·32	Smaltite	6·4–6·6	6·5
Uraninite	8·0–9·97	9·17	Bismuthinite	6·5–6·7	6·57
			Argentite	7·2–7·36	7·25
			Niccolite	7·3–7·67	7·5
			Galena	7·4–7·6	7·5
			Cinnabar	8·0–8·2	8·1

(b) *Density from underground measurements.* One method becomes feasible where it is possible to make gravity measurements underground. If readings are taken at points directly below one another (at surface and at one or more underground openings), then the difference between two of these values is given by

$$\delta g = (0 \cdot 094 - 0 \cdot 02554\sigma)h + \varepsilon_T,$$

where h is the vertical distance in feet and ε_T the difference in terrain corrections between the stations. Hence the average bulk density in the contained slab is

$$\sigma = 3 \cdot 68 - (\delta g - \varepsilon_T)/0 \cdot 02554h. \qquad (2.22)$$

Since ε_T depends upon σ, eq. (2.22) can be solved by successive approximations.

Although this is a good density measurement, the application is clearly limited. Measurements could also be made with a borehole gravity meter; although several models of this type of instrument have been developed, they are not used to any extent.

(c) *Nettleton's method.* A reasonably satisfactory estimate of the near-surface density, due to Nettleton, may be obtained from a representative gravity profile over the survey area. The field readings are reduced to produce Bouguer gravity profiles, assuming several different values of σ for the Bouguer and terrain corrections. The smoothest of these profiles, that is, the one which reflects the topography least, is the one with the correct density. This method is illustrated in fig. 2.7, where it can be seen that some of the profiles follow the topography and others invert it; the one which shows no particular correlation corresponds to the correct density.

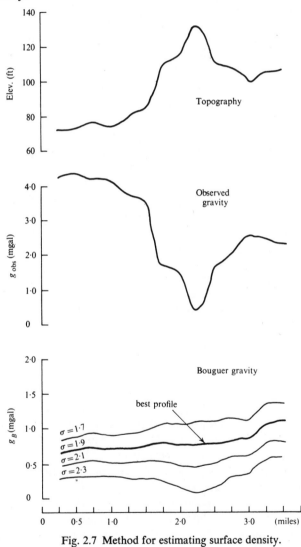

Fig. 2.7 Method for estimating surface density.

On flat ground it is difficult to choose the proper profile, since they may all be quite smooth. However, in this case one can tolerate a larger density error. In some situations it might be necessary to assume different values of σ for different sections of the profile, but this obviously requires more geological information than would normally be available.

In applying this method we are considering only the mass contained between the lowest and highest stations. Thus the datum plane should be the elevation of the lowest station. This procedure provides little control for density changes below this plane, nor is it reliable if the surface terrain is structurally controlled.

(d) *Parasnis' method.* An analytical approach somewhat similar to Nettleton's graphical method has been developed by Parasnis (1962, p. 40). Rearranging the terms in eq. (2.21*b*), §2.3.2*h*, we obtain

$$(g_{obs} - g_r \pm dg_L + 0.094h) - \delta g_B = (0.01277h - T)\sigma. \qquad (2.21c)$$

Since we are attempting to determine an average bulk density over the survey area, the Bouguer anomaly δg_B is considered to be a random error of mean value zero. In particular, for a single profile, eq. (2.21*c*) resembles a straight line of the form $y = mx - b$. If we plot $(g_{obs} - g_r \pm dg_L + 0.094h)$ versus $(0.01277h - T)$ and draw the best-fit straight line through the points and through the origin, its slope will be σ. Obviously all the points will not lie on this line unless the ground is uniform and the Bouguer anomaly everywhere zero; the best fit must be obtained by some method such as least squares.

As in the graphical determination, larger scale or regional variations in gravity are effectively eliminated. Presumably the analytical method would give better results over uniform ground. Also it can be extended to the two-dimensional case by developing the grid areas as a low-order polynomial and solving for σ by least squares.

2.4 Gravity instruments

2.4.1 *General*

Since the detection of an anomaly in gravity prospecting requires that we measure changes in g at least as small as 0.1 mgal, it is not possible, as mentioned earlier, to determine δg and absolute g with the same instrument. The absolute measurement is carried out at a fixed installation and involves the accurate timing of a swinging pendulum or of a falling weight.

Relative measurements may be made in various ways. Three types of instruments have been widely used at different times for gravity exploration. These are the torsion balance, the pendulum and the gravity meter (also known as the gravimeter). The latter is the sole instrument now used for prospecting; the other two have only historical interest in applied geophysics. For this reason only brief descriptions are included of the measurement of absolute g and of the torsion balance and pendulum, while the gravimeter will be discussed in some detail.

2.4.2 *Absolute measurement of gravity*

(a) *Falling body*. Although the timing of a freely falling body was the first method of measuring g, the accuracy was very poor because of the difficulty in measuring small time intervals. Within the last fifteen years this method has been revived as a result of great improvements in instrumentation. Elaborate installations for this purpose are now located at several national laboratories (e.g., France, Canada, England).

The acceleration of gravity can be found from Newton's equation of motion by noting the time interval between two points in a vertical fall. If the falling body, which starts with an unknown initial velocity, falls distances s_1 and s_2 in time intervals of t_1 and t_2 respectively (all quantities being measured from the starting point), then

$$g = 2(s_2 t_1 - s_1 t_2)/(t_2 - t_1) t_1 t_2. \tag{2.23}$$

In order to obtain an accuracy of 1 mgal with a fall of one or two metres, it is necessary to measure time to about 10^{-8} seconds and distance to less than $\frac{1}{2}$ micron.

(b) *Pendulum*. Until recently the standard method for measuring g employed a modified form of the reversible pendulum originally developed by Kater in 1818. Installations of this type exist at Potsdam, Washington and Teddington. The value of 981·274 gals obtained at Potsdam in 1906 is still used as the nominal value for comparison of g at other stations, although it is now known to be too large by about 14 mgals.

The value of g is obtained by timing a large number of oscillations, then using the simplified formula

$$g = 4\pi^2 I/T^2 m h, \tag{2.24}$$

where I is the moment of inertia, T the period, m the mass and h the distance from the pivot to the centre of mass of the pendulum. In the reversible pendulum the factor I/mh, which cannot be determined with high precision, is replaced by ℓ, the length of the equivalent simple pendulum – in effect a weightless, perfectly rigid connection between pivot and point mass. The accuracy required for T and ℓ here is similar to that of the falling weight apparatus.

2.4.3 *Relative measurement of gravity*

(a) *General*. Any method of determining changes in g (or, in fact, of comparing g-values from station-to-station) is a relative measurement in that it has not been obtained by a specific determination of time and length. Obviously the great majority of measurements are of this type. In the following sections we shall describe various techniques for making relative measurements of g.

(b) *Portable pendulum*. The pendulum has been used both for geodetic and prospecting purposes. For example, Bouguer compared the periods of a pendulum at various locations in South America, Lapland and Paris about 1750.

If we differentiate eq. (2.24) we get

$$\frac{dg}{dT} = \frac{-8\pi^2}{T^3} \frac{I}{mh} = \frac{-2g}{T},$$

so that the difference in gravity (to the first order) is given by

$$dg = -2g \, dT/T = -2g(T_2 - T_1)/T_1. \qquad (2.25)$$

Thus if we can measure the periods at two stations to about 1 μsec, the gravity difference is accurate to 1 mgal. This is not difficult with the precise clocks (quartz crystal, caesium, etc.) now available.

The pendulum has been used extensively for geodetic work in determining relative g over the earth surface, both on land and at sea; the three-pendulum apparatus of Vening Meinesz was designed specifically for submarine operation.

Portable pendulums were used in oil exploration during the early 1930s. The usual arrangement employed two instruments and compared the periods at a base and a movable station. In this case the relation for dg is more complicated than in eq. (2.25), since the constants of the two pendulums are not necessarily equal. The oscillations were recorded on light-sensitive paper and the records correlated by radio time signals. Sensitivity was said to be about 0·25 mgal.

Pendulum apparatus is complex and bulky. Two pendulums, swinging in opposite phase to reduce sway of the mounting, are required at each station. These are enclosed in an evacuated, thermostatically controlled chamber to eliminate pressure and temperature effects. To get the required accuracy in T_1 and T_2 it is necessary to record for about half an hour.

(c) *Torsion balance.* Strictly speaking, the instrument used by Cavendish in 1791 to determine the gravitational constant (§2.2.1) is a form of torsion balance, although it was not used as such. Baron Roland von Eötvös, a Hungarian physicist, developed the torsion balance for geodetic purposes about 1880. Modified forms of this instrument were employed in gravity prospecting from 1915 to about 1950. It has been replaced by the gravity meter which, although its sensitivity is no greater, has all the advantages of portability, ruggedness and speed of operation.

Thus the torsion balance now has only historic interest in exploration; its former extensive use, however, requires that it be discussed at least briefly. More detailed treatment may be found in Nettleton (1940), Heiland (1940), and Jakosky (1950).

A simplified schematic of the torsion balance is shown in fig. 2.8. The equal masses m are separated by the rigid bar of length 2ℓ (this is the type of balance used by Cavendish) and vertically by a height h (the modification of Eötvös). The whole is suspended at the centre of the bar by a torsion fibre with a small mirror attached to measure rotation of a light beam. The masses m are of the order of 30 g, the horizontal beam about 40 cm long, the vertical separation 50 cm and the torsion constant of the suspension 0·5 dyne-cm/radian. Non-magnetic metals have been generally used for the balance parts, the masses being small gold spheres, the arms aluminium and the torsion wire tungsten or an alloy of low temperature coefficient such as platinum-iridium.

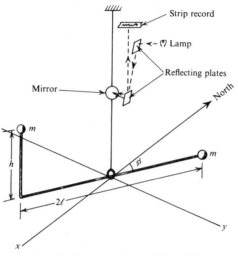

Fig. 2.8 Torsion balance (schematic).

For good sensitivity the torsion balance period may be as large as 30 minutes and the distance between mirror and recording strip 50 cm or more. In some instruments the light is reflected twice to multiply the angular deflection four times.

To minimize temperature and air convection effects the balance is contained in an inner chamber well insulated from at least one outer container. To reduce effect of sway in the support, the balance is double, that is, with two complete parallel beam systems. Three azimuth positions of the beam systems, normally 120° apart, are required to make a station measurement. The whole apparatus is housed in a small hut. Obviously this assembly is bulky and the readings take considerable time. During a full day's operation perhaps eight or ten stations might be occupied.

As mentioned in §2.2.5c and §2.2.5d, the torsion balance measures the horizontal gravity gradient and the differential curvature, two quantities which are intimately related to the distortion of the gravitational field caused by local density variations and terrain irregularities. It does not measure vertical gravity g_z. That is to say, there is no tilting of the horizontal beam due to changes in g_z; the motion is entirely rotational, caused by slight differences in the horizontal components of g acting on the masses.

The effects – greatly exaggerated – of these tiny forces of horizontal gravity on the torsion balance are illustrated in fig. 2.9. A section of uniform ground containing a spherical mass of positive density contrast is shown in fig. 2.9a. The equipotential surfaces above this anomaly are warped upward, the distortion

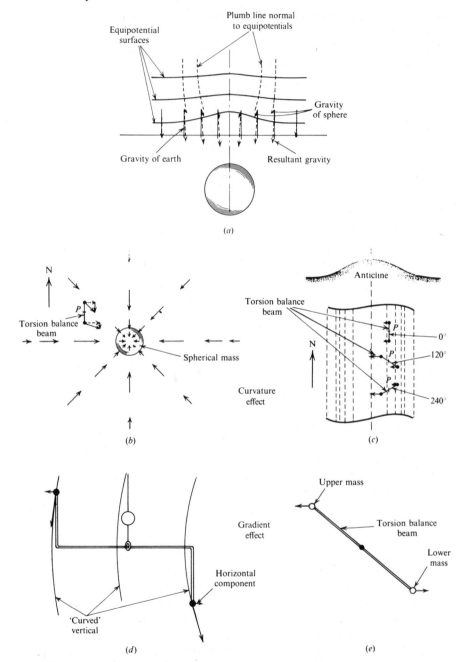

Fig. 2.9 Origin of the forces acting on the torsion balance. (*a*) Distortion of gravity field over a sphere; (*b*) curvature effect – horizontal forces due to a sphere; (*c*) curvature effect – horizontal forces due to an anticline; (*d*) gradient effect – elevation view; (*e*) gradient effect – plan view.

becoming smaller with increasing altitude. Since the plumb line must be everywhere normal to these surfaces, a vertical line will be curved rather than straight. The total gravity effect will be a vector combination of the component due to the uniform surroundings and that due to the sphere; clearly there is a resultant horizontal component directed toward the vertical diameter of the sphere.

Figure 2.9*b* is a plan view of the horizontal *g* components over the sphere. If the torsion balance beam were hung at point *P*, initially pointing north (i.e., it would point north if no anomalous mass were present), the net moment would rotate it counterclockwise; for the other two positions, 120° and 240° (see fig. 2.9*b*), the tendency is to rotate clockwise. When the balance is located directly over the centre of the sphere, there is no net force to rotate it either way.

The effect of a two-dimensional structure is illustrated in fig. 2.9*c*. Since the curvature is zero parallel to the long axis, the horizontal force components all point E–W and there is zero rotation of the beam at the point *P*, when it is pointing north; at 120° and 240° the rotation is counterclockwise and clockwise respectively.

These two examples show the effect of curvature of the equipotential surfaces on the torsion balance. To respond to curvature forces it is only necessary that the two masses be at points of different curvature; they need not be at different elevations. Thus the Cavendish balance measures differential curvature.

The difference in elevation between the masses permits measurement of horizontal gravity gradient (U_{xz}, U_{yz}, as in eq. (2.13)), which is related to the curvature of the vertical. (Saying this in another way, the HGG depends on the density or convergence of the equipotential surfaces: as fig. 2.9*a* shows, these are crowded together where the distortion is a maximum over the sphere.) Figures 2.9*d* and 2.9*e* illustrate the horizontal components acting on the masses in vertical section and plan.

From the diagrams in fig. 2.9 it can be seen that the moments due to curvature and gradient do not necessarily tend to rotate the balance in the same sense. For an arbitrary orientation there will generally be a net moment which will rotate the beam to an equilibrium azimuth where this moment is equal and opposite to the torsion of the suspension. This equilibrium condition is described by the equation

$$n - n_0 = \frac{2NDm\ell^2}{\tau} (2U_{xy} \cos 2\phi + U_\Delta \sin 2\phi) +$$
$$+ \frac{2NDm\ell h}{\tau} (U_{yz} \cos \phi - U_{xz} \sin \phi), \qquad (2.26)$$

where $U_\Delta = |U_{yy} - U_{xx}|$ (these quantities, along with U_{xy}, U_{xz} and U_{yz}, are defined by eqs. (2.13) and (2.14)), τ = torsion constant of fibre (dyne-cm/radian), N = number of reflections (mirror to reflecting plate), D = mirror-plate distance (cm), m = mass of each weight (g), 2ℓ = beam length (cm), h = vertical separation of weights (cm), ϕ = angle between balance beam and *x*-axis, n = deflection on plate (cm), n_0 = deflection on plate for zero torque (cm).

All the mechanical dimensions are illustrated in fig. 2.8. There are five unknown quantities in this equation: n_0, U_{xy}, U_Δ, U_{yz}, U_{zz}. Since the instrument has two beams (the second is reversed so that $\phi_2 = \pi - \phi_1$), there are also two values of n_0, one for each. Thus with three set-ups per station it is possible to solve for all the unknown quantities. This is generally done by successive orientations of $\phi = 0°$ (north), 120° and 240°.

The various components of gradient and curvature are measured in *Eötvös units*: 1 EU $= 10^{-6}$ mgal/cm. Thus if the horizontal gradient is 1 EU, the gravity difference between two points 1 metre apart is 0·0001 mgal. For curvatures, it can be shown that

$$\sqrt{(U_\Delta^2 + 4U_{xy}^2)} = g(1/\rho_1 - 1/\rho_2), \tag{2.27}$$

where g is the gravity value at the station and ρ_1, ρ_2 are the radii of maximum and minimum curvature respectively ($\rho_1 < \rho_2$). For any curved surface these two radii lie on mutually perpendicular planes. Thus curvature is also measured in Eötvös units. The sensitivity of torsion balances used in the field was of the order of a few *EU*, equivalent to a deflection n on the recording plate of a fraction of a millimetre.

(d) *Gravity metres – stable type.* The development and application of *gravity meters* (gravimeters) for field measurements of Δg_z dates from the early 1930s. They may properly be considered in the order of their development, the early models being of the stable type. These have been superseded entirely by the more sensitive unstable meters and so are of historic interest only. However, the principles involved are similar to those in modern instruments.

All gravity meters are essentially extremely sensitive mechanical balances in which a mass is supported by a spring. Small changes in gravity move the weight against the restoring force of the spring. The stable type of instrument, which has a linear dependence on gravity over a large range, requires a considerable amplification of the minute changes in length of the spring. This amplification may be mechanical, optical or electrical, or a combination of these.

The basic elements of a stable gravity meter are shown in fig. 2.10. Since the displacement of the spring is small, Hooke's law applies, i.e., the force is proportional to the change in length, hence

$$F = M\delta g = k\delta s, \quad \text{or} \quad \delta g = k\delta s/M, \tag{2.28}$$

where k is the spring constant in dynes/cm.

In order to measure δg to 0·1 mgal or better, we must detect a fractional change in spring length of $1/10^7$ (since $\delta g/g = \delta s/s$), hence the need for some form of magnification. Mechanically we can make k/M small by using a large mass and a weak spring, but obviously this enhancement of sensitivity is quite limited. The period of oscillation of this system is

$$T = 2\pi\sqrt{(M/k)}. \tag{2.29}$$

Combining (2.28) and (2.29) we get

$$\delta g = 4\pi^2 \delta s / T^2. \qquad (2.30)$$

Thus the period is very large for good sensitivity and a measurement of δg requires considerable time.

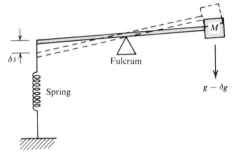

Fig. 2.10 Basic principle of the stable gravity meter.

Several versions of the stable gravity meter were developed between 1932 and 1938, employing ingenious means of magnification. Brief descriptions are given in the following sections of two of the better-known versions.

(e) *Gulf gravimeter*. This instrument measured the rotation of a spring, wound from a flat ribbon, rather than its elongation. For this flat spiral the rotation is, in fact, greater than the relative elongation. The essential parts are shown schematically in fig. 2.11a. The mirror underneath the mass (≈ 100 g) on the end of the spring, in conjunction with a fixed mirror or set of mirrors, the latter partially silvered to permit multiple reflections, produces optical amplification of the original small rotation angle. The sensitivity was said to be better than 0·1 mgal. This instrument was used extensively in the U.S.A. for petroleum exploration. The weight was about 100 lb, later reduced to 25 lb.

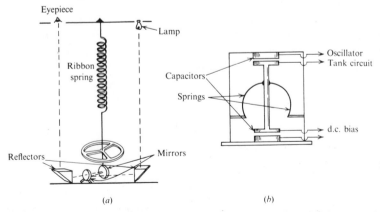

Fig. 2.11 Typical stable gravity meters. (*a*) Gulf gravity meter; (*b*) Boliden gravity meter. (After Garland, 1965.)

(f) *Boliden gravimeter*. Developed in Sweden in 1938, this instrument employed an electronic detector and electrical balancing device. Figure 2.11b is a schematic view. The mass is in the form of a bobbin suspended by bowed springs; the flat end discs are concentric with a pair of fixed discs to form two electrical capacitances. The upper plates form the capacitor in a tuned oscillator circuit so that small changes in capacitance vary the frequency of oscillation. The lower pair are connected to a d.c. supply which produces an electrostatic balance. Measurement of δg is achieved by adjustment of the d.c. voltage to restore the plates to a fixed reference or null position. The sensitivity was about 0.1 mgal.

Some further development on this type of meter has since been carried out, in an effort to make it into an airborne instrument which would measure vertical gradient or dg/dz. The parts were all extremely small, the end discs being the size of a dime. Since it is now possible to measure capacitance variations to an accuracy of $1/10^8$, this electrical approach seems promising.

(g) *Extraneous effects on gravity meters*. Apart from the difficulties of achieving good sensitivity in measuring δg, the instruments are extremely sensitive to other physical effects, such as changes in pressure, temperature and small magnetic and seismic variations.

The effect of temperature changes on the older instruments, which were generally made with rather bulky metal parts, was enormous – about 10 mgal/°C. To maintain the temperature constant within 0·01°C, the working parts of the meter are mounted in a well-insulated thermostatically controlled inner box, surrounded by one or more outer containers. Although this regulation of temperature is equivalent to only 0·1 mgal, it is sufficient, since sudden fluctuations of temperature could not occur in the sealed inner chamber. The thermal control requires 10–20 watts of power, generally supplied by a storage battery. With these additions the equipment becomes relatively heavy and bulky. Furthermore, the temperature regulator must be in operation for some hours to reach equilibrium.

(h) *Gravity meters – unstable types*. Also known as *labilized or astatized gravimeters*, these instruments have an additional negative restoring force, operating in the same sense as gravity against the restoring spring. They are essentially in a state of unstable equilibrium and thus have greater sensitivity than the stable meters. The range over which the readings vary linearly with gravity is less than for stable gravimeters, so they are usually operated as null instruments.

(i) *Thyssen gravimeter*. Although now obsolete, this instrument illustrates the instability effect particularly well and is worth discussing for this reason. It is illustrated in fig. 2.12.

The addition of the mass m above the pivot raises the centre of gravity and produces the instability condition. If g increases, the beam tilts to the right and the moment of m about the pivot enhances the rotation: the converse is true for a decrease in gravity. The increase in sensitivity can be shown by taking moments about the pivot for the position shown in fig. 2.12 and when the beam is horizontal (in the latter case mass m has no effect). Then we have

$$M(g + \delta g)\ell \cos \theta + m(g + \delta g)h \sin \theta = k(s + \delta s)\ell \cos \theta, \quad \text{and} \quad Mg\ell = ksl.$$

Dividing the first expression through by $\cos \theta$, subtracting the second from the result, substituting $g = ks/M$ and $\tan \theta \approx \delta s/\ell$, we obtain

$$\delta g = \frac{k}{M} \left\{ \frac{1 - (mhs)/M\ell^2}{1 + (mh\delta s)/M\ell^2} \right\} \delta s \approx \frac{k}{M} \left(1 - \frac{m}{M} \frac{h}{\ell} \frac{s}{\ell} \right) \delta s. \tag{2.31}$$

Since the term in the bracket is less than unity, the inherent sensitivity of this arrangement is greater than the stable condition of eq. (2.28).

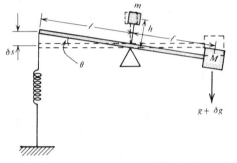

Fig. 2.12 Thyssen gravity meter. (After Dobrin, 1960.)

The Thyssen meter employed two parallel beams with the end weights reversed It was rather big and heavy (beam length 20 cm) and had a sensitivity of 0·25 mgal.

(j) *LaCoste-Romberg gravimeter*. This type of meter, which has been manufactured widely under various names (Askania, Frost, Magnolia, North American), is a modified vertical seismograph of long period, originally designed by L. J. B. LaCoste about 1934.

In connection with this seismograph LaCoste introduced the zero-length

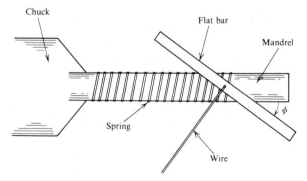

Fig. 2.13 Construction of the zero-length spring.

spring, which has been incorporated in almost all gravity meters since that time. The arrangement for winding this spring is shown in fig. 2.13. The wire is fed through a hole in a flat bar which is oriented at an angle ϕ with respect to the mandrel on which the spring is wound. Thus a pre-tension, proportional to ϕ and the wire tension, is put into the spring during the winding. When it is removed from the mandrel the spring shrinks to a minimum length.

A *zero-length spring* is defined as one in which the tension is proportional to the actual length of the spring, that is, if all external forces were removed the spring would collapse to zero length. Of course, this is physically impossible because of the thickness and weight of the wire. The advantage of the zero-length spring is that if it supports the beam and mass M (see fig. 2.14) in the horizontal position, it will support them in any position.

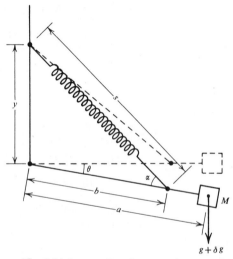

Fig. 2.14 Lacoste-Romberg gravimeter.

To derive the expression for the sensitivity of the LaCoste-Romberg gravimeter, we write $k(s - z)$ for the tension in the spring when its length is s; thus, z is a small correction for the fact that the spring is not truly zero-length. Taking moments about the pivot in fig. 2.14, we get

$$Mga \cos \theta = k(s - z)b \sin \alpha$$
$$= k(s - z)b(y \cos \theta)/s,$$

using the sine law. Thus,

$$g = \left(\frac{k}{M}\right)\left(\frac{b}{a}\right)\left(1 - \frac{z}{s}\right)y.$$

When g increases by δg, the spring length increases by δs where

$$\delta g = \left(\frac{k}{M}\right)\left(\frac{b}{a}\right)\left(\frac{z}{s}\right)\left(\frac{y}{s}\right)\delta s. \tag{2.32}$$

For a given change δg, we can make δs as large as we wish by decreasing one or more of the factors on the right-hand side; moreover, the closer the spring is to the zero-length spring, the smaller z is and the larger δs becomes.

In operation this is used as a null instrument, a second spring being used which can be adjusted to restore the beam to the horizontal position. The sensitivity is about 0·01 mgal.

Early models of this meter were quite bulky and sensitive to temperature changes, since the working parts were metal. Like the stable types, the meter required an elaborate temperature regulating system and the moving parts had to be clamped for transportation. In recent versions these disadvantages have been overcome.

(k) *Worden gravimeter.* The original model of this instrument, which appeared in 1948, incorporated several design changes which enormously improved the portability and speed of operation as well as reducing the chances of mechanical damage. This was achieved by using small, very light parts of quartz, for example, the mass M weighs only 5 milligrams. Thus it is not necessary to clamp the movement between stations. Sensitivity to temperature and pressure changes was greatly reduced by enclosing the system in a vacuum flask. In addition there is an automatic temperature compensating arrangement. As a result of these features the Worden meter is small (10 in. high, 5 in. in diameter) and weighs about 6 lb. Since the temperature compensation is automatic, the only power requirement is two penlight cells for illuminating the scale.

A simplified schematic is shown in fig. 2.15. The moving system is similar to the LaCoste-Romberg meter. The arm OP' and beam OM are rigidly connected and pivot about O, changing the length of the main spring $P'C$, which is fixed at C (see also fig. 2.16b). From fig. 2.15 we have the following relations:

$$s = CP, \quad \delta s = CP' - CP \approx b\theta \sin\left(\frac{\pi}{2} - \alpha\right), \quad \theta \approx \delta s/(b\cos\alpha).$$

The correction factor z, which appeared in the treatment of the LaCoste-Romberg meter, is negligible for the Worden meter. Taking moments about the pivot for the case where $\theta = 0$, we get

$$Mga = ksb\cos\alpha.$$

When g increases to $(g + \delta g)$, P moves along the circle to P' and

$$M(g + \delta g)a\cos\theta = kb(s + \delta s)\cos\left(\alpha + \frac{\theta}{2}\right).$$

To the first approximation, this becomes

$$M(g + \delta g)a = kb(s + \delta s)\left(\cos \alpha - \frac{\theta}{2}\sin \alpha\right)$$

$$= kb(s + \delta s)\left(\cos \alpha - \frac{\delta s}{2b}\tan \alpha\right)$$

$$= kb\left\{s \cos \alpha - \delta s\left(\frac{s}{2b}\right)\tan \alpha + \delta s \cos \alpha\right\}.$$

Subtracting the first moment equation to eliminate g, we get

$$Ma\delta g = kb\left(\cos \alpha - \frac{s}{2b}\tan \alpha\right)\delta s.$$

Using the relation $\sin \alpha = s/2b$, we get finally

$$\delta g = \left(\frac{k}{M}\right)\left(\frac{b}{a}\right)\left(\frac{\cos 2\alpha}{\cos \alpha}\right)\delta s. \tag{2.33}$$

As in the LaCoste-Romberg meter the sensitivity can be increased by decreasing the factors (k/M) and (b/a); in addition the factor $(\cos 2\alpha/\cos \alpha)$ approaches zero when α approaches 45°, thus furnishing another method of obtaining high sensitivity. In practise the sensitivity is 0·01 mgal and the range is about 60 mgal.

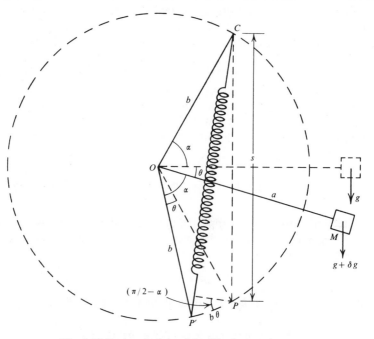

Fig. 2.15 Basic principle of the Worden gravimeter.

The temperature compensating system, which contains the only metal parts in the instrument, is shown in fig. 2.16a. The two long arms are made of materials having different expansion coefficients, and expand or contract in such a way as to raise or lower the upper end of the spring in the proper sense to correct for temperature effects in the spring.

Like the LaCoste-Romberg instrument, the Worden meter is read by measuring the force required to restore the beam to the horizontal position. Figure 2.16b, which is a more realistic schematic diagram, shows the auxiliary spring used for nulling, as well as a coarse spring which can be adjusted to change the range.

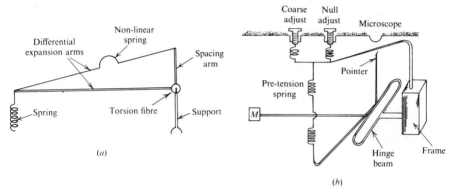

Fig. 2.16 The Worden gravimeter. (After Dobrin, 1960.) (a) Temperature compensation device; (b) schematic (omitting temperature compensation).

2.4.4 *Calibration of gravity meters*

At present the Worden and LaCoste-Romberg meters are the only types used for gravity exploration. Both are null instruments and changes in gravity are shown as arbitrary scale divisions on a micrometer dial on the outside of the housing. There are several methods for converting these scale readings to gravity units.

Calibration theoretically can be carried out by tilting, since a precise geometrical system is involved, but this is not the usual procedure. Generally readings are taken at two or more stations where absolute or relative values of g are already known. If the value of δg between two stations is large enough to cover a reasonable fraction of the total instrument range, it is customary to assume a linear response between them. However, one should occupy several additional stations, if possible.

2.5 Field operations

2.5.1 *General*

Gravity exploration is carried out both on land and at sea. Although some attempts have been made to develop an airborne instrument, this mode of operation is not yet practical. The distinction between reconnaissance and detailed field work is based on the objective, that is, whether the purpose is to find features of interest or to map them. Gravity surveys for oil exploration typically involve one or two stations per square mile on the average, although the distribution of stations may be far from uniform.

In mineral exploration, on the other hand, gravity reconnaissance is practically non-existent. Here it is normally employed as a secondary detailed method for confirmation and further analysis of anomalies already outlined by magnetic and electrical techniques.

2.5.2 *Torsion balance surveys*

In §2.4.3c the determination of gradient and curvature terms by three orientations of a double beam system at each station was discussed briefly. These operations were generally controlled automatically, each azimuth set-up requiring 20 minutes or more. Thus six or eight stations per day were about the maximum and the equipment, including the hut enclosure was transported by truck.

The standard method of displaying results is shown in fig. 2.17. The horizontal gradient, fig. 2.17a, consists of the terms U_{xz} (generally assumed north) and U_{yz} (east). The vector resultant of U_{xz} and U_{yz} represents the magnitude of the total horizontal gravity gradient in the direction of maximum increasing gradient. One can find the difference in gravity, δg, between two stations by multiplying the appropriate component of the gradient by the station spacing.

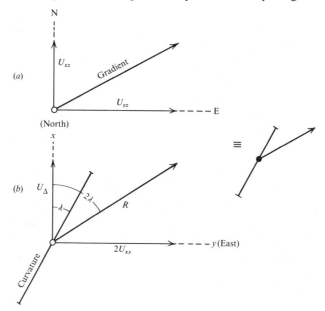

Fig. 2.17 Plotting torsion-balance results. (*a*) Plotting gradients; (*b*) plotting curvatures.

The curvature vector seen in fig. 2.17b, is more complicated. Usually the magnitude of $(U_{yy} - U_{xx})$ (U_{Δ} in eq. (2.26)), is plotted in the north or x-direction and $2U_{xy}$ towards the east. The resultant would have the correct magnitude, but twice the proper angle with respect to north direction, since $2U_{xy}$ has been plotted E-W. Therefore the curvature symbol is drawn as shown – a vector of length $\sqrt{(U_\Delta{}^2 + 4U_{xy}{}^2)} = g(1/\rho_1 - 1/\rho_2)$, centred at the station and making an angle $\lambda = \frac{1}{2}\tan^{-1} 2U_{xy}/U_\Delta$ with the north. This vector is in the direction of the plane of minimum curvature.

2.5.3 *Gravity meter surveys*

(a) *General.* Station spacings in field-work with the gravity meter vary from 10 miles to as little as 20 ft for some detailed work. The station interval is usually selected on the basis of known or assumed values of the anomaly depth and size. For oil exploration, one or two stations per square mile are commonly used because structures associated with oil accumulation are usually larger than this and hence their expression would not be lost with such a spacing. While a more-or-less uniform grid of stations is desirable, stations are sometimes run on loops which are operationally easier. Stations half a mile apart on loops roughly 4 × 4 miles in size might be typical.

In mineral exploration the spacing is determined mainly by earlier detailed magnetic and electrical surveys which have located relatively small and shallow anomalies. Normally the gravity readings are made at the same stations used in the previous surveys, commonly 50 or 100 ft apart.

Occasionally regional gravity surveys are carried out to establish large-scale geological structure, in which the stations might be anything from $\frac{1}{8}$ mile to 10 miles apart. However, such work is usually done by government or academic agencies and the immediate purpose is not mineral exploration.

With modern gravity meters there are generally no insurmountable problems arising from the actual field measurements. In swampy ground and in wooded areas when the wind is strong, it is often difficult to get a stable null on the instrument; however, a little extra care and time will generally give an acceptable measurement. In marine gravity work using instruments which rest on the bottom, strong and variable currents present problems of a similar nature when the water depth is 200 ft or greater, because of the pull on the cable; in severe cases it may not be possible to level the meter. In high latitudes when working in muskeg or in areas covered by shallow inland waters, it is an advantage to do the survey in winter. Obviously, for reasonable speed of operation, some sort of vehicle – truck, snow sled, etc. – ought to be used, if possible, for large area surveys.

It is most important that the gravity meter be precisely levelled for each reading. The levels themselves should be checked regularly because the meter becomes unstable and drifts erratically when they are out of alignment.

(b) *Drift correction.* All gravimeters change null reading with time, even when set up at a fixed station. This drift is the result mainly of creep in the springs and under ideal static conditions the change is unidirectional. In addition, if the movement is not clamped between readings, or is subjected to sudden motion or jarring during transport, the change may be somewhat erratic. A recent LaCoste-Romberg portable instrument, however, has very little drift under any conditions.

The net result of drift is that, over a period of days or even hours, repeated readings at one station will give a series of different gravity values. Consequently it is necessary to reoccupy some of the stations periodically during a gravity survey in order to produce a drift curve for the instrument. The maximum time allowable between repeat readings depends on the accuracy desired in the survey, but would seldom be greater than two or three hours; if the instrument is bumped it is wise to check a known station immediately.

Two examples of drift curves are shown in figs. 2.18a and 2.18b. The first is for a fixed installation and varies quite smoothly with time. The second is a portion of the drift during an actual field survey. It is not necessary to use the same station throughout for checking drift, since this would waste considerable travel time returning to the same point. A reasonable sequence for a grid survey is shown in fig. 2.18c, in which the original station 1 is occupied three times early in the day. Loops are then made on the other traverse lines and station 1 is near enough to be checked conveniently as the grid is completed at the end of the day (note that figs. 2.18b and 2.18c are unrelated).

The intermediate gravity stations, which are occupied only once, can then be corrected for the drift which has occurred during the appropriate fraction of the time interval between repeat stations. These corrections can be taken directly off the drift curve, keeping in mind that positive drift requires negative correction and vice-versa.

Since there is no way of allowing for possible erratic drift between check stations, we can only draw straight lines joining these points on the drift curve and trust that the variation was linear with time. Non-linear changes may be caused by earth tides (see fig. 2.5 and §2.3.2f). By repeating stations within two hours or less it is possible to remove the tidal variation reasonably well.

The importance of establishing a detailed drift curve, particularly in the early stages of an extensive survey, cannot be over emphasized. With good control it is relatively easy to repeat station readings a month later within 0·04 mgal, even though the meter may have drifted several milligals in the interval.

(c) *Surveying.* The precision required in surveying stations (0·1 ft in elevation and about 30 ft in location) may double the cost of field-work in a gravity survey. In open ground of low relief, the surveying is quite simple and can be done faster than the gravity measurements; even so, at least two men are required. However, in heavily wooded areas and particularly in rugged terrain the gravity work will proceed much faster than the surveying. The latter requires multiple set-ups and considerable pruning of the line, so that three or four men are necessary.

In large-scale gravity work it may be possible occasionally to save the time and expense of precise transit surveying by obtaining elevations with a barometer, or preferably with two barometers, one located at a base station and the other measuring elevations at the gravity stations. Even with two instruments the accur-

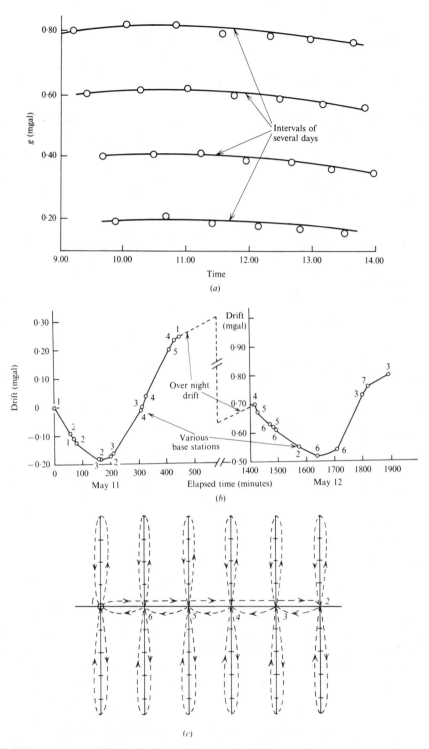

Fig. 2.18 Gravimeter drift. (a) Drift curve for fixed location; (b) drift curve during field survey; (c) field survey loop for drift correction.

acy is probably no better than ±1 ft, which means the reduced gravity readings may be in error by about 0·1 mgal.

(d) *Field reductions.* With the time involved in detailed surveying and the multiple reductions which must be applied to obtain Bouguer gravity (particularly if terrain corrections are necessary), it is difficult to keep up with the field work. This problem applies to gravity work more than other techniques. Thus the interpreter may have little day-to-day control over the progress of the survey. Sometimes it is preferable to curtail gravity measurements, if circumstances permit, in order to concentrate on processing the data already accumulated. This procedure is advisable specially in detailed surveys over rough terrain.

The end result of the gravity survey is a set of profiles and a contour map of Bouguer gravity. Further analysis can be carried out to assist in the interpretation of the gravity data; this will be discussed in §2.6.

2.5.4 *Marine surveys*

(a) *Locating marine stations.* Considerable gravity work has been done over water-covered areas and also under water. There are many difficulties in marine operations peculiar to the environment, the most serious being the location of the station. Stabilization of the instrument against wave motion is also a problem. Although stations occasionally can be located by sighting on two or more known shore points, it is usually necessary to employ sophisticated electronic navigation systems such as Shoran, Raydist, Loran, etc. Recently the communication satellites have provided accurate fixes for position. Offshore navigation is discussed in more detail in §4.5.5e. The accuracy in offshore location is inherently lower than on land and the method is very much more expensive.

(b) *Deep sea operations.* Several types of gravity measuring instruments have been modified for use at sea. About 40 years ago Vening Meinesz developed a three-pendulum system which was installed in a submarine and used for geodetic purposes. More recently shipborne gravity meters have been employed in large-scale sea reconnaissance. An example of this type of equipment is the Graf Askania meter; mounted on an elaborate gyro-stabilized platform, it is located in a region of maximum stability against ship roll. Gravity readings are in the form of a continuous record, the sensitivity being about 2 mgal. Measurements with a shipborne LaCoste-Romberg meter, repeatable within 1 mgal, have also been reported. This is about the same sensitivity as the submarine pendulum apparatus.

If a gravity meter has an eastward component of velocity during a gravity measurement, the ship's velocity adds to the rotational velocity resulting from the rotation of the earth, hence increases the centrifugal force and decreases the gravity reading. Conversely, for a westward component of velocity, the gravity reading is increased. Thus, a correction, the *Eötvös correction*, has to be made for the east or west component of velocity. This correction, dg_E, is given by

$$dg_E = 7\cdot503V \cos \phi \sin \alpha + 0\cdot004154V^2,$$

where dg_E is in milligals when V is in knots, ϕ being the latitude and α the course direction with respect to true north. The accuracy with which the Eötvös correction can be made is usually the factor limiting the accuracy of gravity data obtained on a moving ship.

(c) *Remote control systems.* Standard portable gravity meters have been adapted for operation on the water bottom to depths of at least 750 ft. This method of measurement is suitable for most inland waters and coastal areas and under favourable conditions the sensitivity is almost as good as would be obtained in a ground survey.

The meter is enclosed in a pressure housing which is supported on a squat tripod with disc feet. About half the total weight of the assembly is in the tripod to provide maximum stability when resting on bottom. The overall weight of one model is 650 lb. The complete assembly is connected by cable to a boat from which it is lowered to sit on the bottom.

Levelling to within 10″ of arc is obtained with small motors driving a gimbal system, powered and controlled from the boat. Alternatively the levelling control may be maintained with a servo system. The readings are recorded photographically on a slow-moving film in the housing; in another version, they are transmitted to the boat by a remote reading device.

Although the high sensitivity of this equipment is an advantage, the operation is quite slow in deep water because of levelling problems and because the assembly must be raised to surface between station set-ups. In these conditions the average daily coverage is about 6–8 stations. In shallow (<100 ft) water, however, the speed of operation is nearly the same as for land surveys.

(d) *Shallow water surveys.* In areas covered by water to a depth of about 20 ft, gravity surveys have been carried out on platforms supported by a tripod. Since the operator is on the platform as well, the meter needs no modification. This type of operation is no faster than the underwater instrument, because the heavy tripod must be loaded on a barge for transport between stations. In addition the tripod is sensitive to waves and wind.

2.5.5 *Airborne gravity*

Although considerably more difficult than ground gravity surveys, marine operations are quite practical. The problems involved in an airborne installation are so far insurmountable, although considerable test work has been carried out.

The main difficulty arises from huge and very rapid changes in g caused by the high speed of the aircraft. Variations in g are associated with changes in the aircraft altitude, linear acceleration, roll and heading. Although the last three effects are also quite significant in shipborne gravity work, they can be corrected reasonably well because of the low velocity and slow changes in velocity and course.

In the air the situation is much worse. An altitude change of 50 ft or a variation in heading of 0·16° are equivalent to 5 mgal. A correction of 1000 mgal is required for variation in centrifugal acceleration (the *Eötvös effect*) resulting from E–W flight paths at 200 mph. Clearly one must have exceptional control and smooth, constant velocity flight to obtain useful results.

Nettleton (1960) has described airborne tests with a LaCoste-Romberg meter mounted in a B-17 flying at 12,000 ft over the Imperial Valley in California. About 500 line-miles were flown and records made with aerial cameras, radar altimeter and a hypsometer for small, sudden changes in altitude, in addition to the actual gravity readings. As noted, the flight was made at high altitude and during a time when air turbulence was not troublesome. The airborne data were compared with an earlier regional free-air gravity ground survey.

Following corrections, a contour map was prepared with intervals of 20 mgal, covering an area about 50 by 70 miles. It appears that the sensitivity is about 10 mgal under good conditions. Coupled with the high cost of the operation, this would indicate that airborne gravity could be used for geodetic purposes and possibly for large-scale oil reconnaissance surveys.

In §2.4.3*f*, in connection with the Boliden gravity meter, mention was made of a similar type of meter adapted to measure vertical gravity gradient from an aircraft. The presence state of this instrument, a Swedish development, is not known.

2.6 Interpretation

2.6.1 *General*

The Bouguer gravity map is similar in appearance to a map of total field or vertical component in magnetics, although the gravity contours are usually smoother than contours on magnetic maps. This, of course, is due to the small variations in density, compared to magnetic susceptibility, and to the fact that the gravity field changes less rapidly in space than the dipolar magnetic field. Although this relative simplicity of the gravity map would appear to make the interpretation easier, the reverse may also be true, since the interesting features do not generally stand out as boldly as on a magnetic map.

Taking a specific example, consider a spherical mass 400 ft in diameter whose top is 200 ft below surface. Situated directly below it is a vertical step of 1000 ft in the basement with an average depth of 5000 ft. If the magnetization is approximately vertical and the susceptibility contrast of both features with respect to the surroundings the same, the near-surface magnetic anomaly is about twice the size of that caused by the basement step. In the equivalent gravity situation, assuming equal density contrasts, the effect of the sphere is only a few per cent that of the basement step.

Thus the large-scale deep-seated structures predominate on the gravity map to such an extent that it may be very difficult to recognize smaller or shallower features. Nevertheless, careful detailed analysis is well worth the effort in gravity, not only because the field-work is expensive, but also because it is possible to obtain more specific information than in any of the other geophysical methods except seismic.

2.6.2 *Regionals and residuals*

(a) *Regional gravity*. As noted above, the interesting anomalies on the gravity map frequently are masked by deep-seated structures. The removal of the so-called *regional* resulting from these deep structures is a more serious problem in gravity than in other geophysical methods.

There is a close analogy between this operation and a filtering process (optical, electrical). The regional effects correspond to low frequencies, while the residuals correspond to high frequencies. However, the solution is not as simple as inserting a highpass filter in an electrical circuit to remove the low frequencies; it is often more realistic to consider the desired residuals as random frequencies, or noise.

Thus the passband is not well defined except in the simplest cases and there is inevitably some bias involved in the process of removing the regional effect. As in any geophysical technique, the most useful factor in interpretation is some knowledge of the local geology.

It should be emphasized that, in describing the regional as the effect of 'large-scale deep structures', the latter phrase must be considered in terms of the scale of the survey; the same consideration should apply to the residual (e.g., the size of a residual feature in petroleum surveying may be about the same as the regional in mineral search). It would be quite correct to define the regional as the effect in which we are not interested.

There are several methods of removing the unwanted regional. One approach is entirely graphical, another is analytical (although in some cases graphical methods are also incorporated). We consider these techniques below.

(b) *Graphical and smoothing techniques*. The simplest example of removing the regional by smoothing is illustrated in figs. 2.19a and 2.19b. In the first we have a profile which obviously reflects two disturbances of quite different size; even if no other data were available, the separation into two effects as shown in the diagram seems quite reasonable. Similarly in fig. 2.19b the regional contours are so regular that one would have no hesitation in developing the residual by subtracting the smoothed contours from the originals.

Clearly this method will be satisfactory when the residuals are fairly evident to begin with, which implies that their general trend is distinctly different from that of the regional. In other cases the two trends may be roughly parallel, or the residuals are located in closures, saddles or other complex irregularities in the regional field. Then the simple smoothing process usually becomes ineffective.

Such situations call for additional graphical procedure. In one method several profiles are plotted for lines roughly at right angles to the regional contours, smooth regionals are drawn on the profiles and these smoothed profiles are used

52   Gravity method

to modify the regionals in their vicinity. This approach is suitable when the regional trend is mainly unidirectional but not uniform.

A further refinement in graphical smoothing requires plotting profiles in two directions and smoothing both sets. For example, as an extension of the procedure described in the previous paragraph, smoothing might be applied approximately in the direction of the regional contours as well as normal to them. Alternatively, if the survey has been carried out in detail with close, uniform spacing of stations and lines, the stations themselves could be used. Although this two-dimensional technique requires considerably more time, partly because of the extra lines, partly because the values at the points of intersection of the profiles must be adjusted to be equal, the extra control may justify the additional work. In fact, when the regionals are so irregular that the directional trend is not immediately apparent, two-dimensional smoothing may be the only possible graphical method.

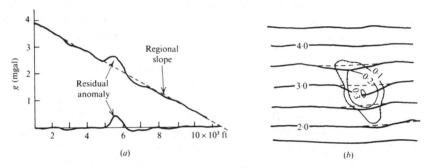

(a)   (b)

Fig. 2.19 Graphical separation of residual gravity from regional trend. (a) Using a profile; (b) using contours.

Having contoured the regional, the residual is now obtained by subtracting the regional from the original Bouguer map, either graphically or numerically – the former by drawing contours of constant difference, the latter by subtraction and drawing a new map.

The result obtained by smoothing contours is inevitably biased by the interpreter. This is not necessarily bad. If he is experienced and has additional information, particularly geology, to guide him, it may be a decided advantage. However, the method requires considerable work by experienced people and this is expensive.

(c) *Empirical gridding method.* This technique, originally described by Griffin (1949), is really a simple way of removing the regional by second-derivative analysis. The regional is considered to be the average value of gravity in the vicinity of the station, and is obtained by averaging observed values on the circumference of a circle centred at the station. To give this concept of the regional some mathematical justification, it can be written

$$\bar{g}(r) = \frac{1}{2\pi} \int_0^{2\pi} g(r,\theta) \, d\theta,$$

where the integral is generally replaced by a sum of n discrete values:

$$\bar{g}(r) = \{g(r,\theta_1) + g(r, \theta_2) + \ldots g(r, \theta_n)\}/n.$$

In practise, the various $g(r, \theta)$ terms are obtained by interpolation from the gravity map contours, or if the survey has been done in sufficiently regular detail, by using the actual station values (fig. 2.20a).

Having obtained the regional, the *residual* is then the difference between $\bar{g}(r)$ and g_s, the station gravity at the centre of the circle:

$$g_{res} = g_s - \bar{g}(r). \tag{2.34}$$

This approach is basically the same as that used in graphical smoothing, except that the latter is on a larger scale. Moreover, it bears some resemblance to the second-derivative methods. The result obtained depends on the choice of the number of points on the circle and particularly on the radius selected. Clearly if the radius is very small the residual is zero, if it is too large, the residual is approximately g_s. Also relatively large circles may overlap other nearby residuals, which would make the results meaningless.

(d) *Second derivative and residual.* Since the second vertical derivative enhances near-surface effects at the expense of deeper sources (see §2.2.5b), there should be a connection between a second-derivative map and a residual map. In fact the

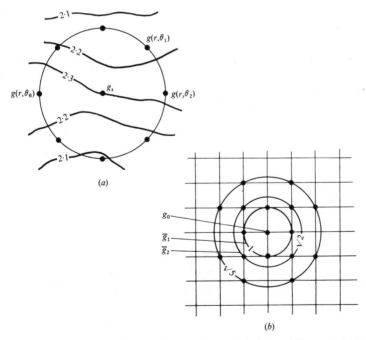

Fig. 2.20 Analytical separation of residual and regional gravity. (a) Griffin method; (b) second-derivative method.

second derivatives are a measure of curvature, as mentioned before; large curvatures are associated with shallow or residual anomalies. Furthermore it is possible to obtain $\partial^2 g/\partial z^2$ from the horizontal second derivatives, since the gravity field satisfies Laplace's equation. The following is a treatment due to Elkins (1951) to show how this may be done.

We assume a function $g(x, y, z)$ harmonic within a region, that is, it has continuous second derivatives and satisfies Laplace's equation. In the plane $z = 0$, considered to be the datum plane of the gravity map, we define another function:

$$\bar{g}(r) = \frac{1}{2\pi} \int_0^{2\pi} g(r\cos\theta, r\sin\theta, 0) \, d\theta. \tag{2.35}$$

As in the previous section, this is the average value of $g(x, y, z)$ around a circle of radius r in the plane $z = 0$, with the origin (any particular gravity station) as its centre. This function can be written as a power series, convergent about $r = 0$,

$$\bar{g}(r) = a_0 + a_2 r^2 + a_4 r^4 + \ldots, \tag{2.36}$$

since such a series is also harmonic. The odd powers of r are absent, because in eq. (2.35) we would have terms of the form $\sin^m \theta \cos^n \theta$ multiplied by r^{m+n}; when integrated between 0 and 2π, these are zero if $m + n$ is odd. Writing Laplace's equation in the form

$$\frac{\partial^2 g}{\partial z^2} = -\left(\frac{\partial^2 g}{\partial x^2} + \frac{\partial^2 g}{\partial y^2}\right),$$

transferring to polar coordinates in the xy-plane, i.e., $x = r\cos\theta$, $y = r\sin\theta$ and integrating both sides for θ from 0 to 2π, we have

$$\int_0^{2\pi} \frac{\partial^2 g}{\partial z^2} \, d\theta = -\left(\frac{\partial^2}{\partial r^2} + \frac{1}{r}\frac{\partial}{\partial r}\right) \int_0^{2\pi} g \, d\theta - \frac{1}{r^2} \int_0^{2\pi} \frac{\partial^2 g}{\partial \theta^2} \, d\theta.$$

The second term on the right gives

$$\frac{1}{r^2}\left(\frac{\partial g}{\partial \theta}\right)\Bigg|_0^{2\pi}.$$

This is zero because the derivative is equal to a series having terms in θ of the form $\sin^m \theta \cos^n \theta$, where m and n are integers, neither of which may be zero. Put $z = 0$ and divide both sides by 2π to get

$$\frac{1}{2\pi} \int_0^{2\pi} \left(\frac{\partial^2 g(r\cos\theta, r\sin\theta, z)}{\partial z^2}\right)_{z=0} d\theta = -\left(\frac{\partial^2}{\partial r^2} + \frac{1}{r}\frac{\partial}{\partial r}\right) \bar{g}(r). \tag{2.37}$$

With $r \to 0$ and substituting for $\bar{g}(r)$ from (2.36) we have

$$\frac{1}{2\pi}\int_0^{2\pi}\left\{\frac{\partial^2 g(r\cos\theta\, r\sin\theta, z)}{\partial z^2}\right\}\Bigg|_{\substack{r=0\\z=0}}d\theta = \lim_{r\to 0}\frac{\partial^2}{\partial z^2}\left\{\frac{1}{2\pi}\int_0^{2\pi}g(r\cos\theta, r\sin\theta)\,d\theta\right\}\Bigg|_{z=0}$$

$$= \lim_{r\to 0}\left(\frac{\partial^2\bar{g}}{\partial z^2}\right)_{z=0} \approx \left(\frac{\partial^2\bar{g}}{\partial z^2}\right)_{\substack{r=0\\z=0}} \approx \left[-\left\{\frac{\partial^2}{\partial r^2} + \frac{1}{r}\frac{\partial}{\partial r}\right\}(a_0 + a_2 r^2 + \ldots)\right]\Bigg|_{r=0}$$

$$= -4a_2. \qquad (2.38)$$

Now if we differentiate eq. (2.36) with respect to r^2 in the vicinity of $r = 0$, we get

$$\frac{\partial\bar{g}}{\partial(r^2)} = a_2,$$

which suggests a graphical method of obtaining $\partial^2\bar{g}/\partial z^2$ for the station at $r = z = 0$. If we determine $g(r)$ for several radii and plot the values against r^2, the slope of the resulting curve at the origin is a_2. Multiplying this slope by -4 gives $\partial^2\bar{g}/\partial z^2$. For other stations we merely shift the origin of coordinates, $r = 0$, on the plane $z = 0$.

This result, obtained partly by graphical means, is oversimplified. More accurate determinations of $\partial^2 g/\partial z^2$ can be obtained by methods which are entirely analytical, using several concentric circles (generally 3) of different radii. The effect of each circle is weighted by a coefficient term. This technique has been developed independently by several people; their results are all of the form

$$\frac{\partial^2 g}{\partial z^2} = \frac{K}{s^2}(k_0\,g_0 + k_1\,\bar{g}_1 + k_2\,\bar{g}_2 + \ldots), \qquad (2.39a)$$

where g_0 is the station gravity (at the centre of the circles), $\bar{g}_1, \bar{g}_2, \ldots$ are average values on successive circles, $k_0, k_1 \ldots$ are weighting coefficients ($\sum_0^n k_i = 0$), K is a numerical factor and s is a distance, generally the station spacing or some simple fraction of it. The circle radii are normally selected on the basis of the grid dimensions. For example, if the survey had been done on a square grid with spacing s, the successive radii generally are s, $\sqrt{2}s$, $\sqrt{5}s$, etc, with $r_0 = 0$ for the first term which represents the common centre (see fig. 2.20b). A convenient formula of this type, which works very well, is due to Henderson and Zietz (1949a). It is

$$\frac{\partial^2 g}{\partial z^2} = 2(3g_0 - 4\bar{g}_1 + \bar{g}_2). \qquad (2.39b)$$

Since the coefficients employed in the general formula (2.39a) vary from one author to another, some bias obviously still remains in this type of residual enhancement.

(e) *Polynomial fitting.* This is a purely analytical method, in which matching of the regional by a polynomial surface of low order exposes the residual features as random errors. The treatment is based on statistical theory. In practice the polynomial is rarely extended beyond the second order. In any case the operation requires a computer and is somewhat formidable.

The residuals left by graphical smoothing have one fundamental difference from those obtained by analytical methods. In the former the interpreter normally selects an arbitrary background value of gravity in smoothing the contours or profiles such that each residual appears entirely as a maximum or minimum, but not both. On the other hand, the analytical approach assumes the residuals to be random errors whose sum is zero. Thus the analytical residual is composed of positive and negative parts. This difference is illustrated in fig. 2.21 for a profile.

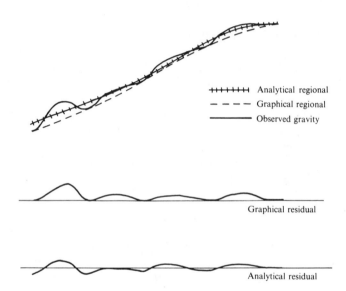

Fig. 2.21 Comparison of graphical and analytical methods of removing regional gravity.

The extensive literature on the subject emphasizes the significance of the problem of regional-residual separation. The techniques above are listed in order of increasing sophistication and complexity. The choice of method for removing the regional depends upon many factors, the most important being the total labour involved, the complexity of the gravity map, the density and distribution of the stations and the quality of the data. With large quantities of data and complex regional effects, analytical methods are clearly indicated. With small amounts of data and/or simple regionals, graphical methods would be preferred. In between the two extremes, the choice of technique must be based upon a careful balancing of the advantages and limitations of each method.

2.6.3 *Gravity effects of simple shapes*

(a) *General.* Only after the camouflaged residual gravity has been exposed can the usual geophysical interpretation, that of matching the field anomalies with various simple geometrical shapes, be attempted. In gravity this technique is better developed than in other methods because the theoretical analysis is straightforward. In the following examples dimensions are in grams, centimeters and milligals unless otherwise noted, the density symbol σ is actually the density contrast with respect to the surroundings, the symbol g refers to g_z, the vertical gravity anomaly, and the minus sign in eq. (2.2) has been dropped (see the note following eq. (2.15)).

(b) *The sphere.* The gravity effect at a point P (fig. 2.22), directed along r, is $g_r = \gamma M/r^2$. Calculating the vertical component, we get

$$g = g_r \cos \theta = \gamma \frac{Mz}{r^3} = \frac{4\pi\gamma\sigma a^3}{3} \frac{z}{(x^2 + z^2)^{\frac{3}{2}}}. \tag{2.40a}$$

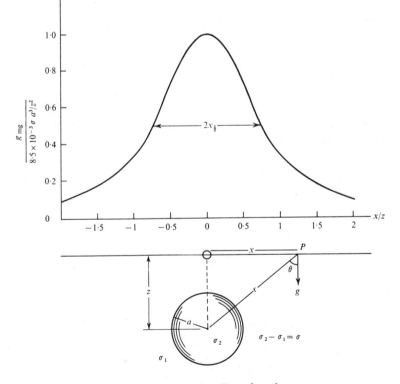

Fig. 2.22 Gravity effect of a sphere.

In practical units of milligals and feet, the units of γ and σ remaining the same, this can be modified to a convenient numerical relation:

$$g_{mg} = 8{\cdot}5 \times 10^{-3} \frac{\sigma a^3}{z^2(1 + x^2/z^2)^{\frac{3}{2}}}. \qquad (2.40b)$$

Note that z is the depth to the sphere centre, that the profile is symmetrical about the origin directly above the centre and, of course, is the same for any horizontal traverse through the origin. The maximum value of g, which occurs at the origin, is

$$g_{max} = 8{\cdot}5 \times 10^{-3}\sigma a^3/z^2.$$

The value of z can be found from the principal profile. When $g = g_{max}/2$, $z = 1{\cdot}3x_{\frac{1}{2}}$, where $x_{\frac{1}{2}}$ is the so-called *half-width* of the profile, that is, half its width at half-maximum value, We can also express the mass of the sphere in terms of $x_{\frac{1}{2}}$ and g_{max}. It is $M = 26 \, g_{max}(x_{\frac{1}{2}})^2$ in tons, when the units are milligals and feet.

Because of the relatively sharp profile ($g \propto 1/x^3$), it is unlikely that the spherical anomaly could be detected at all if the spacing between traverse lines were greater than twice the depth to its centre.

The spherical shape is particularly useful as a first approximation in the interpretation of three-dimensional anomalies which are approximately symmetrical.

(c) *Thin rod.* The geometry is shown in fig. 2.23, the rod having an inclination α and cross-section A. Then the gravity effect at P due to an element $d\ell$ is given by: $\delta g_r = \gamma\sigma A d\ell/r^2$. The vertical component is $\delta g = \gamma\sigma A \, d\ell \sin\theta/r^2$. Now we have the following relations in $\triangle OPB$: $r^2 = (\ell + z \cosec \alpha)^2 + x^2 + 2x$ $(\ell + z \cosec \alpha)\cos\alpha$, $\sin\theta = (\ell + z \cosec \alpha) \sin\alpha/r$, and the limits of ℓ are $0, L$. Then the total effect of the rod is

$$g = \gamma\sigma A \sin\alpha \int_0^L \frac{(\ell + z \cosec \alpha) \, d\ell}{\{(\ell + z \cosec \alpha)^2 + 2x(\ell + z \cosec \alpha) \cos\alpha + x^2\}^{\frac{3}{2}}}.$$

Thus

$$g = \gamma\sigma A \sin\alpha \left[\frac{\cos\alpha(\ell + z \cosec \alpha) + x}{x \sin^2\alpha\{(\ell + z \cosec \alpha)^2 + 2x(\ell + z \cosec \alpha) \cos\alpha + x^2\}^{\frac{1}{2}}} \right]_L^0.$$

With a little manipulation, this becomes

$$g = \frac{\gamma\sigma A}{x \sin\alpha} \left[\frac{x + z \cot\alpha}{(z^2 \cosec^2\alpha + 2xz \cot\alpha + x^2)^{\frac{1}{2}}} - \right.$$

$$\left. - \frac{x + z \cot\alpha + L \cos\alpha}{\{(L + z \cosec \alpha)^2 + x^2 + 2x(L \cos\alpha + z \cot\alpha)\}^{\frac{1}{2}}} \right]. \qquad (2.41a)$$

In practical units of milligals and feet, the multiplier in front of the main bracket is

$$2{\cdot}03 \times 10^{-3}\sigma A / x \sin \alpha. \tag{2.41b}$$

When the rod is vertical, this simplifies to

$$g = 2{\cdot}03 \times 10^{-3}\sigma A \left[\frac{1}{(z^2 + x^2)^{\frac{1}{2}}} - \frac{1}{\{(z + L)^2 + x^2\}^{\frac{1}{2}}}\right]. \tag{2.41c}$$

If L is very large, eq. (2.41a) becomes

$$g = \frac{2{\cdot}03 \times 10^{-3}\sigma A}{x \sin \alpha} \left[\frac{x + z \cot \alpha}{(z^2 \operatorname{cosec}^2 \alpha + 2xz \cot \alpha + x^2)^{\frac{1}{2}}} - \cos \alpha\right], \tag{2.41d}$$

and under the same conditions, eq. (2.41c) is

$$g = \frac{2{\cdot}03 \times 10^{-3}\sigma A}{(z^2 + x^2)^{\frac{1}{2}}}. \tag{2.41e}$$

The profiles in fig. 2.23 correspond to a thin rod dipping 45° and to a vertical rod. In both cases the flanks of the profiles slope less steeply than for the sphere. As one would expect, the curve for the dipping rod has the steeper slope on the up-dip side and the peak is displaced downdip.

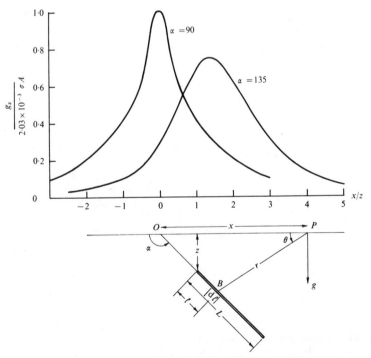

Fig. 2.23 Gravity effect of a thin rod dipping 45° ($\alpha = 135°$) and 90°.

If the rod dips steeply (i.e., practically vertical) we can get some idea of the depth z from the width of the profile. Assuming a vertical rod and L very large, $z = 0.58x_{\frac{1}{2}}$, where $x_{\frac{1}{2}}$ is the half-width of the profile at half-maximum, as defined previously. When $L = z$, $z \approx 1.05x_{\frac{1}{2}}$. Taking an average value of $0.75x_{\frac{1}{2}}$, the depth estimate would be in error by less than $\pm 30\%$, provided $L \geqslant z$. If the rod has a shallow dip, the depth cannot be estimated from the profile shape.

As we shall see later, the thin rod is considerably easier to analyse than a thick cylinder. As a rough rule, the thin rod may be used for a cylinder ($\leqslant 6\%$ error) when the diameter is less than the depth to the top. Clearly if the body outcrops it cannot be considered a thin rod, since g varies as $1/z$ in the equations above; in this case the true cylinder must be used.

(d) *Horizontal rod.* Figure 2.24a shows a horizontal rod of radius R, parallel to the y-axis and centred under the x-axis at a depth z. The rod is a horizontal distance x from the y-axis and a slant distance r_1. We calculate the vertical gravity at any point P on the y-axis by finding the total attraction of the rod at P, taking the

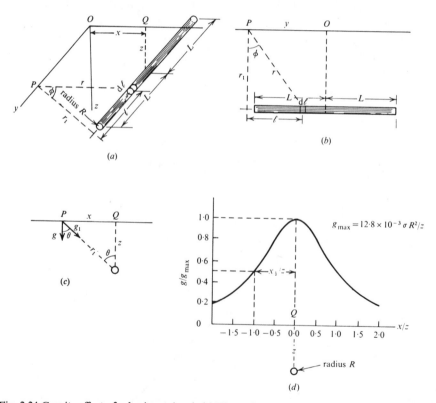

Fig. 2.24 Gravity effect of a horizontal rod. (a) Three-dimensional view; (b) projection on the plane containing the rod and the y-axis; (c) projection on the plane perpendicular to the rod (parallel to the xz-plane); (d) gravity profile along the x-axis.

component of this perpendicular to the rod and then finding the vertical component of this to get g (the component of the attraction parallel to the rod is horizontal and hence can be ignored).

Figure 2.24b shows the geometry in the plane containing the rod and the y-axis. The gravity effect at P due to the element $d\ell$ is

$$\delta g_r = \gamma \delta M/r^2 = \gamma \pi R^2 \sigma \, d\ell/r^2 = \gamma \pi R^2 \sigma(r_1 \, d\phi/\cos^2\phi)/r^2 = \gamma \pi R^2 \sigma \, d\phi/r_1.$$

The component along r_1 then becomes

$$\delta g_1 = \delta g_r \cos \phi = \gamma \pi R^2 \sigma \cos \phi \, d\phi/r_1,$$

and integrating between the limits $\tan^{-1}\{(y + L)/r_1\}$ and $\tan^{-1}\{(y - L)/r_1\}$, we get

$$g_1 = \frac{\gamma \pi R^2 \sigma}{r_1} \left[\frac{y + L}{\{(y + L)^2 + r_1^2\}^{\frac{1}{2}}} - \frac{y - L}{\{(y - L)^2 + r_1^2\}^{\frac{1}{2}}} \right].$$

Referring now to fig. 2.24c which represents a plane perpendicular to the rod, hence parallel to the xz-plane, we get for the vertical component

$$g = g_1 \cos \theta = \frac{\gamma \pi R^2 \sigma z}{r_1^2} \left[\frac{y + L}{\{(y + L)^2 + r_1^2\}^{\frac{1}{2}}} - \frac{y - L}{\{(y - L)^2 + r_1^2\}^{\frac{1}{2}}} \right]$$

$$= \frac{\gamma \pi R^2 \sigma}{z(1 + x^2/z^2)} \left[\frac{1}{\{1 + (x^2 + z^2)/(y + L)^2\}^{\frac{1}{2}}} - \frac{1}{\{1 + (x^2 + z^2)/(y - L)^2\}^{\frac{1}{2}}} \right],$$

$$(2.42a)$$

This can be expressed in milligals and feet exactly as was eq. (2.41b).

If the rod is of infinite length (practially, $L > 10z$), the original limits of integration would be $\pm\pi/2$. The expression for g becomes

$$g = \frac{2\gamma \pi R^2 \sigma}{z(1 + x^2/z^2)}, \qquad (2.42b)$$

which is the more familiar formula for gravity over a long horizontal cylinder. The profile is shown in fig. 2.24d. Note that z is the depth to the centre of the section and is given by $z = x_{\frac{1}{2}}$, if $L \geqslant 10z$. Thus there is no approximation involved in assuming the rod to be thin when it lies horizontal, since $z \geqslant R$. Because g varies as $1/x^2$, this profile is sharper than that of the dipping rod ($g \propto 1/x$) and broader than the sphere.

(e) *Thick vertical cylinder.* The gravity effect off-axis is not easily calculated except in special circumstances. However, the value on the axis, which is also the maximum value, is of considerable importance and is easily calculated.

First we find vertical g on the axis for a disc $d\ell$ thick (fig. 2.25a), by starting with an elementary ring of width dr. The mass of this ring is $\delta m = 2\pi \sigma r \, dr \, d\ell$, so that the gravity effect is $\delta g = 2\pi\gamma\sigma\,d\ell\sin\phi\,d\phi$. Integrating from 0 to $\tan^{-1}R/\ell$, we get for the disc

$$g_d = 2\pi\gamma\sigma\,d\ell\{1 - \ell/\sqrt{(\ell^2 + R^2)}\}.$$

For the whole cylinder we integrate with respect to ℓ from z to $z + L$:

$$g = 2\pi\gamma\sigma\int_z^{z+L}\{1 - \ell/\sqrt{(\ell^2 + R^2)}\}\,d\ell = 2\pi\gamma\sigma[L + \sqrt{(z^2 + R^2)} -$$

$$- \sqrt{\{(z + L)^2 + R^2\}}]. \qquad (2.43a)$$

In units of milligals and feet, this becomes

$$g = 12{\cdot}77 \times 10^{-3}\sigma[L + \sqrt{(z^2 + R^2)} - \sqrt{\{(z + L)^2 + R^2\}}]. \qquad (2.43b)$$

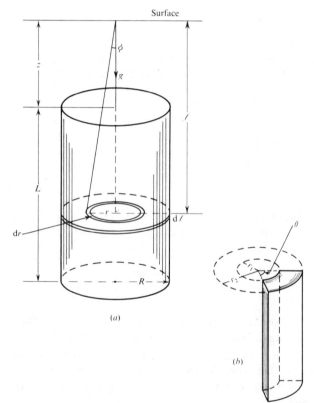

Fig. 2.25 Gravity effect of a vertical cylinder. (*a*) Calculation of gravity on the axis; (*b*) geometry of cylindrical slice.

There are several significant limiting cases of this formula.
(i) If $R \to \infty$, we have an infinite horizontal slab:

$$g = 12{\cdot}77 \times 10^{-3}\sigma L, \quad \text{or} \quad g/L = 0{\cdot}01277\sigma \text{ mgal/ft.} \qquad (2.43c)$$

This is the Bouguer correction, given in §2.3.2*d*. At first sight it seems surprising that *g* is independent of the depth of the slab and varies only with its thickness.

(ii) If we consider only a sector of the cylinder (see fig. 2.25*b*), the limits of integration for the disc are 0 to θ and $\tan^{-1} r_1/\ell$ to $\tan^{-1} r_2/\ell$. This modifies the expression for g_d:

$$g_d = \gamma\sigma\theta \, d\ell\{\ell/\sqrt{(r_1^2 + \ell^2)} - \ell/\sqrt{(r_2^2 + \ell^2)}\}.$$

As a result eq. (2.43*a*) becomes

$$g = \gamma\sigma\theta[\sqrt{(z^2 + r_2^2)} - \sqrt{(z^2 + r_1^2)} + \sqrt{\{(z + L)^2 + r_1^2\}} -$$
$$- \sqrt{\{(z + L)^2 + r_2^2\}}].$$

Now putting $z = 0$, we have

$$g = \gamma\sigma\theta\{(r_2 - r_1) + \sqrt{(r_1^2 + L^2)} - \sqrt{(r_2^2 + L^2)}\}. \qquad (2.43d)$$

This is the terrain correction (see eq. (2.19*a*) in §2.3.2*e*), where *L*, the depth of the sector, corresponds to the difference between the height of the station ($z = r = 0$) and the average elevation in the sector.

(iii) When $z = 0$, the cylinder outcrops and eq. (2.43*b*) becomes

$$g = 12 \cdot 77 \times 10^{-3}\sigma\{L + R - \sqrt{(L^2 + R^2)}\}. \qquad (2.43e)$$

(iv) If $L \rightarrow \infty$, eq. (2.43*b*) becomes

$$g = 12 \cdot 77 \times 10^{-3}\sigma\{\sqrt{(z^2 + R^2)} - z\}, \qquad (2.43f)$$

and if, in addition, $z = 0$, we have

$$g = 12 \cdot 77 \times 10^{-3}\sigma R, \qquad (2.43g)$$

which gets around the difficulty posed by the thin rod.

When the cylinder length is considerably larger than the depth to the top ($L \gg z$), we can used eq. (2.43*f*) to compute the gravity off-axis using well-known methods of solving Laplace's equation (see, for example, Pipes (1958, pp. 484–6) and MacRobert (1948, pp. 151–5)). Since *g* satisfies Laplace's equation, we can express it for $r > z > R$ in a series of Legendre polynomials of the form

$$g(r, \theta) = k \sum_{n=0}^{\infty} b_n r^{-(n+1)} P_n(\cos \theta),$$

where $k = 12 \cdot 77 \times 10^{-3}\sigma$, b_n are coefficients, $P_n(\cos \theta)$ is the Legendre polynomial, $r^2 = x^2 + z^2$, and $\tan \theta = x/z$ (see fig. 2.26). On the axis $\theta = 0$, $r = z$, and the series reduces to

$$g = k(b_0 P_0/z + b_1 P_1/z^2 + b_2 P_2/z^3 + b_3 P_3/z^4 + \ldots)$$
$$= k(b_0/z + b_1/z^2 + b_2/z^3 + b_3/z^4 + \ldots),$$

where P_0, P_1, etc., are the Legendre polynomials for $\theta = 0$, all having the value $+1$.

This result must be the same as that given by eq. (2.43*f*); expanding this equation in terms of R/z, we get

$$g = k(R^2/2z - R^4/8z^3 + R^6/16z^5 - 5R^8/128z^7 + \ldots).$$

Equating the coefficients of the last two series we see that $b_n = 0$ when n is odd, also $b_0 = R^2/2$, $b_2 = -R^4/8$, $b_4 = R^6/16$, $b_6 = -5R^8/128, \ldots$ Thus the expression for $g(r, \theta)$ for off-axis points becomes

$$g(r, \theta) = 12\cdot77 \times 10^{-3}\sigma R \left\{ \frac{1}{2}\left(\frac{R}{r}\right) - \frac{1}{8}\left(\frac{R}{r}\right)^3 P_2(\mu) + \frac{1}{16}\left(\frac{R}{r}\right)^5 P_4(\mu) - \right.$$
$$\left. - \frac{5}{128}\left(\frac{R}{r}\right)^7 P_6(\mu) + \ldots \right\}, \qquad (2.44a)$$

where $\mu = \cos\theta$.

The above can be rewritten to resemble eq. (2.41*c*) for the long thin rod. By inserting $r = \sqrt{(x^2 + z^2)}$, we get

$$g(r, \theta) = 6\cdot4 \times 10^{-3}\sigma R^2 \left\{ \frac{1}{(x^2 + z^2)^{\frac{1}{2}}} - \frac{R^2 P_2(\mu)}{4(x^2 + z^2)^{\frac{3}{2}}} + \right.$$
$$\left. + \frac{R^4 P_4(\mu)}{16(x^2 + z^2)^{\frac{5}{2}}} - \ldots \right\} \qquad (2.44b)$$

This is a more precise result than eq. (2.41*c*), although the difference between the two is negligible if $z \geqslant 2R$. Obviously the thin rod relation is much simpler to apply.

A more useful result for the same thick cylinder can be developed when $z < R$, since in this case the rod approximation is not valid. We expand eq. (2.43*f*) in terms of z/R rather than R/z, to get

$$g = 12\cdot77 \times 10^{-3}\sigma R(1 - z/R + z^2/2R^2 - z^4/8R^4 + z^6/16R^6 - \ldots).$$

Within the limits $z \leqslant r \leqslant R$ we can equate this series to an off-axis series:

$$g(r, \theta) = k \sum_{m=0}^{\infty} a_m r^m P_m(\mu) = k\{a_0 + a_1 rP_1(\mu) + a_2 r^2 P_2(\mu) + $$
$$+ a_3 r^3 P_3(\mu) + \ldots\}.$$

Equating coefficients on the axis ($r = z$, $\theta = 0$), we find that

$$a_0 = R, a_1 = -1, a_2 = 1/2R, a_3 = a_5 = \ldots$$
$$= a_{2n+1} = 0, a_4 = -1/8R^3, \ldots$$

Thus, for points off the z-axis the expression becomes for $z \leqslant r \leqslant R$

$$g(r, \theta) = 2\pi\gamma\sigma R \left\{ 1 - \left(\frac{r}{R}\right) P_1(\mu) + \frac{1}{2}\left(\frac{r}{R}\right)^2 P_2(\mu) - \frac{1}{8}\left(\frac{r}{R}\right)^4 P_4(\mu) + \ldots \right\}.$$

$$(2.45a)$$

which reduces to (2.43*f*) when $r = z$.

If R is intermediate between r and z, that is, $r > R > z$, we must use a different series which turns out to be identical in form with eq. (2.44a); writing $g'(r, \theta)$ to avoid confusion with $g(r, \theta)$ of eq. (2.45a), we have

$$g'(r, \theta) = 2\pi\gamma\sigma R \left\{ \frac{1}{2} \left(\frac{R}{r}\right) - \frac{1}{8} \left(\frac{R}{r}\right)^3 P_2(\mu) + \frac{1}{16} \left(\frac{R}{r}\right)^5 P_4(\mu) - \ldots \right\}.$$

$$(2.45b)$$

The two expressions must match for $r = R$, $z < R$; for $z = R/2$, $r = R$, $\theta = \tan^{-1}x/z = \tan^{-1}[\sqrt{(r^2 - z^2)}/z] = \tan^{-1}\sqrt{3} = 60°$, we find on carrying out the expansions to $P_{12}(\cos 60°)$ that $g(r, \theta) = g'(r, \theta) = 2\pi\gamma\sigma R \times 0.486$. The profile is shown in fig. 2.26, the quantity plotted being the normalized value, that is, $g'(r, \theta)$ in eq. (2.45b) divided by the value of $g'(r, \theta)$ for $\theta = 0$, or $g'(r, \theta)/2\pi\gamma\sigma R$.

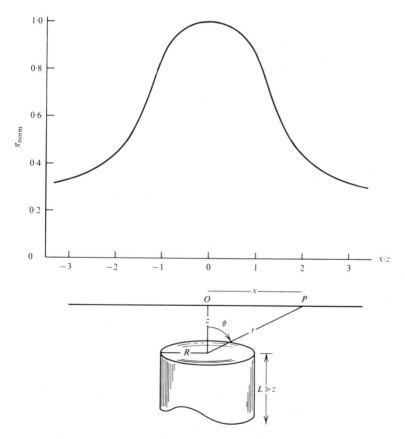

Fig. 2.26 Gravity effect off the axis of a thick vertical cylinder. $g_{\text{norm}} = g'(r, \theta)/2\pi\gamma\sigma R$ (see eq. (2.45b)).

66   Gravity method

(f) *Two-dimensional bodies.* The geometrical bodies considered this far have been three-dimensional. Considerable simplification is possible when a body can be considered two-dimensional; in general a body can be considered two-dimensional when its strike length is about twenty times all the other dimensions, including the depth below the surface, and the cross-section is the same at all points along the strike. If a body has a uniform cross-section in the direction of strike but the strike length is finite, say, $2L$, instead of being effectively infinite, the two-dimensional formulae can be modified as in §2.2.3c to give the correct results for the finite length along strike.

(g) *Thin dipping sheet.* As in the approximation of a thin rod for a true cylinder, it is practical to substitute a thin sheet for a true prism, since the expression is less complicated. Referring to fig. 2.27, we have the following geometrical relations:

$$p = (x - h \cot \alpha) \sin \alpha = x \sin \alpha - h \cos \alpha, \quad r = p \sec \theta,$$

$$z = r \sin \left(\alpha + \theta - \frac{\pi}{2} \right) = p(\sin \alpha \tan \theta - \cos \alpha), \quad r_1 = \sqrt{(x^2 + h^2)},$$

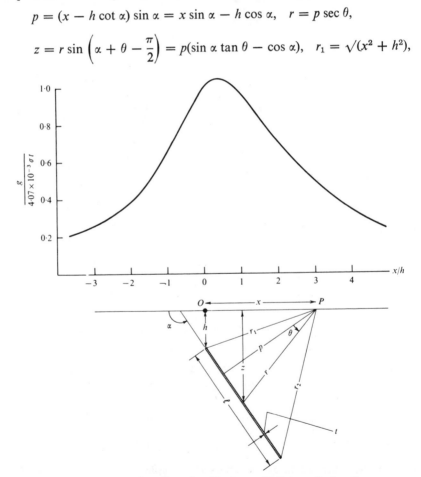

Fig. 2.27 Gravity effect of a thin sheet of infinite strike length.

$$dz = p \sin \alpha \sec^2 \theta \, d\theta \, , \quad r_2 = \sqrt{\{(x + \ell \cos \alpha)^2 + (h + \ell \sin \alpha)^2\}},$$

$$dx = \operatorname{cosec} \alpha \, dt, \quad \tan \theta_1 = \sqrt{(r_1{}^2 - p^2)}/p = (x \cos \alpha + h \sin \alpha)/p,$$

$$\tan \theta_2 = \sqrt{(r_2{}^2 - p^2)}/p = (x \cos \alpha + \ell + h \sin \alpha)/p.$$

Now using eq. (2.7), §2.2.3c, for a two-dimensional structure (dropping the minus sign as explained in the note following eq. (2.15)) we get

$$g = 2\gamma\sigma \int_x \int_z \frac{z}{r^2} \, dx \, dz = 2\gamma\sigma \int_t \int_\theta \frac{(\sin \alpha \tan \theta - \cos \alpha) \operatorname{cosec} \alpha \sin \alpha \sec^2 \theta}{\sec^2 \theta} \, dt \, d\theta$$

$$= 2\gamma\sigma t \int_{\theta_1}^{\theta_2} (\sin \alpha \tan \theta - \cos \alpha) \, d\theta = 2\gamma\sigma t \left| \sin \alpha \log \sec \theta - \theta \cos \alpha \right|_{\theta_1}^{\theta_2}.$$

Inserting the appropriate values of θ_1 and θ_2, this becomes

$$g = 2\gamma\sigma t \left[\tfrac{1}{2} \sin \alpha \log \left\{ \frac{(h + \ell \sin \alpha)^2 + (x + \ell \cos \alpha)^2}{(x^2 + h^2)} \right\} \right.$$

$$\left. - \cos \alpha \left\{ \tan^{-1} \left(\frac{h \sin \alpha + \ell + x \cos \alpha}{x \sin \alpha - h \cos \alpha} \right) - \tan^{-1} \left(\frac{h \sin \alpha + x \cos \alpha}{x \sin \alpha - h \cos \alpha} \right) \right\} \right].$$

$$(2.46a)$$

To convert to milligals and feet, we replace 2γ with $4 \cdot 07 \times 10^{-3}$.

If the sheet is vertical, eq. (2.46a) simplifies to

$$g = 2 \cdot 03 \times 10^{-3} \sigma t \log \left\{ \frac{(h + \ell)^2 + x^2}{x^2 + h^2} \right\}. \qquad (2.46b)$$

When the sheet is not sufficiently elongated to be considered two-dimensional, let the strike length be $2Y$. It is necessary to put finite limits on the integral in the y-axis (see §2.2.3c dealing with logarithmic potential); the limits $\pm \infty$ become $\pm Y$. The equation for a profile at $y = Y$, perpendicular to strike, then becomes

$$g = 4 \cdot 07 \times 10^{-3} \sigma t \left[\tfrac{1}{2} \sin \alpha \log \left\{ \frac{\{(h + \ell \sin \alpha)^2 + (x + \ell \cos \alpha)^2 + Y^2\}^{\frac{1}{2}} - Y}{\{(h + \ell \sin \alpha)^2 + (x + \ell \cos \alpha)^2 + Y^2\}^{\frac{1}{2}} + Y} \times \right. \right.$$

$$\left. \times \frac{(x^2 + h^2 + Y^2)^{\frac{1}{2}} + Y}{(x^2 + h^2 + Y^2)^{\frac{1}{2}} - Y} \right\}$$

$$- \cos \alpha \tan^{-1} \left\{ \frac{Y(h \sin \alpha + \ell + x \cos \alpha)}{\{(h + \ell \sin \alpha)^2 + (x + \ell \cos \alpha)^2 + Y^2\}^{\frac{1}{2}}(x \sin \alpha - h \cos \alpha)} \right\}$$

$$\left. + \cos \alpha \tan^{-1} \left\{ \frac{Y(h \sin \alpha + x \cos \alpha)}{(x^2 + h^2 + Y^2)^{\frac{1}{2}}(x \sin \alpha - h \cos \alpha)} \right\} \right].$$

$$(2.46c)$$

The thin sheet is a very good approximation to a prism section unless the thickness is somewhat greater than h, the depth to the top. When the dip is steep ($\geqslant 60°$) the depth h can be roughly estimated from the half-width of the profile in limiting cases. For example, when $h \approx \ell$, $h \approx 0.7x_{\frac{1}{2}}$ and when $h \gg \ell$, $h \approx x_{\frac{1}{2}}$. However, when ℓ is large or when the dip angle is small, it is not possible to get a reliable estimate.

(h) *Horizontal thin sheet.* When the sheet lies horizontal, eq. (2.46a) reduces to

$$g = 4.07 \times 10^{-3}\sigma t\{\tan^{-1}(\ell - x)/h + \tan^{-1}(x/h)\}, \qquad (2.47a)$$

where h is now the depth to the central axis of the sheet. If, in addition, $\ell \to \infty$, we have, for a semi-infinite horizontal sheet,

$$g = 4.07 \times 10^{-3}\sigma t\{\pi/2 + \tan^{-1}(x/h)\}, \qquad (2.47b)$$

and if the slab extends to infinity in the other direction (i.e., x goes to infinity as well) we have the Bouguer correction, as in eq. (2.43c), with t replacing h.

The profile for a semi-infinite horizontal sheet is shown in fig. 2.28. As one would

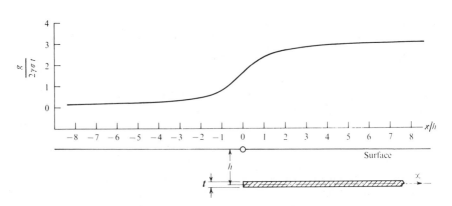

Fig. 2.28 Gravity effect of a semi-infinite horizontal sheet.

expect, this is an asymmetrical profile. From eq. (2.47b) we can obtain the following relations:

$$\text{as } x \to -\infty, g_{-\infty} \to 0; \quad \text{as } x \to +\infty, g_{+\infty} \to 12.77 \times 10^{-3}\sigma t;$$
$$\text{when } x = 0, g_0 \approx 6.38 \times 10^{-3}\sigma t.$$

Also,

$$dg/dx = 4.07 \times 10^{-3}\sigma th/(h^2 + x^2).$$

When $x = 0$,

$$(dg/dx)_0 = 4.07 \times 10^{-3}\sigma t/h.$$

Hence we can calculate σt and h as follows:

$$\Delta g_{max} = g_{-\infty} + g_{+\infty} = 12{\cdot}77 \times 10^{-3}\sigma t,$$

$$\sigma t = \Delta g_{max}/12.77 \times 10^{-3}, \qquad (2.47c)$$

$$h = \Delta g_{max}/\pi(dg/dx)_0. \qquad (2.47d)$$

The thin sheet approximation is correct within 2% if $h \geqslant 2t$.

(i) *Fault approximation.* A fault structure can be approximated by two semi-infinite horizontal sheets, one displaced vertically from the other, as in figs. 2.29a, and 2.29b. The geometrical relations for the two sheets are:

$$r_1 = h_1 \operatorname{cosec} \theta_1, \qquad\qquad r_2 = h_2 \cos \theta_2,$$

$$x - (\ell_1 - h_1 \cot \alpha) = h_1 \cot \theta_1, \qquad x + (\ell_2 + h_2 \cot \alpha) = h_2 \cot \theta_2,$$

$$d\ell_1 = h_1 \operatorname{cosec}^2 \theta_1 \, d\theta_1, \qquad\qquad d\ell_2 = -h_2 \operatorname{cosec}^2 \theta_2 \, d\theta_2,$$

$$\cot^{-1}(x + h_1 \cot \alpha)/h_1 \leqslant \theta_1 \leqslant \pi, \qquad \cot^{-1}(x + h_2 \cot \alpha)/h_2 \leqslant \theta_2 \leqslant 0.$$

Then the vertical gravity effect of the two elements $d\ell_1$ and $d\ell_2$ at P is

$$\delta g = 2\gamma\sigma t(\sin \theta_1 \, d\ell_1/r_1 + \sin \theta_2 \, d\ell_2/r_2) = 2\gamma\sigma t(d\theta_1 - d\theta_2),$$

and integrating with respect to θ_1 and θ_2, we have

$$g = 2\gamma\sigma t \left\{ \int_{\theta_a}^{\pi} d\theta_1 - \int_{\theta_b}^{0} d\theta_2 \right\} = 2\gamma\sigma t\{\pi - \cot^{-1}(x + h_1 \cot \alpha)/h_1 +$$

$$+ \cot^{-1}(x + h_2 \cot \alpha)h_2\}, \quad (2.48a)$$

where $\theta_a = \cot^{-1}(x + h_1 \cot \alpha)/h_1$, $\theta_b = \cot^{-1}(x + h_2 \cot \alpha)/h_2$. Writing this in a slightly different form by means of the identity $(\tan^{-1} A + \cot^{-1} A) = \pi/2$,

$$g = 2\gamma\sigma t[\pi + \tan^{-1}\{(x/h_1) + \cot \alpha\} - \tan^{-1}\{(x/h_2) + \cot \alpha\}]. \quad (2.48b)$$

In milligals and feet, the multiplier becomes $4{\cdot}07 \times 10^{-3}\sigma t$.

If the fault is vertical, this simplifies to give

$$g = 4{\cdot}07 \times 10^{-3}\sigma t(\pi + \tan^{-1} x/h_1 - \tan^{-1} x/h_2). \quad (2.48c)$$

Three profiles are shown in figs. 2.29a and 2.29b, two for $\alpha = \pm 30°$, the other 90°, with $\sigma = 1$ g/cm³, $t = 1000$ ft, $h_1 = 2500$ ft, $h_2 = 4500$ ft. As in the case of the single sheet, the profiles are asymmetrical; in addition, they are increasingly lopsided as the dip departs from 90°. From eq. (2.48b) we have:

As $x \to \pm \infty$, $g \to 2\pi\gamma\sigma t$; when $x = 0$, $g_0 = 2\pi\gamma\sigma t$,

that is, the background gravity and the value at the origin are the same, which could be predicted from consideration of the infinite slab used for Bouguer correction. (It should be noted that the vertical scales for g in figs. 2.29a and 2.29b are

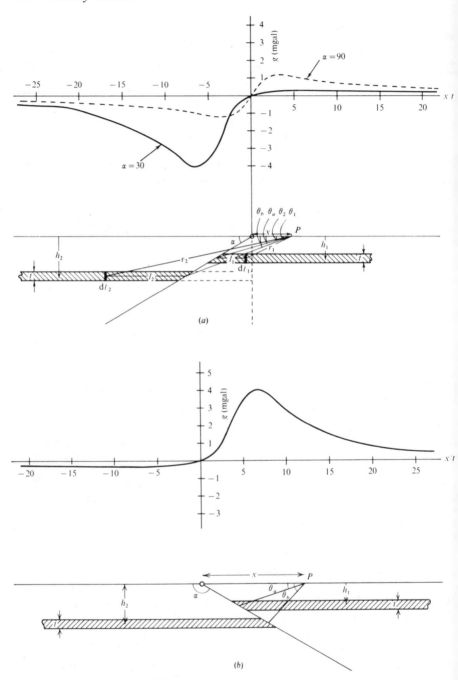

Fig. 2.29 Gravity effect of a faulted horizontal sheet: $t = 1000$ ft, $h_1 = 2500$ ft, $h_2 = 4500$ ft, $\sigma = 1$ g/cm^3. (*a*) Normal fault, $\alpha = 30°$, $90°$; (*b*) reverse fault, $\alpha = 150°$ ($-30°$).

actually $(g - 2\pi\gamma\sigma t)$; the constant term $2\pi\gamma\sigma t$ cannot be observed in a field measurement.) The maximum and minimum are obtained from the zero slope:

$$\frac{dg}{dx} = 2\gamma\sigma t \left\{ \frac{h_1}{(x + h_1 \cot \alpha)^2 + h_1^2} - \frac{h_2}{(x + h_2 \cot \alpha)^2 + h_2^2} \right\} = 0.$$

Thus, $x_m = \pm\sqrt{(h_1 h_2)} \operatorname{cosec} \alpha$ where x_m = horizontal distance from the origin to the maximum and minimum values. The slope at the origin is

$$\left(\frac{dg}{dx}\right)_0 = 2\gamma\sigma t \left(\frac{1}{h_1 \operatorname{cosec}^2 \alpha} - \frac{1}{h_2 \operatorname{cosec}^2 \alpha}\right) = \frac{2\gamma\sigma t (h_2 - h_1)}{x_m^2},$$

which allows us to calculate σt and $(h_2 - h_1)$ from the relations:

$$\sigma t = g_0/12 \cdot 77 \times 10^{-3}, \tag{2.48d}$$

$$(h_2 - h_1) = (\pi x_m^2/g_0)(dg/dx)_0. \tag{2.48e}$$

It is not possible in general to determine α uniquely, although the ratio g_{max}/g_{min}, the maximum positive to maximum negative gravity, is diagnostic of the dip angle. This ratio varies from 1 for $\alpha = 90°$ to about $0\cdot06$ when $\alpha = 15°$, for a normal fault, and from 1 to 16 over the same dip angles for a reverse fault. Having estimated α in this crude fashion, one can then obtain rough values for h_1 and h_2. Of course, when the fault is vertical, $|g_{max}| = |g_{min}|$ and h_1, h_2 may be determined easily, at least in theory. Typical field situations, however, are very rarely as simple as these theoretical expressions would indicate.

The actual maximum and minimum values are:

$$g_{max} = 4\cdot07 \times 10^{-3}\sigma t \left\{ \pi + \tan^{-1} \left(\frac{\sqrt{h_2} \operatorname{cosec} \alpha + \sqrt{h_1} \cot \alpha}{\sqrt{h_1}}\right) \right.$$
$$\left. - \tan^{-1} \left(\frac{\sqrt{h_1} \operatorname{cosec} \alpha + \sqrt{h_2} \cot \alpha}{\sqrt{h_2}}\right) \right\}$$

$$g_{min} = 4\cdot07 \times 10^{-3}\sigma t \left\{ \pi - \tan^{-1} \left(\frac{\sqrt{h_2} \operatorname{cosec} \alpha - \sqrt{h_1} \cot \alpha}{\sqrt{h_1}}\right) \right.$$
$$\left. + \tan^{-1} \left(\frac{\sqrt{h_1} \operatorname{cosec} \alpha - \sqrt{h_2} \cot \alpha}{\sqrt{h_2}}\right) \right\}$$

$$\Delta g = g_{max} - g_{min} = 4\cdot07 \times 10^{-3}\sigma t \left(\tan^{-1} \frac{\sqrt{h_2} \operatorname{cosec} \alpha + \sqrt{h_1} \cot \alpha}{\sqrt{h_1}} + \right.$$
$$+ \tan^{-1} \frac{\sqrt{h_2} \operatorname{cosec} \alpha - \sqrt{h_1} \cot \alpha}{\sqrt{h_1}} - \tan^{-1} \frac{\sqrt{h_1} \operatorname{cosec} \alpha + \sqrt{h_2} \cot \alpha}{\sqrt{h_2}} -$$
$$\left. - \tan^{-1} \frac{\sqrt{h_1} \operatorname{cosec} \alpha - \sqrt{h_2} \cot \alpha}{\sqrt{h_2}} \right). \tag{2.48f}$$

For $\alpha = 90°$, this becomes

$$\Delta g = 8\cdot14 \times 10^{-3}\sigma t\{\tan^{-1}\sqrt{(h_2/h_1)} - \tan^{-1}\sqrt{(h_1/h_2)}\}. \qquad (2.48g)$$

(j) *Thick prism*. The thin sheet approximation breaks down when the depth to the top is small compared to the sheet thickness; for such bodies we must consider a true prism shape. The geometry is illustrated in fig. 2.30. We can use an expression similar to eq. (2.47a) for the horizontal thin sheet at depth z, provided the origin 0

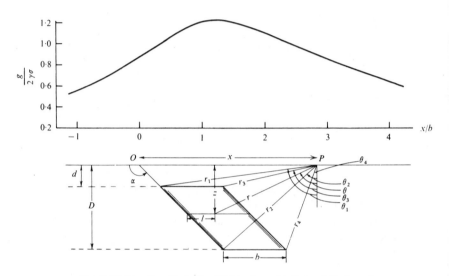

Fig. 2.30 Gravity effect of a thick prism of infinite strike length.

in fig. 2.27 is moved to point 0 in fig. 2.30, the width ℓ in fig. 2.27 is changed to b, and h and t are replaced by z and dz. Then,

$$\delta g = 2\gamma\sigma \, \mathrm{d}z \left\{\cot^{-1}\left(\frac{x - b + z\cot\alpha}{z}\right) - \cot^{-1}\left(\frac{x + z\cot\alpha}{z}\right)\right\}.$$

The gravity for the complete prism is given by

$$g = 2\gamma\sigma \int_d^D \left\{\cot^{-1}\left(\frac{x - b + z\cot\alpha}{z}\right) - \cot^{-1}\left(\frac{x + z\cot\alpha}{z}\right)\right\}\mathrm{d}z.$$

The integration can be carried out by making a change of variable and intregating by parts. The result is

$$g = 2\gamma\sigma \left[\tfrac{1}{2}x\sin^2\alpha \, . \, \log\left\{\frac{D^2 + (x + D\cot\alpha)^2}{D^2 + (x - b + D\cot\alpha)^2} \, \frac{d^2 + (x - b + d\cot\alpha)^2}{d^2 + (x + d\cot\alpha)^2}\right\}\right.$$

$$+ \tfrac{1}{2}b\sin^2\alpha \cdot \log\left\{\frac{D^2 + (x - b + D\cot\alpha)^2}{d^2 + (x - b + d\cot\alpha)^2}\right\} - x\sin\alpha\cos\alpha\left\{\tan^{-1}\left(\frac{x - b}{D} + \right.\right.$$

$$\left.+ \cot\alpha\right) - \tan^{-1}\left(\frac{x}{D} + \cot\alpha\right)$$

$$- \tan^{-1}\left(\frac{x - b}{d} + \cot\alpha\right) + \tan^{-1}\left(\frac{x}{d} + \cot\alpha\right)\right\} +$$

$$+ b\sin\alpha\cos\alpha\left\{\tan^{-1}\left(\frac{x - b}{D} + \cot\alpha\right)\right.$$

$$\left.- \tan^{-1}\left(\frac{x - b}{d} + \cot\alpha\right)\right\} - D\left\{\tan^{-1}\left(\frac{x - b}{D} + \cot\alpha\right) - \right.$$

$$\left.- \tan^{-1}\left(\frac{x}{D} + \cot\alpha\right)\right\}$$

$$- d\left\{\tan^{-1}\left(\frac{x - b}{d} + \cot\alpha\right) - \tan^{-1}\left(\frac{x}{d} + \cot\alpha\right)\right\}\bigg]. \qquad (2.49a)$$

By using the following relations,

$$r_1 = \sqrt{\{d^2 + (x + d\cot\alpha)^2\}}, \quad r_2 = \sqrt{\{D^2 + (x + D\cot\alpha)^2\}},$$

$$r_3 = \sqrt{\{d^2 + (x - b + d\cot\alpha)^2\}} \quad r_4 = \sqrt{\{D^2 + (x - b + D\cot\alpha)^2\}},$$

$$\tan\theta_1 = (x + d\cot\alpha)/d, \quad \tan\theta_2 = (x + D\cot\alpha)/D,$$

$$\tan\theta_3 = (x - b + d\cot\alpha)/d, \quad \tan\theta_4 = (x - b + D\cot\alpha)/D,$$

and noting that the angles, $\alpha, \theta_1, \ldots, \theta_4$, are measured in the clockwise sense, the expression can be simplified somewhat to give

$$g = 4 \cdot 07 \times 10^{-3}\sigma\left[\sin^2\alpha\left\{x\log\left(\frac{r_2 r_3}{r_1 r_4}\right) + b\log\left(\frac{r_4}{r_3}\right)\right\} - \right.$$

$$- \sin\alpha\cos\alpha\{x(\overline{\theta_4 - \theta_3} - \overline{\theta_2 - \theta_1}) - b(\theta_4 - \theta_3)\} -$$

$$\left.- D(\theta_4 - \theta_2) + d(\theta_3 - \theta_1)\right]. \qquad (2.49b)$$

If the sides of the prism are vertical, $\alpha = \pi/2$, and (2.49a) becomes

$$g = 2\gamma\sigma\left[\frac{x}{2}\log\left\{\frac{D^2 + x^2}{d^2 + x^2}\frac{d^2 + (x - b)^2}{D^2 + (x - b)^2}\right\} + \frac{b}{2}\log\left\{\frac{D^2 + (x - b)^2}{d^2 + (x - b)^2}\right\} - \right.$$

$$\left.- D\left\{\tan^{-1}\left(\frac{x - b}{D}\right) - \tan^{-1}\left(\frac{x}{D}\right)\right\} + d\left\{\tan^{-1}\left(\frac{x - b}{d}\right) - \tan^{-1}\frac{x}{d}\right\}\right]$$

$$(2.49c)$$

$$= 4 \cdot 07 \times 10^{-3} \sigma \left[x \log \left(\frac{r_2 \, r_3}{r_1 \, r_4} \right) + b \log \left(\frac{r_4}{r_3} \right) - D(\theta_4 - \theta_2) + d(\theta_3 - \theta_1) \right].$$

$$(2.49d)$$

If the prism outcrops, $d = 0$ and eq. (2.49a) is

$$g = 2\gamma\sigma \left[\frac{x}{2} \sin^2 \alpha \cdot \log \left\{ \frac{(x - b)^2}{x^2} \frac{D^2 + (x + D \cot \alpha)^2}{D^2 + (x - b + D \cot \alpha)^2} \right\} + \right.$$

$$+ \frac{b}{2} \sin^2 \alpha \cdot \log \left\{ \frac{D^2 + (x - b + D \cot \alpha)^2}{(x - b)^2} \right\} -$$

$$- x \sin \alpha \cos \alpha \left\{ \tan^{-1} \left(\frac{x - b}{D} + \cot \alpha \right) - \tan^{-1} \left(\frac{x}{D} + \cot \alpha \right) \right\} -$$

$$- b \sin \alpha \cos \alpha \left\{ \frac{\pi}{2} - \tan^{-1} \left(\frac{x - b}{D} + \cot \alpha \right) \right\} -$$

$$\left. - D \left\{ \tan^{-1} \left(\frac{x - b}{D} + \cot \alpha \right) - \tan^{-1} \left(\frac{x}{D} + \cot \alpha \right) \right\} \right]. \quad (2.49e)$$

When $\alpha = 90°$, this is further simplified to:

$$g = 2\gamma\sigma \left[\frac{x}{2} \log \left\{ \frac{(x - b)^2}{x^2} \frac{D^2 + x^2}{D^2 + (x - b)^2} \right\} + \frac{b}{2} \log \left\{ \frac{D^2 + (x - b)^2}{(x - b)^2} \right\} - \right.$$

$$\left. - D \left\{ \tan^{-1} \left(\frac{x - b}{D} \right) - \tan^{-1} \left(\frac{x}{D} \right) \right\} \right]. \quad (2.49f)$$

An estimate of d, the depth to the top of the body, is not very satisfactory for the thick prism. When $d = b$, we find that $d = 0 \cdot 67x_{\frac{1}{2}}$ for $D = 2b$; whereas $d = 0 \cdot 33x_{\frac{1}{2}}$ when $D = 10b$, that is, a factor of two depending on the depth extent. As the depth of cover becomes still smaller the half-value criterion becomes quite unreliable. Moreover, it is impossible to make a good estimate of the width of the prism from the shape of the curve. In general the profiles become sharper as both D and d get smaller.

Various additional structures can be derived from limiting cases of the thick prism. These are considered in the following sections.

(k) *Semi-infinite horizontal slab.* If the width, b, of the prism is very great, $r_3 \approx r_4 \approx \infty$, $\theta_3 \approx \theta_4 \approx -\pi/2$. Equation (2.49b) now becomes

$$g = 4 \cdot 07 \times 10^{-3} \sigma \{ x \sin^2 \alpha \log (r_2/r_1) + x(\theta_2 - \theta_1) \sin \alpha \cos \alpha +$$

$$+ D(\pi/2 + \theta_2) - d(\pi/2 + \theta_1) \}. \quad (2.50a)$$

If, in addition, $\alpha = 90°$, this can be written

$$g = 4\cdot07 \times 10^{-3}\sigma \left[\frac{x}{2} \log \left(\frac{D^2 + x^2}{d^2 + x^2}\right) + D\left\{\frac{\pi}{2} + \tan^{-1}\left(\frac{x}{D}\right)\right\} - \right.$$

$$\left. - d\left\{\frac{\pi}{2} + \tan^{-1}\left(\frac{x}{d}\right)\right\}\right]. \qquad (2.50b)$$

Again, if the slab outcrops, $d = 0$, $r_1 = x$, $\theta_1 = \pi/2$, and eq. (2.50a) becomes

$$g = 4\cdot07 \times 10^{-3}\sigma \left[\frac{x}{2} \sin^2 \alpha \log \left\{\frac{D^2 + (x + D \cot \alpha)^2}{x^2}\right\} - \right.$$

$$- x \sin \alpha \cos \alpha \left\{\frac{\pi}{2} - \tan^{-1}\left(\frac{x}{D} + \cot \alpha\right)\right\} +$$

$$\left. + D\left\{\frac{\pi}{2} + \tan^{-1}\left(\frac{x}{D} + \cot \alpha\right)\right\}\right]. \qquad (2.50c)$$

If $\alpha = 90°$, eq. (2.50c) is further simplified to give

$$g = 4\cdot07 \times 10^{-3}\sigma \left[\frac{x}{2} \log \left(\frac{D^2 + x^2}{x^2}\right) + D\left\{\frac{\pi}{2} + \tan^{-1}\left(\frac{x}{D}\right)\right\}\right]. \qquad (2.50d)$$

When thick structures are not truly two-dimensional, the formulae can be modified exactly as in eq. (2.46c) for the thin sheet. Also, it is possible to calculate values of depth and thickness as in eqs. (2.47c) and (2.47d). For example, from (2.50b) we have

$$\Delta g_{max} = g_{+\infty} - g_{-\infty} = 4\cdot07 \times 10^{-3}\sigma\pi(D - d),$$

$$\left(\frac{dg}{dx}\right)_{x=0} = 4\cdot07 \times 10^{-3}\sigma \log(D/d).$$

From these relations we obtain the following:

$$\sigma(D - d) = \Delta g_{max}/12\cdot77 \times 10^{-3}, \qquad (2.50e)$$

$$\log(D/d) = \frac{1}{4\cdot07 \times 10^{-3}\sigma}\left(\frac{dx}{dg}\right)_{x=0} \qquad (2.50f)$$

By assuming a value for σ, we can solve for D and d.

(l) *The fault*. Obviously one can combine a pair of the above slabs to make a true fault. This results in a long cumbersome formula, but it can be shortened to the more convenient form (Geldart *et al.* 1966):

$$g = 4\cdot07 \times 10^{-3}\sigma[\pi t + x \cos^2 \beta\{(F_2 - F_1) - (F_4 - F_3)\}], \qquad (2.51a)$$

where $F_i = \psi_i \cot \psi_i - \log \sin \psi_i$; $i = 1, 2, 3, 4$; $t = $ thickness of slab; $\psi_i = (\theta_i - \beta)$; $\tan \theta_i = (x/z_i) + \tan \beta$; $\beta = (\pi/2 - \alpha)$. The geometry is shown in fig. 2.31, together with a typical profile.

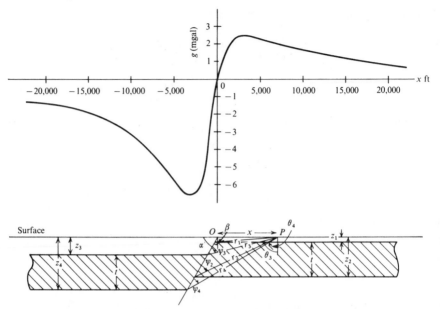

Fig. 2.31 Gravity effect of a faulted thick horizontal bed: $t = 4000$ ft, $z_1 = 500$ ft, $z_2 = 4500$ ft, $z_3 = 2000$ ft, $z_4 = 6000$ ft, $\alpha = 60°$, $\sigma = 1$ g/cm^3. (From Geldart *et al.*, 1966.)

Although the formulae for thick structures are complex compared to the thin sheet approximation, they must be used in situations where the depth of cover is less than about twice the thickness of the slab, as mentioned before. In particular, when the slab outcrops, there is no solution possible for the thin sheet.

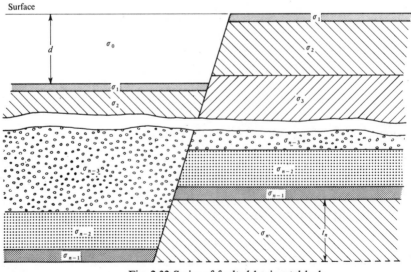

Fig. 2.32 Series of faulted horizontal beds.

Obviously one can extend eq. (2.51a) to include a series of beds at increasing depth. Figure (2.32) illustrates a fault extending downward from surface to a uniform basement (assumed to be infinite in thickness). The upper bed on the left is not present on the right because of erosion or nondeposition. We can calculate the fault anomaly by summing the effects of the successive beds of density $\sigma_1, \sigma_2, \ldots, \sigma_{n-1}$, plus the effect of the surface slab σ_0 and the basement step σ_n (the crosshatched section, bottom right).

Suppose for simplicity that the fault is vertical (actually the same result holds for any dip angle) and extends from surface down to the nth bed. Then the value of g at $x = \pm\infty$ is given by

$$g_{+\infty} = 2\pi\gamma \sum_{i=1}^{n} \sigma_i t_i, \quad g_{-\infty} = 2\pi\gamma \sum_{i=0}^{n-1} \sigma_i t_i.$$

Then the resultant gravity relief will be the difference, that is,

$$\Delta g = g_{+\infty} - g_{-\infty} = 2\pi\gamma(\sigma_n t_n - \sigma_0 d)$$
$$= 12\cdot77 \times 10^{-3}(\sigma_n t_n - \sigma_0 d). \qquad (2.51b)$$

The significance of this result is that the overall gravity effect of the fault is controlled by the top and bottom half-slabs, hence the shape of the profile is primarily dependent on their thickness and density. Furthermore the complete profile will tend to resemble that of a single slab rather than a fault, unless there are strong density contrasts within the intermediate beds.

(m) *Dipping beds.* If the thick prism extends to great depth as well as being effectively of infinite strike length, no analytical solution is possible, since the maximum gravity effect approaches infinity. The same situation clearly holds for dipping beds on the downthrown side of a fault. The problem may be solved only by maintaining a finite downdip length.

2.6.4 *Complex shapes*

(a) *Graphical methods.* When the field profile cannot reasonably be approximated by an analytical solution of a simple geometrical shape, such as those discussed above, it may be necessary to resort to a graphical approach. Several templates have been employed to calculate the effect of a two-dimensional structure of arbitrary cross-section. The template is superimposed on the irregular section of the structure to be analysed, to divide it into elementary areas. The integrated effect at a surface station is then obtained by summation.

A template of this type, due to Hubbert (1948), is illustrated in fig. 2.33. The total gravity effect in milligals at a point on the surface is

$$g \approx 7\cdot1 \times 10^{-5}\sigma N\phi z, \qquad (2.52)$$

where N = number of segments covering the cross-section, ϕ = angular separation of radial lines, z = separation of horizontal lines (ft).

Frequently the cross-section of the structure should be drawn with one scale expanded or compressed, to accommodate flat plates, lenses, steeply dipping thin veins, etc. For such situations several transparent templates containing only the radial lines for angles may be constructed to fit scales of, say, 2 : 1, 5 : 1, 10 : 1, or whatever. It is also convenient to draw the horizontal lines for constant z directly on the cross-section.

A *dot chart*, used by Gulf Research, is divided into sectors with radial lines, as in fig. 2.33, but uses semicircles instead of horizontal lines. Each sector contains a number of dots proportional to the gravity effect at the origin for that particular area.

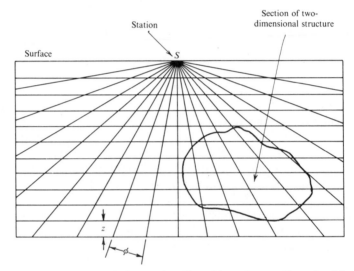

Fig. 2.33 Template for calculating gravity effect of irregular two-dimensional bodies.

When the structure is not really two-dimensional the finite strike length can be allowed for by applying an end-effect correction. For a point in the plane of a cross-section of the finite structure, at a distance r from the section's centre of gravity, the correction is obtained from the relation

$$\frac{g}{g_\infty} = \frac{1}{2}\left\{\frac{1}{\sqrt{(1 + r^2/Y_1^2)}} + \frac{1}{\sqrt{(1 + r^2/Y_2^2)}}\right\}, \qquad (2.53)$$

where g = actual gravity of the finite-length body, g_∞ = gravity for a body of the same cross-section but infinite in length, Y_1, Y_2 = distance from the cross-section to the ends of the body.

Graphical methods have also been employed on three-dimensional bodies by placing templates over contours of the body in the horizontal plane. In effect the body is broken up into a stack of horizontal slabs whose thickness is determined by the contour interval. This approach is more difficult than in the two-dimensional case because the chart must have a variable scale parameter to allow for the different depths of the slabs.

(b) *Analytical methods*. By using an n-sided polygon to approximate the outline of the vertical section of a two-dimensional body, one can calculate the gravity effect by hand or by digital computer (Talwani *et al.*, 1959). A simple section is illustrated in fig. 2.34. It can be shown that the gravity effect of this section is equal to a line integral around the perimeter (Hubbert, 1948). The relation is

$$g = 2\gamma\sigma \oint z\,\mathrm{d}\theta.$$

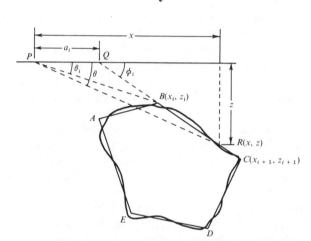

Fig. 2.34 Polygon approximation of irregular vertical section of two-dimensional body.

From the geometry of fig. 2.34 we have the following relations:

$$z = x \tan\theta = (x - a_i)\tan\phi_i, \quad \text{or} \quad z = (a_i \tan\theta \tan\phi_i)/(\tan\phi_i - \tan\theta).$$

The line integral for the side BC is

$$\int_{BC} z\,\mathrm{d}\theta = \int_B^C \frac{a_i \tan\theta \tan\phi_i}{\tan\phi_i - \tan\theta}\,\mathrm{d}\theta = Z_i.$$

Thus,
$$g = 2\gamma\sigma \sum_{i=1}^{n} Z_i. \tag{2.54a}$$

In the most general case, Z_i is given by

$$Z_i = a_i \sin\phi_i \cos\phi_i \left[(\theta_i - \theta_{i+1}) + \tan\phi_i . \log\left\{ \frac{\cos\theta_i\,(\tan\theta_i - \tan\phi_i)}{\cos\theta_{i+1}(\tan\theta_{i+1} - \tan\phi_i)} \right\} \right],$$

$$\tag{2.54b}$$

$$\theta_i = \tan^{-1}\left(\frac{z_i}{x_i}\right), \quad \phi_i = \tan^{-1}\left(\frac{z_{i+1} - z_i}{x_{i+1} - x_i}\right), \quad a_i = x_{i+1} - z_{i+1}\cot\phi_i$$

$$= x_{i+1} + z_{i+1}\left(\frac{x_{i+1} - x_i}{z_i - z_{i+1}}\right).$$

This technique has also been used for three-dimensional bodies by replacing the contours in the horizontal plane with n-sided polygons. The solution, from line integrals of the polygons, is essentially a more complicated version of eq. (2.54a).

Terrain corrections may also be computed by this method. For example, in the three-dimensional case, polygons are fitted to the topographic contours.

2.6.5 *Characteristic curves*

Most of the formulae for simple gravity shapes are far from simple to apply. Even when we can assume with some confidence that a field result may be reasonably matched by a specific geometry, it is still a tedious task to plot theoretical profiles from an expression, such as that in eq. (2.46c) for the thin sheet, which contains four unknowns, α, ℓ, h and Y, apart from the density contrast. A collection of *characteristic curves* greatly reduces the labour involved in this operation.

In order to prepare such curves we must establish some significant features associated with the profiles. It is also customary to reduce the number of variables by measuring in terms of one of them, preferably a dimension which has little influence on the characteristic features. Consider, for example, the thin dipping sheet, dealt with in Grant and West (1965, pp. 273–80); it is clear that the dip angle affects the profile symmetry more than its sharpness or maximum value while the depth h has more influence on the sharpness than on the symmetry. Furthermore the depth extent ℓ has less effect on the sharpness than h, while not controlling the symmetry particularly; consequently it seems reasonable to use the ratio h/ℓ and dip angle α as the characteristic features of the profile.

For the symmetry factor, we consider the ratio of the maximum slopes, s_1 and s_2, on either limb of the profile. Sharpness is estimated from a ratio of profile width at two points, $x_{\frac{2}{3}}$ and $x_{\frac{1}{3}}$, corresponding to $0\cdot67g_{max}$ and $0\cdot33g_{max}$. The principal profile in fig. 2.35 illustrates these parameters.

If we plot $|s_2/s_1|$ versus $x_{\frac{2}{3}}/x_{\frac{1}{3}}$, for a variety of dip angles and ratios of h/ℓ from eq. (2.46c), we have a set of curves which determine α and h/ℓ for anomalies of this shape. From another characteristic set, obtained by plotting $g_{max}/4\cdot0$ $\times 10^{-3}\sigma t$ vs $x_{\frac{1}{3}}/\ell$ for the same range of dip angles and h/ℓ ratios as in the first plot, we then obtain $x_{\frac{1}{3}}/\ell$ and hence ℓ, h and σt. These quantities, together with α, determine the geometry of the anomaly very well, although it is not possible to separate the product σt except by assuming a value for one of them.

For a field anomaly which is effectively two-dimensional, only two sets of curves are required. If, as is usually the case, the strike length is finite, it is necessary to prepare more curves of the same type for a number of values of Y/ℓ (i.e., a range of $0\cdot5 \leqslant Y/\ell \leqslant \infty$). In these circumstances an extra set of curves for first

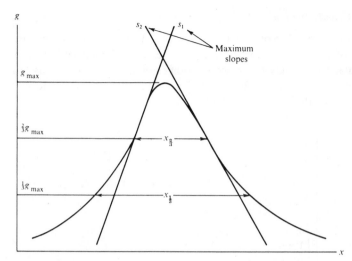

Fig. 2.35 Parameters for characteristic curves of a dipping sheet.

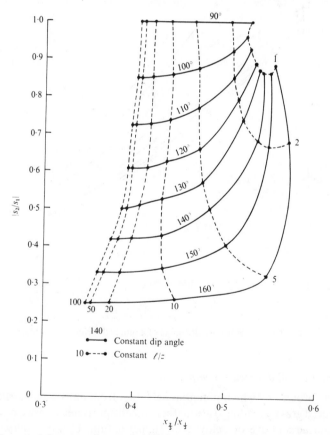

Fig. 2.36 Characteristic curves for gravity effect of a thin rod (principal profile, symmetry versus sharpness).

estimating the value of Y/ℓ is also needed. For example, a comparison of the widths of the profiles along the x- and y-axes would be suitable $(x_{\frac{1}{3}}/y_{\frac{1}{3}}$ versus $x_{\frac{2}{3}}/x_{\frac{1}{3}})$.

Similar curves may be drawn for other gravity features such as faults, prisms and cylinders. Figures 2.36 and 2.37 are characteristic curves for the principal profile of the thin rod, eq. (2.41a).

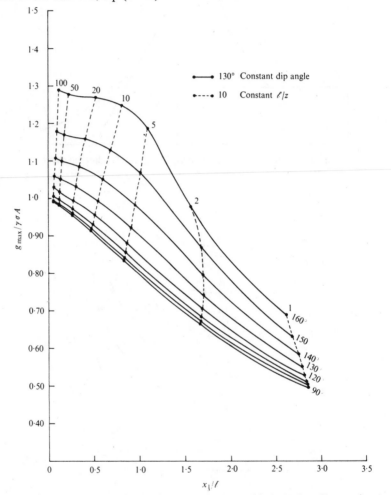

Fig. 2.37 Characteristic curves for gravity effect of a thin rod (principal profile, $g_{\text{max}}/\gamma\sigma A$ versus $x_{1/3}/\ell$).

2.6.6 *Upward and downward continuation*

This is the process by which potential field data from one datum surface (in this case, reduced gravity readings) are mathematically projected upward or downward to level surfaces above or below the original datum. *Upward continuation* is a straightforward operation, since the surfaces are in field-free space; in *downward continuation*, they may not be. The latter process, which is of considerable interest in interpretation, is more dangerous, because of the inherent uncertainty in the location and size of the structures represented by the Bouguer gravity at the datum plane. In projecting to a higher plane, we are effectively smoothing, in going downward we are sharpening the anomalies obtained at ground surface.

It is possible that we may complicate the results as well. Referring to the combined gravity effects of a sphere and basement step, previously described in §2.6.1, it was noted that the sphere was barely discernible at a depth of 200 ft. However, if we were able to make measurements 200 ft below surface, so that the top of the sphere outcropped – which is essentially the process of downward continuation – the maximum gravity effect of the sphere would be enhanced four times. Assuming that we were interested only in the basement relief, this projection would get us into trouble. Admittedly this is rather a trivial case, but the point is that anomalous masses lying between surface and the desired depth of projection may cause violent fluctuations in the gravity values at intermediate levels.

For this reason the downward continuation process is perhaps best suited to interpretation in oil exploration when the principal features of interest are largely controlled by the basement and the overlying relatively uniform sedimentary beds produce smaller gravity effects. For mineral prospecting the second derivative calculation is probably more suitable. Obviously these procedures may be applied in magnetics as well as gravity.

The following is a simplified treatment of continuation for a two-dimensional structure (Roy, 1966) which employs a grid in the xz-plane. This is illustrated in Fig. 2.38.

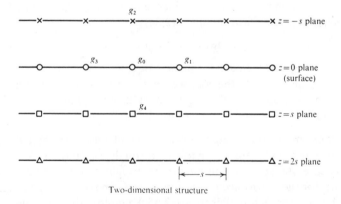

Fig. 2.38 Simple graphical downward continuation.

Given a set of Bouguer gravity readings on the x-axis on the datum surface $(z = 0)$, it is possible to project these downwards to the level $z = s$ (the unit vertical distance corresponds to the station spacing s), by first determining the upward continuation values on the plane $z = -s$. The latter operation is known as solving the Dirichlet problem. If we have a gravity map on the datum plane

$S_1(x_0, y_0, 0)$, then the values on a plane above it, $S_2(x, y, z)$ where $z < 0$, are given by

$$g(x, y, z) = \frac{1}{2\pi} \int_{-\infty}^{\infty} \int_{-\infty}^{\infty} \frac{z\, g(x_0, y_0)\, dx_0\, dy_0}{\{z^2 + (x_0 - x)^2 + (y_0 - y)^2\}^{\frac{3}{2}}}. \qquad (2.55a)$$

In the two-dimensional case the expression is simplified to

$$g(x, z) = \frac{1}{\pi} \int_{-\infty}^{\infty} \frac{z\, g(x_0)\, dx_0}{z^2 + (x_0 - x)^2}. \qquad (2.55b)$$

The variables (x_0, y_0) are the coordinates of stations in the datum plane; in addition we assume $z = -s$. The two-dimensional integral may be replaced by a sum:

$$g(m, -s) = \frac{1}{\pi} \sum_{-\infty}^{+\infty} g(n, 0)\{(\tan^{-1}(n - m + \tfrac{1}{2}) - \tan^{-1}(n - m - \tfrac{1}{2})\}, \qquad (2.55c)$$

where n refers to station coordinates in the datum plane and m is the x-coordinate of the point of calculation in the plane $z = -s$. It is necessary to carry the summation to station values 10 or more units away from the point of calculation. The first few terms of this summation for $m = 0$, $s = 1$, are

$$g(0, -1) = 0{\cdot}295g(0, 0) + 0{\cdot}165\{g(1, 0) + g(-1, 0)\} +$$

$$+\ 0{\cdot}066\{g(2, 0) + g(-2, 0)\} + 0{\cdot}033\{g(3, 0) + g(-3, 0)\} + \ldots$$
$$(2.55d)$$

This gives us a set of gravity values in the upper plane. We can then find the corresponding station values on the plane $z = s$ below the datum from the following relation (a modification of eq. (2.34)):

$$g_4 = 4g_0 - (g_1 + g_2 + g_3), \qquad (2.55e)$$

where g_0, g_1, and g_3 are located in the plane $z = 0$, g_2 in the plane $z = -s = -1$, and g_4 in the plane $z = 1$. In this way we get a row of points in the plane $z = 1$. These in turn may be used with the original data from the plane $z = 0$ to continue down to the plane $z = 2$, and so on (see fig. 2.38).

A similar, but more complicated, procedure can be used for three dimensions in which concentric circles are drawn to pass through grid stations. As in the two-dimensional case, the upward continuation is straightforward; there are various arrangements for the downward projection, at least one of which gives very poor results. A simple approach similar to eq. (2.55e), employs the following relation:

$$g_6 = 6g_0 - (g_1 + g_2 + g_3 + g_4 + g_5), \qquad (2.56)$$

where g_6 is the downward continued value in the plane $z = s$, g_0 is the station reading directly above it in the datum plane, $g_1, \ldots, g_4$ are values at the corners of a square or on a circle surrounding g_0, also in the plane $z = 0$, and g_5 is the upward continued value directly above g_0 in the plane $z = -s$.

One of the requirements for the numerical multipliers, such as those in eq. (2.55c) (known as *weighting coefficients*), is that their sum should equal unity. If not, it is necessary to insert a remainder term with an extra coefficient, which makes up the difference.

The mathematics required for more sophisticated treatment of downward continuation is beyond the level of this book. One method employs Fourier transform theory. Generally a digital computer is necessary.

2.6.7 *Excess mass*

As frequently remarked, there is no unique solution to a set of potential field data. However, in gravity work it is possible to determine uniquely the total anomalous mass, regardless of its geometrical distribution. This is a most significant calculation, particularly useful (although potentially dangerous in the wrong hands) in estimating ore tonnage in mineral exploration.

To find the magnitude of this *excess mass* we refer to Gauss' Theorem, §2.2.4, in which it was shown that the total flux across a closed surface in a gravitational field was proportional to the total mass enclosed by the surface. From eq. (2.8b) (dropping the minus sign – see the note following eq. (2.15)), we have the expression

$$\int_s g_n \, dS = 4\pi\gamma M.$$

Assume the mass (or masses) to be surrounded by a hemisphere of radius R whose flat face is the datum plane $z = 0$. Then the surface integral may be separated into two parts, one for the flux through the half-sphere, the other through the plane. From fig. 2.39 we have

$$\int_s g_n \, dS = \int_{z=0}\int g_n \, dx \, dy + \int_H\int g_n R^2 \sin\theta \, d\theta \, d\phi = 4\pi\gamma M,$$

where g_n in the integral over the datum plane $z = 0$ can be replaced with $g(x,y)$, the residual anomaly. Moreover, R can be made as large as we wish, hence we take it large enough that M appears to be a point mass and the distance $|\mathbf{R} - \mathbf{r}|$ can be taken equal to R. We now have

$$\int_{z=0}\int g(x, y) \, dx \, dy + \int_{\phi=0}^{2\pi}\int_{\theta=\pi/2}^{\pi} \gamma \frac{M}{R^2} R^2 \sin\theta \, d\theta \, d\phi = 4\pi\gamma M.$$

On evaluating the second integral we get

$$\int_{z=0}\int g(x, y) \, dx \, dy = 2\pi\gamma M,$$

hence the anomalous mass M is given by

$$M = \frac{1}{2\pi\gamma} \int_{-\infty}^{+\infty} \int_{-\infty}^{+\infty} g(x, y) \, dx \, dy. \tag{2.57a}$$

Practically, the mass M is obtained by dividing the survey area into suitably sized elements, estimating the average gravity in each element, multiplying each average gravity value by the element of area and summing. In units of milligals, tons and feet, the numerical expression is

$$M = 2 \cdot 44 \sum \bar{g}(x, y) \, \Delta x \, \Delta y. \tag{2.58a}$$

In polar coordinates, eq. (2.57a) becomes

$$M = \frac{1}{2\pi\gamma} \int_0^\infty \int_0^{2\pi} g(r, \theta) r \, dr \, d\theta. \tag{2.57b}$$

The elementary areas in this expression are similar to the terrain correction charts. Replacing the second integral in (2.57b) with a summation and using a procedure similar to that in §2.6.2c, we get

$$M = \frac{1}{\gamma} \sum_0^R \bar{g}(r_n, r_{n+1}) \int_{r_n}^{r_{n+1}} r \, dr = \frac{1}{2\gamma} \sum_0^R \bar{g}(r_n, r_{n+1})(r_{n+1}^2 - r_n^2)$$

$$= 7 \cdot 66 \sum_0^R \bar{g}(r_n, r_{n+1})(r_{n+1}^2 - r_n^2), \tag{2.58b}$$

where $\bar{g}(r_n, r_{n+1}) = \Delta g$ and $\pi(r_{n+1}^2 - r_n^2) = \Delta x \Delta y$ in eq. (2.58a), the units being tons, milligals and feet as before. Physically this mathematical operation

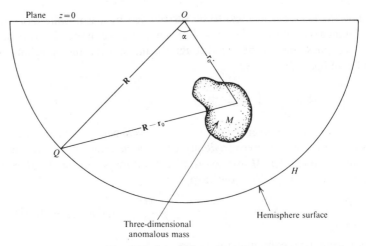

Fig. 2.39 Calculation of excess mass.

amounts to spreading the anomaly over the enclosing surface in a thin sheet, known as *Green's Equivalent Layer*. The actual mass producing the anomaly can only be determined if we know its density σ_a and the density contrast $\Delta\sigma$. Then eq. (2.58a) is modified to

$$M = 2 \cdot 44 \frac{\sigma_a}{\Delta\sigma} \sum \Delta g \, \Delta x \, \Delta y. \tag{2.58c}$$

Clearly it is necessary to carry out the summation far enough to reach very small (background) values of $g(x, y)$. If the regional has not been properly removed, or if there are other residual anomalies in the vicinity, the estimate obviously will be in error.

2.6.8 *Overburden effects*

In most field situations we know very little about the depth of overburden. Since the density contrast between overburden and bedrock is likely to be larger than the contrast between different rocks, variations in thickness of overburden can produce significant gravity anomalies. Table 2.2 shows an average density of 1·95 g/cm^3 for an assortment of overburden material when wet, 1.55 when dry. From the same table, the average for wet and dry sedimentary rocks is 2·50 and 2·20 respectively. Thus a contrast of 0·6 g/cm^3 would not be unusual.

As a rough estimate, we would expect the overburden to be thicker in valleys and low lying flat land than on steep hillsides and elevated plateaus. Abrupt changes in overburden thickness, however, are common enough. In any gravity survey, and particularly in mineral exploration, it is worthwhile to consider the extent to which gravity anomalies may be caused by variations in overburden thickness.

From the Bouguer correction given in eq. (2.18), and the effect of a semi-infinite horizontal slab for which $d = 0$, as given by eq. (2.50c), we can get some idea of the magnitude of the overburden effect. The maximum gravity variation resulting from a sudden change D in overburden thickness, where the density contrast is $\Delta\sigma$, is given by

$$\Delta g_{max} = 12{\cdot}77 \times 10^{-3}\Delta\sigma D.$$

Thus the thickness of overburden is

$$D \approx 78\Delta g_{max}/\Delta\sigma. \tag{2.59}$$

For example, if $\Delta g_{max} = 0{\cdot}5$ mgal and $\Delta\sigma = 0{\cdot}6$ g/cm^3, $D = 65$ ft.

The maximum horizontal gradient of gravity will, of course, be large if overburden irregularity is the source. For abrupt changes in depth of 20 ft or more, assuming $\Delta\sigma = 0{\cdot}6$ as above, the value of $(dg/dx)_{max}$ will be of the order of 0·1 mgal in 10 ft. In fact this steep gradient on the profile is probably more diagnostic, if it occurs, than the magnitude of g_{max}.

It is clear, therefore, that the depth of overburden should be measured in areas of shallow gravity anomalies. This is best done by small-scale seismic refraction or possibly surface resistivity measurements.

2.6.9 *Maximum-depth rules*

Smith (1959) has given several formulae for maximum depths of gravity distributions whose shapes are not known. The only restriction is that the anomalous bodies have a density contrast with the host rock which is either entirely positive or entirely negative. Then, if $|g_{max}|$ and $|\partial g/\partial x_{max}|$ are the maximum

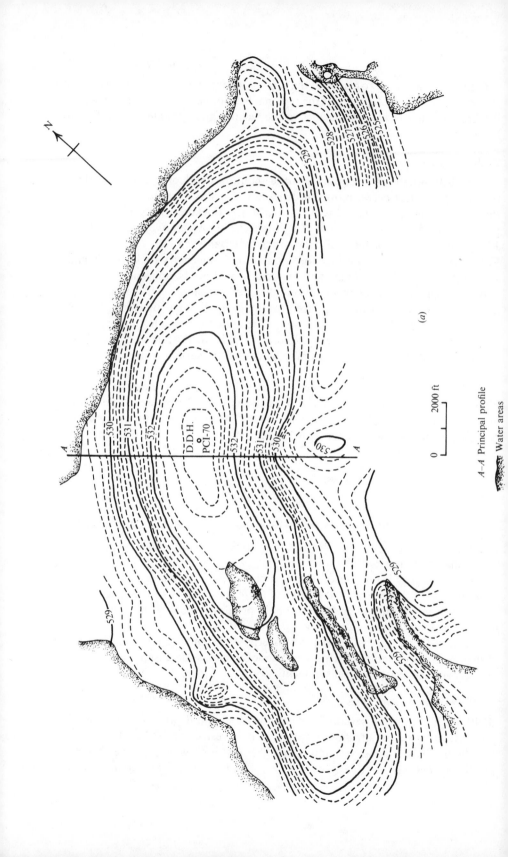

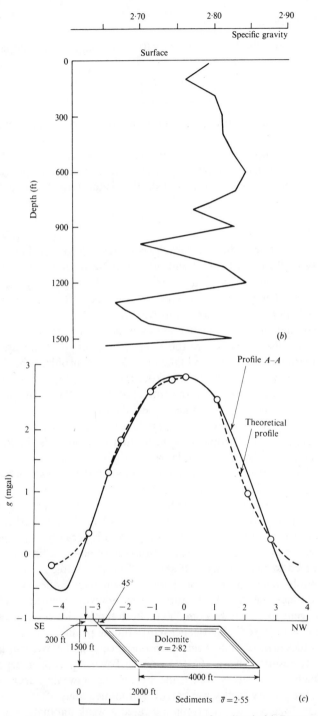

Fig. 2.40 Gravity survey, Portland Creek Pond, Newfoundland. (a) Bouguer gravity map: contour interval 0·2 mgal; (b) density log, D.D.H. PC1-70; (c) comparison with calculated profile for a thick dolomite slab.

values of gravity and the horizontal derivative, the depth to the upper surface has a limiting value, given by

$$z \leqslant \frac{0 \cdot 86 |g_{max}|}{|(\partial g/\partial x)_{max}|}.$$ (2.60)

If the anomaly is reasonably two-dimensional, the factor $0 \cdot 86$ becomes $0 \cdot 65$ in eq. (2.60).

This expression is not particularly accurate, as can easily be verified for simple geometries. In the case of a sphere, eq. (2.60) gives the depth to the centre; for a long thin vertical rod and a horizontal sheet, it is too large by about 70% and 100% respectively.

2.7 Field examples

(1) Figure 2.40a shows a Bouguer gravity contour map compiled from a survey in the vicinity of Portland Creek Pond, northern Newfoundland. This was an exploration programme for oil and gas in an area of sedimentary rocks whose thickness, a few miles south, is known to be over 5000 ft. The topography is reasonably flat and no terrain corrections were required.

It is evident that the large positive anomaly is not a reflection of basement structure because the gradients are too steep. For example, if we use the thin slab approximation of eq. (2.47) and take out a SE–NW profile, as at *A–A* in the diagram, it is a simple exercise to show from the values of g_{max} and $(\partial g/\partial x)_{max}$ that *h* is not greater than 700–800 ft. In contrast to the thin slab we might consider an intrusive body in the form of a dike of great depth extent, using, for example, eq. (2.49b). Such a structure produces a profile whose flanks are not nearly steep enough to match the field profile. It appears that the source is shallow and of limited depth extent – that is, it must lie within the sediments.

A vertical diamond drill hole was put down to 1500 ft in the centre of this gravity anomaly. Its location is shown in fig. 2.40a and a density log, obtained from measurement of core samples at 100 ft intervals, is shown in fig. 2.40b. The presence of dolomite, from near surface to a depth of 900 ft and interbedded with dark shales from 900 to 1500 ft, accounts for the positive gravity. Average density of the dolomite samples was $2 \cdot 82$ g/cm³. If the surrounding sedimentary formations are assumed to have a density of about $2 \cdot 55$ it is possible to match the field profile reasonably well with the dipping prism shown in fig. 2.40c. This analysis is oversimplified since the actual structure is neither two-dimensional ($Y \approx 9b$), nor is it homogeneous in the bottom 500 ft. Both factors, however, should have the desirable effect of producing steeper flanks on the profile.

(2) The profiles in fig. 2.41 illustrate the pronounced effect of overburden thickness on gravity results. This is the Louvicourt Township copper deposit near Val d'Or, Quebec. Discovery was made by drilling a weak Turam anomaly (see §7.5.5c); the gravity survey was carried out immediately after.

The original Bouguer gravity profile indicated a weak anomaly of $0 \cdot 15$ mgal directly over the conductor and a much broader and larger magnitude anomaly about 250 ft to the north. Obviously the small peak would not have aroused any great enthusiam. Later, when it had been established that the overburden thickness increased appreciably immediately over the sulphide zone, it was possible to correct for this variable thickness as discussed in §2.6.8, using a density contrast of about 0·8 g/cm³ between the host rock and the overburden. This is equivalent to 0·01 mgal per foot of overburden thickness. In the corrected field profile the larger anomaly to the north has practically disappeared and the small peak has been enhanced to 0·3 mgals. A third profile calculated from density measurements of diamond drill cores is also shown.

This example clearly indicates the importance of measuring the overburden thickness in conjunction with gravity applied to small-scale mineral exploration. Particularly is this true in surveys for vein-type base-metal deposits which respond to EM methods. The overburden effect would be less pronounced in regions favourable for IP, that is, large-area low-grade disseminated mineralization.

(3) The Delson fault is a well-documented structural feature in the St Lawrence Lowlands. Striking roughly E–W, it is located east of the St Lawrence River several miles southeast of Montreal. Although the area is generally covered by about 50 ft of overburden, there are exposures of Utica shales and Chazy limestones in river beds to indicate the location and direction of the fault. The sedimentary beds of the Lowlands are flat-lying shales, limestones and dolomites of Paleozoic age underlain by Precambrian basement rocks at a depth usually greater than 2500 ft.

Figure 2.42 shows a Bouguer gravity profile taken across the Delson fault in a S–N direction, together with a geologic section. The north end of the profile terminates at the St Lawrence River about 7 miles east of the Mercier bridge. A linear regional trend of 0·74 mgal/mile, positive to the south, has been removed.

The profile in fig. 2.42 resembles the gravity effect of a horizontal slab rather than a fault (compare with figs. 2.28, 2.29 and 2.31, and see discussion of multiple beds in §2.6.3*l*). The only appreciable gravity effect from the sedimentary beds would be provided by the juxtaposition of the Chazy and Utica formations near surface and the displaced Potsdam layer (whose thickness is in some doubt) at greater depth. The first pair produces a gravity profile of the proper shape with a total variation of 0·57 mgal and maximum slope of 1·14 mgal/1000 ft; thus the total anomaly is too small and the slope too large to fit the field profile. The low density Potsdam section, on the other hand, would tend to reduce the anomaly, since the bed nearer surface lies on the south side of the fault; the total effect, however, is only about −0·1 mgal and maximum slope −0·045 mgal/1000 ft.

By postulating a density of 2·96 in the Precambrian rocks and a step of 900 ft on the south side of the fault, we obtain a total anomaly of 2·1 mgal with maximum slope of 0·36 mgal/1000 ft. The theoretical profile in fig. 2.42 is the result.

The theoretical profile corresponds to a fault trace 1000 ft south of the mapped location. There are two explanations for this. First, the Delson fault is not vertical, but dips north about 80°. Second, faults very rarely show single clear-cut faces, that is, there is a faulted region of some width.

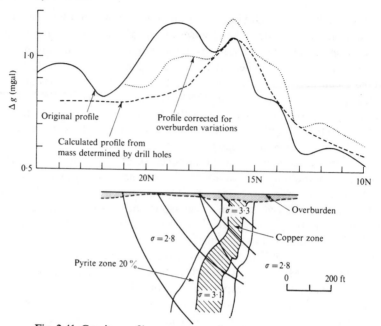

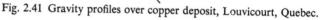

Fig. 2.41 Gravity profiles over copper deposit, Louvicourt, Quebec.

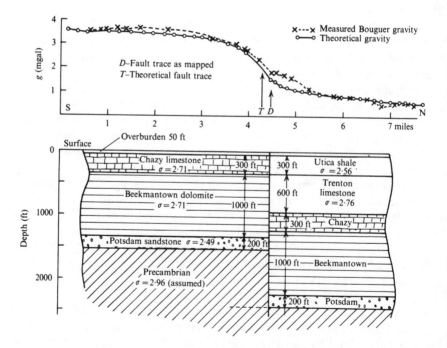

Fig. 2.42 Gravity profile and geologic section across the Delson fault, St Lawrence Lowlands.

The field profile also shows a small anomaly about $2\frac{1}{2}$ miles north of Delson fault, although there is no supporting geological evidence.

2.8 Problems

1. The following data were obtained over a two-day period of gravity follow-up on a base-metal prospect.

	Aug. 31				Line 0			
Stn	Time (hr)	g_{obs} (mgal)	Stn	Time (hr)	g_{obs} (mgal)	Stn	Time (hr)	g_{obs} (mgal)
°53	13.55	182·78	42	14.29	182·95	31	15.07	184·48
52	14.00	182·77	41	14.31	183·06	32	15.10	184·81
51	14.04	182·73	40	14.35	183·15	33	15.12	184·53
50	14.07	182·92	°39	14.37	183·13	34	15.14	184·33
49	14.10	183·05	25	14.50	183·82	35	15.16	184·17
48	14.15	183·19	26	14.52	183·97	36	15.19	184·03
47	14.17	182·99	27	14.55	183·99	37	15.22	183·73
46	14.20	182·88	28	14.59	183·96	38	15.25	183·38
45	14.22	182·89	29	15.01	184·25	°39	15.27	183·35
44	14.25	182·85	30	15.05	184·48	°53	15.45	183·01
43	14.27	182·91						
	Sept. 6				Line 2S			
0°53	9.10	185·02	33	10.56	185·39	44	11.44	184·53
2S°53	9.20	185·11	34	11.00	185·39	45	11.47	184·56
2S°39	9.40	184·93	35	11.05	185·52	46	11.52	184·64
24	10.19	185·86	36	11.09	185·37	47	11.55	184·67
25	10.23	185·66	37	11.15	185·00	48	11.59	184·67
26	10.27	185·65	38	11.20	184·86	49	12.02	184·76
27	10.31	185·66	°39	11.25	184·86	50	12.05	184·76
28	10.34	185·59	40	11.29	184·79	51	12.08	184·80
29	10.38	185·47	41	11.32	184·69	52	12.12	184·90
30	10.41	185·51	42	11.35	184·67	2S°53	12.16	185·09
31	10.45	185·46	43	11.39	184·60	0°53	12.28	185·03
32	10.50	185·39						

Stations marked ° have been visited at least twice for drift correction. For example, station 53, line 0 has been occupied at the beginning and end of each of the two days, station 53, line 2S at the beginning and end of the second day's work and the stations 39 on both lines were used as base stations for checking drift at intermediate times.

Draw-up a drift curve for the instrument for the two-day period and correct all stations accordingly. (Note that each of the four stations 53 and 39 must have the same gravity readings – or very nearly so, allowing for slight changes in the instrument height – as a result of the drift corrections.)

2. The gravity data tabled below were obtained during a detailed follow-up on a small sphalerite showing in Eastern Ontario. Reduce the raw gravity readings by taking out the drift, free-air (including height of instrument), Bouguer (assuming an average density of 2.67 g/cm^3) and latitude corrections (lines are N–S). Plot the two profiles. Are there

any indications of a small high-grade (10–15%) or larger low-grade (2–5%) sphalerite deposit? Can you suggest any reason for the general shape of the profiles?

Stn L 8W	g_{obs} (mgal)	Time (pm)	H.I. (ft)	Elev. (ft)	Stn L 8W	g_{obs} (mgal)	Time (pm)	H.I. (ft)	Elev. (ft)
°3S	37·04	2.20	1·42	0·00	1 + 75	36·86	3.33	1·42	1·60
3 + 25	36·82	2.30	1·25	3·69	1 + 50	36·88	3.37	1·58	1·84
3 + 50	36·87	2.33	1·33	3·63	1 + 25	37·49	3.42	1·33	−5·81
3 + 75	36·84	2.36	1·33	4·78	1S	37·77	3.45	1·08	−9·68
4S	36·80	2.40	1·33	5·85	0 + 75	38·00	3.49	1·10	−13·99
4 + 25	36·68	2.45	1·33	7·11	0 + 50	38·03	3.53	0·83	−14·93
4 + 50	36·63	2.48	1·50	8·26	0 + 25	38·07	3.59	0·83	−15·06
4 + 75	36·57	2.52	1·17	10·03	B.L.	38·03	4.02	1·17	−15·30
5S	36·47	2.55	1·33	11·42	L 10W				
5 + 25	36·56	2.57	1·25	10·19	B.L.	37·62	4.23	1·42	−3·06
5 + 50	36·67	3.00	1·33	8·91	1S	37·94	4.34	1·08	−7·41
5 + 75	36·67	3.04	1·33	8·21	2S	37·60	4.38	1·23	−9·14
6S	36·73	3.06	1·42	7·46	3S	37·55	4.40	1·42	−8·20
°3S	37·06	3.13	1·33	0·00	4S	37·27	4.46	1·42	−4·85
2 + 75	36·96	3.17	1·33	−0·76	5S	37·45	4.50	1·17	1·04
2 + 50	36·74	3.23	1·42	1·18	6S		4.53	1·42	−1·04
2 + 25	36·89	3.28	1·25	1·23	L 8W				
2S	36·86	3.30	1·42	1·39	°3S	37·10	5.00	1·17	0·00

L = line; B.L. = base line = 0 ft on each line; 3S = 300 ft south of base line, 3 + 75 = 375 ft south of base line, etc.; H.I. = height of instrument (gravimeter) above the surface whose elevation is given.

3. Show that the gravity anomaly produced by a cone at its apex is

$$\Delta g_{max} = 2\pi\gamma\sigma h(1 - \cos\alpha),$$

where h is the vertical height and α half the apex angle of the cone. Hence show that the terrain correction at the apex is

$$\delta g_t = 2\pi\gamma\sigma h \cos\alpha.$$

If $h = 1000$ ft, $\alpha = 68°$, $\sigma = 2·67$ g/cm³, determine with the aid of a template and table 2.1 the terrain correction required at various points on the sloping sides and on flat ground surrounding the cone (obviously the correction is the same at any particular elevation because of symmetry). How close to the base can gravity measurements be made without a significant terrain effect?

4. The topographic map in fig. 2.43 was prepared in considerable detail to take out a terrain correction for a gravity survey. Make a template of appropriate scale for the first four or five zones of table 2.1 (zones B–F) and calculate the terrain correction at several stations, such as (2N, 3W), (8N, 12W), (0, 9W), etc. Assuming that the topography is reasonably flat in the area surrounding the section illustrated, how many additional zones would be necessary to make a complete terrain correction?

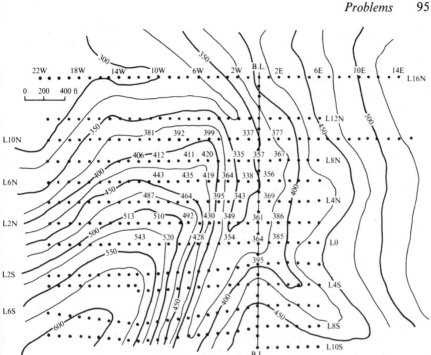

Fig. 2.43 Topographic map for gravity terrain corrections.

In fact, there is a steep ridge to the northwest, striking SW–NE; at a distance of about 2 miles from the centre of the area in fig. 2.43 the elevations increase from 300 ft to 1100 ft within one-half mile. There is also a lake, which is roughly one mile across and a maximum of 100 ft in depth, situated about 2 miles to the northeast. To what extent would these large topographic features affect the overall terrain correction?

5. The following reduced gravity readings (Bouguer) were obtained from an E–W traverse with stations at 100 ft intervals.

			Line 81N				
Stn	g_B (mgal)	Stn	g_B (mgal)	Stn	g_B (mgal)	Stn	g_B (mgal)
85E	1·35	89E	1·20	93E	0·77	97E	0·51
86	1·30	90	1·15	94	0·75	98	0·50
87	1·25	91	1·07	95	0·66	99	0·40
88	1·22	92	0·87	96	0·55		

An electromagnetic survey carried out earlier had outlined a conducting zone about 2400 ft long, striking roughly N–S, with a maximum width of 160 ft in the central part of the area. The gravity work was done as an attempt to assess the metallic content of the conductor. Four additional parallel gravity traverses, on lines adjacent to line 81N, produced essentially similar results. No information is available on the depth of over-burden. Plot the profile and make a qualitative interpretation of the nature of the conducting zone based on the gravity survey.

6. The Bouguer gravity readings tabled below are taken from a survey in the sedimentary Pine Point area of the North-West Territories. Station spacing is 100 ft on a N–S line.

Stn	g_B (mgal)	Stn	g_B (mgal)	Stn	g_B (mgal)	Stn	g_B (mgal)	Stn	g_B (mgal)
8S	0·02	4S	0·08	B.L.	0·19	4N	0·82	8N	0·22
7	0·06	3	0·06	1N	0·41	5	0·82	9	0·20
6	0·08	2	0·01	2	0·62	6	0·67	10	0·18
5	0·09	1	0·04	3	0·79	7	0·33	11	0·16

Plot the profile and interpret the gravity anomaly, assuming it to be approximately two-dimensional. Make two interpretations, assuming first that (a) below is valid, then assuming that (b) (but not (a)) is valid: (a) a Turam survey has not located any conductors in the area, while soil geochemistry shows minor lead and zinc; (b) the gravity anomaly coincides with a strong IP anomaly.

Attempt to match the gravity profile with a simple geometrical cross-section with particular regard to its depth, depth extent and width.

7. Residual gravity contours obtained from a survey over a base-metal area are shown in fig. 2.44. A regional trend of about 0·8 mgal/1000 ft was removed to produce this map. A profile along the S–N line has also been plotted. The gravity anomaly is obviously caused by a plug-type of structure of considerable positive density contrast. Make an interpretation of the source as precisely as possible with this data.

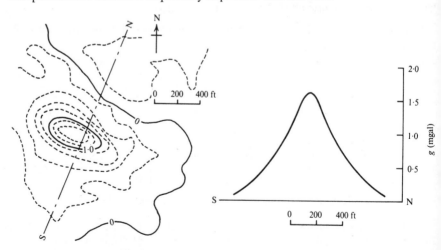

Fig. 2.44 Residual gravity contours and principal profile, Northwest Quebec. (Contour interval 0·2 mgal). (After Seigel, 1957.)

8. Tabled below are Bouguer gravity readings from a survey made in the Bathurst area of northern New Brunswick. Station spacing is 100 ft and the line spacing is as noted.

Because of limited time and money only four lines were selected for a follow-up of an earlier combined magnetic-electrical-geochemical survey. (Although this type of spot gravity work is used to some extent in base-metal exploration, it has obvious limitations in this case.) There is a pronounced regional gradient, positive to the east, but it is not uniform, being stronger in the north lines and weakest on line 10N. Furthermore the large and irregular spacing between the lines makes it difficult to plot well-defined contours. On the other hand, if we attempt to remove the gradients from each line independently it is not clear what background level should be selected; since we are looking for massive sulphides the tendency is to overemphasize positive gravity areas. No measurements of overburden depth were made.

Stn	L20N (mgal)	L16N (mgal)	L10N (mgal)	L2N (mgal)	Stn	L20N (mgal)	L16N (mgal)	L10N (mgal)	L2N (mgal)
4W	45·38	45·71	46·10	—	2E	46·21	46·42	46·55	47·09
3W	45·47	45·94	46·10	46·43	3E	46·08	46·40	46·72	47·00
2W	45·68	45·90	46·40	46·48	4E	46·54	46·53	46·80	47·12
1W	45·79	45·97	46·22	46·60	5E	46·62	46·75	46·66	47·50
BL	45·91	45·93	46·27	46·68	6E	46·91	46·87	46·61	47·61
1E	46·09	46·14	46·31	47·00					

Try to remove the regional by graphical smoothing and by the Griffin point-and-circle method (more sophisticated techniques are not warranted). Interpret any residuals remaining after the separation process.

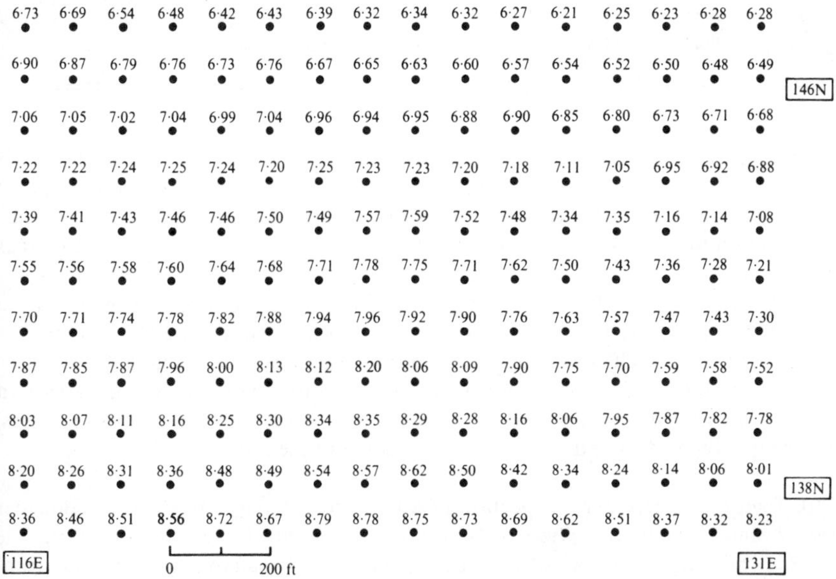

6·73 6·69 6·54 6·48 6·42 6·43 6·39 6·32 6·34 6·32 6·27 6·21 6·25 6·23 6·28 6·28

6·90 6·87 6·79 6·76 6·73 6·76 6·67 6·65 6·63 6·60 6·57 6·54 6·52 6·50 6·48 6·49 [146N]

7·06 7·05 7·02 7·04 6·99 7·04 6·96 6·94 6·95 6·88 6·90 6·85 6·80 6·73 6·71 6·68

7·22 7·22 7·24 7·25 7·24 7·20 7·25 7·23 7·23 7·20 7·18 7·11 7·05 6·95 6·92 6·88

7·39 7·41 7·43 7·46 7·46 7·50 7·49 7·57 7·59 7·52 7·48 7·34 7·35 7·16 7·14 7·08

7·55 7·56 7·58 7·60 7·64 7·68 7·71 7·78 7·75 7·71 7·62 7·50 7·43 7·36 7·28 7·21

7·70 7·71 7·74 7·78 7·82 7·88 7·94 7·96 7·92 7·90 7·76 7·63 7·57 7·47 7·43 7·30

7·87 7·85 7·87 7·96 8·00 8·13 8·12 8·20 8·06 8·09 7·90 7·75 7·70 7·59 7·58 7·52

8·03 8·07 8·11 8·16 8·25 8·30 8·34 8·35 8·29 8·28 8·16 8·06 7·95 7·87 7·82 7·78

8·20 8·26 8·31 8·36 8·48 8·49 8·54 8·57 8·62 8·50 8·42 8·34 8·24 8·14 8·06 8·01 [138N]

8·36 8·46 8·51 8·56 8·72 8·67 8·79 8·78 8·75 8·73 8·69 8·62 8·51 8·37 8·32 8·23

[116E]   0 ——— 200 ft   [131E]

Fig. 2.45 Bouguer gravity data digitized on 100-ft uniform grid (values in milligals).

9. Additional data obtained in the gravity survey of problem 1 are tabled below.

Line 0

Stn	H.I. (m)	Elev. (m)	Stn	H.I. (m)	Elev. (m)	Stn	H.I. (m)	Elev. (m)
53	0·46	6·76	42	0·50	6·64	31	0·40	2·22
52	0·43	6·99	41	0·46	5·82	32	0·43	0·00
51	0·49	7·08	40	0·39	5·96	33	0·42	1·04
50	0·47	5·95	39	0·45	6·21	34	0·48	1·55
49	0·51	4·73	25	0·50	2·59	35	0·47	2·50
48	0·53	4·68	26	0·46	3·15	36	0·47	3·00
47	0·53	6·00	27	0·40	3·78	37	0·49	4·06
46	0·48	6·42	28	0·50	3·17	38	0·47	5·38
45	0·50	6·40	29	0·43	1·96	39	0·48	6·21
44	0·48	6·75	30	0·45	1·31	53	0·45	6·76
43	0·47	7·20						

Line 2S

Stn	H.I. (m)	Elev. (m)	Stn	H.I. (m)	Elev. (m)	Stn	H.I. (m)	Elev. (m)
53	0·46	8·03	42	0·42	8·41	31	0·42	6·51
52	0·44	7·85	41	0·45	8·02	32	0·42	7·01
51	0·43	8·20	40	0·44	7·65	33	0·45	6·89
50	0·43	8·50	39	0·48	7·62	34	0·45	6·34
49	0·46	8·75	24	0·44	3·84	35	0·44	5·39
48	0·46	8·91	25	0·46	4·74	36	0·45	6·58
47	0·47	8·61	26	0·41	4·88	37	0·46	7·80
46	0·46	8·36	27	0·42	4·69	38	0·46	7·67
45	0·40	8·64	28	0·42	5·69	39	0·45	7·62
44	0·41	9·03	29	0·45	6·15	53	0·43	8·03
43	0·43	9·06	30	0·46	6·03			

Lines are east–west, 200 metres apart, stations are 50 metres apart with the larger numbers to the west. For some reason the base line, station zero, was cut at an angle of 20° east of true north, so that each station on line 0 is displaced about 73 metres east of its equivalent on line 2S. Obviously there is a latitude correction between the two lines as well.

Using the data in problem 1 (corrected for drift), apply the appropriate reductions to obtain Bouguer gravity for all stations, using an average density of 2·67 g/cm³ for the local formations, which consist of gneiss on the western portion extending roughly to station 25 and ultrabasic rocks to the east. There are frequent large outcrops in the eastern region, while the gneiss is covered by a fairly uniform thin overburden of 1–2 metres. Plot the gravity profiles and make an interpretation of the results.

10. Figure 2.45 represents a portion of a large gravity map on which the Bouguer gravity values have been digitized at 100-ft station intervals. Presented in this form, the data may be analysed in several ways but are particularly suitable for numerical techniques. The following methods are suggested.

(i) Remove the regional by drawing contours and employing graphical smoothing.
(ii) Remove the regional by the Griffin method.
(iii) Remove the regional by calculating the second derivative (see eq. (2.39*a*)).
(iv) Carry out downward continuation by any method you know.

Compare the results achieved with the different methods and consider their relative advantages and limitations. Interpret the residual anomaly or anomalies, if any, and calculate the excess mass. Is the digitized section actually large enough to give reasonable results?

11. A topographic section and Bouguer gravity profile over a long east–west ridge are shown in fig. 2.46. With no other relevant information available is it possible to answer the following questions with assurance?

(a) Was a terrain correction taken out in reducing the gravity readings?
(b) Is a terrain correction necessary?
(c) Is the gravity profile essentially a reflection of the topography?
(d) Is a regional gravity effect present which is independent of the topography?
(e) Assuming there is a definite gravity anomaly caused by a subsurface structure, can you locate it and estimate its approximate section? (As an aid to making this interpretation it would be very helpful to replot the gravity profile on an expanded vertical scale.)

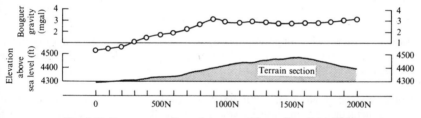

Fig. 2.46 Bouguer gravity and topographic profiles, East Africa.

12. Figure 2.47 is a Bouguer gravity contour map of the area whose topography is shown in fig. 2.43. The geology is sedimentary, with sandstone and limestone beds which are known to be over 5000 ft thick a few miles to the west. Within the survey area the limestone-sandstone contact can be seen at the bottom of the hill just west of the base line. The limestone bed extends east from this contact at surface and appears to be less than 50 ft thick; since it does not continue west under the hill, a fault is indicated. Under normal circumstances, the sandstone would have the lower density; but in this area, since the limestone is thin and since there are underground streams flowing in it, the density contrast is probably insignificant and may even be in the opposite sense.

Problem 4, with which fig. 2.43 is connected, was an exercise in making a few terrain corrections in rugged ground. Preparation of the map in fig. 2.47 required more than 15 man-days of work, most of it spent on terrain corrections for about 400 gravity stations. The end result is a gravity map which appears to be a fair reproduction of the topography.

Do the two maps, figs. 2.43 and 2.47, resemble each other in spite of the elaborate terrain correction or because of it? If the latter is true, would this be a fundamental

100   Gravity method

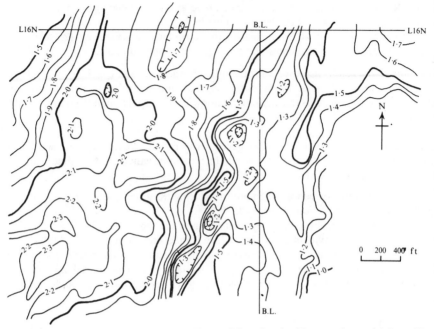

Fig. 2.47 Bouguer gravity contours, Eastern Nova Scotia. (Contour interval 0·1 mgal.)

argument against carrying out gravity surveys in regions where the topography is highly irregular? On the other hand, if the anomalous gravity is not due to topography, how do you explain it?

13. The reduced gravity contours in fig. 2.48 show a portion of a large survey made over a base-metal property in southern New Brunswick. Remove the regional. Given that there are ore-grade sulphides in the area, interpret the residual and estimate the excess mass.

14. Figure 2.49 represents a principal profile of residual Bouguer gravity across a three-dimensional structure of positive density contrast in the St Lawrence lowlands east of Montreal. The contours of residual gravity are approximately circular. Drill logs from gas and oil exploration holes in the vicinity have indicated flat-lying sedimentary beds to a depth of over 4000 ft. The maximum density contrast among the different sediments is not greater than 0·2 g/cm^3 and generally is closer to 0·1. Density of the Precambrian basement rocks is not known, but they are probably denser than the average sediments by 0·25–0·30.

To interpret this anomaly, first consider the maximum gravity combined with the maximum slope of the flanks, with respect to the known depths and density contrasts in the area. This will indicate an approximate depth to the source. Then attempt to match the profile with a simple shape, such as the sphere, rod, cylinder, etc. (Note that it is possible to simulate a pill-box type of structure with the aid of eqs. (2.45a) and (2.45b) by taking the difference between the gravity effects of two long cylinders at different depths.)

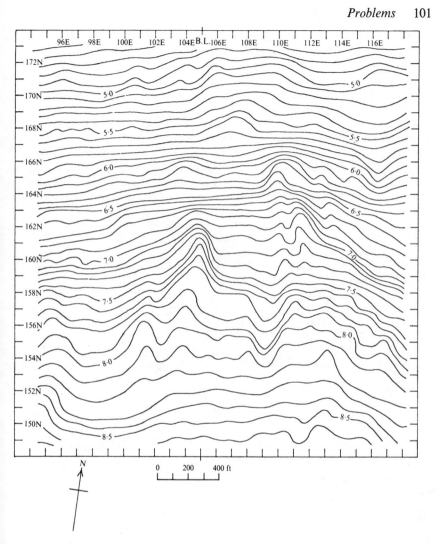

Fig. 2.48 Bouguer gravity contours, Southern New Brunswick. (Contour interval 0·1 mgal.)

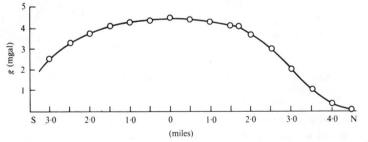

Fig. 2.49 Residual gravity profile, three-dimensional structure, St Lawrence Lowlands.

15. On the portion of the Bouguer anomaly map shown in fig. 2.50, the most negative values are found in the lower left corner. (The numerical values shown are with respect to an arbitrary datum.) (1) A large fault strikes N 20° W just east of AUS. Examine the shape of the fault anomaly by drawing the profile $A'A$. (2) On the simple fault model (fig. 2.28), the point where the fault's gravity expression is half its maximum value locates the fault; what difficulties are encountered in the practical application of this rule? Another rule is that the fault is located at the inflection point on the fault profile. Locate the fault on this profile by applying these rules. What does the asymmetry of the fault profile indicate? (3) The half-width rule states that the depth to the midpoint on a fault is equal to the distance between the points on the fault's gravity profile where the fault's gravity expression is $\frac{1}{4}$ and $\frac{1}{2}$ (or $\frac{1}{2}$ and $\frac{3}{4}$) of its maximum expression. What depth does this give for this fault? What difficulties are encountered in the practical application

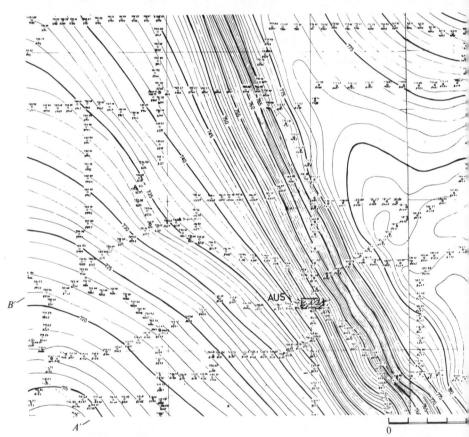

Fig. 2.50 Bouguer gravity map, Western Australia. (Contour interval 1·0 mgal.) (Courtesy West Australian Petroleum.)

of this rule? (4) What is the magnitude of this fault in milligals? Assume that a uniform density contrast across the fault exists from the surface to twice the depth given in part (3); how much density contrast is implied?

16. On the Bouguer anomaly map of fig. 2.50, the bow in the contours about 9 km NW of AUS indicates a structure. (1) Draw the profile $B'B$ to separate this anomaly from the fault anomaly examined in problem 15. Where is the centre of the anomaly? (2) Draw a profile at right angles to $B'B$ through the centre of the anomaly; is the anomaly easier to see on this profile? Are the residual anomalies consistent with each other? What is the magnitude of the anomaly in milligals? (3) Assume that the anomaly can be approximated by a buried spherical mass (see fig. 2.22); how deep must the centre of the sphere be? If a density contrast of 0.1 g/cm^3 is assumed, what would be the radius of the sphere? (4) The bow in the contours about 5 km SSW of AUS might be interpreted as a similar anomaly. What is unreliable about this anomaly?

17. A grid residual is one way of isolating anomalies from a regional background. Read the contour values on a 2 km grid over the Bouguer anomaly map of fig. 2.50. Assuming that the expected value at each grid station is the average of the four values which are 2 km away, the 'residual' is the difference between the observed and the expected values; this is a variation of the methods shown in fig. 2.20. Determine the residuals for all the grid points and contour the resulting map. Note how the fault studied in problem 15 and the anomaly studied in problem 16 are emphasized by this process.

References

F – field methods; G – general; I – instruments; T – theoretical

Algermissen, S. T. (1961). Underground and surface gravity survey, Leadwood, Missouri. *Geophysics*, **26**, 158–68. (F).

Bickel, H. C. (1948). A note on terrain corrections. *Geophysics*, **13**, 255–8. (T)

Bott, M. H. P. and Smith, R. A. (1958). The estimation of the limiting depth of gravitating bodies. *Geophys. Prosp.* **6**, 1–10. (T)

Bullard, E. C. and Cooper, R. I. B. (1948). Determination of the masses required to produce a given gravitational field. *Royal Soc. Proc.* A, **194**, 332–47. (T)

Dana, E. S. (1955). *A textbook of mineralogy* (revised by W. E. Ford, 4th ed.) New York, Wiley. (G)

Dean, W. C. (1958). Frequency analysis for gravity and magnetic interpretation. *Geophysics*, **23**, 97–127. (T)

Elkins, T. A. (1951). The second derivative method of gravity interpretation. *Geophysics*, **16**, 29–50. (T)

Garland, G. D. (1965). *The earth's shape and gravity*. London, Pergamon. (G)

Geldart, L. P., Gill, D. E. and Sharma, B. (1966). Gravity anomalies of two-dimensional faults. *Geophysics*. **31**, 372–97. (T)

Goetz, J. F. (1958). A gravity investigation of a sulphide deposit. *Geophysics*, **23**, 606–23. (F)

Grant, F. S. (1954). A theory for the regional correction of potential field data. *Geophysics*, **19**, 23–45. (T)

Grant, F. S. and West, G. F. (1965). *Interpretation theory in applied geophysics*. New York, McGraw-Hill. (T)

Griffin, W. R. (1949). Residual gravity in theory and practise. *Geophysics*, **14**, 39–56. (T)

Hammer, S. (1939). Terrain corrections for gravimeter stations. *Geophysics*, **4**, 184–94. (T)

104   Gravity method

Hammer, S. (1945). Estimating ore masses in gravity prospecting. *Geophysics*, **10**, 50–62. (T)

Heiland, C. A. (1940). *Geophysical exploration*. New York, Prentice-Hall. (G)

Henderson, R. G. and Zietz, I. (1949a). The computation of second vertical derivatives of geomagnetic fields. *Geophysics*, **14**, 508–16. (T)

Henderson, R. G. and Zietz, I. (1949b). The upward continuation of anomalies in total magnetic intensity fields. *Geophysics*, **14**, 517–34. (T)

Hubbert, M. K. (1948). A line integral method of computing the gravimetric effects of two-dimensional masses. *Geophysics*, **13**, 215–25. (T)

LaCoste, L. J. B. Jr. (1934). A new type long period vertical seismograph. *Physics*, **5**, 178–80. (I)

LaCoste, Lucien, Clarkson, N. and Hamilton, G. (1967). LaCoste and Romberg stabilized platform shipboard gravity meter. *Geophysics*, **32**, 99–109. (I)

LaFehr, T. R. and Nettleton, L. L. (1967). Quantitative evaluation of a stabilized platform shipboard gravity meter. *Geophysics*, **32**, 110–18. (F)

Legge, J. A. Jr. (1944). A proposed least squares method for the determination of the elevation factor. *Geophysics*, **9**, 175–9. (T)

Mesko, A. (1965). Some notes concerning the frequency analysis for gravity interpretation. *Geophys. Prosp.* **13**, 475–88. (T)

MacMillan, W. D. (1958). *Theory of the potential*. New York, Dover. (G)

MacRobert, T. M. (1948). *Spherical harmonics*. New York, Dover. (G)

Nettleton, L. L. (1954). Regionals, residuals and structures. *Geophysics*, **19**, 1–22. (T)

Nettleton, L. L., LaCoste, L. and Harrison, J. C. (1960). Tests of an airborne gravity meter. *Geophysics*, **25**, 181–202. (F)

Oldham, C. H. G. and Sutherland, D. B. (1955). Orthogonal polynomials: their use in estimating the regional effect. *Geophysics*, **20**, 295–306. (T)

Parasnis, D. S. (1962). *Principles of applied geophysics*. London, Methuen. (G)

Pepper, T. B. (1941). The Gulf underwater gravity meter. *Geophysics*, **6**, 34–44. (I)

Peters, L. J. (1949). The direct approach to magnetic interpretation and its practical application. *Geophysics*, **14**, 290–320. (T)

Pipes, L. A. (1958). *Applied mathematics for engineers and physicists*, 2nd ed. New York, McGraw-Hill. (G)

Ramsey, A. S. (1940). *Introduction to the theory of Newtonian attraction*. London, Cambridge Univ. Press. (G)

Roy, A. (1966). Downward continuation and its application to electromagnetic data interpretation. *Geophysics*, **31**, 167–84. (T)

Seigel, H. O. (1957). In *Methods and Case Histories in Mining Geophysics, 6th Commonwealth Mining & Metallurgical Congress*. Montreal, Mercury Press.

Siegert, A. J. F. (1942). Determination of the Bouguer correction constant. *Geophysics*, **7**, 29–34. (T)

Smith, R. A. (1959). Some depth formulae for local magnetic and gravity anomalies. *Geophys. Prosp.* **7**, 55–63. (T)

Talwani, M., Worzel, J. L. and Landisman, M. (1959). Rapid gravity computations for two-dimensional bodies with application to the Mendocino submarine fracture zones. *Jour. Geophys. Res.* **64**, 49–59. (T)

Talwani, M. and Ewing, M. (1960). Rapid computation of gravitational attraction of three-dimensional bodies of arbitrary shape. *Geophysics*, **25**, 203–25. (T)

3. Magnetic methods

3.1 Introduction

Study of the earth's magnetism is the oldest branch of the subject of geophysics. It has been known for more than three centuries that the earth behaves as a large and somewhat irregular magnet. The fact that a splinter of magnetite, hanging by a thread, takes up a definite position resulted in it being called a *lodestone* or *leading stone*. Sir William Gilbert (1540–1603) made the first scientific investigation of terrestrial magnetism and recorded in his book *de Magnete* that knowledge of this property of magnetite was brought to Europe from China by Marco Polo. Gilbert showed that the earth's magnetic field was equivalent to that of a permanent magnet, lying in a general north-south direction, near the earth's rotational axis.

The properties of the terrestrial magnetic field have been studied ever since Gilbert's time, but it was not until 1843 that von Wrede first used variations in the field to locate deposits of magnetic ore. Just as the work by Gilbert and Newton marked the beginning of geophysics, the publication of *The Examination of Iron Ore Deposits by Magnetic Measurements* by Thalén in 1879 marked the beginning of applied geophysics. Since that publication and especially in the last few years, there have been great advances in instrumentation and interpretation of measurements in this, the oldest of the geophysical methods of locating both hidden ores and structures associated with deposits of oil and gas.

As discussed in §2.1, magnetic and gravity methods have much in common. The magnetic map (usually total field or vertical component), however, is generally more complex and the variations in field more erratic and localized than the gravity map. This is partly due to the difference between the dipole magnetic field and the polar gravity field, the former having magnitude and variable direction, the latter magnitude and vertical direction only. Where the gravity map shows mainly regional effects, the magnetic map appears to be a multitude of residual anomalies, the result of large variations in the fraction of magnetic minerals contained in the near-surface rocks.

Thus the precise interpretation of magnetic field data is much more difficult than for gravity. On the other hand, the field measurements are easily made, cheap and simple compared to most geophysical techniques and corrections to readings are practically unnecessary. Partly because of this and also because magnetic field variations are very often diagnostic of mineral structures, as well as regional structures characteristic of favourable petroleum areas, the magnetic method is by far the most versatile of all geophysical prospecting techniques.

One can hardly imagine a geophysical programme in which magnetics would not be employed at least in reconnaissance.

3.2 Principles and elementary theory

3.2.1 *Definitions*

(a) *Magnetic force.* The expression for magnetic force is obtained from Coulomb's law for magnetic poles and symbolically is almost identical to Newton's law of gravity force. The expression is

$$\mathbf{F} = (m_1 m_2/\mu r^2)\mathbf{r}_1, \tag{3.1}$$

where, using *cgs* electromagnetic units (which are convenient in magnetostatics, henceforth referred to merely as *emu*), $\mathbf{F}$ is the force in dynes on m_2, the poles are r cm apart and $\mathbf{r}_1$ is a unit vector directed from m_1 towards m_2. The poles themselves are somewhat of a fiction, since they cannot exist isolated but only in pairs; however, if we assume two very long bar magnets with two poles close together and the other two far apart, the situation is fulfilled in practise. Finally, μ is the *permeability* of the medium surrounding the magnets; it is a dimensionless quantity whose value is precisely 1 in a vacuum and is practically 1 in air.

If two poles m_1 and m_2, each having strength of 1 *emu*, are placed 1 cm apart in a vacuum (or in air in practise), the force between them is 1 dyne. Unlike the gravity case, in which the force is always attractive, the magnetostatic force is attractive if the poles are opposite sign, repulsive if they are the same. The sign convention adopted is that a *positive pole* is one which is attracted towards the earth's north magnetic pole, a *negative pole* is one which is attracted towards the earth's south magnetic pole. The terms *north-seeking* and *south-seeking* are also used in place of *positive* and *negative*.

(b) *Magnetic field strength.* A more practical quantity than the force is the strength of magnetic field existing at a point in space, as a result of a pole of strength m located at a distance r from it. The *magnetic field strength* $\mathbf{H}$ is defined as the force on a unit pole:

$$\mathbf{H} = \mathbf{F}/m' = (m/\mu r^2)\mathbf{r}_1, \tag{3.2}$$

where m' is a fictitious pole at the point in space and is in effect the instrument of measurement. It is assumed that m' is not large enough to disturb the field $\mathbf{H}$ at the point of measurement, i.e., $m' \ll m$.

The field $\mathbf{H}$ may also be produced by current flowing in a wire (e.g. eqs. (7.4) to (7.9), §7.2.2b to §7.2.2f) rather than by the pole or poles of magnetized material. Furthermore, the current may be unidirectional or alternating in time; in the latter case the magnetic field also changes at the same rate as the current.

In *emu* $\mathbf{H}$ is in oersteds, that is, dynes/unit pole.

(c) *Magnetic moment.* Since magnetic poles always exist in pairs, the fundamental magnetic entity is the *magnetic dipole*, two poles of strength $+m$ and $-m$ separated by a distance ℓ. Then the *magnetic moment* is defined as

$$\mathbf{M} = m\ell\mathbf{r}_1 = \mathcal{M}\mathbf{r}_1, \tag{3.3}$$

$\mathbf{M}$ being a vector in the direction of the unit vector $\mathbf{r}_1$ extending from the negative pole towards the positive pole.

(d) *Intensity of magnetization.* A magnetic body placed in an external magnetic field becomes magnetized by induction. The *intensity of magnetization* is proportional to the strength of the field and its direction is in the direction of that field. It is defined as the magnetic moment per unit volume, that is,

$$\mathbf{I} = \mathbf{M}/v = \mathcal{I}\mathbf{r}_1. \tag{3.4}$$

Practically this magnetization by induction amounts to lining up the dipoles of the magnetic material; for this reason $\mathbf{I}$ is often referred to as the *magnetic polarization*. If $\mathbf{I}$ is constant and has the same direction throughout, the body is said to be *uniformly magnetized*.

(e) *Magnetic susceptibility.* The degree to which the body is magnetized is determined by its *magnetic susceptibility*, defined as

$$k = \mathcal{I}/H, \quad \text{or} \quad \mathbf{I} = k\mathbf{H}. \tag{3.5}$$

Susceptibility is the fundamental parameter in magnetic prospecting, since the magnetic response of rocks and minerals is determined by the amount of magnetic materials in them and the latter have k values much larger than the rocks and minerals themselves. The susceptibilities of various materials are discussed in §3.3.7.

(f) *Magnetic induction.* The magnetic body which we are considering, when placed in the external field $\mathbf{H}$, has its internal poles more or less lined up by the field $\mathbf{H}$ to produce a field of its own, $\mathbf{H}'$, which increases the total field within the body. This extra field is related to the intensity of magnetization. The *magnetic induction*, $\mathbf{B}$, is defined as the total field within the body and can be written

$$\mathbf{B} = \mathbf{H} + \mathbf{H}' = \mathbf{H} + 4\pi\mathbf{I}.$$

Combining this with eq. (3.5), we get

$$\mathbf{B} = (1 + 4\pi k)\mathbf{H}. \tag{3.6a}$$

Now by definition the ratio of induction $\mathbf{B}$ to magnetizing force $\mathbf{H}$ is the *magnetic permeability*, μ, which has already appeared in eq. (3.1). Thus (3.6a) may be written

$$\mathbf{B} = (1 + 4\pi k)\mathbf{H} = \mu\mathbf{H}, \tag{3.6b}$$

and $\mu = (1 + 4\pi k)$ is the relation between magnetic susceptibility and permeability. In *cgs* magnetic units, **B** is measured in gauss, so that the permeability μ becomes gauss/oersted. However, since μ is dimensionless in these units, the gauss is dimensionally the same as the oersted. This is unfortunate, because it has led to an indiscriminate interchanging of the quantities **B** and **H**. In fact they are not equivalent, since **H**, as stated previously, is an intrinsic magnetic field strength, while **B** represents an induced magnetic field plus the intrinsic magnetic field; in some magnetic materials, for instance, **B** may be quite large as a result of previous magnetization having no relation to the present external field **H**.

(g) **B-H** *relations: the hysteresis loop.* The relation between **B** and **H** discussed above can be quite complex in ferromagnetic materials (see §3.3.5*d*). This is illustrated in the cycle of magnetization, a familiar elementary physics experiment performed with a steel rod in a solenoid. On plotting *B* versus *H*, we obtain the well-known *hysteresis loop* shown in fig. 3.1.

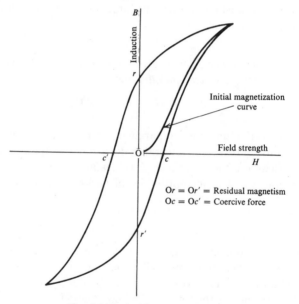

Fig. 3.1 Typical hysteresis loop.

If the sample is initially demagnetized we obtain the first portion of the curve in which **B** increases with **H** until it begins to flatten off due to saturation. On decreasing the external field, the curve does not follow the same path and shows a positive value of **B** when **H** = 0. This is known as the *residual magnetism* in the sample. When **H** is reversed it is found that **B** finally becomes zero at some negative value of **H**, known as the *coercive force*. The other half of the hysteresis loop is then obtained by making **H** still more negative until reverse saturation is reached and then returning **H** to the original positive saturation value.

The intercepts on the *B*-axis are a measure of the induced magnetic polarization remaining in the sample when the external field is zero, while those on the *H*-axis indicate how much reverse field is necessary to remove the induced magnetization. The area under the curve represents the energy loss per cycle per unit volume of the magnetic material as a result of hysteresis. Residual effects in various magnetic materials will be discussed in more detail in §3.3.6.

(h) *Magnetic units.* Although MKS units are used generally in this book they are not very suitable in magnetostatics. The equivalent relations for eqs. (3.2) and (3.6*b*) in MKS units are

$$\mathbf{H} = (m/\mu\mu_0 r^2)\mathbf{r}_1 \quad \text{and} \quad \mathbf{B} = \mu\mu_0\mathbf{H},$$

where $\mathbf{H}$ is in ampere-turns/m, m in webers, $\mathbf{B}$ in webers/m² and μ_0 in henrys/m, the latter being the permeability of free space. Converting $\mathbf{H}$ and $\mathbf{B}$ from MKS to em units, we get

$$1 \text{ ampere-turn/m} = 4\pi \times 10^{-3} \text{ oersteds,} \quad \text{and} \quad 1 \text{ weber/m}^2 = 10^4 \text{ gauss.}$$

In studying magnetic properties of the earth, these MKS units are rather confusing because they are not the right order of magnitude and because they describe magnetic effects in terms of current flow.

In magnetic prospecting we measure variations of the order of $1/10^4$ of the earth's main field, which is about 0·5 oersted. A new unit of magnetic intensity or field strength, the *gamma* (γ), is introduced:

$$1\gamma = 10^{-5} \text{ oersteds.}$$

This unit is a convenient size for geophysical work.

3.2.2 *Basic theory*

(a) *Magnetostatic potential: dipole field.* As in gravity, the vector magnetic field may be derived from a scalar potential function:

$$\mathbf{F}(\mathbf{r}) = -\nabla A(\mathbf{r}), \tag{3.7a}$$

and this potential may also be defined as the work done in moving a unit pole against the magnetic field, from the relation

$$A(\mathbf{r}) = -\int_\infty^r \mathbf{F}(\mathbf{r}) \cdot d\mathbf{r} = \frac{m}{\mu r}. \tag{3.7b}$$

Since the single magnetic pole is a pure fiction, the scalar magnetic potential is somewhat nebulous, although mathematically useful. A realistic entity, as mentioned previously, is the magnetic dipole. Referring to fig. 3.2, in which $\mu = 1$ for the surrounding medium, we can calculate A at an external point P, since

$$A = \frac{m}{r_1} - \frac{m}{r_2} = m \left\{ \frac{1}{\sqrt{(r^2 + l^2 - 2rl\cos\theta)}} - \frac{1}{\sqrt{(r^2 + l^2 + 2rl\cos\theta)}} \right\}. \tag{3.8a}$$

If $r \gg l$, this is simplified to

$$A \approx 2ml \cos \theta / r^2 \approx \mathscr{M} \cos \theta / r^2. \tag{3.8b}$$

From eq. (3.7a), we can derive the vector magnetic field, which has a radial component along r and an angular component normal to r. These are given by

$$
\left.
\begin{aligned}
F_r &= -\frac{\partial A}{\partial r} = -m \left\{ \frac{r + l \cos \theta}{(r^2 + l^2 + 2rl \cos \theta)^{\frac{3}{2}}} - \frac{r - l \cos \theta}{(r^2 + l^2 - 2rl \cos \theta)^{\frac{3}{2}}} \right\}, \\
F_\theta &= -\frac{1}{r}\frac{\partial A}{\partial \theta} = m \left\{ \frac{l \sin \theta}{(r^2 + l^2 + 2rl \cos \theta)^{\frac{3}{2}}} + \frac{l \sin \theta}{(r^2 + l^2 - 2rl \cos \theta)^{\frac{3}{2}}} \right\},
\end{aligned}
\right\} \tag{3.9a}
$$

and again, if $r \gg l$ these expressions become

$$F_r \approx 2\mathscr{M} \cos \theta / r^3, \quad \text{and} \quad F_\theta \approx \mathscr{M} \sin \theta / r^3. \tag{3.9b}$$

Two special cases, $\theta = 0$ and $\theta = \pi/2$, representing the Gauss-A (or end-on) and Gauss-B (or side-on) positions, are given by

$$F_r = 2\mathscr{M}r/(r^2 - l^2)^2, \quad \text{and} \quad F_\theta = 0, \quad (\theta = 0) \tag{3.9c}$$

$$F_r = 0, \quad \text{and} \quad F_\theta = \mathscr{M}/(r^2 + l^2)^{\frac{3}{2}}. \quad (\theta = \pi/2) \tag{3.9d}$$

If $\mathbf{r} \gg l$, these are further simplified to

$$F_r \approx 2\mathscr{M}/r^3 \quad (\theta = 0), \quad \text{and} \quad F_\theta \approx \mathscr{M}/r^3. \quad (\theta = \pi/2) \tag{3.9e}$$

The total or resultant magnitude from (3.9b) is

$$F = \frac{\mathscr{M}}{r^3} \sqrt{(4 \cos^2 \theta + \sin^2 \theta)}, \tag{3.9f}$$

and the direction with respect to F_r is given by

$$\tan \alpha = F_\theta / F_r = \tfrac{1}{2} \tan \theta. \tag{3.9g}$$

In vector notation, the total field corresponding to (3.9b) becomes

$$\mathbf{F} = \left(\frac{2\mathscr{M} \cos \theta}{r^3} \right) \mathbf{r}_1 + \left(\frac{\mathscr{M} \sin \theta}{r^3} \right) \boldsymbol{\theta}_1, \tag{3.9h}$$

where unit vectors $\mathbf{r}_1$ and $\boldsymbol{\theta}_1$ are measured in the direction of increasing r and θ (counterclockwise in fig. 3.2).

(b) *The general magnetic anomaly.* A volume of magnetic material can be considered to be an assortment of magnetic dipoles. Whether these are initially lined up to any extent so that they exhibit residual magnetism, depends on the previous magnetic history of the body and its surroundings; they will, however, be aligned by induction, due to the presence of an external field. In any case we may assume that the body has a continuous distribution of dipoles resulting in a vector dipole moment per unit volume which we will call $\mathbf{M(r)}$. Equation (3.8*b*) gives the scalar potential at a point some distance from a dipole as

$$A = \mathcal{M} \cos \theta / r^2 = -\mathbf{M} \cdot \nabla (1/r).$$

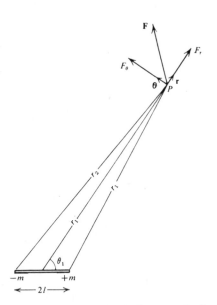

Fig. 3.2 Calculating the field of a magnetic dipole.

Then the potential for the whole body is

$$A(\mathbf{r}_0) = -\int_v \mathbf{M(r)} \cdot \nabla \left(\frac{1}{|\mathbf{r}_0 - \mathbf{r}|} \right) dV, \qquad (3.10a)$$

where the coordinate system is shown in fig. 3.3. If $\mathbf{M}$ is constant and has a constant direction denoted by $\boldsymbol{\alpha}_1 = \ell\mathbf{i} + m\mathbf{j} + n\mathbf{k}$, then

$$\mathbf{M} \cdot \nabla = \mathcal{M} \frac{\partial}{\partial \alpha} = \mathcal{M} \left(\ell \frac{\partial}{\delta x} + m \frac{\partial}{\partial y} + n \frac{\partial}{\partial z} \right),$$

and

$$A(\mathbf{r}_0) = -\mathcal{M} \frac{\partial}{\partial \alpha} \int_v \frac{dV}{|\mathbf{r}_0 - \mathbf{r}|} \right). \qquad (3.10b)$$

The resultant magnetic field of this body can be obtained by employing eq. (3.7*a*) with (3.10*a*); this gives

$$\mathbf{F}(\mathbf{r}_0) = \nabla \int_v \mathbf{M(r)} \cdot \nabla \left(\frac{1}{|\mathbf{r}_0 - \mathbf{r}|} \right) dV. \qquad (3.11)$$

This field exists in the presence of the main earth's field $\mathbf{F}_0$ at the location, i.e., the total field is given by

$$\mathbf{F}_t = \mathbf{F}_0 + \mathbf{F}(\mathbf{r}_0),$$

where the directions of $\mathbf{F}_0$ and $\mathbf{F}(\mathbf{r}_0)$ are not necessarily the same. Only if the magnitude of $\mathbf{F}(\mathbf{r}_0)$ is much smaller than $\mathbf{F}_0$, or if the body has no residual magnetism at

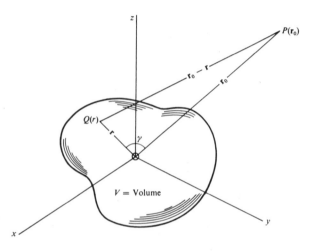

Fig. 3.3 General magnetic anomaly.

all, can $\mathbf{F}_t$ and $\mathbf{F}_0$ be assumed approximately equidirectional. Thus in the general situation, where $\mathbf{F}(\mathbf{r}_0)$ is an appreciable fraction (say 25% or more) of $\mathbf{F}_0$, and $\mathbf{M}(\mathbf{r})$ has a different direction than $\mathbf{F}_0$, the component of $\mathbf{F}(\mathbf{r}_0)$ in the direction of the field $\mathbf{F}_0$ becomes

$$F_\beta(\mathbf{r}_0) = -\boldsymbol{\beta}_1 \cdot \nabla A(\mathbf{r}_0) = -\frac{\partial A(\mathbf{r}_0)}{\partial \beta} = \mathscr{M}\frac{\partial^2}{\partial\alpha\partial\beta}\int_v \frac{dV}{|\mathbf{r}_0 - \mathbf{r}|}, \qquad (3.12a)$$

where $\boldsymbol{\beta}_1$ signifies the direction of $\mathbf{F}_0$. If the magnetic moment of the body has not been misoriented by residual effects, then the magnetization will be mainly or entirely induced by $\mathbf{F}_0$ in the direction $\boldsymbol{\beta}_1$. Then using (3.4) and (3.5), eq. (3.12a) becomes

$$F_\beta(\mathbf{r}_0) = \mathscr{M}\frac{\partial^2}{\partial\beta^2}\int_v \frac{dV}{|\mathbf{r}_0 - \mathbf{r}|} = kF_0\frac{\partial^2}{\partial\beta^2}\int_v \frac{dV}{|\mathbf{r}_0 - \mathbf{r}|}. \qquad (3.12b)$$

As in gravity, we can also define the logarithmic potential for two-dimensional features, in which eq. (3.12a) becomes

$$F_\beta(\mathbf{r}_0) = 2\mathcal{M} \frac{\partial^2}{\partial\alpha\partial\beta} \int_s \log|\mathbf{r}_0 - \mathbf{r}| \, dS, \qquad (3.13a)$$

or,

$$F_\beta(\mathbf{r}_0) = 2kF_0 \frac{\partial^2}{\partial\beta^2} \int_s \log|\mathbf{r}_0 - \mathbf{r}| \, dS, \qquad (3.13b)$$

where S is the cross-section of the body.

Obviously the magnetic problem is inherently more complicated than its equivalent in gravity. Even for the simplest geometrical shapes it is necessary to make additional assumptions (because of the dipole nature of the magnetic field) about the direction of magnetization.

(c) *Poisson's relation.* There is a relation, due to Poisson, between gravity potential U and magnetic potential A of a body, when both density, σ, and dipole moment are constant (Garland, 1951). It can then be shown that

$$A = -\frac{\mathcal{I}}{\gamma\sigma} \frac{\partial U}{\partial\alpha} = -\frac{\mathcal{I}}{\gamma\sigma} \nabla U \cdot \boldsymbol{\alpha}_1 = -\frac{\mathcal{I}}{\gamma\sigma} g_\alpha, \qquad (3.14a)$$

where α and the unit vector $\boldsymbol{\alpha}_1$ are in the direction of the polarization and g_α is the component of gravity in the same direction.

In terms of fields rather than potentials, we can write

$$F_\beta(\mathbf{r}) = \text{component of } \mathbf{F}(\mathbf{r}) \text{ in the direction } \boldsymbol{\beta}_1,$$

$$= -\frac{\partial A}{\partial\beta} = \frac{\mathcal{I}}{\gamma\sigma} \frac{\partial g_\alpha}{\partial\beta}. \qquad (3.14b)$$

In the special case where the polarization is vertical, we can write

$$Z = \frac{\mathcal{I}}{\gamma\sigma} \frac{\partial g_z}{\partial z}, \qquad (3.14c)$$

where Z is the vertical component of the magnetic field. This is a most useful relation, because, as mentioned above, the gravity field is generally easier to solve for than the magnetic field.

(d) *Field equations.* Clearly in the homogeneous region outside the volume V of fig. 3.3, the magnetic potential, like the gravity potential, satisfies Laplace's equation. Thus by analogy with eq. (2.9), §2.2.4, we can write

$$\nabla^2 A = 0. \qquad (3.15)$$

Similarly the magnetic potential everywhere within a region containing magnetic material, like the gravity potential within a mass anomaly, satisfies Poisson's equation, so that we have

$$\nabla^2 A = 4\pi \nabla \cdot \mathbf{M(r)}. \tag{3.16}$$

As one would expect, the magnetic expression for Poisson's equation is more complicated than for gravity.

3.3 Magnetism of the earth

3.3.1 *Nature of the geomagnetic field*

As mentioned in the introductory section of this chapter, studies of the earth's magnetic field have been made for several hundred years. Measurements have been made at sea for navigational purposes, on land in connection with prospecting and in magnetic observatories to establish variations in the field. In comparison with other geophysical data relating to the earth, the amount of information accumulated for the geomagnetic field is relatively enormous.

As a result of these studies it is well established that the geomagnetic field is composed of three parts, so far as exploration geophysics is concerned:

(i) the *main field*, which although not constant in time, varies relatively slowly and is of internal origin,

(ii) the *external field*, a small fraction of the main field, which varies rather rapidly, partly cyclically and partly randomly and which originates outside the earth,

(iii) *variations of the main field*, usually but not always much smaller than the main field, relatively constant in time and place and caused by local magnetic anomalies in the near-surface crust of the earth; these are the targets in magnetic prospecting.

3.3.2 *The main field*

(a) *Elements of the earth's magnetic field.* If a steel needle, not previously magnetized, could be hung at its centre by a thread so that it were free to orient itself in any direction in space, at most points on the earth's surface it would assume a direction neither horizontal nor in line with the geographical meridian. This orientation is the direction of the earth's total magnetic field at this point. The *magnitude* of this field, F, the *inclination* of the needle from the horizontal, I, and its *declination*, D, the angle it makes with geographic north, completely define the magnetic field.

The various magnetic elements are illustrated in fig. 3.4. In addition to F, I and D, they are Z (or V), the vertical component reckoned positive downwards, H, the horizontal component which is always positive and X, Y, the components of H

onsidered positive to the north and east respectively. From the diagram we have
the following relations among these elements:

$$F^2 = H^2 + Z^2 = X^2 + Y^2 + Z^2,$$
$$H = F \cos I, \quad Z = F \sin I, \quad \tan I = Z/H,$$
$$X = H \cos D, \quad Y = H \sin D, \quad \tan D = Y/X, \tag{3.17}$$

As defined previously, the end of the needle which dips downward in northern
latitudes is called the north-seeking or positive pole; in the southern hemisphere
the end dipping downward is the south-seeking or negative pole.

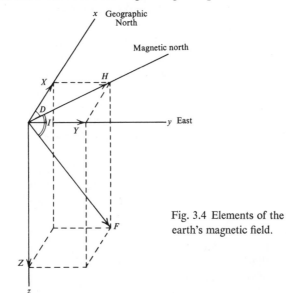

Fig. 3.4 Elements of the earth's magnetic field.

Lines of equal declination, inclination, horizontal intensity, etc., when plotted
on maps, are called *isomagnetic charts*. They show the variations in the geomagnetic field over the earth's surface. Oddly enough, the magnetic field reflects
little or nothing of the variation in surface geology and geography such as
mountain ranges, submarine ridges, earthquake belts. This indicates that the
source of the field lies deep within the earth or far outside it.

In fact the geomagnetic field resembles that of a dipole whose north and south
magnetic poles are located approximately at $78\frac{1}{2}°$ N, 69° W and $78\frac{1}{2}°$ S, 111° E,
the axis being inclined $11\frac{1}{2}°$ to the polar diameter and its centre displaced some
300 km towards Indonesia. Since this dipole is merely a close approximation
to the total magnetic field of the earth, its poles do not coincide with the surface
points where the dip needle stands vertical, i.e. where $H = 0$. Known as the
dip poles, or magnetic north and south poles, these are presently located at
75° N, 101° W and 67° S, 143° E.

The magnitudes of F at the north and south magnetic poles are 0·6 and 0·7
oersteds respectively; the minimum value, about 0·25 oersteds, occurs in the

Pacific, off Antofagasta in northern Chile. In a few locations F is larger than 3 oersteds, because of abnormal near-surface magnetic features. The line of zero inclination (where $Z = 0$) is never more than 15° from the equator, the largest deviation being to the south, in South America and in the eastern Pacific; in Africa and Asia, it is slightly north of the equator. Contours of equal declination are considerably more complex.

The various components of eq. (3.17) have been measured extensively over the earth's surface, both on land and at sea. World charts are compiled from such data: for example, *isogonic*, *isoclinic* and *isodynamic maps* show lines of equal declination, dip and equal values of F, H or Z respectively.

(b) *Origin of the main field.* The main geomagnetic field theoretically could be caused by an internal or external source, either of permanent magnetism or of unidirectional current flow, or it could be the result of current flowing in and out of the earth's surface. The latter possibility can be ruled out because the observed air-to-earth currents are much too small to account for the existing magnetic field of the earth. Spherical harmonic analysis of the observed surface magnetic fields shows that at least 99% is due to sources inside, the remaining 1% to sources outside the earth.

Several hypotheses have been put forward to account for the mechanism of the internal source. We are not concerned with details in this book, but some of the possibilities can be easily dispatched. For instance, the assumptions that the field is (i) a fundamental property associated with rotating bodies, or (ii) caused by the rotation of electric charges, break down because (1) magnetization arising from the angular velocity of the earth is much too small, and (2) potential gradient in the earth would be impossibly large. The suggestion that the earth is uniformly magnetized requires a value of $\mathscr{I}$ far too large for known surface rocks. Finally the inevitably high temperatures in the earth's interior ($\sim 2000°$ C) probably rule out the possibility of permanent magnetism at depth, since all known magnetic materials lose their magnetic properties at much lower temperatures (Fe- 750°C, Ni- 360°C, magnetite- 575°C, etc.), although it is not clear what effect the huge interior pressures may have on the temperature of demagnetization, known as the *Curie point*.

The present theory is that the main field is caused by electric currents circulating in the outer core – known to be liquid from seismic evidence – which extends from a radius of 1300 km to 3500 km. For several reasons the earth's core is assumed to be a combination of iron and nickel, both good electrical conductors. Even if the internal material were a nonconductor, however, the enormous pressures in the core ($\sim 10^6$ bars) may conceivably squeeze out electrons to form some sort of free-electron gas of suitable conductivity. Whatever the constituents of the core, the magnetic source is thought to be a type of self-excited dynamo, in which a highly conductive fluid moves about in complex mechanical motion, while electric currents, possibly caused by chemical or thermal variations, flow through it. The combination of motion and current creates a magnetic field. Since little is known, or is likely to be known, of the earth's core, the theoretical development is difficult. Laboratory tests have shown, however, that the dynamo source may be a valid explanation and also that it may account for certain slow or secular variations which are known to have occurred in the earth's magnetic field.

(c) *Secular variations.* As a result of nearly 400 years of continuous study, it has been established that the geomagnetic field is far from permanent. A striking example of this is seen in records of *I* and *D* from magnetic observatories in London and Paris extending back to about 1580. The inclination has changed some 10° (75° to 65°) and the declination about 35° (10° E to 25° W and back to 10° W) during this period. Although the latter variation appears to be cyclic, as if the magnetic poles were precessing at a very rapid (geologically speaking) time rate about the geographic polar axis, records from other localities are quite different.

Similar wandering of the supposedly fixed field is evident from long-term records of other magnetic elements. In all cases these secular variations appear to be regional rather than world wide. Although their source is not understood, it is thought to be internal, possibly connected with changes in convection currents in the core, in the core-mantle coupling and in the rotational speed of the earth.

3.3.3 *The external magnetic field*

Most of the remaining 1% of the geomagnetic field, which originates outside the earth, appears to be associated with electric currents in the ionized layers of the outer atmosphere. The variation in time is much more rapid than for the so-called permanent field. Several well-documented effects are listed below.

(i) A cycle of 11 years duration, correlated with sunspot activity, has a latitude distribution which indicates an external origin.

(ii) Solar diurnal variations, with a period of 24 hours and range of 30 γ, vary with latitude and season and are probably controlled by action of the sun on ionospheric currents.

(iii) Lunar diurnal variations of 25-hour period and amplitude about 2γ vary cyclically through the month and seem to be associated with a moon-ionosphere interaction.

(iv) Magnetic storms are transient disturbances with amplitudes as great as 1000γ in most latitudes and even larger in polar regions, where they are usually associated with the aurora. Although rather erratic, they often occur at intervals of about 27 days, the period correlating with sunspot activity. At the height of a magnetic storm, which may last several days, long-range radio reception is considerably affected and magnetic prospecting may be impractical.

3.3.4 *Local magnetic anomalies*

Aside from the occasional effects of magnetic storms, none of these variations in the geomagnetic field has any appreciable significance in magnetic prospecting. Nor do the latitude variations of *F* or $Z(\sim 7\gamma/\text{mile})$ require corrections to field data, except in the case of large-scale surveys, such as airborne magnetics employed in oil exploration.

As mentioned previously, important changes occur in the main field as the result of variations in the magnetic mineral content of near-surface rocks. These anomalies are occasionally large enough to double the main field locally. In general they do not persist over very great distances; that is to say, magnetic maps do not exhibit large-scale regional features, such as isostatic anomalies in gravity. Perhaps this is too sweeping a generalization, since a huge area like the Canadian Shield shows a definite magnetic contrast with the Western Plains. On the other hand, there is an enormous number of large, erratic variations within the Shield itself, resulting in an extremely complex magnetic map. The sources of these local magnetic anomalies can hardly be very deep, since temperatures at the bottom of the crust should be well above the Curie point. Thus they are associated with near-surface features.

3.3.5 *Magnetism of rocks and minerals*

(a) *Basic types of magnetism.* Since magnetic anomalies are entirely caused by the amount of magnetic minerals contained in the rocks, it is necessary to discuss these minerals (which are surprisingly few in number), and in particular their magnetic susceptibilities.

All materials – elements, compounds, etc. – can be classified in three groups according to their magnetic properties: diamagnetic, paramagnetic and ferromagnetic (the third group including several subdivisions).

(b) *Diamagnetism.* A *diamagnetic substance* is one which has a negative magnetic susceptibility. This means that in eq. (3.5) the intensity of magnetization induced in the substance by the field H is in the opposite direction to H. All materials are fundamentally diamagnetic, since the orbital motion of the negatively charged electrons in the substance, in the presence of the external field, H, is in such a direction as to oppose H. However, the diamagnetism will prevail only if the net atomic magnetic moment of all the atoms is zero when H is zero. This situation is characteristic of atoms with closed (completely filled) electron shells. This is a weak effect compared to other forms of magnetism described shortly. Many of the elements and compounds exhibit diamagnetism; the most common diamagnetic earth materials are graphite, gypsum, marble, quartz and salt.

(c) *Paramagnetism.* By definition all materials which are not diamagnetic are *paramagnetic*, that is, k is positive. In a paramagnetic substance each atom or molecule has a net magnetic moment in zero external field. Paramagnetism is characteristic of substances whose sub-shells are not filled to the maximum. Examples are the series of elements $_{22}Ca-_{28}Ni$ (including the ferromagnetics), $_{41}Nb-_{45}Rh$, $_{57}La-_{78}Pt$, $_{90}Th-_{92}U$. The effect decreases with temperature.

(d) *Ferromagnetism.* Iron, cobalt and nickel are paramagnetic elements in which the magnetic interaction between atoms and groups of atoms is so strong that there is an alignment of moments within large regions or domains of the substance. While the susceptibilities of diamagnetic and paramagnetic materials are mostly less than $\pm 10^{-3}$ *emu*, the three ferromagnetics have values 10^6 times this. Ferromagnetism also decreases with temperature and disappears entirely at the Curie temperature. Ferromagnetic minerals apparently do not exist in nature.

(e) *Ferrimagnetism*. Materials in which the magnetic domains are subdivided into regions which may be aligned in opposition to one another, but whose net moment is not zero when $\mathbf{H} = 0$, are called *ferrimagnetic*. This explanation presumes that either one set of sub-domains has a stronger magnetic alignment than the other, where equal numbers of both exist, or that there are more sub-domains of one type than the other. Examples of the first type of ferrimagnetic minerals are magnetite, titanomagnetite and ilmenite, oxides of iron or iron and titanium; pyrrhotite is a magnetic mineral of the second type. Practically all magnetic minerals are ferrimagnetic.

(f) *Antiferromagnetism*. If the net magnetic moments of parallel and antiparallel sub-domains cancel each other in a material which would otherwise be considered ferromagnetic, the resultant susceptibility is very small, of the order of paramagnetic substances. Such a material is called *antiferromagnetic* for obvious reasons; hematite is the most common example.

3.3.6 *Residual magnetism*

Although in many cases the magnetization of rocks depends mainly upon the present strength of the ambient geomagnetic field and the magnetic mineral content, in general this not true. In practise residual magnetism often contributes to the total magnetization in rocks, both in amplitude and direction. The effect is very complicated because of the dependence upon the magnetic history of the rocks. This residual is called *normal remanent magnetization* (NRM) and may be' due to one of several causes.

(i) *Chemical remanent magnetization* (CRM) takes place when magnetic grains increase in size or are changed from one form to another as a result of chemical action at moderate temperatures, that is, below the Curie point. This process is particularly significant in sedimentary and metamorphic rocks.

(ii) *Detrital magnetization* (DRM) occurs during the slow settling of fine-grained particles in the presence of an external field. Varved clays exhibit this type of remanence.

(iii) *Isothermal remanent magnetization* (IRM) is the residual left following the removal of an external field (see fig. 3.1). The earth's magnetic field is much too small to produce appreciable amounts of IRM. Lightning strokes produce IRM over very small irregular areas.

(iv) *Thermoremanent magnetization* (TRM) results when magnetic material is cooled from the Curie point in the presence of an external field. The remanence acquired in this fashion is particularly stable. In some cases it may be opposite in direction to the magnetizing field. This is the main mechanism in the magnetization of igneous rocks.

(v) *Viscous remanent magnetization* (VRM) is produced by a long exposure to an external field, the buildup of remanence being a logarithmic function of time; VRM is probably more characteristic of fine-grained rocks than coarse. The remanence is quite stable.

The discussion above shows that residual magnetic effects are quite complex. Studies of the magnetic history of the earth (*paleomagnetism*) indicate that the earth's field has varied in magnitude and has actually reversed its polarity many times in the past. On the other hand, there is no specific evidence that it has disappeared entirely for any appreciable period, or that it has ever existed except as a dipole field. Paleomagnetic tests have also shown that many rocks have remanent magnetization not oriented in the direction of, or opposite to, the present local field; such results support the hypothesis of continental drift.

Paleomagnetism is not a direct concern of the applied geophysicist. However, the information obtained through this discipline concerning residual magnetism in rocks has been of considerable value in the interpretation of magnetic anomalies, particularly since it is impossible to separate residual from induced magnetization in the field.

3.3.7 *Magnetic susceptibilities of rocks and minerals*

Magnetic susceptibility is the significant variable in magnetics, playing the same role as density in gravity interpretation. As in the case of density, it is not generally possible to measure susceptibility directly in the field; although instruments are available for this purpose, they can only be used on outcrop or on rock samples. Measurements of this type do not necessarily give the bulk susceptibility of the formation.

From fig. 3.1 it is obvious that k (hence μ also) is not a constant for a magnetic substance; as H increases, k increases rapidly at first, reaches a maximum and then decreases to zero. Furthermore, although magnetization curves have the same general shape, the value of H for saturation varies greatly with the type of magnetic mineral. Thus it is important, in making susceptibility determinations, to use a value of H about the same as the earth's field.

Since the ferrimagnetic minerals, particularly magnetite, are the main source of local magnetic anomalies, there have been numerous attempts to establish a quantitative relation between rock susceptibility and Fe_3O_4 concentration. For example, an extremely rough linear dependence (k ranging from 10^{-4} to 10^{-1} emu as the per cent of Fe_3O_4 by volume increases from $0 \cdot 05\%$ to 35%) is shown in one report, but the scatter is large and results from other areas are quite different.

Table 3.1 is a list of magnetic susceptibilities for a variety of rocks. Although there is a great variation in the k-values, even for a particular rock, and a wide overlap between different types, sedimentary rocks have the lowest average susceptibility and basic igneous rocks the highest. In every case the susceptibility

Table 3.1. *Magnetic susceptibilities of various rocks*

Type	Susceptibility × 10⁶ emu Range	Average	Type	Susceptibility × 10⁶ emu Range	Average
Sedimentary			Igneous		
Dolomite	0–75	10	Granite	0–4000	200
Limestones	2–280	25	Rhyolite	20–3000	
Sandstones	0–1660	30	Dolerite	100–3000	1400
Shales	5–1480	50	Augite-Syenite	2700–3600	
Av. Var. Sed. (48)	0–4000	75	Olivine-Diabase		2000
Metamorphic			Diabase	80–13,000	4500
Amphibolite		60	Porphyry	20–16,700	5000
Schist	25–240	120	Gabbro	80–7200	6000
Phyllite		130	Basalts	20–14,500	6000
Gneiss	10–2000		Diorite	50–10,000	7000
Quartzite		350	Pyroxenite		10,500
Serpentine	250–1400		Peridotite	7600–15,600	13,000
Slate	0–3000	500	Andesite		13,500
Av. Var. Met (61)	0–5800	350	Av. acid Ign.	3–6530	650
			Av. basic Ign.	44–9710	2600

depends only upon the amount of ferrimagnetic minerals present, mainly magnetite, sometimes ilmenite or pyrrhotite.

Table 3.2 lists susceptibilities for various minerals. The values of chalcopyrite and pyrite are typical of many sulphide minerals which are basically nonmagnetic.

Table 3.2. *Magnetic susceptibilities of various minerals*

Type	Susceptibility × 10⁶ emu Range	Average	Type	Susceptibility × 10⁶ emu Range	Average
Graphite		−8	Siderite	100–310	
Quartz		−1	Pyrite	4–420	130
Rock salt		−1	Limonite		220
Anhydrite, Gypsum		−1	Arsenopyrite		240
Calcite	−0·6– −1		Hematite	40–3000	550
Coal		2	Chromite	240–9400	600
Clays		20	Franklinite		36,000
Chalcopyrite		32	Pyrrhotite	10²–5 × 10⁵	125,000
Sphalerite		60	Ilmenite	2·5 × 10⁴–	
Cassiterite		90		3 × 10⁵	1·5 × 10⁵
			Magnetite	10⁵–1·6 × 10⁶	5 × 10⁵

It is often possible to locate minerals of negative susceptibility, although the negative values are very small, by means of detailed magnetic surveys. It is also worth noting that many iron minerals are only slightly magnetic.

3.3.8 *Magnetic susceptibility measurements*

(a) *General.* There are several laboratory instruments for measuring k. The simplest types involve a comparison of the sample with a standard, the sample generally being mounted in the Gauss-*A* position with respect to the detector. A more common commercial instrument employs an inductance bridge with the sample inserted in a coil; field measurements may be made on outcrop by coupling the bridge coil to the rock itself. Measurements of remanent susceptibility are considerably more complicated. A brief description of equipment for measuring k follows.

(b) *Laboratory and field measurement of* k. The simplest method of measurement is to compare the deflection produced on a tangent magnetometer by a prepared sample (either a drill core or powdered rock contained in a tube) with that of a standard sample of magnetic material (often $FeCl_3$ powder in a test tube) when the sample is in the Gauss-*A* position. The susceptibility of the sample is found from the ratio of deflections:

$$k_s = k_{std} \cdot d_s/d_{std},$$

d_s and d_{std} being the deflections for the sample and standard respectively. Obviously measurements can also be made in the field, but the sample must be of uniform size.

A similar comparison method employs an inductance bridge (Owen or Maxwell type of a.c. bridge) having several air-core coils of different cross-sections to accommodate samples of different sizes. The sample is inserted into one of the coils and the bridge balance condition compared with the balance obtained when a standard sample is in the coil. Alternatively the bridge may be calibrated over a range to give the sample susceptibility directly.

In the latter case the sample need not be shaped to have a particular geometry, although the calibration may not be valid for samples of highly irregular shape. This is the type of instrument used in field measurements on outcrop; the coil is usually of large diameter in this case. The bridge is balanced first with the coil remote from the outcrop and then lying on it. A calibration curve obtained with a standard provides a relation between k and the change in inductance.

(c) *Measurement of remanent magnetism.* One method of determining remanent magnetization uses an *astatic magnetometer*, consisting of two magnets of equal moment, rigidly mounted parallel to each other in the same horizontal plane with the poles opposed. The magnet system is suspended by a torsion fibre. The specimen is placed in various orientations below the astatic system and the angular deflections measured. This device in effect measures a magnetic field gradient produced by the sample, hence it is important to locate it in an area of very uniform field. Consequently the entire assembly is mounted inside a three-component coil system which cancels the earth's field.

An alternative instrument for analysis of residual component is the *spinner magnetometer*. The rock sample is rotated at high speed near a small pickup coil and its magnetic moment generates a.c. in the coil. The phase and intensity of the coil signal is compared with a reference signal of the same frequency and fixed phase generated by the rotating system itself. The total moment of the sample is obtained by rotating it about different axes.

3.4 Field instruments for magnetic measurements

3.4.1 *General*

The sensitivity required in magnetic instruments is between 1γ and 10γ in a total field rarely larger than $50,000\gamma$; thus it is much less than that of gravity meters. Some magnetometers measure both the absolute field and its variations, although this is not a particular advantage and is seldom done in magnetic prospecting surveys.

The earliest devices used specifically for magnetic exploration were modifications of the mariner's compass, such as the *Swedish mining compass*, which measured the dip I, as well as the declination, D. An assortment of instruments has been developed during the past fifty years for determination of the force components F, Z and H. The earlier models, known as *magnetic variometers*, are essentially dip needles of high sensitivity. More modern instruments include the *fluxgate*, *nuclear precession* and *rubidium vapor* (*optical pump*) *magnetometers*. These will be described in historical sequence, with only passing reference to obsolete types.

3.4.2 *Magnetic variometers*

(a) *General*. As the name implies, these instruments measure variations in magnetic field rather than absolute field. Variometers were widely employed in ground magnetics until about fifteen years ago. Although there has been considerable variety in the design, all have a sensitive magnetic dipole as the detecting element. Some variometers were designed to measure horizontal component, a few total field and/or inclination. However, most prospecting was done with vertical-component models.

One might specify two general requirements for this type of instrument. First, in order to eliminate the effect of the component which is not being measured, the axis of rotation of the dipole must be parallel to this component. Although the best of the vertical-component models fulfilled this requirement, the horizontal-component types did not, because of an inherent lack of sensitivity. The second desirable condition is that the dipole position should be maintained normal to the component that is being measured to achieve maximum sensitivity. Again, the z-component instruments were easily arranged to meet this condition by employing the force of gravity, while the equivalent situation was not so simple when measuring H. These points will be clarified by considering a few magnetometers of this type.

(b) *Sensitivity of vertical balance.* The dipole shown in fig. 3.5 is loaded with a small mass m below the centre of rotation, which lowers the centre of gravity of the system, and a counterweight of mass M, which maintains the dipole approximately horizontal in the earth's field. It is essentially a dip needle. If the needle is

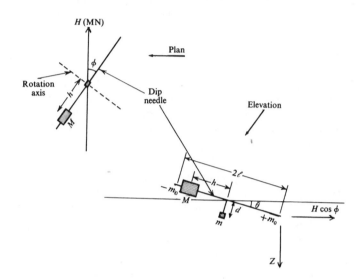

Fig. 3.5 Using a dipole to measure Z: M = latitude counterweight, m = sensitivity counterweight, 2ℓ = dipole length ($<$geometrical length).

free to move in a vertical plane with azimuth angle ϕ, and comes to rest at an angle θ with the horizontal, there are components of H, plus the weights mg and Mg, which tend to rotate the needle counterclockwise. These are balanced by the Z-component clockwise. The equilibrium is defined by the expression

$$2Zm_0\ell \cos \theta = g(Mh \cos \theta + md \sin \theta) + 2Hm_0\ell \cos \phi \sin \theta,$$

Hence, writing $\mathcal{M} = 2m_0\ell$,

$$Z = \frac{g}{\mathcal{M}}(Mh + md \tan \theta) + H \cos \phi \tan \theta \approx \frac{Mgh}{\mathcal{M}} + \left(\frac{mgd}{\mathcal{M}} + H \cos \phi\right)\theta,$$

$$(3.18a)$$

since θ is a small angle. Then the sensitivity relation is

$$\Delta Z = (mgd/\mathcal{M} + H \cos \phi)\Delta\theta. \qquad (3.18b)$$

Clearly the sensitivity could be made very great by choosing ϕ to make the bracketed term nearly zero. The position, however, would be inherently unstable; furthermore the value of H varies from place to place.

The ordinary dip needle, having low sensitivity, is often oriented in the magnetic meridian so that $\phi = 0$ or π. Magnetometers of good sensitivity are set up so that the dipole is across the meridian, $\phi = \pi/2$ and we have

$$\Delta Z = mgd \, \Delta\theta/\mathcal{M}.$$ (3.18c)

This relation is similar to that for the stable gravity meter (see eq. (2.28), §2.4.3*d*). To increase the sensitivity, it is necessary to make *m*, *d* small and $\mathcal{M}$ large. Obviously, the latter condition is limited by the size of the magnet.

(c) *Schmidt vertical balance.* This was the standard ground instrument for measuring *Z* for many years and was manufactured widely in Germany, England, U.S.A. and Canada. The complete magnetometer is shown in fig. 3.6 and schematics of the parts in fig. 3.7. The dipole consists of two thin tapered magnets of tungsten steel, connected together by a cube of aluminium. This cube serves to support the

Fig. 3.6 Schmidt vertical balance.

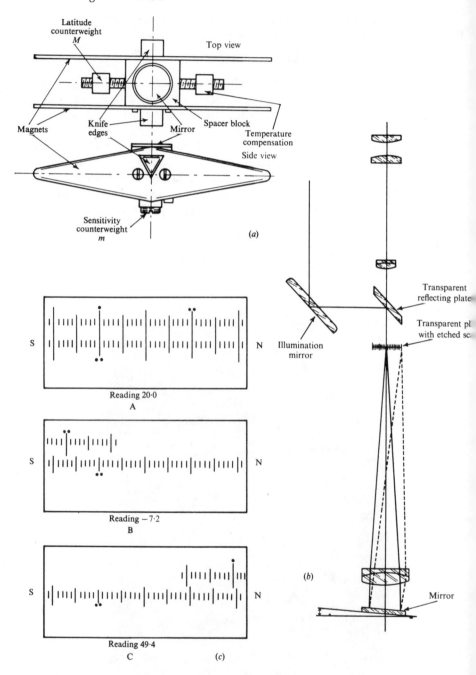

Fig. 3.7 Schmidt vertical balance (schematic). (*a*) Magnet system (after Heiland, 1940); (*b*) optical system; (*c*) scale.

quartz knife edges (fig. 3.7*a*) so that the system somewhat resembles the beam of a chemical balance. The magnets may be made horizontal by adjustment of the brass screw which holds the counterweight, usually called the *latitude counterweight*, because its position must be adjusted for a particular magnetic latitude. This is, in effect, a control of the position *h* of the mass *M* in eq. (3.18*a*) (see also fig. 3.7*a*). The sensitivity is altered by adjusting a large brass screw located below the knife edges on the lower face of the aluminum block (this varies *d* in eq. (3.18*c*)).

A circular mirror, attached to the upper side of the block, reflects a beam of light, thus indicating slight variations in the tilt of the needle. The optical arrangement is shown in fig. 3.7*b*. A small plane mirror outside the eyepiece may be adjusted to reflect daylight on the clear glass reflecting plate, arranged at 45° so as to redirect the light through the plate with the etched scale and down to the mirror, from which the light is reflected back to the eyepiece. When the balance is horizontal, the image of the scale reflected by the mirror will appear to coincide with the direct image. If the balance tilts, one scale appears to move over the other, as shown in fig. 3.7*c*.

A second counterweight, whose position with respect to the fulcrum can also be adjusted by a screw, is mounted opposite the latitude weight, as shown in fig. 3.7*a*. It provides limited automatic temperature compensation for the dipole system. When the temperature increases, the magnetic moment decreases and the local field appears to decrease because the couple $Z\mathcal{M}$ is smaller. At the same time the temperature sensitivity bar expands sufficiently that its couple, acting in the same sense as $Z\mathcal{M}$, corrects for the decrease in $\mathcal{M}$. The resultant sensitivity of the instrument to temperature change is generally reduced to less than $1\gamma/°C$ over a 20° range.

The magnetometer is contained in an insulated box which is mounted with levelling screws on a tripod head. The latter may be rotated to place the dipole in any azimuth. Projecting from the inner walls of the box and meshing with the magnets is a set of copper vanes for damping the free oscillations of the magnet system. Two levels, mounted on top of the box, indicate the horizontal position of the magnets. A thermometer is provided for use when temperature variations are very large. Mounted on the side of the box is a clamp control allowing the magnet system to be taken off its bearings for transportation.

Since it is necessary to orient the moving system perpendicular to the magnetic meridian, a compass is provided with the equipment. In some models this is a separate unit, mounted on the tripod head before setting up the magnetometer. Otherwise it is permanently attached to the tripod and its magnetic effect on the sensitive element is compensated.

One of the disadvantages of the older models was the small scale range of about 2000 γ. Although a set of calibrated bar magnets, which could be mounted in Gauss-*A* position on an adjustable hanger below the instrument, provided adequate extension of the range, it was not a convenient design.

A sensitivity of 10γ/division was not difficult to achieve in these instruments; in this regard they were perfectly adequate for ground prospecting and were as good as more modern instruments. Furthermore they were comparatively cheap and rugged. However, the fact that the element required careful orientation at each station and clamping between readings, in addition to the cumbersome auxiliary magnet system, made the field work slow. Consequently magnetometers of this type are not much used at present.

(d) *Torsion-head magnetometer*. Instruments in which the rotational couple of the magnetic field component is opposed by the torsion of a thin fibre supporting the dipole element have been used in various forms for about forty years. Recent models are a considerable improvement on the Schmidt balance, although not as convenient to use as the fluxgate magnetometer.

The general sensitivity relation of eq. (3.18c) becomes

$$\Delta Z = \tau \Delta \theta / \mathcal{M}, \tag{3.18d}$$

where τ is the torsion constant of the supporting fibre. The sensitivity is about the same as the Schmidt balance.

A model of this type of instrument made in Finland, which is quite inexpensive and functionally attractive, is illustrated in fig. 3.8. The sensing magnet, which is about the size of a needle, is suspended on a fine quartz fibre so that it is free to

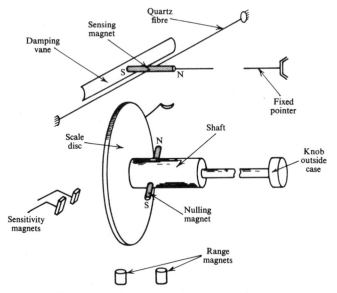

Fig. 3.8 Arvela magnetometer (schematic).

rotate in a vertical plane. Large oscillations are damped out by the thin sheet vane, attached to one end of the dipole, which acts as an air brake. A thin pointer, coaxially mounted on the other end of the dipole, lines up with a fixed wire when the dipole is horizontal. The relative positions of the two pointers may be viewed in an aperture on top of the case, where the moving pointer appears to oscillate from side to side, rather than vertically.

This is a null-reading instrument. The dipole is levelled by rotating a nulling magnet, mounted on a horizontal shaft below the balance magnet. A light disc with a 500-division scale on its edge is attached to the shaft and rotates with it. This scale reading may be observed in a second aperture on the top of the case.

The full rotation of the nulling magnet is 180°, that is, it provides a variable field at the balance dipole whose vertical component is given by

$$Z \approx 2\mathcal{M} \cos \theta / z^3, \quad 0 \leqslant \theta \leqslant \pi,$$

(see eq. (3.9*b*), §3.2.2*a*) where *m* is the nulling magnet dipole moment and *z* the distance between it and the sensing magnet. This cosine half-cycle is the scale on the disc. The full-scale range is nominally 10,000γ for a sensitivity of 20γ per division; the sensitivity may be adjusted by a pair of magnets on the side of the case. In addition, two auxiliary magnets in the base extend the total range to approximately 30,000γ. Some models also have a calibrating coil wound round the bottom of the case, which may be energized by an external d.c. source.

The detector magnet may be clamped for rough handling of the instrument, although this is generally unnecessary between station readings. Azimuth orientation is not critical. A test carried out on one instrument gave a maximum variation of 40γ for 360° rotation in azimuth and less than 20γ over a 90° sector in which the needle was mainly N–S. The chief disadvantage of this type of magneto-meter is that a tripod is required, since this increases reading time and is an added bulk in transport.

(e) *Horizontal-component instruments.* It was mentioned in §3.4.2*a* that the axis of rotation should be parallel to the unwanted component in making variometer readings. A device arranged to measure *H* in this fashion would thus resemble a compass. Several instruments of this type, such as the Kohlrausch, Thalén-Tiberg and Schmidt compensation magnetometers, were formerly used for this purpose. Since it was impossible to balance the needle with the force of gravity, they required compensating magnets. The Thalén-Tiberg model, for instance, used a magnet in the Gauss-*A* position on a calibrated arm at right angles to the magnetic meridian; the ratio of this field to *H* varied with the tangent of an angle. Such instruments, while simple in principle, had a low sensitivity and are now obsolete.

(f) *Schmidt horizontal balance.* This instrument is very similar to the vertical balance, the difference being that the twin magnets are mounted vertically and placed in the magnetic meridian so that the axis of rotation is perpendicular to the meridian. The centre of gravity is displaced by a weight below the rotational axis. The resultant equilibrium condition involves both the horizontal and vertical components, the former acting approximately at right angles to the dipoles.

The sensitivity relation then becomes

$$\Delta H \approx (mgh/\mathcal{M} + Z)\Delta\theta, \tag{3.18e}$$

where $\Delta\theta$ is the deflection from the vertical. Since it is not possible to dispose of the vertical component unless $\Delta\theta$ is zero, the instrument is highly sensitive to misalignment in azimuth. Because Z must be known at every station, it was customary to measure both components, using two instruments.

Apart from the difference in mounting the balance magnet and auxiliary magnet (the latter being in Gauss-B position below the head) the description of the vertical balance covers this model as well and the sensitivity is the same. In addition to the inherent disadvantages in these instruments, it is quite unusual to measure horizontal component nowadays; as a result they are very seldom used.

3.4.3 *Fluxgate magnetometer*

This device was originally developed during World War II as a submarine detector. Several designs of such instruments have been used for recording diurnal variations in the earth's field, for airborne geomagnetics and as portable ground magnetometers. The principle is the same in all applications.

The fluxgate detector consists essentially of a core of magnetic material, such as mu-metal, permalloy, ferrite, etc., having a very high permeability at low magnetic fields. In the most common design, two cores are each wound with primary and secondary coils, the two assemblies being as nearly as possible identical and mounted parallel so that the windings are in opposition. The two primary windings are connected in series and energized by a low frequency (50–1000 Hz) current produced by a constant current source. The maximum current is sufficient to magnetize the cores to saturation, in opposite polarity, twice each cycle. The secondary coils, which consist of many turns of fine wire, are connected to the inputs of a *differential amplifier*, that is, an amplifier whose output is proportional to the difference between two input signals.

The effect of saturation in the fluxgate elements is illustrated in fig. 3.9. In the absence of an external d.c. magnetic field, the saturation of the cores is symmetrical and of opposite sign near the peak of each half-cycle. As a result, the outputs from the two secondary windings cancel. The presence of an external field component parallel to the cores causes saturation to occur earlier for one half-cycle than the other. This produces an unbalance because the core fluxes do not cancel. Since the differential output voltage from the secondary windings is proportional to dB/dt, a series of voltage pulses is fed into the amplifier, as shown in fig. 3.9d. The pulse height is proportional to the amplitude of the biasing field of the earth. Obviously any component can be measured by suitable orientation of the cores.

The original problem with this type of magnetometer – a lack of sensitivity in the core – has been solved by the development and use of materials having sufficient initial permeability to saturate in fields of a few oersteds. Clearly the hysteresis loop should be as thin as possible. There remains a relatively high noise level, caused by hysteresis effects in the core. As mentioned above, the fluxgate elements should be identical. They should also be long and thin, or at least have a large ratio of length to diameter, to reduce eddy currents. Several further improvements have been introduced to increase the signal-to-noise ratio. The most significant are the following:

(i) by deliberately unbalancing the two elements, voltage spikes are present with or without an ambient field; the presence of the earth's field then increases the voltage of one polarity more than the other and this difference is amplified,

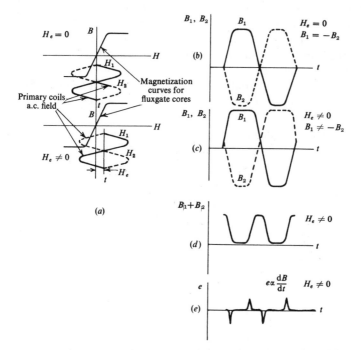

Fig. 3.9 Basic principle of the fluxgate magnetometer. (From Whitham, 1960.) (*a*) Magnetization of the cores, (*b*) graph of flux in the cores for $H_e = 0$, (*c*) graph of flux in the cores for $H_e \neq 0$; (*d*) sum of fluxes in the two cores for $H_e \neq 0$; (*e*) output voltage for $H_e \neq 0$.

(ii) since the odd harmonics of the a.c. component are cancelled fairly well in a reasonably matched set of cores, the even harmonics (generally only the second is significant) are amplified to appear as positive or negative signals, depending on the polarity of the earth's field,

(iii) most of the ambient field is cancelled and variations in the remainder are detected with an extra secondary winding,

(iv) negative feedback of the d.c. output of the amplifier can be used to reduce the effect of the earth's field,

(v) by tuning the output of the secondary windings with a capacitance, the second harmonic is greatly increased; in some models a phase-sensitive detector, rather than the difference amplifier, is used with this arrangement.

With one or more of these modifications, the signal-to-noise ratio can be increased by a factor of nearly 10. For instance, 100 $\mu v/\gamma$ signal versus 20 $\mu v/\gamma$ noise may be improved to 40 $\mu v/\gamma$ signal versus 1 $\mu v/\gamma$ noise. The tuned output in (v) gives a signal of about 1 mv/γ which is equivalent to 1° phase shift in a phase-sensitive detector.

Figure 3.10 is a block diagram of a fluxgate magnetometer used in a fixed installation for recording variations of the earth's field. Although more elaborate than a portable prospecting meter, particularly at the output end, the components are typical of the models using the second harmonic of the signal.

Another version of the fluxgate detector requires only one core from which even harmonics are amplified. This technique is employed in at least one portable ground instrument for measuring Z. The output is adjusted to read zero on the most sensitive scale of the meter by a current control which cancels the vertical field at a selected station. Variations are then read at other stations. Four or five scale ranges are included, from $\pm 1000\gamma$ to a top scale of $\pm 100{,}000\gamma$, with a maximum sensitivity of better than 10γ.

There are several fundamental sources of error in the fluxgate instrument. These include inherent unbalance in the two cores (if two are used), thermal and shock noise in cores, drift in biasing circuits, temperature sensitivity of $1\gamma/°C$ or

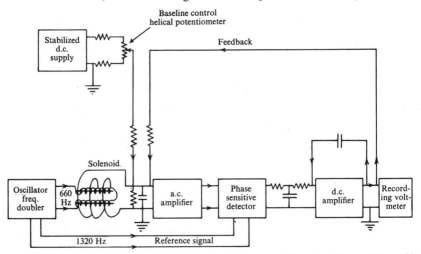

Fig. 3.10 Block diagram of recording fluxgate magnetometer. (Courtesy Serson, Dom. Obs., Ottawa.)

less. These disadvantages are minor, however, compared to the obvious advantages—direct readout, no azimuth orientation, rather coarse levelling requirements, light weight (5–7 lb) and small size. The price is about the same as that of the best variometers and the sensitivity quite as good. Another attractive feature is that any component of the magnetic field may be measured. Since no elaborate tripod is required, readings may be made very quickly, generally in about 15 seconds. A portable fluxgate instrument is shown in fig. 3.11.

3.4.4 *Total-field instruments*

(a) *Nuclear precession magnetometer.* This instrument, which makes use of an entirely different physical principle than the magnetometers considered so far, grew out of the discovery of nuclear magnetic resonance about 1945. In the original work it was found that some nuclei have a net magnetic moment which, coupled with their spin, cause them to precess about an axial magnetic field. The free precession of protons in the earth's magnetic field was detected in 1950 and was later used as the basis of a magnetometer (Packard and Varian, 1954).

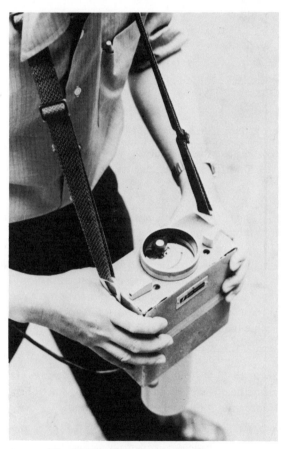

Fig. 3.11 Portable fluxgate magnetometer.

The nuclear precession magnetometer depends on the measurement of the free-precession frequency of protons (hydrogen nuclei) which have been polarized in a direction approximately normal to the direction of the terrestrial field. When the polarizing field is suddenly removed, the protons precess like a spinning top, the earth's field supplying the precessing force corresponding to that of gravity in the case of the top. The analogy is illustrated in fig. 3.12.

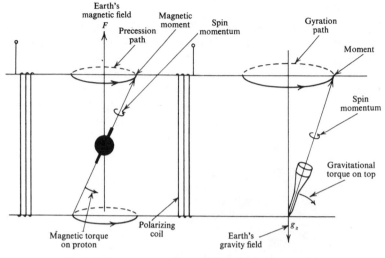

Fig. 3.12 Proton precession and the spinning-top analogy.

The proton precesses at an angular velocity ω, known as the *Larmor precession frequency*, which is proportional to the magnetic field strength F, so that

$$\omega = \gamma_p F. \qquad (3.19a)$$

The constant γ_p is the *gyromagnetic ratio of the proton*, that is, the ratio of its magnetic moment to its spin angular momentum. The value of γ_p is known to an accuracy of $\frac{1}{4} \times 10^{-4}$. Since precise frequency measurements are relatively easy, it is clear that the magnetic field can be determined to the same accuracy if it is possible to detect a signal derived from the precession of the proton. The detection is possible by means of a coil surrounding the sample; the proton, being a moving charge, induces in the coil a voltage which varies at the precession frequency. Thus we can determine the magnetic field from the relation

$$F = \omega/\gamma_p = 2\pi f/\gamma_p, \qquad (3.19b)$$

where the factor $2\pi/\gamma_p = 23\cdot4874 \pm 0\cdot0018$ in units of gammas/Hz.

When the polarizing field is cut off, the protons precess about the direction of the earth's field. Hence only total field may be measured.

The essential components of a magnetometer of this type include a source of protons, a polarizing magnetic field considerably stronger than that of the earth and directed roughly normal to it, a pickup coil coupled tightly to the source, an amplifier to boost the minute voltage induced in the pickup coil and a frequency measuring device. The latter operates in the audio range since, from eq. (3.19b), $f = 2130$ Hz for $F = 0\cdot5$ oersted; it must also be capable of indicating frequency differences of about $0\cdot4$ Hz for an instrument sensitivity of 10γ.

The proton source is usually a small bottle of water (the nuclear moment of ordinary oxygen is zero), or some organic fluid rich in hydrogen, such as methanol, ethyl alcohol, benzene, etc. The polarizing field of 50–100 oersteds is obtained by passing direct current through a solenoid wound round the bottle – which is oriented roughly E–W for the measurement. When the solenoid current is abruptly cut off, the proton precession about the earth's field is detected by a second coil as a transient voltage building up and decaying over an interval of about 3 seconds duration, modulated by the precession frequency. In some models the same coil is used for both polarization and detection. The modulation signal is amplified to a suitable level and the frequency measured. The last operation may be performed in several ways. A block diagram of the magnetometer circuits is shown in fig. 3.13.

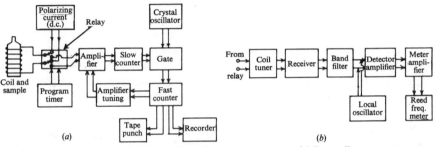

Fig. 3.13 Block diagram of nuclear precession magnetometer. (*a*) Recording magnetometer (from Dobrin, 1960); (*b*) portable direct-readout magnetometer.

The actual precession signal is worth considering in more detail. First, the polarizing field H_p aligns the protons in a volume V of the sample to produce a net nuclear magnetic moment $\mathcal{M}$, where

$$\mathcal{M} = k_n V H_p,$$

k_n being the nuclear susceptibility, which is about 3×10^{-10} emu/cm³ for water. For $H_p = 100$ oersteds and $V = 500$ cm³, the moment is 1.5×10^{-5}.

The polarizing field is cut off abruptly in a time which is short compared to the period of precession (50 μsec versus 500 μsec) and the protons precess about the axis of the earth's field. This effect requires a finite time to build up to a maximum, after which it decays as the precession becomes random; clearly it must persist for many precession cycles to make the frequency measurement possible. The decay time is determined by the spin-spin relaxation time of the nuclei, the moment decreasing at an exponential rate, according to the formula

$$\mathcal{M}_t = \mathcal{M} e^{-t/\tau},$$

where τ is about 2 seconds for water.

The voltage induced in the detecting coil, having N turns and cross-sectional area A, is proportional to $\mathcal{M}_t$ and f. Its maximum value is

$$V_1 = 2\pi f N A \cdot 4\pi \alpha \mathcal{M} \times 10^{-8} \text{ (volts)},$$

where α is a *filling factor* (depending upon the coupling between sample and coil) whose value is $0.01 - 0.05$. For the sample considered above, with $A = 50$ cm²,

$N = 1000$, $f = 2$ kHz, $\alpha = 0.04$, $\mathscr{M}_t = 0.3 \times 10^{-5}$ the signal is roughly $10\,\mu\text{V}$, sufficiently above the noise level to be detected.

The measurement of frequency may be carried out electronically by actually counting precession cycles in an exact time interval, or by comparing them with a very stable frequency generator. In one ground model the precession signal is mixed with a signal from a local oscillator of very high precision to produce low frequency ($\sim$100 Hz) beats, which are used to drive a vibrating reed frequency meter. Whatever method is used, the frequency must be measured to an accuracy of about $\frac{1}{4} \times 10^{-4}$ to realize the capabilities of the method. Although this is not particularly difficult in a laboratory or fixed installation, it poses some problems in small portable equipment.

The important advantages of the proton magnetometer are that it measures absolute field and that its sensitivity ($\sim 1\gamma$) is higher than any of the instruments considered so far. The fact that it requires no orientation or levelling makes it very attractive for marine and even more for airborne operations. Furthermore, there are essentially no mechanical parts in the detector element to cause trouble, although the electronic components are relatively complex.

Probably the main disadvantage is that only the total field can be measured. In addition the instrument cannot record continuously, requiring a second or more between readings. This is no problem in ground or marine prospecting; in an aircraft travelling at 180 mph, the distance interval is several hundred feet. Portable instruments of this type are somewhat heavier than other magnetometers.

(b) *The optical-pump magnetometer.* A variety of scientific instruments and techniques has been developed by using the energy involved in transferring atomic electrons from one energy level to another (ionization, spin, etc.). For example, by irradiating a gas with light or radio frequency waves of the proper frequency, electrons may be raised to a higher energy level; if they can be accumulated in such a state and then suddenly returned to a lower level, they release some of the previously absorbed energy in the process. This energy may be used for amplification, as in the case of *masers*, or for an intense light beam, such as produced by a *laser*.

An application of this type in geophysical instruments is the *optical pump magnetometer*. The principle of operation may be understood from an explanation of the diagram in fig. 3.14a, which shows three possible energy levels, A_1, A_2 and B for a hypothetical atom. Under normal conditions of pressure, temperature, etc., the atoms occupy ground state levels, A_1 and A_2. The energy difference between A_1 and A_2 is very small ($\sim 10^{-8}$ eV), representing essentially a magnetic fine structure due to atomic electron spins which are normally not all aligned in the same direction. Even thermal energies ($kT \sim 10^{-2}$ eV) are much larger than this, so that the atoms are as likely to be in level A_1 as A_2.

Level B represents a much higher energy, the transitions from A_1 or A_2 to B corresponding to infrared or visible spectral lines. If we irradiate a sample with a beam from which spectral line A_2B has been removed, atoms in level A_1 can absorb energy and rise to B, but atoms in A_2 will not be excited. When the excited atoms fall back to ground state, they may return to either level, but if they fall to A_1, they can be removed by photon excitation to B again. Clearly there is an accumulation of atoms in level A_2.

The technique of overpopulating one energy level in this fashion is known as optical pumping. As the atoms are moved from level A_1 to A_2 by this selective process, the sample becomes increasingly transparent to the irradiating beam; when all atoms are in the A_2 state, a suitable photosensitive detector will register a maximum current, as shown in fig. (3.14b). If now we apply an RF signal, having energy corresponding to the transition between A_1 and A_2, the pumping effect is nullified and the sample transparency drops to a minimum again. The proper frequency for this signal is given by $f = E/h$, were E is the energy difference between A_1 and A_2 and h is Planck's constant ($6 \cdot 62 \times 10^{-34}$ joule-sec).

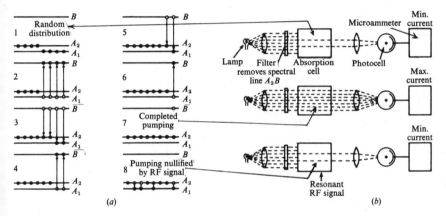

Fig. 3.14 Optical pumping. (*a*) Energy level transitions; (*b*) effect of pumping on light transmission.

To make this device into a magnetometer, it is necessary to select atoms having magnetic energy sub-levels which are suitably spaced to give a measure of the weak magnetic field of the earth. Elements which have been used for this purpose include caesium, rubidium, sodium and helium. The first three each have a single electron in the outer shell whose spin axis lies either parallel or antiparallel to an external magnetic field. These two orientations correspond to the energy levels A_1 and A_2 (actually the sub-levels are more complicated than this, but the simplification serves to illustrate the pumping action adequately) and there is a difference of one quantum of angular momentum between the parallel and antiparallel states. Instead of filtering to eliminate the optical energy transition A_2 to B, the irradiating beam is circularly polarized so that the photons in the light beam have a single spin axis. Atoms in sub-level A_1 then can be pumped to B, gaining one energy quantum by absorption, while those in A_2 already have the same momentum as B and cannot make the transition.

Figure 3.15 is a schematic diagram of the rubidium vapour magnetometer. Light from the Rb lamp is circularly polarized to illuminate the Rb vapour cell after which it is refocussed on a photocell. The axis of this beam is inclined approximately 45° to the earth's total field, causing the electrons to precess about the

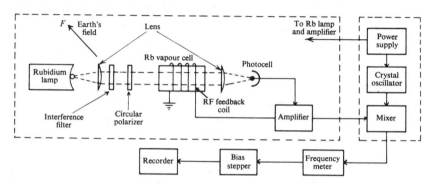

Fig. 3.15 Rubidium vapour magnetometer (schematic).

axis of the field at the Larmor frequency, exactly as in the case of protons in the nuclear precession magnetometer. At one point in the precession cycle the atoms will be most nearly parallel to the light beam direction, one half cycle later they will be more antiparallel. In the first position more light is transmitted through the cell than in the second. Thus the precession produces a variable light intensity which flickers at the Larmor frequency. If the photocell signal is amplified and fed back to a coil wound on the cell, the coil-amplifier system becomes an oscillator, whose frequency is given by

$$F = 2\pi f/g, \tag{3.20}$$

where g is the *gyromagnetic ratio* of the electron and F is the earth's field. It is a straightforward operation to measure the frequency of this oscillator.

For Rb, Na and He the values of $g/2\pi$ are approximately 4·67, 7 and 28 Hz/γ respectively, while the corresponding frequencies for $F = 0·5$ oersted are 233, 350 and 1400 kHz respectively. Because the g-factor for the electron is known to a precision of about 1 part in 10^7 and because of the relatively high frequencies involved, it is not difficult to measure magnetic field variations as small as 0·01γ with a magnetometer of this type; obviously it is capable of measuring absolute field as well.

The sensitivity of this instrument is considerably greater than is normally required in prospecting. Since about 1965 such magnetometers using rubidium or caesium vapour have been increasingly employed. Test surveys have been flown using two detectors separated vertically by about 120 ft. Such an arrangement measures dF/dz, the total field vertical gradient; theoretically the ultimate sensitivity is increased from about 0·02γ to 0·005γ by reducing random instrumental errors due to drift and motion of the single detector. However, independent pitch and yaw of the two birds produces noise which partially reduces the gain in sensitivity. A solid-state magnetometer of 0.001γ sensitivity is in the development state at Massachusetts Institute of Technology. With this device a gradiometer could be constructed having about 1 ft vertical separation on a rigid mounting.

Results from a high resolution test survey in the Timmins area of northern Ontario are shown in §3.7.2. In another application lightweight (2–5 lb) models of the rubidium vapour magnetometer have been installed and operated successfully in satellites. The cost of airborne surveys with the optical pump magnetometer is considerably greater than with the fluxgate meter.

Of the magnetometers described in the preceding pages the fluxgate type is most suitable for ground prospecting. It has also been used more in airborne work than the proton precession and optical pump instruments, partly because of its earlier development. For these reasons, as well as the moderate price, it is the most versatile instrument available at present. Although the nuclear precession and optical pump types have inherently greater sensitivity, this may not always be necessary or even desirable. Basically the latter two measure total field only, although they could be modified to measure H or Z by approximately cancelling the unwanted component; this procedure, however, is too elaborate for a portable ground instrument and impractical in the airborne model.

3.4.5 *Calibration of magnetometers*

All magnetometers may be calibrated by placing them in a suitably oriented, variable magnetic field of known value. In the earlier variometer models, the calibration could be made with the range magnets which are mounted in Gauss-*A* or Gauss-*B* position below the detector. Some instruments came equipped with calibrating coils.

The most dependable and direct calibration employs a *Helmholtz coil* large enough to surround the instrument. This is a pair of identical coils (same area, number of turns) coaxially spaced a distance apart equal to the radius. The resultant magnetic field, for a current i flowing through the pair when connected series-aiding, is directed along the axis and given by

$$H = \frac{3 \cdot 2\pi Ni}{\sqrt{(125)a}} \approx \frac{0 \cdot 9Ni}{a}, \tag{3.21a}$$

where i is in amperes, H in oersteds, a = the mean radius of either coil and also the coil separation (in centimetres) and N is the number of turns on either coil.

Since H varies directly with current, this can be written more conveniently for calibration purposes, in the form

$$\Delta H = 0 \cdot 09N\Delta i/a, \tag{3.21b}$$

where H is in gammas and i in microamperes.

The great advantages in using a coil arrangement of this type is that the field is uniform within about 6% over a cylinder of diameter a and length $0 \cdot 75a$, concentric with the coils. Thus it is a simple matter to determine the scale constant for any magnetometer in gammas per division. Also the Helmholtz pair may be oriented for any desired field component.

3.5 Field operations

3.5.1 *General*

Magnetic exploration is carried out on land, at sea and in the air. For areas of appreciable extent, reconnaissance over both land and sea is very conveniently done with the airborne magnetometer. Although it is not uncommon to use a portable instrument for ground reconnaissance (since the readings can be taken very quickly), most of this work is done by air.

In oil exploration magnetics is exclusively a reconnaissance tool, generally in the form of airborne surveys. Along with gravity it is done as a preliminary to seismic work to establish approximate depth, topography and character of the basement rocks. Since the susceptibilities of sedimentary rocks are relatively small, the main response is due to igneous rocks below the sediments. Consequently we would expect rather low magnetic relief in such areas and this is usually the case.

Mineral exploration reconnaissance is also done extensively from the air, frequently combined with airborne EM. In almost all cases of ground follow-up, detailed magnetic surveys are included as well. As in oil search, the method is usually indirect, that is, the primary interest is in geologic mapping and not in the magnetic concentration *per se*. Frequently the association of characteristic magnetic anomalies with base-metal sulphides, gold, asbestos, etc., has been used as a marker in mineral exploration. There is also, of course, an application for magnetics in the direct search for certain iron and titanium ores.

3.5.2 *Airborne magnetic surveys*

(a) *General.* The use of airborne magnetics is so widespread that it will be considered first as the leading field method. In Canada, government agencies had completely mapped well over half of the country by 1971; aeromagnetic maps on a scale of 1 mile/inch are available at a nominal sum. In other countries this approach has not been used to the same extent by local governments, but large areas in all parts of the world have been surveyed by private interests in the course of oil and mineral exploration programmes.

The fluxgate, nuclear precession and optical pump magnetometers have been adapted for airborne work. In fact, the fluxgate airborne model originally developed for submarine detection preceded the portable fluxgate ground instrument by several years.

The sensitivity of airborne magnetometers is generally greater than those used in ground exploration – about $1-5\gamma$ as compared with $10-20\gamma$. There are two reasons for this: first, because of the initial large cost of the aircraft and availability of space for the airborne set, it is practical to instal more sophisticated equipment than could be handled in a portable instrument; second, greater sensitivity is an advantage in making measurements several hundred feet above ground surface whereas the same sensitivity is usually unnecessary and may even be undesirable in ground surveys.

Airborne magnetometers measure the total field; this is no problem for the proton precession and optical pump instruments, since essentially they can measure nothing else. Although total field measurement is not always desirable, stability problems have so far prohibited detection of the vertical component with the fluxgate instrument.

(b) *Instrument mounting.* Aside from stabilization, there are certain problems in mounting the sensitive magnetic detector in an airplane, since the latter has a complicated magnetic field of its own, the result of permanently magnetized iron, induced magnetization, electrical currents in various circuits in the plane plus eddy currents in nonmagnetic metals, such as the aluminium fuselage and wing. One obvious way to eliminate these effects is to tow the sensing element some distance behind the aircraft. This was the original mounting arrangement and is still used regularly. The detector is housed in a streamlined cylindrical container known as the *bird* connected by cable to the airplane. The cable, which carries the low-level detector signal as well as power for preamplifier, stability controls, etc., is 100–500 ft long. Thus the bird may be as much as 200 ft nearer the ground than the aircraft and sufficient altitude must be maintained to prevent fouling it in tree tops. A photograph of a bird mounting is shown in fig. 3.16*a*.

An alternative scheme is to mount the detector on a wing tip or slightly behind the tail. The stray magnetic effects of the plane are minimized by permanent magnets and soft iron or permalloy shielding strips suitably arranged in and around the housing, by currents in compensating coils and by an assortment of metallic sheets for electric shielding of the eddy currents. The shielding is a cut-and-try process, since the magnetic effects vary with the aircraft and mounting location. Figure 3.16*b* shows a typical installation, with the magnetometer head in the tail.

(c) *Stabilization.* Since the nuclear and atomic type magnetometers measure total field, the problem of stable orientation of the sensing element is minor. Although the polarizing field in the proton precession instrument must not be parallel to the total field direction, practically any other orientation will do; the signal amplitude becomes inadequate only within a cone of about 5° around the axis of the field.

Stabilization of the fluxgate magnetometer is considerably more difficult, since the sensing element must be maintained accurately in the *F*-axis. This is accomplished with two additional fluxgate detectors which are rigidly oriented orthogonally with the first; that is, the three elements form a three-dimensional orthogonal coordinate system. The whole set is mounted on a small platform which rotates freely in all directions. When the sensing fluxgate is accurately aligned in the total field axis, there is zero signal in the other two. Any tilt away from this axis produces a signal in one or both of the control elements which,

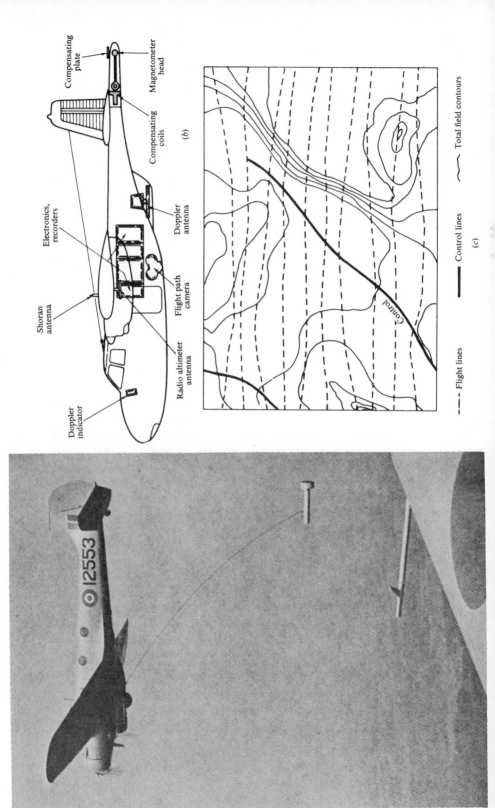

Compensating
plate

Magnetometer
head

Compensating
coils

(b)

Electronics,
recorders

Doppler
antenna

Shoran
antenna

Flight path
camera

Radio altimeter
antenna

Doppler
indicator

Control

Total field contours

Control lines

Flight lines

(c)

after amplification, drives a servo motor (one for each axis) to restore the system to proper orientation.

An alternative stabilization device employs a magnetic vane mounted radially on a disc which spins at 3600 rpm about an axis parallel to the total field. Any misalignment of this axis produces in the detector coil a 60-Hz component whose amplitude and phase are proportional to the tilt angle and its direction respectively. This signal controls two servo motors to correct the axial error.

(d) *Flight pattern.* For the actual survey, the magnetometer is flown along a preselected set of lines, the altitude (above the surface for land surveys, above sea-level for marine surveys) being kept as nearly constant as possible. The height is continuously recorded by an altimeter, either radio or barometric. The flight pattern, an example of which is shown in fig. 3.16c, consists of parallel lines spaced anywhere from $\frac{1}{8}$ to several miles apart, the heading where possible being normal to the main geological trend in the area. Since both the magnetometer sensitivity and the earth's field may change with time, due to instrument drift and diurnal or more sudden variations, it is customary to make repeat readings at several stations during a day's flight. This is sometimes done by a recording magnetometer on the ground, or by ground-station checks on a baseline roughly in the centre of the survey area and parallel to strike. A further check is generally obtained by flying several control lines which cross the regular flight pattern to record second readings at a number of intersections.

(e) *Effect of variations in flight path.* Needless to say, flight lines are normally neither parallel nor of constant altitude. Errors arising from deviations in flight path over a magnetic prism in northern latitudes have been discussed by Bhattacharyya (1970). A very brief summary is included here.

(i) Altitude: in flying a N–S path passing 500 ft west of the prism centre and maintaining an altitude of 1000 ft, the value of F_{max} is 2000γ; if the altitude varies smoothly from 1000 to 1100 ft and back again to 1000 ft over a horizontal distance of 1000 ft, starting 50 ft south and ending 950 ft north of the peak, the maximum decrease in total field is 135γ, all on the north flank of the profile. Thus the slope of the north flank is changed considerably. For a 300-ft increase in altitude over a distance of 2500 ft, the peak field is reduced 36% and, of course, the slope near the peak is also reduced.

(ii) Heading: if, during the same N–S traverse, the horizontal heading changes 100 ft east over 1000 ft immediately south of the peak, the maximum change in F is 68γ (or 3% of F_{max}) and occurs at the maximum. For a 200 ft easterly deviation over a distance of 2000 ft straddling the peak, the maximum change is -8% of F_{max}, occurring on the north flank and changing its slope.

These deviations are particularly significant in the analysis of high-resolution aeromagnetic data.

(f) *Aircraft location.* One of the problems in airborne surveying is to locate the aircraft at all times with respect to ground position. In the simplest method the pilot controls his flight path by using a plan marked on maps or aerial photographs, while an aerial camera takes photos on a strip film. The film and the magnetic record are tagged simultaneously at intervals as well. Over featureless terrain, such as snow or water, location by aerial photography must be replaced by more elaborate and expensive radio navigation controls, such as Shoran, in which aircraft position is determined by signals from two or more ground stations. An alternate type of fix may be obtained with Doppler radar equipment carried by the aircraft. As the name implies, the Doppler effect creates a shift in frequency between radar beams reflected fore and aft of the plane, which is a measure of ground speed. Two other beams directed to the right and left of the plane measure cross drift. This information, when combined with good flight direction data, automatically provides a precise flight path. A more detailed discussion of radio positioning devices may be found in §4.5.5e.

As mentioned previously, the magnetic information consists of profiles plotted on continuous strip chart or tape records. The fluxgate instrument produces a continuous record. The proton magnetometer cannot do this, because of the periodic process of polarization and precession; readings are made at the rate of about 1 per second, recorded in digital form on punched tape and then reproduced on a strip chart.

(g) *Advantages and disadvantages of airborne magnetics.* The additional parts required to convert a portable ground magnetometer to airborne surveying are nowhere near as elaborate as the necessary auxiliary equipment, which is complex, bulky and expensive. Nevertheless, airborne magnetic surveying is extremely attractive for reconnaissance because of the low cost per mile and the high speed of surveying. Worldwide prices during the last few years averaged $14 per line-mile for approximately one million line-miles covered. The equivalent price for ground magnetics during the same period was about $100 per line-mile.

The main advantages in airborne magnetics result from the great speed of surveying, which is an advantage in itself. This speed reduces the cost per line-mile, as we have seen, but also decreases the effect due to time variations in the magnetic field and instrument drift. Erratic magnetic features near surface, frequently a nuisance in ground work, are reduced or even eliminated by the height of the aircraft. In some cases the elevation may be chosen purposely to favour geologic structures of a certain size and depth. There are in general no operational problems associated with irregular terrain, sometimes a source of difficulty in ground magnetics (see §3.5.4b). (There are, of course, exceptional areas, such as the Rocky Mountains, Andes, etc., where airborne operations may be very difficult or even impossible.) As a consequence of the height of flight, the data are smoother, which may make the analysis easier. Finally, airborne magnetics can be done over water and in regions which are inaccessible for ground work.

The disadvantages in airborne magnetometer work apply mainly to mineral exploration. One is the high initial cost, which for surveys over small areas may be prohibitive. Another is caused by limited accuracy in locating the aircraft; under the best conditions horizontal position can be found only within 50 ft at an altitude of 1000 ft. In fact the advantages due to the height of the aircraft, cited above, may become limitations in mineral search where the magnetic anomalies are shallow and of limited extent.

3.5.3 *Shipborne magnetic surveys*

Both the fluxgate and proton magnetometers have been used in marine operations. There are no major problems in ship installation. The sensing element must be towed some distance (500–1000 ft) astern, to reduce magnetic effects of the vessel, in a watertight housing appropriately known as the *fish*, which usually rides about 50 ft below surface. Stabilization is necessary and is similar to that employed in the airborne bird.

Clearly there is no particular advantage and a considerable increase in cost using a ship instead of an aircraft in marine exploration. The operation becomes more practical, however, when carried out in conjunction with other surveys such as shipborne gravity. The main application has been in large-scale oceanographic surveying related to earth physics and petroleum search.

3.5.4 *Ground magnetic surveys*

(a) *General.* Magnetic surveying on the ground with portable instruments is a relatively old and well-established technique. In oil exploration, since the method is used for reconnaissance, the station spacing varies from $\frac{1}{4}$ mile to several miles. Airborne magnetics, however, has almost entirely replaced the ground survey in this connection. If for some reason the surface operation is necessary, it is convenient to combine it with gravity reconnaissance, since the station intervals and survey areas are normally about the same.

Mineral exploration with the ground magnetometer, on the other hand, is very widely used. Although the main application is in detailed surveys, ground magnetics may also be employed as a reconnaissance method in base-metal search to follow up geochemical reconnaissance before establishing a grid location and line cutting. In either case, stations are normally taken at close intervals, since the portable fluxgate meter (now almost exclusively used for ground surveys) can be read in a few seconds. Thus the station spacing is usually 50–200 ft, occasionally as small as 20 ft.

Most ground magnetic surveys measure the vertical component, Z. Thirty or more years ago, horizontal-component instruments were also used to some extent, although generally as an addition to measurement of Z. Nowadays the H-component is very rarely a subject of interest. Some total field ground surveys are done with the proton magnetometer. Occasionally the vertical gradient is measured in detailed work, that is, $\partial Z/\partial z$, using a vertical-component instrument at two or more elevations.

(b) *Corrections*. In precise work, as in gravity surveys, repeat readings should be made every two hours or so at a previously occupied station. This will provide corrections for diurnal and possible erratic variations of the magnetic field and changes in the instrument caused by temperature and drift. Alternately a second magnetometer may be installed at a base station to correct for the field variations. It must be admitted that neither of these procedures is followed very often in mineral prospecting. In many cases such precautions are unnecessary, because the anomalies of interest are larger than 500γ.

Since most ground magnetometers have a sensitivity of about 10γ, stations should not be located near railroad tracks, wire fences, drill-hole casings, culverts, that is, any sizable objects containing iron. The instrument operator should also divest himself of iron articles, or at least confine them to rear pockets. Belt buckles, compasses, knives, iron rings, even steel spectacle frames, can produce significant variations in magnetometer readings.

Apart from the effects discussed above – which are usually ignored or avoided, rather than corrected – the reductions required for ground magnetic data are insignificant, especially when compared to gravity. The vertical gradient of total field varies from a maximum of approximately 0.01γ/ft at the poles to a minimum -0.005γ/ft at the magnetic equator. The horizontal or latitude variation, while not uniform, is rarely larger than 10γ/mile between equator and poles. Thus free-air and latitude magnetic corrections are generally unnecessary.

The influence of topography on ground magnetics, on the other hand, can be very significant. This is quite apparent when making measurements in stream gorges, for example, where the neighbouring rock walls above the station frequently produce abnormal magnetic lows. It has been found that terrain anomalies as large as 700 gammas occur at steep ($45°$) slopes of only 30 ft in formations containing 2% magnetite ($k = 0.002$); the effect appears to increase linearly with susceptibility. For details the reader is referred to a paper by Gupta and Fitzpatrick (1971). In such cases a form of terrain correction is required, although it cannot be applied merely as a function of topography alone, as is done in gravity (eq. (2.19), §2.3.2e), since there are situations, for example, sedimentary formations of very low and uniform susceptibility, in which no terrain distortion of the local field is observed and no correction is needed.

A terrain-effect smoothing correction may be carried out, if necessary, by reducing measurements (usually of Z) from an irregular surface $z = h(x, y)$ to a horizontal plane, say $z = 0$, above it. To a first approximation, we can write this relation in the form of a Taylor's series of two terms:

$$Z(x, y, 0) = Z(x, y, h) - h(\partial Z/\partial z)_{z=h}. \qquad (3.22)$$

The gradient of Z may be obtained by direct measurement at the station, but usually it is calculated from the magnetic contour map with the aid of templates. The discussion of this calculation is deferred to the following section.

3.6 Interpretation

3.6.1 *General*

The end result of a magnetic survey, either ground or airborne, is a set of profiles or a magnetic contour map. In sedimentary areas there may be some similarity between magnetic and gravity maps. In general, however, the magnetic anomalies are more numerous, more erratic, less persistent and of larger magnitude than in gravity. Consequently the regional-residual separation is very difficult and often is not even attempted. Matching of field anomalies with simple geometrical shapes is the most common method of interpretation. Downward continuation and second-derivative techniques are also used to some extent. The former is not very suitable in areas of complex, shallow magnetics, characteristic of mineral exploration regions, but would be attractive for estimating the thickness of sedimentary formations encountered in petroleum surveys. Conversely, second-derivative analysis is more useful for interpretation in mineral prospecting to enhance small-scale features near surface, while upward continuation may be applied to suppress them. The latter technique, as mentioned in the previous section, may also be required to reduce topographic effects in ground magnetic work, since eq. (3.22) is a crude form of upward continuation. An elementary discussion follows.

3.6.2 *Smoothing by upward continuation*

Suppose we have a set of vertical-component magnetic data, $Z(x', y')$ on an irregular surface $z = h(x', y')$. We want to calculate the equivalent $Z(x, y)$ values on a plane above the surface h, say the plane $z = 0$, z being positive downwards. As mentioned in §2.6.6, this is known as the Dirichlet problem in mathematical physics which can be solved by the use of Green's Theorem. The procedure involves replacing the volume integral for Z (eq. (3.11)) with a surface integral. The relation then is

$$Z(x, y, z) = -\frac{1}{2\pi} \int_{-\infty}^{\infty} \int_{-\infty}^{\infty} \frac{Z(x', y')z \, dx' \, dy'}{\{(x' - x)^2 + (y' - y)^2 + z^2\}^{\frac{3}{2}}}, \quad (3.23)$$

where the left-hand side is the value of Z at coordinates (x, y) in a plane above or at the surface, and $z = h(x', y')$ on the right-hand side. This solution holds for all $z < h(x', y')$. If we convert to polar coordinates by putting $x' = r \cos \theta$, $y' = r \sin \theta$ and shift the origin to make $x = y = 0$, the expression becomes

$$Z(x, y, z) = -\frac{1}{2\pi} \int_0^{\infty} \int_0^{2\pi} \frac{Z(r, \theta) \, zr \, dr \, d\theta}{(r^2 + z^2)^{\frac{3}{2}}}. \quad (3.24)$$

Since the surface for the coordinates x', y' is in fact irregular, the conversion to plane polar coordinates is clearly not rigorous and is justified only because we are going to use a first-order approximation. If we define

$$\overline{Z(r)} = \frac{1}{2\pi} \int_0^{2\pi} Z(r, \theta) \, d\theta$$

as the average value of $Z(r, \theta)$ around a circle of radius r (see eq. (2.35), §2.6.2d) we have

$$Z(x, y, z) = -\int_0^\infty \frac{\overline{Z(r)} \cdot zr \cdot dr}{(r^2 + z^2)^{\frac{3}{2}}}.$$

Differentiating both sides of this expression with respect to z, we obtain

$$\frac{\partial Z}{\partial z} = -\int_0^\infty \frac{\overline{Z(r)}(r^2 - 2z^2)r \, dr}{(r^2 + r^2)^{\frac{5}{2}}}. \tag{3.25}$$

Part of this expression may be integrated directly but, since $\overline{Z(r)}$ is a function of r, the complete integration can only be done numerically, that is,

$$\frac{\partial Z}{\partial z} = \sum \overline{Z(r)} \int_0^\infty \frac{(2z^2 r - r^3) \, dr}{(r^2 + z^2)^{\frac{5}{2}}} = \sum_{r_n = 0}^\infty \left[\overline{Z(r_n)} \left\{ \frac{r^2_{n+1}}{(r_{n+1}{}^2 + z^2)^{\frac{3}{2}}} - \frac{r^2_n}{(r_n{}^2 + z^2)^{\frac{3}{2}}} \right\} \right]$$

$$= \frac{Z(0)r_1{}^2}{(r_1{}^2 + z^2)^{\frac{3}{2}}} - \overline{Z(r_1)} \left\{ \frac{r_1{}^2}{(r_1{}^2 + z^2)^{\frac{3}{2}}} - \frac{r_2{}^2}{(r_2{}^2 + z^2)^{\frac{3}{2}}} \right\} - \overline{Z(r_2)} \left\{ \frac{r_2{}^2}{(r_2{}^2 + z^2)^{\frac{3}{2}}} - \right.$$

$$\left. - \frac{r_3{}^2}{(r_3{}^2 + z^2)^{\frac{3}{2}}} \right\} - \cdots \tag{3.26}$$

$Z(0)$ is the measured value at the station whose coordinates on the irregular surface are (x, y, h). Then,

$$\left(\frac{\partial Z}{\partial z} \right)_{z=h} = \frac{Z(x, y, h)}{r_1(1 + h^2/r_1{}^2)^{\frac{3}{2}}} - \overline{Z(r_1)} \left\{ \frac{1}{r_1(1 + h^2/r_1{}^2)^{\frac{3}{2}}} - \frac{1}{r_2(1 + h^2/r_2{}^2)^{\frac{3}{2}}} \right\} -$$

$$- \overline{Z(r_2)} \left\{ \frac{1}{r_2(1 + h^2/r_2{}^2)^{\frac{3}{2}}} - \frac{1}{r_3(1 + h^2/r_3{}^2)^{\frac{3}{2}}} \right\} - \cdots$$

Using this expression for the vertical gradient we can find the value of $Z(x, y, 0)$ at any point in the plane $z = 0$ above the irregular surface. From eq. (3.22), this is

$$Z(x, y, 0) = Z(x, y, h) \left\{ 1 - \frac{h}{r_1(1 + h^2/r_1{}^2)^{\frac{3}{2}}} \right\} + h \left[\overline{Z(r_1)} \left\{ \frac{1}{r_1(1 + h^2/r_1{}^2)^{\frac{3}{2}}} - \right. \right.$$

$$\left. \left. - \frac{1}{r_2(1 + h^2/r_2{}^2)^{\frac{3}{2}}} \right\} + \overline{Z(r_2)} \left\{ \frac{1}{r_2(1 + h^2/r_2{}^2)^{\frac{3}{2}}} - \frac{1}{r_3(1 + h^2/r_3{}^2)^{\frac{3}{2}}} \right\} + \cdots \right]. \tag{3.27a}$$

The procedure is to draw concentric circles around the station (x, y, h) of successively larger radii $r_1, r_2, \ldots r_n$, average the values of Z around each circle and multiply each average by the appropriate square-bracket term. Note that h varies from station to station.

For two-dimensional features, where the gradient in the strike direction

(y-axis) is very small, we can replace r by x in eq. (3.27a) and the one-dimensional effect becomes

$$Z(x, 0) = Z(x, h) \left\{ 1 - \frac{2h/\pi}{x_1(1 + h^2/x_1^2)} \right\} + \frac{2h}{\pi} \left[\overline{Z(x_1)} \left\{ \frac{1}{x_1(1 + h^2/x_1^2)} - \right. \right.$$

$$\left. \left. - \frac{1}{x_2(1 + h^2/x_2^2)} \right\} + \overline{Z(x_2)} \left\{ \frac{1}{x_2(1 + h^2/x_2^2)} - \frac{1}{x_3(1 + h^2/x_3^2)} \right\} + \cdots \right],$$

$$(3.27b)$$

where

$$\overline{Z(x_n)} = Z(x_n) - Z(x_{-n}).$$

Actually this development is merely an approximation of the upward-continuation technique, since we used only the first expansion term of the Taylor's series in eq. (3.22); it is adequate for a topographic correction in magnetics, where neither the elevations nor the lateral susceptibility variations are accurately known.

For a more precise upward continuation of three-dimensional anomalies, we can employ the formulae of Henderson (1960), given below for the vertical magnetic component:

$$Z(-h) = \sum_{i=0}^{10} \overline{Z(r_i)} K(r_i, h), \qquad (3.28a)$$

where $Z(-h) =$ vertical component at height h above $z = 0$, $\overline{Z(r_i)} =$ average value on $z = 0$ for circle radius r_i, $K(r_i, h) =$ upward continuation coefficients.

Table 3.3 lists the values of coefficients for $h = 1, 2 \ldots 5$ in units of grid spacing, together with the various circle radii.

Table 3.3. *Coefficients for upward continuation*

i	r_i	$K(r_i, 1)$	$K(r_i, 2)$	$K(r_i, 3)$	$K(r_i, 4)$	$K(r_i, 5)$
0	0	0·11193	0·04034	0·01961	0·01141	0·00742
1	1	0·32193	0·12988	0·06592	0·03908	0·02566
2	$\sqrt{2}$	0·06062	0·07588	0·05260	0·03566	0·02509
3	$\sqrt{5}$	0·15206	0·14559	0·10563	0·07450	0·05377
4	$\sqrt{8}$	0·05335	0·07651	0·07146	0·05841	0·04611
5	$\sqrt{13}$	0·06586	0·09902	0·10226	0·09173	0·07784
6	5	0·06650	0·11100	0·12921	0·12915	0·11986
7	$\sqrt{50}$	0·05635	0·10351	0·13635	0·15474	0·16159
8	$\sqrt{136}$	0·03855	0·07379	0·10322	0·12565	0·14106
9	$\sqrt{274}$	0·02273	0·04464	0·06500	0·08323	0·09897
10	25	0·03015	0·05998	0·08917	0·11744	0·14458

Multiplying these coefficients by the appropriate values of $\overline{Z(r_i)}$ and adding the ten products we get quite precise values for the upward-continued field (within 2% of the theoretical value) over simple geometric shapes, such as a sphere, thin rod, etc.

For two-dimensional anomalies we need only consider a profile and eq. (3.28*a*) is replaced by the following relation:

$$Z(-h) = \sum_{i=0}^{n} \overline{Z(x_i)} K(x_i, h),$$ (3.28*b*)

where

$$K(x_i, h) = \frac{1}{2\pi} \frac{h(x_{i+1} - x_{i-1})}{h^2 + (x_i - x)^2}, \, i = 0, 1, 2, 3, \ldots$$

Expanding this summation for clarity and putting $x = 0$, we have

$$Z(-h) = Z(x_0) \frac{h}{2\pi} \frac{(x_1 - x_{-1})}{h^2} + \{Z(x_1) - Z(x_{-1})\} \frac{h(x_2 - x_0)}{2\pi(h^2 + x_1^2)} +$$

$$+ \{Z(x_2) - Z(x_{-2})\} \frac{h(x_3 - x_1)}{2\pi(h^2 + x_2^2)} + \ldots \quad (3.28c)$$

Obviously these formulae are equally suitable for gravity applications. Furthermore an approximate downward continuation can be carried out by first calculating the field on the plane $z = -1$, say, then proceeding as in §2.6.6.

3.6.3 *Crude interpretation*

Because of the erratic and complex character of magnetic maps, interpretation is quite often only qualitative. Indeed the technique is something of a fine art; the interpreter experienced in magnetics can usually see geological structure merely by looking at a magnetic map, much as one can visualize surface features from the contours of a topographic map. Frequently there is a connection between magnetics and topography, as well as with buried geologic structures, particularly in mineral exploration areas.

In sedimentary regions, particularly where the basement depth is roughly 5000–10,000 ft, the magnetic contours are normally smooth and the variations quite small. Here the magnetic anomalies are reflections of the basement rocks rather than of near-surface features. Also the larger anomalies are usually caused by susceptibility variations, rather than topographic relief in the basement. This was discussed in §2.6.1 to illustrate the difference between deep-seated gravity and magnetic anomalies caused by basement topography. Had we considered a lateral change in susceptibility (e.g., a vertical contact between two slabs of different susceptibility) instead of a vertical step in the basement, the magnetic anomaly could have been quite large, since it depends only on the magnitude of Δk and is independent of the depth. Consequently, anomaly magnitude is not of much value in finding basement depth; instead, depth calculations are usually based upon anomaly shape, especially sharpness.

Regions like the Precambrian, Appalachians, etc., on the other hand, in which metamorphic and igneous rocks predominate, usually exhibit complex magnetic variations. In such areas basement features are frequently camouflaged by higher frequency magnetic effects originating nearer surface. Several techniques for separating these anomalies have been mentioned in the previous section. For crude preliminary interpretation a visual study of the magnetic map can be quite fruitful; in this regard practice and experience are essential.

A few rough ground rules for large-scale magnetic anomalies can be cited. From a study of aeromagnetic maps over huge areas in Western and Central North America and Venezuela, Affleck (1963) has concluded that there are predominant trends for magnetic anomaly systems, mainly NE–SW, NW–SE, E–W and N–S, and that the spacing between parallel features is about four miles. These systematic trends appear to fall into two groups. Within a single magnetic-tectonic province the dominant direction is usually NE–SW or NW–SE and normally terminates at the province boundary. Superimposed almost everywhere on these are moderate to weak features, trending E–W and N–S, which frequently extend across the province boundaries and are probably of more recent origin than the dominant trends.

The areas considered in this statistical analysis presumably are mainly sedimentary. A cursory study of the Canadian Appalachian region (the Maritime provinces and Eastern Quebec south of the St Lawrence) as well as Northern Saskatchewan and Alberta appears to confirm the trends. Parts of the Canadian Shield are so complex that they would require considerable study; even here, however, large-scale northeast trends are quite obvious on the east and west flanks.

Thus, qualitative analysis of magnetic data can be a valuable tool in structural mapping. Needless to say, it should only become a substitute for the latter if no geological information is available. The effect of remanent magnetization, in particular, can lead to incorrect interpretation when it is overlooked in this type of work.

3.6.4 *Magnetic effects of simple shapes*

(a) *General.* As in gravity and EM, it is the custom to match field anomalies with models. The problem is considerably more difficult because of the dipole character of magnetism and because of the possibility of remanence in the magnetic source. Consequently only the most elementary shapes are considered.

This section is divided in two parts. In the first the vertical component is derived for several simple geometrical shapes; these results are suitable for curve matching of anomalies obtained in ground surveys, where the vertical component is normally measured. In the second, total field response is obtained for comparison with airborne results.

(b) *The isolated pole.* Strictly speaking this magnetic feature is a fiction, but practically it may be used to represent a steeply dipping dipole of cross-section small compared to its length, whose lower pole has negligible effect at surface – a chimney or pipe-like structure. The geometry and typical profiles are shown in fig. 3.17*a*.

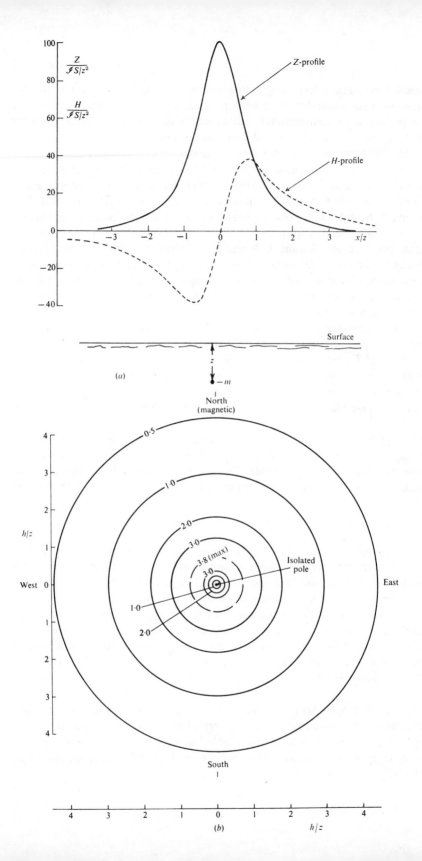

$\dfrac{Z}{\mathscr{I}S/z^2}$

$\dfrac{H}{\mathscr{I}S/z^2}$

Z-profile

H-profile

Surface

z

$-m$

(a)

North
(magnetic)

h/z

0.5

1.0

2.0

3.0

3.8 (max)

3.0

Isolated
pole

West

East

1.0

2.0

South

(b)

h/z

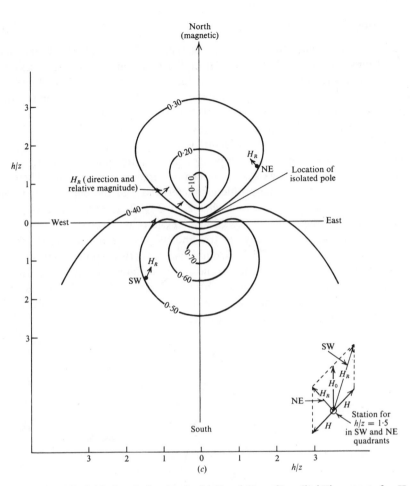

Fig. 3.17 Magnetic field of an isolated pole. (*a*) Z and H profiles; (*b*) $|H|$ contours for $H_{\max}$ $= 3 \cdot 8 \gg H_0$ (note that the direction of H is towards the pole at all points); (*c*) $|H_R|$ contours for $H_{\max} = H_0 =$ horizontal component of the earth's field $= 0 \cdot 38$.

The magnetic potential at P is given by $A = m/r$, where m is the pole strength. If the body is magnetized by induction only (and assuming the magnetization is uniform) we can put $m = \mathscr{I}S$, where $\mathscr{I}$ is the intensity of magnetization and S the cross-section area. Thus the vertical and horizontal components are

$$Z = -\partial A/\partial z = \mathscr{I}Sz/r^3, \quad H = -\partial A/\partial h = \mathscr{I}Sh/r^3 = \mathscr{I}S\sqrt{(x^2 + y^2)}/r^3,$$
$$(3.29)$$

where

$$h^2 = x^2 + y^2, r^2 = h^2 + z^2.$$

The horizontal component is included here because, although it is now seldom measured, it has some significance in this case.

If the remanent magnetization of the pipe is appreciable the induction magnetization I must be modified to $I + I_r$. Since the remanence may not necessarily
be in the same direction as the induction field, unless one of them is predominant,
the anomaly may be considerably different from that of either I or I_r alone.
However, in a long rod having appreciable dip, the remanence will (over about
80% of the earth's surface area) be mainly along the rod axis. Thus the effect of
random remanence in this type of feature is usually small.

The vertical- and horizontal-component profiles are illustrated in fig. 3.17a
for a traverse along the x-axis. Z_{max} is located directly over the pole. There are
several relations between the profile characteristics and the depth of the pole.
When $Z = Z_{max}/2$, $x_\frac{1}{2} = 0.75z$ and when $Z = Z_{max}/3$, $x_\frac{1}{3} = z$, where $x_\frac{1}{2}$, $x_\frac{1}{3}$ are
the half-widths of the profile at $Z_{max}/2$ and $Z_{max}/3$ respectively.

The H-profile is perfectly asymmetric and its positive half intersects the Z-
profile very nearly at $Z_{max}/3$. The horizontal distance between positive and negative
peaks is approximately $3z/2$. This profile would be independent of the traverse
direction only in the absence of the earth's field or if the effect of the pole is vastly
larger than the horizontal component of the earth's field, as shown in fig. 3.17b;
clearly such a situation would be very rare.

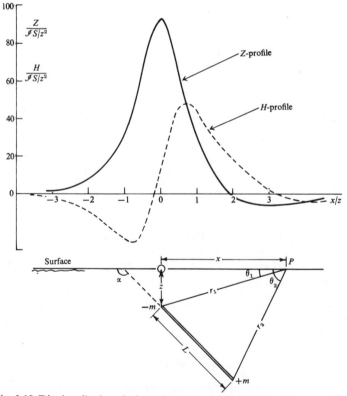

Fig. 3.18 Dipping dipole polarized along its axis, Z and H profiles, $\alpha = 135°$.

A more realistic condition is shown in fig. 3.17c, where the horizontal component of the earth's field, H_0, is equal to H_{max} for the pole. The contours of the resultant horizontal component are considerably distorted in comparison with those of fig. 3.17b, and it is clear that the H-profile of fig. 3.17a is not reproduced on any traverse across the pole under these conditions.

(c) *The dipole.* A three-dimensional magnetic structure which occurs quite frequently is the dipole. Although an expression for the dipole field has already been derived (see eq. (3.9)), it is convenient to derive it again using a different geometry as shown in fig. 3.18.

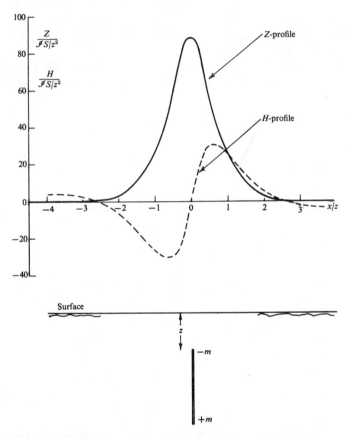

Fig. 3.19 Vertical dipole polarized along its axis, Z and H profiles.

If we write the pole strength m in the form $m = \mathscr{I}S = kF_0S$, k being the susceptibility and F_0 the earth's total field, the vertical field being positive downward, we have for the vertical component of the dipole field for a traverse in a vertical plane containing the dipole

$Z = $ (Vert. comp. of field due to $-m$) $-$ (Vert. comp. of field due to $+m$)

$$= kF_0S\left\{\left(\frac{1}{r_1{}^2}\right)\left(\frac{z}{r_1}\right) - \left(\frac{1}{r_2{}^2}\right)\left(\frac{z + L\sin\alpha}{r_2}\right)\right\}$$

$$= kF_0S\{(z/r_1{}^3) - (z + L\sin\alpha)/r_2{}^3\}. \tag{3.30a}$$

Similarly the horizontal component is given by

$$H = kF_0S\{(x/r_1{}^3) - (x + L\cos\alpha)/r_2{}^3\}. \tag{3.30b}$$

Since we have assumed that the rod is magnetized along its axis, these expressions are valid only under one or both of the following conditions: (i) the intrinsic field of the rod is very much larger than the external field; (ii) the rod is oriented along the external field direction.

The first assumption is quite possible, the second highly unlikely. The general derivation – arbitrary orientation of the dipole in an external field – is very

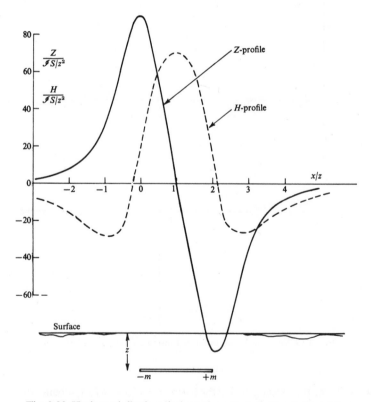

Fig. 3.20 Horizontal dipole polarized along its axis, Z and H profiles.

difficult but can be approximated by using a thin sheet of finite length. This derivation will be discussed in the next section. Two special orientations may be mentioned in connection with eqs. (3.30a) and (3.30b). First, when the rod is vertical and vertically polarized, $\alpha = \pi/2$, $F_0 = Z_0$ and we have

$$Z = kZ_0S \left(\frac{z}{r_1^3} - \frac{z+L}{r_2^3} \right), \tag{3.30c}$$

$$H = kZ_0Sx \left(\frac{1}{r_1^3} - \frac{1}{r_2^3} \right). \tag{3.30d}$$

In the second case, if the rod is horizontal and horizontally polarized, $\alpha = \pi$, $F_0 = H_0$ and

$$Z = kH_0Sz \left(\frac{1}{r_1^3} - \frac{1}{r_2^3} \right), \tag{3.30e}$$

$$H = kH_0S \left(\frac{x}{r_1^3} - \frac{x-L}{r_2^3} \right). \tag{3.30f}$$

The vertical orientation is a fair approximation in high latitudes and is often used in aeromagnetic interpretation. Presumably the horizontally polarized dipole might be equally common in equatorial regions, but one would expect it to be the result of remanence rather than induction.

Profiles for the various Z- and H-components in eqs. (3.30a) to (3.30f) are shown in figs. 3.18 to 3.20. Compared to the isolated pole, the vertical-component profiles are somewhat sharper and the maxima slightly smaller, due to the effect of the N-seeking pole at finite depth. Analytically it can be shown that the value of x corresponding to the depth to the top of the rod occurs at about $0.23Z_{max}$, when it is very short, and at about $0.35Z_{max}$, when it is very long.

Dip may be estimated roughly by the symmetry of the Z-profile, which has a steeper slope and negative tail on the downdip side; this tail is not at all pronounced, however, unless the dip is less than 30°. In fig. 3.20 the Z-profile is completely asymmetrical, with the peaks almost exactly above the ends of the rod for $z/L \ll 2$; when $z/L > 1$, the peaks occur beyond the ends. In this orientation it is not possible to estimate depth although obviously a steep slope at the zero crossover would be caused by a shallow source.

(d) *Dipole – arbitrary orientation.* It is possible to get an approximate solution for the arbitrarily oriented dipping dipole, magnetized by induction, from eq. (3.36j), §3.6.4k, for the thin dipping sheet of finite length (see also fig. 3.26a).

Putting $Y = t$, where $2Y$ is the strike length, t the thickness, the sheet becomes a dipole of cross-section $S = 2t^2$ and the vertical component is

$$Z = kS \left[\frac{H_0 \sin \beta \sin (\alpha + \theta_2) + Z_0 \cos (\alpha + \theta_2)}{r_2 \sqrt{(r_2{}^2 + t^2)}} - \right.$$

$$- \frac{H_0 \sin \beta \sin (\alpha + \theta_1) + Z_0 \cos (\alpha + \theta_1)}{r_1 \sqrt{(r_1{}^2 + t^2)}} +$$

$$+ \left\{ \frac{x \sin \alpha \cos \alpha (H_0 \sin \beta \sin \alpha + Z_0 \cos \alpha)}{x^2 \sin^2 \alpha + t^2} \right\}$$

$$\left. \times \left\{ \frac{\cot (\alpha - \theta_2)}{\sqrt{(r_2{}^2 + t^2)}} - \frac{\cot (\alpha - \theta_1)}{\sqrt{(r_1{}^2 + t^2)}} \right\} \right]. \qquad (3.31a)$$

When $r_2 \to \infty$, we have an isolated pole and the expression is somewhat shorter:

$$Z = -kS \left\{ \frac{H_0 \sin \beta \sin (\alpha + \theta_1) + Z_0 \cos (\alpha + \theta_1)}{r_1 \sqrt{(r_1{}^2 + t^2)}} + \right.$$

$$\left. + \frac{x \sin \alpha \cos \alpha \cot (\alpha - \theta_1)(H_0 \sin \beta \sin \alpha + Z_0 \cos \alpha)}{(x^2 \sin^2 \alpha + t^2) \sqrt{(r_1{}^2 + t^2)}} \right\} \qquad (3.31b)$$

Profiles for dipoles dipping $45° W$ and $45° S$ are shown in figs. 3.21a and 3.21b. In each case the earth's field inclination is $45°$ N. The surprising feature of these profiles is the large negative values when the depth extent is very great.

The depth d is approximately half the horizontal distance between positive and negative peaks of the profiles. The depth extent appears to be related to the amplitude of the negative tail. Estimate of dip would be difficult.

(e) *Polarized sphere.* As mentioned previously it is possible to use gravity results (actually the torsion balance parameters U_{zz}, U_{zz}, etc.) to determine magnetic effects for the equivalent geometrical shapes. The conversion is made possible by Poisson's Relation, eq. (3.14). Two-dimensional features are generally amenable to this derivation; three-dimensional structures, on the other hand, usually are not, except for the sphere.

The coordinate system in relation to the sphere is shown in fig. 3.22. The sphere centre is the origin of coordinates and it is assumed that the sphere is uniformly magnetized, by induction only, in the field $\mathbf{F}_0$. Since $\mathbf{I}$ is parallel to $\mathbf{F}_0$, we take $\boldsymbol{\alpha}_1$ in eq. (3.14a) in the direction of $\mathbf{F}_0$, hence

$$A = - \frac{\mathscr{I}}{\gamma \sigma} \frac{\partial U}{\partial \alpha} = - \frac{\mathscr{I}}{\gamma \sigma} \nabla U . \boldsymbol{\alpha}_1 = - \frac{1}{\gamma \sigma} \nabla U \cdot \mathbf{I}$$

$$= - \frac{1}{\gamma \sigma} \left(\mathscr{I}_z \frac{\partial U}{\partial x} + \mathscr{I}_y \frac{\partial U}{\partial y} + \mathscr{I}_z \frac{\partial U}{\partial z} \right),$$

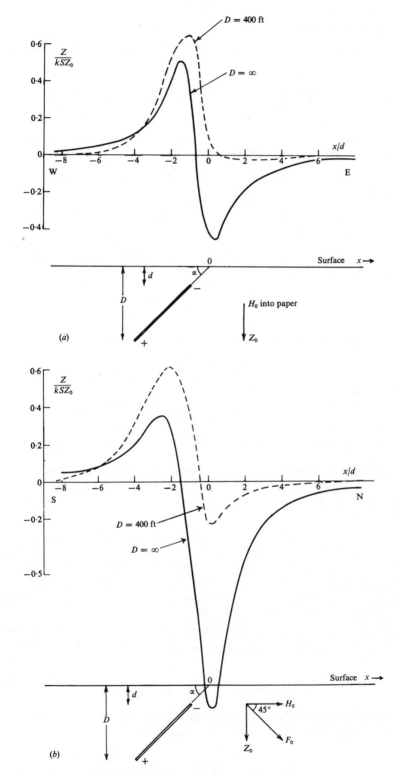

Fig. 3.21 Dipping dipole polarized by the earth's field, Z profiles. $\alpha = 45°$, $d = 100$ ft.
(a) Dipole dipping 45°W; (b) dipole dipping 45°S.

where $\mathscr{I}_x = kH_0 \sin \beta$, $\mathscr{I}_y = -kH_0 \cos \beta$, $\mathscr{I}_z = kZ_0$. Then the vertical component of the field is

$$Z = -\frac{\partial A}{\partial z} = \left(\frac{k}{\gamma\sigma}\right)\frac{\partial}{\partial z}\left(H_0 \sin \beta \frac{\partial U}{\partial x} - H_0 \cos \beta \frac{\partial U}{\partial y} + Z_0 \frac{\partial U}{\partial z}\right). \tag{3.32a}$$

The various gradient and curvature values for the sphere are now easily derived. Since $U = \gamma M/r = \gamma\sigma V/r$, we have

$$U_{xz} = \partial^2 U/\partial x\,\partial z = \gamma\sigma V(3xz/r^5),\ U_{yz} = \gamma\sigma V(3yz/r^5),\ U_{zz} = \gamma\sigma V(3z^2 - r^2)/r^5,$$

where V = volume of the sphere and $r^2 = x^2 + y^2 + z^2$.

If we make the x-axis coincide with the magnetic meridian, $\beta = \pi/2$ and $r^2 = x^2 + z^2$ for a N–S traverse. Then we have

$$Z = \frac{k}{\gamma\sigma}\left(H_0\frac{\partial^2 U}{\partial x\,\partial z} + Z_0\frac{\partial^2 U}{\partial z^2}\right) = kVZ_0\left\{\frac{H_0 3xz}{Z_0\ r^5} + \frac{(2z^2 - x^2)}{r^5}\right\}$$

$$= \frac{kVZ_0}{z^3}\left\{\frac{2 + 3(x/z)\cot I - x^2/z^2}{(1 + x^2/z^2)^{\frac{5}{2}}}\right\}. \tag{3.32b}$$

Similarly we can obtain the horizontal component; thus,

$$H = \frac{kVZ_0}{z^3}\left\{\frac{(2x^2/z^2 - 1)\cot I + 3x/z}{(1 + x^2/z^2)^{\frac{5}{2}}}\right\}. \tag{3.32c}$$

The resultant N–S profiles are shown in fig. 3.22 for a dip angle of 45°. They are similar to the dipping dipole of fig. 3.18 which indicates that the sphere, when uniformly magnetized by induction, behaves as a dipole magnetized along its axis.

Depth of the sphere may be estimated from the width of the profile: at $Z_{max}/2$ the total width (rather than the half-width) is roughly equal to the depth to the centre of the sphere. However, this is not a very useful estimate in a field situation.

(f) *Demagnetization.* In the preceding example we have assumed that the induced magnetization is exactly the product of k, the volume susceptibility of the body, and the external field $\mathbf{F}_0$. In fact this is true only for rod-like shapes magnetized along the axis and having a cross-section small compared to their length, such as the dipole of eq. (3.30). In general there is a demagnetization, so that the resultant field inside the body is different from $\mathbf{F}_0$.

The demagnetization effect results in a reduction of the magnetic moment $\mathbf{M}$ by a factor which varies with the shape of the body. The effective moment is given by

$$\mathbf{M}' = kV\mathbf{F}_0/(1 + Nk), \quad (0 \leqslant N \leqslant 4\pi). \tag{3.33}$$

Maximum demagnetization occurs in thin sheets magnetized normal to the face; in this case, $N = 4\pi$. For the sphere $N = 4\pi/3$. Obviously the effect is still quite

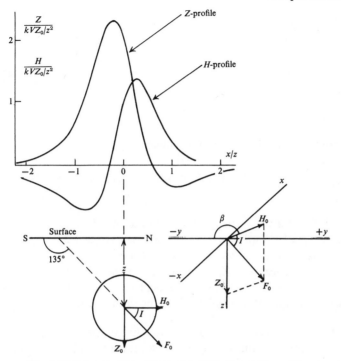

Fig. 3.22 Sphere polarized in earth's field, Z and H curves.

small unless $k \geqslant 0{\cdot}01$ emu. Hence the demagnetization is significant only in massive pyrrhotite and in rocks containing more than 5–10% of magnetite.

(g) *Two-dimensional shapes.* As in gravity, the magnetic body is not accurately two-dimensional unless its strike length is at least ten, and preferably twenty, times larger than any other dimension. This is even less likely to be so in magnetics than in gravity; however, the formulae may be modified for finite strike length.

The two-dimensional magnetic features may be derived with relative ease from the corresponding gravity shapes, by the method already used for the polarized sphere.

(h) *Horizontal cylinder.* The geometry is shown in fig. 3.23. If the cylinder is considered to have infinite length in the direction of the y-axis, the gradient and curvature terms involving y disappear and we have from eq. (3.32a)

$$Z = -\frac{\partial A}{\partial z} = \frac{k}{\gamma \sigma}\left(H_0 \sin \beta \frac{\partial^2 U}{\partial x\,\partial z} + Z_0 \frac{\partial^2 U}{\partial z^2}\right),$$

and since

$$U = 2\gamma \sigma S \log\left(\frac{1}{r}\right),\quad \frac{\partial^2 U}{\partial x\,\partial z} = \frac{4\gamma \sigma S x z}{r^4},\quad \frac{\partial^2 U}{\partial z^2} = 2\gamma \sigma S \frac{(z^2 - x^2)}{r^4},$$

therefore

$$Z = \frac{2kS}{r^4}\left\{2H_0 xz \sin\beta + Z_0(z^2 - x^2)\right\}. \tag{3.34a}$$

If the cylinder length is $2Y$, this expression is modified since we start with a potential

$$U = \gamma\sigma S \log\left\{\frac{\sqrt{(r^2 + Y^2)} + Y}{\sqrt{(r^2 + Y^2)} - Y}\right\},$$

and hence the vertical component becomes

$$Z = \frac{2kSY}{r^4(r^2 + Y^2)^{\frac{3}{2}}}[H_0(3r^2 + 2Y^2)xz \sin\beta + Z_0\{Y^2(z^2 - x^2) + r^2(2z^2 - x^2)\}]. \tag{3.34b}$$

Two profiles for the vertical component are shown in fig. 3.23, one where the cylinder strikes E–W ($\beta = \pi/2$) and one for a strike N30W ($\beta = 30°$). If the cylinder axis is in the direction of the magnetic meridian, the curve would be perfectly symmetrical with Z_{max} directly above the centre of the cylinder.

As in the case of the sphere, the depth to the centre of the cylinder is approximately equal to the full width of the profile at $Z_{max}/2$.

(i) *Thick prism.* Magnetic anomalies caused by igneous instrusions in the form of dikes are common features in regions favourable for mineral exploration, since there is frequently a contrast in the magnetic mineral content of such intrusions with respect to the host rock. Such structures may be simulated by a prism, whose strike length is considerably larger than its thickness. The geometry is illustrated in fig. 3.24a, from which we have the following relations:

$$r_1{}^2 = d^2 + (x + d \cot\alpha)^2, \quad r_2{}^2 = D^2 + (x + D \cot\alpha)^2,$$
$$r_3{}^2 = d^2 + (x + d \cot\alpha - b)^2, \quad r_4{}^2 = D^2 + (x + D \cot\alpha - b)^2,$$
$$\phi_1 = \tan^{-1} d/(x + d \cot\alpha,) \text{ etc.}$$

Using the same approach as in the previous examples, we develop the gravity gradient and curvature expressions and apply Poisson's Relation. From eq. (2.13b), §2.2.5c, the component of the gradient of g along the x-axis normal to the infinite strike length is given by (x', z' being coordinates of a point in the prism)

$$U_{xz} = 4\gamma\sigma \iint \frac{x'z'}{r^4}\, dx'\, dz' = 4\gamma\sigma \int z'\, dz' \int \frac{x'\, dx'}{(z'^2 + x'^2)^2}$$

$$= 4\gamma\sigma \int z'\, dz' \left\{\frac{-1}{2(z'^2 + x'^2)}\right\}\Bigg|_{x + z'\cot\alpha}^{x + z'\cot\alpha - b}$$

$$U_{xz} = 2\gamma\sigma \iint \left\{\frac{z'}{z'^2 \cosec^2\alpha + 2z'x \cot\alpha + x^2}\right.$$
$$\left. - \frac{z'}{z'^2 \cosec^2\alpha + 2z'(x - b) \cot\alpha + (x - b)^2}\right\} dz',$$

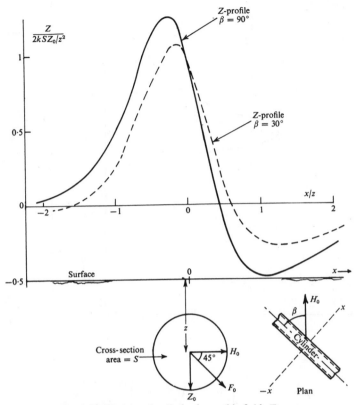

Fig. 3.23 Horizontal cylinder in earth's field, Z curves.

where the limits of integration are D and d. After some manipulation this becomes

$$U_{xz} = 2\gamma\sigma \left\{ \sin^2 \alpha \log \left(\frac{r_2 r_3}{r_1 r_4} \right) + \sin \alpha \cos \alpha (\phi_1 - \phi_2 - \phi_3 + \phi_4) \right\}.$$

From eq. (2.14) we can obtain U_{xx}, which in the two-dimensional case is equal to $-U_{zz}$. In this case,

$$U_{xx} = -U_{zz} = -2\gamma\sigma \left\{ \sin \alpha \cos \alpha \log \left(\frac{r_2 r_3}{r_1 r_4} \right) - \sin^2 \alpha (\phi_1 - \phi_2 - \phi_3 + \phi_4) \right\}.$$

Applying the Poisson Relation as in eq. (3.34a), the vertical component is

$$Z = 2k \sin \alpha \left\{ (H_0 \sin \beta \sin \alpha + Z_0 \cos \alpha) \log \left(\frac{r_2 r_3}{r_1 r_4} \right) + \right.$$

$$\left. + (H_0 \sin \beta \cos \alpha - Z_0 \sin \alpha)(\phi_1 - \phi_2 - \phi_3 + \phi_4) \right\}. \quad (3.35a)$$

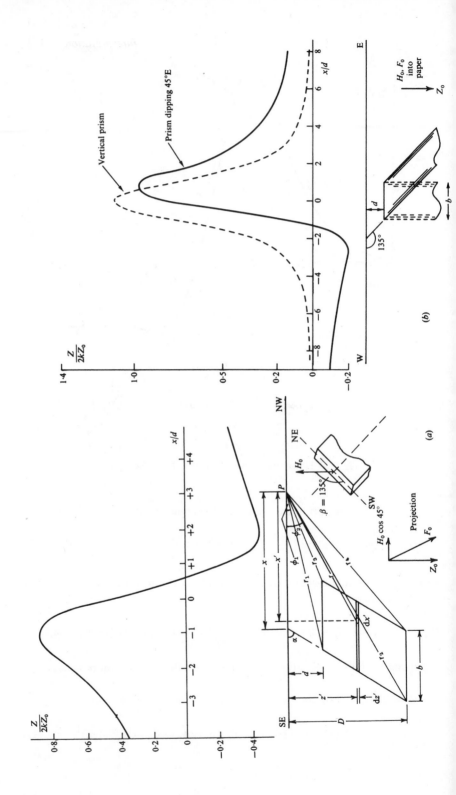

$\dfrac{Z}{2kZ_0}$

Vertical prism

Prism dipping 45°E

x/d

W   E

H_0, F_0 into paper

Z_0

$135°$

d

b

(b)

$\dfrac{Z}{2kZ_0}$

x/d

SE   NW

$\beta = 135°$

NE

SW

H_0

$H_0 \cos 45°$

F_0

Z_0

Projection

P

ϕ_2

ϕ_1

r_0

r

x

x'

α

d

z'

dz'

D

dx'

r_1

r_2

b

(a)

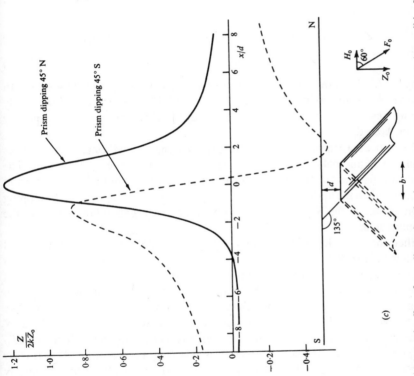

Fig. 3.24 Magnetic effect of two-dimensional prism, Z curves, $D = \infty$. (a) Prism striking NE–SW, dipping 45° to SE, $I = 60°$, profile normal to strike; (b) prism striking N–S, dipping 45°E, 90°; (c) prism striking E–W, dipping 45°N and 45°S.

When the dike strikes E–W, $\sin \beta = 1$, and the relation is

$$Z = 2k \sin \alpha \left\{ (H_0 \sin \alpha + Z_0 \cos \alpha) \log \left(\frac{r_2 \, r_3}{r_1 \, r_4} \right) + \right.$$

$$\left. + (H_0 \cos \alpha - Z_0 \sin \alpha)(\phi_1 - \phi_2 - \phi_3 + \phi_4) \right\}. \tag{3.35b}$$

If the strike is N–S, $\beta = 0$, and

$$Z = 2k \sin \alpha \, Z_0 \left\{ \cos \alpha \log \left(\frac{r_2 \, r_3}{r_1 \, r_4} \right) - \sin \alpha \, (\phi_1 - \phi_2 - \phi_3 + \phi_4) \right\}. \tag{3.35c}$$

If in addition the prism has vertical sides, $\alpha = \pi/2$, and the last two expressions are further simplified to

$$Z = 2k\{H_0 \log (r_2 \, r_3/r_1 \, r_4) - Z_0(\phi_1 - \phi_2 - \phi_3 + \phi_4)\}, \tag{3.35d}$$

$$Z = -2kZ_0(\phi_1 - \phi_2 - \phi_3 + \phi_4). \tag{3.35e}$$

Figure 3.24b shows a set of profiles for two-dimensional prisms of great depth striking N–S and having dip angles of 90° and 135°. The profile for a body dipping west 45° is a reflection in the z-axis of the 135° profile and consequently has not been drawn.

The vertical prism whose long axis is parallel to the magnetic meridian produces a symmetric anomaly, since the horizontal component does not affect it. The significant features of the profiles for the dipping prism are the steep slope and negative tail on the updip side. This configuration is just the opposite of the dipping dipole profile in fig. 3.18, where the steep slope and negative tail appear on the downdip side.

When the prism lies E–W as in fig. 3.24c, the profiles are of the same general character as for N–S strike, but the asymmetry is less pronounced for north dips and more pronounced when the dip is to the south. In fact, when the dip of a prism striking E–W is parallel to the total field direction, the profile is symmetrical.

An illustration of a more general situation is shown in fig. 3.24a. Here the dike strikes SW–NE and dips SE. The resultant profile is similar to that for the E–W prism dipping south in fig. 3.24c. In high magnetic latitudes, where the H_0-component is small, the strike direction is relatively unimportant. This is clear from eqs. (3.35d) and (3.35e); if $H_0/Z_0 \ll 1$ the expressions for Z are practically the same.

It is useful here to provide a qualitative explanation for the character of the profiles in figs. 3.24, based on pole distribution. Consider a dike dipping east and striking N–S as in fig. 3.25. The normal pole distribution would result in S-poles across the upper face, N-poles along the bottom. If the magnetization is mainly induced by the earth's field, however, the vertical component will produce N- and S-poles along the footwall and hanging wall respectively. This can be seen by resolving Z_0 into components Z_1 and Z_2, parallel and normal to the dip axis, as shown in fig. 3.25. The latter component is responsible for this additional pole distribution. Then the footwall N–poles produce a negative component, super-imposed on the normal profile which would be due to Z_1 alone, on the west flank. Similarly the S–poles on the hanging wall face reinforce the Z_1 profile to the east. This effect is large enough to reverse the profile (compare figs. 3.18 and 3.24b).

A similar explanation accounts for the more pronounced asymmetry in the profile of the E–W dike dipping south in fig. 3.24c. In this case the N–poles on the footwall or north side are provided by components of both H_0 and Z_0. In fact the dike tends to be magnetized transversely, since F_0 is practically normal to its dip axis.

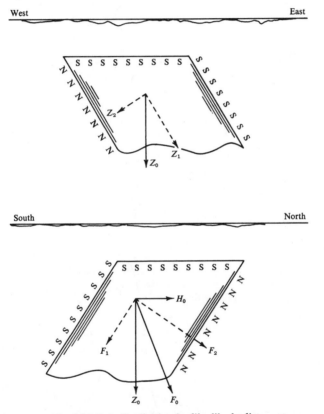

Fig. 3.25 Pole distribution in dike-like bodies.

Depth estimates for magnetic dikes are difficult unless the profiles are quite symmetrical and the width b is no greater than the depth to the top face. Under these restrictions the half-width at half-maximum rule holds within about 20%, that is, $x_{\frac{1}{2}} = d$ at $Z = Z_{max}/2$. However, for asymmetric profiles and/or for large widths, this rule will indicate a depth greater than the actual value.

Direction of dip for such features would appear to be fairly obvious from the profiles, provided we know the total-field direction. The situation is complicated,

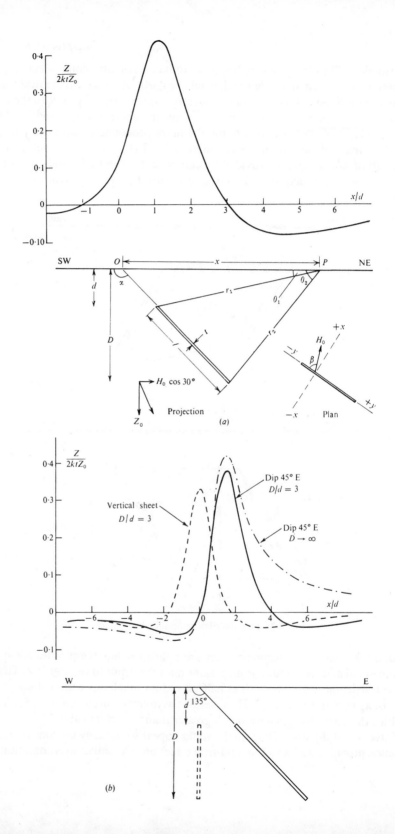

$\frac{Z}{2ktZ_0}$

x/d

SW   O x P   NE

d

α

r_1

θ_2

θ_1

D

l

t

r_2

$H_0 \cos 30°$

Projection

Z_0

$+x$

H_0

β

$-y$

$+y$

$-x$

Plan

(a)

$\frac{Z}{2ktZ_0}$

Dip 45° E
$D/d = 3$

Vertical sheet
$D/d = 3$

Dip 45° E
$D \rightarrow \infty$

x/d

W   E

d 135°

D

(b)

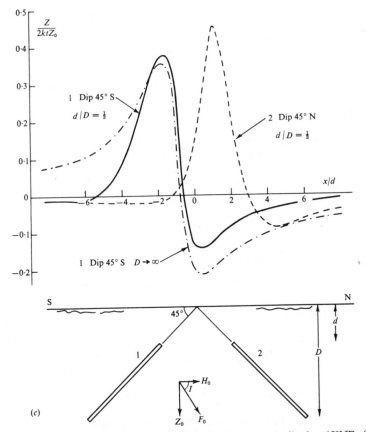

Fig. 3.26 Thin sheet, Z curves, $I = 60°$. (a) Sheet striking NW–SE, dipping 45°NE; (b) sheet striking N–S, dipping 45°E, 90°; (c) sheet striking E–W, dipping 45°S, 45°N.

however, by two factors, one being the effect of remanent magnetization. The other arises from the difficulty in determining the zero line or background reading for a field profile, that is, the problem of isolating a single anomaly. This problem, which is present to some degree in all geophysical work, is particularly troublesome in magnetics.

The prism profiles illustrated here are simplified in that they are all of infinite depth extent. They are not much changed unless the depth extent is less than about five times the width of the top face. In such cases the positive tails (the flanks of smaller slope) may be pulled down slightly negative; for very shallow dip and short length (effectively a flat-lying plate magnetized transversely) the profile approaches a symmetrical shape again, with a broad maximum of small magnitude above the plate and negative tails at the flanks.

(j) *Dipping sheet.* The expression for the vertical-component profile over a thin sheet may be derived in a manner similar to the above, by replacing the horizontal width b by t cosec α, where t is the thickness of the sheet (see fig. 3.26a). The result is

$$Z = 2kt \left\{ \frac{H_0 \sin \beta \sin (\alpha + \theta_2) + Z_0 \cos (\alpha + \theta_2)}{r_2} - \right.$$
$$\left. - \frac{H_0 \sin \beta \sin (\alpha + \theta_1) + Z_0 \cos (\alpha + \theta_1)}{r_1} \right\}. \tag{3.36a}$$

As before, if the sheet strikes E–W, $\sin \beta = 1$, and

$$Z = 2kt \left\{ \frac{H_0 \sin (\alpha + \theta_2) + Z_0 \cos (\alpha + \theta_2)}{r_2} - \right.$$
$$\left. - \frac{H_0 \sin (\alpha + \theta_1) + Z_0 \cos (\alpha + \theta_1)}{r_1} \right\}, \tag{3.36b}$$

while if the strike is N–S, $\beta = 0$, and

$$Z = 2ktZ_0 \left\{ \frac{\cos (\alpha + \theta_2)}{r_2} - \frac{\cos (\alpha + \theta_1)}{r_1} \right\}. \tag{3.36c}$$

If the sheet is vertical, the three expressions above are further simplified to

$$Z = 2kt \left(\frac{H_0 x \sin \beta - Z_0 D}{r_2^2} - \frac{H_0 x \sin \beta - Z_0 d}{r_1^2} \right), \tag{3.36d}$$

$$Z = 2kt \left\{ H_0 x \left(\frac{1}{r_2^2} - \frac{1}{r_1^2} \right) - Z_0 \left(\frac{D}{r_2^2} - \frac{d}{r_1^2} \right) \right\}, \tag{3.36e}$$

$$Z = 2ktZ_0 \left(\frac{d}{r_1^2} - \frac{D}{r_2^2} \right). \tag{3.36f}$$

The profiles of figs. 3.26 have somewhat the same general character as those for the prism. The principal reason for considering the thin sheet is that the formulae are considerably simpler than for the prism and they can be used with good accuracy provided the thickness t is not greater than the depth to the top, d. In addition the thin-sheet geometry is common in areas favourable for mineral exploration.

As mentioned in the discussion of prism profiles, depth estimates are reasonably good when the curves for thin sheets are fairly symmetrical. For example, from fig. 3.26b, the half-width at $Z_{max}/2$ for the vertical and dipping sheets gives $x_{\frac{1}{2}} = 0.8d$ and $x_{\frac{1}{2}} = 0.9d$ respectively. The values in fig. 3.26c are not appreciably different. Rough dip estimates are also possible when the strike of the body and the total field direction are known.

When the depth extent is very great, r_2 approaches infinity in eq. (3.36a) and the sheet is effectively a half-plane. Then we have

$$Z = -2kt \times$$
$$\left[\frac{H_0 \sin \beta \{\sin \alpha (x + d \cot \alpha) + d \cos \alpha\} + Z_0 \{\cos \alpha (x + d \cot \alpha) - d \sin \alpha\}}{d^2 + (x + d \cot \alpha)^2} \right].$$
$$\tag{3.36g}$$

For this limiting case we can determine the depth and dip angle uniquely. By putting $dZ/dx = 0$ in (3.36g) we can find the maximum and minimum values for Z and hence obtain the following relations:

$$\frac{x_m}{d} = \frac{x_{max} - x_{min}}{d} = \frac{2\sqrt{(H_0^2 \sin^2 \beta + Z_0^2)}}{H_0 \sin \beta \sin \alpha + Z_0 \cos \alpha},$$

$$\frac{w_{\frac{1}{2}}}{d} = \frac{2\sqrt{(H_0^2 \sin^2 \beta + Z_0^2)}}{H_0 \sin \beta \cos \alpha + Z_0 \sin \alpha},$$

where $x_{max} - x_{min}$ = horizontal distance between Z_{max} and Z_{min}, and $w_{\frac{1}{2}}$ = full width of profile at $Z_{max}/2$.

Combining these expressions we obtain for the dip angle

$$\tan \alpha = \frac{H_0 w_{\frac{1}{2}} \sin \beta - Z_0 x_m}{H_0 x_m \sin \beta + Z_0 w_{\frac{1}{2}}} \tag{3.36h}$$

while the depth is given by

$$d = x_m \, w_{\frac{1}{2}}/2\sqrt{(x_m^2 + w_{\frac{1}{2}}^2)}. \tag{3.36i}$$

(k) *Effect of finite strike length.* As in the case of gravity shapes (§2.6.3g, eq. (2.46c)), the expressions for truly two-dimensional magnetic shapes may be modified to allow for finite strike length. For the dipping sheet with a length $2Y$, the principal Z-profile – that is, a traverse across the centre – is obtained from eq. (3.36a) as follows:

$$Z = 2kt\left[\frac{Y}{\sqrt{(r_2^2 + Y^2)}}\left\{\frac{H_0 \sin \beta \sin (\alpha + \theta_2) + Z_0 \cos (\alpha + \theta_2)}{r_2}\right\} - \right.$$
$$- \frac{Y}{\sqrt{(r_1^2 + Y^2)}}\left\{\frac{H_0 \sin \beta \sin (\alpha + \theta_1) + Z_0 \cos (\alpha + \theta_1)}{r_1}\right\} +$$
$$+ Y\left\{\frac{H_0 \sin \beta \sin \alpha + Z_0 \cos \alpha}{x^2 \sin^2 \alpha + Y^2}\right\}\left\{\frac{x \cos^2 \alpha + D \cot \alpha}{\sqrt{(r_2^2 + Y^2)}} - \right.$$
$$\left. \left. - \frac{x \cos^2 \alpha + d \cot \alpha}{\sqrt{(r_1^2 + Y^2)}}\right\}\right].$$

Employing the relation $x \cos \alpha + D \operatorname{cosec} \alpha = x \sin \alpha \cot (\alpha - \theta_2)$, this becomes

$$Z = 2ktY\left[\frac{H_0 \sin \beta \sin (\alpha + \theta_2) + Z_0 \cos (\alpha + \theta_2)}{r_2\sqrt{(r_2^2 + Y^2)}} - \right.$$
$$- \frac{H_0 \sin \beta \sin (\alpha + \theta_1) + Z_0 \cos (\alpha + \theta_1)}{r_1\sqrt{(r_1^2 + Y^2)}} +$$
$$+ \frac{x \sin \alpha \cos \alpha(H_0 \sin \beta \sin \alpha + Z_0 \cos \alpha)}{x^2 \sin^2 \alpha + Y^2}\left\{\frac{\cot (\alpha - \theta_2)}{\sqrt{(r_2^2 + Y^2)}} - \right.$$
$$\left. \left. - \frac{\cot (\alpha - \theta_1)}{\sqrt{(r_1^2 + Y^2)}}\right\}\right], \tag{3.36j}$$

with corresponding modifications for eqs. (3.36b) to (3.36f). For example (3.36f) becomes

$$Z = 2ktZ_0 \left\{ \frac{d}{r_1^2 \sqrt{(1 + r_1^2/Y^2)}} - \frac{D}{r_2^2 \sqrt{(1 + r_2^2/Y^2)}} \right\}, \qquad (3.36k)$$

which shows that the profile is reduced in magnitude but otherwise unchanged in shape.

Similar corrections for finite length may be made for the thick prism, fault and other two-dimensional features.

(l) *Effect of demagnetization.* As mentioned previously, demagnetization is most pronounced in the transverse direction in plate-like bodies, that is, the apparent susceptibility is anisotropic, being smaller transverse to the sheet than parallel to it. The resultant direction of magnetization tends to be deflected away from the direction of F_0 towards the plane of the plate.

To allow for this change of magnetization direction it may be necessary to modify the multiplier in the two-dimensional formulae. The susceptibility k is replaced by

$$k' = \frac{k}{1 + Nk} \sqrt{(1 - \cos^2\lambda \cos^2 i)}, \qquad (3.37)$$

where the factor $k/(1 + Nk)$ is the *apparent susceptibility* (the result of demagnetization, as in eq. (3.33)), the term in brackets allows for the change in direction of magnetization, i = dip of resultant magnetization, λ = declination of resultant magnetization, measured with respect to the y-axis and the horizontal projection of the resultant magnetization. We have assumed in the previous formulae that $\lambda = \beta$ and $i = I = \tan^{-1}(Z_0/H_0)$. The effect of this modification may be considerable, particularly with high magnetite content. However, the correction is generally difficult to estimate.

(m) *Horizontal thin sheet; fault approximation.* When $\alpha = \pi$, the tabular magnetic body of §3.6.4j is horizontal, d becomes the depth to the central axis and eq. (3.36a) gives

$$Z = 2kt\{(H_0 \sin \beta \sin \theta_1 + Z_0 \cos \theta_1)/r_1 - (H_0 \sin \beta \sin \theta_2 + Z_0 \cos \theta_2)/r_2\},$$

where $r_1^2 = (x^2 + d^2)$, $r_2^2 = (\ell - x)^2 + d^2$, $\sin \theta_1 = d/r_1$, $\sin \theta_2 = d/r_2$, $\cos \theta_1 = x/r_1$, $\cos \theta_2 = (x - \ell)/r_2$. Rearranging this expression, we get

$$Z = 2ktZ_0 \left\{ \frac{d(H_0/Z_0) \sin \beta + x}{r_1^2} - \frac{d(H_0/Z_0) \sin \beta + (x - \ell)}{r_2^2} \right\}, \qquad (3.38a)$$

with modifications for N–S and E–W strike corresponding to (3.36b) and (3.36c).

Figure 3.27 shows two profiles over a horizontal sheet whose width is four times its depth. When the sheet strikes N–S the profile is perfectly symmetrical (compare with the *H*-profile in fig. 3.20). For E–W strike the characteristic negative tail appears to the north. In neither case is it possible to estimate depth; the width of the profile is somewhat less than the horizontal extent of the sheet. The effect of demagnetization would be to make the E–W profile even more asymmetric (see the *Z*-profile in fig. 3.20).

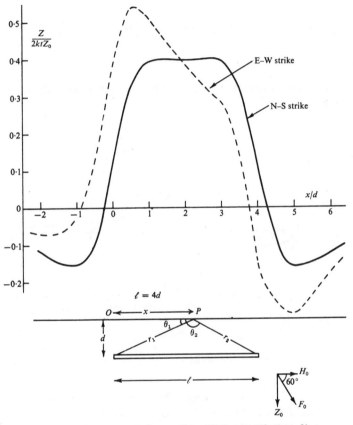

Fig. 3.27 Horizontal sheet striking N–S or E–W, *Z* profiles.

If ℓ becomes very large, we have a semi-infinite sheet and r_2 approaches infinity. The second term in eq. (3.38a) goes to zero and we get the relation

$$Z = 2kt(H_0 d \sin \beta + Z_0 x)/r_1^2. \qquad (3.38b)$$

Profiles for N–S and E–W strike of a single semi-infinite horizontal sheet are shown in fig. 3.28. Here the profile is perfectly asymmetrical when the strike is N–S and lopsided when the strike is E–W. In both instances the depth *d* is one-half the horizontal distance between Z_{max} and Z_{min}. It is necessary to traverse a considerable distance before the magnetic background is reached; for example, when $d = 100$ ft in fig. 3.28 the traverse should extend at least 2500 ft either way from the edge of the sheet. In a practical situation the anomaly would be difficult to extract, since there are likely to be other magnetic features in the vicinity.

If we introduce another semi-infinite sheet at a different depth, lying to the left of the origin in fig. 3.28, we have an approximation to the fault. The geometry is shown in fig. 3.29. For the general case when the fault face is not vertical, eq. (3.38b) becomes

$$Z = 2kt \left\{ \frac{H_0 d \sin \beta + Z_0(x + d \cot \alpha)}{r_1^2} - \frac{H_0 D \sin \beta + Z_0(x + D \cot \alpha)}{r_2^2} \right\}$$

(3.39)

with the usual simplifications when the fault is vertical ($\alpha = \pi/2$) and when the strike is N–S or E–W.

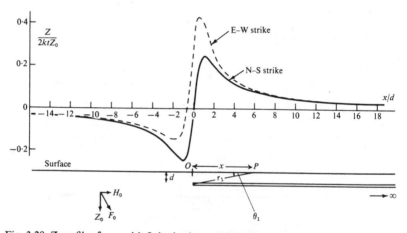

Fig. 3.28 Z profiles for semi-infinite horizontal thin sheet with edge striking N–S or E–W.

Three typical profiles are shown in fig. 3.29 corresponding to a vertical fault and normal and reverse faults dipping 45° and 135° respectively; the vertical fault profile is shown for N–S strike while the others are given for both N–S and E–W strike. In general the same remarks apply here as for the single horizontal sheet. However, the E–W profile for reverse faulting does not have the characteristic asymmetry of the others, hence it would be very difficult to distinguish it from a dipping sheet.

A depth estimate to the upper bed may be made if it is possible to establish that we are dealing with the fault or semi-infinite flat plate (that is, a fault with very large displacement). Within approximate limits it is given by

$$0 \cdot 7 \leqslant x_{\frac{1}{2}}/d \leqslant 1,$$

where $x_{\frac{1}{2}}$ is one-half the distance between Z_{max} and Z_{min}. The lower limit applies when the lower bed is only slightly displaced, the upper when D is very large.

When the angle α is fairly steep, D can be estimated from the relation

$$D \approx x_{\frac{1}{2}} \sqrt{\left\{ \frac{Z_t(1 + x_{\frac{1}{2}}^2/d^2)}{2x_{\frac{1}{2}}(dZ/dx)_{\max} - Z_t(1 + x_{\frac{1}{2}}^2/d^2)} \right\}},$$

where $Z_t = Z_{\max} - Z_{\min}$, $(dZ/dx)_{\max}$ = maximum slope of profile.

(n) *Thick horizontal slab; fault.* As in gravity calculations the fault approximation is accurate to a few per cent provided d is larger than $2t$. When this assumption is not valid it is necessary to use the thick semi-infinite slab. From eq. (3.35a), with $r_3 = r_4 = \infty$ and $\phi_3 = \phi_4 = \pi$, the relation is

$$Z = 2k \sin \alpha \{ (H_0 \sin \beta \sin \alpha + Z_0 \cos \alpha) \log (r_2/r_1) +$$
$$+ (H_0 \sin \beta \cos \alpha - Z_0 \sin \alpha)(\phi_1 - \phi_2) \} \quad (3.40a)$$

with corresponding modifications for eqs. (3.35b) to (3.35e).

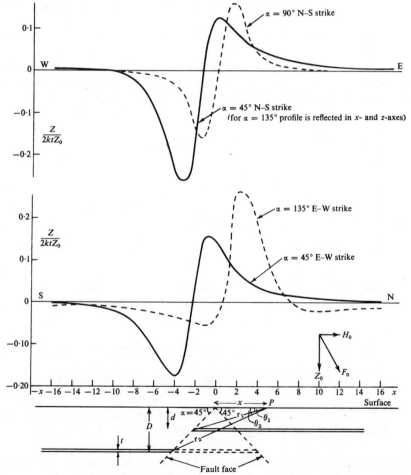

Fig. 3.29 Thin-sheet approximation for a fault, Z profiles.

To simulate the fault we add a similar semi-infinite slab to the left at a different depth, which produces the anomaly

$$Z' = 2k \sin \alpha \left\{ (H_0 \sin \beta \sin \alpha + Z_0 \cos \alpha) \log \left(\frac{r_7}{r_8} \right) + \right.$$

$$\left. + (H_0 \sin \beta \cos \alpha - Z_0 \sin \alpha)(\phi_8 - \phi_7) \right\},$$

where $r_7{}^2 = d_1{}^2 + (x + d_1 \cot \alpha)^2$, $r_8{}^2 = D_1{}^2 + (x + D_1 \cot \alpha)^2$, $\phi_7 = \cot^{-1} \{(x + d_1 \cot \alpha)/d_1\}$, $\phi_8 = \cot^{-1}\{(x + D_1 \cot \alpha)/D_1\}$, and d_1 and D_1 are the depths to the top and bottom of the faulted bed.

The total effect is obtained by adding these:

$$Z'' = 2k \sin\alpha \left\{ (H_0 \sin \beta \sin \alpha + Z_0 \cos \alpha) \log\left(\frac{r_2 r_7}{r_1 r_8} \right) + \right.$$

$$\left. + (H_0 \sin \beta \cos \alpha - Z_0 \sin \alpha)(\phi_1 - \phi_2 - \phi_7 + \phi_8) \right\}. \quad (3.41)$$

Modifications of this formula are easily derived for E–W and N–S strike and for vertical dip.

Two profiles for the single horizontal slab, one for E–W strike, the other N–S, are shown in fig. 3.30. The first is more symmetrical than the second, the profile being determined (for the selected values of α and H_0/Z_0) entirely by the log term.

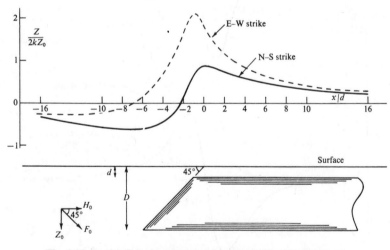

Fig. 3.30 Semi-infinite horizontal slab, edge striking N–S or E–W.

The N–S strike gives a relatively asymmetric profile. Curves for the fault have not been plotted; they are generally similar to those shown in fig. 3.29.

(o) *Contact between slabs of different susceptibilities.* A rather common magnetic feature results from the contact of two slabs with contrasting k values (see fig. 3.31).

The expression for the slab to the left becomes

$$Z' = -2k' \sin \alpha \left\{ (H_0 \sin \beta \sin \alpha + Z_0 \cos \alpha) \log \left(\frac{r_2}{r_1} \right) + \right.$$

$$\left. + (H_0 \sin \beta \cos \alpha - Z_0 \sin \alpha)(\phi_1 - \phi_2) \right\}.$$

Combining with equation (3.40a) the resultant magnetic effect is

$$Z'' = 2\Delta k \sin \alpha \left\{ (H_0 \sin \beta \sin \alpha + Z_0 \cos \alpha) \log \left(\frac{r_2}{r_1} \right) + \right.$$

$$\left. + (H_0 \sin \beta \cos \alpha - Z_0 \sin \alpha)(\phi_1 - \phi_2) \right\}. \qquad (3.40b)$$

If we replace $\Delta k = (k - k')$ by k, this expression is identical to (3.40a).
Figure 3.31 shows three profiles over vertical contacts of considerable depth

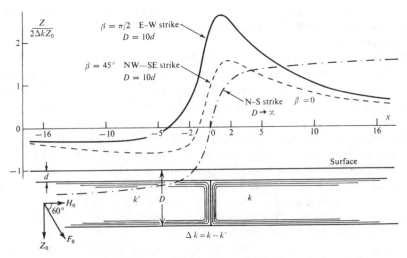

Fig. 3.31 Two semi-infinite horizontal slabs in vertical contact.

extent ($D \geqslant 10d$). It is worth noting that when the vertical contact strikes N–S
and the depth extent of the slabs is very great, the relation in (3.40b) becomes

$$Z'' = 2\Delta k Z_0 (\pi/2 - \cot^{-1} x/d) = 2\Delta k Z_0 \tan^{-1} x/d. \qquad (3.40c)$$

In this case the difference between the limiting values of Z at $x = \pm \infty$ is

$$\Delta Z'' = Z''_{\max} - Z''_{\min} = 2\pi \Delta k Z_0.$$

In addition, the maximum slope occurs directly over the contact and is given by

$$(dZ''/dx)_{x=0} = 2\Delta k Z_0/d.$$

Thus the susceptibility contrast and depth to the bed can be found from the relations

$$\Delta k = \Delta Z''/2\pi Z_0, \quad \text{and} \quad d = \Delta Z''/\pi (dZ''/dx)_{x=0}. \tag{3.40d}$$

3.6.5 *Gradient measurements; discrimination*

The direct measurement of $\partial Z/\partial z$ or $\partial F/\partial z$, that is, the vertical gradient of the vertical field or of the total field, was mentioned briefly in §3.5.4*b* in connection with magnetic terrain corrections. In fact there is considerable merit in carrying out this measurement in general survey work as well because of the additional information obtained.

It is merely necessary to record two readings of the same meter at each station, one directly above the other. With instrument sensitivity about 10γ a change in elevation of 2 or 3 ft will normally be sufficient. Then the vertical gradient is given by

$$\partial Z/\partial z = (Z_2 - Z_1)/\delta z,$$

where Z_2, Z_1 are the readings at the higher and lower elevations respectively and δz the vertical separation. A similar expression, of course, holds for the total field measurement.

Although the older types of magnetic balances were not convenient for a gradient measurement, since they required rather precise levelling, the fluxgate and proton precession magnetometers are well suited in this regard. Of the two the fluxgate is superior, as the proton precession signal decays rather rapidly in high gradient fields and probably could not be used at all if $\partial F/\partial z \geqslant 200\gamma/\text{ft}$.

For three of the geometrical shapes considered in the last section – isolated pole, finite length dipole and vertical contact of great depth extent – the first vertical-derivative profile provides a better depth estimate than does the Z- or F-profile. For the first two features, the width of the profile at $\frac{1}{2}(\partial Z/\partial z)_{\text{max}}$ is equal to the depth of cover within a few per cent. Furthermore the half-maximum value is relatively easier to locate, since the profiles are steeper than the corresponding Z- or F-profiles, although of the same general shape.

In the case of the vertical contact, the depth is equal to half the separation of maximum and minimum values, while the location of the contact itself is better defined than in the Z- and F-profiles because of the steeper slopes directly over it.

The possibility of discriminating between neighbouring magnetic anomalies is also enhanced by the vertical gradient measurement. For example, consider two isolated poles at the same depth z, separated by a distance $x = z$. These can be identified as separate peaks on the $(\partial Z/\partial z)$-profile, but the separation must be $1 \cdot 4z$ to give the same contrast on the Z-profile. For vertical sheets of great depth extent, the corresponding limits are $1 \cdot 1z$ and $1 \cdot 8z$ respectively.

3.6.6 *Dike of great depth extent; master curves*

As noted previously, it is often difficult to establish a background or datum level for magnetic measurements made in the field. Thus the matching of field results with corresponding profiles obtained from simple shapes is highly uncertain. A recent analysis (Koulomzine *et al.*, 1970) for prisms and dikes of infinite depth extent solves this problem and presents master curves which give depth, dip and width of the dike.

The general prism formula consists of one term involving angles and the other a logarithmic function. Writing eq. (3.35a) in a different form to shift the origin to a point midway above the top of the dike, changing b to $2b$ and assuming $r_2 = r_4 = \infty$, $\phi_2 = \phi_4$, we have

$$Z = 2k \sin \alpha \{(H_0 \sin \beta \sin \alpha + Z_0 \cos \alpha) \log (r_3/r_1) + \\ + (H_0 \sin \beta \cos \alpha - Z_0 \sin \alpha)(\phi_1 - \phi_3)\},$$

where

$$r_1{}^2 = d^2 + (x + b)^2, \quad r_3{}^2 = d^2 + (x - b)^2,$$
$$\phi_1 = \cot^{-1}(x + b)/d, \quad \phi_3 = \cot^{-1}(x - b)/d.$$

Dividing through by d to obtain dimensionless parameters we have

$$Z = M\{\cot^{-1}(X + B) - \cot^{-1}(X - B)\} + \\ - (N/2)[\log \{(X - B)^2 + 1\} - \log \{(X + B)^2 + 1\}], \quad (3.42)$$

where $X = x/d$, $B = b/d$. The first term is known as the *symmetric component of the profile*, the second as the *antisymmetric component*, for obvious reasons. Over the centre of the dike the former has a maximum and the latter is zero.

Now if we choose two conjugate points, X_1 and X_2, on the dike profile such that the sum of the Z values is equal to the value of Z at $X = 0$, that is,

$$Z(X_1) + Z(X_2) = Z(0), \quad (3.43a)$$

then the corresponding relations for the separate symmetric and antisymmetric components will be

$$S(X_1) + S(X_2) = S(0), \quad A(X_1) + A(X_2) = 0. \quad (3.43b)$$

Considering the symmetric component first, we have

$$\cot^{-1}(X_1 + B) - \cot^{-1}(X_1 - B) + \cot^{-1}(X_2 + B) - \cot^{-1}(X_2 - B) = 2 \cot^{-1}B.$$

Applying the formula for summing arc cotangents twice, this becomes

$$\cot^{-1}\left(\frac{1 - B^2 + X_1{}^2}{2B}\right) + \cot^{-1}\left(\frac{1 - B^2 + X_2{}^2}{2B}\right)$$
$$= \cot^{-1}\left\{\frac{(1 - B^2 + X_1{}^2)(1 - B^2 + X_2{}^2) - 4B^2}{2B(2 - 2B^2 + X_1{}^2 + X_2{}^2)}\right\} = \cot^{-1}\left(\frac{1 - B^2}{2B}\right).$$

The arguments must be equal. After some manipulation we find that

$$X_1 X_2 = \pm (1 + B^2).$$

Similarly, for the antisymmetric component,

$$\log\left\{\frac{(X_1 - B)^2 + 1}{(X_1 + B)^2 + 1}\right\} + \log\left\{\frac{(X_2 - B)^2 + 1}{(X_2 + B)^2 + 1}\right\} = 0,$$

$$\{(X_1 - B)^2 + 1\}\{(X_2 - B)^2 + 1\} = \{(X_1 + B)^2 + 1\}\{(X_2 + B)^2 + 1\}.$$

Solving again for $X_1 X_2$, this gives

$$X_1 X_2 = -(1 + B^2), \tag{3.43c}$$

which resolves the ambiguity in sign for the symmetric component. Clearly we are free to choose any number of pairs of conjugate points X_{n-1}, X_n in order to locate the point $X = 0$, provided they satisfy eq. (3.43a). Two pairs are sufficient to give a unique solution.

From the profile in fig. 3.32a we can see how to find the point $x = X = 0$, the origin above the dike centre. Points $Z_1, \ldots, Z_4$ are located on the profile such that $Z_{max} - Z_2 = Z_1 - Z_{min} = e$, and $Z_{max} - Z_4 = Z_3 - Z_{min} = E$. The corresponding distances between the abscissae are

$$X_3 - X_4 = \ell, \; X_3 - X_2 = m, \; X_1 - X_3 = n;$$

also, since

$$X_1 X_2 = X_3 X_4,$$

hence,

$$X_3 = mn/(\ell - m + n), \; X_2 = m(m - \ell)/(\ell - m + n), \text{ etc.} \tag{3.43d}$$

Note that $Z(0)$ is the point on the profile located a horizontal distance X_3 from the point Z_3, X_2 from Z_2, etc. For most accurate results it is best to locate Z_3 and Z_4 as close as possible to the midpoint of the profile and Z_2, Z_1 near the maximum and minimum.

Thus the datum or background line $Z = 0$ can be drawn at a distance above Z_{min} equal to the vertical distance between Z_{max} and Z_0. This follows from eq. (3.43a) if we put $Z(X_1) = Z_{max}$ and $Z(X_2) = Z_{min}$, that is, $-Z_{min} = Z_{max} - Z(0)$.

The establishment of the background value of Z is of considerable significance in itself because the dike structure is now located laterally with respect to the field profile. The analysis may now be carried further to establish the dike parameters. First, making use of the relations $S(X) = S(-X)$ and $A(X) = -A(-X)$ for the symmetric and antisymmetric components, it can be shown that

$$S(X) = \tfrac{1}{2}\{Z(X) + Z(-X)\}, \text{ and } A(X) = \tfrac{1}{2}\{Z(X) - Z(-X)\}. \tag{3.44}$$

Thus we can plot $S(X)$ and $A(X)$ from the original profile by taking off points which are equidistant either side of the line $X = 0$. On these two component-profiles we mark points $S_{\frac{3}{4}}$, $S_{\frac{1}{2}}$ and $A_{\frac{1}{2}}$ with corresponding abscissae $X_{\frac{3}{4}}$, $X_{\frac{1}{2}}$ and

$X_{e/2}$ as well as X_e (see fig. 3.32*b*). By a development similar to that used for eq. (3.43*d*), it can be shown that

$$d = x_{\frac{1}{2}}(\phi^2 - 1)/2 = 2x_{\frac{1}{2}}D, \qquad 2b = x_{\frac{1}{2}}\sqrt{\{4 - (\phi^2 - 1)^2\}} = 2x_{\frac{1}{2}}W,$$

$$d = x_e(1 - \mu)^2/2\mu = 2x_e\mathscr{D} \qquad 2b = x_e\sqrt{\{4\mu^2 - (1 - \mu)^4\}}/\mu = 2x_e\mathscr{W},$$

where

$$\phi = x_{\frac{1}{2}}/x_{\frac{1}{2}}, \quad \mu = x_e/x_{e/2}, \quad D = (\phi^2 - 1)/4, \quad \mathscr{D} = (1 - \mu)^2/4\mu$$

$$W = \tfrac{1}{2}\sqrt{\{4 - (\phi^2 - 1)^2\}}, \mathscr{W} = \frac{1}{2\mu}\sqrt{\{4\mu^2 - (1 - \mu)^4\}}, x_{\frac{1}{2}} = X_{\frac{1}{2}}d, \text{ etc.}$$

$$(3.45)$$

Finally, we can find the dip angle α from the relation

$$\alpha = \pi - \cot^{-1}\left(\frac{H_0 \sin\beta}{Z_0}\right) + \tan^{-1}\left\{\frac{A(X)_{\max}}{S(X)_{\max}} \cdot \frac{4\tan^{-1}B}{\log\{(X_e - B)/(X_e + B)\}}\right\}$$

or,

$$\alpha = \pi - \cot^{-1}\left(\frac{H_0 \sin\beta}{Z_0}\right) + \tan^{-1}\left\{\frac{A(X)_{\max}}{S(X)_{\max}} \cdot \frac{4\tan^{-1}B}{\log\{(X_{\frac{1}{2}} - B)/(X_{\frac{1}{2}} + B)\}}\right\}.$$

$$(3.46)$$

Six master curves involving functions of d, b and α for the symmetric and anti-symmetric components of the dike profile are shown in fig. 3.32*c*. The functions for d and b are clear enough; the dip-angle functions P and $\mathscr{P}$ are related to known quantities in the following expansions of eq. (3.46):

$$\alpha - \{\pi - \cot^{-1}(H_0 \sin\beta/Z_0)\} = \tan^{-1}\xi P \quad \text{or} \quad \tan^{-1}\xi\mathscr{P},$$

where

$$\xi = \frac{A(X)_{\max}}{S(X)_{\max}}, P = \frac{4\tan^{-1}[\sqrt{\{4 - (\phi^2 - 1)^2\}}/(\phi^2 - 1)]}{\log[2 - \sqrt{\{4 - (\phi^2 - 1)^2\}}] - \log[2 + \sqrt{\{4 - (\phi^2 - 1)^2\}}]},$$

$$\mathscr{P} = \frac{4\tan^{-1}[\sqrt{\{4\mu^2 - (1 - \mu)^4\}}/(1 - \mu)^2]}{\log[2\mu - \sqrt{\{4\mu^2 - (1 - \mu)^4\}}] - \log[2\mu + \sqrt{\{4\mu^2 - (1 - \mu)^4\}}]}.$$

The analysis for the dike of infinite depth extent has been extended to cover the prism of finite length and depth extent. Because of the extra terms involving Y, r_3, r_4, ϕ_3, ϕ_4, it is necessary to provide more master curves, but the procedure is similar to the above.

Although we have considered the vertical component only, it is possible to solve the total-field anomaly as well. Clearly this development can also be extended to dikes or prisms in gravity.

3.6.7 *Maximum depth estimates; Smith rules*

As in §2.6.9 dealing with depth estimates for gravity anomalies, there are corresponding limiting values in magnetics, derived by Smith (1961). If $|\mathscr{I}|_{\max}$ is the maximum magnetization of a body in which the magnetization is parallel throughout, though not necessarily uniform or even in the same sense, and if

$|\partial Z/\partial x|_{max}$ and $|\partial^2 Z/\partial x^2|_{max}$ are the absolute maxima of first and second derivatives along the x-profile, then the depth to the upper surface is given by

$$z \leqslant \frac{5 \cdot 45 |\mathscr{I}|_{max}}{|\partial Z/\partial x|_{max}}, \quad \text{and} \quad z^2 \leqslant \frac{26 \cdot 6 |\mathscr{I}|_{max}}{|\partial^2 Z/\partial x^2|_{max}}. \tag{3.47}$$

When $\mathscr{I}$ is everywhere vertical and in the same direction (down or up), the numerical factors are reduced to 2·59 and 3·14 respectively.

For two-dimensional magnetic features having infinite length in the y-direction, in which the total magnetization is parallel throughout, the equivalent expressions become

$$z \leqslant \frac{4 |\mathscr{I}|_{max}}{|\partial Z/\partial x|_{max}}, \quad \text{and} \quad z^2 \leqslant \frac{4 \cdot 7 |\mathscr{I}|_{max}}{|\partial^2 Z/\partial x^2|_{max}}. \tag{3.48}$$

When the body is uniformly magnetized by induction we may replace $\mathscr{I}_{max}$ by kF_0 or $kF_0/(1 + Nk)$ as in eq. (3.33).

Since we do not normally have any value for $\mathscr{I}$, these estimates, obtained by combining the two limits, are even cruder than the equivalent relations for

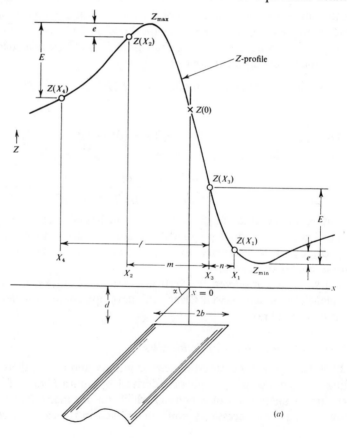

(a)

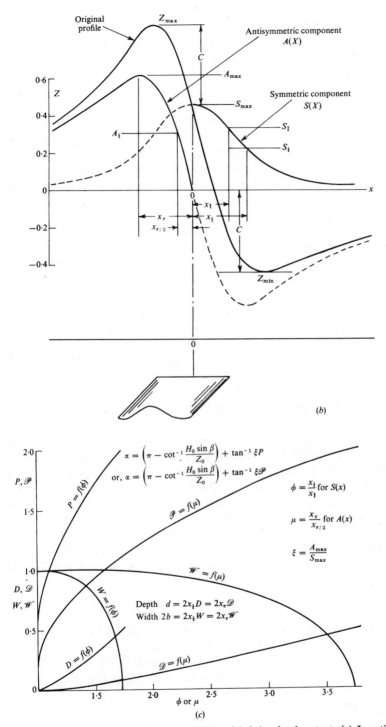

Fig. 3.32 Magnetic effect of two-dimensional dike of infinite depth extent. (a) Location of conjugate points and the point $X = 0$; (b) illustrating antisymmetric and symmetric components; (c) master curves.

gravity. For a semi-infinite vertical thin sheet the result is within 50%, but for three-dimensional features it appears to be very poor.

3.6.8 *Interpretation of airborne anomalies*

(a) *General.* Because of the enormous areas which have been surveyed by the airborne magnetometer, interpretation of aeromagnetics has become almost a subject in itself. The approach is normally different than in ground magnetics because the data are less detailed. Depth determinations are most important, lateral extent less so (airborne anomalies are mainly three-dimensional) and reasonable dip estimates quite difficult. In this regard aeromagnetic and airborne EM interpretation are similar.

In view of these limitations very simple geometrical models are employed: isolated pole, dipole, line of poles and dipoles, bottomless prism, thin plate, dike and vertical contact. There is also a fairly clear demarcation between petroleum and mineral exploration interpretation. In the former, depth to basement is the prime concern since the sediments provide low magnetic relief, while in the latter more detail is desirable but not always possible.

The potentialities of high resolution aeromagnetics, such as may be obtained with the optical pump instruments, measuring to $0 \cdot 01\gamma$, are only now being exploited to a limited extent. With such high sensitivity the results are greatly affected by fluctuations in the earth's main field. Measurements of vertical gradient may be a solution to this problem. Whatever type of instrument is used, as mentioned previously, only the total field is measured in aeromagnetic surveys.

(b) *Isolated pole.* The induced magnetization in a body is usually along the direction of the earth's field. However, for a long slender body with appreciable dip, the direction of magnetization will tend to be along the long axis, except in the region of the magnetic equator. Provided the length is great enough, we have in effect a single pole, $-m$ (in the northern hemisphere). Moreover, since the field of the pole, $\mathbf{F}_m$, is much smaller than the field of the earth, $\mathbf{F}_0$, the component of $\mathbf{F}_m$ normal to $\mathbf{F}_0$ will make a negligible contribution to the total field. Thus, the anomaly in the total field resulting from the single pole is the component of $\mathbf{F}_m$ in the direction of $\mathbf{F}_0$. Writing F for the magnitude of this component, we get from eq. (3.12a),

$$F = -\boldsymbol{\alpha}_1 \cdot \nabla A,$$

where $A = -m/r$, $\boldsymbol{\alpha}_1 =$ unit vector in the direction of $\mathbf{F}_0$,

$$= \mathbf{F}_0/F_0 = (X\mathbf{i} + Y\mathbf{j} + Z\mathbf{k})/F_0$$
$$= (F_0 \cos I \cos D\,\mathbf{i} + F_0 \cos I \sin D\,\mathbf{j} + F_0 \sin I\,\mathbf{k})/F_0$$
$$= (\cos I \cos D)\mathbf{i} + (\cos I \sin D)\mathbf{j} + (\sin I)\mathbf{k},$$

(see fig. 3.33). Since

$$\mathbf{r} = x\mathbf{i} + y\mathbf{j} + (z - z_m)\mathbf{k},$$

and

$$\nabla A = \frac{\mathrm{d}A}{\mathrm{d}r}\frac{\partial r}{\partial x}\mathbf{i} + \frac{\mathrm{d}A}{\mathrm{d}r}\frac{\partial r}{\partial y}\mathbf{j} + \frac{\mathrm{d}A}{\mathrm{d}r}\frac{\partial r}{\partial z}\mathbf{k},$$

$$= \frac{m}{r^3}\{x\mathbf{i} + y\mathbf{j} + (z - z_m)\mathbf{k}\},$$

we obtain

$$F = -\frac{m}{r^3}\{x \cos I \cos D + y \cos I \sin D + (z - z_m) \sin I\}.$$

If we make our observations in the xy-plane by means of a traverse along the x-axis, this axis being along the magnetic meridian, this equation simplifies to

$$F = -\frac{m(x \cos I - z_m \sin I)}{(x^2 + z_m^2)^{3/2}}. \tag{3.49a}$$

The maximum and minimum points of F occur at points on the x-axis given by

$$x_{\max} = z_m\left\{\frac{3 - \sqrt{(9 + 8 \cot^2 I)}}{4 \cot I}\right\}, \; x_{\min} = z_m\left\{\frac{3 + \sqrt{(9 + 8 \cot^2 I)}}{4 \cot I}\right\}, \tag{3.49b}$$

while the maximum and minimum values are

$$F_{\max} = \left(\frac{m \sin I}{4z_m^2}\right)\left\{\frac{1 + \sqrt{(9 + 8 \cot^2 I)}}{(1 + x_{\max}^2/z_m^2)^{3/2}}\right\},$$

$$F_{\min} = \left(\frac{m \sin I}{4z_m^2}\right)\left\{\frac{1 - \sqrt{(9 + 8 \cot^2 I)}}{(1 + x_{\min}^2/z_m^2)^{3/2}}\right\}. \tag{3.49c}$$

Since the assumption is that we are in the northern hemisphere (S-seeking pole near the surface), $F_{\max}$ is south of the origin and $F_{\min}$ is north of the origin. Also, F is zero at $x_0 = z_m \tan I$, that is, north of the origin. Profiles for various values of I are shown in fig. 3.33; in southern latitudes the curves are reflected in the F-axis.

These curves illustrate clearly that, as the inclination decreases (that is, as we move toward the magnetic equator in north latitudes), the total-field profile on a magnetic meridian becomes progressively more asymmetric. At the same time the maximum decreases and the minimum increases and both are displaced progressively southward. Obviously the statement applies to the southern hemisphere as well if we interchange the signs of the maxima and minima and proceed from south to north. When

$$I = 0, \; |x_{\max}| = |x_{\min}| \approx 0{\cdot}7z_m.$$

Pole depth may be estimated from the relations in eq. (3.49b). For example,

$$z_m = \frac{2(x_{\min} - x_{\max})}{\sqrt{(8 + 9 \tan^2 I)}} = \frac{4(x_0 - x_{\max})}{\tan I + \sqrt{(8 + 9 \tan^2 I)}}.$$

For high latitudes, however, it may be difficult to locate x_{min} and x_0, since the curves are practically symmetrical when $I \geqslant 60°$. An alternative estimate, good within 10%, is then given by $z_m \approx 1.4x_{\frac{1}{4}}$, where $x_{\frac{1}{2}}$ is the usual half-width of the profile at $F_{max}/2$.

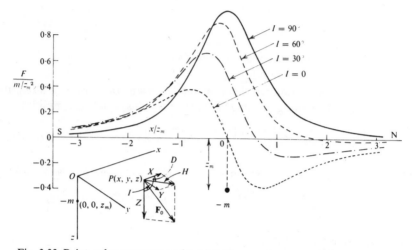

Fig. 3.33 Point pole: geometry and total field anomaly curves. (After Smellie, 1967.)

(c) *Dipole.* For a rod-like feature, magnetized in the direction of the earth's field, with length small compared to the height of the aircraft, the magnetic potential is given by eq. (3.8b). Here $\theta = I + \tan^{-1} (z_c/x)$ as illustrated in fig. 3.34. Then the potential is

$$A = \mathcal{M} \cos \theta / r^2 = \mathcal{M} (x \cos I - (z - z_c) \sin I)/(x^2 + z_c^2)^{3/2},$$

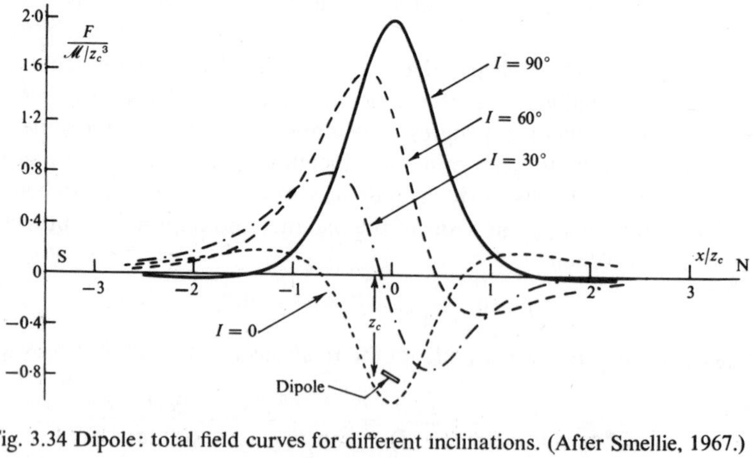

Fig. 3.34 Dipole: total field curves for different inclinations. (After Smellie, 1967.)

where z_c is the depth to the dipole centre. By the same analysis as for the point pole, we find the total field along the meridian in the plane $z = 0$ to be

$$F(x) = \frac{\mathscr{M}}{z_c^3} \left\{ \frac{(3 \sin^2 I - 1) - 6(x/z_c) \sin I \cos I + (3 \cos^2 I - 1)(x/z_c)^2}{(1 + x^2/z_c^2)^{5/2}} \right\}.$$

(3.50)

A set of total-field dipole profiles for various inclinations is shown in fig. 3.34. The asymmetry is more pronounced at steep dip angles than in the point-pole examples and in fact the curve is symmetrical and inverted at $I = 0°$. The same shift of the maximum and minimum to the south is evident.

Although it is possible to obtain exact values for x_{max}, x_{min}, etc., as for the point pole, the expressions are rather cumbersome. Depth to the dipole centre can be estimated roughly from the profiles themselves. For values of $I \geqslant 45°$ and when $I = 0°$, the full width of the profile at $F_{max}/2$ is about equal to z_c. For shallower inclinations the estimate is not as good, the depth being roughly the horizontal distance between F_{max} and F_{min}. Note that in all cases we have assumed the dipole length to be small compared to its distance from the point of measurement.

(d) *Line of poles.* Although two-dimensional features are quite uncommon in airborne magnetics, the line of poles serves as a simple shape for initial depth estimates. This is an approximation to the long shear or fracture zone, thin dike, etc., with an appreciable susceptibility contrast and extending to considerable depth. The magnetic potential is given by the logarithmic relation:

$$A = 2m_\ell \log r,$$

where $r^2 = x^2 + (z - z_0)^2$ and m_ℓ is the pole intensity per unit length (assumed to be uniform). By an entirely similar development to that used for the point pole, we find the total field profile along the x-axis, normal to strike, to be

$$F(x) = -2m_\ell \sin I (x \cot I \sin \beta - z_0)/(x^2 + z_0^2).$$

(3.51)

Positive x-values are considered to be in the half-plane north of the strike axis. A set of profiles for dip angles of $90°$, $60°$, $30°$ and $0°$, traversing normal to a line of poles striking NW (or SW) is shown in fig. 3.35. These have the same character as for the single pole, although somewhat broader. The half-width of the profile at $Z_{max}/2$ is about equal to, or slightly larger than, the depth z_0 for dip angles larger than $30°$; when I is smaller the depth is roughly equal to half the horizontal distance between F_{max} and F_{min}.

(e) *Line of dipoles.* The opposite extreme to the line of poles is the magnetic stringer of limited depth extent, which can be approximated by a line of dipoles. If $\mathscr{M}_\ell$ is the dipole moment per unit length in the total-field direction, the potential is

$$A = 2\mathscr{M}_\ell (x \cos I \sin \beta - z_c \sin I)/(x^2 + z_c^2),$$

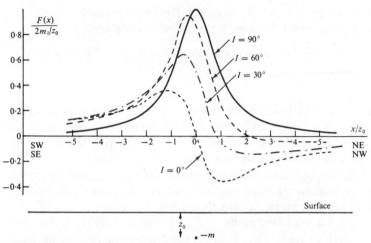

Fig. 3.35 Total field curves over line of poles striking NW–SE or NE–SW.

with the same notation as in the previous examples. Then the total field is found in the usual fashion, the result being

$$F(x) = 2\mathcal{M}_\ell \cos^2 I \sin^2 \beta \left\{ \frac{(x^2 - z_c^2)(1 - \tan^2 I \cosec^2 \beta) - 4x \tan I \cosec \beta}{(x^2 + z_c^2)^2} \right\},$$

(3.52)

$$= \frac{2\mathcal{M}_\ell \cos^2 I \sin^2 \beta}{z_c^2} \times$$

$$\left[\frac{\{(x^2/z_c^2) - 1\}(1 - \tan^2 I \cosec^2 \beta) - 4(x/z_c) \tan I \cosec \beta}{(1 + x^2/z_c^2)^2} \right].$$

Profiles for the dipole line are found in fig. 3.36. The depth to the centre, z_c, for

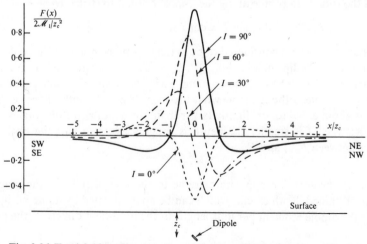

Fig. 3.36 Total-field profiles over line of dipoles striking NW–SE or NE–SW.

$I \geqslant 45°$ and $I \leqslant 15°$, is approximately equal to the full width of the profile at $F_{max}/2$, while for values of I between $15°$ and $45°$ the depth is roughly equal to the horizontal distance between F_{max} and F_{min}.

(f) *Vertical prism.* The simple pole and dipole approximations are mainly applicable to airborne magnetic anomalies associated with mineral exploration areas. In oil reconnaissance the basement magnetics of prime importance are so far removed from the aircraft that the discrimination between sources of small lateral extent is generally impossible. Thus a number of closely spaced lines of poles or dipoles would resemble a prism or plate anomaly. For estimating depth of basement, the vertical prism has been widely used in airborne interpretation. The Z-component over a vertical prism of great depth extent can be obtained from eq. (3.35a):

$$Z = 2kZ_0\{(H_0/Z_0) \sin \beta \log (r_3/r_1) + (\phi_3 - \phi_1)\},$$

where

$$r_1{}^2 = z^2 + x^2, r_3{}^2 = z^2 + (x - b)^2, \phi_3 = \cot^{-1}\{(x - b)/z\}, \phi_1 = \cot^{-1}(x/z).$$

The horizontal component can also be found from the Poisson Relation. For this geometry it becomes

$$H = 2kZ_0\{\log (r_3/r_1) - (H_0/Z_0) \sin \beta(\phi_3 - \phi_1)\}.$$

Then, since $F = H \sin \beta \cos I + Z \sin I$, the total field is

$$F = 2kZ_0\{2\cos I \sin \beta \log (r_3/r_1) + (\sin I - \cot I \cos I \sin^2 \beta)(\phi_3 - \phi_1)\},$$
$$= 2kF_0\{2\sin I \cos I \sin \beta \log (r_3/r_1) + (\sin^2 I - \cos^2 I \sin^2 \beta)(\phi_3 - \phi_1)\}. \tag{3.53}$$

Anomalies of this type are very rarely two-dimensional, however, and a more realistic approach is to assume a strike length $2Y$. Equation (3.53) must then be modified. For a principal profile, we obtain

$$F = 2kF_0\left[\sin I \cos I \sin \beta \log \left\{ \frac{\sqrt{(r_1{}^2 + Y^2)} + Y}{\sqrt{(r_1{}^2 + Y^2)} - Y} \times \frac{\sqrt{(r_2{}^2 + Y^2)} - Y}{\sqrt{(r_2{}^2 + Y^2)} + Y} \right\} \right.$$

$$+ (\sin^2 I - \cos^2 I \sin^2 \beta) \left\{ \tan^{-1}\left(\frac{Y}{x}\right) - \tan^{-1}\left(\frac{Y}{x - b}\right) \right.$$

$$\left. \left. - \tan^{-1}\frac{Yz}{x\sqrt{(r_1{}^2 + Y^2)}} + \tan^{-1}\frac{Yz}{(x - b)\sqrt{(r_3{}^2 + Y^2)}} \right\} \right] \tag{3.54}$$

Two profiles for total field over the vertical prism are shown in fig. 3.37. When the strike length is infinite ($2Y/b > 10$), it is possible to use the dike analysis of §3.6.6 to find the depth and width. This procedure is not valid for a finite strike length, however. This is clear from the second profile in which $2Y = 4b$; if we attempt to find the datum or background value, it appears to be at $F/kF_0 = -0.45$

rather than zero, because of the strong negative tail to the NE. In such situations it is necessary to refer to characteristic curves (Koulomzine *et al.*, 1970; Grant & West, 1965; Vacquier *et al.*, 1951). Of these, the first two methods require 30 or 40 master curves and are simpler in this respect than the third. The method of Koulomzine has the added advantage that it is unnecessary to consider the *y*-axis profile or to estimate the average slope.

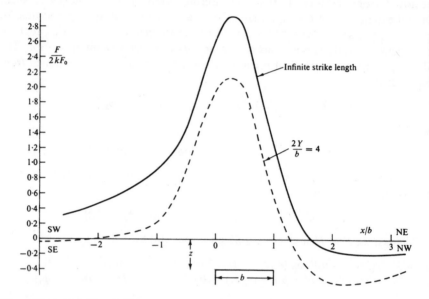

Fig. 3.37 Total-field profiles over vertical prism of great depth extent. $I = 75°$, $\beta = 45°$, $z/b = 0.5$, $2Y =$ strike length.

(g) *Horizontal thin plate.* The opposite extreme to the vertical prism of great depth extent is the geometry of the thin plate, for which the Z-component is given by eq. (3.38*a*). The horizontal component for this feature is

$$H = 2ktZ_0 \left\{ \frac{z - x \cot I \sin \beta}{r_1^2} - \frac{z - (x - b) \cot I \sin \beta}{r_3^2} \right\}.$$

Thus the total field becomes

$$F = 2ktF_0 \left\{ \frac{2z \cos I \sin I \, \sin \beta + x(\sin^2 I - \cos^2 I \sin^2 \beta)}{r_1^2} \right.$$
$$\left. - \frac{2z \cos I \sin I \sin \beta + (x - b)(\sin^2 I - \cos^2 I \sin^2 \beta)}{r_3^2} \right\}. \quad (3.55)$$

In the above z is the depth to the centre of the plate and t its thickness. When the plate has a finite length $2Y$, (3.55) becomes

$$F = 2ktF_0 \left[\frac{x - b\,(\cos^2 I \sin^2 \beta - \sin^2 I) - 2z \sin I \cos I \sin \beta}{r_2^2 \sqrt{(1 + r_2^2)/Y^2}} \right.$$

$$\left. - \frac{x\,(\cos^2 I \sin^2 \beta - \sin^2 I) - 2z \sin I \cos I \sin \beta}{r_1^2 \sqrt{(1 + r_1^2)/Y^2}} \right]. \quad (3.56)$$

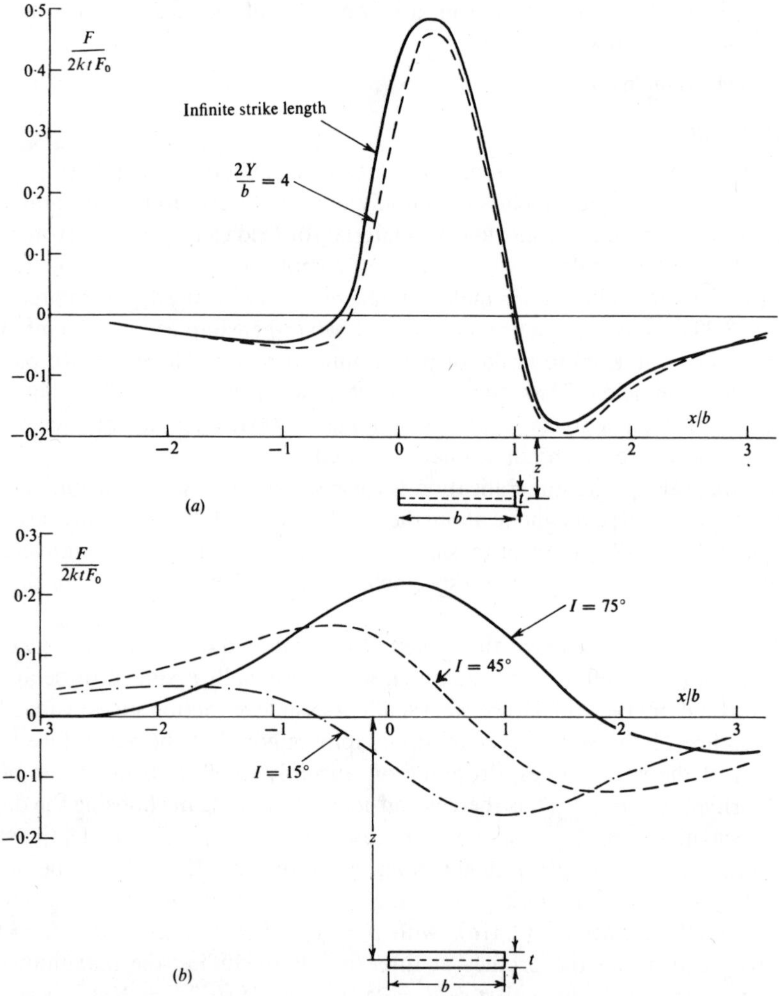

Fig. 3.38 Total-field profiles over thin horizontal plate of finite width. $\beta = 45°$, $t/b = 0.125$. (a) Effect of strike length on principal profiles. $I = 75°$, $(z/b) = 0.5$; (b) effect of inclination of the earth's field on principal profile. $I = 15°$, $45°$, $75°$, $(z/b) = 2$.

Total-field principal profiles for the thin plate are illustrated in fig. 3.38a, one for infinite strike length, the other for $2Y = 4b$. They are remarkably similar; this is also true for gravity over a thin plate. Unless $Y < b$ the effect of finite length does not lower the curve more than about 20%.

Again it is not possible to use the procedure of §3.6.6 to find the zero line without the addition of master curves. Although the number of curves required is less than for the prism of great depth extent, because of the smaller effect of finite strike length, the preparation of a complete set is still a considerable task. However, it is not usually possible to get a reasonable depth estimate by any other method, since the width of the principal profile is controlled to a large extent by the width of the plate. This is apparent from a comparison of figs. 3.38a and 3.38b, the latter for $z/b = 2$ rather than 0·5.

3.7 Field examples

3.7.1 *Ground surveys*

As noted previously in this chapter, the magnetic method is the only geophysical prospecting technique which is commonly used in all exploration programmes. Consequently there is no shortage of material for field examples and problems.

(1) The first example shows the inherent complexity of ground magnetic data and the resultant difficulties in making interpretation with any degree of precision. Figure 3.39 displays magnetic contours and two rather simple vertical component profiles, taken normal to strike of pyrite mineralization. There are two parallel pyrite zones in acidic flows, near a contact between the latter and rhyolite porphyry. Both have a strike length greater than 1000 ft and the zone nearer the contact appears to pinch out at line 75 + 00SE.

Although the pyrite mineralization is clearly associated with a magnetic trend in the area, the large magnetic anomalies on lines 69 and 75 could only be due to magnetite or possibly pyrrhotite, since the susceptibility of pyrite is relatively low (table 3.2). However, there is no specific indication of these minerals in the drill logs of holes 1–4.

Since the overburden near the collars of the diamond drill holes was generally quite thick (82 ft at T−1, for example), it was originally assumed to be at least 50 ft throughout the grid. However, a shallow seismic refraction survey carried out later on line 75 showed a bedrock topography only 5–10 ft below surface in the vicinity of the pyrite zones, dropping off abruptly to 50–75 ft northeast of the acidic-rhyolite contact. Thus there is considerable latitude in choosing the depths of the magnetic sources, which must be close to surface and of small depth extent.

Considering the single 13,000 gamma peak on line 75 + 00SE, the source would appear to be a dipole at very shallow depth, with steep dip and limited strike length. Using eq. (3.31a), with $\beta = 45°$, $I = 70°$, $\alpha = 90°$, $Z_0 = 0·36$ oersteds and fitting the profiles at three points (including the maximum), we obtain a reasonable fit with $t \approx d = 25$ ft, $D = 75$ ft, $2Y = 100$ ft, $k = 0·25$ (see fig. 3.39b).

If we try to match the double peak profile on line 69 + 00 by assuming two

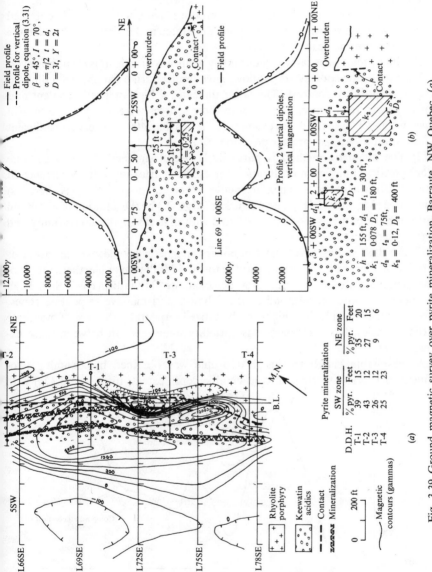

Fig. 3.39 Ground magnetic survey over pyrite mineralization, Barraute, NW Quebec. (a) Magnetic map; (b) attempt to match profiles on lines 75 + 00SE and 69 + 00SE.

vertical dipoles of identical cross-section separated by 160 ft and inductively magnetized in the earth's field, eq. (3.31a) produces the following parameters: $d \approx t = 75$ ft, $D = 225$ ft, $2Y = 300$ ft, $k_1 = 0\cdot1$, $k_2 = 0\cdot15$. The central trough and the northeast flank correspond to the field profile, but the southwest flank is much too large. This curve has not been reproduced here. A better fit was obtained with the two vertically magnetized dipoles illustrated in the diagram, although the trough between the peaks is too deep.

This interpretation is rather crude and by no means unique. It is clear, however, that the magnetic sources are shallow, have limited strike length, steep dip and large susceptibility contrast. The latter indicates a high magnetic content and possibly large remanence. Furthermore these sources were not detected in drill holes T-1 and T-3, so the depth extent must be less than 400 ft. There is insufficient information to warrant greater precision.

(2) The magnetic method is particularly useful (in fact, the only suitable geophysical technique) in exploration for asbestos, because of its occurrence in ultrabasic intrusive rocks rich in magnetite. When olivine (Mg_2SiO_4) is altered to serpentine $(Mg_3SiO_5(OH)_4)$ and magnesite $(MgCO_3)$ by the addition of water and carbon dioxide, the asbestos is associated with high magnetic susceptibility and massive serpentinite.

Figure 3.40 shows a vertical component magnetic profile over an asbestos prospect in Garrison Township, near the town of Matheson in Northern Ontario. The geologic section, under an overburden of about 50 ft, is displayed below. High magnetic responses correspond to the asbestos and massive serpentine zones, lows over the volcanics and highly carbonatized serpentine; the area of moderate carbonatization has a magnetic susceptibility somewhere in between these extremes.

A reasonable match to this field profile was obtained by assuming two-dimensional prism sections of considerable depth extent, using eq. (3.35a) with

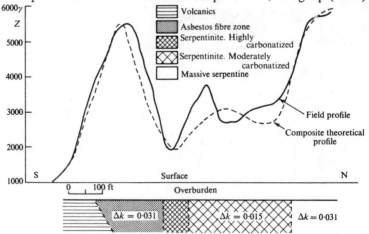

Fig. 3.40 Vertical-component ground magnetic profile in area of asbestos mineralization, Garrison Township, Ontario.

$r_2 = r_4$, $\phi_2 = \phi_4$, $\beta = \pi/2$, $\alpha = \pi/3$, $\alpha = \pi/2$. Obviously the presence of asbestos in the massive serpentine zones can only be established by drilling.

3.7.2 *Airborne surveys*

(1) The Monteregian hills are very familiar topographic features of the St Lawrence Lowland region near Montreal. They were formed by igneous intrusions of the sedimentary basin rocks. These hills are magnetic as well as topographic anomalies because of their pronounced contrast with the uniformly low susceptibility background in the sediments. The aeromagnetic maps (Canadian Government Aeromagnetic Series, St Jean and Beloeil, P.Q.) show this very clearly for Mts Bruno, St Hilaire, Rougemont and St Gregoire. On the same sheets we also see two well-defined magnetic highs, one about 3 miles west of St Gregoire, the larger 7 miles northwest of Bruno. Neither of these is a topographic feature, although the one near St Gregoire may just break the surface. One would assume that they were igneous plugs which failed to reach the eminence of the Monteregian hills.

These two magnetic features, plus Mt St Gregoire, provide three excellent examples of the vertical-prism model commonly employed in aeromagnetic interpretation. Figure 3.41a shows the total-field contours from the original maps, with the larger subsurface anomaly reduced to a scale of 1 in = 2 miles. The principal profiles are displayed in fig. 3.41b.

Two methods have been employed to assess the magnetic characteristics of these three intrusions. Equation (3.54) for total-field response of a vertical prism allows us to calculate k, z, $2Y$ and b, the susceptibility contrast, depth, strike length and width, by matching the principal profiles. The results are shown in fig. 3.41b. The models of Vacquier *et al.* (1951) will also give the same parameters, although we are restricted to the discrete shapes provided. In both cases it is necessary to know or assume the inclination I and magnitude F_0 of the local total field. These values are not precisely known; although I is about 60° and $F_0 \approx 0.5$ in the Montreal area, a measurement made at Mt Yamaska (another of the Monteregian hills about 7 miles east of Rougement) gave $I \approx 75°$ and $F_0 \approx 0.6$.

	I	k	z	b	2Y	Source
			ft	ft	ft	
St Gregoire	60°	0·0033	370	2900	2900	Vacquier, fig. A60
	75	0·004	370	2900	2200	Vacquier, fig. A70
	60	0·005	500	2600	2600	Equation (3.54)
Anomaly near	60°	0·002	750	5500	5500	Vacquier, fig. A60
St Gregoire	75	0·0025	750	5500	4100	Vacquier, fig. A70
	60	0·0035	1000	4000	4000	Equation (3.54)
Anomaly	75°	0·0067	3700	15,000	30,000	Vacquier, fig. A75
near Bruno	60	0·0045	4000	16,000	40,000	Equation (3.54)

The theoretical and field profiles fit reasonably well. If we assume I to be 75° rather than 60°, the theoretical curves have steeper slopes on the left (south or southeast) flanks; it would be necessary to increase the lateral dimensions of the sources considerably to match the field profiles.

In practical interpretation of this type of anomaly the depth to the top of the prism, z, is the most significant dimension. In the first two examples, we know that St Gregoire rises 600 ft above the surrounding plain and that the anomaly near

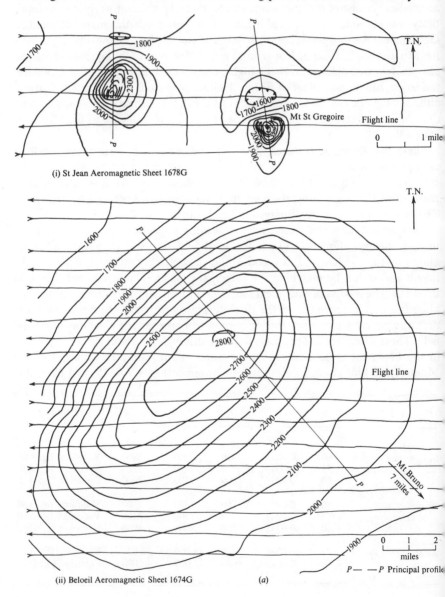

(i) St Jean Aeromagnetic Sheet 1678G

(ii) Beloeil Aeromagnetic Sheet 1674G   (a)

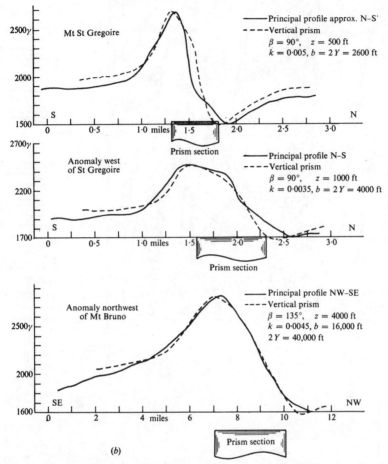

Fig. 3.41 Total-field magnetic data for three anomalous areas in the St Lawrence Lowlands. (*a*) Contour maps (contour values in gammas); (*b*) principal profiles. $I = 60°$, $F_0 = 0.6$ oersteds.

it has practically zero elevation. The height of flight of the aeromagnetic survey is given as 1000 ft. One east–west flight line crosses directly over the Mt St Gregoire magnetic maximum, while the adjacent line passes over the maximum of the surface anomaly. It is not known whether the aircraft climbed on approaching Mt St Gregoire, but the 1000 ft altitude must surely have been maintained on the adjacent line since the ground is flat everywhere. In any case the z-values obtained by curve fitting are larger than those from the Vacquier models.

In the case of the subsurface plug near Mt Bruno, the depth estimates agree very well and locate the source about 2800 ft below ground level.

(2) The mapping of high resolution aeromagnetic data has been described by B. K. Bhattacharyya (1971). In 1969 the Geological Survey of Canada arranged an experimental survey of this type in the Precambrian shield of northern Ontario near Timmins, using a caesium vapour magnetometer having a sensitivity of 0.02γ.

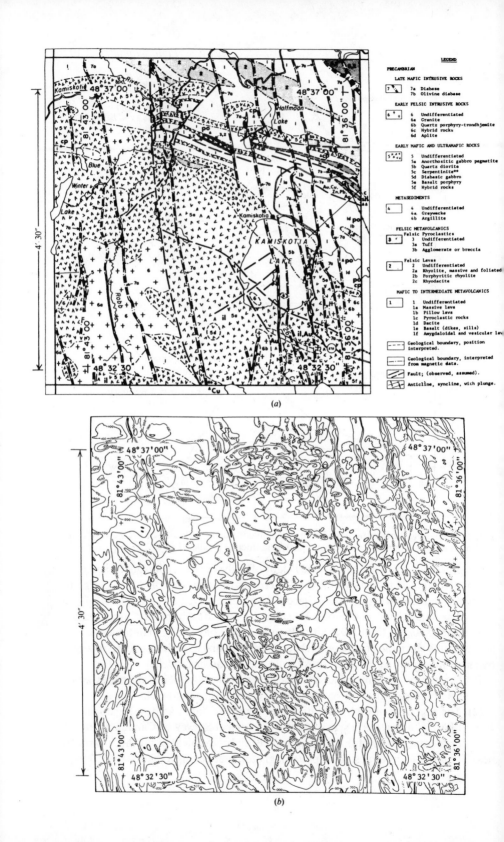

(a)

(b)

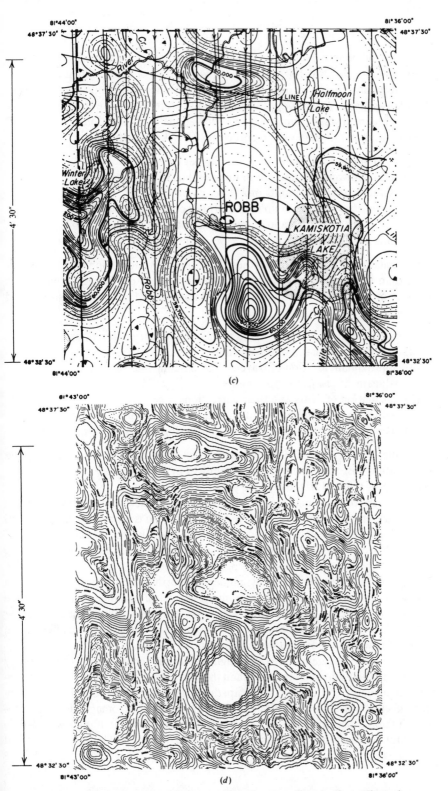

Fig. 3.42 High-resolution aeromagnetic survey, Timmins area, Ontario (from Bhattacharyya,

Control of the actual survey was much tighter than in conventional aeromagnetic work. Line spacing was 1000 ft at an average altitude of 820 ft. Flight paths were straight within ± 300 ft over any 15-mile distance. Perpendicular to these, double baselines, flown in opposite directions, were required every 5 miles. The earth's total field was continuously recorded at one ground station. In the aircraft the following parameters were recorded on 7-track magnetic tapes:

(1) total magnetic field in units of 0.02γ per count,
(2) total field vertical gradient in units of 0.005γ per count,
(3) terrain clearance in units of 2 ft per count,
(4) barometric altitude in units of 10 ft per count,
(5) Doppler along-track and across-track distances in units of 150 ft per count,
(6) time in seconds.

Compilation of the data also required considerably more sophisticated techniques than the manual method normally employed in producing aeromagnetic maps in order that the maximum errors did not exceed 1γ. This process involved the following steps:

(1) initial check of in-flight digital data and necessary corrections,
(2) calculation of data point coordinates,
(3) approximate location of traverse and baseline intersections,
(4) adjustment of intersection points,
(5) calculation of drifts and their corrections,
(6) reduction of magnetic field data to a common datum,
(7) reduction of corrected values for contouring and quantitative analysis.

Needless to say, most of these operations were done on a computer; the compilation cost was about $1 per line mile. For details on the data processing, the reader is referred to the original paper.

Four maps of the same area, a section of the total survey about 6×6 miles in extent, are shown in figs. 3.42a to 3.42d. Figure 3.42a is a provisional geological map, prepared with some help from the vertical intensity ground magnetic map of fig. 3.42b; a conventional aeromagnetic map and the high resolution aeromagnetic map are shown in fig. 3.42c and fig. 3.42d respectively.

The bedrock, which is cut by numerous north-south diabase dikes, is an Archean complex of gabbro, granite, mafic and felsic volcanics. There are three major fault systems: one, striking N30° W, is the main control for the diabase dikes while the other two, trending W–NW and NE, appear to have affected the dikes somewhat as indicated by shearing and deflections.

Of the three magnetic maps, the high-resolution version reflects the geology more accurately and in more detail than the other two. Although there is considerable detail in the ground magnetic map, the large geologic trends are broken up by the effect of variable susceptibility. The conventional aeromagnetic map (produced from a proton-precession instrument survey flown at 1000 ft with $\frac{1}{2}$-mile spacing), however, has much less detail than the other two. This is due to several factors. In comparison with the high-resolution survey, the magnetometer is less sensitive by a factor of 50, the line spacing is $2\frac{1}{2}$ times greater and the compilation method is relatively crude. Compared to the ground survey, the extra sensitivity (a factor of about 3) is nullified by the large line spacing and the vertical separation of 1000 ft.

Apart from the general geologic trends, there are several pronounced magnetic anomalies in the section which are probably due to gabbro. These are very obvious on both aeromagnetic maps, although they are more numerous and show greater detail on fig. 3.42*d* for the same reasons given above. The ground magnetic map, on the other hand, does not expose these features nearly as well; they seem lost in the detailed clutter.

Clearly the high-resolution aeromagnetic survey provides much the best picture of the general geology of the area. One might say that this technique produces an excellent compromise between the ground magnetic survey, which is too close to the source, and the usual type of airborne magnetics, which is too far away from it. Although the cost of the high-resolution survey is larger, by a factor of about 6, than for conventional airborne magnetic work, it is by no means excessive compared to a detailed ground survey; if the latter were done on 400-ft line spacing, the price would be roughly the same.

3.8 Problems

1. Chromite is found in serpentine in a certain district. A magnetic analysis of several specimens indicated that there was less magnetite in the chromite-bearing serpentine than in the barren serpentine. The following vertical component readings were taken on a N–S line at 25-ft stations. Assuming a two-dimensional zone of mineralization, estimate the depth and cross-section of the magnetic anomaly from this profile.

Stn	Z	Stn	Z	Stn	Z	Stn	Z	Stn	Z
0N	275γ	3N	230γ	6N	155γ	9N	-10γ	12N	150
1	220	4	185	7	35	10	-15	13	220
2	224	5	185	8	-40	11	$+100$	14	220

2. Figure 3.43 shows vertical-component contours obtained from a detailed ground magnetic survey carried out in the Noranda district. The area, of course, is rich in sulphides and graphite. The rocks to the south are rhyolites, rhyolitic breccias and tuffs; to the north we find basic volcanics and tuffs with occasional rhyolite. These formations strike roughly E–W with practically vertical dip. If there are any anomalous sulphides and/or graphite in the vicinity, where would you locate them? Are they shallow or deep? If they are sulphides, what varieties could definitely be ruled out? Do you think the fence and power line have affected the magnetometer readings to any extent?

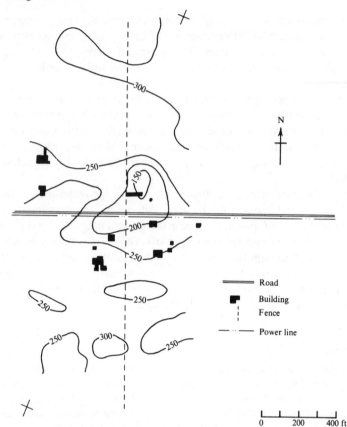

Fig. 3.43 Vertical magnetic field ground survey, Noranda Area. (Contour interval 50γ.) (After Seigel, 1957.)

3. The following vertical-component magnetometer readings were made on a detailed traverse during a field school held in the Quebec Eastern Townships region.

Stn	Z	Stn	Z	Stn	Z
17W	2040γ	11W	9200γ	5W	5140γ
16	2320	10	3400	4	4260
15	2080	9	−9500	3	2680
14	2080	8	−7000	2	2220
13	1800	7	−1060	1	2240
12	3280	6	2720	0	1940

Stations are 10 ft apart on a line approximately E–W. This large anomaly was originally detected during a pace-and-compass traverse exercise, when the magnetic declination suddenly changed by about 110°. The effect did not persist for any appreciable distance.

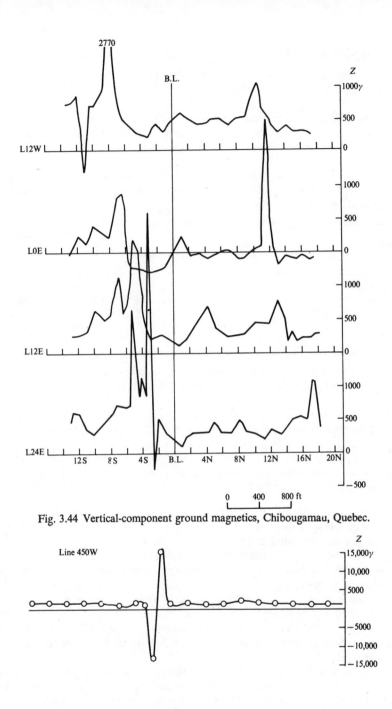

Fig. 3.44 Vertical-component ground magnetics, Chibougamau, Quebec.

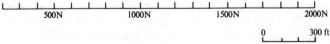

Fig. 3.45 Vertical-component ground magnetics, East Africa.

Make an interpretation of the source with regard to its location, depth, dip, lateral extent and possible mineral character.

4. The vertical-component ground magnetic profiles shown in fig. 3.44 are part of a base-metal survey in an area west of Chibougamau, P.Q. Using the same criteria as in Problems 2 and 3, make an interpretation of these results.

5. Vertical-component magnetometer readings tabulated below are taken from a detailed survey in the vicinity of an old mining property. The primary metal was zinc, with low-grade copper and minor silver.

Stn (ft)	Z (gammas)	Stn (ft)	Z (gammas)	Stn (ft)	Z (gammas)
6 + 00S	45	3 + 00S	50	0 + 50N	75
5 + 00	50	2 + 00	90	1 + 00	40
4 + 50	52	1 + 00	135	2 + 00	8
4 + 00	50	0 + 50	130	3 + 00	0
3 + 50	45	0 + 00	95	4 + 00	−5

Assuming the source of the weak magnetic anomaly to be a dike of large strike length and depth extent, use the method of §3.6.6 to determine its parameters.

6. The vertical-component ground magnetic profile shown in fig. 3.45 is from a large-scale multiple-method survey for base metals in Tanzania. The general geology is similar to that described in Problem 20, chapter 7. Components of the earth's magnetic field in this region are: $F_0 \approx 0.35$ oersted, $D = 4°$ W, $I \approx -31°$ N. There are known sulphide bodies in the vicinity with very large E–W strike extent. Make an interpretation of this anomaly.

7. The magnetic contours in fig. 3.46 provide a striking illustration of the effect of line spacing on geophysical data. A small section, taken from what would normally be considered a detailed ground magnetic survey, is shown in fig. 3.46a. The total area surveyed was about $2\frac{3}{4} \times 3\frac{1}{2}$ miles, with lines spaced 200 ft apart N–S and station readings every 50 ft, reduced to 20 ft in the vicinity of anomalous values. As a result the Z-contours are greatly elongated in the N–S direction, that is, between grid lines.

In the course of more detailed follow-up work, some of the magnetic zones were resurveyed on 50 ft grids, 400 ft N–S by 200 ft E–W. Even with this uniform and smaller grid spacing it was necessary at times to make readings 5–10 ft apart because of the small lateral extent of the magnetic anomalies. Three of these detailed grids are shown in figs. 3.46b, 3.46c and 3.46d.

The differences between the contours on the two grid scales are very apparent. These differences are particularly related to lateral extent, strike axis, magnitude and location of anomaly maxima. What are they? Other geophysical surveys have indicated the presence of massive sulphides at the centres of the grids shown in figs. 3.46b and 3.46d but not at 164N, 103E. Given this additional information, would you conclude that the magnetic anomalies are directly or indirectly related to the sulphides? Do they have any association with the sulphides? Would you reach the same conclusion if only fig.

3.46*a* were available? Estimate the depth, lateral extent, attitude, susceptibility and probable content of a few of the anomalies in all four diagrams. Do they show evidence of strong remanent magnetization?

8. The vertical-component ground magnetic contours in fig. 3.47 were taken from a survey over a nickel prospect in northern Manitoba. Zone C is approximately $2\frac{1}{2}$ miles NE of Zone A. There is considerable overburden throughout the area. Electromagnetic surveys showed that both zones were good conductors. One of them contained ore-grade nickel sulphides, the other was mainly barren sulphides and graphite. It it possible to distinguish the economic mineralization solely from the magnetic results? Estimate the depths of the main magnetic anomalies in the two zones.

9. The single ground magnetic vertical-component profile in fig. 3.48 is from an area near Senneterre, Quebec. Extensive bands of massive pyrite and pyrrhotite, along with some graphite, are known to occur in meta-sedimentary breccias and tuffs, interbedded with lava flows, in the vicinity. Make an interpretation of the magnetic profile assuming either that: (*a*) the magnetic anomalies are associated with sulphide mineralization, or (*b*) there is no pyrrhotite present.

10. Figure 3.49 shows four vertical-component magnetic profiles from the Manitoba Nickel Belt, obtained during large-scale base-metal exploration programmes. Sulphides and graphite occur in the Precambrian rocks below Paleozoic sediments and thick overburden; the mineralized zones frequently extend for miles.

Diamond drilling has established that the mineralization associated with two of these profiles is pyrite and pyrrhotite, a third is graphite and pyrite and the fourth graphite and pyrrhotite, and that they are located at four different depths. With this information use the magnetic data to locate the mineralized sections as precisely as you can.

11. The two ground magnetic contour maps in fig. 3.50 illustrate the effect of irregular topography on magnetic measurements. The areas surveyed in fig. 3.50*a* and fig. 3.50*b* are only 60 miles apart. The terrain is quite rugged in both as can be seen from the dotted contours, but the geology is entirely different. In the former the rocks are sedimentary throughout the area and to great depth; at the latter site there are granites in most of the north and west parts of the map and sediments in the lower ground at the bottom of the diagram. It is thought that these sedimentary beds extend for some distance up the hill in the lower left corner and there is assumed to be a contact between the granites and volcanic formations somewhere in the upper left portion of the map.

The uniform magnetic response over the sediments in fig. 3.50*a* (about 300γ maximum contrast) is to be expected because of the low susceptibility of sandstone and limestones. There is no particular correlation between the topographic and magnetic contours, hence there is no need to apply a topographic correction.

In fig. 3.50*b* the situation is quite different. A definite magnetic contrast exists between the granites and the sedimentary area, although the map does not extend far enough south and southeast to indicate this very clearly. However, there is a pronounced magnetic low which follows the topography from north to south, then west to east, starting in the upper left area of the map. This is a very clear reflection of the terrain effect on ground magnetics.

Using the method of §3.6.2, apply the terrain correction at a few selected points on the

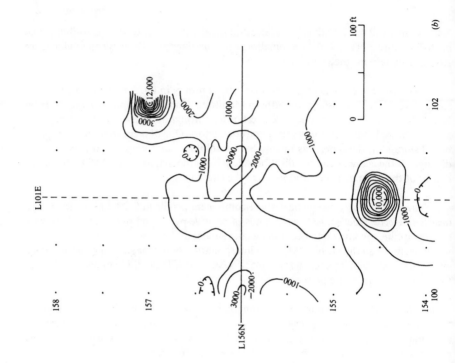

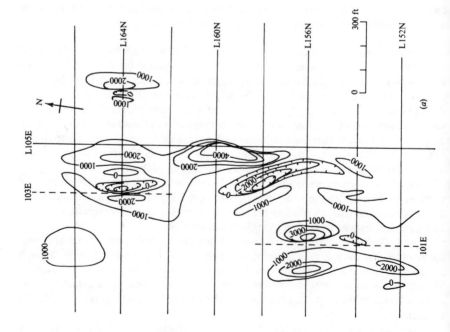

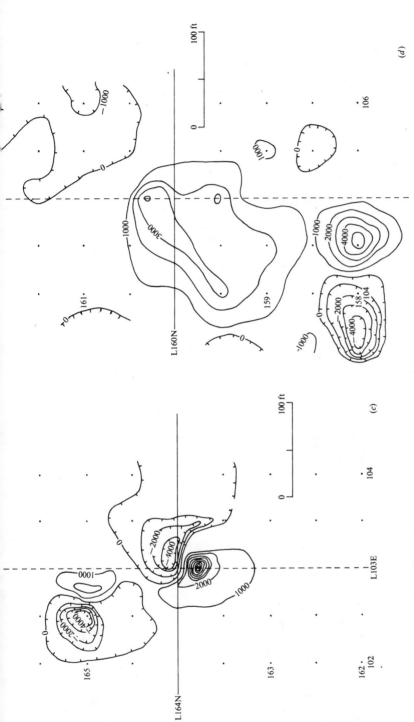

Fig. 3.46 Effect of line spacing on magnetic data. (a) Conventional magnetic map, contour interval 1000γ; (b) detailed magnetic map centred at 156N, 101E; (c) detailed magnetic map centred at 164N, 103E; (d) detailed magnetic map centred at 160N, 105E.

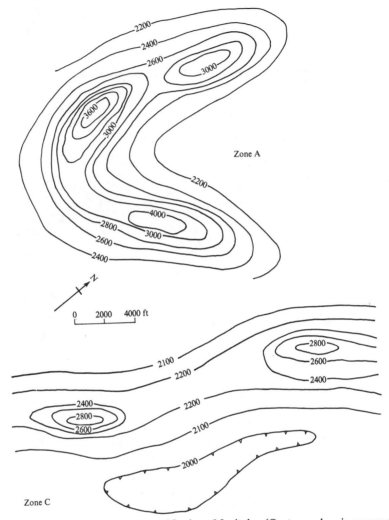

Fig. 3.47 Ground magnetic contours, Northern Manitoba. (Contour values in gammas.)

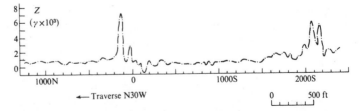

Fig. 3.48 Vertical-component magnetic profile near Senneterre, Quebec.

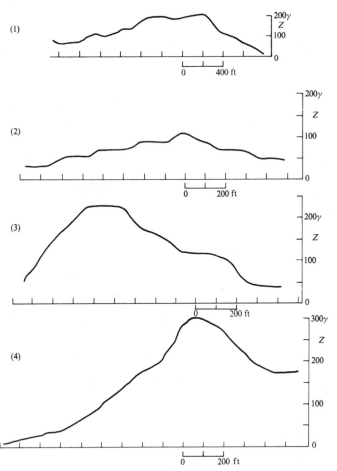

Fig. 3.49 Vertical-component magnetic profiles, Manitoba Nickel Belt.

map of fig. 3.50*a*. For example, the 100γ low on the steep slope near the bottom of the map, midway between the east–west boundaries, lies on the 425 ft contour; if we choose the 600-ft elevation for $z = 0$, the value of h will be 175 ft. Reasonable values for circle radii would be $r_1 = 100$ ft, $r_2 = 200$ ft, etc. For other stations one might select one at the top of the hill, one in the northwest corner and one to the southeast. Do these modified Z-values aid the magnetic interpretation in any way?

Repeat the procedure for several strategically located stations on the map of fig. 3.50*b*, particularly in the area of the magnetic lows following the stream gorge. (Obviously the best method for handling analysis of this type would be to digitize the contoured data and use a computer, if this were possible.) Are the terrain corrections significant in this case? Would these corrections be more or less reliable if the vertical gradients had been measured in the course of the survey? Why?

12. The two sets of contours shown in fig. 3.51 illustrate the differences between total-field airborne and vertical-component ground magnetic survey results. Only the relative

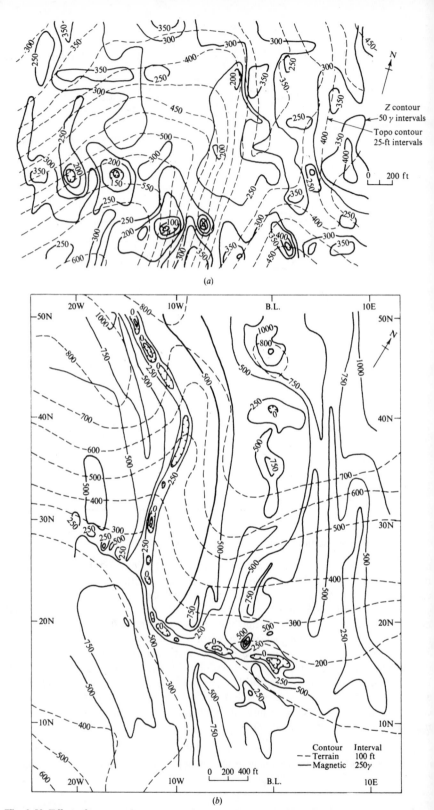

Fig. 3.50 Effect of topography on magnetic measurements. (a) Elevation contours and vertical-component contours in an area of sedimentary rocks; (b) elevation contours and vertical-component contours in an area of granitic and sedimentary rocks.

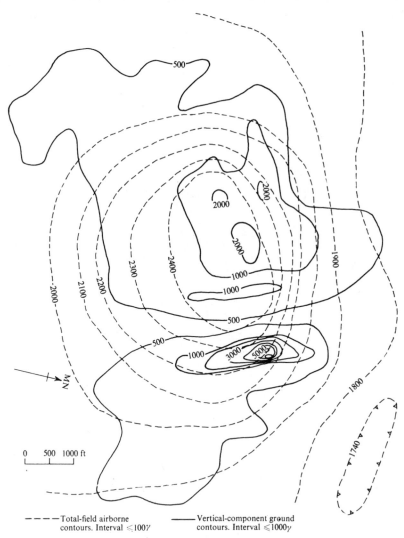

————— Vertical-component ground
contours. Interval ≤1000γ

Fig. 3.51 Comparison of ground vertical-component data with airborne total-field data.

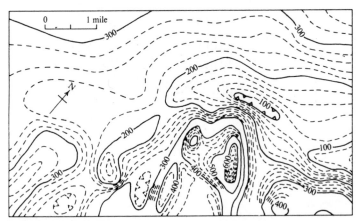

Fig. 3.52 Total-field aeromagnetic contours, Northwest Newfoundland. Contour interval 20γ, flight altitude 1000 ft.

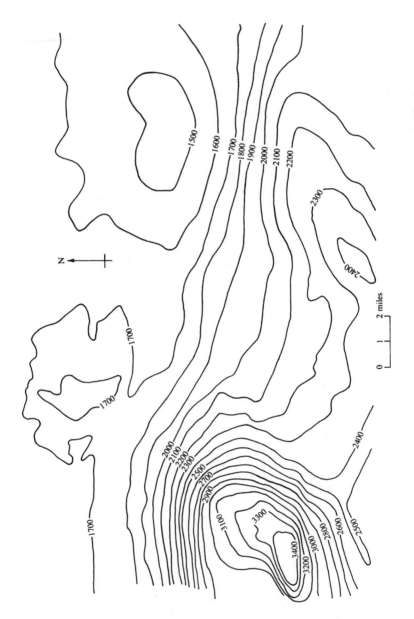

Fig. 3.53 Total-field aeromagnetic contours, St Lawrence Lowlands. (Contour interval 100 γ.)

values in gammas are significant in each; there is no relation between the absolute magnitudes of the fields. Although the ground survey has not been done in detail, the coverage is sufficient to indicate at least five distinct maxima, compared to a single symmetrical anomaly in the airborne contours. Furthermore, the largest of these ground maxima is displaced about 2200 ft from the centre of the total field anomaly.

The altitude of the aircraft was 1000 ft and the ground is flat. The magnetic inclination in the area is 75°. Calculate the depths and approximate lateral extents of the 6000γ and the larger 1000γ anomalies from the ground contours as well as the airborne anomaly.

With the aid of eq. (3.28a) and table 3.3, carry out an upward continuation of the ground data to 1000 ft by choosing $r_1 = 200$ ft and $h = 5$. The remarks in problem 11 apply here as well; one may perform the operation at several selected points to obtain a very rough idea of the matching between ground and airborne data, but it would be preferable to use a computer.

13. Figure 3.52 shows a 6 × 4 mile section from a Canadian government aeromagnetic survey in northwest Newfoundland. The rocks in the area are sedimentary, consisting of sandstones, shales and limestones with some dolomite. In the upper half and the lower left quarter of the section the topography is flat, the average elevation being about 350 ft. A steep escarpment, in the shape of an inverted U with the apex to the north, occupies the lower middle portion of the figure. It follows the closed 200γ contour on the left, continues north and east to overlap the east half of the 100γ low, then turns southeast between the 300γ contours on the lower right. This scarp rises about 700 ft, in places having a slope of nearly 30°; as a result the magnetic high in the lower part of the diagram is on a 1000-ft plateau. Flight lines were east–west, 1000 ft above ground level.

With this information make an interpretation of the magnetic anomaly in the lower central part of the section. Could it be entirely or partly the result of topography? Is it the reflection of a single magnetic structure? Is it possible that the larger magnetic low area, contained in the 200γ contour striking roughly east–west, might represent a distinctly different structure?

14. The data tabled below are vertical magnetic component readings on traverses across an obviously strong magnetic structure of great length, striking roughly E–W. A heavily damped vertical-component instrument, flying at 300–400 ft above ground surface, was used in the airborne traverse; the ground measurements, made approximately along the flight line, were obtained with a conventional fluxgate magnetometer.

Airborne		Ground			
Stn	Z	Stn	Z	Stn	Z
0N	600γ	0N	4600γ	600N	36000γ
400	2600	100	7400	700	28600
600	4100	200	13700	800	16000
800	5700	300	28600	900	8000
1000	4050	400	40000	1000	4600
1200	2750	500	40000		
1500	750				

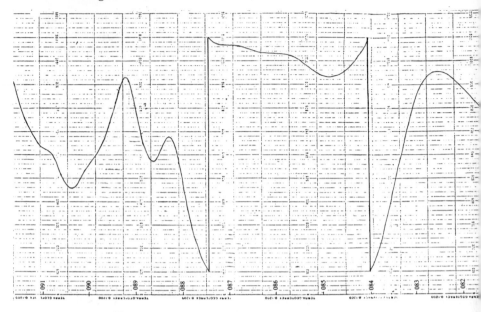

Fig. 3.54 East-west aeromagnetic

Station intervals are in feet and airborne station 800N corresponds approximately to a point between 400N and 500N on the ground traverse. Determine the depth, cross-section, approximate susceptibility and direction of magnetization of the source from each traverse. As a check on the results, continue the ground profile upwards to find out if it matches the airborne profile.

15. A section from Canadian government aeromagnetic maps of the St Lawrence Lowlands sedimentary region is reproduced in fig. 3.53. In the area shown, the general low magnetic relief to the north appears considerably disturbed in the southern portion. There are at least three, and possibly four, large structural features producing the magnetic anomalies. Two are well documented geologically.

Can you distinguish any fault zones? Any domes or plugs? Are these anomalies produced by structures in the sediments or in the underlying basement rocks? Analyse this aeromagnetic section as precisely as possible with particular emphasis on depths to the sources. Altitude of the aircraft was 1000 ft above ground level and the flight lines were E–W. The topography is essentially flat throughout.

16. Figure 3.54 shows a portion of an east–west aeromagnetic profile. The fiducial marks (numbers on the bottom of the profile) are 2 km apart and the aircraft was flown at a constant barometric elevation of 750 m. A semiquantitative measure of the depth to magnetic basement is given by the horizontal distance over which the magnetic curve is closely approximated by a straight line (the distance *A* in fig. 3.55). Another such measure is half the distance between the points where the slope is half the maximum intervening slope (the distance *B* in fig. 3.55). Analyse the profile of fig. 3.54 using these methods.

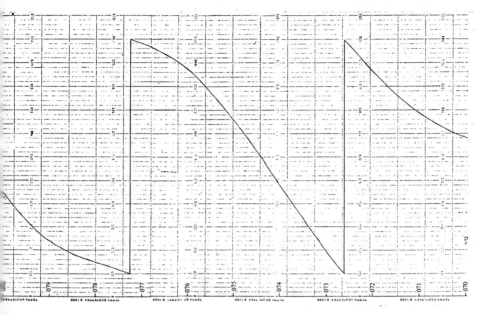

le = 600γ. (Courtesy Chevron Oil.)

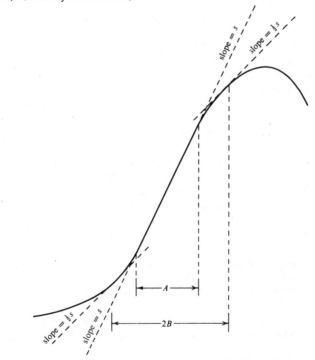

Fig. 3.55 Determining anomaly depth from the slope of a magnetic profile.

References

F – field methods; G – general; I – instruments; T – theoretical

Affleck, J. (1963). Magnetic anomaly trend and spacing patterns. *Geophysics*, **28** 379–95. (T)

Balsey, J. R. and Buddington, A. F. (1958). Iron-titanium oxide minerals, rocks and aeromagnetic anomalies of the Adirondack area, New York. *Econ. Geol.* **53**, 777–805. (G)

Bhattacharyya, B. K. (1964). Magnetic anomalies due to prism-shaped bodies with arbitrary polarization. *Geophysics*, **29**, 517–31. (T)

Bhattacharyya, B. K. (1970). Some important considerations in the acquisition and treatment of high-resolution aeromagnetic data. *Bolletino di Geofisica Teorica ea Applicata*, **12**, 45–6, 21–44. (F, T)

Bhattacharyya, B. K. (1971). An automatic method of compilation and mapping of high-resolution aeromagnetic data. *Geophysics*, **36**, 695–716. (T)

Chapman, S. and Bartels, J. (1940). *Geomagnetism*. Oxford, Clarendon. (G)

Cook, K. L. (1950). Quantitative interpretation of vertical magnetic anomalies over veins. *Geophysics*, **15**, 667–86. (T)

Evjen, H. M. (1936). The place of the vertical gradient in gravitational interpretations. *Geophysics*, **1**, 127–36. (T)

Fleming, J. A. (ed.), (1939). Terrestrial magnetism and electricity. *Physics of the earth* Series 8, N.R.C. New York, McGraw-Hill. (G)

Garland, G. D. (1951). Combined analysis of gravity and magnetic anomalies. *Geophysics*, **16**, 51–62. (T)

Gay, S. P. Jr. (1966). Standard curves for interpretation of magnetic anomalies over long tabular bodies. *Mining Geophysics*, **2**, 512–48. Tulsa, Soc. of Explor. Geophys. (T)

Green, R. (1960). Remanent magnetization and the interpretation of magnetic anomalies. *Geophys. Prosp.* **8**, 98–110. (T)

Gupta, V. K. and Fitzpatrick, M. M. (1971). Evaluation of terrrain effects in ground magnetic surveys. *Geophysics*, **36**, 582–89. (F)

Henderson, R. G. and Zeitz, I. (1948). Analysis of total magnetic-intensity anomalies produced by point and line sources. *Geophysics*, **13**, 428–36. (T)

Henderson, R. G. and Zeitz, I. (1957). Graphical calculation of total intensity-anomalies of three-dimensional bodies. *Geophysics*, **22**, 887–904. (T)

Henderson, R. (1960). A comprehensive system of automatic computation in magnetic and gravity interpretation. *Geophysics*, **25**, 569–85. (T)

Henderson, R. G. and Zeitz, I. (1966). Magnetic doublet theory in the analysis of total intensity anomalies. *Mining Geophysics*, **2**, 490–511. Tulsa, Soc. of Explor. Geophys. (T)

Hood, Peter and McClure, D. J. (1965). Gradient measurements in ground magnetic prospecting. *Geophysics*, **30**, 403–10. (F)

Hutchinson, R. D. (1958). Magnetic analysis by logarithmic curves. *Geophysics*, **23**, 749–69. (T)

Jacobs, J. A. (1963). *The earth's core and geomagnetism*. London, Pergamon. (G)

Jahren, C. E. (1963). Magnetic susceptibility of bedded iron formation. *Geophysics*, **28**, 756–66. (F)

Koulomzine, Th. Lamontagne, Y. and Nadeau, A. (1970). New methods for direct interpretation of magnetic anomalies caused by inclined dikes of infinite length. *Geophysics*, **35**, 812–30. (T)

Methods and Case Histories in Mining Geophysics, 6th Commonwealth Congress. Montreal, Mercury Press.

Muffly, Gary (1946). The airborne magnetometer. *Geophysics*, **11**, 321–34. (I)

Packard, M. E. and Varian, R. H. (1954). Free nuclear induction in the earth's magnetic field. *Phys. Rev.*, **93**, 941. (I)

Peters, L. J. (1949). The direct approach to magnetic interpretation and its practical application. *Geophysics*, **14**, 290–320. (T)

Runcorn, S. K. (1956). The magnetism of the earth's body. *Handbuch der Physik* (ed. by J. Bartels), **47**, 493–533. Berlin, Springer-Verlag. (G)

Seigel, H. O. (1957). In *Methods and Case Histories in Mining Geophysics, 6th Commonwealth Congress.* Montreal, Mercury Press.

Smellie, D. W. (1967). Elementary approximations in aeromagnetic interpretation. *Mining Geophysics*, **2**, 474–89. Tulsa, Soc. of Explor. Geophys. (T)

Smith, R. A. (1961). Some theorems concerning local magnetic anomalies. *Geophys. Prosp.* **9**, 399–410. (T)

Steenland, N. C. and Brod, R. J. (1960). Basement mapping with aeromagnetic data, Blind River Basin. *Geophysics*, **25**, 586–601. (T)

Vacquier, V., Steenland, N. C. Henderson, R. G. and Zeitz, I. (1951). Interpretation of aeromagnetic maps. *Geol. Soc. Am., Memoir*, **47**. (T)

Whitham, K. (1960). Measurement of the geomagnetic elements. *Methods and Techniques in Geophysics* (ed. by S. K. Runcorn), **1**, 134–48. New York, Interscience. (G)

4. Seismic methods

4.1 Introduction

4.1.1 *Importance of seismic work*

The seismic method is by far the most important geophysical technique in terms of expenditures (see table 1.1) and number of geophysicists involved. The predominance of the seismic method over other geophysical methods is due to various factors, the most important of which are the high accuracy, high resolution and great penetration of which the method is capable.

The widespread use of seismic methods is principally in exploring for petroleum, the locations for exploratory wells rarely being made without seismic information. Seismic methods are also important in groundwater searches and in civil engineering, especially to measure the depth to bedrock in connection with the construction of large buildings, dams, highways and harbour surveys. Seismic techniques have found little application in direct exploration for minerals because they do not produce good definition where interfaces between different rock types are highly irregular. However, they are useful in locating features such as buried channels in which heavy minerals may be accumulated.

Exploration seismology is an offspring of earthquake seismology. When an earthquake occurs, the earth is fractured and the rocks on opposite sides of the fracture move relative to one another. Such a rupture generates seismic waves which travel outward from the fracture surface. These waves are recorded at various sites using seismographs. Seismologists use the data to deduce information about the nature of the rocks through which the earthquake waves travelled.

Exploration seismic methods involve basically the same type of measurements as earthquake seismology. However, the energy sources are controlled and movable and the distances between the source and the recording points are relatively small. Much seismic work consists of *continuous coverage* where the response of successive portions of earth is sampled along lines of profile. Explosives and other energy sources are used to generate the seismic waves and arrays of seismometers or geophones are used to detect the resulting motion of the earth. The data usually are recorded on magnetic tape so that computer processing can be used to enhance the signals with respect to the noise, extract the significant information and display the data in such a form that a geological interpretation can be carried out readily.

The basic technique of seismic exploration consists of generating seismic waves and measuring the time required for the waves to travel from the sources to a series of geophones, usually disposed along a straight line directed towards the source. From a knowledge of traveltimes to the various geophones and the velocity of the waves, one attempts to reconstruct the paths of the seismic waves. Structural information is derived principally from paths which fall into two main categories: *head-wave* or *refracted* paths in which the principal portion of the path is along the interface between two rock layers and hence is approximately horizontal, and *reflected* paths in which the wave travels downward initially and at some point is reflected back to the surface, the overall path being essentially vertical. For both types of path, the traveltimes depend upon the physical properties of the rocks and the attitudes of the beds. The objective of seismic exploration is to deduce information about the rocks, especially about the attitudes of the beds, from the observed arrival times and (to a very limited extent) from variations in amplitude and frequency.

The importance of seismic work in the exploration for petroleum is evidenced by its extensive application. Almost all of the major oil companies rely on seismic interpretation for selecting the sites for exploratory oil wells. Despite the indirectness of the method – most seismic work results in the mapping of geological structure rather than finding petroleum directly – the likelihood of a successful venture is improved more than enough to pay for the seismic work. Likewise, engineering surveys, mapping of water resources and other studies requiring accurate knowledge of subsurface structure derive valuable information from seismic data.

We shall first give a brief outline of the history of seismic exploration and of the field methods used for acquiring seismic data. This will be followed by a brief discussion of seismic methods to provide a background for the following sections. The subsequent sections will then discuss the theory of seismic wave propagation, the geometry of seismic ray paths and the characteristics of seismic events. We shall then examine in more detail the recording instrumentation and field techniques used for land and marine reflection and refraction surveys. Finally we shall describe the processing of seismic data and conclude with a discussion of interpretation techniques.

4.1.2 *History of seismic exploration*

Much of seismic theory was developed before instruments were available which were capable of sufficient sensitivity to permit significant measurements. Earthquake seismology preceded exploration applications. Mallet experimented with 'artificial earthquakes' in 1845 in attempts to measure seismic velocities. Knott developed the theory of reflection and refraction at interfaces in a paper in 1899 and Zoeppritz and Wichert published on wave theory in 1907. During the First World War, both the Allies and Germany carried out research directed toward the location of heavy guns by means of recording the arrival of seismic waves generated by the recoil. This work was fundamental in the development of exploration seismology and several workers engaged in this research later pioneered in the development of seismic prospecting techniques and instruments; among these, Mintrop in Germany and Karcher, McCollum and Eckhardt in the United States were outstanding.

In 1919 Mintrop applied for a patent on the refraction method and in 1922 Mintrop's Seismos Company furnished two crews to do refraction seismic prospecting in Mexico and the Gulf Coast area of the United States using a mechanical seismograph of rather low sensitivity. The discovery of the Orchard Salt dome in Texas in 1924 led to an extensive campaign of refraction shooting during the next six years, the emphasis being principally on the location of salt domes. By 1930 most of the shallow domes had been discovered and the refraction method began to give way to the reflection method. While refraction techniques were ideal for locating salt domes, reflection techniques are more suitable for mapping other types of geologic structures commonly encountered.

Reflection seismic prospecting stemmed principally from the pioneering work of Reginald Fessenden about 1913. This work was directed towards measuring water depths and detecting icebergs using sound waves. In the early 1920s Karcher developed a reflection seismograph which saw field use in Oklahoma. It was not until 1927, however, that commercial utilization of the method began with a reflection survey of the Maud Field in Oklahoma by the Geophysical Research Corporation using a vacuum tube amplifier. Oklahoma proved to be particularly suitable for the application of reflection methods, just as the Gulf Coast had been for refraction techniques, and the reflection method rapidly grew in popularity until it eventually displaced the refraction method to a large extent. Although the reflection method has continued to be the principal seismic method, there are certain areas and certain types of problems where refraction techniques enjoy advantages over reflection shooting and so they continue to be used to a modest degree.

A distinctive reflection was characteristic of the first reflection application in Oklahoma. The low-velocity Sylvan shale overlies the high-velocity Viola limestone, providing a high-contrast reflector which could be recognized on an isolated seismic trace. Hence the first reflection work utilized the *correlation method* whereby a map (of the top of the Viola) can be constructed by recognizing the same event on isolated individual records. However, most areas are not characterized by such a distinctive reflector and so the correlation method has little application in general.

In 1929 the calculation of dip from the time differences across several traces of a seismic record permitted the successful application of reflection exploration in the Gulf Coast area where reflections were not distinctive of a particular lithologic break and could not be followed for long distances. This method proved to be much more widely applicable than correlation shooting and so led to rapid expansion of seismic exploration.

Rieber in 1936 published the idea of processing seismic data using variable-density records and photocells for reproducing; however, widespread use of playback processing did not begin until magnetic tape became commercially available in 1953. Magnetic tape recording spread rapidly in the next few years. The compositing of magnetic tape data made weaker energy sources feasible while the use of a dropped weight as a source of seismic energy in 1955 was the forerunner of a series of different kinds of impulsive sources.

The application of information theory and computer processing to seismic signal extraction began with the Geophysical Analysis Group Project at the Massachusets Institute of Technology in 1953. The common-depth-point method (also called common-reflection-point) was patented in 1956. The redundancy of data achieved with this method made practicable several schemes for the attenuation of noise (including multiple reflections) and so improved data quality that most areas have been remapped with the new techniques.

4.1.3 *Outline of the seismic reflection method*

To provide a background for the following sections, a brief outline of the seismic reflection technique will be given at this point (although the reasons for the various steps will only be given later). Seismic techniques have changed considerably within the last few years. The word 'conventional' often is used to distinguish yesterday's techniques from present-day methods. Thus the conventional seismic source is the explosion of a charge of dynamite in a drilled shothole, although many other energy sources are now in use. Conventional reflection work employs continuous-coverage *split-dip spreads* (arrangements of geophones), although almost all present-day work uses *common-depth-point* (CDP) redundancy. Conventional records (one photographic paper record resulting from each explosion from which the interpretation is to be made) were replaced by *analog recording* on *magnetic tape* which has itself been superseded by *digital recording*. Even though conventional techniques have been largely superseded in actual practice, they form the basis from which modern practice has developed. At the same time they are more readily understood and provide a simpler starting point, so we shall begin by describing them. Modern variations and refinements will be described later.

Assume a land crew using conventional reflection methods with an explosive charge as the energy source. The first step in the field work is the drilling of a vertical hole in the earth at the *shotpoint*, the hole diameter being perhaps 10 or 12 cm and the depth usually between 6 and 30 m. A *charge* of 2 to 25 kg of explosive is armed with an electric blasting *cap* and then placed near the bottom of the hole. Two wires extend from the cap to the surface where they are connected to a *blaster* which is used to send an electrical current through the wires to the cap which then explodes, initiating the explosion of the dynamite (the *shot*).

Shotpoints are usually spaced at equal intervals of about 1/4 mile (400 meters). Two *cables* are laid out in a straight line extending each way from the hole about to be fired to the adjacent holes. The cables contain many pairs of electrical conductors, each pair terminating in an electrical connector at both ends of the cable. In addition, each pair of wires is connected to one of several outlets spaced at intervals of 30 to 70 m along the cable. Several *geophones* (*seismometers*) are connected to each of these outlets so that each pair of wires in the cable carries the output energy of a *group* of geophones back to the recording instruments. Because of the small spacing between the geophones in the group attached to one pair of wires, the whole group is approximately equivalent to a single fictitious geophone located at the centre of the group. Usually 24 or more geophone groups are located at equal intervals along the cable. When the dynamite charge is exploded, each geophone group generates a signal which depends upon the motion of the ground in the vicinity of the group. The net result is the generation of signals furnishing information about the ground motion at a number of regularly spaced points (the *group centres*) along a straight line passing through the shotpoint. This information constitutes a *profile*.

The electrical signals from the geophone groups go to an equal number of amplifiers. These amplifiers increase the overall signal strength and partially eliminate (*filter out*) parts of the input deemed to be undesirable. The outputs from the amplifiers along with accurate timing signals are recorded on magnetic tape and on paper records. Thus the recorded data consist of several *traces*, each trace showing how the motion of one geophone group varies with time after the shot.

Events, that is, arrivals of energy which vary systematically from trace to trace and which are believed to represent reflected energy, are identified on the records. The *arrival times* (the interval between the shot instant and the arrival of the energy at a geophone group, also known as the *traveltime*) of these events are measured for various geophone groups. The location and attitude of the interface which gave rise to each reflection event are then calculated from the arrival times. The results for various shot locations are combined into cross-sections and contour maps which represent the structure of the geological interfaces responsible for the events. The presence or absence of hydrocarbons or other minerals is inferred from the structural information, since the seismic data are not appreciably affected by such accumulations (see, however, §4.8.2g and fig. 4.116).

We have introduced above a number of terms used in a specialized sense in seismic work (*indicated by italics*), for example, shotpoint, group, profile, trace, events, arrival time. Applied seismology abounds in such technical terms. We shall follow the definitions given for such terms by Sheriff in the *Encyclopedic Dictionary of Exploration Geophysics* (1973).

4.2 Seismic theory

4.2.1 *Theory of elasticity*

(a) *General.* The seismic method utilizes the propagation of waves through the earth; since this propagation depends upon the elastic properties of the rocks, we shall discuss briefly the basic concepts of elasticity.

The size and shape of a solid body can be changed by applying forces to the external surface of the body. These external forces are opposed by internal forces which resist the changes in size and shape. As a result the body tends to return to its original condition when the external forces are removed. Similarly, a fluid resists changes in size (volume) but not changes in shape. This property of resisting changes in size or shape and of returning to the undeformed condition when the external forces are removed is called *elasticity*. A perfectly elastic body is one which recovers completely after being deformed. Many substances including rocks can be considered perfectly elastic without appreciable error provided the deformations are small.

The theory of elasticity relates the forces which are applied to the external surface of a body to the resulting changes in size and shape. The relations between the applied forces and the deformations are most conveniently expressed in terms of the concepts, stress and strain.

(b) *Stress. Stress* is defined as force per unit area. Thus, when a force is applied to a body, the stress is the ratio of the force to the area on which the force is applied. If the force varies from point to point, the stress also varies and its value at any point is found by taking an infinitesimally small element of area centred at the point and dividing the total force acting on this area by the magnitude of the area. If the force is perpendicular to the area, the stress is said to be a *normal stress* (or *pressure*). In this book, positive values correspond to tensile stresses (the opposite convention of signs is also used frequently). When the force is tangential to the element of area, the stress is a *shearing stress*. When the force is neither parallel nor perpendicular to the element of area, it can be resolved into components parallel and perpendicular to the element; hence any stress can be resolved into component normal and shearing stresses.

If we consider a small element of volume inside a stressed body, the stresses acting upon each of the six faces of the element can be resolved into components, as shown in fig. 4.1 for the two faces perpendicular to the x-axis. Subscripts denote the x-, y- and z-axes respectively and σ_{xy} denotes a stress parallel to the x-axis acting upon a surface perpendicular to the y-axis. When the two subscripts are the same (as with σ_{xx}), the stress is a normal stress; when the subscripts are different (as with σ_{xy}), the stress is a shearing stress.

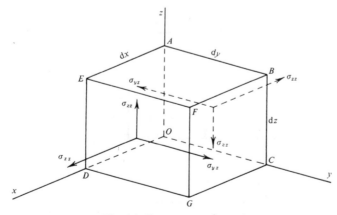

Fig. 4.1 Components of stress.

When the medium is in static equilibrium, the stresses must be balanced. This means that the three stresses σ_{xx}, σ_{yx} and σ_{zx} acting upon the face OABC must be equal and opposite to the corresponding stresses shown on the opposite face DEFG, with similar relations for the remaining four faces. In addition, a pair of shearing stresses, such as σ_{yx}, constitute a *couple* tending to rotate the element about the z-axis, the magnitude of the couple being

$$(\text{Force} \times \text{Lever arm}) = (\sigma_{yx}\, dy\, dz)\, dx.$$

If we consider the stresses on the other four faces, we find that this couple is opposed solely by the couple due to the pair of stresses σ_{xy} with magnitude $(\sigma_{xy}\, dx\, dz)\, dy$. Since the element is in equilibrium, the total moment must be zero; hence $\sigma_{xy} = \sigma_{yx}$. In general, we must have

$$\sigma_{ij} = \sigma_{ji}.$$

(c) Strain. When an elastic body is subjected to stresses, changes in shape and dimensions occur. These changes, which are called *strains*, can be resolved into certain fundamental types.

Consider a rectangle PQRS in the xy-plane (see fig. 4.2). When the stresses are applied, let P move to P', PP' having components u and v. If the other vertices Q, R and S have the same displacement as P, the rectangle is merely displaced as a whole by the amounts u and v; in this case there is no change in size or shape and no strain exists. However, if u and v are different for the different vertices, the rectangle will undergo changes in size and shape and strains will exist.

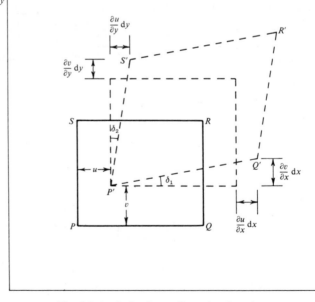

Fig. 4.2 Analysis of two-dimensional strain.

Let us assume that $u = u(x, y)$, $v = v(x, y)$. Then the coordinates of the vertices of $PQRS$ and $P'Q'R'S'$ are as follows:

$P(x, y)$: $P'(x + u, y + v)$;

$Q(x + dx, y)$: $Q'\left(x + dx + u + \dfrac{\partial u}{\partial x}dx, y + v + \dfrac{\partial v}{\partial x}dx\right)$;

$S(x, y + dy)$: $S'\left(x + u + \dfrac{\partial u}{\partial y}dy, y + dy + v + \dfrac{\partial v}{\partial y}dy\right)$;

$R(x + dx, y + dy)$: $R'\left(x + dx + u + \dfrac{\partial u}{\partial x}dx + \dfrac{\partial u}{\partial y}dy, y + dy + v + \dfrac{\partial v}{\partial x}dx + \right.$

$\left. + \dfrac{\partial v}{\partial y}dy\right)$.

In general, the changes in u and v are much smaller than the quantities dx and dy; accordingly we shall assume that the terms $(\partial u/\partial x)$, $(\partial u/\partial y)$, and so on are small enough that powers and products can be neglected. With this assumption, we see that:

(1) PQ increases in length by the amount $(\partial u/\partial x)dx$ and PS by the amount $(\partial v/\partial y)dy$; hence $\partial u/\partial x$ and $\partial v/\partial y$ are the fractional increases in length in the direction of the axes;

(2) the infinitesimal angles δ_1 and δ_2 are equal to $\partial v/\partial x$ and $\partial u/\partial y$ respectively;

(3) the right angle at P decreases by the amount $(\delta_1 + \delta_2) = (\partial v/\partial x + \partial u/\partial y)$;

(4) the rectangle as a whole has been rotated counter-clockwise through the angle $(\delta_1 - \delta_2) = (\partial v/\partial x - \partial u/\partial y)$.

Strain is defined as the relative change (that is, the fractional change) in a dimension or shape of a body. The quantities $\partial u/\partial x$ and $\partial v/\partial y$ are the relative increases in length in the directions of the x- and y-axes and are referred to as *normal strains*. The quantity $(\partial v/\partial x + \partial u/\partial y)$ is the amount by which a right angle in the xy plane is reduced when the stresses are applied, hence is a measure of the change in shape of the medium; it is known as a *shearing strain* and will be denoted by the symbol ε_{xy}. The quantity $(\partial v/\partial x - \partial u/\partial y)$, which represents a rotation of the body about the z-axis, does not involve change in size or shape and hence is not a strain; we shall denote it by the symbol θ_z.

Extending the above analysis to three dimensions, we write (u, v, w) as the components of displacement of a point $P(x, y, z)$. The elementary strains thus are:

$$\text{Normal strains} \quad \left. \begin{aligned} \varepsilon_{xx} &= \frac{\partial u}{\partial x}, \\[4pt] \varepsilon_{yy} &= \frac{\partial v}{\partial y}, \\[4pt] \varepsilon_{zz} &= \frac{\partial w}{\partial z}; \end{aligned} \right\} \qquad (4.1a)$$

$$\text{Shearing strains} \quad \varepsilon_{xy} = \varepsilon_{yx} = \frac{\partial v}{\partial x} + \frac{\partial u}{\partial y},$$

$$\varepsilon_{yz} = \varepsilon_{zy} = \frac{\partial w}{\partial y} + \frac{\partial v}{\partial z}, \qquad (4.1b)$$

$$\varepsilon_{zx} = \varepsilon_{xz} = \frac{\partial u}{\partial z} + \frac{\partial w}{\partial x}.$$

In addition to these strains, the body is subjected to simple rotation about the three axes given by

$$\theta_x = \frac{\partial w}{\partial y} - \frac{\partial v}{\partial z},$$

$$\theta_y = \frac{\partial u}{\partial z} - \frac{\partial w}{\partial x}, \qquad (4.1c)$$

$$\theta_z = \frac{\partial v}{\partial x} - \frac{\partial u}{\partial y}.$$

The changes in dimensions given by the normal strains result in volume changes when a body is strained. The change in volume per unit volume is called the *dilatation* and represented by Δ. If we start with a rectangular parallelepiped with edges dx, dy, and dz in the unstrained medium, in the strained medium the dimensions are $dx(1 + \varepsilon_{xx})$, $dy(1 + \varepsilon_{yy})$, $dz(1 + \varepsilon_{zz})$; hence the increase in volume is approximately $(\varepsilon_{xx} + \varepsilon_{yy} + \varepsilon_{zz})\, dx\, dy\, dz$. Since the original volume was $(dx\, dy\, dz)$, we see that

$$\Delta = \varepsilon_{xx} + \varepsilon_{yy} + \varepsilon_{zz} = \frac{\partial u}{\partial x} + \frac{\partial v}{\partial y} + \frac{\partial w}{\partial z}. \qquad (4.2)$$

(d) *Hooke's law.* In order to calculate the strains when the stresses are known, we must know the relationship between stress and strain. When the strains are small, this relation is given by *Hooke's law* which state‧ that a given strain is directly proportional to the stress producing it. When several stresses exist, each produces strains independently of the others; hence the total strain is the sum of the strains produced by the individual stresses. This means that each strain is a linear function of all of the stresses and vice versa. In general, Hooke's law leads to complicated relations but when the medium is *isotropic*, that is, when properties do not depend upon direction, it can be expressed in the following relatively simple form:

$$\sigma_{ii} = \lambda\Delta + 2\mu\varepsilon_{ii}, \quad i = x, y, z; \qquad (4.3a)$$

$$\sigma_{ij} = \mu\varepsilon_{ij}, \quad i, j = x, y, z, \quad i \neq j. \qquad (4.3b)$$

The quantities λ and μ are known as *Lamé's constants*. If we write $\varepsilon_{ij} = (\sigma_{ij}/\mu)$, it is evident that ε_{ij} is smaller the larger μ is. Hence μ is a measure of the resistance to shearing strain and is often referred to as the *modulus of rigidity* or *shear modulus*.

Although Hooke's law has wide application, it does not hold for large stresses.

For a given substance when the stress is increased beyond an *elastic limit*, Hooke's law no longer holds and strains increase more rapidly. Strains resulting from stresses which exceed this limit do not entirely disappear when the stresses are removed.

(e) *Elastic constants.* Although Lamé's constants are convenient when we are using (eqs. (4.3a) and (4.3b). other elastic constants are also used. The most common are *Young's modulus (E)*, *Poisson's ratio (σ)* and the *bulk modulus (k)* (the symbol σ is more-or-less standard for Poisson's ratio – the subscripts should prevent any confusion with the stress σ_{ij}). To define the first two we consider a medium in which all stresses are zero except σ_{xx}. Assuming σ_{xx} is positive (that is, a tensile stress), dimensions parallel to σ_{xx} will increase while dimensions normal to σ_{xx} will decrease; this means that ε_{xx} is positive (elongation in the x-direction) while ε_{yy} and ε_{zz} are negative. Also, we can show (see problem 2) that $\varepsilon_{yy} = \varepsilon_{zz}$. We now define E and σ by the relations

$$E = \sigma_{xx}/\varepsilon_{xx}, \tag{4.4a}$$

$$\sigma = -\varepsilon_{yy}/\varepsilon_{xx} = -\varepsilon_{zz}/\varepsilon_{xx}, \tag{4.4b}$$

the minus signs being inserted to make σ positive.

To define k, we consider a medium subjected only to a hydrostatic pressure p; this is equivalent to the statements

$$\sigma_{xx} = \sigma_{yy} = \sigma_{zz} = -p; \sigma_{xy} = \sigma_{yz} = \sigma_{zx} = 0.$$

Then, k is defined as the ratio of the pressure to the dilatation,

$$k = -p/\Delta, \tag{4.4c}$$

the minus sign being inserted to make k positive.

By substituting the above values in Hooke's law we can obtain the following relations between E, σ and k and Lamé's constants, λ and μ (see problem 2):

$$E = \frac{\mu(3\lambda + 2\mu)}{(\lambda + \mu)}, \qquad \lambda = \frac{E\sigma}{(1+\sigma)(1-2\sigma)} \tag{4.4d}$$

$$\sigma = \frac{\lambda}{2(\lambda + \mu)}, \qquad \mu = \frac{E}{2(1+\sigma)} \tag{4.4e}$$

$$k = \frac{3\lambda + 2\mu}{3}. \tag{4.4f}$$

By eliminating different pairs of constants among the three equations many different relations can be derived expressing one of the five constants in terms of two others.

The preceding theory assumes an isotropic medium. Rocks, especially sedimentary and metamorphic rocks, are frequently not isotropic. Measurements of the elastic constants of sedimentary rocks often yield values which depend upon the direction of measurement and differences of 20 to 25% have been reported between measurements parallel and normal to the bedding planes. In spite of this, in discussing wave propagation we generally ignore such differences and treat sedimentary rocks as isotropic media; the results when one does so are useful and to do otherwise leads to extremely complex and cumbersome mathematical equations.

The elastic constants are defined in such a way that they are positive numbers. As a consequence of this, σ must have values between 0 and 0·5 (this follows from eq. (4.4e) since both λ and μ are positive and hence $\lambda/(\lambda + \mu)$ is less than unity). Values range from 0·05 for very hard, rigid rocks to about 0·45 for soft, poorly consolidated materials. Liquids have no resistance to shear and hence for them $\mu = 0$ and $\sigma = 0·5$ For most rocks, E, k, and μ lie in the range from 0·2 to 1·2 megabars (2×10^{10} to 12×10^{10} newton/m²), E generally being the largest and μ the smallest of the three. Extensive tables of elastic constants of rocks have been given by Birch (1966).

4.2.2 *Wave motion*

(a) *Wave equation.* Up to this point we have been discussing a medium in static equilibrium. We shall now remove this restriction and consider what happens when the stresses are not in equilibrium. In fig. 4.1 we now assume that the stresses on the rear face of the element of volume are as shown in the diagram but that the stresses on the front face are respectively

$$\sigma_{xx} + \frac{\partial \sigma_{xx}}{\partial x}\, dx,\ \sigma_{yx} + \frac{\partial \sigma_{yx}}{\partial x}\, dx,\ \sigma_{zx} + \frac{\partial \sigma_{zx}}{\partial x}\, dx.$$

Since these stresses are opposite to those acting on the rear face, the net (unbalanced) stresses are

$$\frac{\partial \sigma_{xx}}{\partial x}\, dx,\ \frac{\partial \sigma_{yx}}{\partial x}\, dx,\ \frac{\partial \sigma_{zx}}{\partial x}\, dx.$$

These stresses act on a face having an area $(dy\, dz)$ and affect the volume $(dx\, dy\, dz)$; hence we get for the net forces per unit volume in the directions of the x-, y- and z-axes the values

$$\frac{\partial \sigma_{xx}}{\partial x},\ \frac{\partial \sigma_{yx}}{\partial x},\ \frac{\partial \sigma_{zx}}{\partial x},\ \text{respectively.} \qquad \left(\frac{\text{force}}{\text{unit volume}}\right)$$

Similar expressions hold for the other faces; hence we find for the total force in the direction of the x-axis the expression

$$\left(\frac{\partial \sigma_{xx}}{\partial x} + \frac{\partial \sigma_{xy}}{\partial y} + \frac{\partial \sigma_{xz}}{\partial z} \right).$$

Newton's second law of motion states that the unbalanced force equals the mass times the acceleration; thus we obtain the equation of motion along the x-axis,

$$\rho\frac{\partial^2 u}{\partial t^2} = \text{Unbalanced force in the } x\text{-direction,}$$

$$= \frac{\partial\sigma_{xx}}{\partial x} + \frac{\partial\sigma_{xy}}{\partial y} + \frac{\partial\sigma_{xz}}{\partial z}, \tag{4.5a}$$

where ρ is the density. Similar equations can be written for the motion along the y- and z-axes.

Equation (4.5a) relates the displacements to the stresses. We can obtain an equation involving only displacements by using Hooke's law to replace the stresses with strains and then expressing the strains in terms of the displacements, using eqs. (4.1a), (4.1b), (4.2), (4.3a) and (4.3b). Thus,

$$\rho\frac{\partial^2 u}{\partial t^2} = \frac{\partial\sigma_{xx}}{\partial x} + \frac{\partial\sigma_{xy}}{\partial y} + \frac{\partial\sigma_{xz}}{\partial z},$$

$$= \lambda\frac{\partial\Delta}{\partial x} + 2\mu\frac{\partial\varepsilon_{xx}}{\partial x} + \mu\frac{\partial\varepsilon_{xy}}{\partial y} + \mu\frac{\partial\varepsilon_{xz}}{\partial z},$$

$$= \lambda\frac{\partial\Delta}{\partial x} + \mu\left\{2\frac{\partial^2 u}{\partial x^2} + \left(\frac{\partial^2 v}{\partial x\,\partial y} + \frac{\partial^2 u}{\partial y^2}\right) + \left(\frac{\partial^2 w}{\partial x\,\partial z} + \frac{\partial^2 u}{\partial z^2}\right)\right\},$$

$$= \lambda\frac{\partial\Delta}{\partial x} + \mu\nabla^2 u + \mu\frac{\partial}{\partial x}\left(\frac{\partial u}{\partial x} + \frac{\partial v}{\partial y} + \frac{\partial w}{\partial z}\right),$$

$$= (\lambda + \mu)\frac{\partial\Delta}{\partial x} + \mu\nabla^2 u, \tag{4.5b}$$

where $\nabla^2 u = $ *Laplacian of* $u = (\partial^2 u/\partial x^2 + \partial^2 u/\partial y^2 + \partial^2 u/\partial z^2)$. By analogy we can write the equations for v and w

$$\rho\frac{\partial^2 v}{\partial t^2} = (\lambda + \mu)\frac{\partial\Delta}{\partial y} + \mu\nabla^2 v, \tag{4.5c}$$

$$\rho\frac{\partial^2 w}{\partial t^2} = (\lambda + \mu)\frac{\partial\Delta}{\partial z} + \mu\nabla^2 w. \tag{4.5d}$$

To obtain the wave equation, we differentiate these three equations with respect to x, y and z respectively and add the results together. This gives

$$\rho\frac{\partial^2}{\partial t^2}\left(\frac{\partial u}{\partial x} + \frac{\partial v}{\partial y} + \frac{\partial w}{\partial z}\right) = (\lambda + \mu)\left(\frac{\partial^2\Delta}{\partial x^2} + \frac{\partial^2\Delta}{\partial y^2} + \frac{\partial^2\Delta}{\partial z^2}\right) +$$

$$+ \mu\nabla^2\left(\frac{\partial u}{\partial x} + \frac{\partial v}{\partial y} + \frac{\partial w}{\partial z}\right),$$

that is,

$$\rho\frac{\partial^2\Delta}{\partial t^2} = (\lambda + 2\mu)\nabla^2\Delta,$$

or

$$\frac{1}{\alpha^2} \frac{\partial^2 \Delta}{\partial t^2} = \nabla^2 \Delta,$$

where

$$\alpha^2 = \frac{\lambda + 2\mu}{\rho}.$$

$$\left.\vphantom{\frac{1}{\alpha^2}\frac{\partial^2\Delta}{\partial t^2}}\right\} \quad (4.6)$$

By subtracting the derivative of eq. (4.5c) with respect to z from the derivative of eq. (4.5d) with respect to y, we get

$$\rho \frac{\partial^2}{\partial t^2} \left(\frac{\partial w}{\partial y} - \frac{\partial v}{\partial z} \right) = \mu \nabla^2 \left(\frac{\partial w}{\partial y} - \frac{\partial v}{\partial z} \right),$$

that is,

$$\frac{1}{\beta^2} \frac{\partial^2 \theta_x}{\partial t^2} = \nabla^2 \theta_x,$$

where

$$\beta^2 = \frac{\mu}{\rho}.$$

$$\left.\vphantom{\frac{1}{\beta^2}\frac{\partial^2\theta_x}{\partial t^2}}\right\} \quad (4.7)$$

By subtracting appropriate derivatives, we obtain similar results for θ_y and θ_z. These equations are different examples of the *wave equation* which we can write in the general form

$$\frac{1}{V^2} \frac{\partial^2 \psi}{\partial t^2} = \nabla^2 \psi, \qquad (4.8a)$$

where V is a constant.

(b) *Plane-wave solutions.* Let us consider first the case where ψ is a function only of x and t, so that eq. (4.8a) reduces to

$$\frac{1}{V^2} \frac{\partial^2 \psi}{\partial t^2} = \frac{\partial^2 \psi}{\partial x^2}. \qquad (4.8b)$$

Any function of $(x - Vt)$,

$$\psi = f(x - Vt), \qquad (4.9a)$$

is a solution of eq. (4.8b) (see problem 3) provided that ψ and its first two derivatives have no discontinuities. This solution (known as *D'Alembert's solution*), furnishes an infinite number of particular solutions (for example, $e^{k(x-Vt)}$ $\sin(x - Vt)$, $(x - Vt)^3$, where we must exclude points at which these functions and their first two derivatives cease to exist). The answer to a specific problem consists of selecting the appropriate combination of solutions which also satisfies the boundary conditions for the problem.

A *wave* is defined as a 'disturbance' which travels through the medium. In our notation, the disturbance ψ is a volume change when $\psi = \Delta$ and a rotation when

$\psi = \theta_x$. Obviously the disturbance in eq. (4.9a) is travelling along the x-axis. We shall now show that it travels with a speed equal to the quantity V.

In fig. 4.3, a certain part of the wave has reached the point P_0 at the time t_0. If the coordinate of P_0 is x_0, then the value of ψ at P_0 is $\psi_0 = f(x_0 - Vt_0)$. If this same portion of the wave reaches P_1 at the time $t_0 + \Delta t$, then we have for the value of ψ at P_1

$$\psi_1 = f\{x_0 + \Delta x - V(t_0 + \Delta t)\}.$$

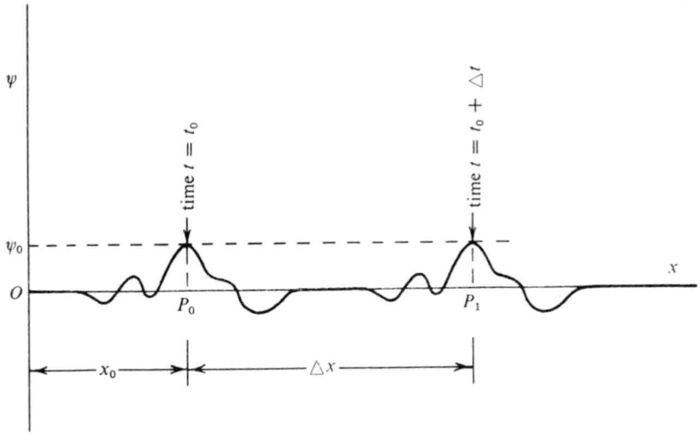

Fig. 4.3 Illustrating the velocity of a wave.

But, since this is the same portion of the wave which was at P_0 at time t_0, we must have $\psi_0 = \psi_1$, that is,

$$x_0 - Vt_0 = x_0 + \Delta x - V(t_0 + \Delta t).$$

Thus, the quantity V is equal to $\Delta x/\Delta t$ and is therefore the speed with which the disturbance travels.

A function of $(x + Vt)$, for example, $\psi = g(x + Vt)$, is also a solution of eq. (4.8b); it denotes a wave travelling in the negative x-direction. The general solution of eq. (4.8b)

$$\psi = f(x - Vt) + g(x + Vt), \tag{4.9b}$$

represents two waves travelling along the x-axis in opposite directions with the velocity V.

Since the value of ψ is independent of y and z, the disturbance must be the same everywhere in a plane perpendicular to the x-axis. This type of wave is called a *plane wave*.

The quantity $(x - Vt)$ (or $(x + Vt)$) is known as the *phase*. The surfaces on which the wave motion is the same, that is, surfaces on which the phase has the same value, are known as *wavefronts*. In the case which we are considering, the wavefronts are planes perpendicular to the x-axis. Note that the wave is travelling in the direction normal to the wavefront; this holds for all waves in isotropic media. A line denoting the direction of travel of the wave energy is called a *raypath.*

It is convenient at times to have an expression for a plane wave travelling along a straight line inclined at an angle to each of the axes. Assume that the wave is travelling along the x'-axis which has direction cosines (l, m, n) relative to the x-, y- and z-axes (fig. 4.4). Then, at a point P on the x'-axis at a distance x' from the origin, we have

$$x' = lx + my + nz,$$

where the coordinates of P are (x, y, z). Then,

$$\psi = f(lx + my + nz - Vt) + g(lx + my + nz + Vt). \tag{4.10}$$

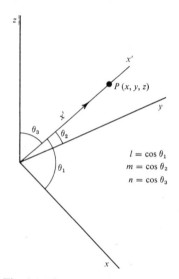

Fig. 4.4 Wave direction not along an axis.

(c) *Spherical-wave solutions.* In addition to plane waves, we shall have occasion to use another important type of wave, the *spherical wave* where the wavefronts are a series of concentric spherical surfaces. We express eq. (4.8*a*) in spherical coordinates (r, θ, ϕ), where θ is the colatitude and ϕ the longitude:

$$\frac{1}{V^2}\frac{\partial^2\psi}{\partial t^2} = \frac{1}{r^2}\left\{\frac{\partial}{\partial r}\left(r^2\frac{\partial\psi}{\partial r}\right) + \frac{1}{\sin\theta}\frac{\partial}{\partial\theta}\left(\sin\theta\frac{\partial\psi}{\partial\theta}\right) + \frac{1}{\sin^2\theta}\frac{\partial^2\psi}{\partial\phi^2}\right\}. \tag{4.11}$$

We consider only the special case when the wave motion is independent of θ and ϕ, hence is a function only of r and t. Then we get the simplified equation,

$$\frac{1}{V^2}\frac{\partial^2\psi}{\partial t^2} = \frac{1}{r^2}\frac{\partial}{\partial r}\left(r^2\frac{\partial\psi}{\partial r}\right). \tag{4.12}$$

A solution of the above equation is

$$\psi = \frac{1}{r} f(r - Vt)$$

(see problem 3, part (c)). Obviously,

$$\psi = \frac{1}{r} g(r + Vt)$$

is also a solution and the general solution of eq. (4.12) is

$$\psi = \frac{1}{r} f(r - Vt) + \frac{1}{r} g(r + Vt), \tag{4.13}$$

in which the first term represents a wave expanding outward from a central point and the second term a wave collapsing toward the central point.

When r and t are fixed, $(r - Vt)$ is constant and hence ψ is constant. Thus, at the instant t the wave has the same value at all points on the spherical surface of radius r. The spherical surfaces are therefore wavefronts and the radii are rays. Obviously the rays are normal to the wavefront as in the case of plane waves.

As the wave progresses outward from the centre, the radius increases by the amount V during each unit of time. Eventually the radius becomes very large and the portion of the wavefront near any particular point will be approximately plane. If we consider fig. 4.5, we see that the error which we introduce when we

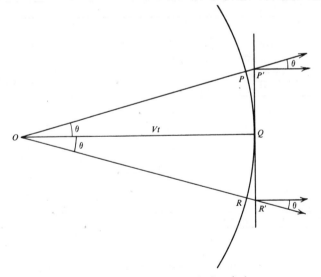

Fig. 4.5 Relation between spherical and plane waves.

replace the spherical wavefront PQR by the plane wavefront $P'QR'$ is due to the divergence between the true direction of propagation given by the direction of the radius and the assumed direction normal to the plane. By taking OQ very large or PR very small (or both), we can make the error as small as desired. Since plane waves are easy to visualize and also the simplest to handle mathematically, we generally assume that conditions are such that the plane-wave assumption is valid.

(d) *Harmonic waves.* So far we have discussed only the geometrical aspect of waves, that is, the way in which the wave depends upon the space coordinates. However, ψ is a function of t also, and so we must consider the time dependence of waves as well.

The simplest form of time variation is that of the *harmonic wave*, that is, waves involving sine or cosine expressions such as

$$\psi = A \cos 2\pi k(x - Vt), \tag{4.14a}$$

$$\psi = A \sin 2\pi k(lx + my + nz - Vt), \tag{4.14b}$$

$$\psi = \frac{B}{r} \cos 2\pi k(r + Vt). \tag{4.15}$$

At a fixed point, ψ varies as the sine or cosine of the time; hence the motion is simple harmonic. The values of ψ range from $+A$ to $-A$ for the plane wave of eqs. (4.14a) and (4.14b) and from $+B/r$ to $-B/r$ for the spherical wave of eq. (4.15). The limiting value, A or B/r, is known as the *amplitude* of the wave ψ.

For a fixed value of t, whenever x in eq. (4.14a) increases by the amount $1/k$, the argument of the cosine increases by 2π and hence the value of ψ repeats. This distance, $1/k$, is called the *wavelength*, usually represented by the symbol, λ. The quantity k is called the *wave number* (some writers use $(2\pi/\lambda)$ as the wave number). In eqs. (4.14b) and (4.15), $(lx + my + nz)$ and r represent the distance the wave has travelled from the origin, hence are equivalent to x in eq. (4.14a); therefore, k has the same significance here as in eq. (4.14a).

If the space coordinates in eqs. (4.14a), (4.14b) and (4.15) are kept fixed and t allowed to increase, the value of ψ repeats each time that t increases by the amount T where $kVT = 1 = (VT)/\lambda$; consequently,

$$\left. \begin{aligned} T &= \lambda/V, \\ \nu &= \frac{1}{T} = \frac{V}{\lambda}, \\ V &= \nu\lambda, \end{aligned} \right\} \tag{4.16}$$

where T is called the *period* and ν the *frequency* of the wave. Another frequently used quantity is the *angular frequency*, ω, defined as $\omega = 2\pi\nu = 2\pi kV$. Using the above symbols, we can write eq. (4.14a) in the following equivalent forms:

$$\left. \begin{aligned} \psi &= A \cos 2\pi k(x - Vt) = A \cos \frac{2\pi}{\lambda}(x - Vt) \\ &= A \cos(2\pi kx - \omega t) = A \cos \omega \left(\frac{x}{V} - t \right) \\ &= A \cos 2\pi \left(\frac{x}{\lambda} - \nu t \right) = A \cos 2\pi(kx - \nu t). \end{aligned} \right\} \tag{4.14c}$$

(e) *P-waves and S-waves.* Up to this point our discussion of wave motion has been based upon eq. (4.8a). The quantity ψ has not been defined; we have merely inferred that it is some disturbance which is propagated from one point to another with the speed V. However, in a homogeneous isotropic medium, eqs. (4.6) and (4.7) must be satisfied. We can identify the functions Δ and θ_x with ψ and conclude that two types of waves can be propagated in a homogeneous isotropic medium, one corresponding to changes in the dilatation, Δ, the other to changes in one or more components of the rotation given in eq. (4.1c).

The first type is variously known as a *dilatational, longitudinal, irrotational, compressional* or *P-wave*, the latter name being due to the fact that this type is usually the first (primary) event on an earthquake recording. The second type is referred to as the *shear, transverse, rotational* or *S-wave* (since it is usually the second event observed on earthquake records). The *P*-wave has the velocity α in eq. (4.6) and the *S*-wave the velocity β in eq. (4.7) where

$$\left.\begin{array}{l} \alpha = \{(\lambda + 2\mu)/\rho\}^{\frac{1}{2}}, \\ \beta = (\mu/\rho)^{\frac{1}{2}}. \end{array}\right\} \tag{4.17}$$

Since the elastic constants are positive, α is always greater than β. Writing γ for the ratio β/α, we see that

$$\gamma^2 = \frac{\beta^2}{\alpha^2} = \frac{\mu}{\lambda + 2\mu} = \frac{\frac{1}{2} - \sigma}{1 - \sigma}, \tag{4.18}$$

using eq. (4.4e). As σ decreases from 0·5 to zero, γ increases from zero to its maximum value, $1/\sqrt{2}$; thus, the velocity of the *S*-wave ranges from zero up to 70% of the velocity of the *P*-wave.

For fluids μ is zero and hence β and γ are also zero; therefore *S*-waves do not propagate through fluids.

Let us investigate the nature of the motion of the medium corresponding to the two types of wave motion. Consider a spherical *P*-wave of the type given by eq. (4.13). Figure 4.6 shows wavefronts drawn at quarter-wavelength intervals, t being chosen so that kVt is an integer. The arrows represent the direction of motion of the medium at the wavefront. The medium is undergoing maximum compression at B (that is, the dilatation Δ is a minimum) and minimum compression (maximum Δ) at the wavefront D.

We can visualize the plane-wave situation by imagining that the radius in fig. 4.6 has become very large so that the wavefronts are practically plane surfaces. The displacements will everywhere be perpendicular to these planes so that there will no longer be convergence or divergence of the particles of the medium as they move back and forth parallel to the direction of propagation of the wave. Such a displacement is longitudinal which explains why *P*-waves are sometimes called longitudinal waves. *P*-waves are the dominant waves involved in seismic exploration.

To determine the motion of a medium during the passage of an S-wave, we return to eq. (4.7) and consider the case where a rotation θ_z which is a function of x and t only is being propagated along the x-axis. We have

$$\frac{1}{\beta^2}\frac{\partial^2 \theta_z}{\partial t^2} = \frac{\partial^2 \theta_z}{\partial x^2}.$$

Since

$$\theta_z = \frac{\partial v}{\partial x} - \frac{\partial u}{\partial y} = \frac{\partial v}{\partial x}$$

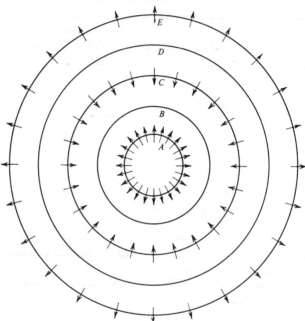

Fig. 4.6 Displacements for a spherical P-wave.

from eq. (4.1c), we see that the wave motion consists solely of a displacement v of the medium in the y-direction, v being a function of both x and t. Since v is independent of y and z, the motion is everywhere the same in a plane perpendicular to the x-axis; thus the case we are discussing is that of a plane S-wave travelling along the x-axis. Therefore, by eq. (4.9b) the displacement v must have the form

$$v = f(x - \beta t) + g(x + \beta t).$$

We can visualize the above relations using fig. 4.7. When the wave arrives at P, it causes the medium in the vicinity of P to rotate about the axis $Z'Z''$ (parallel to the z-axis) through an angle ε. Since we are dealing with infinitesimal strains, ε must be infinitesimal and we can ignore the curvature of the displacements and consider that points such as P' and P'' are displaced parallel to the y-axis to the points Q' and Q''. Thus, as the wave travels along the x-axis, the medium is displaced transversely to the direction of propagation, hence the name transverse wave. Moreover, since the rotation varies from point to point at any given instant, the medium is subjected to varying shearing stresses as the wave moves along; this accounts for the name shear wave.

Since we might have chosen to illustrate θ_y in fig. 4.7 instead of θ_z, it is clear that shear waves have two degrees of freedom, unlike P-waves which have only one – along the radial direction. (A rotation about the x-axis propagated along the same axis is not easily generated and so we ignore this possibility.) In practice S-wave motion is usually resolved into components parallel and perpendicular to the surface of the ground; these are known respectively as SH and SV *waves*. (When the wave is travelling neither horizontally nor vertically, the motion is resolved into a horizontal (SH) component and a component in the vertical plane through the direction of propagation.)

Because the two degrees of freedom of S-waves are independent, we can have an S-wave which involves motion in only one plane, for example, SH or SV motion; such a wave is said to be *plane polarized*. We can also have a wave in which the SH and SV motion have the same frequency and a fixed phase difference; such a wave is *elliptically polarized*. However, polarization of S-waves is not important in seismic exploration.

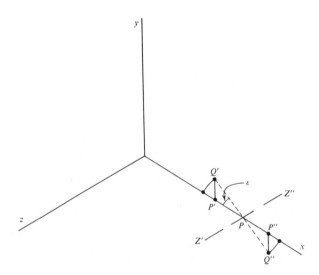

Fig. 4.7 Motion during passage of an S-wave.

In the case of a medium which is not homogeneous and isotropic, it may not be possible to resolve wave motion into separate P-waves and S-waves. However, inhomogeneities and anisotropy in the earth are small enough that assumption of separate P- and S-waves is valid for practical purposes.

(f) *Surface waves.* In an infinite homogeneous isotropic medium, only *P-* and *S*-waves exist. However, when the medium does not extend to infinity in all directions but is bounded by a surface, other types of waves can be generated. These waves are called *surface waves* since they are confined to the vicinity of one of the surfaces which bound the medium.

In exploration seismology the only type of surface wave of importance is the *Rayleigh wave.* This wave travels along the surface of the earth; it involves a combination of longitudinal and transverse motion with a definite phase relation to each other. The amplitude of this wave motion decreases exponentially with depth. The particle motion is confined to the vertical plane which includes the direction of propagation of the wave. During the passage of the wave, a particle traverses an elliptical path, the major axis of the ellipse being vertical. The direction of particle motion around the ellipse is called *retrograde* (see fig. 4.8) because it is opposite to the more familiar direction of motion of particles in waves on the surface of water. The velocity of Rayleigh waves depends upon the elastic constants in the vicinity of the surface and is always less than the *S*-wave velocity, β;

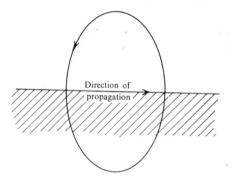

Fig. 4.8 Motion during passage of a Rayleigh wave.

when $\sigma = \frac{1}{4}$, the Rayleigh wave velocity is 0.92β. The exponential decrease in amplitude with depth depends on the wavelength of the waves; since the elastic constants change with depth, the velocity of Rayleigh waves varies with wavelength. A variation of velocity with wavelength (or frequency) is called *dispersion* and it results in a change of the shape of the wave train with distance (see §4.2.2i).

In earthquake seismology another type of surface wave called *Love wave* is observed. A Love wave involves transverse motion parallel to the surface of the ground and sometimes is called an *SH* wave. Love waves have velocities intermediate between the *S*-wave velocity at the surface and that in deeper layers, and exhibit dispersion. Energy sources used in seismic work do not generate Love waves to a significant degree and hence Love waves are unimportant in seismic exploration. Also, modern geophones designed to respond only to vertical motion of the surface would not detect any Love waves which might exist.

(g) *Energy density; intensity.* Probably the single most important feature of any wave is the energy associated with the motion of the medium as the wave passes through it. Usually we are not concerned with the total energy of a wave but rather with the energy in the vicinity of the point where we observe it; the *energy density* is the energy per unit volume in the neighbourhood of a point.

Consider a spherical harmonic P-wave for which the radial displacement for a fixed value of r is given by

$$u = A \cos (\omega t + \phi),$$

where ϕ is a *phase angle*. The displacement u ranges from $-A$ to $+A$. Since the displacement varies with time, each element of the medium has a velocity, $\dot{u} = \partial u / \partial t$, and an associated kinetic energy. The kinetic energy δL contained within each element of volume δv is

$$\delta L = \tfrac{1}{2}(\rho \delta v)\dot{u}^2.$$

The kinetic energy per unit volume is

$$\frac{\delta L}{\delta v} = \tfrac{1}{2}\rho\dot{u}^2 = \tfrac{1}{2}\rho\omega^2 A^2 \sin^2 (\omega t + \phi).$$

This expression varies from zero to a maximum of $\rho\omega^2 A^2/2$.

The wave also involves potential energy resulting from the elastic strains created during the passage of the wave. As the medium oscillates back and forth, the energy is converted back and forth from kinetic to potential form, the total energy remaining fixed. When a particle is at zero displacement, the potential energy is zero and the kinetic energy is a maximum, and when the particle is at its extreme displacement the energy is all potential. Since the total energy equals the maximum value of the kinetic energy, the energy density E for a harmonic wave is

$$E = \tfrac{1}{2}\rho\omega^2 A^2 = 2\pi^2\rho\nu^2 A^2. \tag{4.19}$$

Thus we see that the energy density is proportional to the first power of the density of the medium and to the second power of the frequency and amplitude of the wave.

We are interested also in the rate of flow of energy and we define the *intensity* as the quantity of energy which flows through a unit area normal to the direction of wave propagation in unit time. Take a cylinder of infinitesimal cross-section area, δA, whose axis is parallel to the direction of propagation and whose length is equal to the distance travelled in the time, δt. The total energy inside the cylinder at any instant t is $EV \delta t \, \delta A$; at the time $t + \delta t$ all of this energy has left the cylinder through one of the ends. Dividing by the area of the end of the cylinder, δA, and by the time interval, δt, we get I, the amount of energy passing through unit area in unit time:

← velocity

$$I = EV. \tag{4.20a}$$

For a harmonic wave, this becomes

$$I = \tfrac{1}{2}\rho V \omega^2 A^2. \tag{4.20b}$$

In fig. 4.9 we show a spherical wavefront diverging from a centre O. By drawing sufficient radii we can define two portions of wavefronts, A_1 and A_2, of radii r_1 and

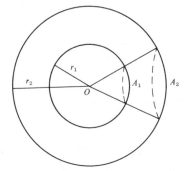

Fig. 4.9 Dependance of intensity upon distance.

r_2, such that the energy which flows outward through the spherical cap A_1 in one second must be equal to that passing outward through the spherical cap A_2 in one second (since the energy is moving only in the radial direction). The flow of energy per second is the product of the intensity and the area; hence

$$I_1 A_1 = I_2 A_2.$$

Since the areas A_1 and A_2 are proportional to the square of their radii, we get

$$\frac{I_2}{I_1} = \frac{A_1}{A_2} = \left(\frac{r_1}{r_2}\right)^2.$$

Moreover, it follows from eq. (4.20a) that E is proportional to I and hence

$$\frac{I_2}{I_1} = \frac{E_2}{E_1} = \left(\frac{r_1}{r_2}\right)^2. \tag{4.21}$$

Thus, geometrical spreading causes the intensity and the energy density of spherical waves to decrease inversely as the square of the distance from the source (Newman, 1973). This is called *spherical divergence*.

(h) *Absorption.* We shall also consider two other mechanisms, absorption and partition at interfaces, which cause the energy density of a wave to decrease. In the preceding section we considered variations of the energy distribution as a function of geometry. Implicit in the discussion was the assumption that none of the wave energy disappeared, that is, was transformed into other forms of energy. In reality this assumption is always incorrect since, as the wave passes through the medium, the elastic energy associated with the wave motion is gradually absorbed by the medium, reappearing ultimately in the form of heat. This process is called *absorption* and is responsible for the eventual complete disappearance of the wave motion.

The mechanisms by which the elastic energy is transformed into heat are not understood clearly. During the passage of a wave, heat is generated during the compressive phase and absorbed during the expansive phase. The process is not perfectly reversible since the heat conducted away during the compression is not equal to the heat flowing back during the expansion. Internal friction is undoubtedly involved and many other mechanisms may contribute, such as loss of energy involved in the creation of new surfaces (fracturing near an explosion), piezoelectric and thermoelectric effects, viscous losses in the fluids filling the rock pores, etc.

The measurement of absorption is very difficult. Absorption varies with frequency and laboratory measurements, which are invariably made at high frequencies, may have little application to actual field conditions. Field measurements must be corrected for reflection or refraction effects and the entire path should be through the same homogeneous medium. Measurement difficulties have resulted in wide divergence in absorption measurements.

In many physical phenomena, the loss of energy by absorption is exponential with distance and this appears to be approximately true for elastic waves in rocks. Thus, we can write

$$I = I_0 e^{-\alpha x}, \tag{4.22}$$

where I and I_0 are values of the intensity at two points a distance x apart and α is the *absorption coefficient* (not to be confused with the velocity of a *P*-wave). Experimental evidence suggests that the absorption coefficient is approximately proportional to frequency (that is, $\alpha\lambda$ is roughly constant for a particular rock containing certain fluids). While geophysicists do not all agree that this is true, absorption certainly increases with frequency and provides one mechanism responsible for the rapid loss of high frequencies with distance.

One important measurement reported by McDonal *et al.* (1958) in the Pierre Shale, a massive shale formation in Colorado about 4000 ft (1200 m) thick with average velocity 7100 ft/sec (2330 m/sec), found a value for α of approximately 0·85 db/wavelength. Thus, a 10-Hz wave loses 0·85 db every 233 m while a 30-Hz wave loses the same amount every 79 m. These values are probably higher than typical values for sedimentary rocks (see Tullos and Reid, 1969).

To compare the loss by absorption with the loss of intensity by geometrical spreading, we have calculated the losses in going from a point 200 m from the source to various distances from the source assuming $\alpha = 0·25$ db/λ (an arbitrary value thought to be more typical than the value from the Pierre Shale). The results shown in table 4.1 were calculated using the following relations:

Absorption: Loss in db $= 10 \log_{10} (I_0/I) = 4{\cdot}3\alpha x = 1{\cdot}1(x/\lambda)$
$= 1{\cdot}1(x_s - 200)/\lambda = 1{\cdot}1\nu(x_s - 200)/2000;$

Spreading: Loss in db $= 10 \log_{10} (I_0/I) = 20 \log_{10} (x_s/200),$

where $x_s =$ distance to the shotpoint.

Table 4.1. *Energy losses by absorption and spreading.* ($\alpha = 0.25$ db/wavelength; $V = 2000$ m/sec)

	Frequency (ν)	Distance from shotpoint (x_s)			
		1200 m	2200 m	4200 m	8200 m
Absorption	1 Hz	0·55 db	1·1 db	2·2 db	4·4 db
	3	1·6	3·3	6·6	13
	10	5·5	11	22	44
	30	16	33	66	130
	100	55	110	220	440
Spreading	All	16	21	26	32

The table shows that losses by spreading are more important than losses by absorption for low frequencies and short distances. As the frequency and distance increase, absorption losses increase and eventually become dominant. The more rapid loss of higher frequencies results in change of waveshape with distance.

In addition to absorption and spreading, the partitioning of energy at interfaces is also responsible for the decrease in the energy of a wave with distance. This is discussed in §4.2.2*m*.

(i) *Dispersion; group velocity.* The velocity, V, α or β, which appears in eqs. (4.8) to (4.18) is known as the *phase velocity* because it is the distance travelled per unit time by a point of constant phase, such as a peak or trough. This is not necessarily the same as the speed with which the energy travels, which is the *group velocity* and will be denoted by U. Consider, for example, the wave train shown in fig. 4.10*a*; we could determine the group velocity U by drawing the envelope of the pulse (the double curve ABC, $AB'C$) and measuring the distance which the envelope travels in unit time. The relation between U and V is shown in fig. 4.10*b*, where V is given by the rate of advance of a certain phase (such as a trough) while U is measured by the speed of the maximum amplitude of the envelope.

If we decompose a pulse into its component frequencies by Fourier analysis, we find a spectrum of frequencies. If the velocity is the same for all frequencies the pulse shape will remain the same and the group velocity will be the same as the phase velocity. However, if the velocity varies with frequency, the pulse changes shape as it travels and the group velocity is different from the phase velocity, that is, the medium is dispersive. It can be shown (see problem 5) that the group velocity U is

$$U = V - \lambda \frac{dV}{d\lambda} = V + \omega \frac{dV}{d\omega}, \tag{4.23}$$

where V, λ, ω, $dV/d\lambda$ and $dV/d\omega$ are average values for the range of frequencies which make up the principal part of the pulse.

When V decreases with frequency, V is larger than U as illustrated in fig. 4.10 where the envelope travels slower than the individual cycles which overtake and pass through the envelope and disappear as they reach the leading edge. When V increases with frequency, the opposite is true.

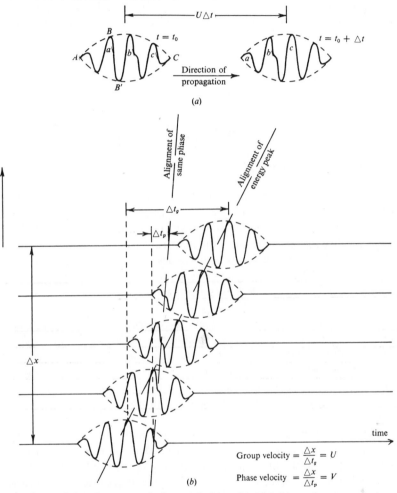

Fig. 4.10 Comparison of group and phase velocities. (*a*) Definition of group velocity, U; (*b*) arrival of a dispersive wave at successive geophones.

Dispersion is not a dominant feature of exploration seismology since most rocks exhibit little variation of velocity with frequency in the seismic frequency range. However, dispersion is important in connection with surface waves and certain other phenomena.

(j) *Huygens' principle.* This principle, which most readers will have encountered in elementary physics, is important in understanding wave travel and is frequently useful in drawing successive positions of wavefronts. *Huygens' principle* states that every point on a wavefront can be regarded as a new source of waves. Given the location of a wavefront at a certain instant, future positions of the wavefront can be found by considering each point on the first wavefront as a new wave source. In fig. 4.11, AB is the wavefront at the time t_0 and we wish to find the wavefront at a later time $(t_0 + \Delta t)$. During the interval Δt, the wave will advance a distance

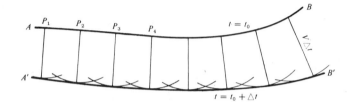

Fig. 4.11 Using Huygens' principle to locate new wavefronts.

$V\Delta t$, V being the velocity (which may vary from point to point). We select points on the wavefront, P_1, P_2, P_3 and so on, from which we draw arcs of radius $V\Delta t$. Provided we select enough points, the envelope of the arcs $(A'B')$ will define as accurately as we wish the position of the wavefront at the time $(t + \Delta t)$. Except on the envelope, the elemental waves interfere destructively with each other so that their effects cancel. When AB is plane, we need draw only two arcs and the straight line tangent to the two arcs defines the new wavefront.

(k) *Reflection and refraction.* Whenever a wave encounters an abrupt change in the elastic properties, as when it arrives at a surface separating two beds, part of the energy is *reflected* and remains in the same medium as the original energy; the balance of the energy is *refracted* into the other medium with an abrupt change in the direction of propagation occurring at the interface. Reflection and refraction are fundamental in exploration seismology and we shall discuss these in some detail.

We can derive the familiar laws of reflection and refraction using Huygens' principle. Consider a plane wavefront AB incident on a plane interface as in fig. 4.12 (if the wavefront is curved, we merely take A and B sufficiently close together that AB is a plane to the required degree of accuracy). AB occupies the position $A'B'$ when A arrives at the surface; at this instant the energy at B' still must travel the distance $B'R$ before arriving at the interface. If $B'R = V_1\Delta t$, then Δt is the time interval between the arrival of the energy at A' and at R. By Huygens' principle, during the time Δt the energy which reached A' will have travelled either upward a distance $V_1\Delta t$ or downward a distance $V_2\Delta t$. By drawing arcs with centre A' and lengths equal to $V_1\Delta t$ and $V_2\Delta t$, and then drawing the tangents from R to these arcs, we locate the new wavefronts, RS and RT in the upper and lower media. The angle at S is a right angle and $A'S = V_1\Delta t = B'R$; therefore the triangles $A'B'R$ and $A'SR$ are equal with the result that the *angle of* *incidence θ_1* is equal to the *angle of reflection θ_1'*; this is the *law of reflection.* For the refracted wave, the angle at T is a right angle and we have

$$V_2 \Delta t = A'R \sin \theta_2.$$

and

$$V_1 \Delta t = A'R \sin \theta_1;$$

hence

$$\frac{\sin \theta_1}{V_1} = \frac{\sin \theta_2}{V_2}. \tag{4.24}$$

The angle θ_2 is called the *angle of refraction* and eq. (4.24) is the *law of refraction,* also known as *Snell's law.* The angles are usually measured between the ray paths and a normal to the interface but these angles are the same as those between the interface and the wavefronts in isotropic media. The laws of reflection and refraction can be combined in the single statement: at an interface the quantity $p = (\sin \theta_i)/v_i$ has the same value for the incident, reflected and refracted waves. This generalized form of Snell's law will be understood in future references to Snell's law. The quantity p is called the *raypath parameter.*

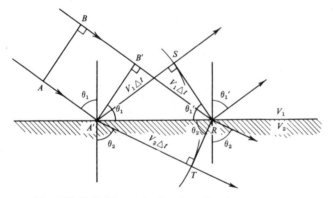

Fig. 4.12 Reflection and refraction of a plane wave.

When the medium consists of a number of parallel beds, Snell's law requires that the quantity p have the same value everywhere for all reflected and refracted rays resulting from a given initial ray.

When V_2 is less than V_1, θ_2 is less than θ_1. However, when V_2 is greater than V_1, θ_2 reaches $90°$ when $\theta_1 = \sin^{-1}(V_1/V_2)$. For this value of θ_1, the refracted ray is travelling along the interface. The angle of incidence for which $\theta_2 = 90°$ is the *critical angle*, θ_c; obviously, $\sin \theta_c = V_1/V_2$. For angles of incidence greater than θ_c, it is impossible to satisfy Snell's law (since $\sin \theta_2$ cannot exceed unity) and *total reflection* occurs (that is, the refracted ray does not exist). This situation is discussed in §4.2.2m.

Snell's law is very useful in determining raypaths and arrival times and in deriving reflector position from observed arrival times, but it does not give information about the amplitudes of the reflected and transmitted waves. This subject is taken up in §4.2.2*m*.

(l) *Diffraction*. Seismic energy travels along other paths besides those given by Snell's law. Whenever a wave encounters a feature whose radius of curvature is comparable to or smaller than the wavelength, the ordinary laws of reflection and refraction no longer apply. In such cases the energy is *diffracted* rather than re-flected or refracted. Since seismic wavelengths are large (often 100 metres or more) compared with some geologic dimensions, diffraction is an important process. The laws of diffraction are complex but at distances greater than several wave-lengths from the diffracting source the diffracted wavefront is essentially that given by Huygens' construction (Trorey, 1970).

Figure 4.13 illustrates the method of constructing diffracted wavefronts pro-duced by a faulted bed. We assume a plane wavefront AB incident normally on

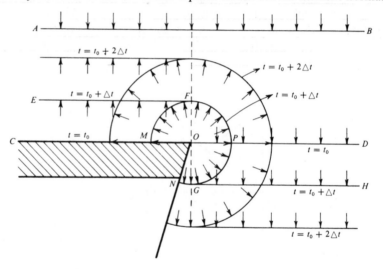

Fig. 4.13 Diffracted wavefronts.

the faulted bed CO, the position of the wavefront when it reaches the surface of the bed at $t = t_0$ being COD. At $t = t_0 + \Delta t$, the portion to the right of O has ad-vanced to the position GH while the portion to the left of O has been reflected and has reached the position EF. We might have constructed the wavefronts EF and GH by selecting a large number of centres in CO and OD and drawing arcs of length $V \Delta t$; EF and GH would then be determined by the envelopes of these arcs. However, for the portion EF there would be no centres to the right of O to define the envelope while for the portion GH there would be no centres to the left of O to define the envelope. Thus, O marks the transition point between centres which give rise to the upward-travelling wavefront EF and centres which give rise to the downward-travelling wavefront GH; the arc FPG with centre O is the diffracted wavefront originating at *O* and connecting the two wavefronts, *EF* and *GH*. The diffracted wavefront also extends into the geometrical shadow area *GN* and into the region *FM*.

The intensity of the reflected wave will be affected by the termination of the reflector as shown by the model results presented in fig. 4.14. As the point of reflection approaches the end of the reflector the intensity of the reflected wave

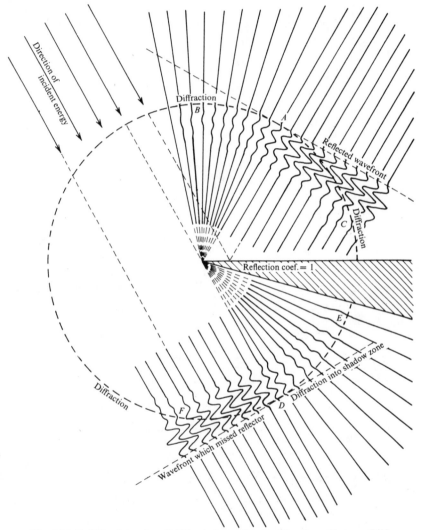

Fig. 4.14 Relative intensity of diffracted wavefronts. (Courtesy Chevron Oil.)

decreases until it is equal to, for the ray *A*, half its normal value (the amplitude decreases by the factor $1/\sqrt{2}$), the decrease in reflected energy being accounted for by the diffracted energy. For ray *A* there is no distinction between reflected and

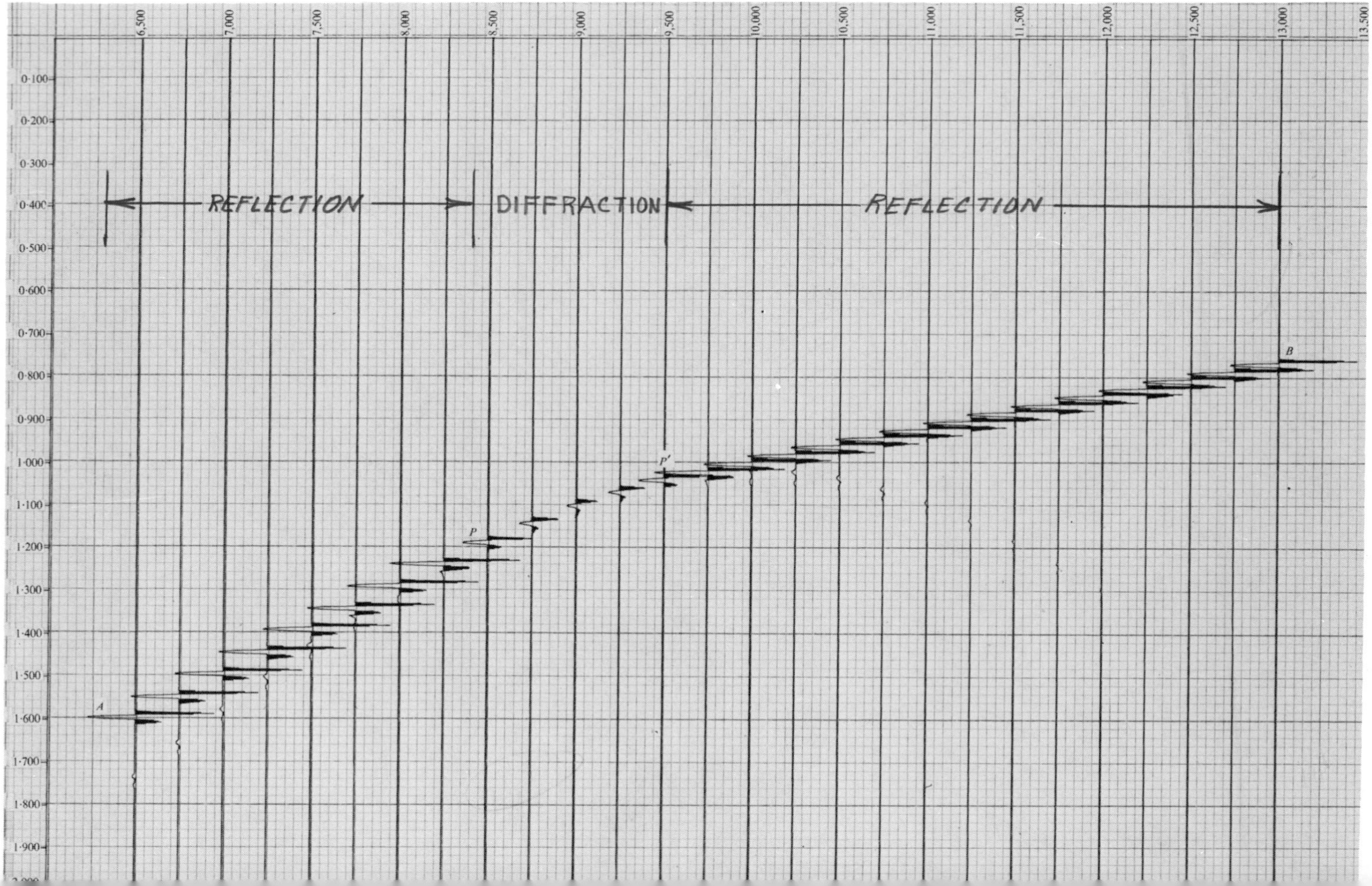

REFLECTION
DIFFRACTION
REFLECTION
A
P
P′
B
6,500
7,000
7,500
8,000
8,500
9,000
9,500
10,000
10,500
11,000
11,500
12,000
12,500
13,000
13,500
0·100
0·200
0·300
0·400
0·500
0·600
0·700
0·800
0·900
1·000
1·100
1·200
1·300
1·400
1·500
1·600
1·700
1·800
1·900

diffracted energy; the reflection is continuous with the diffraction, which decreases in intensity in the region *B*. Hence one cannot identify the point at which the reflection ends on the basis of waveshape or abrupt change of amplitude. In the region *B* the diffraction arrives progressively later and with diminished amplitude as the distance from ray *A* increases. If the instrumentation were to build up the amplitude, as may happen with *AGC* (see §4.5.4*e*), one might attribute the delay in arrival of the wave to a change in the depth of the reflector. Energy is also diffracted into the region *C* where it arrives slightly after the reflection and adds a slight tail to the reflection. The diffraction in the region *C* is 180° out-of-phase with the diffraction in the region *B*; however the diffraction in the region *C* usually is obscured by the reflection and is rarely observed although it contains as much energy as the diffraction in the region *B*. Diffractions *E* and *F* are also present on either side of the raypath *D* which just missed the reflector; relationships between *E* and *F* and between them and the wavefront which missed the reflector are similar to those between *B* and *C* and between them and the reflection. The diffraction *E* provides a mechanism for energy to appear behind an obstruction. Diffractions are discussed further in §4.4.3.

Diffractions are very important in determining the appearance of reflections where the reflectors are not continuous or plane. Figure 4.15 shows the events which a sharply-bent reflector causes. The reflection to the right of $x = 10,000$ gives rise to *BP'*, the reflection to the left of $x = 10,000$ gives *AP*. Diffraction fills in the gap *PP'* and makes the seismic event continuous without a sharp break in slope.

(m) *Partitioning of energy at an interface*. When a wave arrives at a surface separating two media having different elastic properties, it gives rise to reflected and refracted waves as described above. At the boundary the stresses and displacements must be continuous. Two neighbouring points *R* and *S*, which lie on opposite sides of the boundary as shown in fig. 4.16, will in general have different values of

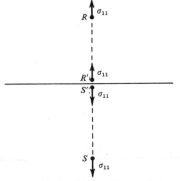

Fig. 4.16 Continuity of normal stress.

normal stress. This difference results in a net force which accelerates the layer between them. However, if we choose points closer and closer together, the stress values must approach each other and in the limit when the two points coincide on the boundary, the two stresses must be equal. If this were not so, the infinitesimally thin layer at the boundary would be acted upon by a finite force and hence have an acceleration which would approach infinity as the two points approach each other. Since the same reasoning applies to a tangential stress, we see that the normal and tangential components of stress must be continuous (cannot change abruptly) at the boundary. Likewise the normal and tangential components of displacement must be continuous. If the normal displacement were not continuous, one medium would either separate from the other leaving a vacuum in between or else would penetrate into the other so that the two media would occupy the same space. If the tangential displacement were not continuous, the two media would move differently on opposite sides of the boundary and one would slide over the other. Such relative motion is impossible and so displacement must be continuous at the boundary.

The continuity of normal and tangential stresses and displacements at the boundary can be expressed by means of four equations (*boundary conditions*) which the wave motion must obey at the interface. Assume a plane *P*-wave with amplitude A_0 incident on the boundary between two solid media. Snell's law fixes the angles of reflection and refraction while the amplitudes of the reflected and refracted waves are fixed by the four boundary conditions. However, to satisfy four equations we must have four unknown amplitudes; hence four waves must be generated at the boundary. These correspond to reflected and refracted *P*-waves and reflected and refracted *S*-waves. This is illustrated in fig. 4.17 where

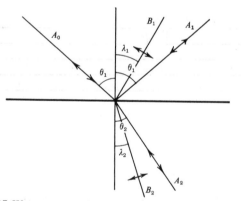

Fig. 4.17 Waves generated at an interface by an incident *P*-wave.

A_1, A_2, θ_1, and θ_2 are the amplitudes and angles of the reflected and refracted *P*-waves and B_1, B_2, λ_1, and λ_2 are the amplitudes and angles of the reflected and refracted *S*-waves.

Snell's law tells us that

$$\frac{\sin \theta_1}{\alpha_1} = \frac{\sin \theta_2}{\alpha_2} = \frac{\sin \lambda_1}{\beta_1} = \frac{\sin \lambda_2}{\beta_2} = p. \qquad (4.25)$$

(This more general statement of Snell's law can be derived following the same reasoning as was used to derive eq. (4.24).) The equations governing the amplitudes were first given by Knott (1899) but he expressed them in terms of potential functions from which the displacements have to be found by differentiation. The corresponding equations in terms of amplitudes were given by Zoeppritz (1919) in the following form:

$$A_1 \cos \theta_1 - B_1 \sin \lambda_1 + A_2 \cos \theta_2 + B_2 \sin \lambda_2 = A_0 \cos \theta_1, \quad (4.26a)$$

$$A_1 \sin \theta_1 + B_1 \cos \lambda_1 - A_2 \sin \theta_2 + B_2 \cos \lambda_2 = - A_0 \sin \theta_1, \quad (4.26b)$$

$$A_1 Z_1 \cos 2\lambda_1 - B_1 W_1 \sin 2\lambda_1 - A_2 Z_2 \cos 2\lambda_2 - B_2 W_2 \sin 2\lambda_2 = - A_0 Z_1 \cos 2\lambda_1, \quad (4.26c)$$

$$A_1 \gamma_1 W_1 \sin 2\theta_1 + B_1 W_1 \cos 2\lambda_1 + A_2 \gamma_2 W_2 \sin 2\theta_2 - B_2 W_2 \cos 2\lambda_2$$
$$= A_0 \gamma_1 W_1 \sin 2\theta_1, \quad (4.26d)$$

where

$$\gamma_i = \beta_i/\alpha_i, \ Z_i = \rho_i \alpha_i, \ W_i = \rho_i \beta_i, \ i = 1, 2.$$

The products of density and velocity (Z_i and W_i) are known as *acoustic impedances*. To apply these equations at an interface, we must know the density and velocities in each of the media, hence $Z_1, Z_2, W_1, W_2, \gamma_1$ and γ_2 are known. For a given A_0 and θ_1, we can calculate θ_2, λ_1 and λ_2 from eq. (4.25) and the four amplitudes, A_1, A_2, B_1 and B_2, from eq. (4.26).

Zoeppritz' equations reduce to a very simple form for normal incidence. Since the curves change slowly for small angles of incidence (say up to 20°), the results for normal incidence have wide application. For a *P*-wave at normal incidence, the tangential stresses and displacements are zero; eqs. (4.26b) and (4.26d), which express the continuity of tangential displacement and stress, show that $B_1 = B_2 = 0$; thus, eqs. (4.26) reduce to the following:

$$A_1 + A_2 = A_0,$$
$$Z_1 A_1 - Z_2 A_2 = -Z_1 A_0.$$

The solution of these equations is

$$\left.\begin{array}{l} \dfrac{A_1}{A_0} = \dfrac{Z_2 - Z_1}{Z_2 + Z_1}, \\[2mm] \dfrac{A_2}{A_0} = \dfrac{2Z_1}{Z_2 + Z_1}. \end{array}\right\} \quad (4.27)$$

These ratios are sometimes called the reflection and transmission coefficients, but the fractions of the incident energy which are reflected and refracted are also called by these names. Writing R and T for the *reflection* and *transmission coefficients* given by the fraction of the incident energy reflected and transmitted, and letting $\delta = Z_2/Z_1$, we find from eqs. (4.20b) and (4.27)

$$R = \frac{\frac{1}{2}\alpha_1\rho_1\omega^2 A_1{}^2}{\frac{1}{2}\alpha_1\rho_1\omega^2 A_0{}^2} = \left(\frac{Z_2 - Z_1}{Z_2 + Z_1}\right)^2 = \left(\frac{\delta - 1}{\delta + 1}\right)^2, \qquad (4.28a)$$

$$T = \frac{\frac{1}{2}\alpha_2\rho_2\omega^2 A_2{}^2}{\frac{1}{2}\alpha_1\rho_1\omega^2 A_0{}^2} = \frac{4Z_1 Z_2}{(Z_2 + Z_1)^2} = \frac{4\delta}{(\delta + 1)^2}, \qquad (4.28b)$$

$$R + T = 1. \qquad (4.28c)$$

The quantity δ is known as the *impedance contrast*. If we define δ as the ratio Z_1/Z_2 instead of Z_2/Z_1, the expressions for R and T are unchanged; hence the coefficients do not depend upon which medium contains the incident wave. When $\delta = 1$, $R = 0$ and all the energy is transmitted (note that this does not require that $\rho_1 = \rho_2$ and $\alpha_1 = \alpha_2$). As the impedance contrast approaches 0 or ∞, T approaches zero and R approaches unity; thus, the farther the impedance contrast is from unity, the stronger the reflected energy.

Table 4.2 shows how the reflected energy varies for impedance contrasts such as may be expected within the earth. Since both density and velocity contrasts are small for most of the interfaces encountered, only a small portion of the energy is reflected at any one interface; this is illustrated by the first four lines in table 4.2. The 'sandstone-on-limestone' interface is about as large a contrast as is apt to be encountered, whereas the 'shallow interface' and 'deep interface' figures are much more typical of most interfaces in the earth; hence usually appreciably less than 1% of the energy is reflected at any interface. The major exceptions involve the bottom and surface of the ocean and the base of the weathering (see §4.2.3b). A much larger proportion of the energy can be reflected from these and hence they are especially important in the generation of multiple reflections and other phenomena with which we shall deal later.

Note that, while the energy fractions R and T do not depend on which side of an

Table 4.2. *Energy reflected at interface between two media*

Interface	First medium Velocity	First medium Density	Second medium Velocity	Second medium Density	$\delta = Z_1/Z_2$	R
Sandstone on limestone	2000	2·4	3000	2·4	0·67	0·040
Limestone on sandstone	3000	2·4	2000	2·4	1·5	(−)0·040
Shallow interface	2100	2·4	2300	2·4	0·93	0·0021
Deeper interface	4300	2·4	4500	2·4	0·97	0·0005
'Soft' ocean bottom	1500	1·0	1500	2·0	0·50	0·11
'Hard' ocean bottom	1500	1·0	3000	2·5	0·20	0·44
Surface of ocean	1500	1·0	360	0·0012	3800	(−)0·9995
Base of weathering	500	1·5	2000	2·0	0·19	0·46

All velocities in m/sec, densities in g/cm³; the minus signs in parentheses indicate 180° phase reversal, not that R is negative.

interface the wave is incident, this is not true of the reflected amplitude A_1 since interchanging Z_1 and Z_2 in eq. (4.27) changes the sign of the ratio A_1/A_0. A negative value of A_1 means that the reflected wave is $180°$ out-of-phase with the incident wave; thus, for an incident wave $A_0 \cos \omega t$ the reflected wave is $A_1 \cos (\omega t + \pi)$. In table 4.2 phase reversal occurs for the situations where δ exceeds unity.

Turning now to the general case where the angle of incidence is not necessarily $90°$, we shall illustrate solutions of Zoeppritz' equations by graphs showing the energy partition as functions of the angle of incidence for certain values of parameters. Many curves would be required to show the variations of energy partitioning as a function of incident angle because of the many parameters which can be varied: incident P-, SH-, or SV-wave, P-wave velocity ratio, density ratio, and S-wave velocities in each medium (or the equivalent of defining Poisson's ratio for each medium). Figure 4.18 shows several cases representative of the variety of results possible.

Figure 4.18*a* shows the partitioning of energy as a function of the angle of incidence when a P-wave is incident in the high-velocity medium for a P-wave velocity ratio $\alpha_2/\alpha_1 = 0.5$, a density ratio $\rho_2/\rho_1 = 0.8$, $\sigma_1 = 0.30$, $\sigma_2 = 0.25$. For small incident angles all of the energy is in the reflected or transmitted P-waves, RP and TP respectively, and hence there are essentially no S-waves. As the incident angle increases, some of the energy goes into reflected and transmitted S-waves, RS and TS respectively, mostly at the expense of the reflected P-wave. Note that at intermediate angles of incidence the reflected S-wave carries more energy than the reflected P-wave. Such *converted waves* (waves resulting from the conversion of P-waves to S-waves or vice-versa at an interface) are sometimes recorded at long offsets where they are evidenced by alignments which disappear as one tries to follow them to shorter offsets. As grazing incidence is approached the energy of the reflected P-wave increases until at grazing incidence all of the energy is in the reflected P-wave.

The opposite situation is shown in fig. 4.18*b* where $\alpha_2/\alpha_1 = 2.0$, $\rho_2/\rho_1 = 0.5$, $\sigma_1 = 0.30, \sigma_2 = 0.25$. Since $Z_1 = Z_2$, the P-wave reflection coefficient is essentially zero for small incident angles. As the incident angle increases, S-wave energy increases. As the critical angle for P-waves is approached, the transmitted P-wave energy falls rapidly to zero and no transmitted P-wave exists for larger incident angles. Also as the critical angle for P-waves is approached, both reflected P-wave and reflected S-wave become very strong; such a build-up in reflection strength near the critical angle is called *wide-angle reflection*. Sometimes it is possible to make use of this phenomenon to map reflectors using long offsets where they cannot be followed at short offsets (Meissner, 1967). As the critical angle for S-waves is approached, the transmitted S-wave falls to zero.

If we had not had a density contrast but otherwise the values had been as indicated in fig. 4.18*b*, there would have been a reflected P-wave at small incident angles (as shown by the dotted curve) whose fractional energy would have decreased slightly as the incident angle increased.

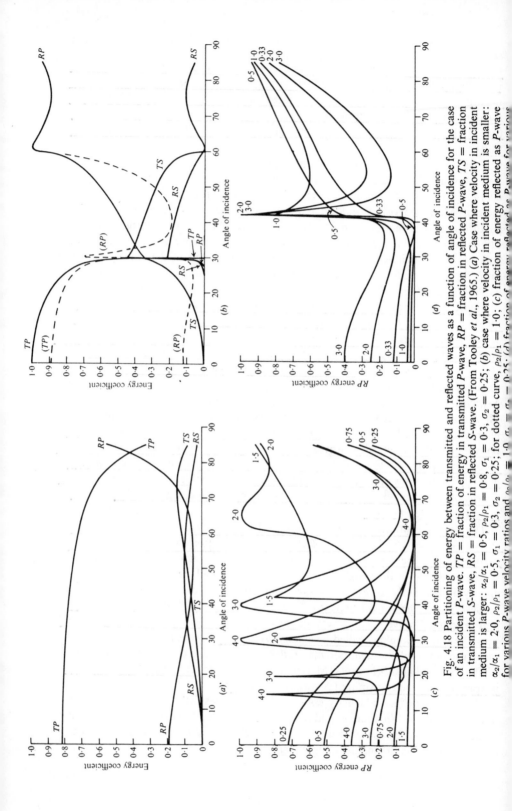

Fig. 4.18 Partitioning of energy between transmitted and reflected waves as a function of angle of incidence for the case of an incident P-wave. TP = fraction of energy in transmitted P-wave, RP = fraction in reflected P-wave, TS = fraction in transmitted S-wave, RS = fraction in reflected S-wave. (From Tooley et al., 1965.) (a) Case where velocity in incident medium is larger: $\alpha_2/\alpha_1 = 0.5$, $\rho_2/\rho_1 = 0.8$, $\sigma_1 = 0.3$, $\sigma_2 = 0.25$; (b) case where velocity in incident medium is smaller: $\alpha_2/\alpha_1 = 2.0$, $\rho_2/\rho_1 = 0.5$, $\sigma_1 = 0.3$, $\sigma_2 = 0.25$; for dotted curve, $\rho_2/\rho_1 = 1.0$; (c) fraction of energy reflected as P-wave for various P-wave velocity ratios and $\rho_2/\rho_1 = 1.0$, $\sigma_1 = \sigma_2 = 0.25$; (d) fraction of energy reflected as P-wave for various

Figure 4.18c shows the *P*-wave reflection coefficient for various *P*-wave velocity ratios when $\rho_1 = \rho_2$ and $\sigma_1 = \sigma_2 = 0\cdot25$. The reflected energy is zero for a velocity ratio of one (no impedance contrast) and increases both as the ratio becomes larger than 1 and as it becomes smaller than 1. The two peaks for $\alpha_2/\alpha_1 > 1$ occur at the critical angles for *P*- and *S*-waves respectively. Figure 4.18d shows the energy of the reflected *P*-wave for various density contrasts when $\alpha_2/\alpha_1 = 1\cdot5$ and $\sigma_1 = \sigma_2 = 0\cdot25$.

(n) *Multiples.* Multiples are events which have undergone more than one reflection. Since the energy of multiples is the product of the energy reflection coefficients for each of the reflectors involved and since *R* is very small for most interfaces, only the strongest impedance contrasts will generate multiples strong enough to be recognized as events.

We may distinguish between two classes of multiples which we call long-path and short-path. A *long-path multiple* is one whose travel path is long compared with primary reflections from the same deep interfaces and hence long-path multiples appear as separate events on a seismic record. A *short-path multiple*, on the other hand, arrives so soon after the associated primary reflection from the same deep interface that it interferes with and adds tail to the primary reflection; hence its effect is that of changing waveshape rather than producing a separate event. Possible raypaths for these two classes are shown in fig. 4.19.

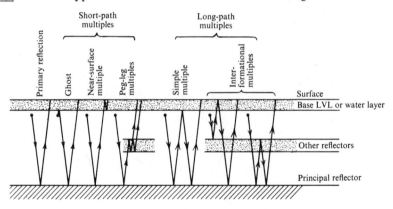

Fig. 4.19 Types of multiples.

The only important long-path multiples are those which have been reflected once at the surface or base of the LVL (see §4.2.3b) and twice at interfaces with relatively large acoustic impedance contrasts. Since *R* is about 50% at the base of the LVL and perhaps 5% for the strongest interfaces at depth, the maximum effective *R* for such multiples will be of the order of $0\cdot05 \times 0\cdot5 \times 0\cdot05 = 0\cdot001$. This value is in the range of typical reflection coefficients so that such multiples may have sufficient energy to be confused with primary events. The principal situation where weaker long-path multiples may be observable is where primary energy is nearly absent at the time of arrival of the multiple energy so that the gain of the recording system is very high.

Short-path multiples which have been reflected successively from the top and base of thin reflectors (fig. 4.20a) on their way to or from the principal reflecting interface with which they are associated (often called *peg-leg multiples*) are important in determining the wave-forms of the events recorded on a seismogram. These peg-leg multiples have the effect of delaying part of the energy and therefore lengthening the wavelet. The stronger peg-leg multiples will often have the same

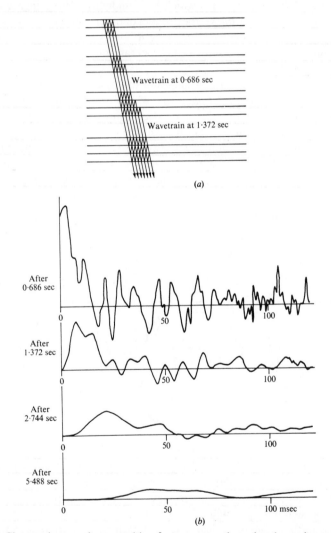

Fig. 4.20 Changes in waveshape resulting from passage through a layered sequence. (After O'Doherty & Anstey, 1971.) (a) Schematic diagram showing peg-leg multiples; (b) wavetrains after different traveltimes.

sign as the primary since successive large impedance contrasts tend to be in opposite directions (otherwise the successive large changes in velocity would cause the velocity to exceed its allowable range). This effectively lowers the signal frequency as time increases (see O'Doherty and Anstey, 1971). Figure 4.20*b* shows how a simple impulse (such as an explosion might generate) becomes modified as a result of passing it through a sequence of interfaces.

4.2.3 *Seismic velocity*

(a) *Factors affecting velocity.* Equation (4.17) shows that the velocity of *P*-waves in a homogeneous solid is a function only of the elastic constants and the density. One might expect that the elastic constants, which are properties of the intermolecular forces, would be relatively insensitive to pressure, whereas the density should increase with pressure because rocks are moderately compressible. This would lead one to expect that the numerator in the expression for velocity would not change very much with increasing pressure whereas the denominator would get larger so that velocity would decrease with depth of burial in the earth. In fact, this is contrary to actual observations.

Birch (1966) shows wide ranges in the velocity of any given rock type, as illustrated in fig. 4.21. While most rocks are mixtures of different minerals, even if we

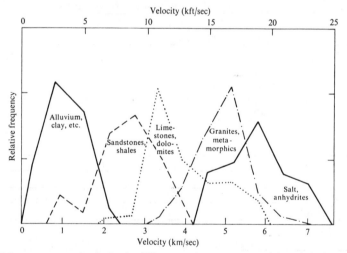

Fig. 4.21 Measurements of velocity in different rock types reported by Birch (1966). (After Grant and West, 1965.)

were to consider only relatively 'pure' rocks such as sandstones composed mainly of quartz or limestones which are almost pure calcite, we would encounter a wide range of velocity values, almost all of them lower than the values for quartz or calcite.

The most important aspect in which rocks differ from homogeneous solids is in having granular structure with voids between the grains. These voids are responsible for the porosity of rocks (see §11.1.2) and porosity is the important factor in determining velocity. Gassmann (1951) derived an expression for the velocity in a model consisting of tightly-packed elastic spherical particles under a pressure so that the contacts between the spheres become areas rather than points. The elastic constants of such a pack vary with pressure and the effect is to make the P-wave velocity vary as the 1/6th power of the pressure. Faust (1953) found an empirical formula for velocity in terms of the depth of burial Z and the formation resistivity R:

$$V = 2 \times 10^3 (ZR)^{1/6}, \tag{4.29}$$

V being in ft/sec when Z is in feet and R in ohm-feet. However, the deviation of individual measurements was very large, indicating the presence of other factors which have not been taken into account.

An earlier form of Faust's law (Faust, 1951) also included the age of the rock as a factor in determining velocity. An older rock might be expected to have a higher velocity, having been subjected for a longer time to pressures, cementation, and other factors which might increase its velocity.

In actual rocks the pore spaces are filled with a fluid whose elastic constants and density also affect the seismic velocity. The fluid is under a pressure which usually is different from that resulting from the weight of the overlying rocks. In this situation the effective pressure on the granular matrix is the difference between the overburden and fluid pressures. Where formation fluids are under abnormal pressures approaching the overburden pressure, the seismic velocity is exceptionally low, a fact which is sometimes used to predict abnormal fluid pressure from velocity measurements.

Subjecting a porous rock to pressure results in both reversible and irreversible changes in porosity; that is, when the pressure is removed part of the original porosity is regained while another part is permanently lost, perhaps because of crushing of the grains, alteration of the packing, or other permanent structural changes. Empirical data suggest that the maximum depth to which a rock has been buried is a measure of the irreversible effect on porosity and is therefore an important parameter in determining porosity. In summary, porosity appears to be the dominant variable in determining the velocity in sedimentary rocks and porosity in turn is determined principally by the existing differential pressure and the maximum depth of burial.

The variation of velocity with depth, usually referred to as the *velocity function*, is frequently a reasonably systematic increase as we go to greater depths. Areas of moderately uniform geology, such as the U.S. Gulf Coast, exhibit little variation in the velocity function from area to area. Because of the seaward regional dip in the Gulf Coast area, as one goes seaward younger section is encountered at a given depth but the velocity function does not vary greatly; the maximum pressures to which the rocks have been subjected are the existing pressures which depend mainly on depth, not age. On the other hand, areas subject to recent structural deformation and uplift, such as California, exhibit rapid variation of velocity function from area to area. Many of the California rocks have been buried to greater depths and subjected to greater stresses than exist at present. The result is rapid lateral changes in velocity which profoundly affect seismic interpretation.

The variation of velocity with density is shown in fig. 4.22. The large range of velocity for any given lithology (e.g., shale velocities range from 1600 to

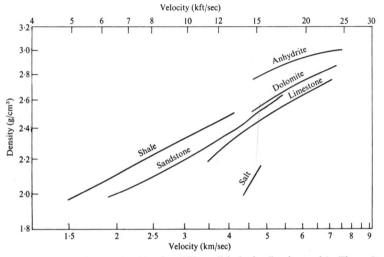

Fig. 4.22 Velocity-density relationships for different lithologies (log-log scale). (From Gardner *et al.*, 1974.)

4000 m/sec) tells the same story as fig. 4.21. The overlap of ranges of velocity makes it impossible to tell the lithology of a sample merely from its velocity. The range of density values results mainly from different porosities and the curves would look very similar if velocity had been plotted against porosity.

The *time-average equation* is often used to relate velocity V and porosity ϕ; it assumes that the traveltime per unit path length in a fluid-filled porous rock is the average of the traveltimes per unit path length in the matrix material, $1/V_m$, and in the fluid, $1/V_f$, the traveltimes being weighted in proportion to their respective volumes:

$$\frac{1}{V} = \frac{\phi}{V_f} + \frac{(1-\phi)}{V_m}. \qquad V = \frac{V_f V_m}{V_f + (V_m - V_f)\phi} \quad (4.30)$$

This relationship is used extensively in well-log interpretation.

The irreversible change in porosity (and consequently in velocity) with depth of burial has been used to determine the maximum depth at which a section formerly lay. If the velocity-depth relationship for a given lithology can be established in an area not subjected to uplift, the maximum depth of burial can be ascertained from the observed velocity-depth relationship and hence the amount of the uplift can be inferred. In fig. 4.23 the shale and limestone regression lines (curves A and B) represent measurements on 'pure' shales and limestones which are believed to be at their maximum depth of burial. Curve C, which is obtained from these curves by interpolation, is the predicted curve based on the relative amounts of shale and limestone actually present and assuming the rocks to be at their maximum depth of burial. The displacement in depth required to fit this curve to the actual measurements is presumed to indicate the amount of uplift which has occurred.

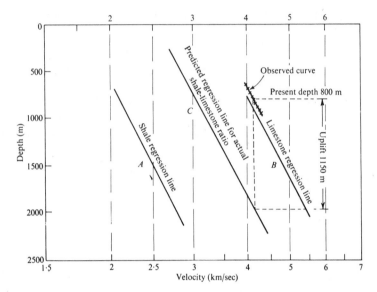

Fig. 4.23 Finding maximum depth of burial from velocity. (From Jankowsky, 1970.)

(b) *The weathered or low-velocity layer.* Seismic velocities which are lower than the velocity in water usually imply that gas (air or methane resulting from the decomposition of vegetation) fills at least some of the pore space (Watkins *et al.*, 1972). Such low velocities are usually seen only near the surface in a zone called the *weathered layer* or the *low-velocity layer*, often abbreviated *LVL*. This layer, which is usually 4 to 50 m thick, is characterized by seismic velocities which are not only low (usually between 250 and 1000 m/sec) but at times highly variable. Frequently the base of the LVL coincides roughly with the water table, indicating that the low-velocity layer corresponds to the aerated zone above the water-saturated zone, but this is not always the case. Obviously the term 'weathering' as used by geophysicists differs from the geologist's 'weathering' which denotes the disintegration of rocks under the influence of the elements.

The importance of the low-velocity layer is four-fold: (1) the absorption of seismic energy is high in this zone, (2) the low velocity and the rapid changes in velocity have a disproportionately large effect on traveltimes, (3) the marked velocity change at the base of the LVL sharply bends seismic rays so that their travel through the LVL is nearly vertical regardless of their direction of travel beneath the LVL and (4) the very high impedance contrast at the base of the LVL makes it an excellent reflector, important in multiple reflections. Because of the first factor, records from shots in this layer often are of poor quality and efforts are made to locate the shot below the LVL. Methods of investigating the low-velocity layer are discussed in §4.5.3*e* and methods of correcting for it in §4.5.7*b*.

4.3 Geometry of seismic wave paths

4.3.1 *Geometry of reflection paths*

(a) *General.* The exact interpretation of reflection data requires a knowledge of the velocity at all points along the reflection paths. However, even if we had such a detailed knowledge of the velocity, the calculations would be tedious and often we would assume a simple distribution of velocity which is close enough to give useable results. The simplest assumption is that the velocity is constant between the surface and the reflecting bed. Although this assumption is rarely even approximately true, it leads to simple formulae which give answers which are within the required accuracy in many instances.

The basic problem in reflection seismic surveying is to determine the position of a bed which gives rise to a reflection on a seismic record. In general this is a problem in three dimensions. However, the dip is often very gentle and the direction of profiling is frequently nearly along either the direction of dip or the direction of strike. In such cases a 2-dimensional solution is generally used. We shall discuss the 2-dimensional problem in the next two sections, then the more general problem.

(b) *Horizontal reflector; normal moveout.* The simplest 2-dimensional problem is that of zero dip illustrated in the lower part of fig. 4.24. The reflecting bed, AB, is at a depth h below the shotpoint S. Energy leaving S along the direction SC will be reflected in such a direction that the angle of reflection equals the angle of incidence.

Although the reflected ray CR can be determined by laying off an angle equal to α at C, it is easier to make use of the image point I which is located on the same normal to the reflector as S and as far below the bed as S is above. If we join I to C and prolong the straight line to R, CR is the reflected ray (since CD is parallel to SI, making all the angles marked α equal).

Denoting the average velocity by V, the traveltime t for the reflected wave is $(SC + CR)/V$. However, $SC = CI$ so that IR is equal in length to the actual path, SCR. Therefore, $t = IR/V$ and in terms of x, the shotpoint-to-geophone distance (*offset*), we can write

$$V^2t^2 = x^2 + 4h^2, \tag{4.31a}$$

or

$$\frac{V^2t^2}{4h^2} - \frac{x^2}{4h^2} = 1. \tag{4.31b}$$

Thus the traveltime curve is a hyperbola as shown in the upper part of fig. 4.24.

The geophone at R will also record the *direct wave* which travels along the path SR. Since SR is always less than $(SC + CR)$, the direct wave arrives first. The traveltime is $t_D = x/V$ and the traveltime curves are the straight lines OM and ON passing through the origin with slopes of $\pm 1/V$.

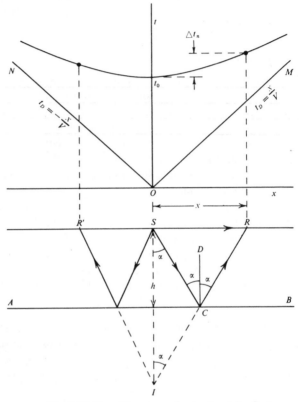

Fig. 4.24 Traveltime curve for horizontal reflector.

When the distance x becomes very large the difference between SR and $(SC + CR)$ becomes small and the reflection traveltime approaches the direct wave traveltime asymptotically.

The location of the reflecting bed is determined by measuring t_0, the traveltime for a geophone at the shotpoint. Setting $x = 0$ in (4.31a), we see that

$$h = \tfrac{1}{2}Vt_0. \tag{4.32}$$

Equation (4.31a) can be written

$$t^2 = \frac{x^2}{V^2} + \frac{4h^2}{V^2} = \frac{x^2}{V^2} + t_0^2. \tag{4.31c}$$

If we plot t^2 against x^2 (instead of t versus x as in fig. 4.24), we obtain a straight line of slope $(1/V^2)$ and intercept $t_0{}^2$. This forms the basis of a well-known scheme for determining V, the '$X^2 - T^2$ method'; this will be described in §4.5.6c.

We can solve eq. (4.31a) for t, the traveltime measured on the seismic record. Generally $2h$ is appreciable larger than x so that we can use a binomial expansion as follows:

$$t = \frac{2h}{V}\{1 + (x/2h)^2\}^{\frac{1}{2}} = t_0\{1 + (x/Vt_0)^2\}^{\frac{1}{2}}$$

$$= t_0\left\{1 + \tfrac{1}{2}\left(\frac{x}{Vt_0}\right)^2 - \tfrac{1}{8}\left(\frac{x}{Vt_0}\right)^4 + \ldots\right\}. \tag{4.33}$$

The difference in traveltime for a given reflection for two geophone locations is known as *moveout* and is represented by Δt. If t_1, t_2, x_1 and x_2 are the traveltimes and offsets, we have to the first approximation

$$t_2 \approx t_0\left[1 + \tfrac{1}{2}\left(\tfrac{x_2}{Vt_0}\right)^2\right]$$

$$\Delta t = t_2 - t_1 \approx \frac{x_2{}^2 - x_1{}^2}{2V^2t_0}. \tag{4.34a}$$

$$t_1 \approx t_0\left[1 + \tfrac{1}{2}\left(\tfrac{x_1}{Vt_0}\right)^2\right]$$

In the special case where one geophone is at the shotpoint, Δt is known as the *normal moveout* which we shall denote by Δt_n. Then,

$$\Delta t_n \approx \frac{x^2}{2V^2t_0}. \tag{4.34b}$$

At times we retain another term in the expansion:

$$\Delta t_n \approx \frac{x^2}{2V^2t_0}\left[1 - \left(\frac{x}{4h}\right)^2\right]$$

$$\Delta t_n \approx \frac{x^2}{2V^2t_0} - \frac{x^4}{8V^4t_0{}^3} = \frac{x^2}{2V^2t_0}\left\{1 - \left(\frac{x}{4h}\right)^2\right\}. \tag{4.34c}$$

From eq. (4.34b) we note that the normal moveout increases as the square of the offset x, inversely as the square of the velocity, inversely as the first power of the traveltime (or depth – see eq. (4.32)). Thus reflection curvature increases rapidly as we go to more distant geophones, at the same time the curvature becoming progressively less with increasing record time.

The concept of normal moveout is extremely important. It is the principal criterion by which we decide whether an event observed on a seismic record is a reflection or not. If the normal moveout differs from the value given by eq. (4.34b) by more than the allowable experimental error, we are not justified in treating the event as a reflection. One of the most important quantities in seismic interpretation is the change in arrival time caused by dip; to find this quantity we must eliminate normal moveout. Normal moveout must also be eliminated before 'stacking' (adding together) common-depth-point records (see §4.5.3a). Finally, eq. (4.34b) can be used to find V by measuring x, t_0 and Δt_n; this forms the basis of the ΔT-method of finding velocity (see §4.5.6d) and also of velocity analysis (§4.7.7a).

(c) *Dipping reflector; dip moveout.* When the bed is dipping in the direction of the profile, we have the situation shown in fig. 4.25, θ being the dip and h the depth (normal to the bed). To draw the ray path for the reflection arriving at the geophone R, we join the image point I to R by a straight line cutting the bed at C. The path is then SCR and t is equal to $(SC + CR)/V$; since $(SC + CR) = IR$, application of the cosine law to the triangle SIR gives

$$V^2t^2 = IR^2$$
$$= x^2 + 4h^2 - 4hx \cos\left(\frac{\pi}{2} + \theta\right)$$
$$= x^2 + 4h^2 + 4hx \sin\theta. \tag{4.35}$$

On completing the squares, we obtain

$$\frac{V^2t^2}{(2h\cos\theta)^2} - \frac{(x + 2h\sin\theta)^2}{(2h\cos\theta)^2} = 1.$$

Thus, as before the traveltime curve is a hyperbola but the axis of symmetry is now the line $x = -2h\sin\theta$ instead of the t-axis. This means that t has different

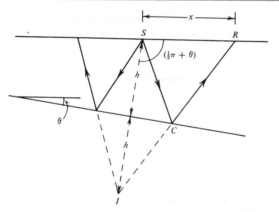

Fig. 4.25 Geometry of reflection path for dipping reflector.

values for geophones symmetrically placed on both sides of the shotpoint, unlike the case for zero dip.

Setting x equal to 0 in eq. (4.35) gives the same value for h as in eq. (4.32); note, however, that h is not the vertical depth here as it was in the earlier result.

To obtain the dip θ, we solve for t in eq. (4.35) by assuming that $2h$ is greater than x and expanding as in the derivation of eq. (4.33). Then

$$t = \frac{2h}{V}\left\{1 + \left(\frac{x^2 + 4hx\sin\theta}{4h^2}\right)\right\}^{\frac{1}{2}}$$
$$\approx t_0\left(1 + \frac{x^2 + 4hx\sin\theta}{8h^2}\right) \tag{4.36}$$

using only the first term of the expansion. The simplest method of finding θ is from the difference in traveltimes for two geophones equally distant from, and on opposite sides of, the shotpoint. Letting x have the values $+s$ for the down-dip geophone and $-s$ for the up-dip geophone and denoting the equivalent traveltimes by t_1 and t_2, we get

$$t_1 \approx t_0 \left(1 + \frac{s^2 + 4hs \sin \theta}{8h^2}\right),$$

$$t_2 \approx t_0 \left(1 + \frac{s^2 - 4hs \sin \theta}{8h^2}\right),$$

$$\Delta t_d = t_1 - t_2 \approx t_0 \left(\frac{s \sin \theta}{h}\right) \approx \frac{2s}{V} \sin \theta.$$

requires a symmetrical spread

The quantity Δt_d is called the *dip moveout*. If we write Δx for the distance $2s$ between the two geophones, the dip becomes

$$\sin \theta \approx V \left(\frac{\Delta t_d}{\Delta x}\right). \tag{4.37a}$$

For small angles θ is approximately equal to $\sin \theta$ so that the dip is directly proportional to Δt_d under these circumstances. Moreover, for a given reflection the dip moveout is directly proportional to Δx. Therefore, to obtain the dip as accurately as possible, we use as large a value of Δx as the data quality permits; for symmetrical spreads, we measure dip moveout between the geophone groups at the opposite ends of the spread, Δx then being the spread length.

It should be noted that normal moveout was eliminated in the derivation of eq. (4.37a). The terms in s^2 which disappeared in the subtraction represent the normal moveout.

Figure 4.26 illustrates diagrammatically the relation between normal moveout and dip moveout. The diagram at the left represents a reflection from a dipping bed; the alignment is curved and unsymmetrical about the shotpoint. Diagram (B) shows what would have been observed if the bed had been horizontal; the alignment is curved symmetrically about the shotpoint position owing to the normal moveout. The latter ranges from 0 to 13 msec (1 millisecond = 10^{-3} sec = 1 msec, the unit of time commonly used in seismic work) at an offset of 400 m. Diagram (C) was obtained by subtracting the normal moveouts shown in (B) from the arrival times in (A). The resulting alignment shows the effect of dip alone; it is straight and has a time difference between the outside curves of 10 msec, that is, $\Delta t_d = 10$ msec when $\Delta x = 800$ m. Thus, we find that the dip is 2500 $(10 \times 10^{-3}/800) = 0.031$ rad $= 1.8°$.

The method of normal-moveout removal illustrated in fig. 4.26 was used solely to demonstrate the difference between normal moveout and dip moveout. If we require only the dip moveout Δt_d we merely subtract the traveltimes for the two outside geophones in (A).

Frequently we do not have a symmetrical spread and we find the dip moveout by removing the effect of normal moveout. As an example, refer to fig. 4.26, curve (D), which shows a reflection observed on a spread extending from

$x = -133$ m to $x = +400$ m. Let $t_0 = 1\cdot225$ sec, $t_1 = 1\cdot223$ sec, $t_2 = 1\cdot242$ sec, $V = 2800$ m/sec. From eq. (4.34b) we get for Δt_n at offsets of 133 and 400 m respectively the values 1 msec and 8 msec (rounded off to the nearest millisecond since this is usually the precision of measurement on seismic records). Subtracting these values we obtain for the corrected arrival times $t_1 = 1\cdot222$, $t_2 = 1\cdot234$; hence the dip moveout is 12 msec for $\Delta x = 533$ m. The corresponding dip is $2800(12 \times 10^{-3}/533) = 0\cdot063$ rad $= 3\cdot6°$.

An alternative to the above method is to use the arrival times at $x = -133$ m and $x = +133$ m, thus obtaining a symmetrical spread and eliminating the need for calculating normal moveout. However, doing this would decrease the effective spread length from 533 m to 267 m and thereby reduce the accuracy of the ratio $(\Delta t_d/\Delta x)$.

Equation (4.37a) is very similar to the equation which gives the 'angle of approach' of a plane wave as it reaches the surface. In fig. 4.27, AC represents an upward-travelling plane wave at the instant when it arrives at the end of the spread, A. After a further interval Δt the same wave reaches the other end of the spread, B. Then, $BC = V\Delta t$ and

$$\sin \theta = \frac{BC}{BA} = V\frac{\Delta t}{\Delta x}. \tag{4.37b}$$

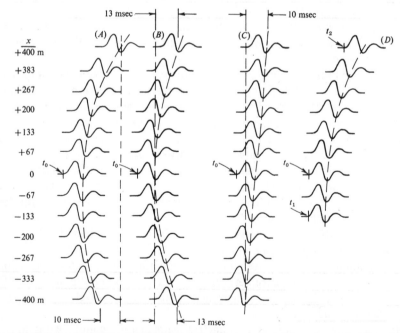

Fig. 4.26 Relation between normal moveout and dip moveout. For curves (A), (B) and (C), $t_0 = 1\cdot000$ sec, $V = 2500$ m/sec. For curve (D), $t_0 = 1\cdot225$ sec, $t_1 = 1\cdot223$ sec, $t_2 = 1\cdot242$ sec, $V = 2800$ m/sec.

This result, although identical in form with eq. (4.37*a*), has a different significance since it gives the direction of travel of a plane wave as it reaches the spread, V being the average velocity between C and the surface. In eq. (4.37*a*) V is the average velocity down to the reflector and θ is the angle of dip. The two results are identical in this instance because we have assumed V constant down to the reflector but often this will not be the case as we shall see presently.

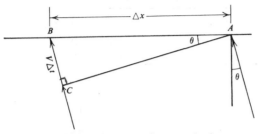

Fig. 4.27 Finding the angle of approach of a wave.

(d) *Cross dip*. When the profile is at an appreciable angle to the direction of dip, the determination of the latter becomes a 3-dimensional problem and we use the methods of solid analytical geometry. In fig. 4.28 we take the xy-plane as horizontal

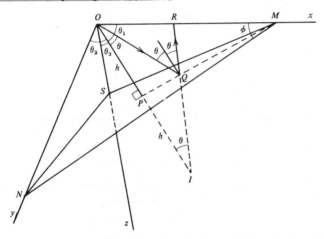

Fig. 4.28 Three-dimensional view of a reflection path for a dipping bed.

with the z-axis extending vertically downward. The line OP of length h is perpendicular to a dipping plane bed which outcrops (that is, intersects the xy-plane) along the line MN if extended sufficiently.

We write θ_1, θ_2, θ_3 for the angles between OP and x-, y- and z-axes. The cosines of these angles (*the direction cosines* of OP) have the values l, m and n. In analytical geometry it is shown that

$$l^2 + m^2 + n^2 = 1.$$

The angle ϕ between MN and the x-axis is the direction of strike of the bed while θ_3 is the angle of dip.

The path of a reflected wave arriving at a geophone R on the x-axis can be found using the image point I. The line joining I to R cuts the reflector at Q, hence OQR is the path. Since $OQ = QI$, the line IR is equal to Vt, t being the traveltime for the geophone at R. The coordinates of I and R are respectively $(2hl, 2hm, 2hn)$ and $(x, 0, 0)$, hence we have

$$
\begin{aligned}
V^2t^2 = IR^2 &= (x - 2hl)^2 + (0 - 2hm)^2 + (0 - 2hn)^2 \\
&= x^2 + 4h^2(l^2 + m^2 + n^2) - 4hlx \\
&= x^2 + 4h^2 - 4hlx.
\end{aligned}
$$

When $x = 0$, we obtain the same relation between h and t_0 as in eq. (4.32). Proceeding as in the derivation of eq. (4.36) we get for the approximate value of t,

$$
t \approx t_0 \left(1 + \frac{x^2 - 4hlx}{8h^2}\right).
$$

By subtracting the arrival times at two geophones located on the x-axis at $x = \pm s$, and letting $2s = \Delta x$ as before, we find

$$
\Delta t_x \approx t_0 \left(\frac{ls}{h}\right)
$$

$$
\approx \frac{l\Delta x}{V},
$$

$$
l = \cos \theta_1 \approx V \left(\frac{\Delta t_x}{\Delta x}\right). \tag{4.38a}
$$

If we also have a spread along the y-axis (*cross spread*), we get

$$
m = \cos \theta_2 \approx V \left(\frac{\Delta t_y}{\Delta y}\right), \tag{4.38b}
$$

where Δt_y is the time difference ('cross dip') between geophones a distance Δy apart and symmetrical about the shotpoint. Since

$$
n = \cos \theta_3 = \{1 - (l^2 + m^2)\}^{\frac{1}{2}},
$$

$$
\sin \theta_3 = (1 - n^2)^{\frac{1}{2}} = (l^2 + m^2)^{\frac{1}{2}}
$$

$$
= V \left\{\left(\frac{\Delta t_x}{\Delta x}\right)^2 + \left(\frac{\Delta t_y}{\Delta y}\right)^2\right\}^{\frac{1}{2}}. \tag{4.39a}
$$

If the spread lengths Δx and Δy are equal, we have

$$
\sin \theta_3 = V \frac{\{(\Delta t_x)^2 + (\Delta t_y)^2\}^{\frac{1}{2}}}{\Delta x}. \tag{4.39b}
$$

When the profile lines are not perpendicular, for example, when one is along the x-axis and the other along the y'-axis at an angle α to the x-axis, the solution is more complicated. Taking the length of a symmetrical spread along the y'-axis as $2s$, the coordinates of the ends of the spread (relative to the x-, y-, z-axes) are $\pm s \cos \alpha, \pm s \sin \alpha, 0$. Then

$$V^2 t_\pm{}^2 = (2hl \pm s \cos \alpha)^2 + (2hm \pm s \sin \alpha)^2 + (2hn)^2$$
$$= s^2 + 4h^2 \pm 4hs(l \cos \alpha + m \sin \alpha).$$

The dip moveout becomes

$$\Delta t' = \frac{\Delta y'}{V} (l \cos \alpha + m \sin \alpha)$$

or

$$(l \cos \alpha + m \sin \alpha) = V \left(\frac{\Delta t'}{\Delta y'}\right). \tag{4.38c}$$

The measured moveouts, Δt_x and $\Delta t'$, divided by the spread lengths give the values of l and $(l \cos \alpha + m \sin \alpha)$. Since α is known, m can be found and the total dip calculated using eqs. (4.38a), (4.38b) and (4.39a). The solution can also be found graphically as shown in fig. 4.29 (see also problem 7).

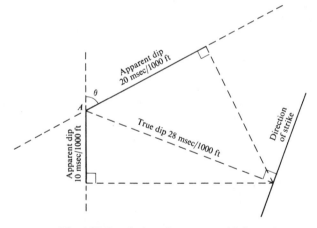

Fig. 4.29 Resolution of cross-spread information.

To find the strike ϕ, we start from the equation of a plane (that is, the reflector) which has a perpendicular from the origin of length h and direction cosines (l, m, n), namely

$$lx + my + nz = h.$$

Setting $z = 0$ gives the equation of the line of intersection of the reflector and the surface; this strike line has the equation

$$lx + my = h.$$

The intercepts of this line on the x- and y-axes are h/l and h/m. Referring to fig. 4.30, we find that

$$\tan \phi = \frac{h}{m} \div \frac{h}{l} = \frac{l}{m}$$

$$= \left(\frac{\Delta t_x}{\Delta x}\right) \left(\frac{\Delta y}{\Delta t_y}\right) \qquad (4.40a)$$

$$= \left(\frac{\Delta t_x}{\Delta t_y}\right) \qquad (4.40b)$$

when $\Delta x = \Delta y$. The strike can also be found using the construction shown in fig. 4.29.

4.3.2 *Velocity gradient and raypath curvature*

(a) *Effect of velocity variation.* The assumption of constant velocity is not valid in general, the velocity usually changing as we go from one point to another. In petroleum exploration we are usually dealing with more-or-less flat-lying bedding and the changes in seismic velocity as we move horizontally are for the most part small, being the result of slow changes in density and elastic properties of the beds.

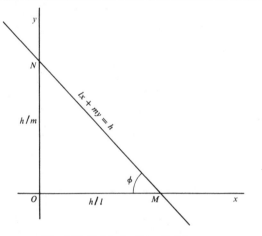

Fig. 4.30 Determination of strike.

These horizontal variations are generally much less rapid than the variations in the vertical direction where we are going from bed to bed with consequent lithological changes and increasing pressure with increasing depth. Because the horizontal changes are gradual they can often be taken into account by dividing the survey area into smaller areas within each of which the horizontal variations can be ignored and the same vertical velocity distribution used. Such areas often are large enough to include several structures of the size of interest in oil exploration so that changes from one velocity function to another do not impose a serious burden upon the interpreter.

(b) *Equivalent average velocity.* Vertical variations in velocity can be taken into account in various ways. One of the simplest is to use a modification of the constant velocity model. We assume that the actual section existing between the surface and a certain reflecting horizon can be replaced with an equivalent single layer of constant velocity V equal to the average velocity between the surface and the reflecting horizon. This velocity is usually given as a function of depth (or of t_0, which is nearly the same except when the dip is large). Thus the section is assigned a different constant velocity for each of the reflectors below it. Despite this inconsistency the method is useful and is extensively applied. The variation of the average velocity with t_0 is found using one of the methods described in §4.5.6. For the observed values of the arrival time t_0 we select the average velocity V corresponding to this reflector; using the values of t_0, the dip moveout Δt_d and V, we calculate the depth h and the dip θ using eqs. (4.32) and (4.37a).

c) *Velocity layering.* A method which is commonly used to take into account velocity variations is to replace the actual velocity distribution with an approximate one corresponding to a number of horizontal layers of different velocities, the velocity being constant within each layer. Simple equations such as (4.32) and (4.37a) are no longer appropriate because rays bend at each layer interface. Often a graphical method is used to find the depth and dip. The method uses a 'wave-front chart'; the preparation and use of these charts will be described in §4.5.7d. In effect this method replaces the actual ray paths with a series of line segments which are straight within each layer but undergo abrupt changes in direction at the boundaries between layers. Dix (1955) shows that the effect of this on eq. (4.31c) is to replace the average velocity V by its rms (root-mean-square) value, V_{rms}, so that

$$t^2 = t_0^2 + x^2 / V^2_{\text{rms}}, \qquad (4.41a)$$

where

$$V_{\text{rms}} = \left(\sum_1^n V_i^2 t_i / \sum_1^n t_i \right)^{\frac{1}{2}}, \qquad (4.41b)$$

V_i and t_i being respectively the average velocity in, and the traveltime through, the ith layer (see also Shah and Levin, 1973).

(d) *Velocity functions.* At times the assumption is made that the velocity varies in a systematic continuous manner and therefore can be represented by a velocity function. The actual velocity usually varies extremely rapidly over short intervals, as shown by sonic logs (see §4.5.6b); however, if we integrate these changes over distances of a wavelength or so (30–100 m), we obtain a function which is generally smooth except for discontinuities at marked lithological changes. If the velocity discontinuities are small, we are often able to represent the velocity distribution with sufficient accuracy by a smooth velocity function. The path of a wave traveling in such a medium is then determined by two integral equations.

To derive the equations, we assume that the medium is divided into a large number of thin beds in each of which the velocity is constant; on letting the number of beds go to infinity, the thickness of each bed becomes infinitesimal and the velocity distribution becomes a continuous function of depth. Referring to fig. 4.31, we have for the nth bed

$$\frac{\sin i_n}{V_n} = \frac{\sin i_0}{V_0} = p,$$

$$V_n = V_n(z),$$

$$\Delta x_n = \Delta z_n \tan i_n,$$

$$\Delta t_n = \frac{\Delta z_n}{V_n \cos i_n}.$$

The raypath parameter p (see §4.2.2k) is a constant which depends upon the direction in which the ray left the shotpoint, that is, upon i_0.

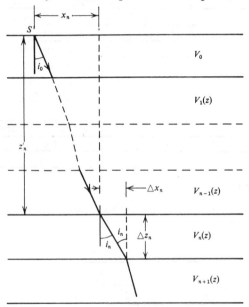

Fig. 4.31 Raypath where velocity varies with depth.

In the limit when n becomes infinite, we get

$$\frac{\sin i}{V} = \frac{\sin i_0}{V_0} = p, \ V = V(z), \tag{4.42}$$

$$\frac{dx}{dz} = \tan i, \frac{dt}{dz} = \frac{1}{V \cos i},$$

$$x = \int_0^z \tan i \, dz, \ t = \int_0^z \frac{dz}{V \cos i};$$

hence

$$x = \int_0^z \frac{pV\,dz}{\{1 - (pV)^2\}^{\frac{1}{2}}},$$

$$t = \int_0^z \frac{dz}{V\{1 - (pV)^2\}^{\frac{1}{2}}}.$$

(4.43)

Since V is a function of z, eq. (4.43) furnishes two integral equations relating x and t to the depth z. These equations can be solved by numerical methods when we have a table of values of V at various depths.

(e) *Linear increase of velocity with depth*. Sometimes we can express V as a continuous function of z and integrate eqs. (4.43). One case of considerable importance is that of a linear increase of velocity with depth, namely

$$V = V_0 + kz,$$

where V_0 is the velocity at the horizontal datum plane, V the velocity at a depth z below the datum plane and k a constant whose value is generally between $0\cdot3$ and $1\cdot3$ per sec.

If we introduce a new variable $u = pV = \sin i$, then $du = pdV = pk\,dz$, and we can solve for x and t as follows:

$$x = \frac{1}{pk} \int_{u_0}^{u} \frac{u\,du}{\sqrt{(1 - u^2)}} = \frac{1}{pk} \sqrt{(1 - u^2)} \Big|_{u}^{u_0} = \frac{1}{pk} \cos i \Big|_{i}^{i_0} =$$

$$= \frac{1}{pk}(\cos i_0 - \cos i), \quad (4.44a)$$

$$t = \frac{1}{k} \int_{u_0}^{u} \frac{du}{u\sqrt{(1 - u^2)}} = \frac{1}{k} \log\left\{\frac{u}{1 + \sqrt{(1 - u^2)}}\right\}\Big|_{u_0}^{u} =$$

$$= \frac{1}{k} \log\left(\frac{\sin i}{\sin i_0} \cdot \frac{1 + \cos i_0}{1 + \cos i}\right) = \frac{1}{k}\log\left(\frac{\tan i/\,2}{\tan i_0/2}\right); \quad (4.44b)$$

hence,

$$i = 2 \tan^{-1}(e^{kt} \tan i_0/2), \quad (4.44c)$$

$$z = \frac{1}{k}(V - V_0) = \frac{1}{pk}(\sin i - \sin i_0). \quad (4.44d)$$

The parametric eqs. (4.44a) and (4.44d) give the coordinates x and z, the parameter i being related to the one-way traveltime t by eq. (4.44b) or (4.44c).

The raypath given by eqs. (4.44a) and (4.44d) is a circle; this can be shown by calculating the radius of curvature ρ which turns out to be a constant:

$$\rho = \frac{(1 + x'^2)^{\frac{3}{2}}}{x''},$$

where

$$x' = \frac{dx}{dz} = \tan i, \text{ using (4.44a) and (4.44d)},$$

$$x'' = \frac{d^2x}{dz^2} = \frac{d}{di}(\tan i)\frac{di}{dz} = \sec^2 i \frac{di}{dz} = pk \sec^3 i, \text{ using (4.44d)}.$$

Hence

$$\rho = \frac{(1 + \tan^2 i)^{\frac{3}{2}}}{pk \sec^3 i} = \frac{1}{pk} = \left(\frac{V_0}{k}\right)\frac{1}{\sin i_0} = \text{constant}.$$

Figure 4.32 shows a ray leaving the shotpoint at the angle i_0. The centre, O, of the circular ray lies above the surface a distance $\rho \sin i_0$, that is, V_0/k. Since this is independent of i_0, the centres of all rays lie on the same horizontal line. This line is located where the velocity would be zero if the velocity function were extrapolated up into the air (since $z = -V_0/k$ at this elevation).

To determine the shape of the wavefront, we make use of fig. 4.33. The raypaths SA and SB are circular arcs with centres O_1 and O_2 respectively. If we continue the

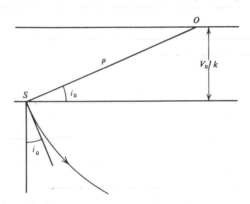

Fig. 4.32 Circular ray leaving the shotpoint at the angle i_0.

arcs upwards to meet the vertical through S at the point S', the line O_1O_2 bisects $S'S$ at right angles. Next we select any point C on the downward extension of $S'S$ and draw the tangents to the two arcs, CA and CB. From plane geometry we know that the square of the length of a tangent to a circle from an external point (for example, CA^2) is equal to the product of the two segments of any chord drawn from the same point (CS . CS' in fig. 4.33). Using both circles we see that

$$CS . CS' = CA^2 = CB^2,$$

hence $CA = CB$. Thus a circle with centre C and radius $R = CA$ cuts the two raypaths at right angles. Since SA and SB can be any raypaths and a wavefront is a surface which meets all rays at right angles, the circle with centre C must be the wavefront which passes through A and B. Even though the arc SA is longer than

SB, the greater path length is exactly compensated for by the higher velocity at the greater depth of the raypath *SA*.

We can draw the wavefront for any value of *t* if we can obtain the values of *h* and *R* in fig. 4.33. Thus, the quantities *h* and *R* are equal to the values of *z* and *x* for a

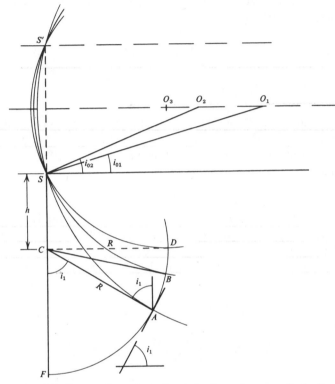

Fig. 4.33 Construction of wavefronts for linear increase of velocity.

ray which has $i = \pi/2$ at the time *t*, that is, *SD* in the diagram. Substitution of $i = \pi/2$ in eqs. (4.44a), (4.44c) and (4.44d) yields

$$\tan i_0/2 = e^{-kt}, \quad \sin i_0 = \operatorname{sech} kt, \quad \cos i_0 = \tanh kt,$$

$$
\left.
\begin{aligned}
h &= \frac{1}{pk}(1 - \sin i_0) = \frac{V_0}{k}\left(\frac{1}{\sin i_0} - 1\right) \\
&= \frac{V_0}{k}(\cosh kt - 1), \\
R &= \frac{1}{pk}\cos i_0 = \frac{V_0}{k}\cot i_0 = \frac{V_0}{k}\sinh kt.
\end{aligned}
\right\}
\tag{4.45}
$$

Equation (4.45) shows that the centre of the wavefront moves downward and the radius becomes larger as time increases.

Field measurements yield values of the arrival time at the shotpoint t_0 and the dip moveout Δt_d. Since the ray which returns to the shotpoint must have encountered a reflecting horizon normal to the raypath and retraced its path back to the point of origin, the dip is equal to the angle i_1 at the time $t = t_0/2$. Thus, to locate the segment of reflecting horizon corresponding to a set of values of t_0 and Δt_d, we make the following calculations:

$$(a)\ t = t_0/2,\ (b)\ i_0 = \sin^{-1}\left(V_0\frac{\Delta t_d}{\Delta x}\right),\ (c)\ i_1 = 2\tan^{-1}(e^{kt}\tan i_0/2),$$

$$(d)\ h = \frac{V_0}{k}(\cosh kt - 1),\ (e)\ R = \frac{V_0}{k}\sinh kt.$$

With these values we find C, lay off the radius R at the angle i_1, and draw the reflecting segment perpendicular to the radius as shown at the point A in fig. 4.33. This method is easily adapted to a simple plotting machine (Daly, 1948) which will be discussed in §4.5.7d.

4.3.3 *Geometry of refraction paths*

(a) *Head waves.* In refraction seismology we make use of waves which have been refracted at the critical angle; these waves are often called *head waves*. In fig. 4.34a we see a P-wave incident on the refracting horizon at the critical angle θ_c. After refraction it travels along the interface in the lower medium. This produces an

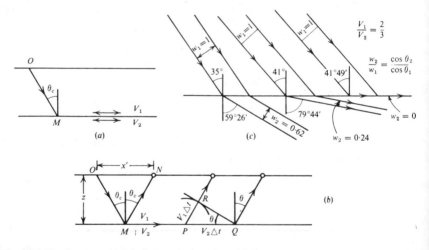

Fig. 4.34 Head waves. (a) Motion at the interface; (b) wavefront emerging from refractor at critical angle; (c) changes in beam width upon refraction.

oscillatory motion parallel to and immediately below the interface (as shown by the double-headed arrow just below the interface). Since relative motion between the two media is not possible, the upper medium is forced to move in phase with the lower medium. The disturbance in the upper medium travels along the interface with the same velocity V_2 as the refracted wave just below the interface. Let us assume that these disturbances represented by the arrows reach the point P in fig. 4.34b at the time t. According to Huygens' principle, P then becomes a centre from which a wave spreads out into the upper medium. After a further time interval Δt, this wave has a radius of $V_1 \Delta t$ while the wave moving along the refractor has reached Q, PQ being equal to $V_2 \Delta t$. Drawing the tangent from Q to the arc of radius $V_1 \Delta t$, we obtain the wavefront RQ. Hence the passage of the refracted wave along the interface in the lower medium generates a plane wave travelling upward in the upper medium at the angle θ, where

$$\sin \theta = \frac{V_1 \Delta t}{V_2 \Delta t} = \frac{V_1}{V_2}.$$

Thus we see that $\theta = \theta_c$, so that the two inclined portions of the path are symmetrically disposed with respect to the normal to the refractor.

The preceding discussion of the geometry of head waves ignores an important point which the alert reader may already have noted, namely that the transmission coefficient T_P in fig. 4.18b is zero at θ_c, hence one would expect that the head wave carries no energy. In general T_P and T_S vanish at the critical angles, not because the amplitude of the refracted wave vanishes, but because the width of the refracted beam approaches zero as the angle of refraction approaches $90°$ (see fig. 3.34c). Since head waves exist and frequently are very strong, we are thus faced with the paradox of a theory which predicts the correct geometry but states that the intensity is zero. Zoeppritz' equations, from which we derived T_P and T_S, assume that the incident wave is plane. A more complete theory based upon curved wavefronts (Grant and West, chap. 6, 1965) predicts the existence of head waves with non-zero intensities and having the same geometrical relations as those predicted by plane-wave theory.

Obviously the head wave will not be observed at offsets less than ON in fig. 4.34b where

$$x' = ON = 2z \tan \theta_c = 2z \tan \left\{ \sin^{-1} \left(\frac{V_1}{V_2} \right) \right\} = 2z \left\{ \left(\frac{V_2}{V_1} \right)^2 - 1 \right\}^{-\frac{1}{2}}. \quad (4.46)$$

As the ratio V_2/V_1 increases, x' decreases. When V_2/V_1 equals $1 \cdot 4$, x' is equal to $2z$. Hence as a rule of thumb, refractions are observed only at offsets greater than twice the depth to the refractor.

(b) *Single horizontal refractor.* In the case of a single horizontal refracting horizon, we can readily derive a formula expressing the arrival time in terms of the offset, the depth and the velocities. In fig. 4.35, the lower part shows a horizontal plane refractor separating two beds of velocities V_1 and V_2, where $V_2 > V_1$. For a geophone at R, the path of the refracted wave is $OMPR$, θ being the critical angle.

The traveltime t can be written

$$t = \frac{OM}{V_1} + \frac{MP}{V_2} + \frac{PR}{V_1} = \frac{MP}{V_2} + 2\frac{OM}{V_1}$$

$$= \frac{x - 2z \tan \theta}{V_2} + \frac{2z}{V_1 \cos \theta}$$

$$= \frac{x}{V_2} + \frac{2z}{V_1 \cos \theta}\left(1 - \frac{V_1}{V_2}\sin \theta\right)$$

$$= \frac{x}{V_2} + \frac{2z \cos \theta}{V_1}, \tag{4.47a}$$

where we have used the relation $\sin \theta = V_1/V_2$ in the last step. This equation can also be written

$$t = \frac{x}{V_2} + t_1, \tag{4.47b}$$

where

$$\left.\begin{array}{l} t_1 = (2z \cos \theta)/V_1, \\[2mm] z = \tfrac{1}{2}V_1 t_1/\cos \theta. \end{array}\right\} \tag{4.47c}$$

or

Equations (4.47a) and (4.47b) represent a straight line of slope $1/V_2$ and *intercept time t_1*. This is illustrated in fig. 4.35 where $OMQ, OMP'R', OMPR$ and

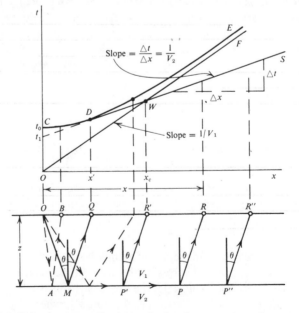

Fig. 4.35 Relation between reflection and refraction raypaths and traveltime curves.

$OMP''R''$ are a series of refraction paths and DWS the corresponding time-distance curve. Note that this straight-line equation does not have physical meaning for offsets less than x' since the refracted wave does not exist for such values of x; nevertheless we can project the line back to the time axis to find t_1.

The problem to be solved usually is to find the depth z and the two velocities V_1 and V_2. The slope of the direct-wave time-distance curve is the reciprocal of V_1 and the same measurement for the refraction event gives V_2. We can then calculate the critical angle θ from the relation $\theta = \sin^{-1}(V_1/V_2)$, and use the intercept time, t_1, to calculate z from eq. (4.47c).

In fig. 4.35 the time-distance curves for the reflection from the interface AP'' and for the direct path are represented by the hyperbola CDE and the straight line OF, respectively. Since the path OMQ can be regarded either as a reflection or as the beginning of the refracted wave, the reflection and refraction time-distance curves must coincide at $x = x'$, that is, at the point D. Moreover, differentiating eq. (4.31a) to obtain the slope of the reflection time-distance curve at $x = x'$, we find

$$\left[\frac{dt}{dx}\right]_{x=x'} = \left[\frac{x}{V_1^2 t}\right]_{x=x'} = \frac{1}{V_1}\left[\frac{OQ}{OM + MQ}\right] = \frac{1}{V_1}\left(\frac{\frac{1}{2}OQ}{OM}\right) = \frac{1}{V_1}\sin\theta = \frac{1}{V_2}.$$

We see therefore that the reflection and refraction curves have the same slope at D, and consequently the refraction curve is tangent to the reflection curve at $x = x'$.

Comparing reflected and refracted waves from the same horizon and arriving at the same geophone, we note that the refraction arrival time is always less than the reflection arrival time (except at D). The intercept time t_1 for the refraction is less than the arrival time t_0 for the reflection at the shotpoint, since

$$t_1 = \frac{2z}{V_1}\cos\theta, \ t_0 = \frac{2z}{V_1}; \text{ hence } t_1 < t_0.$$

Starting at the point Q, we see that the direct wave arrives ahead of the reflected and refracted waves since its path is the shortest of the three. However, part of the refraction path is traversed at velocity V_2, so that as x increases, eventually the refraction wave will overtake the direct wave. In fig. 4.35 these two traveltimes are equal at the point W. If the offset corresponding to W is x_c, we have

$$\frac{x_c}{V_1} = \frac{x_c}{V_2} + \frac{2z}{V_1}\cos\theta$$

$$z = \frac{x_c}{2}\left(1 - \frac{V_1}{V_2}\right)/\cos\theta = \frac{x_c}{2}\left(\frac{V_2 - V_1}{V_2}\right)\frac{V_2}{(V_2^2 - V_1^2)^{\frac{1}{2}}}$$

$$z = \frac{x_c}{2}\left(\frac{V_2 - V_1}{V_2 + V_1}\right)^{\frac{1}{2}}. \tag{4.48}$$

This relation sometimes is used to find z from measurements of the velocities and the *crossover distance* x_c. However, usually we can determine t_1 more accurately than x_c and hence eq. (4.47c) provides a better method of determining z.

(c) *Several horizontal refractors.* Where all beds are horizontal, eqs. (4.47a), (4.47b) and (4.47c) can be generalized to cover the case of more than one refracting horizon. Consider the situation in fig. 4.36 where we have three beds of velocities, V_1, V_2 and V_3. Whenever $V_2 > V_1$, we have the refraction path $OMPR$ and corresponding time-distance curve WS, just as we had in fig. 4.35. If $V_3 > V_2 > V_1$, travel by a refraction path in V_3 will eventually overtake the refraction in V_2.

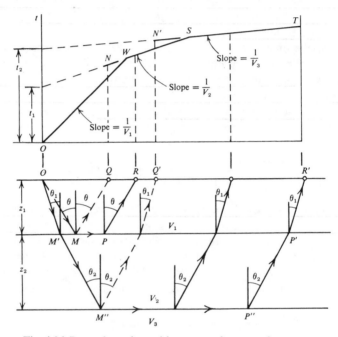

Fig. 4.36 Raypaths and traveltime curves for two-refractor case.

The refraction paths such as $OM'M''P''P'R'$ are fixed by Snell's law:

$$\frac{\sin\theta_1}{V_1} = \frac{\sin\theta_2}{V_2} = \frac{1}{V_3},$$

where θ_2 is the critical angle for the lower horizon while θ_1 is less than the critical angle for the upper horizon. The expression for the traveltime curve ST is obtained as before:

$$
\begin{aligned}
t &= \frac{OM' + R'P'}{V_1} + \frac{M'M'' + P'P''}{V_2} + \frac{M''P''}{V_3} \\
&= \frac{2z_1}{V_1\cos\theta_1} + \frac{2z_2}{V_2\cos\theta_2} + \frac{x - 2z_1\tan\theta_1 - 2z_2\tan\theta_2}{V_3} \\
&= \frac{x}{V_3} + \frac{2z_2}{V_2\cos\theta_2}\left(1 - \frac{V_2}{V_3}\sin\theta_2\right) + \frac{2z_1}{V_1\cos\theta_1}\left(1 - \frac{V_1}{V_3}\sin\theta_1\right)
\end{aligned}
$$

$$= \frac{x}{V_3} + \frac{2z_2}{V_2} \cos \theta_2 + \frac{2z_1}{V_1} \cos \theta_1 \qquad (4.49a)$$

$$= \frac{x}{V_3} + t_2 . \qquad (4.49b)$$

Thus the time-distance curve for this refraction is also a straight line whose slope is the reciprocal of the velocity just below the refracting horizon and whose intercept is the sum of terms of the form $(2z_i/V_i \cos \theta_i)$, each bed above the refracting horizon contributing one term. We can generalize for n beds:

$$t = \frac{x}{V_n} + \sum_i \frac{2z_i}{V_i} \cos \theta_i, \qquad (4.50)$$

where $\theta_i = \sin^{-1} (V_i/V_n)$. This equation can be used to find the velocities and thicknesses of each of a series of horizontal refracting beds, each of constant velocity, provided each bed contributes enough of the time-distance curve to permit it to be analysed correctly. Then we can find all of the velocities (hence the angles θ_i also) by measuring the slopes of the various sections of the time-distance curve and then get the thicknesses of the beds from the intercepts.

(d) *Effect of refractor dip.* The simple situations on which eqs. (4.46) to (4.50) are based are frequently not valid. One of the most serious defects is the neglect of dip

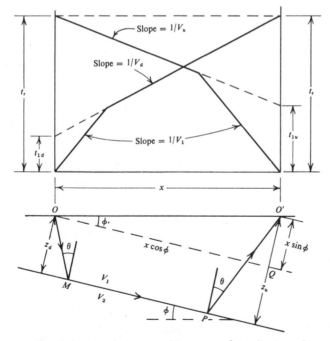

Fig. 4.37 Raypaths and traveltime curves for a dipping refractor.

since dip changes the refraction time-distance curve drastically. The lower part of fig. 4.37 shows a vertical dip-section through a refracting horizon. Let t be the traveltime for the refraction path $OMPO'$. Then, we have

$$
\begin{aligned}
t &= \frac{OM + O'P}{V_1} + \frac{MP}{V_2} \\
&= \frac{z_d + z_u}{V_1 \cos \theta} + \frac{OQ - (z_d + z_u) \tan \theta}{V_2} \\
&= \frac{x \cos \phi}{V_2} + \frac{z_d + z_u}{V_1} \cos \theta.
\end{aligned}
\tag{4.51}
$$

If we place the shotpoint at O and a detector at O', we are 'shooting down-dip'. In this case it is convenient to have t in terms of the distance from the shotpoint to the refractor z_d; hence we eliminate z_u using the relation

$$
z_u = z_d + x \sin \phi.
$$

Writing t_d for the down-dip traveltime, we obtain

$$
\begin{aligned}
t_d &= \frac{x \cos \phi}{V_2} + \frac{x}{V_1} \cos \theta \sin \phi + \frac{2z_d}{V_1} \cos \theta \\
&= \frac{x}{V_1} \sin (\theta + \phi) + \frac{2z_d}{V_1} \cos \theta
\end{aligned}
$$

where

$$
\left.
\begin{aligned}
t_d &= \frac{x}{V_1} \sin (\theta + \phi) + t_{1d}, \\
t_{1d} &= \frac{2z_d}{V_1} \cos \theta.
\end{aligned}
\right\}
\tag{4.52a}
$$

The result for shooting in the up-dip direction is similarly obtained by eliminating z_d:

where

$$
\left.
\begin{aligned}
t_u &= \frac{x}{V_1} \sin (\theta - \phi) + t_{1u}, \\
t_{1u} &= \frac{2z_u}{V_1} \cos \theta.
\end{aligned}
\right\}
\tag{4.52b}
$$

Note that the down-dip traveltime from O to O' is equal to the up-dip traveltime from O' to O; this shotpoint-to-shotpoint traveltime is called the *reciprocal time* and is denoted by t_r.

These equations can be expressed in the same form as (4.47b):

$$t_d = \frac{x}{V_d} + t_{1d}, \tag{4.52c}$$

$$t_u = \frac{x}{V_u} + t_{1u}, \tag{4.52d}$$

where

$$V_d = \frac{V_1}{\sin(\theta + \phi)}, \quad V_u = \frac{V_1}{\sin(\theta - \phi)}. \tag{4.52e}$$

V_d and V_u are known as *apparent velocities* and are given by the reciprocals of the slopes of the time-distance curves.

For reversed profiles such as shown in fig. 4.37, eq. (4.52e) can be solved for the dip ϕ and the critical angle θ (and hence for the refractor velocity V_2):

$$\left. \begin{aligned} \theta &= \tfrac{1}{2}\left\{ \sin^{-1}\left(\frac{V_1}{V_d}\right) + \sin^{-1}\left(\frac{V_1}{V_u}\right) \right\}, \\ \phi &= \tfrac{1}{2}\left\{ \sin^{-1}\left(\frac{V_1}{V_d}\right) - \sin^{-1}\left(\frac{V_1}{V_u}\right) \right\}. \end{aligned} \right\} \tag{4.53}$$

The distances to the refractor, z_d and z_u, can then be found from the intercepts using (4.52a) and (4.52b).

Equation (4.52e) can be simplified where ϕ is small enough that we can approximate by letting $\cos\phi \approx 1$ and $\sin\phi \approx \phi$. With this simplification, (4.52e) becomes

$$\frac{V_1}{V_d} = \sin(\theta + \phi) \approx \sin\theta + \phi\cos\theta,$$

$$\frac{V_1}{V_u} = \sin(\theta - \phi) \approx \sin\theta - \phi\cos\theta;$$

hence

$$\sin\theta = \frac{V_1}{V_2} \approx \frac{V_1}{2}\left(\frac{1}{V_d} + \frac{1}{V_u}\right),$$

so that

$$\frac{1}{V_2} \approx \frac{1}{2}\left(\frac{1}{V_d} + \frac{1}{V_u}\right). \tag{4.54a}$$

An even simpler approximate formula for V_2 (although slightly less accurate) can be obtained by applying the binomial theorem to eq. (4.52e) and assuming that ϕ is small enough that higher powers of ϕ are negligible:

$$V_d = \frac{V_1}{\sin\theta}(\cos\phi + \cot\theta\sin\phi)^{-1} \approx V_2(1 - \phi\cot\theta),$$

$$V_u \approx V_2(1 + \phi\cot\theta),$$

hence

$$V_2 \approx \tfrac{1}{2}(V_d + V_u). \tag{4.54b}$$

Equations similar to (4.52) can be derived for the situation of a number of beds which have the same dip and strike or other specific situations but such equations are of limited value in practice. Not only do they involve a large amount of computation but also one usually is not sure that they are applicable to a specific real situation. Moreover where there are more than two refracting horizons it is often difficult to identify equivalent up-dip and down-dip segments, especially if the refractors are not plane or if the dip and strike change.

4.4 Characteristics of seismic events

4.4.1 *Distinguishing features of events*

The basic task of interpreting reflection records is that of selecting those events on the record which represent primary reflections, translating the arrival times for these reflections into depths and dips and mapping the reflecting horizons. In addition, the interpreter must be alert to other types of events which may yield valuable information, such as multiple reflections and diffractions.

Recognition and identification of seismic events are based upon five characteristics: (*a*) coherence, (*b*) amplitude standout, (*c*) character, (*d*) dip moveout and (*e*) normal moveout. The first of these is by far the most important in recognizing an event. Whenever a wave recognizable as such reaches a spread, it produces approximately the same effect on each geophone. If the wave is strong enough to override other energy arriving at the same time, the traces will look more-or-less alike during the interval in which this wave is arriving; this similarity in appearance from trace to trace is called *coherence* (see fig. 4.38) and is a necessary condition for the recognition of any event. *Amplitude standout* refers to an increase of amplitude such as results from the arrival of coherent energy; it is not always

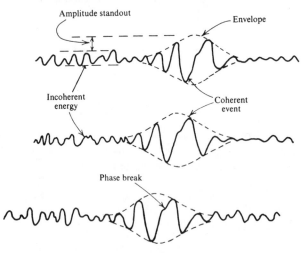

Fig. 4.38 Characteristics of seismic events.

marked, especially when AGC (see §4.5.4e) is used in recording. *Character* refers to a distinctive appearance of the waveform which identifies a particular event; it involves primarily the shape of the envelope, the number of cycles showing amplitude standout and irregularities in phase resulting from interference between components of the event. Moveout, which refers to a systematic difference from trace to trace in the arrival time of an event, has been discussed in §4.3.1b and §4.3.1c.

Coherence and amplitude standout tell us whether or not a strong seismic event is present, but they say nothing about the type of event. The most distinctive criterion for identifying the nature of events is moveout. Character is also often very useful, especially the frequency content and the number of cycles observed.

4.4.2 *Reflections; refractions*

Reflections exhibit normal moveouts which must fall within certain limits set by the velocity distribution. The dip moveout is usually small but occasionally reflections have large dip moveouts (as with fault-plane reflections). Reflection events rarely involve more than two or three cycles and are often rich in frequency components in the range 20–50 Hz; deep reflections at times have considerable energy even below this range.

Refractions are relatively low frequency and usually include more cycles than reflections. Refractions and *reflected refractions* (refracted waves which are reflected back towards the spread, see fig. 4.39a) generally have straight alignments (prior to normal moveout corrections) in contrast with the curved

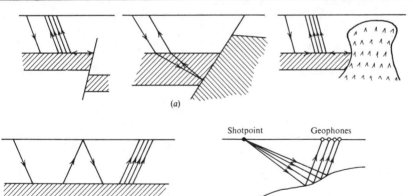

(a)

Shotpoint   Geophones

(b)   (c)

Fig. 4.39 Reflected refractions. (a) Reflected refractions from faults and salt dome; (b) multiply-reflected refraction; (c) broadside reflected refraction (plan view).

alignments of reflections and diffractions. Broadside reflected refractions (fig. 4.39c) have normal moveout appropriate to the refractor velocity. Refractions from deep refractors are not observed on reflection records except when the offsets are unusually long or when the occasional reflected refraction is recorded.

One very powerful technique for distinguishing between reflections, diffractions, reflected refractions and multiples is to play back the magnetic tapes after correcting for (*a*) weathering and elevation (*static corrections*, since the correction is the same for all arrival times on a given trace; see §4.5.7*b*), and (*b*) normal moveout (*dynamic corrections*, since the amount of the correction decreases with increasing arrival time). Provided the correct normal moveout was removed, reflections appear (fig. 4.40) as straight lines while diffractions and multiples will still have some curvature (since their normal moveouts are larger than those of primary reflections) and refractions and other formerly straight alignments will have inverse curvature.

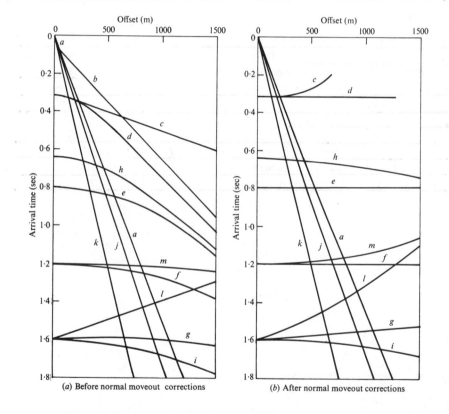

(*a*) Before normal moveout corrections   (*b*) After normal moveout corrections

Fig. 4.40 Types of events on a seismic record. Identities of events are: a = direct wave, V = 650 m/sec; b = refraction at base of weathering, V_H = 1640 m/sec; c = refraction from flat refractor, V_R = 4920 m/sec; d = reflection from refractor in c, V_{av} = 1640 m/sec; e = reflection from flat reflector, V_{av} = 1970 m/sec; f = reflection from flat reflector, V_{av} = 2300 m/sec; g = reflection from dipping reflector, V_{av} = 2630 m/sec; h = multiple of d; i = multiple of e; j = ground roll, V = 575 m/sec; k = air wave, V = 330 m/sec; l = reflected refraction from in-line disruption of refractor of c; m = reflected refraction from broadside disruption of refractor of c. After proper normal-moveout corrections, the primary reflections are straight.

4.4.3 *Diffractions*

Diffractions are indistinguishable from reflections on the basis of character. The amplitude of a diffraction is a maximum at some point along the line of profiling and decreases rapidly as we go away from this point (although this variation of amplitude is usually obscured in recording). Diffractions usually exhibit distinctive moveout. In fig. 4.41, for all shotpoint and geophone positions such that the point

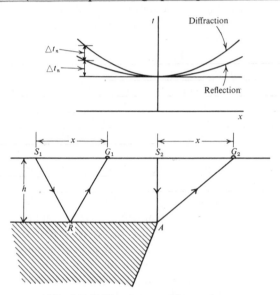

Fig. 4.41 Diffraction traveltime curve.

of reflection R is to the left of A, the reflection traveltime curve is given by eq. (4.31c), that is

$$t_r = \frac{1}{V}(x^2 + 4h^2)^{\frac{1}{2}} \approx \frac{2h}{V} + \frac{x^2}{4Vh} = t_0 + \Delta t_n,$$

assuming that x is smaller than h. The reflection traveltime curve is a hyperbola as shown in fig. 4.24. For the case where the shotpoint is directly above the diffraction source A, the diffraction traveltime curve is given by the equation

$$t_d = \frac{1}{V}\{h + (x^2 + h^2)^{\frac{1}{2}}\} \approx \frac{2h}{V} + \frac{x^2}{2Vh} = t_0 + 2\Delta t_n. \tag{4.55}$$

Thus, the normal moveout for the diffraction shown in fig. 4.41 is twice that of a reflection at the same offset; the reflection corresponds to a virtual source at a depth of $2h$ while the diffraction comes from a source at depth of h. The earliest arrival time on a diffraction curve is for the trace which is recorded directly over the diffracting point (except for unusual velocity-distribution situations) but the diffraction will not necessarily have its peak amplitude on this trace.

4.4.4 *Multiples*

The reflection-seismic technique is based upon the assumption of simple wave paths such as those in fig. 4.24 and 4.25. However, more complex reflection events are recorded, such as the multiples shown in fig. 4.19. The strongest multiples involve reflection at the surface or at the base of the *LVL* (see §4.2.2*n*) where the reflection coefficient is very large because of the large acoustic-impedance contrast. Hence the amplitude of this type of multiple depends primarily upon the reflection coefficients at depth. Since at least two reflections at depth are involved, multiples will be observed as distinctive events when these coefficients are abnormally high. It is important that multiples be recognized as such so that they will not be interpreted as reflections from deeper horizons.

Because velocity generally increases with depth, multiples usually exhibit more normal moveout than primary reflections with the same traveltime. This is the basis of the attenuation of multiples in common-depth-point processing which will be discussed in §4.7.7*b*. However, the difference in normal moveout often is not large enough to identify multiples.

The effect of dip on multiples which involve the surface or the base of the *LVL* can be seen by tracing rays using the method of images. In fig. 4.42, we trace a multiple arriving at symmetrically disposed geophones, G_1 and G_{24}. The first image point, I_1, is on the perpendicular from S to AB as far below AB as S is above. We next draw the perpendicular from I_1 to the surface of the ground where the second reflection occurs and place I_2 as far above the surface as I_1 is below. Finally, we locate I_3 on the perpendicular to AB as far below as I_2 is above. We can now draw the rays from the source S to the geophones (working backwards from the geophones). The dip moveout is the difference between the path lengths I_3G_{24} and I_3G_1; it is about double that of the primary ($I_1G_{24} - I_1G_1$). The multiple at the shotpoint will appear to come from I_3 which is up-dip from I_1, the image point for the primary, and I_3S is slightly less than twice I_1S. Hence we can see that if the reflector dips, the multiple involves a slightly different portion of the reflector than the primary and has a traveltime slightly less than double the traveltime of the primary. The latter fact makes identifying multiples by merely doubling the arrival time of the primary imprecise whenever appreciable dip is present. The arrival time of the multiple will be approximately equal to that of a primary reflection from a bed at the depth of I_1. If the actual dip at I_1 is not double that at AB (and one would not in general expect such a dip), then the multiple will appear to have anomalous dip. If the multiple should be misidentified as a primary, one might incorrectly postulate an unconformity or up-dip thinning which might lead to erroneous geologic conclusions.

Ghosts are the special type of multiple illustrated in fig. 4.19. The energy travelling downward from the shot has superimposed upon it energy which initially travelled upward and was then reflected downward at the base of the *LVL* (in land surveys) or at the surface of the water (in marine surveys). A 180° phase shift, equivalent to half a wavelength, occurs at the additional reflection and hence the effective path difference between the direct wave and the ghost is ($\lambda/2 + 2d$), where d is the depth of the shot below the reflector producing the ghost. The interference between the ghost and the primary depends on the fraction of a wavelength represented by the difference in effective path length; since the seismic wavelet is made up of a range of frequencies, the interference effect will vary for the different components. Thus, the overall effect on the wavelet shape will vary as d is varied. Relatively small changes in shot depth can result in large variations in reflection character, creating serious problems for the interpreter. Therefore the depth of the shot below the base of the weathering or the surface of the water is maintained as nearly constant as possible.

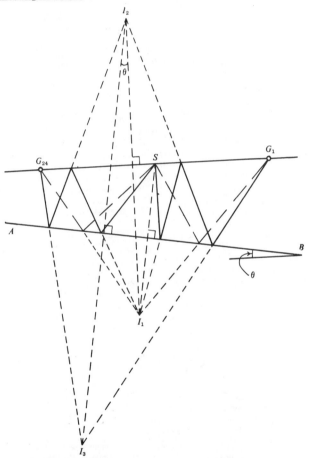

Fig. 4.42 Raypath of a multiple from a dipping bed.

Ghosts are especially important in marine surveys because the surface of the water is almost a perfect reflector and consequently the ghost interference will be strong. If d is small in comparison with the dominant wavelengths, appreciable signal cancellation will occur. At depths of 30 to 50 feet interference is constructive for frequencies of 40 to 25 Hz which is in the usual seismic range. The same effect occurs with the upcoming signal from the reflectors we wish to map. Hence marine sources and marine detectors often are operated at such depths.

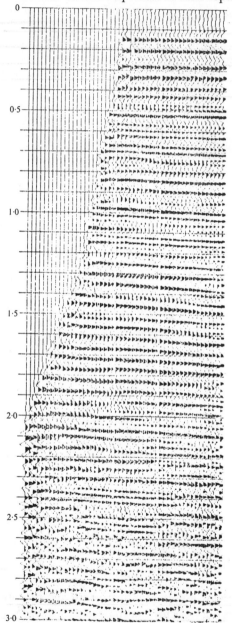

Fig. 4.43 Seismic record showing singing. Compare with fig. 4.101. (Courtesy Petty-Ray Geophysical.)

A particularly troublesome type of multiple produces the coherent noise known as *singing* (also called *ringing* or *water reverberation*) which is frequently encountered in marine work (and occasionally on land). This is due to multiple reflections in the water layer. The large reflection coefficients at the top and bottom of this layer result in considerable energy being reflected back and forth repeatedly, the reverberating energy being reinforced periodically by reflected energy. Depending upon the water depth, certain frequencies are enhanced and as a result the record looks very sinusoidal (see fig. 4.43). Not only is the picking of reflections difficult but measured traveltimes and dip moveouts will probably be in error. This type of noise and its attenuation are discussed in §4.7.3*d*.

4.4.5 *Surface waves*

Surface waves (often called *ground roll*) are usually present on reflection records. For the most part these are Rayleigh waves with velocities ranging from 100 to 1000 m/sec or so. Ground roll frequencies usually are lower than those of reflections and refractions, often with the energy concentrated below 10 Hz. Ground roll alignments are straight, just as in the case of refractions, but they represent much lower velocities. The envelope of ground roll builds up and decays very slowly and often includes many cycles. Surface wave energy generally is high enough even in the reflection band to override all but the strongest reflections; however, because of the low velocity, different geophone groups are affected at different times so that only a few groups are affected at any one time. Sometimes there is more than one ground roll wavetrain, each with different velocities. Occasionally where surface waves are exceptionally strong, in-line offsets are used to permit recording the desired reflections before the surface waves reach the spread.

4.4.6 *Effects of reflector curvature*

Geometrical focusing as a result of curvature of a reflector affects the amplitude of a reflection. Over anticlines the reflected raypaths diverge more than where reflectors are plane, resulting in reduced energy density. Many recording systems change the amplitude in such a way that these variations of amplitude are obscured. However, some effects may remain, as in situations where the reflection from a curved surface interferes with other events whose amplitudes are not similarly affected.

Strengthening of reflections occurs over synclines. Focusing of the energy because of the concave-mirror effect makes more of the reflector surface effective for producing a reflection. If the energy is focused on the recording surface, a situation which occurs when the centre of curvature of the reflector lies on the surface, the amplitude is a maximum.

If the curvature of a convex reflector is great enough, the energy focuses below the surface (see the two deeper reflections in fig. 4.44) and a *buried focus* occurs. This means that for a given surface location reflections may be obtained from more than one part of the reflector; the time-distance plot is no longer a simple curve

but has several 'branches', most commonly three. The two deeper reflections in
fig. 4.44 involve buried foci; each shows branches from each flank of the syncline
plus a *reverse branch* from the curved bottom of the syncline. Obviously the likelihood of a buried focus occurring increases with reflector depth.

The waves producing the reverse branch pass through a focus which results in a
90° phase shift relative to waves which do not pass through the focus (see Wood,
1934, p. 35); however, this phase shift is rarely useful in identifying buried focus
effects. Nevertheless it will affect calculations of reflector depth where picking is

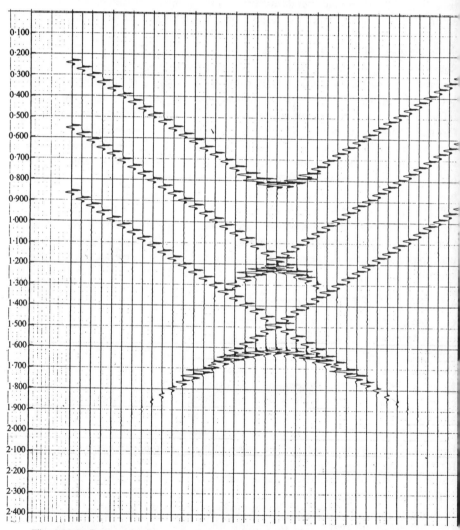

Fig. 4.44 Reflections from curved reflector. In all cases reflector radius of curvature = 1000 m,
$V = 2000$ m/sec. Depths to the bottom of the syncline are 800, 1200 and 1600 m respectively
for the three reflectors. The traces are 100 m apart. (Courtesy Chevron Oil.)

done systematically on the same phase, for example, always in the troughs. The reverse branch is so named because the point of reflection traverses the reflector in the opposite direction from the surface traverse. Thus, in fig. 4.44, as one moves from left to right on the surface the reflection point for the reverse branch moves from right to left.

Just as light can be focused by passing through a lens, seismic waves can also be focused by curved velocity surfaces which result in seismic rays being bent by refraction; such situations are often very complex. Curvature at the base of the weathering can be especially important because of the large velocity contrast usually associated with this surface. Variations in permafrost thickness also cause large effects.

4.4.7 *Types of seismic noise*

The reliability of seismic mapping is strongly dependent upon the quality of the records. However, the quality of seismic data varies tremendously. At one extreme we have areas where excellent reflections (or refractions) are obtained without any special measures being taken; at the other extreme are those areas in which the most modern equipment, extremely complex field techniques and sophisticated data processing methods do not yield usable data (often called *NR areas*, that is, areas of 'no reflections'). In between these extremes lie the vast majority of areas in which useful results are obtained but the quantity and quality of the data could be improved with beneficial results.

We use the term *signal* to denote any event on the seismic record from which we wish to obtain information. Everything else is *noise*, including coherent events which interfere with the observation and measurement of signals. The *signal-to-noise ratio*, abbreviated S/N, is the ratio of the signal energy in a specified portion of the record to the total noise energy in the same portion. Poor records result whenever the signal-to-noise ratio is small; just how small is to some extent a subjective judgment. Nevertheless, when S/N is less than unity, the record quality is usually marginal and deteriorates rapidly as the ratio decreases further.

Seismic noise may be either (*a*) coherent, or (*b*) incoherent. *Coherent noise* can be followed across at least a few traces; *incoherent noise* is dissimilar on all traces and we cannot predict what a trace will be like from a knowledge of nearby traces. Sometimes the difference between coherent and incoherent noise is merely a matter of scale and if we had geophones more closely spaced the incoherent noise would be seen as coherent noise. Nevertheless, incoherent noise is defined with respect to the records being used without regard for what closer spacing might reveal.

Incoherent noise is often referred to as *random noise* (spatially random) which implies not only non-predictability but also certain statistical properties; more often than not the noise is not truly random. (It should be noted that spatial randomness and time randomness may be independent; the usual seismic trace is apt to be random in time since we do not know when a reflection will occur on the basis of what the trace has shown previously, with the exception of multiples.)

Coherent noise is sometimes subdivided into (a) energy which travels essentially horizontally, and (b) energy which reaches the spread more-or-less vertically. Another important distinction is between (a) noise which is repeatable, and (b) noise which is not; in other words, whether the same noise is observed at the same time on the same trace when a shot is repeated. The three properties – coherence, travel direction and repeatability – form the basis of most methods of improving record quality.

Coherent noise includes surface waves, reflections or reflected refractions from near-surface structures such as fault planes or buried stream channels, refractions carried by high-velocity stringers, noise caused by vehicular traffic or farm tractors, multiples, etc. All of the preceding except multiples travel essentially horizontally and all except vehicular noise are repeatable on successive shots.

Incoherent noise which is spatially random and also repeatable is due to scattering from near-surface irregularities and inhomogeneities such as boulders, small-scale faulting, etc.; such noise sources are so small and so near the spread that the outputs of two geophones will only be the same when the geophones are placed almost side-by-side. Non-repeatable random noise may be due to wind shaking a geophone or causing the roots of trees to move generating seismic waves, stones ejected by the shot and falling back to the earth near a geophone, a person walking near a geophone and so on.

4.4.8 *Attenuation of noise*

If the noise has appreciable energy outside the principal frequency range of the signal, frequency filtering can be used to advantage. Very low frequency components (such as high-energy surface waves rich in low frequencies) may be filtered out during the initial recording provided the low frequencies are sufficiently separated from the reflection frequencies. However, the spectrum of the noise often overlaps the signal spectrum and then frequency filtering is of limited value in improving record quality. With modern digital recording, the only low-frequency filtering in the field is often that resulting from the limited low-frequency response of the geophones.

If we add several random noises together, there will be some cancellation because they will be out-of-phase with each other. If they are truly random in the statistical sense, the sum of n random signals will be proportional to $\sqrt{n}$, whereas the sum of n coherent in-phase signals will be proportional to n so that the signal-to-noise ratio will be improved by the factor $\sqrt{n}$. If on the other hand we wish to attenuate coherent noise and distribute (say) 16 geophones uniformly over one wavelength we may achieve a much larger signal-to-noise improvement.

These principles are the bases of the use of multiple geophones or multiple shotpoints (called geophone or shotpoint *arrays*; see §4.5.3c) to cancel noise. If we connect together, for example, 16 geophones which are spaced far enough apart that the noise is spatially random but still close enough together that reflected energy travelling almost vertically is essentially in phase at all 16 geophones, the sum of the 16 outputs will have a signal-to-noise ratio four times greater than the output when the 16 geophones are placed side by side. If on the other hand we are attenuating coherent noise and the 16 geophones are spread evenly over one wavelength of a coherent noise wavetrain (for example, ground roll), then the coherent noise will be greatly reduced.

Noise can also be attenuated by adding together traces shot at different times or different places or both. This forms the basis of several *stacking* techniques including vertical stacking, common-depth-point stacking, uphole stacking and several more complicated methods (see §4.7.7). The gain in record quality often is large because of a reduction in the level of both random and coherent noise. Provided the static and dynamic corrections are accurately made, signal-to-noise improvements for random noise should be about 2·5 for 6-fold and about 5 for 24-fold stacking.

Vertical stacking involves combining together several records for which both the source and geophone locations remain the same. It is extensively used with weak surface energy sources and many marine sources (see §4.5.4c and 4.5.5c). Vertical stacking usually implies that no trace-to-trace corrections are applied but that corresponding traces on separate records are merely added to each other. The effect, therefore, is essentially the same as using multiple shots or multiple source units simultaneously. In difficult areas both multiple source units and vertical stacking may be used. In actual practice the surface source is moved somewhat (3–10 m) between the shots. Up to 20 or more separate records may be vertically stacked, but the stacking of many records becomes expensive both in field time and in processing costs whereas the incremental improvement becomes small after the first few. Vertical stacking is often done in the field, sometimes in subsequent processing. Marine vertical stacking rarely involves more than four records because at normal ship speeds the ship moves so far during the shooting that the data are *smeared* when stacked; smearing means that changes in the reflecting points affect the arrival times so much that the signal may be adversely affected by summing (the effect is similar to using a very large geophone or source array).

The *common-depth-point* technique which is extensively used is very effective in attenuating several kinds of noise. The summation traces comprise energy from several shots using different geophone and shotpoint locations. The field technique will be discussed in §4.5.3a and the processing (which is almost always done in a processing centre rather than in the field) in §4.7.7b.

4.5 Reflection field methods and equipment

4.5.1 *Field-crew organization*

(a) *Clients and contractors.* Most geophysical work is performed by seismic contractors for client companies which utilize the geophysical data obtained in their search for oil. Contracting companies often process data and manufacture geophysical equipment both for their own use and for sale to others. Some client companies operate their own field crews in addition to employing contract crews.

The client company usually designates one of its personnel as *client representative* or 'birddog'; he is the communication channel between his company and the party chief. He often interprets the data and is responsible for the client's interests, which include checking that sound techniques are employed and that data quality and crew efficiency are maintained.

(b) *Land-crew organization.* The organization chart for a land seismic crew (or *party*) is shown in fig. 4.45. The crew usually is headed by a *party chief*, who is a

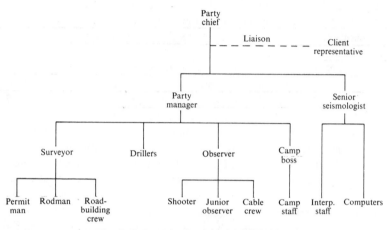

Fig. 4.45 Organization of a land seismic crew.

professional geophysicist. He exercises overall supervision of both field crew and office personnel and is responsible for quality control at all levels, the interpretation of the data and final reports and recommendations based on the work. The party chief also has many non-geophysical responsibilities such as cost control, data security, labour relations, public relations, record keeping, personnel training, safety, etc.

The party chief receives assistance from the supervisor to whom he reports and from contractor specialists in instrument operations, refraction interpretation, etc. The party chief is assisted in the office by a *seismologist* who supervises the office work and does most of the interpretation, and in the field by a party manager who supervises the field work. The party manager's primary responsibility is to obtain maximum production of data of adequate quality at reasonable cost. He employs helpers as the work needs dictate and is responsible for safety, equipment maintenance, maintaining adequate supplies, paying bills and operation of the field camp.

The *observer* is usually the next man in authority in the field. His primary responsibility is to operate the instruments but he is also responsible for the actual field layouts and data acquisition. The observer is usually assisted by a junior observer and by a crew of 'jug hustlers' who actually lay out the cable and the geophones.

The *surveyor* has the responsibility of locating the line and the points along the line in their proper places. As the advance man on the ground, he anticipates difficulties and problems which shooting the line will involve and seeks to avoid or resolve them. This often involves investigating alternative line locations so that the objectives of the survey may be achieved at minimum cost. Frequently he is assisted by a *permit man* who contacts land owners or tenants and secures permission to work on their land. The surveyor is also assisted by rodmen who help measure the seismic line and map the area. In areas of difficult access he may also supervise brush cutters or bulldozer operators who clear the way for the seismic line.

Other members of a field crew vary in numbers depending on the nature of the survey. A crew may have one to four drillers, occasionally more, plus assistants to help drill and haul water for the drilling operation, or one to perhaps four operators of surface source units (see §4.5.4c). A *shooter* is responsible for detonating explosives at the proper time and for cleaning up the shothole area afterwards. Cooks and mechanics may be included where operations are performed out of field camps.

(c) *Marine-crew organization.* A marine seismic crew is usually headed by a party manager who is responsible for the seismic work performed. A marine seismic crew includes several observers and junior observers (who relieve each other during the continuous operations), several technicians who operate the navigation equipment and helpers who handle the streamer (see §4.5.5d) when it is being let out or pulled in.

The seismic ship is under the command of a captain whose authority is final; however, the captain usually follows the instructions of the party manager except where safety is involved. Many supporting personnel are required for a seismic ship: cooks, maintenance engineers, mechanics, etc. Data processing or analysis is usually not done aboard ship.

4.5.2 *Field methods for land surveys*

(a) *The program.* Usually the seismic crew receives the *program* from the client in the form of lines on a map which indicate where data are to be obtained. The seismic crew ordinarily is not responsible for laying out the program, a factor which sometimes contributes to the failure of surveys since it encourages the attitude that the objective is to shoot a certain program rather than to obtain certain information. Without understanding program objectives clearly, the wrong alternatives in operating decisions may be made. Good practice (Agnich and Dunlap, 1959) is to 'shoot the program on paper' before beginning the survey, estimating what the data are likely to show, anticipating problems which may occur, asking what alternatives are available and how data might be obtained which will distinguish between alternative interpretations.

Before beginning a survey the question should be asked, 'Is it probable that the proposed lines will provide the required information?' Data migration (§4.5.7d) may require that lines be located elsewhere than directly on top of features in order to measure critical aspects of a structure. Crestal areas may be so extensively faulted that lines across them may be non-definitive. The structures being sought may be beyond seismic resolving power. Near-surface variations along a proposed line may be so large that the data are difficult to interpret whereas moving the seismic line a short distance may improve data quality. Obstructions to shooting along a proposed line may increase difficulties unnecessarily whereas moving the line slightly may achieve the same objectives at reduced cost. Where the dip is considerable, merely running a seismic line to a wellhead may not tie the seismic data to the well data. Lines may not extend sufficiently beyond faults and other features to establish the existence of such features unambiguously or to determine fault displacements. Lines may cross features such as faults so obliquely that their evidences are not readily interpretable. Lack of cross control may result in features located below the seismic line being confused by features to the side of the line.

(b) *Permitting.* Once the seismic program has been decided upon, it is usually desirable (or necessary) to meet with the owners of the land to be traversed. Permission to enter upon lands to carry out a survey may involve a payment, often a fixed sum per shot-hole, as compensation in advance for 'damages which may be incurred'. Even where the surface owners do not have the right to prevent entry, it is advantageous to explain the nature of the impending operations. Of course, a seismic crew is responsible for damages resulting from their actions whether or not permission is required to carry out the survey.

(c) *Laying out the line.* Once the preliminary operations have been completed, the survey crew lays out the lines to be shot. This is usually done by a transit-and-chain survey which determines the positions and elevations of both the shotpoints and the centres of geophone groups. The *chain* is often a wire equal in length to the geophone group interval. Using this chain the successive group centres are laid out along the line, each centre being marked in a conspicuous manner, commonly by means of brightly coloured plastic ribbon called *flagging*. The transit is used to keep the line straight and to obtain the elevation of each group centre by sighting on a rod carried by the lead chainman.

Many variations from the above procedures are used depending on the sort of terrain being traversed. Plane tables and alidades are sometimes used. Ties to benchmarks and wellheads are often made by transit-theodolite and rod rather than by chaining. Side features are often tied in by triangulation. The surveyor should indicate in his data and maps the locations of all important features such as streams, buildings, roads, fences, etc. The surveyor also plans access routes so that drills, recording trucks, etc., can get to their required locations most expeditiously.

In areas of difficult terrain or heavy vegetation, trail-building or trail-cutting crews may be required. These often precede the survey crew but usually are under the direct supervision of the surveyor who is therefore responsible for the preparation of a straight trail in the proper location.

(d) *Shothole drilling*. The next unit on the scene is the drilling crew (when dynamite is used as the energy source); depending upon the number and depth of holes required and the ease of drilling, a seismic crew will generally have from one to four drilling crews. Whenever conditions permit, the drills are truck-mounted. Water trucks are often required to supply the drills with water for drilling. In areas of rough terrain the drills may be mounted on tractors or portable drilling equipment may be used. In swampy areas the drills are often mounted on amphibious vehicles. Usually the drilling crew places the dynamite in the holes before leaving the site.

(e) *Recording*. The drilling crews are followed by the recording unit. This unit can be divided into three groups on the basis of primary function: the shooting crew who are responsible for loading the shotholes (if the drillers have not already loaded them) and for setting off the dynamite; the jug hustlers who lay out the cables, place the geophones in their proper locations and connect them into the cables; and the recording crew who do the actual recording of the signals.

In the 'conventional' field routine, while the shooting crew are preparing to fire the charge and the jug hustlers are laying out the cables and connecting the geophones, the observer tests and adjusts the amplifiers and other units of the recording system and tests the cables for continuity to ensure that all geophones are connected. Finally, he gives the signal to the shooter via telephone or radio to fire the charge. Just prior to the firing signal, the observer starts the tape transport (see §4.5.4*f*). On receipt of the firing signal the shooter closes two firing switches (two switches are used to reduce the chance of accidental firing) on the *blaster*, the device used to set off the explosive (see §4.5.4*b*). Within a few seconds a micro-switch is closed by the motion of the tape drum, thereby completing the blaster circuit so that the explosion takes place. Immediately after the shot the observer studies a monitor record to see if a second shot is required, perhaps because of improper instrument settings or excessive noise caused by a nearby farm tractor. If a second shot is required, the shooting crew reloads the hole and the process is repeated; if a second shot is not necessary, the shooting and recording equipment are moved to the next location.

With common-depth-point recording shotpoints are close together – of the order of a hundred metres apart versus 400 or 540 metres ($\frac{1}{4}$ or $\frac{1}{3}$ mile) with conventional split-dip recording. The high production and high efficiency needed in order to achieve a low cost per kilometre have altered field procedures. At the same time the redundancy of coverage has lessened the dependence on any individual record so that occasional missed records can be tolerated. Also, the broad dynamic range of digital recording has removed much of the need for filtering in the field and the need to tailor instrument settings to the particular local conditions.

The above considerations dictate that the recording operation must not wait on other units. Contrary to the conventional method, time is no longer spent in repeating shots to improve data quality or on the frequent physical moving of the recording truck. Shotholes may be drilled for the entire line before the recording even begins so that one need never wait on the drills. Extra cables and geophones are laid out and checked out in advance of the recording unit. A *roll-along switch* (see §4.5.3*a*) is used which makes it possible for the recording unit to be located physically at a different place than where it is located electrically. The recording unit connects to the seismic cable at any convenient location, for example, the intersection of a road and the seismic line. The roll-along switch is adjusted so that the proper geophones are connected and the shooters are advised to operate the blaster. Following the shot, the shooters move on to the next shothole (which is not very far away) and the observer adjusts the roll-along switch so that the next geophones are connected. The time between shots may be only a few minutes and the recording truck may not move all day long. Holes where misfires occur are not reloaded and reshot. The shooting unit often walks the line since they need no equipment except the blaster and firing line and perhaps shovels to fill in the shot-hole after the shot. The recording unit does not have to traverse the line and so is subject to less abuse. Damages are reduced because less equipment traverses the line. Thus other benefits accrue besides increased efficiency of recording.

When a seismic crew uses a surface energy source, the source unit moves into place and the signal from the recorder activates the source so that the energy is introduced into the ground at the proper time. Despite the fact that an explosive may not be involved, terms such as 'shot' and 'shotpoint' are still used. The energy from each surface source usually is small compared to the energy from a dynamite explosion so that many records are made for each shotpoint location and vertically stacked to make a single record. Several source units may be used and these may advance a few feet between the component 'sub-shots' which will be combined to make one profile. It is not uncommon to use three or four source trucks and combine thirty or so component sub-shots.

A monitor record is usually made in the field, either in parallel with the recording on magnetic tape or by playback of the magnetic tape. These monitor records are used to make certain that all of the equipment is functioning properly and also to determine weathering corrections (discussed in §4.5.7*b*). However, monitor records usually are not used for interpretation.

The magnetic tapes are shipped to a data processing centre where corrections are applied and various processing techniques are used, for example, velocity analysis, filtering, stacking (see §4.7). The end result of the data processing is usually record sections from which an interpretation is made.

✳ **4.5.3** *Field layouts*

(a) *Split-dip and common-depth-point recording.* Virtually all routine seismic work consists of *continuous profiling*, that is, the cables and shotpoints are arranged so that there are no gaps in the data other than those due to the fact that the geophone groups are spaced at intervals rather than continuously. 'Conventional' coverage implies that each reflecting point is sampled only once except at the ends of each profile; these end points (*tie points*) are sampled again with the adjacent spreads so as to reduce the likelihood of errors in following an event from one record to the next. This is in contrast to common-depth-point, or *redundant, coverage* where each reflecting point is sampled more than once. Areal or cross coverage indicates that the dip components perpendicular to the seismic line have been measured as well as the dip components along the line. Each of these methods can employ various relationships of shotpoints to the geophone groups.

Conventional *split-dip* shooting is illustrated by fig. 4.46. Shotholes are laid out at regular intervals along the line of profiling, often 400 to 540 m apart. A seismic cable which is two shotpoint intervals long is used. Provision is made to connect groups of geophones (often 24 groups) at regular intervals along the cable

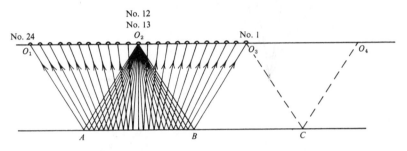

Fig. 4.46 Symmetrical spread with continuous subsurface coverage.

(called the *group interval*). Thus with shotpoints 400 m apart, 24 groups are distributed along 800 m of cable making the group centres $34\frac{3}{4}$ m apart. With the cable stretched from point O_1 to point O_3, shotpoint O_2 is shot; this gives sub-surface control (for flat dip) between A and B. The portion of cable between O_1 and O_2 is then moved between O_3 and O_4 and shotpoint O_3 is shot; this gives subsurface coverage between B and C. The travel path for the last group from shotpoint O_3 is the reversed path for the first group from shotpoint O_2 so that the subsurface coverage is continuous along the line.

Common-depth-point (CDP) or 'roll-along' shooting is illustrated in fig. 4.47a (Mayne, 1962). We have evenly spaced geophone groups which we shall number by their sequence along the seismic line rather than by the trace which they represent on the seismic record. Geophone groups 1 to 24 are connected to the amplifier inputs in the recording truck and shot A is fired. Assuming a horizontal reflector, this gives subsurface coverage from a to g. Geophone groups 3 to 26 are then connected to the amplifier inputs, the change being made by means of the roll-along switch rather than by physically moving the seismic cable. Shot B is then fired giving sub-surface coverage from b to h. Shotpoint C is now fired into geophones 5 to 28 giving coverage from c to i, and so on down the seismic line. Note that the reflecting point for the energy from shot A into geophone group 21 is point f, which is also the reflecting point for the energy from B into geophone group 19, from C into 17, from D into 15, from E into 13 and from F into 11. After removal of normal moveout, these traces will be combined (*stacked*) together in a subsequent data processing operation. Thus reflecting point *f* is sampled six times and the coverage is called '600%' or '6-fold' recording. Obviously, the multiplicity tapers off at the end of the line. Most present-day recording uses at least 6-fold multiplicity and 12 and 24 are common, especially in marine shooting.

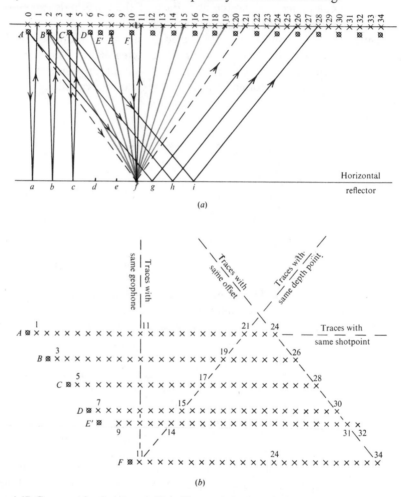

(*a*)

(*b*)

Fig. 4.47 Common-depth-point profiles. The symbols × and ⊗ represent geophone groups and shotpoints respectively. (*a*) Vertical section illustrating common-depth-point shooting; (*b*) common-depth-point stacking chart.

Occasionally one of the regularly-spaced locations will not be a suitable place for a shothole (perhaps because of risk of damage to nearby buildings) and an irregularly-spaced shotpoint will be used. Thus if shotpoint *E* could not be used, a shot might be taken at *E'* instead and then geophone group 14 (instead of 13) would receive the energy reflected at *f*. To help keep track of the many traces involved, *stacking charts* are used (Morgan, 1970). Figure 4.47*b* shows the stacking chart when E' is shot instead of E. Note how the six traces which have the common depth point f line up along a diagonal; points along the opposite diagonal have a common offset while points on a horizontal line have the same shotpoint and points on a vertical line represent traces from a common geophone group. Stacking charts are useful in making static and dynamic corrections and to ensure that the traces are stacked properly.

(b) *Spread types.* By *spread* we mean the relative locations of the shotpoint and the centres of the geophone groups used to record the energy from the shot. Several spread types are shown in fig. 4.48. In split-dip shooting the shotpoint is at the centre of a line of regularly-spaced geophone groups. Since there are 24 groups

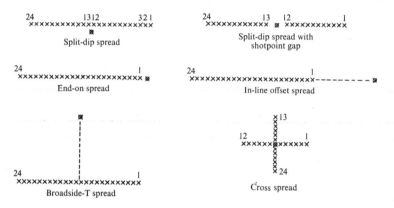

Fig. 4.48 Types of reflection spreads. The symbols × and ⊗ represent geophone groups and shotpoints respectively.

commonly, the shotpoint is usually midway between groups 12 and 13. Such an arrangement does not give an exact time-tie to the next record (since the shotpoint does not coincide with a geophone group), and therefore groups 12 and 13 are sometimes located together at the shotpoint or other minor modifications are made. Placing the shotpoint close to a geophone group often results in a noisy trace (because of gases escaping from the shothole and ejection of tamping material); hence the shotpoint may be moved perpendicular to the seismic line 15–50 m.

Occasionally in split-dip shooting the shothole is located at the end of the line of active geophone groups to produce an *end-on spread.* Sometimes in areas of exceptionally heavy ground roll the shotpoint is removed (*offset*) an appreciable distance (often 500–700 m) along the line from the nearest active geophone group to produce an *in-line offset spread.* Alternatively the shotpoint may be offset in the direction normal to the cable, either at one end of the active part to produce a *broadside-L* or opposite the centre to give a *broadside-T spread.* Both the inline and broadside offsets permit the recording of one to two seconds of reflection energy before the ground-roll energy arrives at the spread. *Cross spreads* consisting of two lines of geophone groups roughly at right angles to each other are used to record three-dimensional dip information.

(c) *Arrays.* The term *array* refers either to the pattern of a group of geophones which feed a single channel or to a distribution of shotholes or surface energy sources which are fired simultaneously; in the latter case it also includes the different locations of a single energy source for which the results are combined by vertical stacking. A wave approaching the surface along the direction of the vertical will affect each geophone of an array simultaneously so that the outputs of the geophones will combine constructively; on the other hand, a wave travelling horizontally will affect the various geophones at different times so that there will be a certain degree of destructive interference. Similarly, the waves travelling vertically downward from an array of shotholes or surface energy sources fired simultaneously will add constructively when they arrive at the geophones whereas the waves travelling horizontally away from the source array will arrive at a geophone with different phases and will be partially cancelled. Thus, arrays provide a means of discriminating between waves arriving at the spread from different directions.

Arrays are classified as *linear* when the elements are spread out in a line, usually along the seismic line, or as *areal* when the group is distributed over an area. The response of a geophone array is usually illustrated by a graph (such as fig. 4.49) showing the output of the array compared to the output of the same number of geophones concentrated at the same location. The response is usually given for the case of a sine-wave input as a function of frequency, apparent velocity, dip moveout or angle of approach. Usually the response is plotted against a dimensionless variable such as the ratio of the apparent wavelength to the element spacing (or some other key dimension of the array).

Response diagrams such as that in fig. 4.49 apply equally to arrays of geophones and arrays of shotpoints (or source units) since theoretically we get the same results by using one shotpoint and 16 geophones as by using one geophone and 16 shotpoints spaced in the same manner and shot simultaneously with $\frac{1}{16}$th the charge in each hole. However, we use multiple geophones much more than multiple shots because the cost is less, although in exceptionally difficult areas both multiple shots and multiple geophones are used at the same time. With some surface sources two to four source units may be used at the same time to provide the effect of multiple shots.

The cancelling of horizontally travelling coherent noise by using geophone and shotpoint arrays presents a more challenging array design problem than does the cancellation of random noise. In the case of random noise the locations of the elements of the array are unimportant provided no two are so close that the noise is identical for both. For coherent noise the size, spacing, and orientation of the array must be selected on the basis of the properties of the noise to be cancelled (Schoenberger, 1970). If the noise is a long sinusoidal wavetrain, an array consisting of n elements spaced along the direction of travel of the wave at intervals of

λ/n, where λ is the apparent wavelength, will provide cancellation. However, the assumption of a single wavelength to be cancelled is not realistic, actual noise often consisting of several types arriving from different directions, each type invariably comprising a range of wavelengths; moreover the nature of the noise

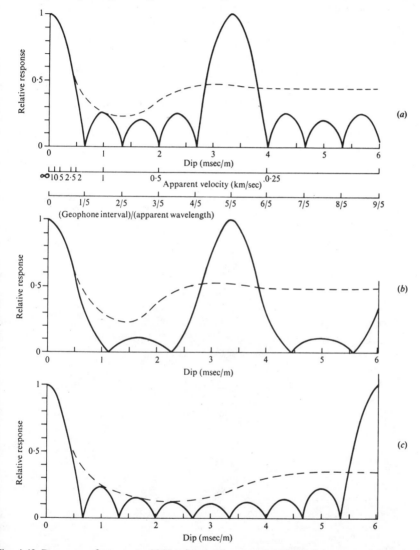

Fig. 4.49 Response of arrays to 30 Hz signal. The overall length of the array, which is the factor controlling the width of the first peak, is the same for all three arrays. The location of the principal secondary (alias) peak is controlled by the element spacing. Weighting increases the attenuation in the reject region. The dotted curves indicate the array response to a bell-shaped frequency spectrum peaked at 30 Hz with a width of 30 Hz. (Courtesy Chevron Oil.) (*a*) Five in-line geophones spaced 10 m apart; (*b*) geophones spaced 10 m apart and weighted 1, 2, 3, 2, 1; (*c*) nine geophones spaced 5·5 m apart.

306 *Seismic methods*

may change from point to point along the line. Therefore, it is usual to resort to two-dimensional arrays in areas of severe noise problems (although the in-line distribution of elements is almost always the most important aspect). One can apply the principles of antenna design to obtain maximum cancellation for a band of frequencies approaching the spread from an arbitrary direction and numerous articles have been written on the subject of geophone and shot arrays; a review paper by McKay (1954) shows examples of the improvement in record quality for different arrays.

In addition to the difficulties in defining the noise wavelengths to be attenuated, actual field layouts rarely correspond with their theoretical design (Newman and Mahoney, 1973). Measuring the location of the individual geophones is not practicable because of the time required. In heavy brush one may have to detour when laying out successive geophones and often one cannot see one geophone from another so that even the orientation of lines of geophones can be very irregular. In rough topography maintaining an array design might require that geophones be at different elevations which may produce far worse effects than those which the array is intended to eliminate. Similar problems arise where the conditions for planting the geophones vary within a group (Lamer, 1970), perhaps as a result of loose sand, mucky soil or scattered rock outcrops. The best rules for array design are often to (1) determine the maximum size which can

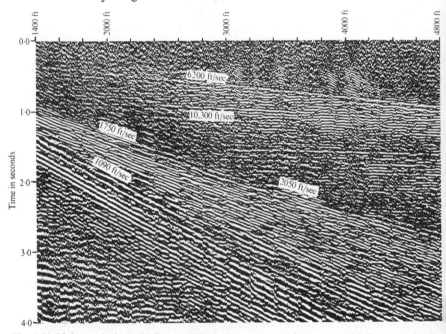

Fig. 4.50 Noise analysis or *Walkaway*. Vibroseis source, geophones spaced 5 ft apart, offset to first geophone 1400 ft. Identification of events: 6200 ft/sec arrival = refraction from base of weathering; 1750 and 2050 ft/sec = ground-roll modes; 1090 ft/sec = air wave; 10,300 ft/sec = refraction event. (From Sheriff, 1973.)

be permitted without discriminating against events with the maximum anticipated dip and (2) distribute as many geophones as field economy will permit more-or-less uniformly over an area a little less than the maximum size permitted, maintaining all geophone plants and elevations as nearly constant as possible even if this requires severe distortion of the layout.

(d) *Noise analysis.* Systematic investigation of coherent noise often begins with shooting a *noise profile* (also called a *micro-spread* or *walkaway*). This is a small-scale profile with a single geophone per trace, the geophones being spaced as closely as 1–3 m over a total spread length of the order of 300 m or more. If the weathering or elevation is variable, corrections should be made for each trace. The corrected data, often in the form of a record section, such as shown in fig. 4.50, are studied to determine the nature of the coherent events on the records, the frequencies and apparent velocities of the coherent noise, *windows* between noise trains where reflection data would not be overridden by such noise, and so on. Once we have some indications of the types of noises present, we can design arrays or other field techniques to attenuate the noise and then field-test our techniques to see if the desired effect is achieved.

(e) *Uphole surveys.* An *uphole survey* is one of the best methods of investigating the near-surface and finding the thickness and velocity of the low-velocity layer, D_W and V_W, and the sub-weathering velocity, V_H. An uphole survey requires a shothole deeper than the base of the LVL. Usually a complete spread of geophones plus an uphole geophone is used. Shots are fired at various depths in the hole, as shown in the lower part of fig. 4.51, beginning at the bottom and continuing until the shot is just below the surface of the ground. Arrival times are plotted against shot depth for the uphole geophone and for several of the distant geophones, including two or more spaced 200 m or more apart, as shown in the upper part of fig. 4.51. The plot for the uphole geophone changes abruptly where the shot enters the LVL; the slope of the portion above the base of the LVL gives V_W and the break in slope usually defines D_W clearly.

For the distant geophones the plot is almost vertical at first since the path length changes very little as long as the shot is in the high-speed layer. However, when the shot enters the LVL there is an abrupt change in slope and the traveltime increases rapidly as the path length in the LVL increases. The refraction velocity at the base of the LVL, V_H, is obtained by dividing the time interval between the vertical portions of the curves for two widely separated geophones (Δt_{17} in fig. 4.51) into the distance between the geophones. This velocity measurement is often different from that given by the slope of the deeper portion of the uphole geophone curve, partly because the latter is less accurate (since the time interval is much less than Δt_{17}), partly because the layering of beds of different velocities has little effect on Δt_{17} but may affect t_{uh} substantially.

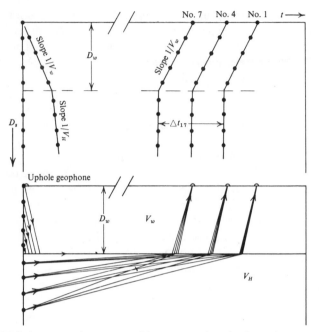

Fig. 4.51 Uphole survey. At top, traveltime versus shot depth; at bottom vertical section showing raypaths.

4.5.4 *Field equipment for land surveys*

(a) *Drilling*. When dynamite is being used as the energy source, holes are drilled so that the explosive can be placed below the low-velocity layer. The holes are usually about 8–10 cm in diameter and 6–30 m in depth, although depths of 80 m or more are used occasionally. Normally the holes are drilled with a rotary drill, usually mounted on a truck bed but sometimes on a tractor or amphibious vehicle for working in difficult areas. Some light drills can be broken down into units small enough that they can be carried. Augers are used occasionally. In work in soft marshes, holes are sometimes jetted down with a hydraulic pump. Typical rotary-drilling equipment is shown in the photograph in fig. 4.52 and in the diagram in fig. 4.53.

Rotary drilling is accomplished with a drill bit at the bottom of a drill pipe, the top of which is turned so as to turn the bit. Fluid is pumped down through the drill pipe, passes out through the bit and returns to the surface in the annular region around the drill pipe. The functions of the drilling fluid are to bring the cuttings to the surface, to cool the bit and to plaster the drill hole to prevent the walls from caving and formation fluids from flowing into the hole. The most common drilling fluid is *mud* which consists of a fine suspension of bentonite, lime and/or barite in water. Sometimes water alone is used, sometimes air is the circulating fluid. Drag bits are used most commonly in soft formations; these tear out pieces of the

Fig. 4.52 Mayhew 1000 drilling rig. (Courtesy Gardner-Denver.)

earth. Hard rock is usually drilled with roller bits or cone bits which cause pieces of rock to chip off because of the pressure exerted by teeth on the bits. In areas of exceptionally hard rock, diamond drill bits are used.

(b) *Explosive energy sources.* Explosives were the sole source of energy used in seismic exploration until weight dropping was introduced in 1953. While explosives are no longer dominant, they continue to be an important seismic energy source in land work.

Two types of explosives have been used principally: gelatin dynamite and ammonium nitrate. The former is a mixture of nitroglycerin and gelatin (which form the explosive component) and an inert material which binds the mixture together and which can be used to vary the 'strength' of the explosive. The velocity of detonation (that is, the velocity with which the explosion travels away from the point of initiation in an extended body of explosive) is high for the explosives used in seismic work, around 6000–7000 m/sec; consequently, the seismic pulses generated have very steep fronts in comparison with other energy sources. This

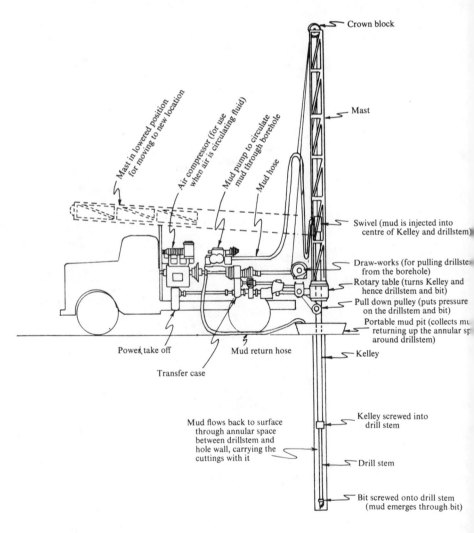

ROTARY   DRILL

Fig. 4.53 Rotary drill. (From Sheriff, 1973.)

high concentration of energy is desirable from the point of view of seismic wave analysis but detrimental from the viewpoint of damage to nearby structures. Ammonium nitrate is cheaper and less dangerous since it is more difficult to detonate than gelatin dynamites. Ammonium nitrate and NCN (nitrocarbo-

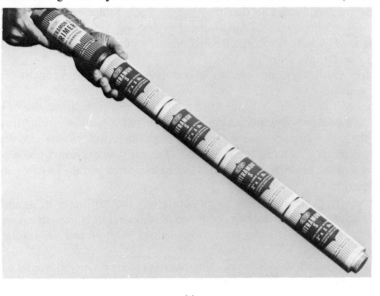

(a)

(b)

Fig. 4.54 Seismic explosives. (Courtesy Dupont.) (a) Cans of Nitramon shothole explosive joined end-to-end; (b) electric blasting cap.

nitrite) are the dominant explosives used today (in such forms as Nitramon*, etc.). Other types of explosives are also used occasionally.

Explosives are packaged in tins or in tubes of cardboard or plastic about 5 cm in diameter which usually contain 1–10 pounds (0·5–5 kg) of explosive. The tubes and tins are constructed so that they can be easily joined together end-to-end (fig. 4.54a) to obtain various quantities of explosives. Ammonium nitrate some-times is used in bulk form, the desired quantity being mixed with fuel oil and poured directly into a dry shothole.

Electric blasting caps are used to initiate an explosion. These consist of small metal cylinders, roughly 0·6 cm in diameter and 4 cm long (see fig. 4.54b). They contain a resistance wire imbedded in a mixture of powder charges, one of which ignites at a relatively low temperature. By means of two wires issuing from the end of the cap, a large current is passed through the resistance wire and the heat generated thereby initiates the explosion. The cap has previously been placed inside one of the explosive charges so that the explosion of the cap detonates the entire charge.

Primers are generally necessary in setting off the explosion in ammonium nitrate explosives. These are cans of more sensitive explosive which are used as one of the elements in making up the total charge. A cap is inserted into a 'well' in the end of the can of primer to set it off.

The current which causes the blasting cap to explode is derived from a *blaster*; this is basically a device for charging a capacitor to a high voltage by means of either batteries or a hand-operated generator and then discharging the capacitor through the cap at the desired time. Incorporated in the blaster is a device which generates an electrical pulse at the instant that the explosion begins. This *time-break* pulse fixes the instant of the explosion, $t = 0$. The time-break pulse is transmitted to the recording equipment by a telephone line or radio where it is recorded along with the seismic data.

Several techniques are used at times to concentrate the energy travelling down-ward from an explosion. The detonating front in an explosive usually travels much faster than the seismic wave in the formation so that the seismic wave originating from the top of a long explosive charge lags behind the wave from the bottom of the charge even where the explosive is detonated at the top (which is the usual method). Explosives with low effective detonating velocity are sometimes used but these are made in long flexible tubes which are difficult to load. Delay units are sometimes used between several concentrated explosive charges to allow the wave in the formation to catch up with the explosive front; these may consist of delay caps (which introduce a fixed delay between the time the detonating shock initiates them and the time they themselves explode) or helically-wound detonating cord (so that the detonating front has to travel a longer distance). Expendable impact blasters have also been used; these detonate when they are actuated by the shock wave from another explosion.

---

* Registered trademark of E.1. Du Pont de Nemours & Co.

While explosives provide the most compact high-energy source, they have many disadvantages which often preclude their use: high cost, the time and expense involved in drilling holes, potential damage to nearby buildings, wells, etc., and most important of all, restrictions about where holes can be drilled and explosives detonated.

(c) *Surface energy sources.* Many alternative energy sources have been developed for use in both land and marine work. Discussion of those which are used primarily at sea and infrequently on land will be postponed until §4.5.5c.

Without exception the surface energy sources are less powerful than explosives and their use on a large scale has been made feasible by vertical stacking methods which permit adding together the effects of a large number of weak impulses to obtain a usable result.

The earliest non-dynamite source to gain wide acceptance was the *thumper* or *weight dropper*. This method was developed largely by the McCullom Geophysical Company. A rectangular steel plate weighing about 3000 kg is dropped from a height of about 3 m. The instant of impact is determined by a sensor on the plate. Weights often are dropped every few metres so that the results of 50 or more drops are composited into a single field record. The time between release of the weight and impact on the ground is not constant enough to permit more than one source to be used simultaneously. Often two or three units are used in succession, one dropping its weight while the others lift their weights into the armed position and move ahead to the next drop point. The use of weight dropping is now largely restricted to desert or semi-desert areas where the massive trucks can move about relatively freely.

The *Dinoseis** method (Godfrey *et al.*, 1968) developed by the Sinclair Oil and Gas Company involves the explosion of a mixture of propane and oxygen within an expandable chamber. The explosion chamber is mounted under a truck and is lowered to the ground when ready for use (see fig. 4.55). The explosion of the gas mixture by means of a spark plug creates a pressure which acts on a moveable

Fig. 4.55 Truck-mounted twin 36-inch Dinoseis gas exploders. (Courtesy Geo Space.)

* Trademark of Atlantic Richfield Co.

plate forming the bottom of the chamber, thus transmitting the pressure pulse into the ground. The weight of the chamber provides the necessary reaction inertia. Several other types of gas exploders have been used in land work. As with weight dropping, the heavy source chamber requires massive field equipment which in turn restricts usage to fairly open areas.

The *Geoflex** method uses explosive detonating cord buried under a few inches of earth. A hundred metres of cord may be buried, often in a source pattern, using a special plough and then exploded by means of caps placed at intervals along the cord.

While the foregoing are primarily surface sources, gas guns, airguns (Brede *et al.*, 1970), and other devices are sometimes used in boreholes, especially in soft marsh where there is little risk of being unable to recover the equipment from the hole. The airguns used on land are modifications of the guns designed for marine use which are discussed in §4.5.5c.

Unlike other energy sources which try to deliver energy to the ground in the shortest time possible, the *Vibroseis†* source passes energy into the ground for 7 seconds or more. A vibrator (usually hydraulic) actuates a steel plate pressed firmly against the ground (see fig. 4.56). The output wave train consists of a sine wave whose frequency increases continuously from about 6 Hz to about 50 Hz during the 7-second 'sweep'. Each returning reflection event is a similar wave train of about 7 seconds duration; since reflections occur much closer together

Fig. 4.56 Vibroseis equipment mounted for off-road survey. (Courtesy Continental Oil.)

than this, the result is a superposition of many wave trains. Subsequent data processing (discussed in §4.7.4e) is necessary to resolve the data; in effect the

* Tradename of Imperial Chemical Industries, Ltd.
† Trademark of Continental Oil Co.

processing compresses each returning wave train into short wavelets, thus removing much of the overlap (see fig. 4.104).

Vibroseis sources produce low energy density; as a result they can be used in cities and other areas where explosives and other sources would cause extensive damage. The use of Vibroseis in land seismic exploration is increasing rapidly.

(d) *Geophones.* Seismic energy arriving at the surface of the ground is detected by *geophones*, frequently referred to as *seismometers, detectors* or *jugs.* Although many types have been used in the past, modern geophones are almost entirely of the moving-coil electromagnetic type for land work and the piezoelectric type for marsh and marine work. The latter will be discussed in §4.5.5*d* in connection with marine equipment.

The moving-coil electromagnetic geophone is shown schematically in fig. 4.57 while fig. 4.58 is a photograph of a cutaway model. The schematic diagram shows a permanent magnet in the form of a cylinder into which a circular slot has been cut,

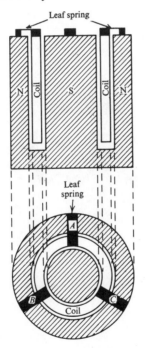

Fig. 4.57 Geophone schematic.

the slot separating the central South Pole from the outer annular North Pole. A coil consisting of a large number of turns of very fine wire is suspended centrally in the slot by means of light leaf springs, *A*, *B* and *C*. The geophone is placed on the ground in an upright position, as shown in the diagram. When the ground moves vertically, the magnet moves with it but the coil, because of its inertia, tends to stay fixed. The relative motion between the coil and magnetic field generates a voltage between the terminals of the coil. The geophone output for horizontal motion is essentially zero since the coil is supported in such a way that it stays fixed relative to the magnet during horizontal motion.

The output voltage of the geophone is directly proportional to the strength of the magnetic field of the permanent magnet, the number of turns in the coil, the radius of the coil and the velocity of the coil relative to the magnet. Modern high sensitivity geophones have an output of 0·5–0·7 volts for a velocity of 1 cm/sec of the ground.

The geophone coil and springs constitute an oscillatory system with natural frequency in the range from 4–15 Hz for reflection work, 1–10 Hz for refraction work. Since the coil tends to continue to oscillate after the ground motion dies away, it is necessary to dampen (attenuate) the motion. This is achieved in part by winding the coil on a metal 'former'; eddy currents induced in the latter when it

Fig. 4.58 Cutaway of digital-grade geophone. (Courtesy Geo Space.)

moves in the magnetic field oppose the motion and hence produce damping. Additional damping is obtained by connecting a shunt resistance across the coil of the geophone; when a current flows through the coil the interaction between the magnetic fields of the current and the permanent magnet further slows down the motion of the coil (the input impedance of the recording system is so large that it does not significantly affect the current and hence the damping). The geophone damping can be adjusted by varying the shunt resistance since this changes the current through the geophone. If a geophone with a high resistance shunt is

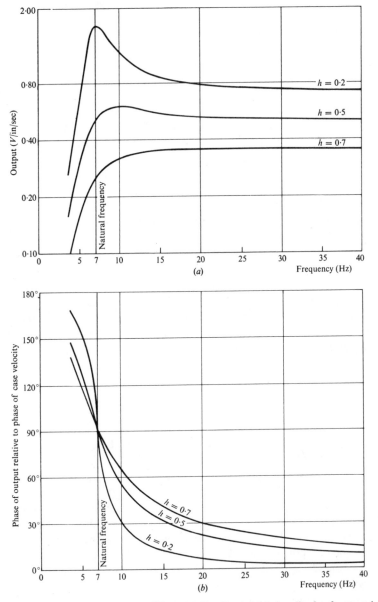

(a)

(b)

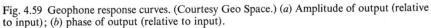

Fig. 4.59 Geophone response curves. (Courtesy Geo Space.) (a) Amplitude of output (relative to input); (b) phase of output (relative to input).

tapped lightly, the coil will oscillate for some time. As the shunt resistance is decreased, the number of oscillations will decrease because of the increased damping until finally a point will be reached where a tap will just fail to produce an oscillation. At this point the geophone is *critically damped*.

The response of a geophone to a harmonic signal depends upon the relation between the frequency of the signal and the natural frequency of the geophone as well as the degree of damping. Figure 4.59*a* shows curves of the amplitude of the output current as a function of the frequency of a harmonic input signal. The maximum velocity of the geophone is the same for all curves. Often the output is normalized with respect to (that is, expressed as a fraction of) the output for high frequencies.

The various curves correspond to different values of the damping, *h* being the ratio of the amount of damping relative to critical damping ($h = 1$). For zero damping ($h = 0$), the output becomes infinite at the natural frequency; obviously this is merely a theoretical result since zero damping can never be achieved. As *h* increases, the output peak decreases in magnitude and moves slowly towards higher frequencies. Somewhere between $h = 0.5$ and $h = 0.7$ the peak disappears and the range of flat response has its maximum extent. As *h* increases beyond this value, the low frequency response gradually falls off. Clearly, therefore, the generally accepted choice of 70% of critical damping for geophones results in more-or-less optimum operating conditions with respect to amplitude distortion in the geophone output.

The output of a geophone is shifted in phase with respect to the input by the amounts shown in fig. 4.59*b*. Phase shift is important since the seismic signal comprises a range of frequencies; if the phase shift does not vary properly with frequency, phase distortion occurs. The optimum phase shift is linear with frequency and with an intercept of $n\pi$; that is, the shift is equal to $(k\omega + n\pi)$ where *k* is a constant and *n* an integer. To show this, we note that an input signal of the form $A \cos \omega t$ will appear in the output as

$$B \cos (\omega t + k\omega + n\pi) = \pm B \cos \omega(t + k).$$

Thus, the net effect is to shift all frequencies in time by the amount *k* and to invert the pulse when *n* is odd, neither of which results in change of waveshape. Referring to fig. 4.59*b*, we see that the phase characteristics for $h = 0.5 - 0.7$ are reasonably straight for the frequency range above 20 Hz.

Usually several closely-spaced geophones are connected in a series-parallel arrangement to produce a single composite output. The entire geophone group is considered to be equivalent to a single geophone located at the centre of the group. However, the damping of each geophone will be affected by the presence of the other geophones because of the change in resistance of the circuit. An exception is an arrangement of *n* parallel branches, each containing *n* identical geophones in series, which has the same resistance as a single geophone and hence the same damping.

(e) *Amplifiers.* Except for very strong signals arriving soon after the shot is fired, the output of the geophone is too weak to be recorded without amplification. Also, the useful range of amplitudes of the geophone output extends from a few tenths of a volt at the beginning of the recording to about 1 μV near the end of the recording several seconds after the shot (signals weaker than about 1 μV are lost in the system noise), a relative change or *dynamic range* of about 10^5 (100 dB). Therefore, besides amplifying weak signals, the amplifier usually is called upon to compress the range of signals as well. In addition, amplifiers are used to filter the geophone output to enhance the signal relative to the noise.

Fig. 4.60 Portable analog seismic instruments. (Courtesy Texas Instruments.)

Seismic amplifiers generally employ solid state circuitry which allows them to be very compact. While they are usually mounted in a recording truck or other vehicle, they can also be carried where necessary. Portable analogue recording equipment is shown in fig. 4.60; this equipment consists of four main 'suitcases': a power supply, the amplifiers, a magnetic tape recorder, and a camera, plus a battery to supply the power and photographic developing tanks. The amplifier suitcase contains 24 complete amplifiers plus test circuitry.

A block diagram of an analogue amplifier is shown in fig. 4.61; the arrangement of circuit elements and the number of amplification stages vary from manufacturer to manufacturer. The cable from the geophones may be connected to a balance circuit which permits adjusting the impedance to ground so as to minimize the coupling with nearby power lines, thus reducing pickup of noise at the power-line frequency (*high-line pickup*). The next circuit element usually is a filter to attenuate the low frequencies which arise from strong ground roll and which otherwise might overdrive the first amplification stage and introduce distortion.

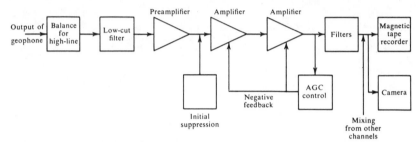

Fig. 4.61 Block diagram of analogue seismic amplifier.

Seismic amplifiers are multi-stage and have very high maximum gain, usually of the order of 10^5 (100 dB), sometimes as much as 10^7 (140 dB); 100 dB means that an input of 5 μV amplitude appears in the output with an amplitude of 0·5 V. Lower amplification can be obtained by means of a multi-position master gain switch which reduces the gain in steps.

The amplifier gain is varied during the recording interval starting with low amplification during the arrival of strong signals at the early part of the record and ending up with the high gain value fixed by the master gain setting. This variation of gain with time (signal compression) can be accomplished with *Automatic Gain (Volume) Control*, usually abbreviated *AGC* or *AVC*. This is accomplished by a negative feedback loop, a circuit which measures the average output signal level over a short interval and adjusts the gain to keep the output more-or-less constant regardless of the input level.

It is important in making corrections for near-surface effects that we be able to observe clearly the *first breaks*, the first arrivals of energy at the different geophones. (For a geophone near the shotpoint, the first arrival travels approximately along the straight line from the shot to the geophone; for a distant geophone the first arrival is a head wave refracted at the base of the low-velocity layer – see fig. 4.86 and the discussion of weathering corrections, §4.5.7*b*.)

If we allow the AGC to determine the gain prior to the first arrivals, the low input level (which is entirely noise) will result in very high gain; the output will then be noise amplified to the point where it becomes difficult to observe the exact instant of arrival of the first breaks. This problem is solved by using *initial suppression* or *presuppression*. A high-frequency oscillator signal (about 3 kHz) is fed into the AGC circuit which reacts by reducing the gain so that the noise is barely perceptible; the high frequency signal is subsequently removed by filtering so that it does not appear in the output. With the reduced gain the relatively strong first breaks stand out clearly. As soon as the first breaks have all been recorded, the oscillator signal is removed, usually by a relay triggered by one of the first breaks. Thereafter, the AGC adjusts the gain in accordance with the seismic signal level.

Seismic amplifiers are intended to reproduce the input with a minimum of distortion and hence the gain (without filters) should be constant for the entire frequency spectrum of interest. For reflection work this range is about 10–100 Hz while for refraction work the range is about 1–50 Hz; accordingly most amplifiers have flat response for all frequencies from about 1 Hz to 200 Hz or more.

Frequency filtering refers to the discrimination against certain frequencies relative to others. Seismic amplifiers have a number of filter circuits which permit us to reduce the range of frequencies which the amplifier passes. While details vary, most permit the selection of the upper and the lower limits of the pass band. Often it is possible to select also the sharpness of the *cut-off* (the rate at which the gain decreases as we leave the pass band). Figure 4.62 shows typical filter response curves. The curves are specified by the frequency values at which the gain has dropped by 3dB (30% of amplitude, 50% of power); the curve marked 'Out' is the response curve of the amplifier without filters.

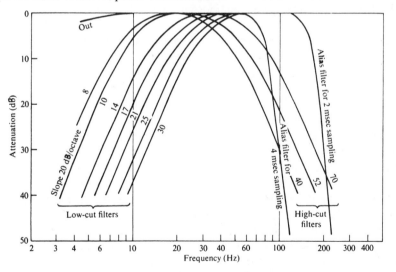

Fig. 4.62 Response of seismic filters.

Seismic amplifiers may include circuitry for *mixing* or *compositing*, that is, combining two or more signals to give a single output. Mixing in effect increases the size of the geophone group and is sometimes used to attenuate certain types of surface waves. The commonest form, called '50% mixing', is the addition equally of the signals from adjacent geophone groups. Magnetic tape recording now has virtually eliminated the need to mix during recording since we can always mix in playback.

The time-break signal often is superimposed on one of the amplifier outputs where it appears as a sharp pulse which marks the point $t = 0$ for the record. When explosives are being used, the output of an *uphole geophone* (a geophone placed near the top of the shothole) is also superimposed on one of the outputs; the interval between the time-break and the uphole geophone signal is called the *uphole time* (t_{uh}); it measures the vertical traveltime from the shot to the surface and is important in correcting for near-surface effects.

High resolution or *HR* amplifiers are used in engineering and mining problems to map the top 200 m or so. To get resolution of a few meters, we must use short wavelengths; accordingly these amplifiers have essentially uniform response up to 300 Hz, sometimes to 500 Hz, and the AGC time constants are correspondingly short. To permit recording very shallow reflections, small offsets are used and the initial suppression permits recording events within 0·050 sec or so after the first break.

⁎ (f) *Analogue data recording.* For the first thirty years or so of seismic exploration, the outputs of the amplifiers were recorded directly on photographic paper by means of a camera. However, about 1952 recording on magnetic tape began. Today few seismic crews are not equipped for magnetic tape recording (analogue or digital).

The feature which originally led to widespread use of magnetic recording was the ability to record in the field with a minimum of filtering, automatic gain control, mixing, etc., and then introduce the optimum amounts of these on playback. Later a more important advantage turned out to be the ability to produce record sections (see §4.5.7e) which proved to be powerful aids in interpretation. However, magnetic tape recording did not develop its full potential until the introduction of digital techniques during the early 1960s.

Analogue magnetic tape recorders usually have heads for recording 26 to 50 channels in parallel. In the early years direct recording was used; the output from the amplifier went directly to the recording head, the intensity of magnetization of the tape being proportional to the current in the recording head and hence proportional also to the signal strength. Later, direct recording was displaced by frequency modulation and pulse-width modulation techniques since these are more noise-free and can accept a wider range of signal strengths.

The data recorded on magnetic tape must be presented in visual form for monitoring and for interpretation. This is done most commonly by a photographic *camera*, although cameras using electrostatic printing are also widely used. Whatever the recording principle, the basic features are much the same for most cameras: (1) a small motor which moves a strip of paper at a constant speed (conventionally about 1 ft/sec) during the recording period, (2) supply and take-up magazines to hold the paper, (3) a series of galvanometers, one for each geophone group, which transform the electrical signals coming out of the amplifiers into intense spots of light moving in accordance with the signals and (4) a device for recording accurate timing marks.

The galvanometers which are generally used are similar to a lead pencil in size and shape. Each contains a small coil suspended in the field of a permanent magnet. A tiny mirror attached to the coil reflects a sharply focused beam of light onto the paper. When the signal current flows through the coil, the interaction between the field of the coil and the permanent magnetic field causes the coil to rotate, thereby deflecting the beam transversely to the direction of movement of the paper. The result is a graph of the motion of the geophone, the time axis being the direction of motion of the record.

Each individual graph representing the motion of a geophone (or the average of a group of geophones) is called a *trace*. In addition to the *wiggly-line* trace just described, variable-area and variable-density traces are used; these three modes of representation are compared in fig. 4.63. The *variable-area* trace can be pictured as the result of passing the light from the galvanometer through a cylindrical lens whose axis would be vertical to give the result shown in the figure. This lens produces a vertical line of light instead of a point as for the wiggly trace. As this line moves up and down with the signal variations, the lower part is cut off by a stop. The *variable-density* trace is obtained by replacing the galvanometer with a light source whose intensity is varied in accordance with the signal variations. For example, the intensity might increase during signal peaks and decrease during signal troughs; the photographic record will then be darker for peaks and lighter for troughs. The upper and lower modes of display in fig. 4.63 show the wiggly trace superimposed on variable area and variable density respectively; these have the advantage of retaining the fine detail afforded by the wiggly trace and at the same time obtaining the graphic effects of the variable-area or variable-density displays.

The heart of the timing system is an oscillating electric circuit whose frequency is accurately controlled by a tuning fork or crystal. In one system, a voltage derived from the oscillating circuit is used to drive a small synchronous motor which turns a hollow cylinder inside a fixed coaxial cylinder. Light issuing from a series of narrow longitudinal slits in the rotating cylinder passes through a slit in the fixed cylinder to make *timing lines* at intervals of 0·010 sec (10 msec). Every tenth slit in the rotating cylinder is slightly wider than the others which causes each tenth timing line to be heavier. More modern timing systems use a light source which is caused to flash every 10 msec, every tenth flash being brighter in order to produce a heavy timing line every tenth second.

(g) *Digital recording.* Digital recording was first introduced into seismic work early in the 1960s and by the end of the decade was in widespread use. Whereas analogue devices represent the signal by a voltage (or other quantity) which varies continuously with time, *digital recording* represents the signal by a series of numbers which denote values of the output of the geophone measured at regular intervals, usually 2 or 4 msec. Digital recording is capable of higher fidelity than analogue recording and permits numerical processing of the data without adding appreciably to the distortion. Digital processing has proven to be so effective in improving seismic data that it has gained widespread acceptance and analogue recording may be completely superseded eventually. However the beginning

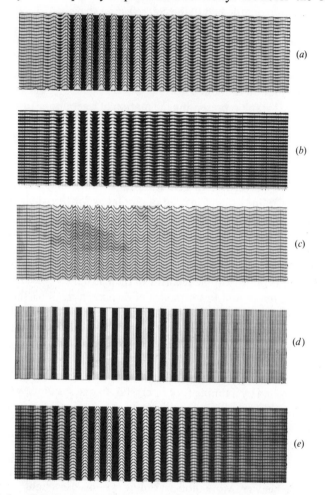

Fig. 4.63 Modes of displaying seismic data. (Courtesy Geo Space.) (*a*) Wiggle superimposed on variable area; (*b*) variable area; (*c*) wiggle; (*d*) variable density; (*e*) wiggle superimposed on variable density.

(geophone response) and end (display) of the recording process continue to be analogue.

Before describing digital recording, we shall discuss digital representations. While we could build equipment to handle data using the scale of ten which forms the basis of our ordinary arithmetic, it is more practical to operate on the *binary scale* of 2. The binary scale uses only two digits, 0 and 1; hence only two different conditions are required to represent binary numbers, for example, a switch opened or closed. Binary arithmetic operations are much like decimal ones. The decimal number 20873 is a shorthand way of saying that the quantity is equal to 3 units plus 7×10 plus 8×10^2 plus 0×10^3 plus 2×10^4. Similarly, the binary number 1011011 is equal to 1 unit plus 1×2 plus 0×2^2 plus 1×2^3 plus 1×2^4 plus 0×2^5 plus 1×2^6, which is the same as the decimal number 91. We can use positive and negative square pulses to represent 1 and 0 or represent them in other ways. Each pulse representing 1 or 0 is called a *bit* and the series of bits which give the value of a quantity is called a *word*.

The most obvious feature of the truck-mounted digital equipment shown in fig. 4.64 which distinguishes it from analogue equipment (such as shown in fig. 4.60) is the digital magnetic tape unit occupying the right-hand rack and the associated power supply occupying the lower half of the centre rack. Since these units are not excessively bulky or heavy, digital equipment can be just as portable as analogue.

The block diagram of a digital recorder in fig. 4.65 can be compared with that of the analogue recorder shown in fig. 4.61. The geophones and amplifiers shown in the diagram could be those normally used for analogue recording; however, to take full advantage of the capabilities of digital recording, we use components of

Fig. 4.64 Digital field equipment mounted in truck. (Courtesy Texas Instruments.)

higher standards of performance than those which suffice for analogue recording. Digital amplifiers usually have enough dynamic range that they do not require low-cut filtering prior to the first amplification stage. Digital amplifiers are usually *binary-gain* amplifiers in which the gain can change only by a factor of two or 6 dB (Siems and Hefer, 1967). We can find the gain at any instant merely by keeping track of the number of times the gain has increased and decreased. Multiplication by two is very simple in the binary number system because it simply involves shifting the 'binary point' (equivalent to 'decimal point').

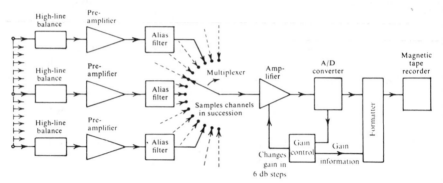

Fig. 4.65 Block diagram of digital seismic amplifier.

Another type of seismic amplifier called 'floating point' has come into use. Binary-gain amplifiers are limited in how quickly they can change the gain, typically only 6 dB (a factor of 2) between successive sampling times. Floating-point amplifiers measure the magnitude of each sample without regard to the magnitude of previous samples. Each value is expressed as a certain number of significant figures times 2 raised to the proper power, where both the significant figures and the power are recorded.

The output of each amplifier passes through an alias filter whose function will be explained in §4.7.3*b*. From the alias filter the signal passes to a *multiplexer* which is in essence a high-speed electronic switch. Starting with the multiplexer we require only one channel – one sample-and-hold unit, one *A/D (analogue-to-digital) converter*, etc., instead of one for each trace.

The multiplexer connects each amplifier in turn to the A/D converter. Here the amplifier is connected to a *sample-and-hold* unit for about 1 μsec which is sufficient time to charge a capacitor to the same voltage as the amplifier output at that instant. The amplifier is then disconnected and during the ensuing 30 μsec or so the A/D converter compares the capacitor voltage with a series of standard voltages. The result of this comparison is a series of pulses giving the polarity and value of the amplifier output voltage (in the binary scale) at the sampling instant. Processing the data for each trace requires about 31 μsec or about 750 μsec for 24 traces. Thus, there is ample time to read each trace even at the fastest sampling rate (1 msec) which is used. The output of the A/D converter is a word, usually with 14 bits, the first of which gives the sign (plus or minus) of the signal and the remaining bits the magnitude.

The output of the A/D converter is used to control the gain in a binary-gain system, the gain being cut in half whenever the measured voltage exceeds a certain value or doubled whenever the voltage falls below a threshold value.

The series of pulses from the A/D converter passes to the *formatter*, the device which controls the sequence in which the various bits and words are recorded on the magnetic tape. In addition to the data from the A/D converter, the formatter also receives data identifying the profile, the time-break, up-hole information, values of the amplifier gain for each sample, a word to mark the end of a record and so on (all of this information being in digital form).

Digital tape recorders use either half-inch or one-inch tape. With half-inch tape the recorder usually has nine recording heads, sometimes seven; with nine heads a word of 14 bits of information is written in two rows (or two *bytes*) across the tape, the remaining four bits being used for additional data. With one-inch tape 21 heads are used. The formatter distributes the various bits among the heads in a fixed pattern known as the *format*. Tape speeds in digital recording range from 20 to 150 in/sec, depending upon the sampling interval and the format arrangement (Northwood *et al.*, 1967; Meiners *et al.*, 1972). The tape speed is adjusted so that the density of data along a single track is constant (800 bits/in. or more).

Digital recording (and processing) involves a long series of operations which take place sequentially on a time scale measured in microseconds. The entire sequence is controlled by an electronic 'clock', a crystal-controlled oscillator operating in the megacycle range which furnishes a continuous series of pulses whose shape and spacing are accurately maintained. Time is measured by counting these pulses and the operating cycles of the component units (such as the multiplexer and the formatter) are controlled by circuits which count the clock pulses and operate electronic switches when the count reaches predetermined values.

(h) *Digital-to-analogue conversion.* While digital data are ideal for versatile data processing, analogue data are more suitable for interpretation and for monitoring the recording system to ensure that all units are functioning properly. Accordingly, digital recording systems provide for reconverting the digital data to analogue form for recording by a camera.

The D/A (digital-to-analogue) conversion is carried out by reversing the recording process. On playing back the digital tape, the bits coming off the tape are rearranged by a deformatter before going to the D/A converter. Usually only the seven or eight 'most significant' bits specifying a signal value are used (for example, 1001100111101 is reduced on playback to 1001100 or 10011001) because the camera unit has a resolution of only about 1% (or 1 part in 1100100 in the binary scale), smaller amplitude changes being undetectable on the paper record.

The D/A converter can be regarded as seven batteries (when seven bits are used) with voltages proportional to 2^6, 2^5, 2^4, 2^3, 2^2, 2^1, 2^0 units. The batteries are connected in series so that the total voltage is 1111111 (that is, 127 in the decimal system). Each 'zero' disconnects the corresponding battery. The bit giving the polarity of the signal is used to control the polarity of the output. The result is a voltage at the output of the D/A converter equal in magnitude and sign to the original signal at the instant of sampling.

The output from the D/A converter goes then to a demultiplexer which connects each voltage to a sample-and-hold circuit in the proper channel. The sample-and-hold voltage has a stair-case form such as shown in fig. 4.66. The output of the sample-and-hold circuit goes through a low-pass filter which smooths out the steps and is then amplified to the voltage level required by the camera.

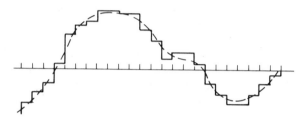

Fig. 4.66 Conversion from digital to analogue form.

4.5.5 *Marine equipment and methods*

(a) *Marine operations.* Marine seismic operations usually imply water which is sufficiently deep to allow freedom of movement for ships which are 30 to 70 m in length (a typical ship is shown in fig. 4.67). Such operations differ from operations on land and in shallow water primarily because of the speed with which they take

Fig. 4.67 A seismic survey ship. (Courtesy Prakla.)

place. Normal production shooting takes place at a speed of the order of 6 knots and can proceed on a 24-hour-per-day basis. Hence with a 2400-m, 24-channel spread, 12-fold CDP coverage of 250 km/day or 2600 profiles/day would be possible if all the time were spent shooting. This much production is never achieved because much time is spent travelling to the line or from the end of one line to the start of the next, waiting for good weather, or because of other factors. Nonetheless production rates are of the order of 2 profiles/minute and each profile may consist of 2 to 4 component *subshots* (separate records which are then vertically stacked).

The monthly cost of a marine crew is large but the high production cuts the unit cost of marine seismic data to about 10% of that of land data. The high production rate requires special emphasis on efficiency in operations. Source and receiver units are towed into place and forward travel does not stop during a recording. While detailed monitoring of data quality is not possible at the pace of operations, the relatively constant water environment surrounding the sources and receivers and the general absence of the low-velocity weathering layer which is usually present on land lessen variations in data quality.

(b) *Bubble effect.* An underwater explosion produces a bubble of gases at high pressure. As long as the gas pressure exceeds the hydrostatic pressure of the surrounding water, the net force will accelerate the water outward away from the shot. The net force decreases as the bubble expands and becomes zero when the bubble expansion reduces the gas pressure to the value of the hydrostatic pressure. However, at this point the water has acquired its maximum outward velocity and so continues to move outwards while decelerating because the net force is now directed inward. Eventually the water comes to rest and the net inward force now causes a collapse of the bubble with a consequent sharp increase in gas pressure, and the process repeats itself. Thus the bubble will oscillate and seismic waves will be generated on each oscillation.

As the bubble loses energy and rises toward the surface its oscillation period decreases (Kramer *et al.*, 1968). For 'conventional' dynamite charges of $16\frac{2}{3}$ lb (7·5 kg) the bubbles produced by the explosion in effect generate additional seismic records every 0·2 to 0·4 sec. These records are superimposed on each other so that one cannot tell from which oscillation a reflection event comes and hence the data cannot be used. The practice when using conventional explosives as the source, therefore, was to shoot within 2 metres or so of the surface so that the bubble would vent to the surface. This produced spectacular plumes of water but was inefficient in generating useful seismic energy.

(c) *Marine energy sources.* Marine seismic reflection work consists mostly of two types, common-depth-point and profiler work; these differ considerably in cost, size of energy source, effective penetration, and various other aspects. We shall discuss here the larger energy sources commonly used in common-depth-point recording: the smaller energy sources used in profiling will be described in §4.5.5*f*.

The most widely used large energy source is the *air gun*, a device which discharges air under very high pressure into the water (Giles, 1968; Schulze-Gattermann, 1972). Pressures up to 10,000 psi are used although 2000 psi is most common. The PAR* air gun is illustrated in fig. 4.68. The gun is shown at the left

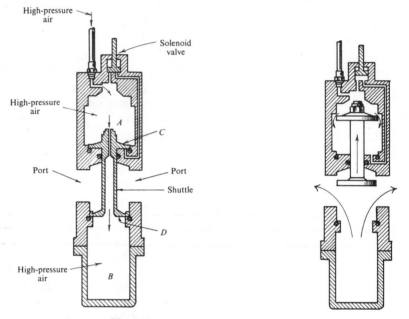

Fig. 4.68 Air gun. (Courtesy Bolt Associates.)

in the armed position, ready for firing. Chambers *A* and *B* are filled with high pressure air which entered *A* at the top left and passed into *B* through an axial opening in the 'shuttle'. The latter is held in the closed position by the air pressure (since flange *C* is larger than flange *D*, resulting in a net downward force). To fire the gun, the solenoid at the top opens a valve which allows high pressure air to reach the underside of flange *C*. This produces an upward force which is large enough to overcome the force holding the shuttle in the closed position and consequently the shuttle opens rapidly. This allows the high pressure air in the lower chamber to rush out through four ports into the water. The bubble of high pressure air then oscillates in the same manner as a bubble of waste gases resulting from an explosion. However, since the energy is smaller, the oscillating frequency is in the seismic range and therefore has the effect of lengthening the original pulse (rather than generating new pulses as with dynamite).

The upward motion of the shuttle is arrested before it strikes the top of chamber *A* because the upward force falls off rapidly as the air enters the water and the

* Registered trademark of Bolt Associates Inc.

downward force of the air in the upper chamber increases. The shuttle then returns to the armed position and the lower chamber again fills with air. The explosive release of the air occurs in 1–4 msec while the entire discharge cycle requires about 25–40 msec. Usually several air guns are used in parallel. Because the dominant frequency of the pulse depends on the energy (that is, on the product of the pressure and volume of air discharged), mixtures of gun sizes (the gun size is the volume of the lower chamber) from 10 to 2000 in^3 are sometimes used to give a broader frequency spectrum.

With the Unipulse* variation of the air gun (Mayne *et al.*, 1971) the flow of air into the bubble continues for some time after the initial discharge to retard the violent collapse of the bubble. This is achieved by dividing the lower chamber into two parts connected by a small orifice; the air in the lowermost chamber is delayed in passing through this orifice before discharging into the water.

The *Vaporchoc*† or *Steam Gun* injects superheated steam into the water. As shown in fig. 4.69, the superheated steam under high pressure passes through an insulated pipe to a submerged tank. When the valve is opened the steam emerges

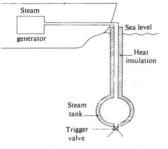

Fig. 4.69 Schematic diagram of Vaporchoc equipment. (Courtesy Compagnie Générale de Géophysique.)

into the water where it forms a bubble. The bubble collapses and disappears because the steam condenses; thus there is no subsequent bubble oscillation. The steam injection time is usually between 10 and 50 msec and shooting rates of 5 to 10 shots per minute are used. The time from valve opening to bubble collapse is not constant, however, so that multiple units are not used simultaneously. The steam injection generates a forerunner seismic pulse when the valve is opened but the main seismic pulse is generated by the bubble collapse. The forerunner precedes the main pulse by about 50 msec and may have 20% of the amplitude of the main pulse. The seismic tape is referenced to the time of valve opening; the time difference between this and the generation of the main pulse is removed in data processing.

The *Flexotir*‡ method utilizes the explosion of a small (2-oz) dynamite charge within a *cage*, a thick-walled cast-iron spherical shell about 2 ft in diameter with

* Registered trademark of Petty-Ray Geophysical Inc.
† Registered trademark of Compagnie Générale de Géophysique.
‡ Registered trademark of Institute Francais de Petrole.

many perforations spaced around the shell (see fig. 4.70). The charge is placed at the centre of the cage by pumping it down a hose leading from the ship; it is then fired electrically. Water is forced out through the perforations in the shell by the expanding gases. The water-flow out of and back into the shell as the bubble of gases oscillates dissipates the energy and dampens subsequent oscillations while having little effect on the initial expansion (Knudsen, 1961; Lavergne, 1970). Flexotir cannot be used in shallow water because, without a head of hydrostatic pressure, the bubble diameter becomes larger than the spherical shell, resulting in its destruction. Cages are sometimes used with large air guns to serve the same function of damping bubble oscillation.

Fig. 4.70 Flexotir sphere and loading hose. (Courtesy Institut Français du Petrole.)

The *sleeve exploder* or *Aquapulse** utilizes the explosion of a mixture of propane and oxygen in a closed flexible chamber. A heavy rubber sleeve (the *boot*) is fastened around a steel frame which is filled with the explosive mixture. The mixture is fired by a spark plug and the products of the explosion expand the rubber boot. A valve opens following the explosion so that the contraction of the boot vents the gases to the surface; this attenuates the bubble effect rapidly, although not completely.

* Registered trademark of Western Geophysical Company of America.

Several other arrangements utilizing explosive gas mixtures have been used. In a marine version of the Dinoseis, the explosion of gas in a metal chamber drives a piston outward against the confining water pressure. Gassp* uses a long neoprene tube in which an explosive gas mixture is fired. Spark plugs in each 20-ft module permit the entire gas mixture to be exploded at once, thus achieving a linear seismic source. The *seismic underwater exploder* (SUE)† and several other types of gas exploders use firing chambers open to the water. The chambers are filled with an explosive gas mixture and fired by a spark plug at the top of the chamber, the waste gases being vented into the water. If placed shallow in the water, such open-chamber exploders are inefficient; if placed deep, they generate severe bubble oscillation.

The *Maxipulse*‡ method makes use of the bubble effect by recording the bubble oscillation and using this information to 'deconvolve' the data (see §4.7.3*d*) in subsequent processing. A special cartridge containing a percussion delay detonator and $\frac{1}{2}$ lb (200 g) of explosive is pumped down a hose hydraulically. At the lower end of the hose the percussion cap strikes a firing wheel and after a one-second delay detonates the charge. During the one-second interval the ship moves about 5 m away from the charge so that the explosion does not destroy the hose and accessory equipment. A pressure detector mounted on the hose records the instant of detonation and the subsequent bubble waveform (*signature*). The detector is a flat-crystal hydrophone which is capable of withstanding large pressures. The source pressure signature is recorded on an auxiliary channel at fixed gain after alias filtering. The first bubble collapse often creates more seismic energy than the initial explosion and the second bubble collapse often 50% as much. The explosion time is not sufficiently predictable to permit the simultaneous use of multiple sources.

The *Aquaseis*§ system uses several hundred feet of detonating cord which trails behind the ship to give a linear charge involving about $\frac{1}{2}$ kg of explosive per 30 m. Secondary bubble pulses are small because of the low explosive content per unit of length.

Several types of imploders (Hydrosein**, Boomer†† and Flexichoc‡‡) are sometimes used. *Imploders* operate by creating a region of very low pressure; the collapse of water into the region generates a seismic shock wave. With the Flexichoc an adjustable-volume chamber is evacuated while the walls of the chamber are kept fixed by a mechanical restraint; upon removal of the restraint the hydrostatic pressure collapses the chamber and so generates a seismic pulse free of spurious bubbles. Air is then pumped into the chamber to expand it again where-

* Registered trademark of Shell Development Company.
† Registered trademark of Geophysical Service Incorporated.
§ Tradename of Imperial Chemical Industries Ltd.
** Registered trademark of Western Geophysical Company of America.
†† Registered trademark of E G & G International.
‡‡ Registered trademark of Compagnie Général de Géophysique.

upon the mechanical restraint holds it open while the air is evacuated, ready for the next collapse. With the Hydrosein, two plates are driven apart suddenly by a pneumatic piston, creating between them a very low pressure region into which the water rushes. With the Boomer, two plates are forced apart suddenly by a heavy surge of electrical current through a coil on one of the plates which generates eddy currents in the other plate, resulting in its being suddenly repelled. The Boomer produces less energy than Flexichoc or Hydrosein.

A marine verion of the Vibroseis has been used, often employing four source units simultaneously. As of 1975 many other marine sources were available and new ones were still being introduced. The use of solid explosives as marine energy sources (except for the Maxipulse and Flexotir methods) had nearly disappeared.

Comparison of the many different energy sources may be made on the basis of energy or waveform shape (*signature*). Rayleigh (1917) while studying the sounds emitted by oscillating steam bubbles related bubble frequency to bubble radius, pressure, and fluid density, and Willis (1941) while studying underwater explosions expressed the relationship in terms of source energy (the Rayleigh-Willis formula). The relationship between source energy and the dominant frequency of the seismic wave resulting therefrom is illustrated in fig. 4.71 for some of the marine seismic sources along with the Rayleigh-Willis curve. In general, large energy involves low frequency and vice-versa.

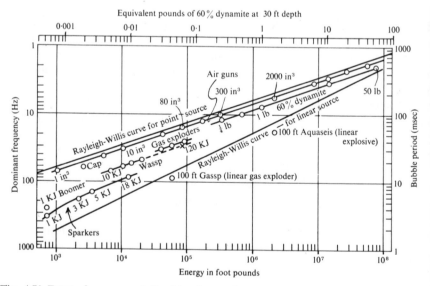

Fig. 4.71 Energy-frequency relationships for marine sources (30 ft depth). (From Kramer *et al.*, 1968.)

(d) *Marine detectors. Hydrophones* or marine pressure geophones are usually of the piezoelectric type, that is, they depend upon the fact that application of pressure to certain substances produces an electric potential difference between two surfaces of the material. Synthetic piezoelectric materials, such as barium zirconate, barium titinate or lead metaniobate are generally used. Hydrophones are often arranged in pairs so that their outputs cancel for translational accelerations but add for pressure pulses. Because hydrophones have high electrical impedance, impedance-matching transformers often are included for each group.

The hydrophones are mounted in a long *streamer* towed behind the seismic ship at a depth often between 10 and 20 m (Bedenbender *et al.* 1970). A streamer is shown diagrammatically in fig. 4.72 and a photograph of a portion in fig. 4.73. Twenty or more hydrophones spaced at intervals of a few feet are connected in series (so that the voltages generated add up) to form a single equivalent geophone 30–100 m in length. The crystals, connecting wires and a stress member (to take the strain of towing) are placed inside a neoprene tube which is then filled with sufficient lighter-than-water oil to make the streamer neutrally buoyant, that is, so that the average density of the tube and contents equals that of the sea water. A lead-in section 100 m or more in length is left between the stern of the ship and the first group of hydrophones to reduce pickup of ship noises. Dead sections are also sometimes included between the different hydrophone groups to give the spread length desired. The last group is often followed by a tail section to which is attached a buoy which floats on the surface; visual or radar sighting on this buoy is used to determine the amount of drift of the streamer away from the track of the seismic ship (caused by water currents). This buoy also helps retrieve the streamer if it should be broken accidentally. The total length of streamer in the water is 1000–2400 m.

When not in use, the streamer is stored on a large motor-driven reel on the stern of the ship. When the streamer is being towed, depth controllers and other devices keep the active part of the cable horizontal at the proper depth. Depth detectors may be included at several places within the streamer to verify that the depth is correct. Water-break detectors are also included at several places along the streamer; these are high-frequency (500–5000 Hz) hydrophones which detect energy from the shot travelling through the water. Knowing the velocity of sound in the water permits converting the water-break traveltime into the offset distance.

When towing a streamer a seismic ship must avoid stopping, making sharp turns, or even drastic reductions in speed, since otherwise the streamer tends to drift which may allow it to get into dangerous locations such as into the ship's propellors. The depth controllers raise or depress the cable by virtue of water flowing by their vanes and thus become ineffective when the streamer is not in motion. Consequently if a shot is missed for any reason it cannot be repeated because the ship will have passed the location. If too many shots are missed the ship must circle with a diameter larger than the streamer length in order to get the streamer in proper location to make up the missed shots. This is very time consuming and consequently very expensive.

(e) *Marine positioning.* Marine seismic navigation involves two aspects: (*a*) placing the ships at a desired position, and (*b*) determining the actual location afterwards

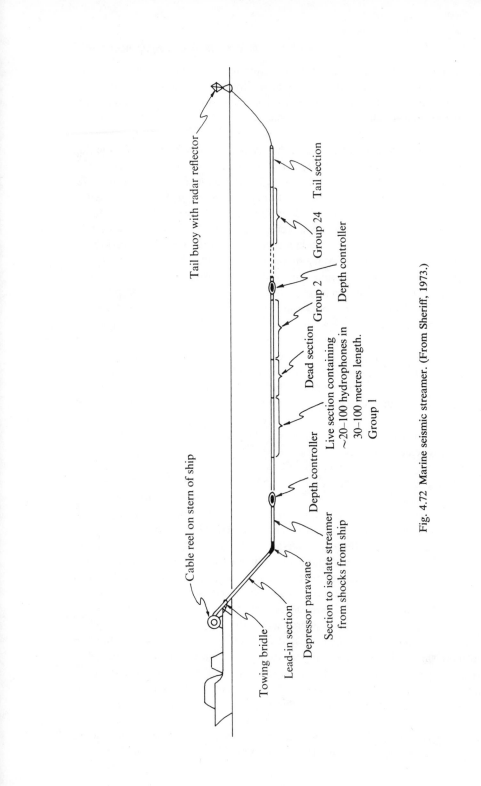

Tail buoy with radar reflector

Cable reel on stern of ship

Towing bridle

Lead-in section

Depressor paravane

Section to isolate streamer
from shocks from ship

Depth controller

Live section containing
~20–100 hydrophones in
30–100 metres length.
Group 1

Dead section

Group 2

Depth controller

Group 24   Tail section

Fig. 4.72 Marine seismic streamer. (From Sheriff, 1973.)

Fig. 4.73 Photo of seismic streamer. Plastic spacers (*a*) are connected by three tensile cables (*b*); a bundle of electrical conductors (*c*) passes through holes in the centres of the spacers. The hydrophone is at (*d*). The streamer covering is a soft plastic which is filled with a liquid to make the streamer neutrally buoyant. A depth controller (*e*) is clamped over the streamer. (Courtesy Seismic Engineering.)

so that the data can be mapped properly. Sometimes (as with reconnaissance surveys) it is not too important that the data be obtained exactly at predetermined locations provided that one can subsequently determine accurately the actual locations which were occupied. In assessing the accuracy of a navigation method, we must distinguish between absolute and relative accuracy. Absolute accuracy is important in tying marine surveys to land surveys and in returning to a certain point later, for example, to locate an offshore well. Relative accuracy is important primarily to ensure the proper location of one seismic profile relative to the next. Relative accuracies of ±20 m are desirable whereas absolute accuracies of ±100 m are sufficient. The actual accuracies obtained in a survey (which are usually very difficult to assess) depend upon the system and equipment used, the configuration of shore stations, the position of the mobile station with respect to the shore stations, variations in the propagation of radio waves, instrument malfunctioning, operator error and so on. Systems capable of giving adequate accuracy under good conditions may not realize such accuracy in geophysical surveys unless considerable care is exercised at all times.

Many types of navigation methods are available, including radiopositioning devices, sonic devices, observations of navigation satellites, etc. Each has advantages and disadvantages, and combinations of systems are often used so that an advantage of one system may compensate for a disadvantage of another.

Radiopositioning, which depends upon the use of radio waves, is used to locate many marine surveys relative to fixed shore stations. Radiopositioning methods can be divided into two basic types depending upon the type of measurement: (*a*) systems which measure the time required for a radio-frequency pulse to travel between a mobile station and a shore station (examples are radar, shoran, the rho-rho mode of Loran-C), and (*b*) systems which measure the difference in traveltime (or phase) of signals from two or more shore stations (these include Raydist, Lorac, Decca, Toran, Loran, and Omega).

Radar and shoran are similar in principle. *Radar* depends upon the reflection of pulses by a target, the distance to the target being equal to one-half the product of the two-way traveltime of the reflected pulse and the velocity of the radio waves. *Shoran* differs from radar in that the target is a shore station which receives the pulse and rebroadcasts it with increased power so that the return pulse is strong. Two or more shore stations are used and the position of the mobile station is found by swinging arcs as in fig. 4.74.

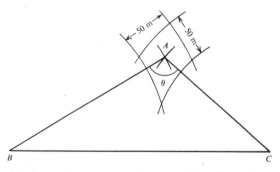

Fig. 4.74 Effect of station angle on errors in shoran position. θ = station angle, A = mobile station, B and C = shore stations. Point A can be anywhere inside the 'parallelogram' formed by the four arcs. [Note: range errors are not to scale.]

Radar and shoran are high-frequency systems, radar frequencies being in the range 3000–10,000 MHz, shoran in the range 225–400 MHz. Since such high frequencies are refracted only very slightly by the atmosphere these methods are basically line-of-sight devices. With normal antennae heights of about 30 m, the range for shoran is roughly 80 km. If the shore stations can be located on hills adjacent to the sea greater ranges can be obtained. By using very sensitive equipment (directional antennae and preamplifiers), ranges of 250 km can be obtained; this variation is called *extended-range* or *XR shoran*. The extension of range beyond the line-of-sight appears to be due to refraction, diffraction and scattering from the troposphere. In some tropical or subtropical regions strong temperature gradients in the atmosphere refract the radio waves so that ranges of 300 km or more can be obtained.

The distance between the ship station and each shore station is normally measured within ±25 m (0·2 μsec). The error in location depends mainly upon the angle between lines joining the shore stations to the mobile station, as shown in fig. 4.74; angles between 30° and 150° usually are considered acceptable.

Several devices utilize the same principles as shoran but use the higher radar frequencies; they 'interrogate' a small *transponder*, a device which emits a signal immediately upon receipt of the interrogating signal. These include RPS and Trisponders, which use frequencies around 9500 MHz and Autotape and Hydrodist, which use frequencies around 3000 MHz. The effective range of these devices is strictly line-of-sight but they are extremely portable.

If two shore stations simultaneously broadcast a radio pulse or coded sequence of pulses, a mobile station can measure the difference in arrival times and so find the difference in distances to the two shore stations. The locus of points with constant difference in distance from two shore stations (*A* and *B* in fig. 4.75, for

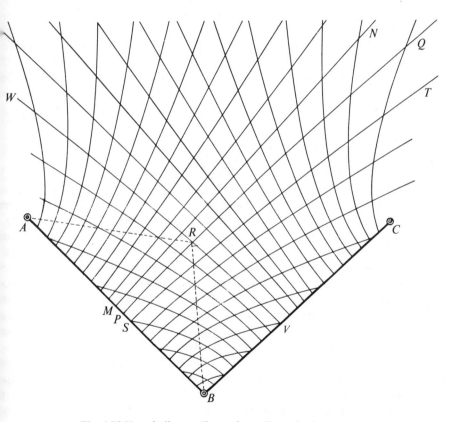

Fig. 4.75 Hyperbolic coordinates for radio-navigation systems.

example) is a hyperbola with foci at the two stations; thus a single measurement determines a hyperbola PQ passing through the location of the mobile station, R. If the difference in arrival times for a second pair of stations (B and C) is measured, the mobile station is located also on the hyperbola VW and hence at the inter-section of the two hyperbolas. This principle forms the basis of Loran and Omega, long-range radio-navigation systems maintained by the U.S. government. *Loran-C* involves the broadcast of a coded sequence of pulses of frequency 100 kHz, the broadcast times being controlled very accurately by atomic clocks. By phase-matching individual cycles, accuracies of 100–300 m can be obtained; however, variations in the ionosphere, atmosphere, ground conductivity and other factors over the long ranges which are involved often make the accuracy only marginally acceptable for seismic surveying. Omega uses frequencies from 10 to 13 kHz but its accuracy is usually considered inadequate for seismic surveying.

Many medium-frequency radio-positioning systems utilize the broadcasting of continuous waves (CW) from several stations, locations being determined by comparison of phase. *Phase-comparison systems* used in seismic exploration gener-ally operate in the frequency band 1·5–4·0 MHz and have ranges up to 400 miles.

Referring again to fig. 4.75, shore stations A and B transmit steady continuous sinusoidal signals which are exactly in phase at M, the midpoint of the baseline AB. A mobile station with a phase-comparison meter will show zero phase difference at M and at all points on the perpendicular bisector MN. If $MP = \lambda/2$ the phase-comparison meter indicates zero phase difference at P also; if the mobile station moves away from P in such a direction that the phase difference remains constant, it traces out the hyperbola PQ. In general, a point R moving in such a way that

$$RA - RB = n\lambda, n = 0, \pm 1, \pm 2, \pm 3, \ldots$$

traces out the family of hyperbolae shown in the diagram.

The zone between two adjacent zero-phase-difference hyperbolae is called a *lane*. If we start from a known point and maintain a continuous record of the phase difference, we know in which lane the mobile station is located at any given time. By using a second pair of stations (one of which can be located at the same point as one of the first pair of stations) transmitting a different frequency we obtain a second family of hyperbolae, hence another hyperbolic coordinate of the mobile station. The accuracy of location decreases with increasing lane width as we go farther from the base stations, also as the angle of intersection of the hyperbolas decreases. Location accuracy is of the order of 30–100 m. If the continuous count of lanes is lost, however, one could be considerably off location as the phase difference meters give only the position within a lane and do not indi-cate in which lane. The major factor governing accuracy in actual system usage therefore is maintaining the lane count accurately. Interference with signals reflected from the ionosphere becomes variable around sunrise and sunset when the ionospheric layering changes because of sunlight-induced ionization, and it is sometimes impossible to maintain accurate lane count during such periods.

Atomic clocks which maintain excellent accuracies have made it possible to maintain highly accurate time at different stations, including mobile stations. Knowing the instant of transmission of a pulse from shore stations thus permits the direct measurement of the traveltime, hence of the distance or *range* to that station. This allows medium- and low-frequency systems to be used as range-measuring devices, locations being determined by measuring two or more ranges. Toran-O and ANA are systems utilizing atomic clocks. Loran-C can also be used in the range-range (or *rho-rho*) mode when the mobile station has an atomic clock.

Doppler-sonar is a *dead-reckoning* system, that is, it determines position with respect to a starting point by measuring and integrating the ship's velocity. The ship's velocity is measured by projecting sonar beams against the ocean floor in four directions from the ship (fig. 4.76). These beams are reflected back to the ship

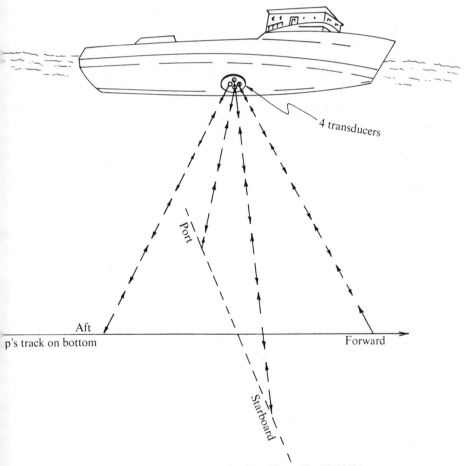

Fig. 4.76 Doppler-sonar navigation. (From Sheriff, 1973.)

but their frequencies undergo a doppler-shift because of motion of the ship with respect to the ocean floor. The frequency shift in each beam thus gives the component of the ship's velocity in that direction. These can be resolved to give the ship's actual velocity (in conjunction with direction information from a gyrocompass) and the velocity can be integrated to give the ship's position. Small errors in velocity measurement accumulate in the integration resulting in position uncertainty of the order of 200 m/hr. The requisite accuracy has to be maintained by periodic *updates*, i.e., periodic determinations of location by independent measurements. In deep water, scatter of the sonar beams by inhomogeneities in the water dominates and the doppler-shifts give a measure of the velocity with respect to the water rather than the ocean floor, resulting in considerable loss of accuracy.

Many seismic ships are equipped to determine their location from observations of *navigation satellites*. The U.S. Navy has placed a number of satellites (six as of 1974) in polar orbits around the world. Each satellite takes about 100 minutes to circle the earth, being in sight of a point under its orbit for about 18 minutes (horizon to horizon). Each satellite transmits continuous waves of frequencies 150 and 400 MHz. The frequencies measured by a receiver on the ship are doppler-shifted because of the relative motion of the satellite with respect to the ship. The differences between the ship's longitude and latitude and the satellite's longitude and latitude at closest approach (see fig. 4.77) are calculated from the doppler-shifts. The satellite transmits information which gives the satellite's location every two minutes. A small computer on the ship combines this information with the doppler-shift measurements and the speed and course of the ship to give the ship's location. Each satellite can be observed on four or more orbits each day, hence 20 or more determinations of position are possible each day. However, the satellites are not uniformly spaced and do not have precisely the same orbital period so that sometimes more than one satellite is visible while at other times several hours may intervene without any satellite being visible. One may expect about two-thirds of the 'passes' to result in satisfactory *fixes* or determinations of position. Satellite fixes may be accurate within ±60 m, provided the ship's velocity is accurately known. The principal disadvantage of satellite navigation for seismic purposes is that it gives no information about position during the interval between fixes.

Most commonly doppler-sonar, gyrocompass and satellite navigation are combined. The satellite gives the periodic updating information needed to maintain the doppler-sonar accuracy while the doppler-sonar along with the gyrocompass gives the velocity information needed for an accurate satellite fix. Satellite fixes may also be combined with radio-positioning systems; the radio system gives velocity information and the satellite fixes remove lane ambiguity from the radio-determined positions. Doppler-sonar and radiopositioning may also be combined, the radiopositioning giving locations in deep water where the doppler-sonar cannot 'see' the ocean floor and the doppler-sonar giving track information during periods of skywave interference.

(f) *Profiling*. The technique of *marine profiling* usually employs a single geophone group, and shooting and recording take place at such short intervals that an essentially continuous record section is obtained. The technique is similar to the continuous recording of water depth using a fathometer (echo sounder). It differs from CDP marine shooting in that it employs smaller ships and weaker energy sources and hence is much cheaper; profilers also have relatively small penetration and cannot discriminate between events on the basis of normal moveout since only

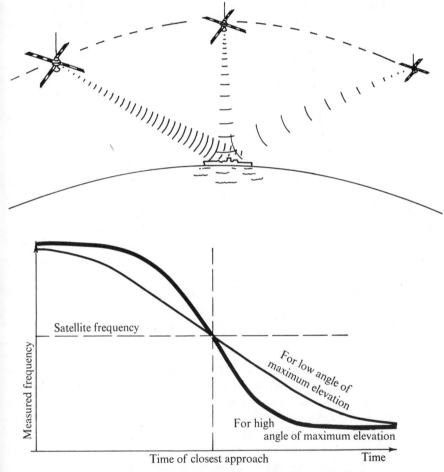

Fig. 4.77 Position-fix from navigation satellite. (From Sheriff, 1973.)

a single geophone group is used. Profiling is extensively used in engineering studies to map the bottom sediments and to locate bed-rock. It is also widely used in oceanographic work to survey large areas cheaply.

The energy sources most commonly used for profiler work are high-powered fathometers, electric arcs, air guns and imploders. The high-powered fathometers are usually piezoelectric devices employing barium titanate or lead zirconate.

Such materials not only generate electric fields when they are compressed (as when they are used as hydrophones) but they also change dimensions when subjected to an electric field. They are thus *transducers* because of their ability to transform electrical energy to acoustic energy and vice-versa. Fathometers used for profiling operate at lower frequencies and higher power levels than fathometers used for water depth measurements; frequencies in the range 2–10 kHz and power levels of roughly 100 watts are commonly used. A repetition rate of once every 2 sec or so is used and penetration of 20–100 m is generally achieved. Sub-bottom reflections usually permit the mapping of the various layers of mud and silt overlying bed-rock, as shown in fig. 4.78. Reflection character can sometimes be interpreted to indicate the nature of the sediments, for example, to find sand layers which can support structures erected on pilings. Surveys for other purposes, such as ones designed to locate pipe-lines buried in the mud, are also made at times.

Electric arcs used as sources for profiler work, usually called *sparkers*, utilize the discharge of a large capacitor to create a spark between two electrodes located in the water. The heat generated by the discharge vaporizes the water creating an effect equivalent to a small explosion. Several sparker units are often used in parallel to give increased penetration. The penetration obtained by earlier models was small but modern sparker arrays deliver as much as 200 kilojoules at 50–2000 Hz and achieve penetrations of 600 m or so (although a 5000-joule source with a penetration of less than 300 m is more common). A variation of the sparker called *Wassp** involves connecting the electrodes by a thin wire which is vaporized by the energy discharge. This increases the duration of the bubble and consequently its low-frequency content.

The air guns used in profiling are similar to those used in CDP marine work except that they are smaller, involving as little as 1 cu. in. of air at 1000 psi and a dominant frequency around 250 Hz. The imploders used in profiling include the Boomer already described and deliver about 200 joules of energy from 50 Hz to a few kilohertz.

Profiler data frequently are recorded on electrosensitive paper using a strip recorder. The paper is moved slowly beneath a rotating cylinder with its axis transverse to the direction of paper movement. A helical wire makes one turn around the cylinder. An electrical current passing from the wire through the paper to the metal plate over which the paper rides produces the image on the paper. The source is fired when the helical wire is touching the paper at the top; as time increases the point on the paper which the wire touches moves down the paper as the cylinder rotates so that the mark produced by energy picked up by the hydrophone is located properly. The interval between source impulses is a small integral multiple of the time of rotation of the cylinder (often 2 sec), the paper moving only a small fraction of an inch during this interval, so that the returning energy is plotted nearly vertically beneath the shotpoint. Reflections with arrival

* Trademark of Teledyne Exploration Co.

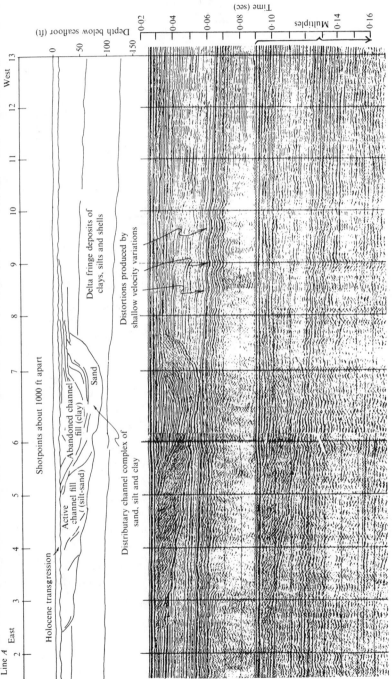

Fig. 4.78 Profiler record showing sub-bottom deposits. (From King, 1973.)

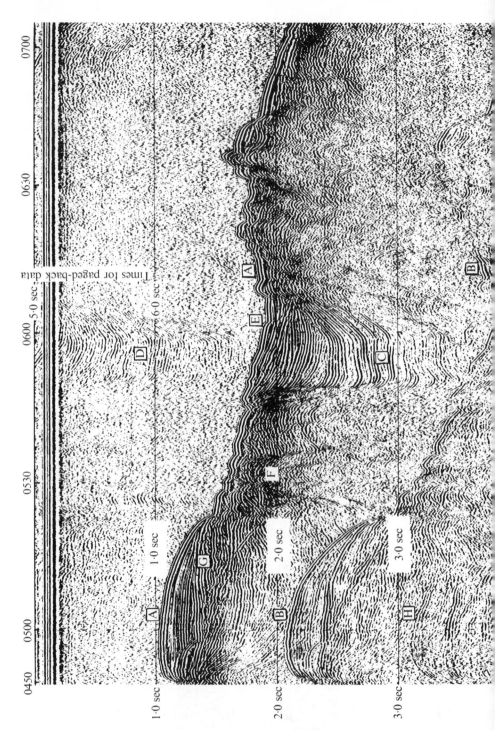

Fig. 4.79 Profiler record offshore Japan. The water-bottom reflection (A) with arrival times of 1·0 to 2·0 sec indicates water depths of 750–1500 m. The ship travelled 8500 m between the 30-minute marks shown at the top of the record. Most primary reflections are obscured below the water-bottom multiple (B) and subsequent multiples. Over 1000 m of sediments are indicated in the vicinity of C; multiples of these appear paged-back at D. E indicates a fault scarp on the ocean floor. F are diffractions, probably from sea-floor relief slightly offset to the line. Note the onlap and thinning above G. H is a second multiple of the ocean floor. (Courtesy Teledyne Exploration.)

times greater than the rotation time of the cylinder will then be written on top of the shallower record for the succeeding impulse, a technique sometimes called paging (see fig. 4.79). Usually it is clear which data belong with which page (or set of arrival times) although occasionally paging creates ambiguity.

Multiples of the seafloor often are very strong and make data arriving after them unusable (Allen, 1972). In engineering work where interest is concentrated in the relatively thin layer of unconsolidated sediments overlying bedrock such as shown in fig. 4.78 this usually does not create a problem; where the unconsolidated sediments are thick the seabottom is apt to be soft (not involving a large acoustic impedance contrast) and hence the seabottom multiple relatively weak. In deep-water oceanographic work a long period of time (a wide *window*) elapses before the ocean bottom multiple arrives (as in fig. 4.79) so that appreciable data are recorded without ambiguity.

4.5.6 *Measurement of velocity*

(a) *Conventional well surveys.* The most accurate methods of determining velocity require the use of a deep borehole. Two types of *well surveys* are used: the 'conventional' method of shooting a well and sonic logging (or continuous velocity survey) which is discussed in the next section.

Shooting a well consists of suspending a geophone or hydrophone in the well by means of a cable and recording the time required for energy to travel from a shot fired near the well down to the geophone (see fig. 4.80). The geophone is specially constructed to withstand immersion under the high temperatures and pressures encountered in deep oil wells. The cable has a threefold role: it supports the geophone, it serves to measure the depth of the geophone and it carries electrical conductors which bring the geophone output to the surface where it is recorded. Shots are fired at one or more points near the wellhead. The geophone is moved between shots so that the results are a set of traveltimes from the surface down to various depths. The geophone depths are chosen to include the most important geological markers, such as tops of formations and unconformities, and also intermediate locations so that the interval between successive measurements is small enough to give reasonable accuracy (often 200 m apart).

Results of a typical well survey are shown in fig. 4.81. The vertical traveltime, t, to the depth, z, is obtained by multiplying the observed time by the factor $(z/\sqrt{\{z^2 + x^2\}})$ to correct for the actual slant distance. The average velocity between the surface and the depth z is then given by the ratio z/t. Figure 4.81 shows the average velocity V and the vertical traveltime t plotted as functions of z. If we subtract the depths and times for two shots, we find the *interval velocity* V_i, the average velocity in the interval $(z_m - z_n)$, by means of the formula

$$V_i = \frac{z_m - z_n}{t_m - t_n}. \tag{4.56}$$

Shooting a well gives the average velocity with good accuracy of measurement. It is, however, expensive since the cost includes not only the one-half to one day's

$$t_{0_1} = t_1 \left(\frac{z}{\sqrt{x^2 + z^2}} \right)$$

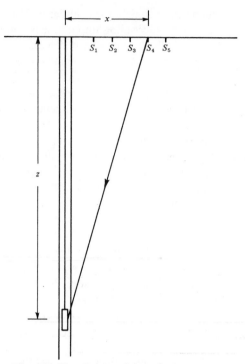

Fig. 4.80 Shooting a well for velocity.

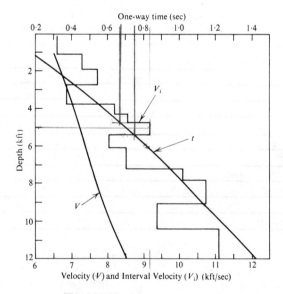

Fig. 4.81 Plot of well-velocity survey.

time of the seismic crew but also the cost of standby time for the well (which often exceeds the seismic cost). Potential damage to the well is another factor which discourages shooting wells; while the survey is being run, the well must stand without drill stem in the hole and hence is vulnerable to cave-in, blow-out, or other serious damage. A further disadvantage in new exploration areas is that seismic surveys are often completed before the first well is drilled.

For marine well surveys, airgun energy sources are used as well as explosives (Kennett and Ireson, 1971). The airguns may be merely hung over the side of the drill platform or drill ship and hence only a small crew is required to obtain data at many depths.

(b) *Velocity logging*. The continuous-velocity survey makes use of one or two pulse generators and two or four detectors, all located in a single unit called a *sonde* which is lowered into the well. Figure 4.82*a* shows the borehole-compensated sonic-logging sonde developed by Schlumberger. It consists of two sources of

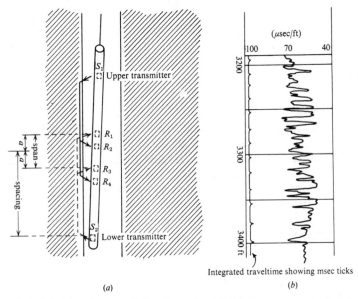

Integrated traveltime showing msec ticks

(*a*)   (*b*)

Fig. 4.82 Sonic logging. (Courtesy Schlumberger.) (*a*) Borehole-compensated logging sonde; (*b*) sonic log.

seismic pulses, S_1 and S_2, and four detectors, R_1 to R_4, the 'span' distances from R_1 to R_3 and from R_2 to R_4 being 2 ft. The velocity is found by measuring the traveltime difference for a pulse travelling from S_1 to R_2 and R_4, similarly for a pulse going from S_2 to R_3 and R_1, then taking the average of the differences. The sonde is run in boreholes filled with drilling mud which has a seismic velocity of roughly 1500 m/sec; however the first energy arrivals are the P-waves which have travelled in the rock surrounding the borehole. Errors arising from variations in borehole size or mud cake thickness near the transmitters are effectively eliminated by measuring the difference in arrival time between two receivers; errors resulting from such variations near the receivers are reduced by averaging the results from the two pairs of receivers. The sonic log (fig. 4.82b) shows as a function of depth the transit time divided by the span (expressed in μsec/ft), the result being the reciprocal of the P-wave velocity in the formation.

The traveltime interval between sonic log receivers is measured by a device which automatically registers the arrival of the signal at each of the two receivers and measures the time interval between the two. Since the signal at the receiver is not a sharp pulse but instead is a wave train, the detector is actuated by the first peak (or trough) which exceeds a certain threshold value. At times the detector is not actuated by the same peak (or trough) at the two receivers and hence the increment of traveltime will be in error. This effect, called *cycle skip*, is readily detected on the sonic log and is easily allowed for since the error is exactly equal to the known interval between successive cycles in the pulse.

The sonic log is automatically integrated to give total traveltime which is then shown as a function of depth by means of ticks at intervals of 1 msec (see fig. 4.82b). However, there is a tendency for small systematic errors to accumulate in the integrated result and check shots are made at the base and top of the sonic log so that the effect of the cumulative error can be reduced by distributing the difference in a linear manner. The sonde often includes a well seismometer of the type used in shooting a well to facilitate taking the check shots.

The instantaneous velocity fluctuates rapidly in many formations, as seen in fig. 4.82b. Nevertheless, while the velocity distribution, if considered in detail, is an extremely irregular function, the wavelengths used in seismic exploration are so long (generally greater than 30 m) that the rapid fluctuations are not significant in determining the path of the wave.

Sonic logs are used for porosity determination (see §11.4.2b) because porosity appears to be the dominant factor in seismic velocity. Although sonic logs are of great value to the geophysicist, they are usually not run with the geophysical uses in mind and hence often do not produce all the information which the geophysicist wants. For example, check shots are not necessary for porosity determination and, therefore, are often omitted; the log usually does not cover the entire hole depth and sonic log data are rarely available for the shallow part of the borehole. Thus the sonic log data are usually incomplete so that using such logs for velocity control involves assumptions where the data are missing. This is especially true in the applications of sonic logs in preparing synthetic seismograms with which we shall deal in §4.8.2c.

Several variations of acoustic borehole logs are used sometimes though they have little application in seismic work. The entire waveform (including other energy arrivals in addition to the first-arrival P-wave) at the receivers is displayed on some logs. Sonic logs are discussed further in §11.4.2.

(c) $X^2 - T^2$ *method.* The arrival time of reflected energy depends not only on the reflection depth and the velocity above the reflector but also on offset distance. Several methods (including the velocity analysis methods which will be described in §4.7.7a) utilize this dependence on offset as a means of measuring the velocity. Two classical methods, $X^2 - T^2$ and $T - \Delta T$ (described in the next section) are central to all surface velocity measurement methods, even though both methods have fallen into disuse.

The $X^2 - T^2$ *method* is based upon eq. (4.41a), that is,

$$t^2 = x^2/V^2_{rms} + t_0^2.$$

When we plot t^2 as a function of x^2, we get a straight line whose slope is $1/V^2_{rms}$ and whose intercept is t_0^2 from which we can determine the corresponding depth.

When the regular seismic profile does not have a sufficiently large range of x-values to enable us to find V with the accuracy required for interpretation purposes, special long-offset profiles are shot, generally using an arrangement described by Dix (1955). This arrangement begins with the symmetrical profile DEF (referring to fig. 4.83a) which gives values from $x = 0$ to $x = \pm s$. Next the

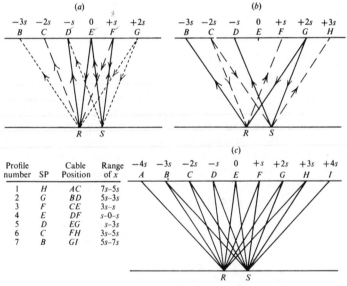

Fig. 4.83 Layout of $X^2 - T^2$ profiles.

shotpoint is moved to the right a distance s (to the point F) and the spread is moved the same distance to the left (to the position CE); values are now obtained from $x = -s$ to $x = -3s$. This process is continued until the maximum desired value of x is reached. This scheme ensures that the reflected energy all comes from the same interval RS of the reflecting bed.

Equation (4.35) shows that the sign of the term which depends on dip $(4hx \sin \theta)$ depends upon the sign of x; thus, averaging values of t^2 for geophones symmetrically disposed with respect to the point E eliminates the effect of dip. (If the dip is small, sufficient accuracy is achieved by averaging the values of t and then squaring the average.) The additional data needed to remove the dependency on dip are obtained by repeating the process described in the preceding paragraph except that the shotpoint is moved to the left and the spread to the right (opposite to that done previously—see fig. 4.83b). The complete set of $X^2 - T^2$ profiles is shown in fig. 4.83c.

An important feature of the Dix arrangement of $X^2 - T^2$ profiles is that time-ties are obtained because reversed paths are involved. These can be used as checks that the picking of the reflection events has not involved a jump of one leg, that is, one cycle. Referring again to fig. 4.83c, we have the following time ties:

Profile	Path	Offset	Reversed path	Tying profile
1	HSC	−5s	CSH	6
6	CRF	+3s	FRC	3
3	FSE	−s	ESF	4
4	ERD	+s	DRE	5
5	DSG	−3s	GSD	2
2	GRB	+5s	BRG	7

An $X^2 - T^2$ survey can give velocities accurate within a few per cent where (1) the records are of good quality and have at least a moderate number of reflections, (2) accurate near-surface corrections are applied, (3) the field work and the interpretation are carefully done, and (4) the velocity distribution is simple (that is, no lateral variation of velocity or complexity of structure).

Once the velocities have been determined to two successive parallel reflectors using eq. (4.41a), the interval velocity, V_n, can be found from the Dix formula. Writing V_L for the rms velocity to the nth reflection and V_U for the rms velocity to the reflection above it, eq. (4.41b) gives

$$\sum_1^n V_i^2 t_i = \sum_1^{n-1} V_i^2 t_i + V_n^2 t_n = V_L^2 \sum_1^n t_i,$$

$$\sum_1^{n-1} V_i^2 t_i = V_U^2 \sum_1^{n-1} t_i.$$

Subtracting and dividing both sides by t_n then gives the Dix formula,

$$V_n^2 = (V_L^2 \sum_1^n t_i - V_U^2 \sum_1^{n-1} t_i)/t_n. \tag{4.57}$$

(d) *T − ΔT method.* The $T - \Delta T$ *method* is based upon eq. (4.34b), which can be written in the form:

$$V = \frac{x}{\sqrt{(2t_0 \Delta t_n)}}.$$

$(4.34b)$ $\Delta t_n = \dfrac{x_2}{2V^2 t_0}$

With symmetrical spreads Δt_n can be calculated from the arrival times of a reflection event at the shotpoint (t_0) and at the outside geophone groups, t_1 and t_l. Dip moveout is eliminated by averaging the moveouts on the opposite sides of the shotpoint:

$$\Delta t_n = \tfrac{1}{2}[(t_1 - t_0) + (t_l - t_0)] = \tfrac{1}{2}(t_1 + t_l) - t_0. \qquad (4.58)$$

The values of Δt_n given by this equation are subject to large errors, mainly because of uncertainties in the near-surface corrections. To get useful results, large numbers of measurements must be averaged in the hope that weathering variations and other uncertainties will be sufficiently reduced; thus, $T - \Delta T$ analyses may involve hundreds of records. Values are usually quantized into groups according to the values of t_0; for example, values of Δt_n in the interval from 1·500 to 1·600 sec might be averaged to give the velocity V corresponding to $t_0 = 1\cdot550$ sec.

4.5.7 Data reduction

(a) *Preliminary processing.* The interpretation of seismic data usually is based on the study of paper records or record sections. The initial step in processing the data is to 'write up' the records. On a label pasted or imprinted on the seismic record or record section one enters (*a*) identifying information such as the name of the company and prospect, (*b*) line and profile numbers, (*c*) survey data (location of the shotpoint and geophone groups), (*d*) description of the spread (number and spacing of geophones per group, in-line and perpendicular offsets of the shot-point), (*e*) source information (size and depth of explosive charge or type of source and source pattern), (*f*) amplifier information (filter and gain settings), (*g*) time of shooting the profile, (*h*) history of any processing which may have been done, etc. Such information may be already printed on computer-generated record sections.

Weathering and elevation corrections are calculated as described in the following section. A time reference, $t = 0$, is marked on the records; this is the shot instant shifted in order to take into account corrections for weathering and elevation so that arrival times measured with respect to this time reference give depths measured from the datum. Timing lines are then labelled at 0·1 sec intervals starting from the time reference.

(b) *Elevation and weathering corrections.* Variations in the elevation of the surface affect traveltimes and it is necessary to correct for such variations as well as for changes in the low-velocity-layer (LVL). Usually a *reference datum* is selected and corrections are calculated so that in effect the shotpoints and geophones are located on the datum surface, it being assumed that conditions are uniform and that there is no LVL material below the datum level.

Many methods exist for correcting for near-surface effects. These schemes are usually based on (1) uphole times, (2) refractions from the base of the LVL, or (3) the smoothing of reflections. We shall describe several of these methods which are simple to apply and adequate to cover most situations. Automatic statics-correction schemes which usually involve statistical methods of smoothing reflections will be discussed in §4.7.4*d*. We shall assume that V_W and V_H, the velocities in the LVL and in the layer just below it, are known; these can be found from an uphole survey or the refraction first breaks, as will be discussed later in the section. In what follows we also assume that the shot is placed below the base of the LVL; if this is not true, modifications have to be made in the following equations in this section (see problem 10).

Figure 4.84 illustrates a method of obtaining the correction for t_0, the shotpoint arrival time. E_d is the elevation of the datum, E_s the elevation of the surface at the

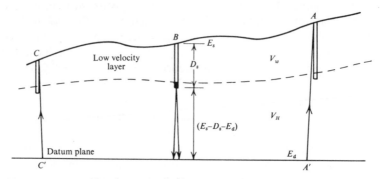

Fig. 4.84 Calculation of weathering corrections.

shotpoint, D_s the depth of the shot below the surface, and t_{uh} is the uphole time, the time required for energy to travel vertically upward from the shot to a geophone on the surface at the shotpoint (in practice the uphole geophone is 3 m or so from the shothole). The deviation of reflection paths from the vertical is usually small enough that we can regard the paths as vertical; therefore the time required for the wave to travel from the shot down to the datum is Δt_s, where

$$\Delta t_s = \frac{(E_s - D_s - E_d)}{V_H}. \tag{4.59}$$

Similarly, the time for the wave to travel up from the datum to a geophone on the surface at B is Δt_g where

$$\Delta t_g = \Delta t_s + t_{uh}. \tag{4.60a}$$

The correction, Δt_0, for the traveltime at the shotpoint is then

$$\Delta t_0 = \Delta t_s + \Delta t_g = 2\Delta t_s + t_{uh} = \frac{2(E_s - D_s - E_d)}{V_H} + t_{uh}. \quad \tag{4.60b}$$

Subtraction of Δt_0 from the arrival time t_0 is equivalent to placing the shot and the shotpoint geophone group on the datum plane, thereby eliminating the effect of the low-velocity-layer if the shot is beneath the LVL. At times the shot may be so far below the datum plane that Δt_s will be negative.

When eq. (4.37*a*) is used to calculate dip, the dip moveout must be corrected for elevation and weathering. The correction to the dip moveout, Δt_c, often called the *differential weathering correction,* is the difference in arrival times at opposite ends of a split-dip spread for a reflection from a horizontal bed. Referring to fig. 4.84, the raypaths from the shot B down to a horizontal bed and back to geophones at A and C have identical traveltimes except for the portions $A'A$ and $C'C$ from the datum to the surface. Assuming as before that $A'A$ and $C'C$ are vertical, we get for Δt_c the expression

$$\Delta t_c = (\Delta t_g)_C - (\Delta t_g)_A$$
$$= (\Delta t_s + t_{uh})_C - (\Delta t_s + t_{uh})_A, \qquad (4.61)$$

where we assume that A and C are shotpoints so that the quantities in brackets are known. If we take the positive direction of dip to be down from A towards C, then Δt_c must be subtracted algebraically from the observed moveout to obtain the true dip moveout.

The following calculation illustrates the effect of the correction. We take as datum a horizontal plane 200 m above sea level; V_H is 2075 m/sec. Data for three successive shotpoints, A, B and C, such as those in fig. 4.84 are as follows:

	Shotpoint C	Shotpoint B	Shotpoint A	
E_s (m)	248	244	257	measured
D_s (m)	15	13	20	
t_{uh} (msec)	48	44	53	
Δt_s (msec)	16	15	18	calculated
Δt_g (msec)	64	59	71	
Δt_0 (msec)	80	74	89	
Δt_c (msec)		−7		

Let us suppose that a reflection on a split profile from shotpoint B gives the following data: $t_0 = 2.421$ sec, $t_A = 2.419$ sec, $t_C = 2.431$ sec. Then, the corrected value of t_0 is $2.421 - 0.074 = 2.347$ sec and the corrected dip moveout is

$$\Delta t_d = 2.431 - 2.419 - (-0.007) = 0.012 + 0.007 = 19 \text{ msec.}$$

If the dip moveout had been negative, for example, −6 msec, the corrected value would have been $-6 - (-7) = +1$ msec so that the correction can change the direction of dip as well as the dip magnitude. Therefore accurate corrections are essential.

Corrections are often required for geophones in between shotpoints where uphole data are not available. The first breaks are frequently used for this purpose. In fig. 4.85, G is a geophone intermediate between adjacent shotpoints A and B for which we have first-break traveltimes. Let t_{AG} and t_{BG} be the first-break times for the paths $A'C'G$ and $B'C''G$. Almost always GC' and GC'' are within 20° of the vertical and $C'C''$ is therefore small. Thus, we can write the approximate relation

$$t_{AG} + t_{BG} \simeq \frac{A'B'}{V_H} + 2t_W \simeq \frac{AB}{V_H} + 2t_W,$$

t_W being the traveltime through the weathered layer at G. Thus,

$$t_W \simeq \{t_{AG} + t_{BG} - (AB/V_H)\}/2. \tag{4.62}$$

Subtracting t_W from the arrival times in effect places the geophone at the base of the LVL; to correct to datum we must subtract the additional amount,

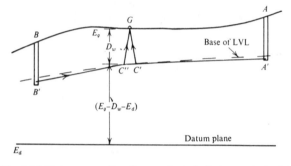

Fig. 4.85 Datum correction for geophone in between shotpoints.

$(E_g - E_d - D_W)/V_H$, where E_g is the mean elevation of the geophone group, D_W being found by multiplying t_W by V_W.

Occasionally special refraction profiles are shot to obtain data for making corrections for intermediate geophones. These profiles may be of the standard type using small charges placed near the surface or a non-dynamite source on the surface; these are interpreted using standard methods such as Wyrobek's (see §4.8.3b) to find the depth and traveltime to the base of the LVL. Alternatively, a shot may be placed just below the LVL as in fig. 4.86; in this event we must modify (4.47a) since the shot is at the base rather than the top of the upper layer. Thus,

$$t = \frac{x - D_W \tan \theta}{V_H} + \frac{D_W}{V_W \cos \theta} = \frac{x}{V_H} + \frac{D_W \cos \theta}{V_W}. \tag{4.63}$$

Most near-surface correction methods require a knowledge of V_H and sometimes of V_W as well. The former can be determined by: (1) an uphole survey, (2) a special refraction survey as described above or (3) analysis of the first breaks for distant geophone groups (since these are equivalent to a refraction profile such as that shown in fig. 4.86). The weathering velocity V_W can be found by (1) measuring the slope of a plot of the first breaks for geophones near the shot-point (correcting distances for obliquity), (2) dividing D_s by t_{uh} for a shot placed near the base of the LVL, (3) an uphole survey, or (4) firing a cap at the surface and measuring the velocity of the direct wave.

(c) *Picking and grading of reflection events.* When the record quality is poor almost any alignment may be mistakenly identified as a reflection. The best criterion in such cases is often the geological picture which results. If this picture does not make sense, we should re-examine the geophysical data with more skepticism. Naturally, 'making sense' does not mean that the result must fit our preconceived ideas but rather that it must be geologically plausible.

When the interpreter decides that an event is a legitimate reflection, he usually marks it and 'times' it. When working with individual records the arrival times at

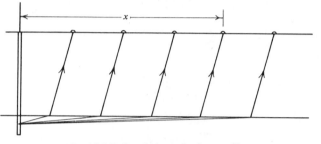

Fig. 4.86 Refraction weathering profile.

the centre of the record and at the two outside traces (or the difference between these outside times, Δt_d), usually corrected for weathering and elevation, are often written directly on the record.

Besides timing reflections, the interpreter often assigns a *grade* to each, for example, VG, G, P, ? (for very good, good, poor, questionable). These grades refer to the certainty that the event is a primary reflection and the accuracy of measurement of the arrival times. Sometimes a two-letter grading system is used to separate the grading of certainty from the grading of timing accuracy.

The process of identifying events on a seismic record and selecting and timing the reflections is referred to as *picking* the records. Figure 4.87 shows a picked record.

(d) *Preparation of cross-sections.* The next stage after the picking of individual records is to prepare a composite representation of all the data for a given line. This can be done by plotting a seismic cross-section on a sheet of graph paper. The shotpoint locations are marked at the top of the sheet on a horizontal line indicating the datum plane. The reflection events for each record are plotted below the corresponding shotpoint according to their arrival times. Plots are often made also of the surface elevation, the depth of shot, the depth of the base of the LVL and the first-break arrival times.

Cross-sections are called *time sections* if the vertical scale is linear with time or *depth sections* if linear with depth. Occasionally the horizontal scale is in units of time (obtained by dividing horizontal distances by V_H, the sub-weathering velocity). A cross-section is *unmigrated* when reflections are plotted vertically below the shotpoint. A *migrated section* (such as fig. 4.88) is one for which we

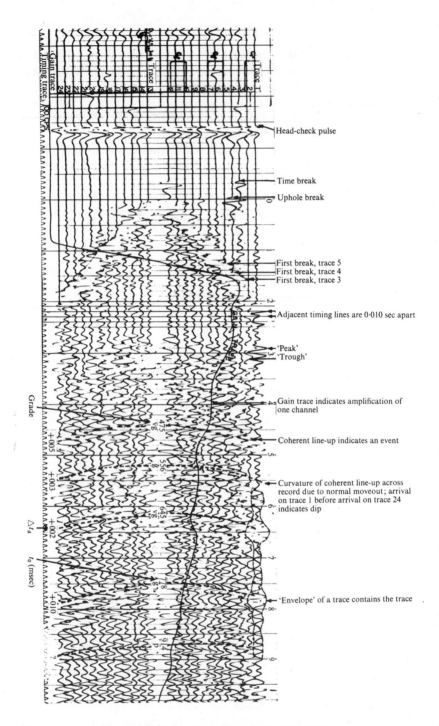

Fig. 4.87 Seismic record. (Courtesy Chevron Oil.)

assume that the seismic line is normal to strike so that the dip moveout indicates true dip and we then attempt to plot in their actual locations the segments of reflecting horizons which produced the recorded events. The scale on migrated sections is usually linear with depth.

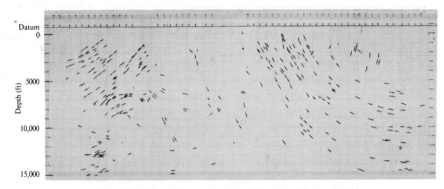

Fig. 4.88 Migrated cross-section. (Courtesy Chevron Oil.)

Unmigrated sections give a distorted picture of the subsurface, the distortion increasing with the amount of dip; they are used for interpretation only in areas of gentle dip or when an accurate structural picture is not required, as for example in a reconnaissance survey. When the dips exceed 5° or so, the distortion inherent in an unmigrated section makes it difficult to determine the subsurface geometry accurately. Figure 4.89 shows the effect of lack of migration on simple structures. The anticline at the left would appear on an unmigrated section as the dotted line $R'ST'$ while the syncline would appear as the dotted line $T'UV'$. Failure to migrate decreases the curvature of an anticline and increases that of a syncline. Since the breadth and height of an anticline affect the volume of oil which would be trapped in the structure, the importance of migration is obvious.

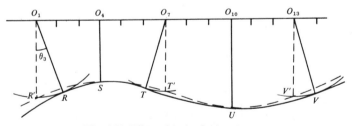

Fig. 4.89 Effect of lack of migration.

Several methods of migration are used. The simplest is to assume a constant velocity down to the reflector and swing an arc whose radius is half the arrival time t_0 multiplied by the average velocity. The radius O_1R in fig. 4.89 makes an angle with the vertical equal to the reflector dip (calculated using (4.37a)) and a straight line segment equal to half the spread length is drawn at R perpendicular to the radius (tangent to the arc) to represent the reflecting segment. This method of migrating has the effect of moving reflector segments from the unmigrated position by too great a distance (*overmigrating*; see problem 13) and hence is not suitable when the dip is large.

Probably the commonest method of migrating reflections uses a wavefront chart, a graph showing wavefronts and raypaths for an assumed vertical distribution of velocity such as shown in fig. 4.81. A wavefront chart is constructed assuming a standard spread length, Δx. The angles of approach at the surface associated with convenient values of dip moveout Δt_d (such as 0, ± 10, . . . ± 50 msec) are calculated using eq. (4.37b). Assuming that different waves left the shotpoint along each of these directions, we trace the paths downward applying Snell's law at each velocity interface. We draw wavefronts through the points reached by the waves travelling along each of the paths at convenient time intervals after the shot is fired (such as 0·4, 0·5 sec, etc.). Each wavefront is labelled with a value of t_0 equal to twice the above time to allow for the return traveltime after reflection.

The simple wavefront chart shown in fig. 4.90 is for seven layers of varying thicknesses with velocities ranging from 6000 to 12,600 ft/sec, assuming $\Delta x = 2000$ ft. To plot a cross-section, the wavefront chart is placed under transparent

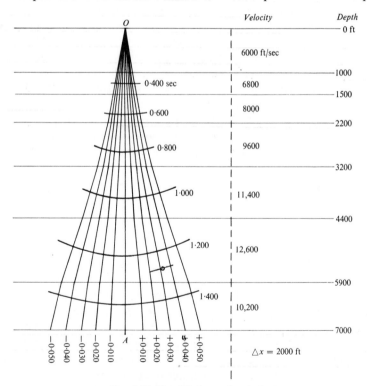

Fig. 4.90 Simplified wavefront chart.

graph paper with the point O at the appropriate shotpoint and the line OA extending vertically downward. A reflection with certain values of t_0 and Δt_d is plotted by interpolating between the wavefronts and rays (actual wavefront charts have more closely spaced wavefronts and rays so that interpolation is more accurate) and drawing a straight line of length equal to half the spread length tangent to the wavefront at the point $(t_0, \Delta t_d)$. The reflection denoted by the symbol $-\circ-$ in the diagram corresponds to $t_0 = 1\cdot270$ sec, $\Delta t_d = 32$ msec.

When reflection data are interpreted using a velocity function which increases linearly with depth, the migrating scheme is simple, especially when use is made of a plotting machine as illustrated in fig. 4.91. The values of h, R, and i_1 are calculated using eqs. (4.45) and (4.44c). Often the plotting arms are labelled directly in arrival time as in fig. 4.91.

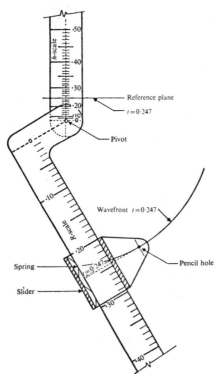

Fig. 4.91 Plotter for linear increase of velocity with depth. (From Rockwell, 1967.)

With an asymmetric spread, we correct the measured moveout for the difference in normal moveout between the two ends of the spread to find the dip moveout, Δt_d. With spread lengths different from the standard length, we multiply the dip moveout by the ratio of the standard spread length to the actual spread length. With in-line offsets we find Δt_d by applying the normal moveout corrections (correcting the arrival time to find t_0 when the offsets are large). On record sections (see next section) we may measure the dip over any distance over which the horizon is dipping uniformly and then correct the value to the standard spread length, allowance being made for the fact that distances on the record section usually represent subsurface coverage while the spread length is twice the subsurface coverage.

Once the cross-section has been prepared, the next step is to identify reflections which are continuous over a large part of the line. Records are compared to identify reflections which appear on both, this *correlation* being based partly upon the similarity of character and partly upon the dip and the agreement in arrival times (*time-tie*). The travel paths for reflections on the outside traces of adjacent profiles which provide continuous subsurface coverage are the same except that they are traversed in opposite directions; hence the traveltimes should be the same provided that adequate weathering and elevation corrections have been applied. For example, in fig. 4.46 trace 1 on the profile shot from O_2 (with path O_2BO_3) should have the same corrected traveltime as trace 24 on the profile sh⌣ from O_3. When two reflections can be correlated, a line is drawn on the cross-section joining the two segments. If the record quality is reasonably good, the result is a series of reflections which can be followed for varying distances.

All too frequently either the record quality is too poor to provide continuous reflections or the continuous reflections which are present are not in the part of the section where information is required. In this event we resort to *phantoms*, lines drawn on the section so that they are parallel to those adjacent dip symbols which the interpreter considers valid. Where data conflict or are absent, the phantom is drawn in the manner which seems most reasonable to the interpreter on the basis of whatever fragmentary evidence may be available.

(e) *Record sections.* Record sections can be regarded as a form of time section obtained by recording successive records side by side. Record sections are usually corrected for elevation, weathering and normal moveout.

Record sections display a large amount of data in a compact form which affords a more comprehensive picture of the geology than can be obtained from the plotted section. They permit one to encompass several kilometres of data at a glance. Correlation of reflections on adjacent records is usually obvious, thus eliminating the time-consuming comparison of individual records. *Jump correlations*, correlations across breaks in the continuity of a reflection, can be made more readily and with more certainty than with individual records or plotted sections. Zones of bad data can be used to deduce structural information, such as the presence of faults. The identification of multiples and diffractions is facilitated. Variable area or variable density record sections can be prepared on a sufficiently reduced scale that a section representing 50 miles or more of line can be viewed at one time, which is useful in presenting the results of a survey.

Record sections also have disadvantages. They frequently are not migrated or are migrated without regard for cross dip and hence give a distorted picture of the structure whenever the dip is appreciable. Some detailed information about phase breaks, envelope, and amplitude is lost, especially with reduced-scale variable-density sections. Not the least of the potential disadvantages occurs because record sections, especially on a reduced scale, are so graphic that people who lack technical understanding of the full significance of the data tend to attach incorrect meanings to the various markings as though the section were a 'photograph' of the rock formations in the ground. It is a short step from this viewpoint to the conclusion that any one in possession of a soft pencil is capable of doing seismic interpretation.

4.6 Refraction field methods and equipment

4.6.1 *Comparison of refraction and reflection methods*

Refraction and reflection work are similar in many aspects, much different in others. The similarities are sufficient that reflection field crews frequently do refraction profiling, though often not with the efficiency of a crew specifically designed for refraction. The differences between reflection and refraction field work mostly result from the long shot-to-geophone distances employed in refraction. The energy input to the ground must be larger for refraction shooting and explosives continue to be the dominant energy source although other seismic sources are also used. The longer travel paths result in the higher frequencies being mostly absorbed so that refraction data are generally of low frequency compared with reflection data. Consequently refraction geophones have lower natural frequencies than reflection geophones, although the response of the latter is often adequate for satisfactory refraction recording. Most digital seismic equipment can be used for refraction but some of the older analogue equipment did not have adequate low-frequency response.

Refraction shooting is usually much slower than reflection shooting because the large offset distances involve more moving time and create problems of communications and logistics. However, refraction profiles are often not as closely spaced as reflection lines and hence the cost of mapping an area is not necessarily greater.

4.6.2 *Refraction profiles*

(a) *In-line refraction profiling.* The basic refraction field method involves shooting reversed refraction profiles, a long linear spread of many geophone groups shot from each end, the distance being great enough that the dominant portion of the travel path is as a head wave in the refractor or refractors being mapped. Usually it is not practical to record simultaneously so many geophone groups spread over such a long distance, and hence refraction profiles are usually shot in segments. Referring to fig. 4.92*a* which shows a single refractor, the spread of geophone groups might be laid out between *C* and *D* and shots at *C* and *G* fired to give two records; the spread would then be moved between *D* and *E* and shots fired at *C* and *G* as before, and 'so on to develop the complete reversed profile *CDEFG*.

The charge size is often varied for the different segments because larger charges are required when the offset becomes greater. Usually one or two groups will be repeated for successive segments to increase the reliability of the time-tie between segments.

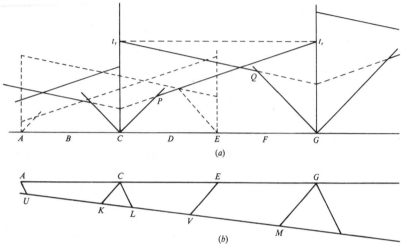

Fig. 4.92 Reversed refraction profiles. (*a*) Time-distance plot for continuous reversed profiling; (*b*) section showing single refractor.

The shothole at C can also be used to record a profile to the left of C and the shothole at G a profile to the right of G. Note that the *reciprocal time* t_r is the same for the reversed profiles and that the intercept times for profiles shot in different directions from the same shotpoint are equal. These equalities are exceedingly valuable in identifying segments of complex time-distance curves where several refractors are present. In simple situations the reversed profile can be constructed without having to actually shoot it by using the reciprocal time and intercept time information. However, usually situations of interest are sufficiently complicated that this procedure cannot be carried out reliably.

The *reversed profiles* shot from C and G allow the mapping of the refractor from L to M. The reversed profile to the left of C permits mapping as far as K but no coverage is obtained for the portion KL. Hence continuous coverage on the refractor requires an overlap of the reversed profiles; a reversed profile between A and E (shown dotted in fig. 4.92*a*) would provide coverage between U and V, thus including the gap KL as well as duplicating the coverage UK and LV. With perfect data the duplicate coverage does not yield new information but in actual profiling it provides valuable checks which increase the reliability of the interpretation.

If we have the two-refractor situation in fig. 4.93, first break coverage on the shallow refractor is obtained from L to K and from M to N when the shots are at C and G; the corresponding coverage on the deeper refractor is from Q to S and from R to P.

If we are able to resolve the refraction events which arrive later than the first breaks, called *second arrivals* or *secondary refractions*, we can increase the coverage obtained with a single profile. However, it is difficult, and sometimes impossible, to adjust the gain to optimize both the first breaks and the second arrivals at the same time; if the gain is too low the first breaks may be weak and ambiguities in timing may result whereas if the gain is too high the secondary refractions may be unpickable. Because of this difficulty, prior to magnetic-tape recording, refraction mapping was generally based on first breaks only. With magnetic tape, recording playbacks can be made at several gains so that each event can be displayed under optimum conditions.

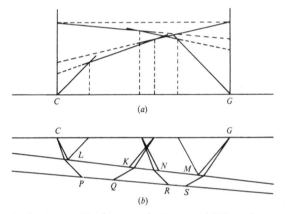

Fig. 4.93 Reversed refraction profiles for two-refractor case. (*a*) Time-distance plot; (*b*) section showing two refractors.

In order to economize on field work, the portions of the time-distance curves which do not add information necessary to map the refractor of interest often are not shot where they can be predicted reasonably accurately. Thus, the portions *CP* and *GQ* of the reversed profile in fig. 4.92*a* often are omitted.

Where a single refractor is being followed, a series of short refraction profiles are often shot rather than a long profile. In fig. 4.94 geophones from *C* to *E* are used with shotpoint *C*, from *D* to *F* with shotpoint *D*, etc. The portions of the time-distance curves attributable to the refractor being mapped are then translated

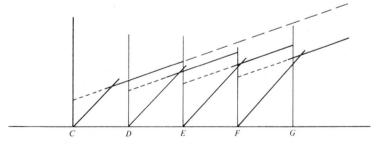

Fig. 4.94 Unreversed refraction profiles for a single refractor.

parallel to themselves until they connect together to make a composite time-distance curve such as that shown by the dashed line. The composite curve may differ from the curve which would actually have been obtained for a long profile from shotpoint *C* because of refraction events from other horizons.

(b) *Broadside refraction and fan shooting.* In *broadside refraction* shooting, shotpoints and spreads are located along two parallel lines (see fig. 4.95) selected so that the desired refraction event can be mapped with a minimum of interference from other events. Where the refraction event can be clearly distinguished from other arrivals, it provides a very economical method of profiling because all the data yield information about the refractor. However, usually the criteria for

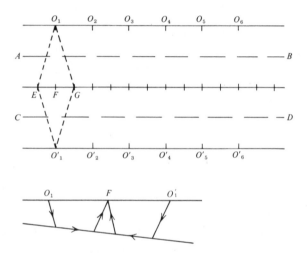

Fig. 4.95 Broadside refraction profiling.

identifying the refraction event are based on in-line measurements (such as the apparent velocity or the relationship to other events) and these criteria are not available on broadside records where the offset distance is essentially constant. Thus if the refractor should unexpectedly change its depth or if another refraction arrival should appear, one might end up mapping the wrong horizon. Consequently broadside refraction shooting is often combined with occasional in-line profiles in order to check on the identity of the horizon being mapped.

The first extensive use of refraction was in searching for salt domes by the *fan-shooting* technique. A salt dome inserts a high-velocity mass into an otherwise low-velocity section so that horizontally travelling energy arrives earlier than if the salt dome were not present; the difference in arrival time between that actually observed and that expected with no salt dome present is called a *lead*. In fan-shooting (fig. 4.96) geophones are located in different directions from the shot-point at roughly the same offset distances, the desire to maintain constant offset distance usually being sacrificed in favour of locations which are more readily accessible. The leads shown by overlapping fans then roughly locate the high-velocity mass. This method is not used for precise shape definition.

4.6.3 *Engineering surveys on land*

The shallow refractions used in engineering applications, such as in determining the depth to bedrock, do not require large energy sources or complex instrumentation. Energy sources are usually very simple, for example, a hammer striking a steel plate on the ground, the instant of impact being determined by an inertial switch on the hammer. Sometimes the energy is obtained by a hand-operated 'tamper', by a weight dropped on the ground or by a small explosion. (Such sources are also used for reflection engineering surveys; see Meidav, 1969.)

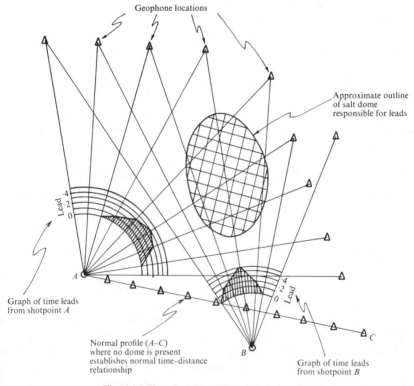

Fig. 4.96 Fan shooting. (From Nettleton, 1940.)

The energy is usually detected by moving-coil geophones similar to those already described. Often only a few channels are used, one to six usually, because otherwise the spread layout becomes complicated. The amplifiers and the camera generally weigh only a few pounds and often are contained in a small metal suitcase. In some systems the recorded data are displayed on a small oscilloscope tube and are photographed with a Polaroid camera so that a permanent record is obtained. In other systems time counters are started at the instant the energy is delivered to the ground and are stopped when the first-break energy arrives at the geophones, thus giving direct readings of traveltimes.

4.6.4 *Marine refraction work*

Because refraction recording requires that there be appreciable distance between the source and the recording locations, two ships have usually been required for marine refraction recording. To shoot a reversed refraction profile in one traverse requires three ships – a shooting ship at each end while the recording ship travels between them. For the shooting ships to travel the considerable distances between shotpoints takes appreciable time because of the relatively low maximum speed of ships and hence the high production rates which make marine reflection work economical are not realized in refraction shooting. Consequently marine refraction work is relatively expensive.

The *sonobuoy* (fig. 4.97) permits recording a refraction profile with only one ship. The sonobuoy is an expendable listening station which radios the information it receives back to the shooting ship. The sonobuoy is merely thrown overboard; the salt water activates batteries in the sonobuoy as well as other devices which cause a radio antenna to be extended upward and one or two hydrophones to be suspended beneath the buoy. As the ship travels away from the buoy, shots are fired and the signals received by the hydrophones are radioed back to the ship where they are recorded. The arrival time of the wave which travels directly through the water from the shot to the hydrophone is used to give the offset distance. After a given length of time the buoy sinks itself and is not recovered. Sonobuoys make it practical to record unreversed refraction profiles while carrying out reflection profiling, the only additional equipment cost being that of the sonobuoy.

4.6.5 *Refraction data reduction*

Refraction data have to be corrected for elevation and weathering variations, as with reflection data. The correction methods are essentially the same except that often geophones are too far from the shotpoint to record the refraction at the base of the LVL and thus there may be no weathering data along much of the line. Additional shots may be taken for special refraction weathering information.

Where complete refraction profiles from zero offset to large offsets are available, playback of the data with judicious selections of filters and AGC may allow one to correlate reflection events with refraction events, thus adding useful information to each type of interpretation. Often the most prominent reflections will not correspond to the most prominent refractions.

Another useful refraction playback technique is to display the data as a *reduced refraction section* where arrival times have been shifted by the amount x/V_R where V_R is near the refractor velocity. If V_R were exactly equal to the refractor velocity the residual times would be the delay times (which will be discussed in §4.8.3b) and relief on the reduced refraction section would correlate with refractor relief (although displaced from the subsurface location of the relief,

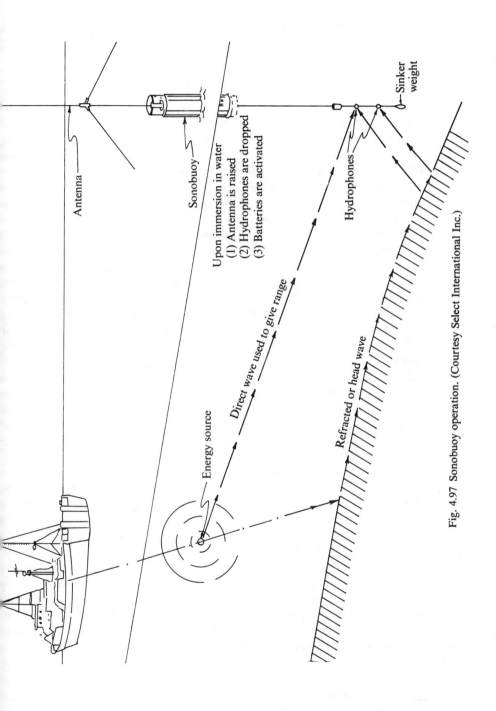

Antenna

Sonobuoy

Upon immersion in water
(1) Antenna is raised
(2) Hydrophones are dropped
(3) Batteries are activated

Sinker weight

Hydrophones

Energy source

Direct wave used to give range

Refracted or head wave

Fig. 4.97 Sonobuoy operation. (Courtesy Select International Inc.)

However, even if V_R is only approximately correct, the use of reduced sections improves considerably the pickability of refraction events, especially secondary refractions.

✳ 4.7 Data processing

4.7.1 *Introduction*

Radar was one of the outstanding technological advances of World War II and was widely used in the detection of aircraft. However, noise frequently interfered with the application of radar and considerable theoretical effort was devoted to the detection of signals in the presence of noise. The result was the birth of a new field of mathematics – Information Theory. Early in the 1950s a research group at the Massachusetts Institute of Technology studied the application of the new field to seismic exploration problems (Flinn *et al*. 1967). Simultaneously with this development, rapid advances in digital computer technology made extensive calculations economically feasible for the first time. These two developments began to have an impact on seismic exploration in the early 1960s and before the end of the decade, data processing (as their application is called) had changed seismic exploration dramatically, so much so that the changes are sometimes referred to as the 'digital revolution'. Most seismic recording is now done in digital form and most data are now subjected to data processing before being interpreted.

At first, information theory was very difficult to understand because it was formulated in complex mathematical expressions and employed an unfamiliar vocabulary; however, as the number of applications has expanded, the basic concepts have been expressed more clearly and simply (Silverman, 1967; Finetti *et al* 1971). Rather than follow the classical development or attempt a rigorous approach, we shall present (without rigorous derivations) a simplified version designed to facilitate the understanding of modern seismic data processing.

Usually we think of seismic data as the variation with time (measured from the shot instant) of the amplitudes of various geophone outputs. When we take this viewpoint we are thinking in the *time domain*, that is, time is the independent variable. We also sometimes find it convenient to regard a seismic wave as the result of the superposition of many sinusoidal waves differing in frequency, amplitude and phase; the relative amplitudes and phases are regarded as functions of the frequency and we are thinking in the *frequency domain*. The frequency domain approach is illustrated by electrical systems which are specified by their effects on the amplitudes and phases of sinusoidal signals of different frequencies. For example, graphs of filter characteristics usually show amplitude ratios or phase shifts as ordinates with frequency along the abscissa.

4.7.2 *Fourier transforms*

Fourier analysis in our context involves transforming functions from the time domain to the frequency domain and *Fourier synthesis* the inverse process of transforming from the frequency domain to the time domain. This distinction is somewhat artificial, however, and analysis and synthesis could be interchanged without making any difference in the final result. The important point with transforms is that no information is lost in transforming. We can, thus, start with a waveform in the time domain, transform it into characteristics in the frequency domain, and then transform the frequency-domain characteristics into a waveform which is identical with the original waveform. This makes it possible to do part of our processing in the time domain and part in the frequency domain, taking advantage of the fact that some processes can be executed more economically in one domain than in the other. In actual transformations we lose a small amount of information because of the truncation of series expansions and round-off errors and such, but we can make these losses as small as we wish by carrying enough terms or enough decimal places.

If we have a reasonably 'well-behaved' periodic function $g(t)$ of period T (that is, $g(t)$ repeats itself every time t increases by T), then the function can be represented by a complex Fourier series

$$g(t) = \sum_{n=-\infty}^{\infty} \alpha_n e^{j2\pi f_n t}, \tag{4.64a}$$

where $f_n = n/T$ and

$$\alpha_n = \frac{1}{T} \int_{-T/2}^{+T/2} g(t) e^{-j2\pi f_n t} dt. \tag{4.65a}$$

Equation (4.64a) can also be written in the real form,

$$g(t) = a_0/2 + \sum_{n=1}^{\infty} (a_n \cos 2\pi f_n t + b_n \sin 2\pi f_n t)$$

$$= \sum_{n=0}^{\infty} c_n \cos (2\pi f_n t - \phi_n) \tag{4.66a}$$

where

$$a_n = (\alpha_n + \alpha_{-n}) = \frac{2}{T} \int_{-T/2}^{+T/2} g(t) \cos 2\pi f_n t \, dt,$$

$$b_n = j(\alpha_n - \alpha_{-n}) = \frac{2}{T} \int_{-T/2}^{+T/2} g(t) \sin 2\pi f_n t \, dt, \tag{4.67a}$$

$$c_0 = a_0/2 = \alpha_0, \phi_0 = 0;$$

for $n \neq 0$,

$$c_n = a_n \cos \phi_n + b_n \sin \phi_n = \frac{2}{T} \int_{-T/2}^{T/2} g(t) \cos (2\pi f_n t - \phi_n) \, dt$$

$$= (a_n^2 + b_n^2)^{\frac{1}{2}}, \tag{4.68}$$

$$\phi_n = \tan^{-1}(b_n/a_n).$$

Equation (4.66a) shows that $g(t)$ can be regarded as the sum of an infinite number

of cosine waves of frequency f_n having amplitude c_n and phase ϕ_n. It thus represents the analysis of the function $g(t)$ into component cosine waves.

As the period T becomes larger, it takes longer for $g(t)$ to repeat; in the limit when
T becomes infinite, $g(t)$ no longer repeats. In this case we get in place of (4.64a) and
(4.65a)

$$g(t) = \int_{-\infty}^{\infty} G(f)\, e^{j2\pi f t}\, df, \tag{4.64b}$$

$$G(f) = \int_{-\infty}^{\infty} g(t)\, e^{-j2\pi f t}\, dt. \tag{4.65b}$$

The function $G(f)$ is the *Fourier transform* of $g(t)$ while $g(t)$ is the *inverse Fourier
transform* of $G(f)$. Using the symbol $\leftrightarrow$ to denote the Fourier-transform and
inverse Fourier-transform operations, we write

$$g(t) \leftrightarrow G(f).$$

We also refer to $g(t)$ and $G(f)$ as a *transform pair*.

Equations (4.64b) and (4.65b) can be written in several ways. In general $G(f)$ is
complex,

$$G(f) = A(f)\, e^{j\phi(f)}, \tag{4.69a}$$

where $A(f)$ and $\phi(f)$ are real and $A(f)$ is also positive. We call $A(f)$ the *amplitude
spectrum* and $\phi(f)$ the *phase spectrum* of $g(t)$. Substitution in (4.64b) gives

$$g(t) = \int_{-\infty}^{\infty} A(f)\, e^{j(2\pi f t + \phi(f))}\, df. \tag{4.64c}$$

For actual waveforms $g(t)$ is real and hence

$$g(t) = \int_{-\infty}^{\infty} A(f) \cos\{2\pi f t + \phi(f)\}\, df. \tag{4.66b}$$

Since $G(f)$ is complex we may separate it into real and imaginary parts,

$$G(f) = a(f) - jb(f), \tag{4.69b}$$

where

$$\left.\begin{aligned} a(f) &= \int_{-\infty}^{\infty} g(t) \cos 2\pi f t\, dt, \\ b(f) &= \int_{-\infty}^{\infty} g(t) \sin 2\pi f t\, dt. \end{aligned}\right\} \tag{4.67b}$$

The integrals, $a(f)$ and $b(f)$, are called the *cosine* and *sine transforms*, respectively.
When $g(t)$ is real, $a(f)$ and $b(f)$ are respectively even (symmetrical) and odd
(antisymmetrical) functions. If $g(t)$ is symmetrical, that is, if $g(t) = g(-t)$,
$b(f) = 0$.

4.7.3 *Convolution*

(a) *The convolution operation.* Let us now consider the time-domain operation called *convolution*. Assume that we feed into a system data sampled at regular intervals, for example, a digital seismic trace. The output of the system can be calculated if we know the *impulse response* of the system, that is, the response of the system when the input is a *unit impulse* (which has zero values everywhere except at the origin where it has the value unity; we write the unit impulse δ_t or $\delta(t)$ depending on whether we are dealing with sampled data or continuous functions). The impulse response of the system will be zero prior to $t = 0$ and then will have the values $g_0, g_1, g_2, \ldots$ at successive sampling intervals. We represent this process diagramatically thus:

$$\delta_t \rightarrow \boxed{\text{System}} \rightarrow g_t = [g_0, g_1, g_2, \ldots].$$

Most systems with which we deal are linear and time-invariant (or very nearly so). A *linear system* is one in which the output is directly proportional to the input while a *time-invariant system* is one in which the output is independent of the time when the input occurred. Writing $\delta_{t-\Delta}$ for a unit impulse which occurs at $t = \Delta$ rather than at $t = 0$, we can illustrate linear and time-invariant systems as follows:

Linear:

$$k\delta_t \rightarrow \boxed{\text{System}} \rightarrow kg_t = [kg_0, kg_1, kg_2, \ldots];$$

Time invariant:

$$\delta_{t-\Delta} \rightarrow \boxed{\text{System}} \rightarrow g_{t-\Delta} = \left[\underbrace{0, 0, 0, \ldots, 0}_{\Delta \text{ zeros}}, g_0, g_1, g_2 \ldots\right].$$

In the last bracket on the right, the first output different from zero is g_0 and occurs at the instant $t = \Delta$.

Obviously any input which consists of a series of sampled values can be represented by a series of unit impulses multiplied by appropriate amplitude factors. We can then use the above two properties to find the output for each impulse and by superimposing these we get the output for the arbitrary input. This process is called *convolution*.

We shall illustrate convolution by considering the output for a filter whose impulse response g_t is $[g_0, g_1, g_2] = [1, -1, \frac{1}{2}]$. When the input x_t is $[x_0, x_1, x_2] = [1, \frac{1}{2}, -\frac{1}{2}]$, we apply to the input the series of impulses $[\delta_t, \frac{1}{2}\delta_{t-1}, -\frac{1}{2}\delta_{t-2}]$ (the last two subscripts meaning that the impulses are delayed by one and two sampling intervals respectively) and obtain the outputs shown below:

$$\delta_t \rightarrow [1, -1, \tfrac{1}{2}],$$
$$\tfrac{1}{2}\delta_{t-1} \rightarrow [0, \tfrac{1}{2}, -\tfrac{1}{2}, \tfrac{1}{4}],$$
$$-\tfrac{1}{2}\delta_{t-2} \rightarrow [0, 0, -\tfrac{1}{2}, \tfrac{1}{2}, -\tfrac{1}{4}].$$

Summing we find the output

$$[\delta_t + \tfrac{1}{2}\delta_{t-1} - \tfrac{1}{2}\delta_{t-2}] \rightarrow [1, -\tfrac{1}{2}, -\tfrac{1}{2}, \tfrac{3}{4}, -\tfrac{1}{4}].$$

Convolution is illustrated in fig. 4.98. This operation is equivalent to replacing each element of the one set by an appropriately scaled version of the other set and then summing elements which occur at the same times. If we call the output z_t and denote the operation of taking the convolution by an asterisk, we may express this as

$$
\begin{aligned}
z_t &= x_t{}^*g_t \\
&= \sum_k x_{t-k}\, g_k \\
&= [x_0 g_0,\ x_0 g_1 + x_1 g_0,\ x_0 g_2 + x_1 g_1 + x_2 g_0, \ldots].
\end{aligned}
\tag{4.70a}
$$

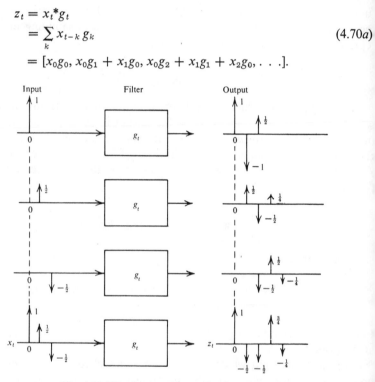

Fig. 4.98 Filtering as an example of convolution.

Note that we would have obtained the same result if we had input g_t into a filter whose impulse response is x_t; in other words, convolution is commutative:

$$x_t{}^*g_t = g_t{}^*x_t = \sum_k x_{t-k}g_k = \sum_k g_{t-k}x_k. \tag{4.70b}$$

While we have been expressing convolution as an operation on sampled data one can also convolve a sample set with a continuous function:

$$z(t) = x_t{}^*g(t) = \sum_k x_k g(t - k). \tag{4.70c}$$

Each term in the summation represents the function $g(t)$ displaced and scaled (displaced to the right k units and multiplied by x_k). A special case of (4.70c) is that of convolving a continuous function $g(t)$ with a unit impulse located at $t = h$:

$$\delta_{t-h}{}^*g(t) = \sum_k \delta_{k-h}g(t-k) = g(t - h)$$

(since δ_{k-h} is zero except for $k = h$, where $\delta_{k-h} = 1$). Hence convolving $g(t)$ with a time-shifted unit impulse displaces the function by the same amount and in the same direction as the unit impulse is displaced.

One can also convolve two continuous functions. In this case the summation becomes an integration:

$$z(t) = x(t)^*g(t) = \int_{-\infty}^{\infty} x(\tau)g(t - \tau)\,d\tau. \tag{4.70d}$$

The *Convolution Theorem* states that the Fourier transform of the convolution of two functions is equal to the product of the transforms of the individual functions; we can state the theorem as follows:

$$x_t \leftrightarrow X(f) = |X(f)|\,e^{j\phi_x(f)},$$
$$g_t \leftrightarrow G(f) = |G(f)|\,e^{j\phi_g(f)},$$
$$x_t{}^*g_t \leftrightarrow X(f)G(f) = [|X(f)|\,e^{j\phi_x(f)}][|G(f)|\,e^{j\phi_g(f)}]$$
$$\leftrightarrow |X(f)||G(f)|\,e^{j[\phi_x(f)+\phi_g(f)]}, \tag{4.71a}$$

where $|X(f)|$ and $|G(f)|$ are the amplitude spectra and $\phi_x(f)$ and $\phi_g(f)$ the phase spectra. This means that if two sets of data are convolved in the time domain, the effect in the frequency domain is to multiply their amplitude spectra and to add their phase spectra. Because of certain symmetry properties of the Fourier transform it can be shown that

$$x_t g_t \leftrightarrow X(f)^*G(f). \tag{4.71b}$$

(b) *Sampling and aliasing.* In the analogue-to-digital conversion, we replace the continuous signal with a series of values at fixed intervals. It would appear that we are losing information by discarding the data between the sampling instants. The transform relationship in eqs. (4.71a) and (4.71b) can be used to understand sampling and the situations in which information is not lost.

We make use of the *comb* or *sampling function*; this consists of an infinite set of regularly-spaced unit impulses (fig. 4.99b). The transform of a comb is also a comb:

$$\text{comb}\,(t) \leftrightarrow k_1\,\text{comb}\,(f), \tag{4.72a}$$

where k_1 depends upon the sampling interval (see problem 15). If the comb in the time domain has elements every 4 msec, the transform has elements every $1/0{\cdot}004 = 250$ Hz. We shall also make use of the *boxcar* (fig. 4.99d), a function which has a constant value between the values $\pm f_0$ and is zero everywhere else.

The transform of a boxcar is a *sinc function*:

$$\text{boxcar}\,(f) \leftrightarrow k_2 \,\text{sinc}\,(2\pi f_0 t) = k_2 \,\frac{\sin 2\pi f_0 t}{2\pi f_0 t}, \qquad (4.72b)$$

where k_2 depends upon the area of the boxcar (see problem 15).

Figure 4.99*a* shows a continuous function $y(t)$ and its transform $Y(f)$:

$$y(t) \leftrightarrow Y(f).$$

The amplitude spectrum, $|Y(f)|$, is symmetric about zero, negative frequencies giving the same values as positive frequencies (we may think of positive frequencies as representing a wave train travelling in one direction, such as from left to right, and negative frequencies as representing a wave train travelling in the opposite direction, from right to left in the above case; both give the same observations at any stationary point).

The sampled data which represent $y(t)$ can be found by multiplying the continuous function by the comb (hence the name 'sampling function'). If we are sampling every 4 msec, we use a comb with elements every 4 msec. According to eq. (4.71*b*),

$$\text{comb}\,(t)y(t) \leftrightarrow k_1 \,\text{comb}\,(f)^* \,Y(f).$$

Convolution is equivalent to replacing each data element (each impulse in comb(f) in this instance) with the other function, $Y(f)$. This is illustrated in fig. 4.99*c*. Note that the frequency spectrum of the sampled function differs from the spectrum of the continuous function in this example by the repetition of the spectrum.

We can recover the spectrum of the original function by multiplying the spectrum of the sampled function by a boxcar. The equivalent time-domain operation (see eq. (4.71*a*)) is to convolve the sample data with the sinc function; as shown in fig. 4.99*e*, this restores the original function in every detail. The sinc function thus provides the precise 'operator' for interpolating between sample values.

In the above instance no information whatsoever was lost in the process of sampling and the interpolating. However, if the continuous function had had a spectrum (shown dotted in fig. 4.99*a*) which included frequency components higher than 125 Hz (in this example), then the time-domain multiplication by the sampling function would have produced an overlap of frequency spectra and no longer would we be able to recover the original spectrum from the spectrum of the sampled data, hence we would not be able to recover the original waveform. Whether or not the original waveform is recoverable depends, therefore, upon whether or not the original waveform contains frequencies higher than half of the sampling frequency.

The relationships demonstrated in the foregoing are summarized by the *sampling theorem*: No information is lost by regular sampling provided that the sampling frequency is greater than twice the highest frequency component in the waveform being sampled. This is equivalent to saying that there must be more than two samples per cycle for the highest frequency. The sampling theorem thus determines the minimum sampling we can use. Since this minimum sampling allows complete recovery of the waveform, we can further conclude that nothing is gained by using a finer sampling. Thus, sampling rates of 2 and 4 msec permit us

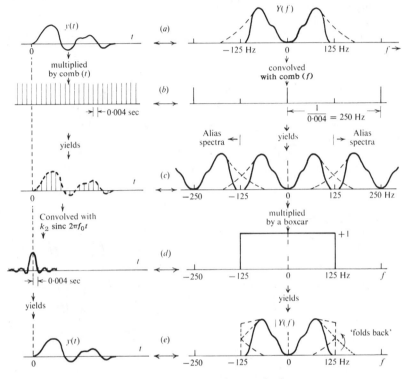

Fig. 4.99 Sampling and reconstituting.

to record faithfully provided none of the signal spectrum lies above 250 and 125 Hz, respectively.

Half the sampling frequency is called the *Nyquist frequency*. Any frequency present in the signal which is greater than the Nyquist frequency f_n by the amount Δf will be indistinguishable from the lower frequency $f_n - \Delta f$. In fig. 4.100 we see that a sampling rate of 4 msec (that is, 250 samples per second) will allow perfect recording of a 75 Hz signal but 175 Hz and 250 Hz signals will appear as (that is, will *alias* as) 75 Hz and 0 Hz (which is the same as a direct current), respectively. Alias signals which fall within the frequency band in which we are primarily interested will appear to be legitimate signals. To avoid this, *aliasing filters* (see fig. 4.65) are used before sampling to remove frequency components higher than the Nyquist frequency. This must be done before sampling because afterwards the alias signals cannot be distinguished. Aliasing is an inherent property of all systems which sample and thus applies not only to digital sampling but also to other situations, such as where we use geophones to sample the earth motion.

Alias filtering also has to be done before any resampling operation which may be performed during data processing.

(c) *Filtering by the earth*. We can think of the earth as a filter of seismic energy. We might consider the wave resulting from an explosion as an impulse $k\delta_t$, that is, the wave motion at the source of the explosion is zero both before and after the explosion and differs from zero only in an extremely short interval (essentially at

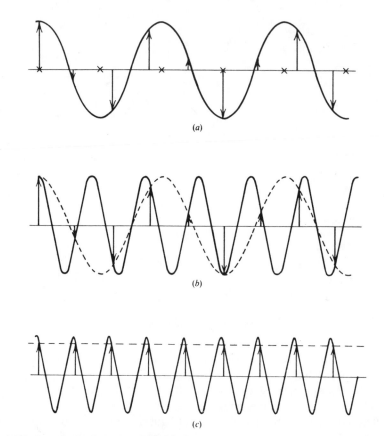

(a)

(b)

(c)

Fig. 4.100 Sampling and aliasing. Different frequencies sampled at 4 msec intervals (250 times per second). (a) 75-Hz signal; (b) 175-Hz signal yields same sample values as 75 Hz; (c) 250-Hz signal yields samples of constant value.

$t = 0$) and during this infinitesimal interval the motion is very large. Ideally, the signal which we record would be simply $k\delta_t$ convolved with the impulse response of the earth. The result would be zero except for sharp pulses corresponding to the arrivals of different reflections. If this were so, we could determine easily from the recorded data the complete solution to the seismic problem. However, in practice we get back not only primary reflections but also multiples, diffractions, surface waves, scattered waves from near-surface irregularities, reflected refractions and so on, all modified by filtering because of absorption and other causes and with random noise always superimposed.

The waveform which we finally record as a seismic record is the result of the successive convolutions of the shot impulse with the impulse response of the various portions of the earth through which the wave travels. We can arrive at an approximate picture by considering the earth to be divided into three zones: (a) the zone near the shot where stress levels and the absorption of the highest frequencies is very severe – we write s_t for the impulse response of this zone; (b) the reflecting sequence of beds whose impulse response e_t is the 'message' information which we are seeking to discover by our seismic exploration; and (c) finally, the near-surface zone which exercises considerable filtering action in changing the waveshape – we write n_t for the impulse response of this last zone. Neglecting additional filtering effects, we thus write the seismic trace g_t as the expression

$$g_t = k\delta_t{}^*s_t{}^*e_t{}^*n_t. \tag{4.73a}$$

When we use a Vibroseis source, the input to the earth is a long wavetrain, v_t. and the seismic trace g'_t which results is

$$g'_t = v_t{}^*s'_t{}^*e_t{}^*n_t, \tag{4.73b}$$

(where we write s'_t rather than s_t because the filtering processes near the Vibroseis source may be different from those near a shot owing to the different magnitude of the stresses involved).

(d) *Water reverberation and deconvolution.* Let us examine the effect of multiples resulting from reflection at the bottom and top of a water layer (Backus, 1959). We write $n\Delta$ for the round trip traveltime from top to bottom and back, n being an integer and Δ the time interval between successive values of our sampled data. We shall assume that the reflection coefficients at the surface and bottom of the water layer are such that the ratio of the reflected to incident amplitudes are -1 and $+r$ respectively, the minus sign denoting phase reversal at the water-air interface. We assume also that the amplitude of a wave returning directly to a hydrophone after reflection at a certain horizon (without a 'bounce' round trip between top and bottom of the water layer) is unity and that its traveltime is t. A wave which is reflected at the same horizon and suffers a bounce either before or after its travel down to the reflector, will arrive at time $t + n\Delta$ with the amplitude $-r$. Since there are two raypaths with the same traveltime for a single-bounce wave, one which bounced before travelling downward and one which bounced after returning from depth, we have in effect a wave arriving at time $t + n\Delta$ with the amplitude $-2r$. There will be three waves which suffer two bounces: one which bounces twice before going downward to the reflector, one which bounces twice upon return to the surface, and one which bounces once before and once after its travel downwards; each of these are of amplitude r^2 so that their sum is a wave of amplitude

$3r^2$ arriving at time $t + 2n\Delta$. Continuing thus, we see that a hydrophone will detect successive signals of amplitudes $1, -2r, 3r^2, -4r^3, 5r^4, \ldots$ arriving at intervals of $n\Delta$. We can therefore write the impulse response of the water layer for various water depths $d = n\Delta/2V$, where V is the velocity in the water:

$$\begin{aligned}
w_t &= [1, -2r, 3r^2, -4r^3, 5r^4, \ldots], & (n = 1) \\
&= [1, 0, -2r, 0, 3r^2, 0, -4r^3, \ldots], & (n = 2) \\
&= [1, 0, 0, -2r, 0, 0, 3r^2, \ldots], & (n = 3)
\end{aligned} \quad (4.74)$$

etc.

If we transform this to the frequency domain we find a large peak (the size of the peak increasing with increasing r) at the frequency $2/n\Delta$ and at multiples of this frequency. These are the frequencies which are reinforced at this water depth (that is, the frequencies for which interference is constructive). The result of passing a wavetrain through a water layer is the same as multiplying the amplitude spectrum of the waveform without the water layer by the spectrum of the impulse response of the water layer. Whenever the reflection coefficient is large (and hence r is large) and the frequency ($2/n\Delta$) (or one of its harmonics) lies within the seismic spectrum, the seismic record will appear very sinusoidal with hardly any variation in amplitude throughout the recording period (see fig. 4.43). Because of the over-riding oscillations, it will be difficult to interpret the primary reflections.

A filter i_t which has the property that

$$w_t * i_t = \delta_t \qquad (4.75)$$

is called the *inverse filter* of w_t. If we pass the reverberatory output from the hydro-phones through the inverse filter (in a data processing centre) we will remove the effect of the water-layer filter. The inverse of the water-layer filter is a simple filter with only three non-zero terms:

$$\begin{aligned}
i_t &= [1, 2r, r^2], & (n = 1) \\
&= [1, 0, 2r, 0, r^2], & (n = 2) \\
&= [1, 0, 0, 2r, 0, 0, r^2], & (n = 3)
\end{aligned} \quad (4.76)$$

etc. (see problem 16). Figure 4.101 shows the result of applying such a filter to the data shown in fig. 4.43.

The process of convolving with an inverse filter is called *deconvolution* and is one of the most important operations in seismic data processing (Middleton and Whittlesey, 1968). While we have illustrated deconvolution as removing the singing effect of a water layer, we could also deconvolve for other filters whose effects we wish to remove if we know enough about the filters and the signal.

4.7.4 Correlation

(a) *Cross-correlation*. The *cross-correlation function* is a measure of the similarity between two data sets. Corresponding values of the two sets are multiplied together and the products summed to give the value of the cross-correlation. Wherever the two sets are nearly the same, the products will usually be positive and hence the

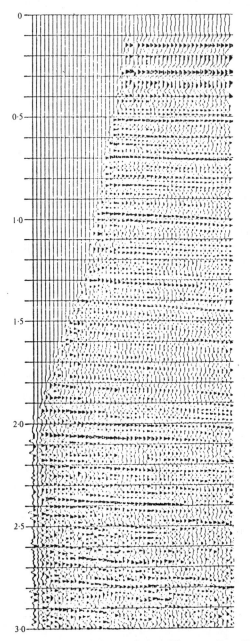

Fig. 4.101 Record of fig. 4.43 after deconvolution. (Courtesy Petty-Ray Geophysical.)

cross-correlation is large; wherever the sets are unlike, some of the products will be positive and some negative and hence the sum will be small. If the cross-correlation function should have a large negative value, it means that the two data sets would be similar if one were inverted (that is, they are similar except that they are out-of-phase). The two data sets might be dissimilar when lined up in one fashion and yet be similar when one set is shifted with respect to the other; thus the cross-correlation is a function of the relative shift between the sets. By convention we call a shift positive if it involves moving the first function to the left with respect to the second function.

We express the cross-correlation of two data sets x_t and y_t as

$$\phi_{xy}(\tau) = \sum_k x_{k+\tau} y_k, \tag{4.77}$$

where τ is the displacement of x_t relative to y_t. (Note that $\phi_{xy}(\tau)$ is a data set rather than a continuous function, because x and y are data sets.) Let us illustrate cross-correlation by correlating the two functions, $x_t = [1, -1, \frac{1}{2}]$ and $y_t = [1, \frac{1}{2}, -\frac{1}{2}]$, shown in fig. 4.102. Diagram (c) shows the two functions in their normal positions. Diagram (a) shows x_t shifted two units to the left; corresponding coordinates are multiplied and summed as shown below the diagram to give

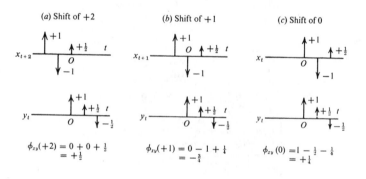

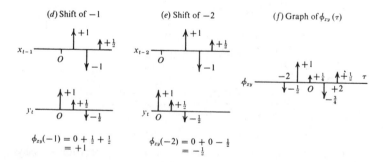

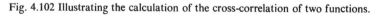

Fig. 4.102 Illustrating the calculation of the cross-correlation of two functions.

$\phi_{xy}(+2)$. Diagrams (*b*) to (*e*) show x_t shifted varying amounts while (*f*) shows the graph of $\phi_{xy}(\tau)$. The cross-correlation has its maximum value (the functions are most similar) when x_t is shifted one unit to the right ($\tau = -1$). Obviously we get the same results if we shift *y* one space to the left. In other words,

$$\phi_{xy}(\tau) = \phi_{yx}(-\tau). \tag{4.78}$$

The similarity between (4.77) and the convolution equation (4.70*a*) should be noted. We may rewrite (4.77) in the form

$$\phi_{xy}(\tau) = \phi_{yx}(-\tau) = \sum_k y_{k-\tau} x_k = \sum_k x_k y_{-(\tau-k)} = x_\tau * y_{-\tau}. \tag{4.79}$$

Hence cross-correlation can be performed by reversing the second data set and convolving.

If two data sets are cross-correlated in the time domain, the effect in the frequency domain is the same as multiplying the complex spectrum of the first data set by the conjugate of the complex spectrum of the second set. Since forming the complex conjugate involves only reversing the sign of the phase, cross-correlation is equivalent to multiplying the amplitude spectra and subtracting the phase spectra. In mathematical terms,

$$x_t \leftrightarrow X(f) = |X(f)|\, e^{j\phi_x(f)},$$
$$y_t \leftrightarrow Y(f) = |Y(f)|\, e^{j\phi_y(f)},$$
$$y_{-t} \leftrightarrow \overline{Y(f)} = |Y(f)|\, e^{-j\phi_y(f)},$$
$$\phi_{xy}(\tau) \leftrightarrow X(f)\overline{Y(f)} = |X(f)||Y(f)|\, e^{j(\phi_x - \phi_y)}. \tag{4.80}$$

We note that changing the sign of a phase spectrum is equivalent to reversing the trace in the time domain.

(*b*) *Autocorrelation.* The special case where a data set is being correlated with itself is called *autocorrelation*. In this case eq. (4.77) becomes:

$$\phi_{xx}(\tau) = \sum_k x_{k+\tau}\, x_k. \tag{4.81a}$$

Autocorrelation functions are symmetrical because a time shift to the right is the same as a shift to the left; from (4.81*a*),

$$\phi_{xx}(\tau) = \phi_{xx}(-\tau). \tag{4.81b}$$

The autocorrelation has its peak value at zero time shift (that is, a data set is most like itself before it is time-shifted). If the autocorrelation should have a large value at some time shift $\Delta t \neq 0$, it indicates that the set tends to be periodic with the period Δt. Hence the autocorrelation function may be thought of as a measure of the repetitiveness of a function.

We can express the preceding concepts in integral form applicable to continuous functions. Equations (4.77) and (4.81*a*) now take the forms

$$\phi_{xy}(\tau) = \int_{-\infty}^{\infty} x(t + \tau)y(t)\, dt, \tag{4.82}$$

$$\phi_{xx}(\tau) = \int_{-\infty}^{\infty} x(t + \tau)x(t)\, dt. \tag{4.83}$$

(c) *Normalized correlation functions.* The autocorrelation value at zero shift is called the *energy* of the trace:

$$\phi_{xx}(0) = \sum_{k} x_k^2. \tag{4.84}$$

(This terminology is justified on the basis that x_t is usually a voltage, current or velocity and hence x_t^2 is proportional to energy.) For the autocorrelation function, (4.80) becomes

$$\phi_{xx}(\tau) \leftrightarrow |X(f)|^2. \tag{4.85}$$

In continuous-function notation,

$$\phi_{xx}(0) = \int_{-\infty}^{\infty} [x(t)]^2\, dt = \int_{-\infty}^{\infty} |X(f)|^2\, df. \tag{4.86}$$

Since the zero-shift value of the autocorrelation function is the energy of the trace, $[x(t)]^2$ is the energy per unit of time or the *power* of the trace and $[X(f)]^2$ is the energy per increment of frequency, usually called the *energy density*.

We often normalize the autocorrelation function by dividing by the energy:

$$\phi_{xx}(\tau)_{\text{norm}} = \frac{\phi_{xx}(\tau)}{\phi_{xx}(0)}. \tag{4.87}$$

The cross-correlation function is normalized in a similar manner by dividing by the geometric mean of the energy of the two traces:

$$\phi_{xy}(\tau)_{\text{norm}} = \frac{\phi_{xy}(\tau)}{[\phi_{xx}(0)\phi_{yy}(0)]^{\frac{1}{2}}}. \tag{4.88}$$

Normalized correlation values must lie between ± 1. A value of $+1$ indicates perfect copy, a value of -1 indicates perfect copy if one of the traces is inverted.

(d) *Automatic statics.* Cross-correlation affords us a means of determining the amount of time shift which will result in the optimum alignment of two seismic traces. If one trace has been delayed with respect to another, for example, in passing through the near-surface layers, the shift which maximizes the cross-correlation is that which produces the optimum alignment (match) of the two traces; the magnitude of the cross-correlation indicates quantitatively how much improvement such a shift will produce. Cross-correlation is a powerful tool and is especially useful when the data quality is poor. It is used in many processes to determine static corrections and the amount of normal moveout to introduce to align traces from different offsets before stacking (Hileman *et al.* 1968; Disher and Naquin, 1970). Criteria can be set which permit such shifts to be determined and applied automatically provided tests are incorporated to ensure that the shifts so introduced are consistent (for example, to ensure that the same corrections are always assumed for the weathering beneath any particular location). Figure 4.103 illustrates the improvement in a CDP stack resulting from use of an automatic statics programme.

(e) *Vibroseis analysis.* The signal g_t' which our geophones record when we use a Vibroseis source (fig. 4.104a) bears little resemblance to e_t, the impulse response of the earth. To obtain a meaningful record, the data are correlated with the Vibroseis sweep signal v_t. The recorded signal g_t' is

$$g_t' = v_t * e_t',$$

where we let $e_t' = s_t' * e_t * n_t$ in (4.73b). Using (4.79) we find for the cross-correlation of the sweep and the recorded signal

$$\begin{aligned}
\phi_{gv}(t) &= g_t' * v_{-t} = (v_t * e_t') * v_{-t} \\
&= e_t' * (v_t * v_{-t}) \\
&= e_t' * \phi_{vv}(t).
\end{aligned} \tag{4.89}$$

(The next to the last step is possible because convolution is commutative.) Hence the overall effect is that of convolving the earth function with the auto-correlation of the Vibroseis sweep signal. The autocorrelation function, $\phi_{vv}(t)$, is quite sharp and has sizeable values only over a very narrow range of time shifts. Therefore, the overlap produced by the passage of a long sweep through the earth has been eliminated almost entirely. This is shown in fig. 4.104 where (a) and (b) are the same Vibroseis record before and after cross-correlation.

4.7.5 *Phase considerations*

The Fourier synthesis of wavetrains according to eqs. (4.66a) or (4.66b) involves adding together sine waves of different frequencies and different phases. If the same components are added together with different phase relations, different waveforms will result. Changing the waveform changes the location of a particular peak or trough and hence measurements of arrival times will be affected by variations in the phase spectra. Because seismic exploration involves primarily determining the arrival times of events, preservation of proper phase relationships during data processing is essential.

Out of all possible wavelets with the same amplitude spectrum, that wavelet whose energy builds up fastest is called the *minimum delay* wavelet; its phase is always less than the other wavelets with the same amplitude spectrum and hence it is also called *minimum phase*. The simplest wavelet (except for an impulse) is a data set which contains only two elements, the set [a, b]. The amplitude spectrum

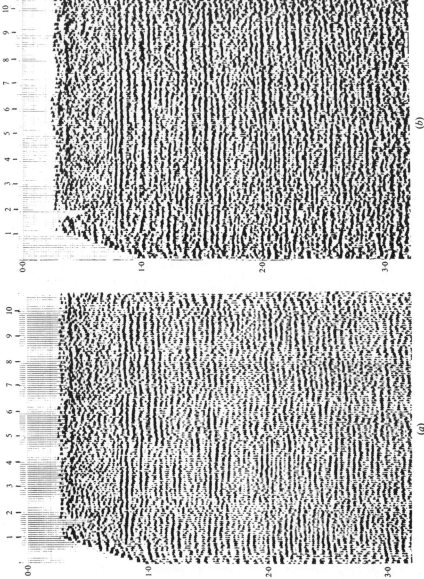

Fig. 4.103 An example of the application of automatic statics. (Courtesy Regional Surveys and

(a)

(b)

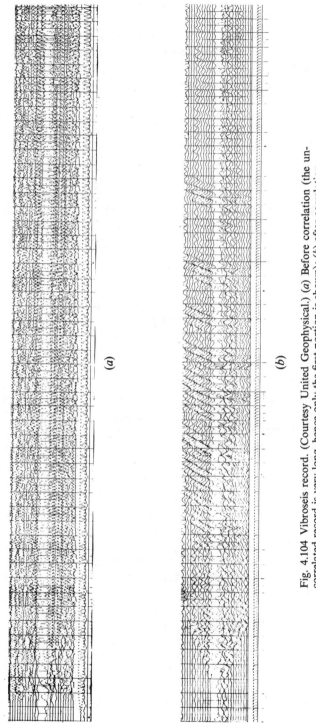

(a)

(b)

Fig. 4.104 Vibroseis record. (Courtesy United Geophysical.) (a) Before correlation (the uncorrelated record is very long, hence only the first portion is shown); (b) after correlation.

of this data set is identical with that of the set [b, a] but no other data set has the same spectrum. If $a > b$, energy is concentrated earlier in the wavelet in the set [a, b] than in the set [b, a] and hence [a, b] is minimum phase (or minimum delay). Larger wavelets can be expressed as the successive convolution of two-element wavelets; a large wavelet is minimum phase if all of its component wavelets are minimum phase. The impulse response of many of the natural filtering processes in the earth are minimum phase. Physical realizability implies minimum phase.

Some filtering processes require that assumption be made about the phase of the signal; generally minimum phase is assumed (Sherwood and Trorey, 1965). Thus deconvolution based upon autocorrelation information has to assume the phase because the phase information of the waveform was lost when its auto-correlation was formed. This can be seen from eq. (4.85) where we note that the auto-correlation function, $\phi_{xx}(t)$, has the transform, $|X(f)|^2$ with zero phase for all values of frequency. Thus all of the phase information present in $X(f)$ has been lost in the autocorrelation.

4.7.6 *Frequency filtering*

(a) *Least-squares (Wiener) filtering.* Sometimes we wish to determine the filter which will do the best job of converting an input into a desired output. The filter which most nearly accomplishes this objective in the least-squares error sense is called the *least-squares filter* or the *Wiener filter*, occasionally the *optimum filter* (Robinson and Treitel, 1967).

Let the input data set be x_t, the filter which we have to determine be f_t, and the desired output set be z_t. The actual result of passing x_t through this filter is x_t*f_t and the 'error' or difference between the actual and the desired outputs is $(z_t - x_t * f_t)$. With the least-squares method we add together the squares of the errors, find the partial derivatives of the sum with respect to the variables f_i (the elements of f_t) and set these derivatives equal to zero. This gives the following simultaneous equations where z_t and x_t are known:

$$\frac{\partial}{\partial f_i} \sum_{t=1}^{m} (z_t - x_t * f_t)^2 = 0, \quad i = 0, 1, 2, \ldots n. \tag{4.90}$$

One such equation is obtained for each of the $n + 1$ elements in f_i; solving for the unknowns f_i we find the filter f_t which minimizes the sum of the errors squared. Further manipulation of (4.90) leads to the so-called *normal equations* which are more convenient than (4.90):

$$\sum_{j=0}^{n} \phi_{xx}(\tau - j)f_j = \phi_{zx}(\tau), \tau = 0, 1, 2, \ldots n. \tag{4.91a}$$

The normal equations for least-squares filtering also have an integral expression for continuous functions,

$$\int_{-\infty}^{\infty} \phi_{xx}(\tau - t)f(t)\,\mathrm{d}t = \phi_{zx}(\tau). \tag{4.91b}$$

These equations can be used to *cross-equalize* traces, that is, to make traces as nearly alike as possible. Suppose we have a group of traces to be stacked such as the components of a common-depth-point stack. After the normal moveout corrections have been made, the traces may still differ from each other because they have passed through different portions of the near surface. The normal equations can be used to find the filters which will make all the traces as nearly as possible like some *pilot trace*, such as the sum of the traces. This procedure will improve the trace-to-trace coherence before the stack and hence improve the quality of the stacked result.

The normal equations are also used to design deconvolution operators. The earth impulse response e_t is assumed to be random, that is, knowledge of the shallow reflections does not help in predicting the deeper reflections. Consequently the autocorrelation of e_t is negligibly small except for zero time shift and we can write

$$\phi_{ee}(\tau) \simeq k\delta_t. \tag{4.92a}$$

The geophone input g_t is regarded (see eq. (4.73a)) as the convolution of e_t with various filters (the most important of which results from near-surface effects), the overall effect being represented by the single equivalent filter n_t:

$$g_t = e_t * n_t.$$

The desired output z_t is the earth's impulse function e_t (which can be shown to be minimum phase); hence using eq. (4.79), we can write

$$
\begin{aligned}
\phi_{zg}(t) &= z_t * g_{-t} \\
&= e_t * (e_{-t} * n_{-t}) \\
&= (e_t * e_{-t}) * n_{-t} \\
&= k\delta_t * n_{-t} \\
&= kn_{-t}.
\end{aligned}
\tag{4.92b}
$$

There can be no output from the filter n_t until after there has been an input to the filter (that is, n_t is physically realizable); this is equivalent to saying that g_t is minimum phase. Hence $n_t = 0$ for $t < 0$. Thus

$$\phi_{zg}(t) = 0 \quad \text{for} \quad t < 0. \tag{4.92c}$$

Therefore if we concern ourselves only with positive values of τ, we have the values required to solve eq. (4.91a) for the deconvolution filter.

(b) *Frequency filtering and deconvolution.* The use of deconvolution to remove the filtering effects of a water layer and the near surface has already been discussed. Although the water-layer filter was presented in a deterministic way, the proper choice of parameters is usually not obvious. Statistical and empirical ways of choosing filter parameters are sometimes used (Kunetz and Fourmann, 1968). Deconvolution by Wiener filtering (using eq. (4.92c)) has also been discussed (Peacock and Treitel, 1969; Robinson, 1972). Another method which is sometimes used involves flattening the frequency spectrum using minimum phase or other assumptions. A deconvolution filter often is based on measurements of such quantities as the autocorrelation of a trace during a certain time interval called the *gate*.

Deeper reflections have a higher per cent of low frequency energy than shallow reflections because of the greater attenuation of the higher frequencies as a result of absorption and other filtering mechanisms; hence time-variant filtering is used in which the pass band moves toward lower frequencies as record time increases. One method of achieving time-variant filtering is to use separate gates (for example, shallow and deep) so that the deconvolution operator varies with record time (Clarke, 1968).

4.7.7 *Multichannel processing*

(a) *Velocity analysis*. The variation of normal moveout with velocity and record time has already been discussed in connection with eq. (4.34b). Several techniques utilize the variation of normal moveout with record time to find velocity (Garotta and Michon, 1967; Cook and Taner, 1969; Schneider and Backus, 1968; Taner and Koehler, 1969). Most assume a stacking velocity (V_{rms}, as discussed in connection with eq. (4.41b)), apply the normal moveouts appropriate for the offsets of the traces being examined as a function of arrival time, and then measure the coherence (degree of match) among all the traces available to be stacked. Several measures of coherence may be used; these are discussed further in §4.7.7f (see eqs. (4.94) and (4.95)). Another stacking velocity is then assumed and the calculation repeated, and so on, until the coherence has been determined as a function of both stacking velocity and arrival time. (Sometimes normal moveout is the variable rather than stacking velocity.)

A velocity-analysis display is shown in fig. 4.105. This is a good analysis because the data involved in fig. 4.105a are good. Peaks on the peak amplitude trace (fig. 4.105b) correspond to events. The locations of the peaks on fig. 4.105c yield the velocities (or normal moveouts) which have to be assumed to optimize the stack (hence the name 'stacking velocity'). Multiples as well as primaries will give rise to peaks and hence the results have to be interpreted to determine the best values to be used to stack the data. In many areas where the velocity increases more-or-less monotonically with depth, the peaks associated with the highest reasonable stacking velocities are assumed to represent primary reflections and peaks associated with lower velocities are attributed to multiples of various sorts. In other areas the relationships are not as obvious and even where the velocity relationships are generally regular, difficulties will be encountered.

The major objective of velocity analysis is to ascertain the amount of normal moveout which should be removed to maximize the stacking of events which are considered to be primaries. This does not necessarily optimize the primary-to-multiple energy ratio and better stacks can be achieved with respect to identifiable multiples. This is not often done, however. An auxiliary objective of velocity analysis is identifying lithology; this is discussed in §4.8.2g.

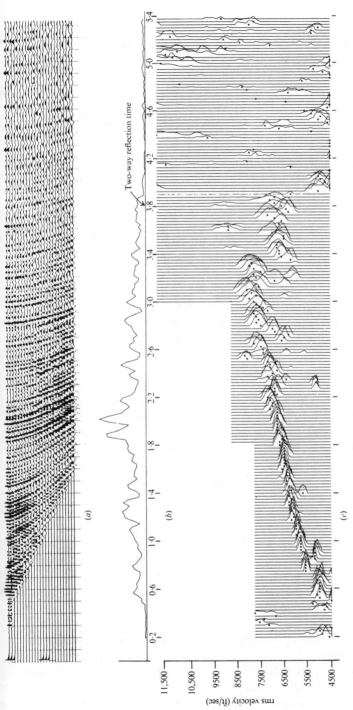

(a)

Two-way reflection time

(b)

(c)

rms velocity (ft/sec)

Fig. 4.105 Velocity analysis. (Courtesy Petty-Ray Geophysical.) (a) Common-depth-point gather showing the data involved in the analysis; (b) maximum amplitude achievable in stacked traces. Peaks indicate the stronger events; (c) amplitude of stacked trace as a function of stacking velocity at 100-msec intervals of t_0. The dots indicate the stacking velocity giving maximum amplitude. The stacking velocity increases with depth from about 5000 ft/sec at 0·6 sec to about 7700 ft/sec at 3·3 sec. The data between 2·7 and 3·9 sec having velocities around 6600 ft/sec are undoubtedly multiples involving the strong reflections at 1·9–2·1 sec. There are probably few primary reflections below 3·3 sec.

(b) *Common-depth-point stacking.* Common-depth-point stacking is probably the most important application of data processing in improving data quality. The principles involved have already been discussed along with the field procedures used to acquire the data. The component data sometimes are displayed as a *gather;* a *common-depth-point gather* (see fig. 4.105a) has the components for the same depth point arranged side-by-side, and a *common-offset gather* has the components for which the shotpoint-to-geophone distance is the same arranged side-by-side.

First-break data and the refraction wavetrains which follow the first breaks usually are so strong that they have to be excluded from the stack to avoid degrading the quality of shallow reflections. This is done by *muting* which involves arbitrarily assigning zero values to traces during the period when the first breaks and following wavetrains are arriving. Thus the multiplicity of a stack increases by steps, the shallowest data often being a twofold stack, slightly deeper data being a fourfold stack, and so on until the full multiplicity of the stack is achieved after the muted events have passed beyond the most distant geophones.

(c) *Apparent-velocity filtering.* The *apparent velocity* of an event, V_A, is found by dividing the distance between two points on the surface of the ground by the difference in arrival time for the same event at geophones located at the points. It is thus the reciprocal of the quantity $\Delta t/\Delta x$ in (4.37b) for a symmetrical spread:

$$V_A = \Delta x/\Delta t = V/\sin \theta, \tag{4.93}$$

where V is the velocity with which a wavefront approaches the spread and θ is the angle between the wavefront and the spread. Apparent velocity is an entirely different quantity than the stacking velocity or the velocity used to convert arrival times to reflector depths. A reflection which arrives from vertically beneath the spread has an infinite apparent velocity (after correction for normal moveout) without regard for the depth of the reflector or the velocity with which the reflection energy has travelled in the earth. Apparent velocity generally decreases for dipping reflectors, becoming smaller as the dip increases but usually it is still much larger than seismic velocities. Horizontally-travelling wavetrains (ground roll, refractions, etc.) have low apparent velocities compared with reflections and so can be discriminated against on this basis. This forms the basis of velocity-filtering methods (Treitel *et al.*, 1967; Sengbush and Foster, 1968). The filtering is achieved by mixing the channels in such a way that events with certain apparent velocities are added out-of-phase and so cancelled.

(d) *Diversity stacking.* Much data processing is far less exotic than is suggested by the mathematical relationships expressed in the foregoing pages. Some of these processes involve merely excluding certain elements of the data, such as the muting operation which has already been discussed.

Diversity stacking is another technique used to achieve improvement by excluding noise. Records in high-noise areas, such as in cities, often show bursts of large-amplitude noise while other portions of the records are relatively little distorted by noise. Under such circumstances amplitude can be used as a discriminant to determine which portions are to be excluded. This can take the form of merely excluding all data where the amplitude exceeds some threshold or perhaps some form of inverse weighting might be used. Such noise bursts often are randomly located on repeated recordings so that sufficient vertical stacking after the weighting tends to produce records free from the high-amplitude noise.

(e) *Automatic migration.* Provided the velocity varies only with depth (that is, not laterally), the traveltime for a diffraction is minimum directly above the diffracting point. The diffraction traveltime at this surface location varies also with shot position and when the shot is also directly above the diffracting point the traveltime has its smallest value, which is the two-way vertical traveltime to the diffracting point. (The situation for diffracted reflections, diffracted refractions and other more complicated paths is more complex.) A *diffraction curve* connects the points on the various traces at which the diffraction energy lies. The energy diffracted by a point can thus be found by summing up along the diffraction curve for which that point is at the apex (Hagedoorn, 1954). If the apex is not a diffracting point, the values along the curve will not be systematic and negatives and positives will tend to cancel.

If the section is searched along all possible diffraction curves, all diffractions will be migrated correctly. The same physical processes are involved in generating reflections and diffractions. One can think of a reflection as the interference product of diffractions from many points on the reflector. Thus if the elements of reflection data are migrated as if they were elements of diffractions, the reflection will be migrated correctly (provided the cross-dip is zero). This principle forms the basis of automatic migration processes (Paturet, 1971). The result of such a process is shown in fig. 4.106.

(f) *Coherency filtering*; *automatic picking.* Trace-to-trace coherence as defined in §4.4.1 can be given a quantitative significance in several ways. For two traces one could use the cross-correlation as a measure of coherence. For a large number of traces we could make use of the fact that when we stack several channels together the resulting amplitude is generally large where the individual channels are similar (coherent) so that they stack in-phase and small where they are unlike (incoherent). The ratio of the energy of the stack compared to the sum of the energies of the individual components would therefore be a measure of the degree of coherence.

If we let x_{ti} be the amplitide of the individual channel i at the time t, then the amplitude of the stack at time t will be $\sum_i x_{ti}$ and the square of this will be the energy. If we call E_t the ratio of the output energy to the sum of the energies of the input traces, we may write

$$E_t = \frac{(\sum_i x_{ti})^2}{\sum_i (x_{ti}^2)}. \qquad (4.94)$$

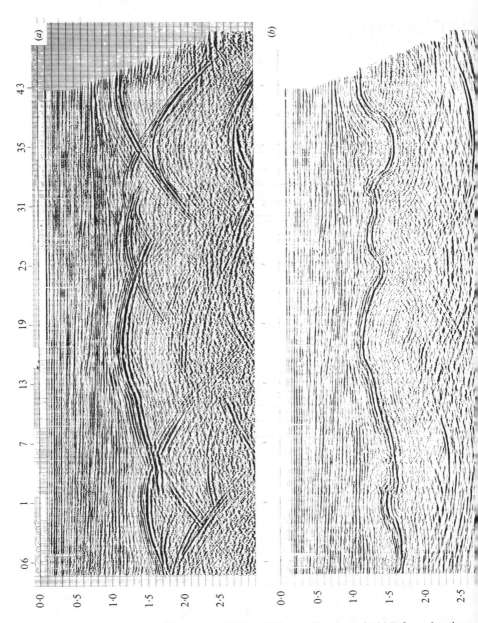

Fig. 4.106 Automigration. (Courtesy A.G.I.P. and Western Geophysical.) (a) Before migration; (b) after migration. Multiple-branch reflection events from sharply-folded basement migrate so as to show a relatively continuous basement reflector. Deeper data, which are probably out-of-the-plane diffractions and other types of events, are not migrated properly.

We expect a coherent event to extend over a time interval; hence a more meaningful quantity than E_t is the *semblance*, S_t (Neidell and Taner, 1971), which denotes the ratio of the total energy of the stack, within a gate of length Δt, to the sum of the energy of the component traces within the same time gate. Using the same terminology as before we can write.

$$S_t = \frac{\sum\limits_{t=t}^{t+\Delta t} (\sum\limits_i x_{ti})^2}{\sum\limits_{t=t}^{t+\Delta t} \sum\limits_i (x_{ti}{}^2)}. \tag{4.95}$$

The semblance will not only tend to be large when a coherent event is present but the magnitude of the semblance will also be sensitive to the amplitude of the event. Thus strong events will exhibit large semblance and weak events will exhibit moderate values of semblance while incoherent data will have very low semblance.

Semblance and other coherence measures are used to determine the values of parameters which will 'optimize' a stack. The semblance is calculated for various combinations of time shifts between the component channels and the optimum time shifts are taken to be those which maximize the semblance. Semblance therefore can be used to determine static corrections or normal-moveout corrections. The determination can be automated although it would be dangerous to automate the determination of normal moveout in this way because no distinction would be made between primary reflections and multiples; when the multiples are strong an automated procedure might optimize the multiples and cancel the primary reflections.

Events can be picked and graded automatically using as criteria coherence measures such as semblance (Paulson and Merdler, 1968; Bois and LaPorte, 1970; Garotta, 1971). At any point on the record where the coherence exceeds a certain threshold value, an event can be marked; the arrival time can be determined by varying the position of the time gate to maximize the coherence. The normal moveout, and hence the stacking velocity, can be found by varying the normal moveout to maximize the coherence. Likewise, the dip moveout can be measured by determining the dip moveout which maximizes the coherence. A grade can be assigned according to some arbitrary scale, for example, one which rates events as grade no. 1 (or 'very good') whenever their coherence exceeds a certain amount. The subsurface distance over which coherence can be maintained can be included as one of the factors in such measurements. The picks can be automatically migrated and plotted on a depth scale, as shown in fig. 4.107.

Automatic picking can be expanded to incorporate intersecting seismic lines, the picks can also be posted on a map automatically and the resulting data contoured automatically. Thus, in principle the output of processing can be contoured depth maps of reflecting horizons and so most of the work usually thought of as 'interpretation' can be automated. However, in the processing many decisions must be made. Criteria have to be specified for determining which events are primary reflections and which multiples, for deciding what to do when events interfere or when events terminate, and so on. The programmer has to define the criteria for each of these decisions in advance and the process will break down if the programmer has not anticipated the need for a certain decision or has not defined the criteria adequately. The value of the final result depends largely on the soundness of the decisions which were made.

Fig. 4.107 Automatically-picked migrated section. (Courtesy Compagnie Générale de Géophysique.)

4.8 Seismic interpretation

4.8.1 *Basic geological concepts*

The interpretation of seismic data in geological terms is the objective and end product of seismic work. However, before discussing this most important and critical phase of interpretation, we shall review briefly some basic geologic concepts which are fundamental in petroleum exploration.

Petroleum is a result of the deposition of plant or animal matter in areas which are slowly subsiding. These areas are usually in the sea or along its margins in coastal lagoons or marshes, occasionally in lakes or inland swamps. Sediments are deposited along with the organic matter and the rate of deposition of the sediments must be sufficiently rapid that at least part of the organic matter is preserved by burial before being destroyed by decay. As time goes on and the area continues to sink slowly (because of the weight of sediments deposited or because of regional (tectonic) forces), the organic material is buried deeper and hence is exposed to higher temperatures and pressures. Eventually chemical changes result in the generation of petroleum, a complex, highly variable mixture of hydrocarbons, including both liquids and gases (part of the gas being in solution because of the high pressure). Ultimately the subsidence will stop and may even reverse.

Sedimentary rocks are porous, *porosity* being the fractional volume of the rock occupied by cavities or pores. Petroleum collects in these cavities, intermingled with the remaining water which was buried with the sediments. When a significant fraction of the pores is interconnected so that fluids can pass through the rock, the rock is *permeable*. Permeability permits the gas, oil and water to separate partially because of their different densities. The oil and gas tend to rise and will eventually reach the surface of the earth and be dissipated unless they encounter a barrier which stops the upward migration. Such a barrier produces a *trap*. Figure 4.108 illustrates the most important types of traps.

The anticline shown in vertical cross-section in fig. 4.108a is a common type of trap and often the easiest to map. In the diagram, bed A is impermeable while the reservoir rock B is permeable. Oil and gas can collect in the reservoir rock of the anticline until the anticline is filled to the *spill point X*. While the diagram is two-dimensional, similar conditions must hold for the third dimension, the structure forming an inverted bowl. If X is the highest point at which oil or gas can escape from the anticline, the contour through X is the *closing contour* and the vertical distance between X and the highest point on the anticline is the *amount of closure*. In fig. 4.108b the closing contour is the -6170 ft contour and the closure is 140 ft. The quantity of oil which can be trapped in the structure depends upon the amount of closure, the area within the closing contour, and the thickness and porosity of the reservoir beds.

Figure 4.108c shows a *fault trap* in which a permeable bed B, overlain by an impermeable bed A, is faulted against impermeable beds C and D. A trap exists if there is also closure in the direction parallel to the fault, for example because of folding as shown by the contours in fig. 4.108d.

Figure 4.108e shows a *pinch-out* in which a reservoir bed B gradually thins and eventually pinches out. A trap is formed when beds A and C are impermeable and closure also exists in the direction at right angles to the diagram, perhaps because of folding or faulting. A *permeability trap* occurs when A, B, and C are separate zones of the same bed differing in permeabilities, B being permeable and A and C impermeable; an example might be a permeable sandstone which becomes *tight* (impermeable) because of fine silt which closes the openings between the grains.

Figure 4.108f represents an *unconformity trap*. Beds B, C, and D became tilted and exposed to erosion, the erosional surface being the wavy line in the diagram. Later the surface sank below sea level and further deposition occurred. If beds A, C, and D are impermeable and B is permeable and closure exists at right angles to the diagram, oil can be trapped under the unconformity.

Figure 4.108g represents a *salt dome* formed when a mass of salt flows upwards under the pressure resulting from the weight of the overlying sediments. The salt dome bows up sedimentary beds and seals off disrupted beds and so provides traps over and around the sides of the dome.

Figure 4.108h shows a limestone *reef* which grew upwards on a slowly subsiding platform (see §4.8.2*f*). The reef is composed of coral or other marine animals with calcareous shells which grow prolifically under the proper conditions of water temperature and depth. As the reef subsides, sediments are deposited around it. Eventually the reef stops growing, perhaps because of a change in the water temperature or the rate of subsidence, and the reef may be buried. The reef material is usually highly porous and often is covered by impermeable sediments.

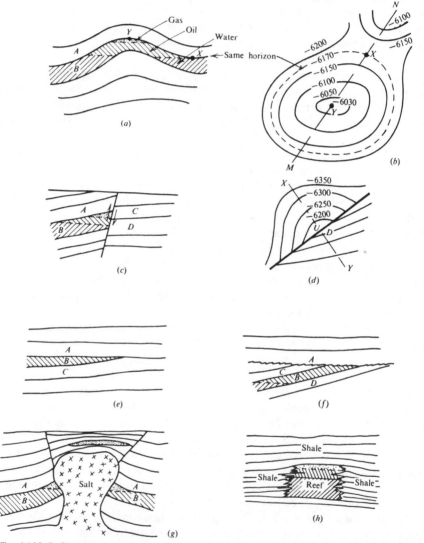

Fig. 4.108 Sedimentary structures which produce oil traps. (*a*) Vertical section through anticline along line *MN* in (*b*); (*b*) map of horizon *A* in (*a*); (*c*) vertical section through fault trap along line *XY* in (*d*); (*d*) map on top of bed *B* in (*c*); (*e*) pinchout; (*f*) unconformity trap; (*g*) salt dome; (*h*) limestone reef.

Hence the reef may form a trap for petroleum generated in the reef itself or flowing into it from another bed.

The primary objective of a seismic survey for hydrocarbons usually is to locate structures such as those shown in fig. 4.108. However, many structures which provide excellent traps do not contain oil or gas in economic quantities. Since drilling wells is very costly, we try to derive from the seismic data as much information as possible about the geological history of the area and about the nature of the rocks in an effort to form an opinion about the probability of encountering petroleum in the structures which we map.

Seismic data are usually interpreted by geophysicists or geologists. The ideal interpreter combines in one person training in both fields. He is fully aware of the processes involved in the generation and transmission of seismic waves, the effects upon the data of the recording equipment and data processing, and the physical significance of the seismic data. At the same time his geological experience helps him assimilate the mass of data, much of it conflicting, and arrive at the most plausible geological picture. Unfortunately, it is rare that we find in the same person the requisite knowledge and experience in both geology and geophysics and often the next best alternative is to have a geophysicist-geologist team working in close cooperation.

4.8.2 *Reflection interpretation*

a) *Mapping reflecting horizons.* Returning now to the geological interpretation of reflection seismic data, we see that the horizons which we draw on seismic sections provide us with a two-dimensional picture only. A three-dimensional picture is necessary to determine whether closure exists, the area within the closing contour, the location of the highest point on the structure, and so on. To obtain three-dimensional information, we usually shoot lines in different directions. Most reflection surveys are carried out along a more-or-less rectangular grid of lines, often with common shotpoints at the intersections of lines to facilitate correlating reflections on the intersecting profiles.

Horizons are first mapped on a cross-section and then this section is compared with the sections for the cross lines in order to identify the same horizons on the cross lines; identification is made on the basis of character and arrival times. The horizons are now 'carried' along the cross-lines and ultimately along all lines in the prospect to the extent that the quality of the data permits.

When a horizon can be carried all the way around a closed loop, we should end up with the same arrival time with which we started. This *closing of loops* provides an important check on reliability. When a loop fails to close within a reasonable error (which depends mainly upon the record quality and the accuracy of the weathering corrections), the cause of the misclosure should be investigated carefully. Migrated sections have to be tied by finding the same reflection on the intersecting sections; such tie points will be displaced from the vertical through the shotpoint by the amount of the migration on each of the lines. Often misclosure is due to an error in correlating from record to record or from line to line, possibly because of inaccurate corrections, a change in reflection character or error in correlating across faults. When the dip is different on the two sides of a fault or the throw varies along the fault, an incorrect correlation across the fault may result in misclosures (but not necessarily). After the sources of misclosure have been carefully examined and the final misclosure reduced to an acceptable level, th remaining misclosure is distributed around the loop.

After horizons have been carried on the sections, maps are prepared. For example, we might map a shallow horizon, an intermediate horizon at roughly the depth at which we expect to encounter oil if any is present, and a deep horizon We map on a *base map* which shows the locations of the seismic lines (usually by means of small circles representing shotpoints) plus other features such as oil wells rivers, shorelines, roads, land and political boundaries, and so on. Values representing the depth of the horizon below the datum plane are placed on the map (*posted*), usually at each shotpoint. Other information relevant to the horizon being mapped (depths in wells, locations of gravity anomalies, relevant geologic information, etc.) is also posted. Faults which have been identified on the record or cross-sections are drawn on the map and the depth values are then contoured

Isopach maps which show the thickness of sediments between two horizons are useful in studying structural growth. They are often prepared by overlaying map of the two horizons and subtracting the contour values wherever the contours of one map cross the contours on the other. The differences are recorded on a blank map and then contoured. If the contours show a trend towards increased thicknes in a certain direction, it may suggest that the region was tilted downwards in thi direction during the period of deposition or that the source of the sediments is in this direction. Uniform thickness of a folded bed indicates that the folding came after the deposition whereas if the thickness increases away from the crest of an anticline, deposition probably was contemporaneous with the growth of the structure. Growth during deposition is usually more favourable for petroleum accumulation since it is more likely for reservoir sands to be deposited on the flanks of structures with even slight relief.

(b) *Drawing conclusions from reflection data.* Structural traps, such as anticline and fault traps, and structural *leads* (possibilities of traps which require more work to define them completely) are usually evident from examination of the maps Traps resulting from pinch-outs and unconformities are more difficult to recognize and nonseismic evidence often must be combined with seismic data to define such features. Nevertheless, careful study of the maps, sections and records plu broad experience and ample imagination will at times disclose variations of dip or other effects which help locate traps of these types.

After the structural information has been extracted, the next step is to work out as much as possible of the geological history of the area. Fundamental in thi connection is the determination of the ages of the different horizons, preferably according to the geological time scale but at least relative to one another. Often seismic lines pass close enough to wells to permit correlating the seismic horizons with geological horizons in the wells. Refraction velocities (if available) may help identify certain horizons. Occasionally a particular reflection has a distinctive character which persists over large areas, permitting not only it to be identified but also other events by their relation to it. Notable examples of persistent identifiable reflections are the low-frequency reflections sometimes associated with massive basement and the prominent reflection from the top of the Ellenburger, a limestone encountered in Northern Texas.

The unravelling of the geological history of the area is important in answering questions such as the following: (*a*) Was the trap formed prior to, during, or subsequent to the generation of the oil and gas? (*b*) Has the trap been tilted sufficiently to allow any trapped oil to escape? (*c*) Did displacement of part of a structure by faulting occur before or after possible emplacement of oil? While the seismic data rarely give unambiguous answers to such questions, often clues can be obtained which, when combined with other information such as surface geology and well data, permit the interpreter to make intelligent guesses which improve the probability of finding oil. Alertness to such clues is the 'art' of seismic interpretation and often the distinction between an 'oil finder' and a routine interpreter.

When the interpretation is finally completed, a report is usually prepared, often both for submission in writing and for oral presentation. In some ways this is the most difficult and most important task of the interpreter. He must present his findings in such a way that the appropriate course of action is defined as clearly as possible. The important aspects should not be obscured by presenting a mass of details nor should they be distorted by presenting carefully selected but nonrepresentative maps and sections. Evidences to support significant conclusions should be given. Alternate interpretations should be presented and an estimate given of the reliability of the results and conclusions. Finally the interpreter should recommend what further action should be undertaken.

(c) *Synthetic seismograms.* If we assume that the shot and near surface produce a certain waveform ($k\delta_t * s_t * n_t$ in eq. (4.73*a*)), we can calculate the reflection waveform (g_t in (4.73*a*)) which would be recorded after modification because of passage through a sequence of layers with given velocities and densities. The wave is assumed to impinge on the first interface where its energy is partitioned among transmitted and reflected waves. Each of the *P*-waves is then followed as it travels to other interfaces where additional *P*-waves are generated, and so on. The resulting seismic record is simply the superposition of those waves which are ultimately reflected back to the geophone station. Since Snell's law determines ray paths and Zoeppritz' equations determine energy relationships, the problem is completely determined and the solution is straightforward. However, actually solving the problem is a formidable task, even for modern computers, because of the tremendous number of waves generated for a realistic sequence of layers; for this reason various simplifications are made. Usually horizontal bedding is assumed and only the ray path normal to the reflecting section is tracked. Density variations are frequently ignored so that the reflection and transmission coefficients are based on velocity changes only. Often small velocity variations are ignored and only the major contrasts are considered, or else the small changes are lumped together into larger steps. Often multiples, especially short-path multiples, are ignored. The result is called a *synthetic seismogram*.

In many areas the synthetic seismogram is a reasonable approximation to actual seismic records and is therefore useful in correlating reflection events with particular horizons. Comparison of the actual and synthetic seismograms may also help to determine which events represent primary reflections and which multiples. Figure 4.109a shows a sonic log plotted to a time scale (rather than the usual depth scale) and fig. 4.109b the synthetic seismogram which would result, assuming a certain initial waveshape and neglecting multiples and changes in density. Figure 4.109c is a field recording made in the same area; some events such as B, D and J correlate well with synthetic events but elsewhere the correlation ranges from fair to poor. Figure 4.109d differs from fig. 4.109b in that simple multiples are included; the principal differences are the appearance of a strong multiple at K and considerable reduction in the amplitude of F, presumably because of interference between multiples. This synthetic seismogram was made without shallow information; sonic logs are often not run in the upper part of a borehole and consequently many of the most important effects cannot be modelled correctly.

A sonic log and a theoretical seismogram fail to have a one-to-one correspondence not only because of the approximations made in the calculation but also because the seismic wavelengths are so much longer than the distances over which the acoustic properties can be assumed to be constant that the actual seismic waveform is the interference composite of many more small events than the computer takes into account. In fig. 4.109a the major changes in the sonic log (such as those corresponding to B, D and J) produce distinct events on the synthetic record; however, in addition weak indistinct events (such as F and G) result from many of the smaller velocity changes.

An important use of synthetic seismograms is in studying the effect of changes in the reflecting layers on the seismic record. One might, for example, assume that a shale passes laterally into a sand and determine how such a change would affect events on the record, possibly altering the number of legs in a reflection event or changing some other characteristic. This might then provide a clue in looking for ancient stream channels or other features of interest. Synthetic seismograms are important in indicating the features which may help to identify stratigraphic traps.

(d) *Evidences of faulting.* Ideally reflection events terminate sharply as the point of reflection reaches the fault plane and resume again in displaced positions on the other side of the fault; in addition, ideally the reflection has a sufficiently distinctive character that the two portions on opposite sides of the fault can be recognized and the fault throw determined. In practice diffractions usually prolong an event so that the location of the fault plane is not clearly evident, although occasionally it is possible to observe sharp terminations. Moreover, although sometimes the

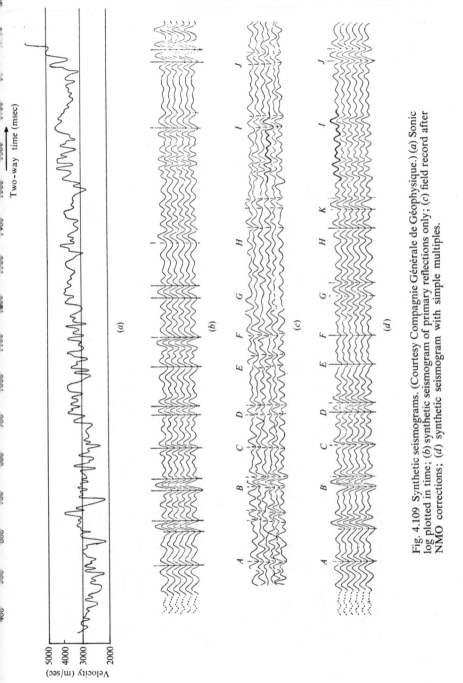

Fig. 4.109 Synthetic seismograms. (Courtesy Compagnie Générale de Géophysique.) (a) Sonic log plotted in time; (b) synthetic seismogram of primary reflections only; (c) field record after NMO corrections; (d) synthetic seismogram with simple multiples.

same reflection can be identified unequivocally on the two sides of a fault, in the majority of cases we can at best make only tentative correlations across faults.

The two record sections in fig. 4.110 join at their north and west ends at right angles. On the N–S section (fig. 4.110*a*) the reflection band consisting of four strong legs marked Σ can be readily correlated across the normal fault which is downthrown to the south by slightly less than 2 cycles (about 65 msec) at the 1·6 sec arrival time of this event. At a velocity of 2300 m/sec this represents a vertical throw of about 75 m. The event near 2·3 sec (marked χ) indicates a throw of about 3 cycles (about 120 msec, the dominant frequency having become slightly

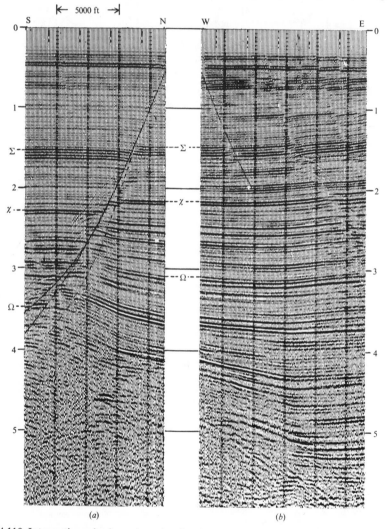

Fig. 4.110 Intersecting seismic sections showing faulting. (Courtesy Geophysical Services Inc.) (*a*) N–S cross-section; (*b*) E–W cross-section.

lower); at a velocity of 3000 m/sec this represents 180 m of throw so that the fault appears to be growing rapidly with depth.

Although the evidence suggests that the fault is a simple break in the shallow section, at greater depths there seems to be a fault zone with subsidiary faults (shown dashed in the figure). If the deeper correlations across the fault(s) are correct, the downthrown event Ω at 3·5 sec is found around 2·9 sec on the upthrown side; assuming a velocity of 4000 m/sec at this depth we get a vertical throw of 1200 m.

The correlation across the fault for the shallow event in fig. 4.110a is based on reflection character; for the deeper event it is based upon intervals between strong reflections, systematic growth of throw with depth and time-ties around loops. Sometimes the displacement of an unconformity or other recognizable feature will indicate the amount of throw. Often, however, the throw cannot be determined clearly from the seismic data.

If the data in fig. 4.110a are transformed into a depth section we get fig. 4.111a; the component of fault-plane dip in this section is around 53°. Note that a fault which is nearly straight on a depth section is concave upwards on a time section because of the increase in velocity with depth. If the fault surface is actually concave upwards, the curvature will be accentuated on a seismic section. Where the

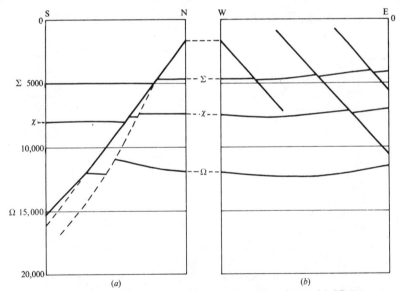

Fig. 4.111 Depth sections of faults and mapped horizons of Fig. 4.110. (a) N–S cross-section; (b) E–W cross-section.

fault was most active (indicated by the most rapid growth in fault throw), the fault surface is most curved.

The fault has not completely died out by the north end of the line and hence the fault trace should appear on the intersecting line (fig. 4.110b). As picked on the

E–W section, the fault offsets the event at 1·6 sec by only about 30 m, indicating that the fault is dying out rapidly toward the east. The fault plane has nearly as much dip in the E–W section so that the strike of the fault plane near the intersection of the two lines is NE–SW and the fault plane dips to the southeast. The true dip of the fault plane is about 60° (the apparent dip on sections is always less than the true dip). Fault indications are not evident below about 2 sec on the E–W section so that the fault appears to have died out at depth towards the east. In poorly consolidated sediments such rapid dying out of faults is common.

Several diffractions can be identified along the fault trace in fig. 4.110a between 1·9 and 2·5 sec and changes in the reflection dip are seen on both sides of the fault trace; these are common evidences of faulting. Other features which are often observed but which are not clearly evident on these sections are various distortions of data from events which passed through the fault plane and therefore experienced velocity changes and consequent bending of the rays at the fault plane; this effect is often accompanied by deterioration in the data quality which sometimes is so

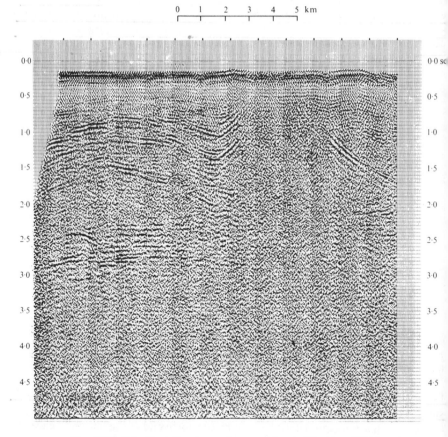

Fig. 4.112 Seismic section across basement horst. (Courtesy Petrobras.)

great that reflections may be almost entirely absent beneath the fault, causing a *shadow zone*. Occasionally the fault plane itself generates a reflection but unless the data are migrated it may not be clear that the fault-plane reflection is associated with the fault. Fault-plane reflections are seen in fig. 4.112 which crosses a basement horst. The variable thickness of the portion of the section around 1·7 to 2·4 sec suggests unconsolidated sediments; faults often die out in such sediments where the fault displacement becomes accommodated by flow (see the following section).

(e) *Folded and flow structures.* When subjected to stresses, rocks may fault, fold or flow, depending on the magnitude and duration of the stresses, the strength of the rocks, the nature of adjacent rocks, etc. The folding of rocks into anticlines and domes provides many of the traps in which oil and gas are found.

Figure 4.113 shows a migrated seismic section across an anticline. Some portions such as *A* which are composed of the more competent rocks (for example limestones and consolidated sandstones) tend to maintain their thickness as they fold. Other portions such as *B* which contain less competent rocks (often shales and evaporites) tend to flow and slip along the bedding, resulting in marked variations in thickness within short distances. Geometry places limits on the amount of folding which is possible and folded structures almost always involve faulting. Note at *C* how a fault is involved with the folding. Arching places sediments under tension so that often they break along normal faults and produce graben-type features on the top. Anticlinal curvature tends to make seismic reflections weaker as well as increase the likelihood of faulting and flowage, so that data quality commonly deteriorates over anticlines.

Salt flow often produces anticlines and domes. In many parts of the world thick salt deposits have been buried fairly rapidly beneath relatively unconsolidated sediments. The sediments compact with depth and so increase their density whereas the salt density remains nearly constant. Thus, below some critical depth the salt is less dense than the overlying sediments. Salt behaves like a very viscous fluid under sufficient pressure, and buoyancy may result in the salt flowing upward to form a salt dome, arching the overlying sediments and sometimes piercing through them. Grabens and radial normal faults (whose throw decreases away from the dome) often result from such arching of the overlying sediments, thus relieving the stretching which accompanies the arching. Salt domes tend to form along zones of weakness in the sediments, such as a large regional fault. The side of a salt dome may itself be thought of as a fault.

Figure 4.114 shows a seismic section across a salt dome. Shallow salt domes are apt to be so evident that they can scarcely be misidentified. Because of the large impedance contrast, the top of the salt dome (or the cap rock on top of the dome) is often a strong reflector. Steep dips may be seen in the sediments adjacent to the salt dome as a result of these having been dragged up with the salt as it flowed upward. The sediments often show rapid thinning toward the dome. The salt itself is devoid of primary reflections, although multiples often obscure this

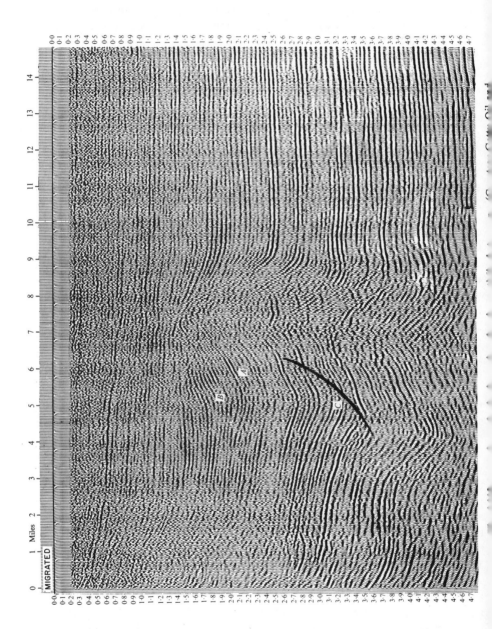

Fig. 4.114 Seismic section across a salt dome of intermediate depth. Note the down-toward-the-dome normal (graben) faulting over the dome. The thinning of section as the dome is approached and the location of the rim syncline with depth can be used to work out the history of the dome's growth. The uplift area and conflicting dip toward the right side of this section probably indicate another dome off to the side of seismic line (dome sideswipe.) (Courtesy Western Geophysical.)

feature, especially if AGC is used. Defining the flank of a salt dome precisely is often very important economically and at the same time very difficult seismically. Frequently the oil is in a narrow belt adjacent to the flank of the dome but since the flank is usually nearly vertical it rarely gives rise to a recognizable reflection. Fortunately the velocity distribution is often only slightly affected by the growth of the dome (except for the velocity in the salt and the cap rock) so that the steep dips of the sediments adjacent to the flanks can be migrated fairly accurately and the flank outlined by the terminations of these reflections. Nevertheless, there remains much art and experience in defining salt-dome flanks.

The salt in a salt dome generally has come from the immediately surrounding region. The removal of the salt from under the sediments around the dome has thus allowed them to subside, producing a rim syncline. The seismic data over such synclines are often very good and aid in mapping the adjacent dome by indicating the volume of salt involved, when movement took place (by sediment thickening), etc. Such synclines may also help provide closure on neighbouring areas where the sediments continue to be supported by residual salt.

Figure 4.115 contains several portions of a seismic line in the North Sea. The horizontal scale has been compressed so as to display a long line on a short section, producing considerable vertical exaggeration. This line shows deep salt swells which have not pierced through the overlying sediments (fig. 4.115a), salt which has pierced through some of the sedimentary section (fig. 4.115b) and also salt which has pierced all the way to the sea floor (fig. 4.115c). The reflection from the base of the salt is generally continuous and unbroken but distortions produced by the variable salt thickness above it at times interrupt this reflection. Because the salt velocity is greater than that of the adjacent sediments, the base of the salt event appears to be pulled up where the salt is thicker. In other areas where the salt velocity is lower than that of the surrounding sediments, flat reflectors beneath the salt may appear to be depressed where the overlying salt is thicker.

Occasionally substances other than salt form flow structures. Poorly consolidated shale may flow, forming structures which strongly resemble salt domes on reflection sections; also, at times shale flows along with salt, producing a salt dome with a sheath of shale.

(f) *Mapping of reefs*. The term 'reef' as used by petroleum geologists comprises a wide variety of types, including both extensive barrier reefs which cover large areas and small isolated pinnacle reefs. It includes carbonate structures built directly by organisms, aggregates comprising limestone and other related carbonate rocks, as well as banks of interstratified carbonate (and sometime also non-carbonate) sediments. Reef dimensions range from a few tens of metres to several kilometres, large reefs being tens of kilometers in length, a few kilometres wide and 200–400 m or more in vertical extent. Because reefs vary so widely, the evidence for reefs shown by seismic data is extremely varied.

We shall describe a model reef so that we may develop the general criteria by which reefs can be recognized in seismic data, keeping in mind that deviations from

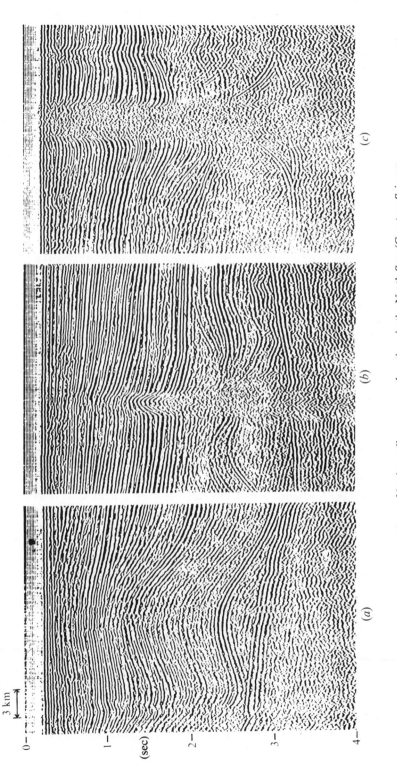

Fig. 4.115 Portions of horizontally compressed sections in the North Sea. (Courtesy Seiscom-Delta.) (a) Deep salt swell; (b) salt dome which has pierced some of the sediments; (c) salt dome which has pierced to the sea floor.

the model may result in large variations from these criteria. Our model reef develops in a tectonically-quiet area characterized by flat-lying bedding which is more-or-less uniform over a large area. The uniformity of the section makes it possible to attribute significance to subtle changes produced by the reef which might go unnoticed in more tectonically-active areas. The reef is the result of the build-up of marine organisms living in the zone of wave action where the water temperature is suitable for sustaining active growth. The site of the reef is usually a topographic high which provides the proper depth. Although the topographic high may be due to a structure in the underlying beds or basement, such as a tilted fault block, more often it is provided by a previous reef; as a result, reefs tend to grow vertically, sometimes achieving thicknesses of 400 m or more and thereby accentuating the effect on the seismic data. In order for the reef to grow upward, the base must subside as the reef builds upward, maintaining its top in the wave zone as the sea transgresses. The reef may provide a barrier between a lagoonal area (the *backreef*) and the ocean basin (the *forereef*), so that sedimentation (and consequently the reflection pattern) may be different on opposite sides of the reef. The surrounding basin may be *starved* (that is, not have sufficient sediments available to keep it filled at the rate at which it is subsiding); at times only one side, more often the ocean side of the reef, may be starved. Alternatively, the reef may not be a barrier to movement of the sediments and in this case it will be surrounded by the same sediments on both sides. Erosion of the reef often provides detritus for deposition adjacent to the reef, resulting in *foreset beds* with dips up to 20°. Eventually the environment for the reef organisms will change so that they can no longer continue to live and build the reef; this might come about because of changes in the water temperature, an increase in the rate of subsidence so that the organic build-up cannot keep pace (called *drowning* of the reef), or various combinations of circumstances. Subsequently the reef may become buried by deep-water shales which may provide both an impermeable cap to the porous reef and sufficient hydrocarbons so that the reef becomes a petroleum reservoir. Additional sediments may continue to be deposited, their weight compacting the sediments which surround the reef more than they compact the relatively rigid reef; thus the overlying sediments which were deposited flat may develop a drape over the reef. The interior of the reef may be more porous and less rigid than the edges so that some differential compaction may occur over the reef itself.

Based on the foregoing model, we might hope to see diffractions from the top and/or flanks of the reef. Abrupt termination of reflections from the surrounding sediments may indicate the location of the reef. If the reef provided a barrier to sedimentation, the entire reflection pattern may differ on the two sides of the reef reflecting the different sedimentary environments. Overlying reflections may show small relief (usually only a few milliseconds in magnitude) because of the differential compaction. The most reliable evidence may result from a velocity difference between the reef materials and the surrounding sediments; as a consequence the traveltime between flat-lying reflections above and below the reef may vary sufficiently between on-reef and off-reef spreads to localize the reef (Davis, 1972). The same velocity difference may produce pseudo-structure on reflecting horizons below the reef. Usually thè velocity in the reef limestone will be greater than that in the surrounding shales so that the reef will be indicated by a time thinning between reflections above and below the reef and by a pseudo-high under the reef. The magnitude of such an anomaly is small, usually less than 20 msec. Sometimes, however, the reef may be surrounded by calcareous shales which have a higher velocity than that of the porous reef limestone so that the time anomaly is reversed.

Reef evidences are often so small and so subtle that seismic mapping of reefs is feasible only in good record areas. Of the first importance is geological information about the nature of the sediments and transgressive periods of deposition, so that one knows beforehand in what portion of the section reefs are more likely to occur.

Similarities between reefs and salt features cause problems at times. The lagoon-al areas behind reefs often provide proper conditions for evaporite deposition so that salt is frequently present in the same portion of the sedimentary column. The amount of salt may not be thick enough to produce *diapiric* (flow) features but differential solution of salt beds followed by the collapse of the overlying sediments into the void thus created may produce seismic features which are similar in many ways to those which indicate reefs.

(g) *Use of velocity, tectonic style and amplitude information.* With common-depth-point data which have a high degree of redundancy (i.e., which sample the same depth point many times) and velocity-analysis programs (§4.7.7a), interval velocities can be calculated from eq. (4.57) for the intervals between parallel reflectors at many points on the section – in fact, almost on a continuous basis. After due allowance for the uncertainties involved in the measurements, systematic variations might be interpreted in stratigraphic terms (Hofer and Varga, 1972). Carbonate and evaporite velocities are sufficiently higher than clastic velocities (especially in Tertiary basins) that they can often be distinguished.

The analysis of velocity data constitutes an important interpretation problem. As with other interpretation problems, some interpretations can be ruled out because they imply impossible or highly improbable situations. From experience we know that velocity does not vary in a 'capricious' manner. Thus it would be unreasonable to expect the velocity to vary from place to place in other than a slow systematic way unless the seismic section shows significant structural or other changes which suggest a reason why the velocity should change rapidly. Thus one might expect two velocity analyses to show some differences. between portions of the section which are separated by a fault whereas one expects little variation in portions which appear to be continuous. Likewise, interpreting an increase in velocity as a reef buildup in an otherwise clastic section might be warranted if the increase is accompanied by changes in reflection character and by structural evidence such as diffractions, or if the increase occurs over deeper structure which could have caused this area to be relatively high during deposition so that a reef might have grown here rather than elsewhere.

The desire to extract stratigraphic information from velocity data sometimes results in interpretations which exceed the limitations of the data. Small errors in normal-moveout measurements can produce sizeable errors in stacking velocities, especially for deep reflections, and these in turn cause large errors in the calculation of interval velocities when the intervals are small. Where the reflectors are not parallel, interval-velocity calculations are meaningless (Taner *et al.*, 1970). Velocity measurements sometimes are severely distorted by various factors, such as interference effects, noise of various kinds, distortions produced by shallow velocity anomalies or weathering variations, and great care must be exercised that these effects are not taken as indications of actual changes in the velocity of the rocks.

'Tectonic style' can also be used as a basis for drawing stratigraphic conclusions. This is a much less tangible aspect of interpretation where broad experience and ingenuity come into play because the various aspects of tectonic style have not been quantified. Certain patterns of reflections often characterize depositional environments so that deltas, foreset bedding, reefs, sandbars, periods of non-deposition or quiet deposition, periods of tectonic activity, etc., may be identified by the aggregate of subtle evidences.

The amplitudes of reflections, especially relative amplitudes, provide information for stratigraphic interpretation. Strong amplitudes result from large changes in acoustic impedance, such as may occur at basement or an unconformity, or because of a change from clastic to carbonate rocks (although strong amplitudes may also be the result of interference, focusing and other causes). Gas accumulations lower the density and the velocity in porous sediments appreciably and hence can sometimes be located by amplitude effects (see fig. 4.116).

Deducing geological significance from the aggregate of many minor observations not only tests the ingenuity of an interpreter, it also tests his in-depth understanding of physical principles. For example, downdip thinning of reflection intervals might result from a normal increase of velocity with depth as well as from thinning of the sediments, and flow of salt or shale may cause illusory structure on deeper horizons. Geometric focusing produced by reflector curvature can produce various effects, especially if not migrated correctly, and energy which comes from a source located off to one side of the line can interfere with the patterns of other reflection events to produce effects which might be interpreted erroneously unless their true nature is recognized. Improper processing likewise can create opportunities for misinterpreting data (Tucker and Yorsten, 1973).

4.8.3 *Refraction interpretation*

(a) *Interpretation of refraction records.* The identification of refraction events is usually simpler than reflection events. Traveltimes are usually available for a relatively long range of offsets and hence it is easy to separate reflections and diffractions with their curved alignments from the direct wave, surface waves

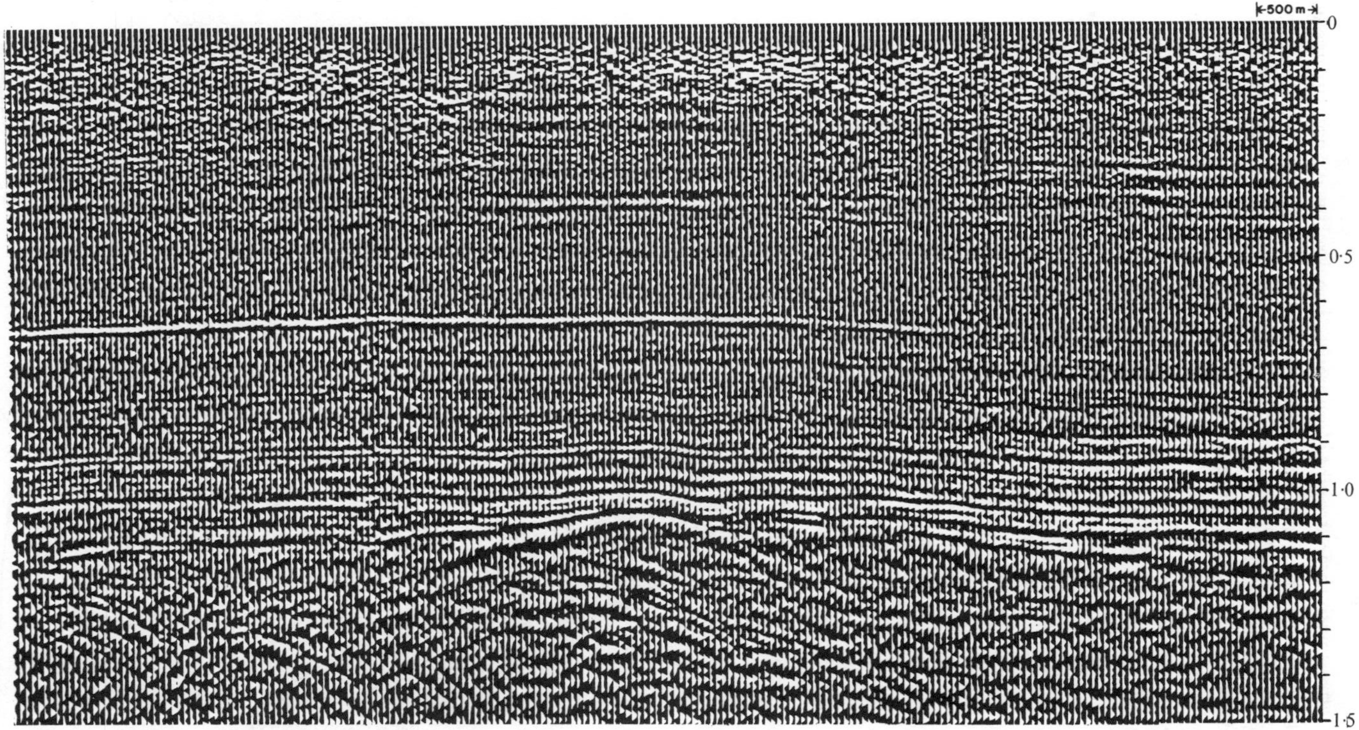

Fig. 4.116 Use of amplitudes in seismic interpretation. Note reflection from gas accumulation just below 0·6 sec. Basement is at about 1·1 sec. (Courtesy Chevron Oil.)

and refractions with their straight alignments. The direct wave and surface waves are easily distinguished from refractions because of the lower velocities of the former. Usually the only problem is in identifying the different refraction events when several refractors are present.

Record sections, while not as widely used as for reflection interpretation, are very useful, especially in studying second arrivals. The refraction profile in fig. 4.117 shows the direct wave as the first arrival near the shotpoint; refractions

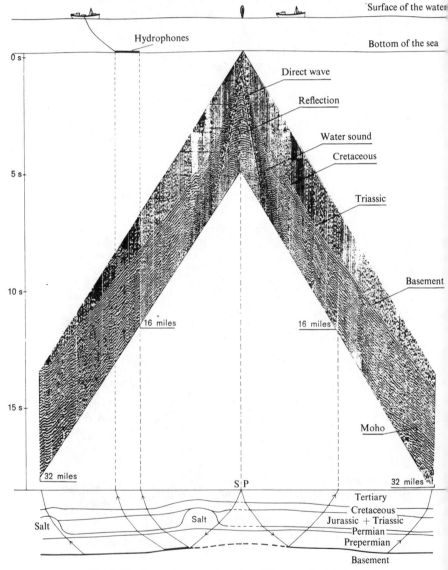

Fig. 4.117 Refraction record section. (Courtesy Prakla-Seismos.)

rom successively deeper refractors become the first arrivals as the offset distance ncreases. Following the first arrivals the continuations of various events are seen fter each has been overtaken by a deeper event. Numerous other events are also een in the zone of second arrivals; most of these are refractions which never ›ecome first arrivals or multiply-reflected refractions (see fig. 4.39*b*).

The simple equations (4.47) through (4.54) can be used when the data are easy ›o interpret and limited in quantity. Often the chief failure of these equations and of most refraction interpretation techniques) is in the assumption of V_1, he velocity of the section above the refractor. Most methods assume straight-line aypaths from the refractor upward to the surface. This is usually not true because he overburden velocity is rarely constant. The biggest improvement in the results ›btained when using the simple equations to calculate refractor depths often is the esult of using a more realistic assumption for V_1 based on information other han that obtainable from the refraction data themselves (Laski, 1973).

Problems sometimes result from a *hidden zone*, a layer whose velocity is lower han that of the overlying bed so that it never carries a head wave. Energy which vould approach it at the critical angle cannot get through the shallower refractors nd hence there is no indication of its presence in the refraction arrivals. The low ·elocity of the hidden layer, however, increases the arrival times of deeper re- ractors relative to what would be observed if the hidden zone had the same ·elocity as the overlying bed, hence results in exaggeration of their depths. Another ituation, which is also referred to at times as a 'hidden zone', is that of a layer vhose velocity is higher than those of the overlying beds but which never produces irst arrivals despite this because the layer is too thin and/or its velocity is not ufficiently greater than those of the overlying beds. Such a bed creates a second rrival but the second arrival may not be recognized as a distinct event.

Refraction interpretation often is based solely on first arrivals, primarily because his permits accurate determination of the traveltimes. When we use second rrivals we usually have to pick a later cycle in the wavetrain and estimate travel- ime from the measured time. However, velocities based on second arrivals will be ccurate and much useful information is available through their study.

Refraction interpretation often involves 'stripping' which is in effect the removal ›f one layer at a time. In this method the problem is solved for the first refractor, fter which the portions of the time-distance curve for the deeper refractors are djusted to give the result which would have been obtained if the shotpoint and eophones had been located on the first refracting horizon. The adjustment con- ists of subtracting the traveltimes along the slant paths from shotpoint down to he refractor and up from the refractor to the geophones, also of decreasing the ›ffsets by the components of the slant paths parallel to the refractor. The new ime-distance curve is now solved for the second refracting layer after which this ayer can be stripped off and the process continued for deeper refractors.

Several additional considerations affect refraction interpretation. Refraction rrival times are corrected to datum in the same manner as reflection arrival times. Vhile the effect of such corrections on the effective shot-to-geophone distance is usually small for reflection data, this is often not so for refraction travel paths above the refractor, since these may have appreciable horizontal components Hence the reference datum should be near the surface to minimize such errors.

If enough data are available, interpretational ambiguities often can be resolved. However, in an effort to keep survey costs down, only the minimum amount of data may be obtained (or less than the minimum) and some of the checks which increase certainty and remove ambiguities may not be possible.

(b) *Delay times.* The concept of delay time, introduced by Gardner (1939), is widely used in routine refraction interpretation, mainly because the various schemes based upon the use of delay times are less susceptible to the difficulties encountered when we attempt to use equations (4.47) to (4.54) with refractors which are curved or irregular. Assuming that the refraction times have been corrected for elevation and weathering, the *delay time* associated with the path $SMNG$ in fig. 4.118 is the observed refraction time at G, t_g, minus the time required for the wave to travel from P to Q (the projection of the path on the refractor) at the velocity V_2. Writing δ for the delay time, we have

$$\delta = t_g - \frac{PQ}{V_2} = \left(\frac{SM + NG}{V_1} + \frac{MN}{V_2}\right) - \frac{PQ}{V_2}$$

$$= \left(\frac{SM + NG}{V_1}\right) - \left(\frac{PM + NQ}{V_2}\right)$$

$$= \left(\frac{SM}{V_1} - \frac{PM}{V_2}\right) + \left(\frac{NG}{V_1} - \frac{NQ}{V_2}\right)$$

$$= \delta_s + \delta_g, \tag{4.96}$$

where δ_s and δ_g are known as the *shotpoint delay time* and the *geophone delay time* since they are associated with the portions of the path down from the shot and up to the geophone.

An approximate value of δ can be found by assuming that the dip is small enough that PQ is approximately equal to the geophone offset x. In this case,

$$\delta = \delta_s + \delta_g \approx t_g - \frac{x}{V_2}. \tag{4.97}$$

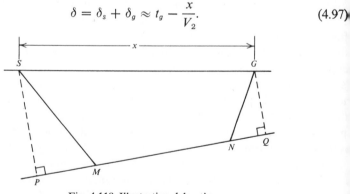

Fig. 4.118 Illustrating delay time.

Provided the dip is less than about 10°, this relation is sufficiently accurate for most purposes. If we substitute the value of t_g obtained from equations (4.47b), (4.52a) and (4.52b), we see that δ is equal to the intercept time for a horizontal refractor but not for a dipping refractor.

Many interpretation schemes using delay time have been given in the literature, for example, Gardner (1939, 1967), Barthelmes (1946), Tarrant (1956), Wyrobek (1956) and Barry (1967). We shall describe only the latter two. The method described by Wyrobek is suitable for unreversed profiles while that of Barry works best with reversed profiles.

The scheme described by Barry, like many based on delay times, requires that we resolve the total delay time δ into its component parts, δ_s and δ_g. In fig. 4.119 we show a geophone R for which data are recorded from shots at A and B. The ray BN is reflected at the critical angle, hence Q is the first geophone to record the head wave from B. Let δ_{AM} be the shotpoint delay time for shot A, δ_{NQ} and δ_{PR} the geophone delay times for geophones at Q and R, δ_{AQ} and δ_{AR} the total delay times for the paths, $AMNQ$ and $AMPR$. Then,

$$\delta_{AQ} = \delta_{AM} + \delta_{NQ}$$
$$\delta_{AR} = \delta_{AM} + \delta_{PR}$$
$$\Delta\delta = \delta_{AQ} - \delta_{AR} = \delta_{NQ} - \delta_{PR}.$$

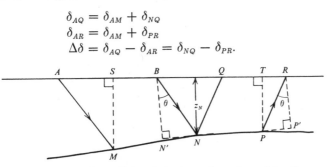

Fig. 4.119 Determining shotpoint and geophone delay times.

For the shot at B, the shot delay time δ_{BN} is approximately equal to δ_{NQ} provided the dip is small. In this case,

$$\delta_{BR} = \delta_{BN} + \delta_{PR} \approx \delta_{NQ} + \delta_{PR}.$$

The geophone delay times are now given by

$$\left.\begin{array}{l}\delta_{NQ} \approx \tfrac{1}{2}(\delta_{BR} + \Delta\delta) \\ \delta_{PR} \approx \tfrac{1}{2}(\delta_{BR} - \Delta\delta)\end{array}\right\}. \qquad (4.98)$$

Thus, it is possible to find the geophone delay time at R provided we have data from two shots on the same side and we can find point Q. If we assume that the bed is horizontal at N and is at a depth z_N we have

$$z_N = V_1\delta_{BN}/\cos\theta, \qquad (4.99a)$$

$$BQ = 2z_N \tan\theta = 2V_1\delta_{BN}\left(\frac{\tan\theta}{\cos\theta}\right) = 2V_2\delta_{BN}\tan^2\theta. \qquad (4.99b)$$

The shot delay time δ_{BN} is assumed to be equal to half the intercept time at B; this allows us to calculate an approximate value of BQ and thus determine the delay times for all geophones to the right of Q for which data from A and B were re corded.

The interpretation involves the following steps which are illustrated in fig. 4.120

(a) the corrected traveltimes are plotted,

(b) the total delay times are calculated and plotted at the geophone positions

(c) the 'geophone offset distances' (PP' in fig. 4.119) are calculated using eq. (4.99b) and the delay times in (b) are then shifted towards the shotpoint by these amounts,

(d) the shifted curves in (c) for the reversed profiles should be parallel; any divergence is due to an incorrect value of V_2, hence the value of V_2 is adjusted and

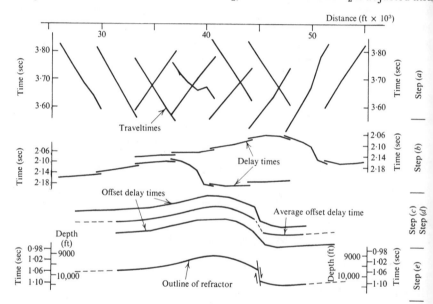

Fig. 4.120 Illustrating the delay-time method of interpreting reversed profiles. (After Barry, 1967.)

steps (b) and (c) repeated until the curves are parallel (with practice only one adjustment is usually necessary),

(e) the total delay times are separated into shotpoint and geophone delay times, the latter being plotted at the points of entry and emergence from the refractor (S and T in fig. 4.119); the delay-time scale can be converted into depth if required using eq. (4.99a).

To illustrate Wyrobek's method we assume a series of unreversed profiles as in the upper part of fig. 4.121. The various steps in the interpretation are as follows:

(a) the corrected traveltimes are plotted and the intercept times measured,

(b) the total delay time δ is calculated for each geophone position for each shot and the values plotted at the geophone position (if necessary, a value of V_2 is assumed); by moving the various segments up or down a composite curve similar to a phantom horizon is obtained.

(c) the intercept times divided by 2 are plotted and compared with the composite delay-time curve; divergence between the two curves indicates an incorrect value of V_2 (see below), hence the value used in step (b) is varied until the two curves are 'parallel', after which the half-intercept time curve is completed by interpolation and extrapolation to cover the same range as the composite delay-time curve,

(d) the half-intercept time curve is changed to a depth curve by using eq. (4.47c), namely

$$z = \tfrac{1}{2}V_1 t_1/\cos \theta$$

(note that we are ignoring the difference between the vertical depth z and the slant depths z_u, z_d in eq. (4.51)).

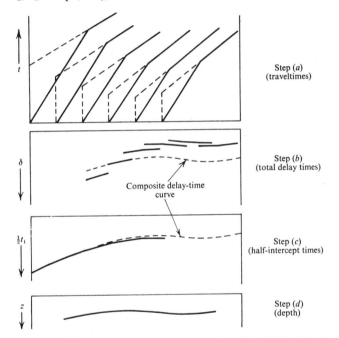

Fig. 4.121 Illustrating Wyrobek's method using unreversed profiles. (After Wyrobek, 1956.)

Wyrobek's method depends upon the fact that the curve of δ is approximately parallel to the half-intercept time curve. For proof of this result, the reader is referred to problem 18. Wyrobek's method does not require reversed profiles because the intercept at a shotpoint does not depend upon the direction in which the cable is laid out.

Delay-time methods are subject to certain errors which must be guarded against. As the shotpoint-to-geophone distance increases, the refraction wavetrain becomes longer and the energy peak shifts to later cycles. There is thus the danger that different cycles will be picked on different profiles and that the error will be interpreted as an increase in shot delay time. If sufficient data are available the error is usually obvious. Variations in refractor velocity manifest themselves in local divergences of the offset total-delay-time curves for pairs of reversed profiles. However, if some data which do not represent refraction travel in the refractor under consideration are accidentally included, the appearance is apt to be the same as if the refractor velocity were varying. In situations where several refractors which have nearly the same velocities are present, unambiguous interpretation may not be possible.

(c) *Wavefront methods.* Wavefront reconstruction, usually by graphical means, forms the basis of several refraction interpretation techniques. The classic paper is one by Thornburgh (1930); other important articles are those by Gardner (1949), Baumgarte (1955), Hales (1958), Hagedoorn (1959), Rockwell (1967) and Schenck (1967).

Figure 4.122 illustrates the basic method of reconstructing wavefronts. The refraction wavefront which reached A at $t = 1{\cdot}600$ sec reached $B, C, \ldots$ at the

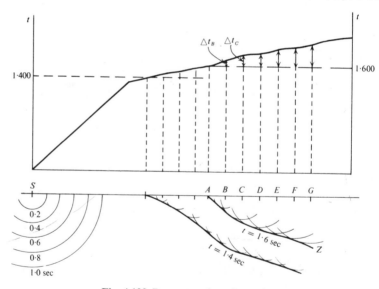

Fig. 4.122 Reconstruction of wavefronts.

times $1{\cdot}600 + \Delta t_B, 1{\cdot}600 + \Delta t_C, \ldots$ By drawing arcs with centres $B, C, \ldots$ and radii $V_1 \Delta t_B, V_1 \Delta t_C, \ldots$, we can establish the wavefront for $t = 1{\cdot}600$ sec (AZ) as accurately as we wish. Similarly, other refraction wavefronts, such as that shown for $t = 1{\cdot}400$ sec, can be constructed at any desired traveltime interval. The direct wavefronts from the shot S are of course the circles shown in the diagram.

In fig. 4.123 we show a series of wavefronts chosen so that only waves which will be first arrivals are shown (all secondary arrivals being eliminated in the interests of simplicity). Between the shotpoint S and the crossover point C (see eq. (4.48)) the direct wave arrives first. To the right of C, the wave refracted at the first horizon arrives first until, to the right of G, the refraction from the deeper horizon overtakes the shallower refraction.

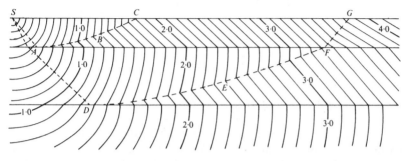

Fig. 4.123 Coincident-time curves. (After Thornburgh, 1930.)

The two systems of wavefronts representing the direct wave and the refracted wave from the shallow horizon intersect along the dotted line ABC; this line, called the coincident-time curve by Thornburgh, passes through the points where the intersecting wavefronts have the same traveltimes. The curve $DEFG$ is a coincident-time curve for the deeper horizon. The coincident-time curves are tangent to the refractors at A and D where the incident ray reaches the critical angle (see problem 19) while the points at which the coincident-time curves meet the surface are marked by abrupt changes in the slopes of the time-distance plot.

Since the coincident-time curve is tangent to the refractor, the latter can be found when we have one profile plus other data, such as the dip, depth, critical angle – or a second profile (not necessarily reversed) since we now have two coincident-time curves and the refractor is the common tangent to the curves.

When reversed profiles are available, the construction of wavefronts provides an elegant method of locating the refractor. The basic principle is illustrated in fig. 4.124 which shows two wavefronts, MCD and PCE, from shots at A and B intersecting at an intermediate point C. Obviously the sum of the two traveltimes from A and B to C is equal to the reciprocal time between A and B, t_r. If we had reconstructed the two wavefronts from the time-distance curve without knowing where the refractor RS was located, we would draw the wavefronts as MCN and PCQ, not MCD and PCE. Therefore if we draw pairs of wavefronts from A and B such that the sum of the traveltimes is t_r, the refractor must pass through the points of intersection of the appropriate pairs of wavefronts in fig. 4.124.

Hagedoorn's *Plus-Minus* method (1959) utilizes a construction similar to that just described. When the refractor is horizontal, the intersecting wavefronts drawn at intervals of Δ milliseconds form diamond-shaped figures (fig. 4.125) whose horizontal and vertical diagonals are equal to $V_2\Delta$ and $V_1\Delta/\cos\theta$ respectively. If we add together the two traveltimes at each intersection and subtract t_r, the resulting 'plus' values equal 0 on the refractor, $+2\Delta$ on the horizontal line

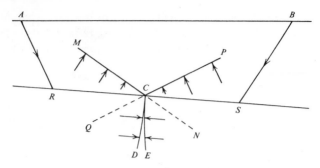

Fig. 4.124 Determining refractor position from wavefront intersections.

through the first set of intersections vertically above those defining the refractor, $+4\Delta$ on the next line up, and so on. Since the distance between each pair of adjacent lines is $V_1\Delta/\cos\theta$, we can use any of the 'plus' lines to plot the refractor. The difference between two traveltimes at an intersection is called the 'minus' value; it is constant along vertical lines passing through the intersections of wavefronts. The distance between successive 'minus' lines as shown in fig. 4.125 is $V_2\Delta$, hence a continuous check on V_2 is possible. Although dip alters the above relations, the changes are small for moderate dip, and the assumption is made that the 'plus' lines are still parallel to the refractor and that the 'minus' lines do not converge or diverge.

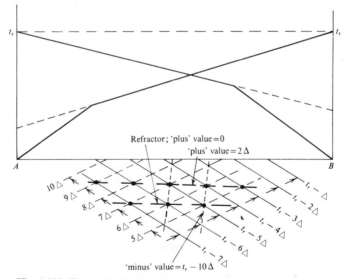

Fig. 4.125 Illustrating the Plus-Minus method. (After Hagedoorn, 1959.)

(d) *Engineering applications.* Refraction methods are commonly applied in mineral-exploration and civil-engineering work to measure depth of bedrock. With the arrangement shown in fig. 4.126, shots are fired from the end points of the spread, A and B, and the midpoint, C. (The 'shot' is usually a hammer blow for shallow overburden or a blasting cap for deeper.) Let t_{AB} be the surface-to-surface travel time from A to B, etc.; then (see problem 20)

$$z_C = \left(\frac{t_{CA} + t_{CB} - t_{AB}}{2} \right) \frac{V_1 V_2}{(V_2{}^2 - V_1{}^2)^{\frac{1}{2}}}, \qquad (4.100a)$$

where V_1 is the overburden and V_2 the bedrock velocity. Frequently $V_2 \gg V_1$ and we can replace the velocity terms by V_1,

$$z_C = \tfrac{1}{2} V_1 (t_{CA} + t_{CB} - t_{AB}), \qquad (4.100b)$$

the error in z_C being less than 6% if $V_2 > 3 V_1$. This method assumes that the overburden is essentially homogeneous, the depth variation is smooth, the dip is small,

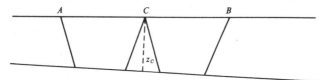

Fig. 4.126 Refraction profile for determining depth to bedrock.

and that the velocity contrast is large so that the perpendicular distance to bedrock is roughly equal to the vertical distance. Depth calculations by this technique are generally good because they depend on the measurement of only one velocity, V_1, and three travel times.

4.9 Problems

1. Consider a cube of elastic isotropic material subjected to a normal stress, σ_{xx}. Equation (4.3a) indicates that the modulus of rigidity enters into the calculation of the strain. Explain why shear is involved.

2. (a) By substituting $\sigma_{xx} > 0$, $\sigma_{yy} = \sigma_{zz} = 0$, in eq. (4.3a), show that $\varepsilon_{yy} = \varepsilon_{zz}$ and verify eq. (4.4e).

(b) By adding the three equations for $\sigma_{xx}, \sigma_{yy}, \sigma_{zz}$, derive eq. (4.4d).
(c) Substituting $\sigma_{xx} = \sigma_{yy} = \sigma_{zz} = -p$ in eq. (4.3a), derive eq. (4.4f).

3. (a) Verify that ψ in eq. (4.9a) is a solution of eq. (4.8b). [Hint: let $\zeta = (x - Vt)$ and show that

$$\frac{\partial \psi}{\partial x} = \frac{df}{d\zeta} \cdot \frac{\partial \zeta}{\partial x} = \frac{df}{d\zeta} = f', \text{ etc.}]$$

(b) Verify that ψ in eq. (4.10) is a solution of eq. (4.8a).
(c) Using the same technique as in parts (a) and (b), show that

$$\psi = \frac{1}{r} f(r - Vt)$$

satisfies eq. (4.12).

4. Verify that eq. (4.11) is the wave equation by substituting the following coordinate transformations (see fig. 4.127) into eq. (4.8a):

$$x = r \sin \theta \cos \phi$$
$$y = r \sin \theta \sin \phi$$
$$z = r \cos \theta.$$

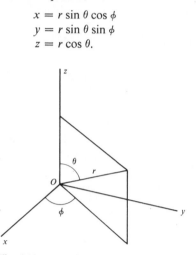

Fig. 4.127 Coordinate transformation.

5. A pulse consists of two frequency components, $\nu_0 \pm \Delta\nu$, of equal amplitudes. We write for the two components

$$A \cos 2\pi(k_1 x - \nu_1 t), \quad A \cos 2\pi(k_2 x - \nu_2 t),$$

where $\nu_1 = \nu_0 + \Delta\nu$, $\nu_2 = \nu_0 - \Delta\nu$, $k_0 = 1/\lambda_0 = \nu_0/V$, $k_1 \approx k_0 + \Delta k \approx (\nu_0 + \Delta\nu)/V$, $k_2 \approx k_0 - \Delta k \approx (\nu_0 - \Delta\nu)/V$.

(a) Show that the pulse is given approximately by the expression

$$B \cos 2\pi(k_0 x - \nu_0 t),$$

where $B = 2A \cos 2\pi\Delta k \left(x - \dfrac{\Delta\nu}{\Delta k} t \right)$.

(b) Why do we regard B as the amplitude? Show that the envelope of the pulse is the graph of B plus its reflection in the x-axis.

(c) Show that the envelope moves with the velocity U where

$$U = \frac{\Delta\nu}{\Delta k} \approx \frac{\mathrm{d}\nu}{\mathrm{d}k} \approx V - \lambda \frac{\mathrm{d}V}{\mathrm{d}\lambda} \approx V + \omega \frac{\mathrm{d}V}{\mathrm{d}\omega}.$$

6. Assume three geophones so oriented that one records only the vertical component of a seismic wave, another records only the horizontal component in the direction of the shot and the third only the horizontal component at right angles to this. Assume a simple wave shape and draw the responses of the three geophones for the following cases: (i) a P-wave travelling directly from the shot to the geophones, (ii) a P-wave reflected from a deep horizon; (iii) an S-wave generated by reflection of a P-wave at an interface, (iv) a Rayleigh wave generated by the shot.

7. Using fig. 4.128 show that the construction of fig. 4.29 gives the same result as eq. (4.38c). [Hint: express OB in terms of $n = \cos\theta_3$ and use eq. (4.38c).]

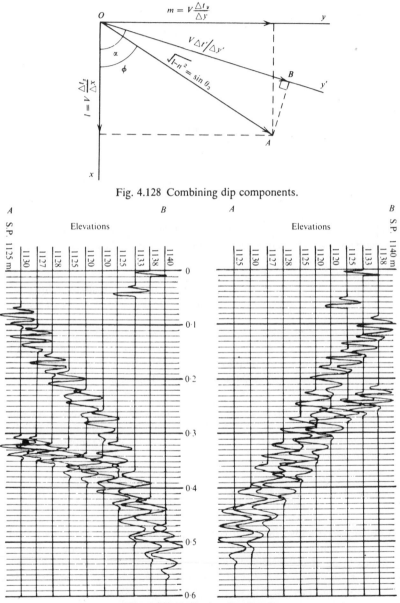

Fig. 4.128 Combining dip components.

Fig. 4.129 Reflection first breaks.

8. (*a*) Given the time-depth information in the following table, calculate interval velocities and average velocity to each depth. Plot velocity against depth and velocity

against time, and determine the equations of straight lines which approximate interval velocity and average velocity against time and depth (4 equations). What are some of the problems involved in making functional fits to data?

Depth	2-way time
1000 m	1·0 sec
2500	2·0
2800	2·1
4800	3·1

(*b*) Using the function derived above for the average velocity as a function of time, calculate the depth of reflections with zero dip moveout and traveltimes of 1·0, 2·0, 2·1 and 3·1 sec. How much error has been introduced by approximating the velocity by a function?

(*c*) Using the function $V(z)$ derived above and the equations in §4.3.2*e*, calculate the depth of flat reflectors corresponding to arrival times of 1·0, 2·0, 2·1 and 3·1 sec.

9. (*a*) Given the velocity-depth relation $V = 1550 + 0.5\ z$(m/sec) and a dip moveout of 50 msec/1000 m, calculate the dip of reflectors for which the arrival times are 1·0, 2·0, 3·0 sec; plot the locations of the reflecting points.

(*b*) Assuming equivalent average velocity as indicated in §4.3.2*b*, the results from part (*a*) of problem 8 and a dip moveout of 50 msec/1000 m, calculate the dip of reflectors corresponding to traveltimes of 1·0, 2·0 and 3·0 sec. Plot the locations of the reflecting points.

(*c*) Assume layers each of constant velocity as follows:

0 to 1000 m: 2000 m/sec
1000 to 2500 m: 3000 m/sec
2500 to 2800 m: 6000 m/sec
2800 to 4800 m: 4000 m/sec.

Verify that this gives the time-depth table of problem 8. By ray tracing through these layers for a dip moveout of 50 msec/1000 m, find the dip and reflecting points for reflectors for which the arrival times are 1·0, 2·0 and 3·0 sec. Compare with the results for parts (*a*) and (*b*).

10. The correction methods discussed in §4.5.7*b* assume that the shot is below the base of the LVL. What changes are required in the equations of this section if this is not the case?

11. Figure 4.129 shows the first arrivals at geophone stations 100 m apart from shots 25 m deep at each end of the spread. (There are actually 11 geophone stations with the shotpoints being at the 1st and 11th stations; however, the geophone group at each shotpoint is not recorded because of hole noise.) The uphole geophone is recorded on the 3rd trace from the right. The weathering velocity is 500 m/sec.

(*a*) Estimate the subweathering velocity V_H by averaging the slopes of lines approximating the first breaks. The valley midway between the shotpoints produces a change in the first-break slopes, as if two reflectors are involved, which is not the case. How can one be sure of the latter?

(b) Determine the weathering thicknesses at the two shotpoints from the uphole times.

(c) What corrections Δt_0 should be applied to reflection times at the two shotpoints for a datum of 1100 m?

(d) Calculate the weathering thickness and the static time correction for each geophone station.

12. (a) Two intersecting seismic spreads have bearings N10°E and N140°E. If the first spread shows an event at $t_0 = 1\cdot760$ sec with dip moveout of 56 msec/1000 m while the same event on the second spread has a dip moveout of 32 msec/1000 m, find the true dip, depth and the strike, assuming that (i) both dips are down to the south and west, (ii) dip on the first spread is down to the south while the other is down to the southeast. Take the average velocity as 3 km/sec.

(b) Calculate the migrated position for each spread as if the cross information had not been available and each had been assumed to be indicating total moveout; compare with the results of part (a). Would the errors be more serious or less serious if the calculations were made for the usual situation where the velocity increases with depth?

13. Consider a reflection with an arrival time at the shotpoint of $1\cdot200$ sec and arrival times at geophones 1500 ft away on opposite sides of the shotpoint equal to $1\cdot162$ and $1\cdot237$ sec respectively, static and dynamic corrections having been applied to all traveltimes.

(a) Using the simplified wavefront chart and velocity distribution shown in fig. 4.90, plot the migrated position of the reflection.

(b) If straight-ray plotting were used at the angle of approach, where would the reflection be plotted?

(c) Using straight-ray plotting and assuming the average velocity at the reflector is 12,600 ft/sec, what is the migrated position?

(d) The average velocity for a vertical travel time of $1\cdot200$ sec is 8660 ft/sec; how can this be determined from the data given in fig. 4.90?

(e) Assuming the average velocity 8660 ft/sec and straight-ray travel, what is the migrated position?

(f) On the adjoining record which has a shotpoint 1500 ft away, what is the arrival time of this reflection for a geophone at the shotpoint?

14. Derive eq. (4.65a) from (4.64a). [Hint: multiply both sides by $e^{-j2\pi f_n t}$ and integrate over one cycle, e.g. from $-T/2$ to $+T/2$. Before substituting the values at the limits, use Euler's theorem, $e^{\pm jx} = \cos x \pm j \sin x$, to express the complex exponentials in terms of sines and cosines.]

15. (a) Since $\delta(t)$ is zero except for $t = 0$ where it equals $+1$, we can apply eq. (4.65b) and find that

$$\delta(t) \leftrightarrow + 1;$$

show that

$$\delta(t - t_0) \leftrightarrow e^{-j2\pi f t_0}.$$

(b) The comb can be written

$$\text{comb}\,(t) = \sum_{n=-\infty}^{+\infty} \delta(t - nt_0);$$

the transform of this expression is clearly

$$\text{comb}\,(t) \leftrightarrow S(f) = \sum_{n=-\infty}^{+\infty} e^{-j2\pi n t_0 f}.$$

Show (see Papoulis, 1962, p. 44) that this represents an infinite series of impulses of height $2\pi/t_0$ spaced $2\pi/t_0$ apart, that is,

$$\sum_{n=-\infty}^{+\infty} \delta(t - n t_0) \leftrightarrow \left(\frac{2\pi}{t_0}\right) \sum_{m=-\infty}^{+\infty} \delta(2\pi f - 2\pi m/t_0),$$

or

$$\text{comb}\,(t) \leftrightarrow \omega_0\,\text{comb}\,(\omega).$$

(c) Show that a boxcar of height h and extending from $-f_0$ to $+f_0$ in the frequency domain has the transform

$$\text{boxcar}\,(f) \leftrightarrow A\,\text{sinc}\,(2\pi f_0 t) = A\,\frac{\sin 2\pi f_0 t}{2\pi f_0 t},$$

where $A = 2h f_0$ = area of the boxcar.

16. (a) Verify that eq. (4.76) is the inverse filter for water reverberation by convolving (4.74) with (4.76), i.e., by substituting the expressions given by eqs. (4.74) and (4.76) in eq. (4.75).

(b) The spectrum of the water-layer filter is shown in fig. 4.130; the large peaks occur at the 'singing frequency'; sketch the amplitude spectrum of the inverse filter. [Hint: time-domain convolution such as shown in eq. (4.75) corresponds to frequency-domain multiplication (see eq. (4.71a)); the frequency spectrum of the unit impulse is $+1$, i.e., flat.]

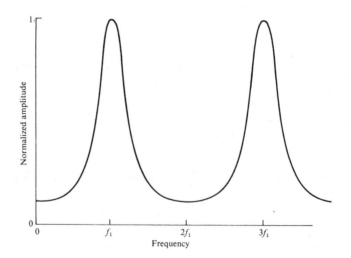

Fig. 4.130 Spectrum of water-layer filter for water-bottom reflection coefficient of 0·5. If h = water depth and V = water velocity, $f_1 = V/2h$.

(c) Verify your sketch of the water-reverberation inverse filter by transforming eq. (4.76):

$$[1, \ 2r, r^2] \leftrightarrow 1 + 2r\,e^{-j2\pi f \triangle} + r^2\,e^{-j4\pi f \triangle}.$$

17. (a) Convolve $[2, 5, -2, 1]$ with $[6, -1, -1]$;

(b) Cross-correlate $[2, 5, -2, 1]$ with $[6, -1, -1]$; for what shift are these functions most nearly alike?

(c) Convolve $[2, 5, -2, 1]$ with $[-1, -1, 6]$; compare with the answer in (b) and explain the difference.

(d) Autocorrelate $[6, -1, -1]$ and $[3, -5, -2]$; the autocorrelation of a function is not unique to that function, for example, other wavelets having the same autocorrelations as the preceding are $[-1, -1, 6]$ and $[-2, -5, 3]$; which member of the set is the minimum delay wavelet?

(e) What is the normalized autocorrelation of $[6, -1, -1]$? What is the normalized cross-correlation in (b)? What do you conclude from the magnitude of the largest value of this normalized cross-correlation?

18. Prove that the half-intercept curve referred to in the discussion of Wyrobek's method in §4.8.3b is parallel to the curve of the total delay time δ (see fig. 4.131). Note that the reciprocal time can be written (see eqs. (4.52c) and (4.52d))

$$t_r = \tfrac{1}{2}\left\{\left(\frac{L}{V_d} + t_{1d}\right) + \left(\frac{L}{V_u} + t_{1u}\right)\right\}.$$

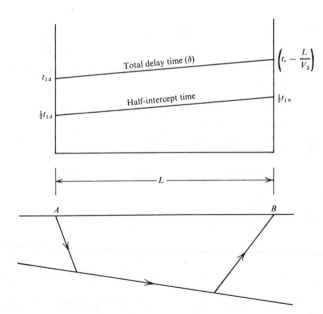

Fig. 4.131 Demonstrating the parallelism of the curves of total delay time and the half-intercept time.

19. Using fig. 4.132 show that: (i) DE, the 'wavefront for $t = 0$', is at a depth of $2h = SD = 2z \cos \theta$ (note that $CD = SA = z/\cos \theta$); (ii) each point on the coincident-time curve is equidistant from S and DE, that is, the curve is a parabola; (iii) taking DE and SD as the x- and y-axes, the coincident-time curve has the equation $4hy = x^2 + 4h^2$; (iv) the slope of the coincident-time curve at A is $\tan \theta$, hence the curve is tangent to the refractor at A.

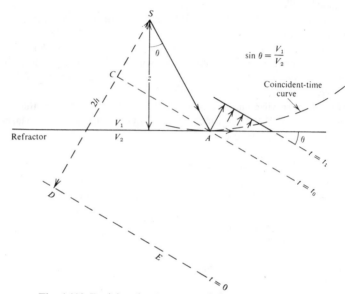

Fig. 4.132 Deriving the properties of the coincident-time curve.

20. Prove eq. (4.100a) assuming that the surface is horizontal and the refractor is plane between the two shotpoints.

21. To find the depth to bedrock in a damsite survey, traveltimes were measured from the shotpoint to 12 geophones laid out at 50-ft intervals along a straight line through the shotpoint. The offsets x range from 50 to 600 ft. Determine the depth of overburden from the data in the following table.

x	t	x	t	x	t
50 ft	19 msec	250 ft	59 msec	450 ft	72 msec
100	29	300	62	500	76
150	39	350	65	550	78
200	50	400	68	600	83

22. To determine the thickness of the surface layer in a certain area, the following readings were obtained from refraction records. Shotpoints A and B are located at the ends of a 600-ft east–west spread of 13 geophones, A being west of B. The ground surface

is flat. No data were obtained from the geophone at the shotpoint. Using the data below, find the velocities in the upper and lower beds, the dip and the vertical depth to the refractor at A and B.

x	t for SP A	t for SP B	x	t for SP A	t for SP B
50 ft	10 msec	10 msec	350 ft	65 msec	57 msec
100	20	20	400	69	63
150	30	30	450	73	69
200	40	40	500	77	74
250	50	46	550	81	80
300	60	51	600	85	85

23. Show that the two geological sections illustrated in fig. 4.133 produce the same refraction time-distance curve. What would be the apparent depth to the lower interface

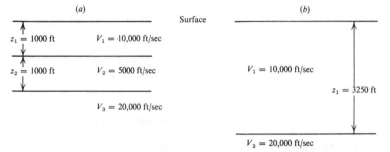

Fig. 4.133 Two different geologic sections which give the same refraction time-distance curve.

(obtained by refraction shooting) if V_3 is approximately equal to V_1 in the section at the left, for example, $V_1 = 10,000$ ft/sec, $V_3 = 10,500$ ft/sec and $V_2 = 5000$ ft/sec as in the diagram?

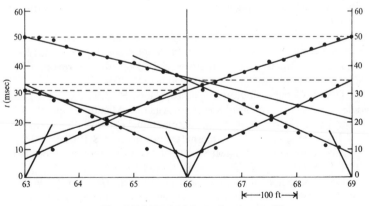

Fig. 4.134 Engineering refraction profile.

434   Seismic methods

24. The time-distance observations in fig. 4.134 constitute an engineering refraction problem.

(a) Using eqs. (4.52) and (4.53), show that the first layer dips about 1° and ranges in thickness from 9 to 11 ft between points 63 and 69.

(b) Show that the second and third layers have velocities of about 11,500 and 18,000 ft/sec respectively, the interface between the two layers dipping about 6°.

(c) Determine the thickness of the second layer by stripping off the shallow layer (see §4.8.3a).

References

Agnich, F. J. and Dunlap, R. C. (1959). Standards of performance in petroleum exploration. *Geophysics*, 24, 916–24.

Agocs, W. B. (1950). Computation charts for linear increase of velocity with depth. *Geophysics*, 15, 227–36.

Allen, F. T. (1972). Some characteristics of marine sparker seismic data. *Geophysics*, 37, 462–70.

Anstey, N. A. (1964). Correlation techniques – a review. *Geophy. Prosp.* 12, 355–82.

Anstey, N. A. (1970). Signal characteristics and instrument specifications: v. 1 of *Seismic prospecting instruments;* Berlin, Gebruder Borntraeger.

Attewell, P. B. and Ramana, Y. V. (1966). Wave attenuation and internal friction as functions of frequency in rocks. *Geophysics*, 31, 1049–56.

Backus, M. M. (1959). Water reverberations – their nature and elimination. *Geophysics*, 24, 233–61.

Balch, A. H. (1971). Color sonagrams – A new dimension in seismic data interpretation. *Geophysics*, 36, 1074–98.

Barbier, M. G. and Viallix, J. R. (1973). Sosie – A new tool for marine seismology. *Geophysics*, 38, 673–83.

Barry, K. M. (1967). Delay time and its application to refraction profile interpretation. In A. W. Musgrave (ed.), *Seismic refraction prospecting*, pp. 348–62. Tulsa, Soc. of Exploration Geophysicists.

Barthelmes, A. J. (1946). Application of continuous profiling to refraction shooting. *Geophysics*, 11, 24–42.

Baumgarte, J. (1955). Konstruktive darstellung von seismischen horizonten unter berucksichtigung der strahlenbrechung im raum: *Geophys. Prosp.* 3, 126–62.

Bedenbender, J. W., Johnston, R. C. and Neitzel, E. B. (1970). Electroacoustic characteristics of marine seismic streamers. *Geophysics*, 35, 1054–72.

Birch, F. (1966). Compressibility; elastic constants. S. P. Clark (ed.), *Handbook of physical constants*, Geological Society of America, Memoir 97, pp. 97–173.

Bois, P. and La Porte, M. (1970). Pointe automatique. *Geophys. Prosp.* 18, 489–504.

Bradley, J. J. and Fort, A. N. (1968). Internal friction in rocks. S. P. Clark (ed.), *Handbook of physical constants*, Geological Society of America, Memoir 97, pp. 175–93.

Brede, E. C., Johnston, R. C. Sullivan, L. B. and Viger, H. L. (1970). A pneumatic seismic energy source for shallow water/marsh areas. *Geophys. Prosp.* 18, 581–99.

Brown, R. J. S. (1969). Normal-moveout and velocity relations for flat and dipping beds and for long offsets. *Geophysics*, 34, 180–95.

Bullen, K. E. (1965). *An introduction to the theory of seismology.* London, Cambridge.

Burg, K. E., Ewing, M., Press, F. and Stulken, E. J. (1951). A seismic wave-guide phenomenon. *Geophysics,* **16**, 594–612.

Bybee, H. H. (1973). Navigation satellites for geophysical exploration. Offshore Technology Conference, paper 1785.

Campbell, F. F. (1965). Fault criteria. *Geophysics,* **30**, 976–97.

Cassand, J., Damotte, B., Fontanel, A., Grau, G., Hemon, C. and Lavergne, M. (1971). *Seismic filtering.* Tulsa, Soc. of Exploration Geophysicists.

Clarke, G. K. C. (1968). Time-varying deconvolution filters. *Geophysics,* **33**, 936–44.

Cole, J. R. (1967). Vibroseis – effective, harmless exploration tool. *Oil & Gas Journal,* October 30, 97–108.

Cook, E. E. and Taner, M. T. (1969). Velocity spectra and their use in stratigraphic and lithologic differentiation. *Geophys. Prosp.* **17**, 433–48.

Coulson, C. A. (1949). *Waves.* Edinburgh, Oliver and Boyd.

Daly, J. W. (1948). An instrument for plotting reflection data on the assumption of a linear increase of velocity. *Geophysics,* **13**, 153–7.

Daly, J. (1953). A universal slide rule for computing the dips of reflecting horizons on the assumption of linear increase in velocity. *Geophysics,* **18**, 820–3.

Davis, T. L. (1972). Velocity variations around Leduc reefs, Alberta. *Geophysics,* **37**, 584–604.

De Bremaecker, J. C., Godson, R. H. and Watkins, J. S. (1966). Attenuation measurements in the field. *Geophysics,* **31**, 562–9.

De Golyer, E. (1947). Notes on the early history of applied geophysics in the petroleum industry. *Early geophysical papers of the Society of Exploration Geophysicists.* Tulsa, Soc. of Exploration Geophysicists, pp. 245–50.

Delaplanche, J., Hagemann, R. F. and Bollard, P. G. C. (1963). An example of the use of synthetic seismograms. *Geophysics,* **28**, 842–54.

Dennison, A. T. (1953). The design of electromagnetic geophones. *Geophys. Prosp.* **1**, 3–28.

Disher, D. A. and Naquin, P. J. (1970). Statistical automatic statics analysis. *Geophysics,* **35**, 574–85.

Dix, C. H. (1952). *Seismic prospecting for oil.* New York, Harper.

Dix, C. H. (1955). Seismic velocities from surface measurements. *Geophysics,* **20**, 68–86.

Dobrin, M. B. (1951). Dispersion in seismic surface waves. *Geophysics,* **16**, 63–80.

Dobrin, M. B. (1976). *Introduction to geophysical prospecting.* New York, McGraw-Hill.

Dunkin, J. W. and Levin, F. K. (1973). Effect of normal moveout on a seismic pulse. *Geophysics,* **38**, 635–42.

Edelmann, H. (1966). New filtering methods with Vibroseis. *Geophy. Prosp.,* **14**, 455–69.

Ergin, K. (1952). Energy ratios of seismic waves reflected and refracted at a rock-water boundary. *Bull. Seis. Soc. Am.* **42**, 349–72.

Evenden, B. S. and Stone, D. R. (1971). Instrument performance and testing. Vol. 2 of *Seismic prospecting instruments.* Berlin, Gebrüder Borntraeger.

Ewing, W. M., Jardetzky, W. S. and Press, F. (1957). *Elastic waves in layered media.* New York, McGraw-Hill.

Faust, L. Y. (1951). Seismic velocity as a function of depth and geologic time. *Geophysics,* **16**, 192–206.

Faust, L. Y. (1953). A velocity function including lithologic variation. *Geophysics,* **18**, 271–88.

Finetti, I., Nicolich, R. and Sancin, S. (1971). Review on the basic theoretical assumptions in seismic digital filtering. *Geophys. Prosp.* **19**, 292–320.

Flinn, E. A., ed., Robinson, E. A. and Treitel, S. (1967). Special issue on the MIT Geophysical Analysis Group reports. *Geophysics*, **32**, 411–525.

Gardner, G. H. F., Gardner, L. W. and Gregory, A. R. (1974). Formation velocity and density – the diagnostic basics for stratigraphic traps. *Geophysics*, **39**, 770–80.

Gardner, L. W. (1939). An areal plan of mapping subsurface structure by refraction shooting. *Geophysics*, **4**, 247–59.

Gardner, L. W. (1947). Vertical velocities from reflection shooting. *Geophysics*, **12**, 221–8.

Gardner, L. W. (1949). Seismograph determination of salt-dome boundary using well detector deep on dome flank. *Geophysics*, **14**, 29–38.

Gardner, L. W. (1967). Refraction seismograph profile interpretation. In A. W. Musgrave, (ed.), *Seismic refraction prospecting*, pp. 338–47. Tulsa, Soc. of Exploration Geophysicists.

Garotta, R. and Michon, D. (1967). Continuous analysis of the velocity function and of the moveout correction. *Geophys. Prosp.* **15**, 584–97.

Garotta, R. (1971). Selection of seismic picking based upon the dip, move-out and amplitude of each event. *Geophys. Prosp.* **19**, 357–70.

Gassmann, F. (1951). Elastic waves through a packing of spheres. *Geophysics*, **16**, 673–85.

Giles, B. F. (1968). Pneumatic acoustic energy source. *Geophys. Prosp.* **16**, 21–53.

Godfrey, L. M., Stewart, J. D. and Schweiger, F. (1968). Application of dinoseis in Canada. *Geophysics*, **33**, 65–77.

Goguel, J. M. (1951). Seismic refraction with variable velocity. *Geophysics*, **16**, 81–101.

Grant, F. S. and West, G. F. (1965). *Interpretation theory in applied geophysics*. New York, McGraw-Hill.

Hagedoorn, J. G. (1954). A process of seismic reflection interpretation. *Geophy. Prosp.* **2**, 85–127.

Hagedoorn, J. G. (1959). The plus-minus method of interpreting seismic refraction sections. *Geophys. Prosp.* **7**, 158–82.

Hales, F. W. (1958). An accurate graphical method for interpreting seismic refraction lines. *Geophys. Prosp.* **6**, 285–314.

Hales, F. W. and Edwards, T. E. (1955). Some theoretical considerations on the use of multiple geophones arranged linearly along the line of traverse. *Geophys. Prosp.* **3**, 65–94.

Harms, J. C. and Tackenberg, P. (1972), Seismic signatures of sedimentation models. *Geophysics*, **37**, 45–58.

Heiland, C. A. (1946). *Geophysical exploration*. New York, Prentice-Hall.

Hileman, J. A., Embree, P. and Pflueger, J. C. (1968). Automated static corrections. *Geophys. Prosp.* **16**, 326–58.

Hilterman, F. J. (1970). Three-dimensional seismic modeling. *Geophysics*, **35**, 1020–37.

Hofer, H. and Varga, W. (1972). Seismogeologic experience in the Beaufort Sea. *Geophysics*, **37**, 605–19.

Holzman, M. (1963). Chebyshev optimized geophone arrays. *Geophysics*, **28**, 145–55.

Jakosky, J. J. (1950). *Exploration geophysics*. Newport Beach, Calif., Trija.

Jankowsky, W. (1970). Empirical investigation of some factors affecting elastic wave velocities in carbonate rocks. *Geophys. Prosp.* **18**, 103–18.

Jeager, J. C. (1958). *Elasticity, fracture and flow*. London, Methuen.

Kennett, P. and Ireson, R. L. (1971). Recent developments in well velocity surveys and the use of calibrated acoustic logs. *Geophys. Prosp.* **19**, 395–411.

King, V. L. (1973). Sea bed geology from sparker profiles, Vermillion Block 321, Offshore Louisiana. Offshore Technology Conference Paper 1802.

Knott, C. G. (1899). Reflexion and refraction of elastic waves, with seismological appplications. *Phil. Mag.*, 5th ser. **48**, 64–97.

Knudsen, W. C. (1961). Elimination of secondary pressure pulses in offshore exploration. *Geophysics*, **26**, 425–36.

Koefoed, O. (1962). Reflection and transmission coefficients for plane longitudinal incident waves. *Geophys. Prosp.* **10**, 304–51.

Kokesh, F. P. and Blizard, R. B. (1959). Geometrical factors in sonic logging. *Geophysics*, **24**, 64–76.

Kramer, F. S., Peterson, R. A. and Walter, W. C. (eds.) (1968). *Seismic energy sources – 1968 handbook*. Pasedena, Bendix United Geophysical.

Kronberger, F. P. and Frye, D. W. (1971). Positioning of marine surveys with an integrated satellite navigation system. *Geophys. Prosp.* **19**, 487–500.

Kunetz, G. and Fourmann, J. M. (1968). Efficient deconvolution of marine seismic records. *Geophysics*, **33**, 412–23.

Lamer, A. (1970). Couplage sol-geophone. *Geophys. Prosp.* **18**, 300–19.

Langstroth, W. T. (1971). Seismic study along a portion of the Devonian salt front in North Dakota. *Geophysics*, **36**, 330–8.

Laski, J. D. (1973). Computation of the time-distance curve for a dipping refractor and velocity increasing with depth in the overburden. *Geophys. Prosp.* **21**, 366–78.

Laster, S. J. and Linville, A. F. (1968). Preferential excitation of refractive interfaces by use of a source array. *Geophysics*, **33**, 49–64.

Lavergne, M. (1970). Emission by underwater explosions. *Geophysics*, **35**, 419–35.

Levin, F. K. (1971). Apparent velocity from dipping interface reflections. *Geophysics*, **36**, 510–6.

Lombardi, L. V. (1955). Notes on the use of multiple geophones. *Geophysics*, **20**, 215–26.

Love, A. E. H. (1944). *A treatise on the mathematical theory of elasticity*. New York, Dover.

Mansfield, R. H. (1947). Universal slide rule for linear velocity vs. depth calculations. *Geophysics*, **12**, 557–75.

Marr, J. D. (1971). Seismic stratigraphic exploration – part I. *Geophysics*, **36**, 311–29; part II. *Geophysics*, **36**, 533–53; part III. *Geophysics*, **36**, 676–89.

Mayne, W. H. (1962). Common-reflection-point horizontal data-stacking techniques. *Geophysics*, **27**, 927–38.

Mayne, W. H. (1967). Practical considerations in the use of common reflection point techniques. *Geophysics*, **32**, 225–9.

Mayne, W. H. and Quay, R. G. (1971). Seismic signatures of large air guns. *Geophysics*, **36**, 1162–73.

McCamy, K., Meyer, R. and Smith, T. J. (1962). Generally applicable solution of Zoeppritz' amplitude equations. *Bull. Seis. Soc. Am.* **52**, 923–55.

McCollum, B. and LaRue, W. W. (1931). Utilization of existing wells in seismograph work: *Bull. A.A.P.G.* **15**, 1409–17. [Reprinted in *Early Geophysical Papers of the Society of Exploration Geophysicists*, 1947. Tulsa, S.E.G., pp. 119–27.]

McDonal, F. J., Angona, F. A., Mills, R. L., Sengbush, R. L., Van Nostrand, R. G. and White, J. E. (1958). Attenuation of shear and compressional waves in Pierre Shale.

Geophysics, **23**, 421–39.

McGee, J. E. and Palmer, R. L. (1957). Early refraction practices: in A. W. Musgrave, ed., *Seismic refraction prospecting*, pp. 3–11. Tulsa, Soc. of Exploration Geophysicists.

McKay, A. E. (1954). Review of pattern shooting. *Geophysics*, **19**, 420–37.

Meidav, T. (1969). Hammer reflection seismics in engineering geophysics. *Geophysics*, **34**, 383–95.

Meiners, E. P., Lenz, L. L., Dalby, A. E. and Hornsby, J. M. (1972). Recommended standards for digital tape formats. *Geophysics*, **37**, 36–44.

Meissner, R. (1967). Exploring deep interfaces by seismic wide-angle measurements. *Geophys. Prosp.* **15**, 598–617.

Meyerhoff, H. J. (1966). Horizontal stacking and multichannel filtering applied to common-depth-point seismic data. *Geophy. Prosp.* **14**, 441–54.

Michon, D., Wlodarczak, R. and Merland, J. (1971). A new method of cancelling multiple reflections – Souston. *Geophys. Prosp.* **19**, 615–25.

Middleton, D. and Wittlesey, J. R. B. (1968). Seismic models and deterministic operators for marine reverberation. *Geophysics*, **33**, 557–83.

Morgan, N. A. (1970). Wavelet maps – a new analysis tool for reflection seismograms. *Geophysics*, **35**, 447–60.

Mossman, R. W., Heim, G. E. and Dalton, F. E. (1973). Vibroseis applications to engineering work in urban area: *Geophysics*, **38**, 489–99.

Moyer, J. R. (1973). The realities of radio positioning: Offshore Technology Conference, paper 1788.

Muskat, M. and Meres, M. W. (1940*a*). Reflection and transmission coefficients for plane waves in elastic media. *Geophysics*, **5**, 115–48.

Muskat, M. and Meres, M. W. (1940*b*). The seismic wave energy reflected from various types of stratified horizons. *Geophysics*, **5**, 149–55.

Neidell, N. S. and Taner, M. T. (1971). Semblance and other coherency measures for multichannel data. *Geophysics*, **36**, 482–97.

Nettleton, L. L. (1940). *Geophysical prospecting for oil*. New York, McGraw-Hill.

Newman, P. (1973). Divergence effects in a layered earth. *Geophysics*, **38**, 481–8.

Newman, P. and Mahoney, J. T. (1973). Patterns – with a pinch of salt. *Geophys. Prosp.* **21**, 197–219.

Northwood, E. J., Weisinger, R. C. and Bradley, J. J. (1967). Recommended standards for digital tape formats. *Geophysics*, **32**, 1073–84.

O'Brien, P. N. S. (1961). A discussion on the nature and magnitude of elastic absorption in seismic prospecting. *Geophys. Prosp.* **9**, 261–75.

O'Brien, P. N. S. (1965). Geophone distortion of seismic pulses and its compensation. *Geophys. Prosp.* **13**, 283–305.

Ocola, L. C. (1972). A nonlinear least-squares method for seismic refraction mapping part I – algorithm and procedure. *Geophysics*, **37**, 260–72; part II – model studies and performance of reframap method. *Geophysics*, **37**, 273–87.

O'Doherty, R. F. and Anstey, N. A. (1971). Reflections on amplitudes. *Geophys. Prosp.* **19**, 430–58.

Officer, C. B., Jr. (1958). *Introduction to the theory of sound transmission*. New York, McGraw-Hill.

Olhovich, V. A. (1964). The causes of noise in seismic reflection and refraction work. *Geophysics*, **29**, 1015–30.

Papoulis, A. (1962). *The Fourier integral and its applications.* New York, McGraw-Hill.

Parasnis, D. S. (1972). *Principles of applied geophysics.* London, Chapman and Hall.

Parr, J. O., Jr. and Mayne, W. H. (1955). A new method of pattern shooting. *Geophysics,* **20,** 539–64.

Paturet, D. (1971). Different methods of time-depth conversion with and without migration. *Geophys. Prosp.* **19,** 27–41.

Paulson, K. V. and Merdler, S. C. (1968). Automatic seismic reflection picking. *Geophysics,* **33,** 431–40.

Peacock, K. L. and Treitel, S. (1969). Predictive deconvolution – theory and practice. *Geophysics,* **34,** 155–69.

Peraldi, R. (1969). Contribution du traitement numerique a l'analyse et a l'interpretation des enregistrements refraction. *Geophys. Prosp.* **17,** 126–64.

Peraldi, R. and Clement, A. (1972). Digital processing of refraction data – study of first arrivals. *Geophys. Prosp.* **20,** 529–48.

Peterson, R. A. and Dobrin, M. B. (1966). *A pictorial digital atlas.* Pasadena, United Geophysical.

Press, F. (1966). Seismic velocities. In S. P. Clark, (ed.), *Handbook of physical constants,* Geological Society of America, Memoir 97, pp. 195–218.

Rackets, H. M. (1971). A low-noise seismic method for use in permafrost regions. *Geophysics,* **36,** 1150–61.

Rayleigh, L. (1917). On the pressure developed in a liquid during the collapse of a spherical cavity. *Philosophical Magazine,* **34,** 94–8.

Rieber, F. (1936). A new reflection system with controlled directional sensitivity. *Geophysics,* **1,** 97–106.

Robinson, E. A. (1967). *Multichannel time series analysis with digital computer programs.* San Francisco, Holden-Day.

Robinson, E. A. and Treitel, S. (1967) Principles of digital Wiener filtering. *Geophys. Prosp.* **15,** 311–33.

Robinson, E. A. and Treitel, S. (1973). *The Robinson-Treitel reader.* Tulsa, Seismograph Service.

Robinson, J. C. (1972). Computer-designed Wiener filters for seismic data. *Geophysics,* **37,** 235–59.

Rockwell, D. W. (1967). A general wavefront method. In A. W. Musgrave, ed., *Seismic refraction prospecting,* pp. 363–415. Tulsa, Soc. of Exploration Geophysicists.

Rosaire, E. E. and Lester, O. C., Jr. (1932). Seismological discovery and partial detail of Vermillion Bay salt dome. *Bull A.A.P.G.* **16,** pp. 51–9. [Reprinted in *Early Geophysical Papers of the Society of Exploration Geophysicists,* 1947. Tulsa, S.E.G., pp. 381–9.]

Rosenkrans, R. R. and Marr, J. D. (1967). Modern seismic exploration of the Gulf Coast Smackover trend. *Geophysics,* **32,** 184–206.

Sattlegger, J. W. (1969). Three-dimensional seismic depth computation using space sampled velocity logs. *Geophysics,* **34,** 7–20.

Schenck, F. L. (1967). Refraction solutions and wavefront targeting. In A. W. Musgrave, (ed.), *Seismic refraction prospecting,* pp. 416–25. Tulsa, Society of Exploration Geophysicists.

Schneider, W. A. (1971). Developments in seismic data processing and analysis (1968–1970). *Geophysics,* **36,** 1043–73.

Schneider, W. A. and Backus, M. M. (1968). Dynamic correlation analysis. *Geophysics,*

33, 105-26.

Schoenberger, M. (1970). Optimization and implementation of marine seismic arrays. *Geophysics*, **35**, 1038-53.

Schulze-Gattermann, R. (1972). Physical aspects of the airpulser as a seismic energy source. *Geophys. Prosp.* **20**, 155-92.

Sengbush, R. L. and Foster, M. R. (1968). Optimum multichannel velocity filters. *Geophysics*, **33**, 11-35.

Shah, P. M. (1973). Use of wavefront curvature to relate seismic data with subsurface parameters. *Geophysics*, **38**, 812-25.

Shah, P. M. and Levin, F. K. (1973). Gross properties of time-distance curves. *Geophysics*, **38**, 643-56.

Sheriff, R. E. (1973). *Encyclopedic dictionary of exploration geophysics*. Tulsa, Soc. of Exploration Geophysicists.

Sherwood, J. W. C. and Poe, P. H. (1972). Continuous velocity estimation and seismic wavelet processing. *Geophysics*, **37**, 769-87.

Sherwood, J. W. C. and Trorey, A. W. (1965). Minimum-phase and related properties of the response of a horizontally stratified absorptive earth to plane acoustic waves. *Geophysics*, **30**, 191-97.

Shetrone, H. A. (1972). Computer mapping of seismic and velocity data in Australia. *Geophysics*, **37**, 313-24.

Siems, L. and Hefer, F. W. (1967). A discussion on seismic binary-gain-switching amplifiers. *Geophys. Prosp.* **15**, 23-34.

Silverman, D. (1967). The digital processing of seismic data. *Geophysics*, **32**, 988-1002.

Slotnick, M. M. (1950). A graphical method for the interpretation of refraction profile data. *Geophysics*, **15**, 163-80.

Slotnick, M. M. (1959). *Lessons in seismic computing*. Tulsa, Soc. of Exploration Geophysicists.

Smith, M. K. (1956). Noise analyses and multiple seismometer theory. *Geophysics*, **21**, 337-60.

Sokolnikoff, Y. (1958). *Mathematical theory of elasticity*. New York, McGraw-Hill.

Stansel, T. A. (1973). Accuracy of geophysical offshore navigation systems. Offshore Technology Conference, paper 1789.

Swan, B. G. and Becker, A. (1952). Comparison of velocities obtained by delta-time analysis and well velocity surveys. *Geophysics*, **17**, 575-85.

Taner, M. T., Cook, E. E. and Neidell, N. S. (1970). Limitations of the reflection seismic method; lessons from computer simulations. *Geophysics*, **35**, 551-73.

Taner, M. T. and Koehler, F. (1969). Velocity spectra – digital computer derivation and applications of velocity functions. *Geophysics*, **34**, 859-81.

Tarrant, L. H. (1956). A rapid method of determining the form of a seismic refractor from line profile results. *Geophys. Prosp.* **4**, 131-9.

Thornburgh, H. R. (1930). Wavefront diagrams in seismic interpretation. *Bull. A.A.P.G.* **14**, 185-200.

Tooley, R. D., Spencer, T. W. and Sagoci, H. F. (1965). Reflection and transmission of plane compressional waves. *Geophysics*, **30**, 552-70.

Treitel, S., Shanks, J. L. and Frasier, C. W. (1967). Some aspects of fan filtering. *Geophysics*, **32**, 789-800.

Trorey, A. W. (1970). A simple theory for seismic diffractions. *Geophysics*, **35**, 762-4.

Tucker, P. M. and Yorston, H. J. (1973). *Pitfalls in seismic interpretation*. Tulsa, Soc. of

Exploration Geophysicists.

Tullos, F. N. and Reid, A. C. (1969). Seismic attenuation of Gulf Coast sediments. *Geophysics*, **34**, 516–28.

Tuman, V. S. (1961). Refraction and reflection of sonic energy in velocity logging. *Geophysics*, **26**, 588–600.

Van Melle, F. A. (1948). Wave-front circles for a linear increase of velocity with depth. *Geophysics*, **13**, 158–62.

Walton, G. G. (1972). Three-dimensional seismic method. *Geophysics*, **37**, 417–30.

Watkins, J. S., Walters, L. A. and Godson, R. H. (1972). Dependence of in-situ compressional-wave velocity on porosity in unsaturated rocks. *Geophysics*, **37**, 29–35.

Weatherby, B. B. (1940). The history and development of seismic prospecting. *Geophysics*, **5**, 215–30.

Weichart, H. F. (1973). Acoustic waves along oil-filled streamer cables. *Geophys. Prosp.* **21**, 281–95.

Werth, G. C., Liu, D. T. and Trorey, A. W. (1959). Offshore singing – field experiments and theoretical interpretation. *Geophysics*, **24**, 220–32.

White, J. E. (1965). *Seismic waves – radiation, transmission and attenuation*. New York, McGraw-Hill.

White, J. E. (1966). Static friction as a source of seismic attenuation. *Geophysics*, **31**, 333–9.

Whitfill, W. A. (1970). The seismic streamer in the marine seismic system. Offshore Technology Conference, paper 1238.

Willis, H. F. (1941). Underwater explosions – time interval between successive explosions. British report WA-47-21.

Wood, R. W. (1934). *Physical optics*. New York, MacMillan.

Wyrobek, S. M. (1956). Application of delay and intercept times in the interpretation of multilayer refraction time distance curves. *Geophys. Prosp.* **4**, 112–30.

Zoeppritz, K. (1919). Über reflexion und durchgang seismischer wellen durch Unstetigkerlsfläschen. *Berlin, Über Erdbebenwellen VII B, Nachrichten der Königlichen Gesellschaft der Wissenschaften zu Göttingen, math-phys. K1.* pp. 57–84.

5. Electrical properties of rocks

5.1 Classification of electrical methods

Electrical prospecting involves the detection of surface effects produced by electric current flow in the ground. There is a much greater variety of techniques available than in the other prospecting methods, where one makes use of a single field of force or anomalous property – gravitation, magnetism, elasticity, radioactivity. Using electrical methods, one may measure potentials, currents and electromagnetic fields which occur naturally – or are introduced artificially – in the earth. Furthermore, the measurements can be made in a variety of ways to determine a variety of results. Basically, however, it is the enormous variation in electrical conductivity found in different rocks and minerals which makes these techniques possible.

Electrical methods include self-potential, telluric currents and magnetotellurics, audio-frequency magnetic fields (AFMAG), resistivity, equipotential point and line and mise-à-la-masse, electromagnetic (EM) and induced polarization (IP). They are often classified by the type of energy source involved, that is, natural or artificial. On this basis the first four above are grouped under natural sources and the remainder as artificial. Such a classification can be made for prospecting methods in general. Hence gravity, magnetics and radioactivity are included in the natural source methods, while seismic requires artificial energy.

In the following chapters we shall study the electrical methods in a slightly different sequence, grouping three natural source methods together but considering AFMAG with EM, since the field techniques are quite similar. For the same reason IP will be considered immediately after resistivity.

5.2 Electrical properties of rocks and minerals

5.2.1 *Electrical potentials*

(a) *General.* Several electrical properties of rocks and minerals are significant in electrical prospecting. They are natural electrical potentials, electrical conductivity (or the inverse, electrical resistivity) and the dielectric constant. Magnetic permeability is also an indirect factor. Of these, electrical conductivity is by far the most important, while the others are of minor significance.

Certain natural or spontaneous potentials occurring in the subsurface are caused by electrochemical or mechanical activity. The controlling factor in all cases is underground water. These potentials are associated with weathering of sulphide mineral bodies, variation in rock properties (mineral content) at geologic contacts, bioelectric activity of organic material, corrosion, thermal and pressure gradients

442

in underground fluids and other phenomena of similar nature. There are four principal mechanisms producing these potentials; the first is mechanical, the latter three chemical, in origin.

(b) *Electrokinetic potential*. Also known as *streaming potential*, this is a well-known effect observed when a solution of electrical resistivity ρ and viscocity η is forced through a capillary or porous medium. The resultant potential difference between the ends of the passage is

$$E_k = -\frac{\phi \Delta P \varepsilon \rho}{4\pi\eta}, \tag{5.1}$$

where ϕ = adsorption (*zeta*) *potential*, ΔP = pressure difference, ε = solution dielectric constant.

The quantity ϕ is the potential of a double layer (solid-liquid) between the solid and solution. Although generally of minor importance, the streaming effect may be the cause of occasional large anomalies associated with topography. It is also observed in self-potential well logging, where the drilling fluid penetrates porous formations (see §11.1.3 and §11.2.3a).

(c) *Liquid-junction* (*diffusion*) *potential*. This is due to the difference in mobilities of various ions in solutions of different concentrations. The value is given by

$$E_d = -\frac{R\theta(I_a - I_c)}{Fn(I_a + I_c)} \log (C_1/C_2), \tag{5.2a}$$

where R = gas constant (8·31 joules/°C), F = Faraday constant (9·65 × 10^4 C/mol), θ = absolute temperature, n = valence, I_a and I_c = mobilities of anions and cations, C_1 and C_2 = solution concentrations. In NaCl solutions, I_a/I_c = 1·49, hence at 25°C,

$$E_d = -11\cdot6 \log_{10}(C_1/C_2), \tag{5.2b}$$

E_d being in millivolts.

(d) *Shale* (*Nernst*) *potential*. When two identical metal electrodes are immersed in a homogeneous solution, there is no potential difference between them. If, however, the concentrations at the two electrodes are different, there is a potential difference given by

$$E_s = -\frac{R\theta}{Fn} \log(C_1/C_2). \tag{5.3a}$$

For $n = 1$, $\theta = 25$°C, this becomes (E_s in millivolts)

$$E_s = -59\cdot1 \log_{10}(C_1/C_2). \tag{5.3b}$$

The combined diffusion and Nernst potentials are known as the *electrochemical*, or *static, self-potential*. For NaCl at 25°C, the electrochemical self-potential (in millivolts) is

$$E_c = -70 \cdot 7 \log_{10}(C_1/C_2).$$
(5.4)

When the concentrations are in the ratio 5:1, $E_c = \pm 50$ mV.

(e) *Mineralization potential*. When two dissimilar metal electrodes are immersed in a homogeneous solution, a potential difference exists between the electrodes. This *electrolytic contact potential*, along with the static self-potential, is undoubtedly among the basic causes of the large potentials associated with certain mineral zones and known as *mineralization potentials*. These potentials, which are especially pronounced in zones containing sulphides, graphite and magnetite, are much larger than those described in the preceding sections; values of several hundred millivolts are common and potentials greater than one volt have been observed in zones of graphite and alunite. Because of the large magnitude, mineralization potentials cannot be attributed solely to the electrochemical potentials described earlier. The presence of metallic conductors in appreciable concentrations appears to be a necessary condition; nevertheless, the exact mechanism is not entirely clear as will be seen in the more detailed discussion of mineralization potentials in §6.1.1 in connection with the self-potential prospecting method.

Other sources of electrical potentials in the earth should be mentioned. From eqs. (5.2a) and (5.3a) it can be seen that the magnitude of the static self-potential depends on temperature; this thermal effect is analogous to the pressure difference in streaming potential and is of minor importance. Obviously metal corrosion – of underground pipes, cables, etc. – is a local source of electrochemical potential. Large-scale earth currents (see §6.2.1) induced from the ionosphere, nuclear blasts, thunderstorms (see AFMAG, §7.5.4e) and the like create small, erratic earth potentials. Currents of bioelectric origin flowing, for instance, in plant roots are also a source of earth potentials. Negative potentials of 100 mV have been reported in this connection, in passing from cleared ground to wooded areas.

Most of the earth potentials discussed above are relatively permanent in time and place. Of the variable types, only telluric and AFMAG sources have been employed in prospecting. When measuring static potentials these fluctuations cause a background noise and may be a nuisance.

5.2.2 *Electrical conductivities*

(a) *General*. Electric current may be propagated in rocks and minerals in three ways: electronic (ohmic), electrolytic and dielectric conduction. The first is the normal type of current flow in materials containing free electrons, such as the metals. In an electrolyte the current is carried by ions at a comparatively slow rate. Dielectric conduction takes place in poor conductors or insulators, which have very few free carriers or none at all. Under the influence of an external varying electric field, the atomic electrons are displaced slightly with respect to their nuclei; this slight relative separation of negative and positive charges is known as dielectric polarization of the material. Ionic and molecular polarization may occur in materials with ionic and molecular bonds. In all these cases dielectric conduction is the result of changing electronic, ionic or molecular polarization caused by the alternating electric field.

(b) *Electronic conduction.* The *electrical resistivity* of a cylindrical solid of length L and cross-section A, having resistance R between the end faces, is given by

$$\rho = RA/L. \tag{5.5}$$

If A is in metres^2, L in metres and R in ohms, the resistivity unit is the ohm-metre (Ωm). For dimensions in centimetres the unit becomes the ohm-centimetre (Ωcm): 1 Ωm = 100 Ωcm.

The resistance R is given in terms of the voltage V applied across the ends of the cylinder and the resultant current I flowing through it, by Ohm's law:

$$R = V/I,$$

where R is in ohms and the units of V and I are volts and amperes.

The reciprocal of resistivity is the *conductivity* σ, the units being mhos/m or mhos/cm. Then

$$\sigma = 1/\rho = L/RA = (I/A)/(V/L) = j/E, \tag{5.6}$$

where j = current density (amperes/m^2), E = electric field (volts/m).

(c) *Electrolytic conduction.* Since most rocks are poor conductors, their resistivities would be extremely large were it not for the fact they are usually porous and the pores filled with fluids, mainly water. As a result the rocks are *electrolytic conductors*, whose effective resistivity may be defined as in eq. (5.5). But it must be kept in mind that the conduction is electrolytic rather than ohmic. That is, the propagation of current is by ionic conduction – by molecules having an excess or deficiency of electrons. Hence the resistivity varies with the mobility, concentration and degree of dissociation of the ions; the latter depends on the dielectric constant of the solvent. As mentioned previously the current flow is slow compared to ohmic conduction and the movement represents an actual transport of material, usually resulting in chemical transformation.

The conductivity of a porous rock varies with the volume and arrangement of the pores and even more with the conductivity and amount of contained water. According to the empirical formula due to Archie (1942),

$$\rho_e = a\phi^{-m}s^{-n}\rho_w, \tag{5.7}$$

where ϕ = fractional pore volume (porosity), s = fraction of the pores containing water, ρ_w = resistivity of water, $n \approx 2$, a, m = constants, $0.5 \leqslant a \leqslant 2.5, 1.3 \leqslant m \leqslant 2.5$.

For example, suppose $s = 1$, $a = 1.5, m = 2$, then $\rho_e/\rho_w = 1.5/\phi^2$ and for values of $\phi = 0.01, 0.1, 0.3, 0.5, \rho_e/\rho_w$ becomes 1.5×10^4, 150, 17 and 6 respectively.

Water conductivity varies considerably (see table 5.2), depending on the amount and conductivity of dissolved chlorides, sulphates and other minerals present.

The geometrical arrangement of the interstices in the rock has a less pronounced effect, but may make the resistivity anisotropic, that is, having different magnitudes for current flow in different directions. Anisotropy is characteristic of stratified rock which is generally more conductive in the bedding plane. The *anisotropy coefficient*, which is the ratio of maximum to minimum resistivity, may be as large as two in some graphitic slates and varies from 1 to 1·2 in rocks such as limestone, shale and rhyolite.

As an example, consider the layered formation shown in fig. 5.1, having resistivities ρ_1 and ρ_2 whose respective fractional volumes are V and $1 - V$.

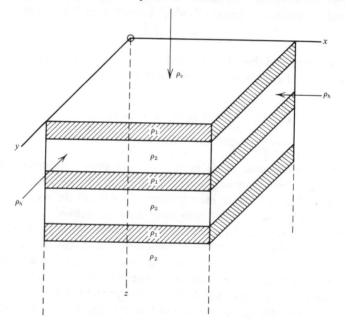

Fig. 5.1 Anisotropic resistivity as a result of horizontal bedding.

Here the resistivity in the horizontal direction – a stack of beds effectively in parallel – is

$$\rho_h = \rho_1\rho_2/\{\rho_1(1 - V) + \rho_2 V\}; \tag{5.8}$$

In the vertical direction, the beds are in series so that

$$\rho_v = \rho_1 V + \rho_2(1 - V). \tag{5.9}$$

Then the ratio is

$$\frac{\rho_v}{\rho_h} = (1 - 2V + 2V^2) + \left(\frac{\rho_1}{\rho_2} + \frac{\rho_2}{\rho_1}\right) V(1 - V).$$

If $V \ll 1$ and $\rho_2/\rho_1 \gg 1$, this simplifies to

$$\frac{\rho_v}{\rho_h} \approx 1 + \frac{\rho_2}{\rho_1} V. \tag{5.10}$$

If the layer of resistivity ρ_1 is for water-saturated beds, this ratio might be quite large.

(d) *Dielectric conduction.* The mechanism of dielectric conduction – the displacement current – was described briefly at the beginning of this section, where it was pointed out that the displacement current flows only in non-conductors when the external electric field changes with time. The significant parameter in dielectric conduction is the *dielectric constant*, k, sometimes called the *specific inductive capacity* of the medium. In analogy with magnetic quantities $\mathscr{I}$, H, k, B and μ (§3.2.1) we have an electrostatic set: *electric polarization* (*electric dipole moment*/ *unit volume*), P, *electric field strength*, E, *electric susceptibility*, η, *electric displacement* (*flux*/*unit area*), D, and *dielectric constant*, k. In electrostatic units, the relations between these are:

$$P = \eta E, \quad D = E + 4\pi P = E(1 + 4\pi\eta) = kE, \tag{5.11}$$

while in MKS units:

$$P = \eta E, \quad D = \varepsilon_0 E + P = E(\varepsilon_0 + \eta) = \varepsilon E, \tag{5.12}$$

and the dielectric constant, $k = 1 + \eta/\varepsilon_0 = \varepsilon/\varepsilon_0$.

In electrostatic units, P, E and D are volts/cm, η and k are dimensionless. In MKS units ε, ε_0, η are farads/m, P, D are coulombs/m^2, E is in volts/m and k is again dimensionless and the same in either system.

The dielectric constant is similar to the conductivity in porous formations in that it varies with the amount of water present (note that water has a very large dielectric constant – see table 5.7). We shall see in §6.2.3 that displacement currents are of secondary importance in earth materials since electrical prospecting methods generally employ low frequencies.

5.2.3 *Magnetic permeability*

Where EM sources are employed, the voltage induced in a subsurface conductor varies not only with the rate of change of magnetic field, but also with the magnetic permeability of the conductor. From Maxwell's equation,

$$\nabla \times \mathbf{E} = -\mu \frac{\partial \mathbf{H}}{\partial t},$$

we see that currents induced in the ground are enhanced by the factor μ. Practically, however, since the maximum susceptibilities, even of magnetic minerals, seldom exceed 0·2 cgsu (see table 5.8), the permeability rarely is appreciably greater than unity. Consequently the magnetic permeability is of no particular significance in electrical work.

5.2.4 *Polarization potentials*

Where a steady current is passed through an electrolytic conductor containing mineral particles it is possible, as described in §5.3.1, to determine the effective resistivity. If a current is suddenly switched on or off in a circuit containing an electrolyte, it will often be observed that a finite time elapses before the potential increases to a fixed value or drops to zero. The delayed build-up or decay of current is characteristic of electrolytic conduction, and is due to accumulation of ions at interfaces between the electrolyte and mineral particles. As a result, a potential opposing the normal current flow is developed across the interface. A similar effect is observed at the contact between electrolytes and clay particles. These are known as *polarization potentials*; the process is called the *induced polarization effect*. The induced polarization (IP) prospecting method involves these interface potentials. They will be considered in more detail in §9.2.

5.3 Measurement of electrical properties of rocks and minerals

5.3.1 *Laboratory measurement of resistivity*

In order to measure directly the true resistivity of a rock, mineral, electrolyte, etc., it is necessary to shape the sample in some regular form, such as a cylinder, cube or bar of regular cross-section. An experimental arrangement is shown in fig. 5.2.

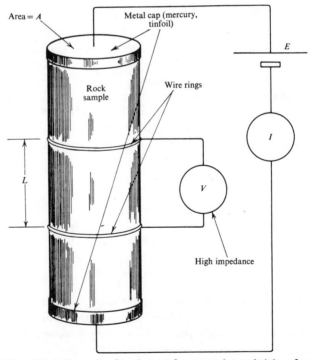

Fig. 5.2 Simplified schematic of equipment for measuring resistivity of core sample.

The main difficulty is in making good electrical contact, particularly for the current electrodes. For this purpose tinfoil or mercury electrodes may be used; sometimes the ends of the sample are dipped in soft solder. From fig. 5.2 and eq. (5.6) the resistivity is given by

$$\rho = AV/LI.$$

The power source may be d.c. or preferably low frequency a.c. (400 Hz or less). The possibility of anisotropy can be checked by measuring the resistivity in two directions, provided the shape is suitable for this.

Obviously one can make these measurements in the field as well on drill core, grab samples, even outcrop if the electrode contact is reasonably good. Estimates of resistivity are frequently made on samples by using an ohmmeter and merely pressing or scraping the terminals of the leads against the surface; such results are not very trustworthy.

5.3.2 *Measurement of dielectric constant*

An a.c. bridge may also be used to measure the resistivity of soils and electrolytes. At audio frequencies any reactive component – normally capacitive – must be accounted for in order to get a good bridge balance. Consequently the measurement determines the effective capacitance, as well as resistivity, of the specimen. Since capacitance varies with the dielectric constant of the material, it is thus possible to determine the latter by substitution.

Figure 5.3 shows a Schering capacitance bridge in which one arm contains an electrolytic cell. The bridge is first balanced with the cell empty, or preferably with

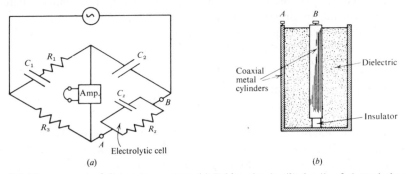

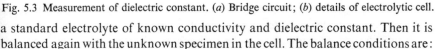

Fig. 5.3 Measurement of dielectric constant. (*a*) Bridge circuit; (*b*) details of electrolytic cell.

a standard electrolyte of known conductivity and dielectric constant. Then it is balanced again with the unknown specimen in the cell. The balance conditions are:

$$R_x = R_3 C_1/C_2, \quad \text{and} \quad C_x = C_2 R_1/R_3, \tag{5.13}$$

and the dielectric constant of the sample is given by

$$k_x = k_s C_x/C_s, \tag{5.14}$$

where C_s and k_s are the capacitance and dielectric constant of the standard.

Normally the sample will be equivalent to a resistance and capacitance in parallel. In some cases it may be necessary to interchange the series and parallel arms of the bridge.

5.4 Typical values of electrical constants of rocks and minerals

5.4.1 *Resistivities of rocks and minerals*

Of all the physical properties of rocks and minerals, electrical resistivity shows the greatest variation. While the range in density, elastic wave velocity and radioactive content is quite small, in magnetic susceptibility it may be as large as 10^5. However, the resistivity of metallic minerals may be as small as 10^{-5} Ωm, that of dry, close-grained rocks like gabbro as large as 10^7 Ωm. The maximum possible range is even greater, from native silver ($1 \cdot 6 \times 10^{-8}$ Ωm) to pure sulphur (10^{16} Ωm).

A *conductor* is usually defined as a material of resistivity less than 10^{-5} Ωm, while an *insulator* is one having a resistivity greater than 10^7 Ωm. Between these limits lie the so-called *semiconductors*. Within this grouping the metals and graphite are all conductors; they contain a large number of free electrons whose mobility is very great. The semiconductors also carry current by mobile electrons but have fewer of them. The insulators are characterized by ionic bonding so that the valence electrons are not free to move; the charge carriers are ions which must overcome larger barrier potentials than exist either in the semiconductors or conductors.

A further difference between conductors and semiconductors is found in their respective variation with temperature. The former vary inversely with temperature and have their highest conductivities in the region of $0°K$. The semiconductors, on the other hand, are practically insulators at low temperatures.

In a looser classification, rocks and minerals are considered to be good, intermediate and poor conductors within the following ranges:

(*a*) minerals of resistivity 10^{-8} to about 1 Ωm,
(*b*) minerals and rocks of resistivity 1 to 10^7 Ωm,
(*c*) minerals and rocks of resistivity above 10^7 Ωm.

Group (*a*) includes the metals, graphite, the sulphides except for sphalerite, cinnabar and stibnite, all the arsenides and sulpho-arsenides except $SbAs_2$, the antimonides except for some lead compounds, the tellurides and some oxides such as magnetite, manganite, pyrolusite and ilmenite. Most oxides, ores and porous rocks containing water are intermediate conductors. The common rock-forming minerals, silicates, phosphates and the carbonates, nitrates, sulphates, borates, etc., are poor conductors.

The following tables list characteristic resistivities for various minerals and rocks. The data are from various sources, including Heiland (1940), Keller (1966), Parasnis (1956, 1966), Jakosky (1950) and Parkhomenko (1967).

The variation in resistivity of particular minerals is enormous, as can be seen from table 5.2. Among the more common minerals, pyrrhotite and graphite

Table 5.1. *Resistivities of metals and elements*

Element	Resistivity (Ωm)		Element	Resistivity (Ωm)	
	Range	Average		Range	Average
Antimony		4.5×10^{-7}	Molybdenum		5.7×10^{-8}
Arsenic		2.2×10^{-7}	Nickel		7.8×10^{-8}
Bismuth		1.2×10^{-6}	Platinum		10^{-7}
Copper		1.7×10^{-8}	Silver		1.6×10^{-8}
Gold		2.4×10^{-8}	Sulphur	10^7–10^{16}	10^{14}
Graphite	5×10^{-7}–10	10^{-3}	Tellurium	10^{-4}–2×10^{-3}	10^{-3}
Iron		10^{-7}	Tin		1.1×10^{-7}
Lead		2.2×10^{-7}	Uranium		3×10^{-7}
Mercury		9.6×10^{-7}	Zinc		5.8×10^{-8}

appear to be the most consistent good conductors, while pyrite, galena and magnetite are often poor conductors in bulk form, although the individual crystals have high conductivity. Hematite and sphalerite, in pure form, are practically insulators, but when combined with impurities may have resistivities as low as 0·1 Ωm. Graphite is often the connecting link in mineral zones which makes them good conductors.

Table 5.2. *Resistivities of minerals*

Mineral	Formula	Resistivity (Ωm)	
		Range	Average
Argentite	Ag_2S	2×10^{-3}–10^4	1.7×10^{-3}
Bismuthinite	Bi_2S_3	18–570	
Covellite	CuS	3×10^{-7}–8×10^{-5}	2×10^{-5}
Chalcocite	Cu_2S	3×10^{-5}–0·6	10^{-4}
Chalcopyrite	$CuFeS_2$	1.2×10^{-5}–0·3	4×10^{-3}
Bornite	Cu_5FeS_4	2.5×10^{-5}–0·5	3×10^{-3}
Marcasite	FeS_2	10^{-3}–3·5	5×10^{-2}
Pyrite	FeS_2	2.9×10^{-5}–1·5	3×10^{-1}
Pyrrhotite	Fe_nS_m	6.5×10^{-6}–5×10^{-2}	10^{-4}
Cinnabar	HgS		2×10^7
Molybdenite	MoS_2	10^{-3}–10^6	10
Galena	PbS	3×10^{-5}–3×10^2	2×10^{-3}
Millerite	NiS		3×10^{-7}
Stannite	Cu_2FeSnS_2	10^{-3}–6×10^3	
Stibnite	Sb_2S_3	10^5–10^{12}	5×10^6
Sphalerite	ZnS	1.5–10^7	10^2
Cobaltite	$CoAsS$	3.5×10^{-4}–10^{-1}	
Smaltite	$CoAs_2$		5×10^{-5}
Arsenopyrite	$FeAsS$	2×10^{-5}–15	10^{-3}
Niccolite	$NiAs$	10^{-7}–2×10^{-3}	2×10^{-5}
Sylvanite	$AgAuTe_4$	4×10^{-6}–2×10^{-5}	
Bauxite	$Al_2O_3 . nH_2O$	2×10^2–6×10^3	

Mineral	Formula	Resistivity (Ωm) Range	Resistivity (Ωm) Average
Braunite	Mn_2O_3	0·16–1·2	
Cuprite	Cu_2O	10^{-3}–300	30
Chromite	$FeCr_2O_4$	1–10^6	
Specularite	Fe_2O_3		6 × 10^{-3}
Hematite	Fe_2O_3	3·5 × 10^{-3}–10^7	
Limonite	$2Fe_2O_3.3H_2O$	10^3–10^7	
Magnetite	Fe_3O_4	5 × 10^{-5}–5·7 × 10^3	
Ilmenite	$FeTiO_3$	10^{-3}–50	
Wolframite	Fe, Mn, WO_4	10–10^5	
Manganite	$MnO(OH)$	10^{-2}–0·3	
Pyrolusite	MnO_2	5 × 10^{-3}–10	
Quartz	SiO_2	4 × 10^{10}–2 × 10^{14}	
Cassiterite	SnO_2	4 × 10^{-4}–10^4	0·2
Rutile	TiO_2	30–1000	500
Uraninite (Pitchblende)	UO_2	1–200	
Anhydrite	$CaSO_4$		10^9
Calcite	$CaCO_3$		2 × 10^{12}
Fluorite	CaF_2		8 × 10^{13}
Siderite	$Fe_2(CO_3)_3$		70
Rock salt	NaCl	30–10^{13}	
Sylvite	KCl	10^{11}–10^{12}	
Diamond	C	10–10^{14}	
Serpentine		2 × 10^2–3 × 10^3	
Hornblende		2 × 10^2–10^6	
Mica		9 × 10^2–10^{14}	
Biotite		2 × 10^2–10^6	
Phlogopite		10^{11}–10^{12}	
Bitum. coal		0·6–10^5	
Coals (various)		10–10^{11}	
Anthracite		10^{-3}–2 × 10^5	
Lignite		9–200	
Fire clay			30
Meteoric waters		30–10^3	
Surface waters (ign. rocks)		0·1–3 × 10^3	
Surface waters (sediments)		10–100	
Soil waters			100
Natural waters (ign. rocks)		0·5–150	9
Natural waters (sediments)		1–100	3
Sea water			0·2
Saline waters, 3%			0·15
Saline waters, 20%			0·05

The range of resistivities of various waters is notably smaller than for solid minerals; the actual resistivities are also lower than those of a great many minerals. Table 5.3 lists resistivities for a variety of ores, from Parkhomenko (1967). In general it appears that pyrrhotite in massive form has the lowest resistivity, that the resistivity of zinc ores is surprisingly low (possibly due to the presence of lead and copper fractions) and that molybdenite, chromite and iron ores have values in the range of many rocks.

Table 5.3. *Resistivities of various ores*

Ore	Other minerals	Gangue	$\rho(\Omega m)$
Pyrite			
18%	2% (chalco)	80%	300
40	20%	40	130
60	5% (ZnS) + 15%	20	0·9
75	10% (ZnS) + 5%	10	0·14
95	5% (ZnS)		1·0
95		5	7·0
Pyrrhotite			
41%		59%	$2·2 \times 10^{-4}$
58		42	$2·3 \times 10^{-4}$
79		21	$1·4 \times 10^{-5}$
82		18	$8·5 \times 10^{-5}$
95		5	$1·4 \times 10^{-5}$
SbS_2 in quartz			$4 \times 10^3 - 3 \times 10^7$
FeAsS 60%	FeS 20%	20% SiO_2	0·39
FeAsS			$10^{-4} - 10^{-2}$
Cu_5FeS_4			3×10^{-3}
Cu_5FeS_4 40%		60% SiO_2	7×10^{-2}
Fe, Mn, WO_4 80%			2×10^4
Fe, Mn, WO_4	CoAsS		$10^3 - 10^7$
PbS, massive			7×10^{-2}
PbS, near massive			0·8
PbS 50–80%			$10^{-2} - 3$
Fe_2O_3			0·1–300
Fe_2O_3, massive			$2·5 \times 10^3$
Iron			
Fe_3O_4 60%			45
Fe_3O_4 from contact met.			$0·5 - 10^2$
Diss. brown iron oxide			$8 \times 10^2 - 3 \times 10^6$
75% brown iron oxide		25%	$2 \times 10^4 - 8 \times 10^5$
Fe_2O_3 fine grained			$2·5 \times 10^3$
Fe_3O_4			$5 \times 10^3 - 8 \times 10^3$
Fe_3O_4 in pegmatite			$7 \times 10^3 - 2 \times 10^5$
Zinc			
30%	5% PbS, 15% FeS	50%	0·75
70%	3% chalco, 17% PbS, 10% FeS		20
80	10% PbS, 10% FeS		$1·7 \times 10^3$
80	2% chalco, 1% PbS, 2% FeS	15%	1·3
90	5% PbS	5%	130

Ore	Other minerals	Gangue	$\rho(\Omega m)$
Graphitic slate			0·13
Graphite, massive			10^{-4}–5×10^{-3}
MoS_2			2×10^2–4×10^3
MnO_2 colloidal ore			1·6
Cu_2S			3×10^{-2}
$CuFeS_2$			10^{-4}–1
$CuFeS_2$ 80%	10% FeS	10%	0·66
$CuFeS_2$ 90%	2% FeS	8% SiO_2	0·65
$FeCr_2O_4$			10^3
$FeCr_2O_4$ 95%		5% Serp.	$1·2 \times 10^4$

Tables 5.4 and 5.5 list typical values for rocks and unconsolidated formations. The ranges here are quite similar to water, which obviously is the controlling factor in many rocks.

Table 5.4. *Resistivities of igneous and metamorphic rocks*

Rock type	Resistivity range (Ωm)
Granite	3×10^2–10^6
Granite porphyry	$4·5 \times 10^3$ (wet)–$1·3 \times 10^6$ (dry)
Feldspar porphyry	4×10^3 (wet)
Albite	3×10^2 (wet)–$3·3 \times 10^3$ (dry)
Syenite	10^2–10^6
Diorite	10^4–10^5
Diorite porphyry	$1·9 \times 10^3$ (wet)–$2·8 \times 10^4$ (dry)
Porphyrite	10–5×10^4 (wet)–$3·3 \times 10^3$ (dry)
Carbonatized porphyry	$2·5 \times 10^3$ (wet)–6×10^4 (dry)
Quartz porphyry	3×10^2–9×10^5
Quartz diorite	2×10^4–2×10^6 (wet)–$1·8 \times 10^5$ (dry)
Porphyry (various)	60–10^4
Dacite	2×10^4 (wet)
Andesite	$4·5 \times 10^4$ (wet)–$1·7 \times 10^2$ (dry)
Diabase porphyry	10^3 (wet)–$1·7 \times 10^5$ (dry)
Diabase (various)	20–5×10^7
Lavas	10^2–5×10^4
Gabbro	10^3–10^6
Basalt	10–$1·3 \times 10^7$ (dry)
Olivine norite	10^3–6×10^4 (wet)
Peridotite	3×10^3 (wet)–$6·5 \times 10^3$ (dry)
Hornfels	8×10^3 (wet)–6×10^7 (dry)
Schists (calcareous and mica)	20–10^4
Tuffs	2×10^3 (wet)–10^5 (dry)
Graphite schist	10–10^2
Slates (various)	6×10^2–4×10^7
Gneiss (various)	$6·8 \times 10^4$ (wet)–3×10^6 (dry)
Marble	10^2–$2·5 \times 10^8$ (dry)
Skarn	$2·5 \times 10^2$ (wet)–$2·5 \times 10^8$ (dry)
Quartzites (various)	10–2×10^8

Table 5.5. *Resistivities of sediments*

Rock type	Resistivity range (Ωm)
Consolidated shales	$20\text{--}2 \times 10^3$
Argillites	$10\text{--}8 \times 10^2$
Conglomerates	$2 \times 10^3\text{--}10^4$
Sandstones	$1\text{--}6\cdot4 \times 10^8$
Limestones	$50\text{--}10^7$
Dolomite	$3\cdot5 \times 10^2\text{--}5 \times 10^3$
Unconsolidated wet clay	20
Marls	3–70
Clays	1–100
Alluvium and sands	10–800
Oil sands	4–800

Very roughly, igneous rocks have the highest resistivity, sediments the lowest, with metamorphic rocks intermediate. However, there is considerable overlapping, as in other physical properties. In addition, the resistivities of particular rock types vary directly with age and lithology, since the porosity of the rock and salinity of the contained water are affected by both. For example, the resistivity range of Precambrian volcanics is 200–5000 Ωm, while for Quaternary rocks of the same kind it is 10–200 Ωm.

The effect of water content on the bulk resistivity of rocks has been frequently mentioned and is evident from table 5.4. Further data are listed in table 5.6, where samples with variable amounts of water are shown. In all cases a small change in the percentage of water effects the resistivity enormously.

Table 5.6. *Variation of rock resistivity with water content*

Rock	% H_2O	$\rho(\Omega$m)	Rock	% H_2O	$\rho(\Omega$m)
Siltstone	0·54	$1\cdot5 \times 10^4$	Pyrophyllite	0·76	6×10^6
Siltstone	0·44	$8\cdot4 \times 10^6$	Pyrophyllite	0·72	5×10^7
Siltstone	0·38	$5\cdot6 \times 10^8$	Pyrophyllite	0·7	2×10^8
Coarse grain SS	0·39	$9\cdot6 \times 10^5$	Pyrophyllite	0	10^{11}
Coarse grain SS	0·18	10^8	Granite	0·31	$4\cdot4 \times 10^3$
Medium grain SS	1·0	$4\cdot2 \times 10^3$	Granite	0·19	$1\cdot8 \times 10^6$
Medium grain SS	1·67	$3\cdot2 \times 10^6$	Granite	0·06	$1\cdot3 \times 10^8$
Medium grain SS	0·1	$1\cdot4 \times 10^8$	Granite	0	10^{10}
Graywacke SS	1·16	$4\cdot7 \times 10^3$	Diorite	0·02	$5\cdot8 \times 10^5$
Graywacke SS	0·45	$5\cdot8 \times 10^4$	Diorite	0	6×10^6
Arkosic SS	1·26	10^3	Basalt	0·95	4×10^4
Arkosic SS	1·0	$1\cdot4 \times 10^3$	Basalt	0·49	9×10^5
Organic limestone	11	$0\cdot6 \times 10^3$	Basalt	0·26	3×10^7
Dolomite	2	$5\cdot3 \times 10^3$	Basalt	0	$1\cdot3 \times 10^8$
Dolomite	1·3	6×10^3	Olivine-pyrox.	0·028	2×10^4
Dolomite	0·96	8×10^3	Olivine-pyrox.	0·014	4×10^5
Peridotite	0·1	3×10^3	Olivine-pyrox.	0	$5\cdot6 \times 10^7$
Peridotite	0·03	2×10^4			
Peridotite	0·016	10^6			
Peridotite	0	$1\cdot8 \times 10^7$			

5.4.2 *Dielectric constants of rocks and minerals*

As mentioned previously, the dielectric constant is a measure of the electrical polarization resulting from an applied electric field. This polarization may be electronic, ionic or molecular. The first type is characteristic of all non-conductors. Ionic displacement occurs in many rock-forming minerals, while water and the hydrocarbons are the only common materials which exhibit molecular polarization.

Because of the relatively slow mobilities of the charge carriers, molecular polarization – the largest of the three effects – and ionic polarization are insignificant at very high frequencies. Thus the dielectric constant, which is proportional to the degree of polarization, varies inversely with frequency. It is also indicative of the amount of water present, since water has a dielectric constant of 80 at low frequencies.

Table 5.7 lists dielectric constants for various minerals and rocks. Most of the measurements have been made at frequencies of 100 kHz and up. For very low frequencies the values would be generally higher by about 30%. In exceptional cases – one example being the measurement of certain ice samples – the results have been larger by several orders of magnitude.

Table 5.7. *Dielectric constants of rocks and minerals*

Rock, mineral	Dielectric const.	Rock, mineral	Dielectric const.
Galena	18	Gypsum	5–11·5
Sphalerite	7·9–69·7	Beryl	5·5–7·8
Corundum	11–13·2	Biotite	4·7–9·3
Cassiterite	23	Epidote	7·6–15·4
Hematite	25	Orthoclase feldspar	3–5·8
Rutile	31–170	Plagioclase feldspar	5·4–7·1
Fluorite	6·2–6·8	Quartz	4·2–5
Calcite	7·8–8·5	Zircon	8·6–12
Apatite	7·4–11·7	Granite (dry)	4·8–18·9
Barite	7–12·2	Gabbro	8·5–40
Peridotite	8·6	Diorite	6·0
Norite	61	Serpentine	6·6
Quartz porphyry	14–49·3	Gneiss	8·5
Diabase	10·5–34·5	Sandstone (dry to moist)	4·7–12
Trap	18·9–39·8	Packed sand (dry to moist)	2·9–105
Dacite	6·8–8·2	Soil (dry to moist)	3·9–29·4
Obsidian	5·8–10·4	Basalt	12
Sulphur	3·6–4·7	Clays (dry to moist)	7–43
Rock salt	5·6	Petroleum	2·07–2·14
Anthracite	5·6–6·3	Water (20°C)	80·36
		Ice	3–4·3

5.4.3 *Magnetic permeability of minerals*

The effect of μ on electrical measurements is very slight except in the case of concentrated magnetite, pyrrhotite and ilmenite. From eq. 3.6*b*, §3.2.1*f*, magnetic permeability is related to susceptibility by the expression

$$\mu = 1 + 4\pi k \text{ in cgs units,}$$
$$\mu/\mu_0 = 1 + k \quad \text{in MKS units.}$$

Generally *k* is too small to change μ appreciably from unity. The following table lists maximum permeabilities of some common minerals.

Table 5.8. *Magnetic permeabilities*

Mineral	Permeability	Mineral	Permeability
Magnetite	5	Rutile	1·0000035
Pyrrhotite	2·55	Calcite	0·999987
Ilmenite	1·55	Quartz	0·999985
Hematite	1·05	Hornblende	1·00015
Pyrite	1·0015		

References

F – field methods; G – general; I – instruments; T – theoretical.

Archie, G. E. (1942). The electrical resistivity log as an aid in determining some reservoir characteristics. *Trans. A.I.M.E.* **146**, 54–62. (T)

Heiland, C. A. (1940). *Geophysical exploration*, chapter 10. New York, Prentice-Hall. (G)

Jakosky, J. J. (1950). *Exploration geophysics*, chapter 5. Newport Beach (Calif.), Trija. (G)

Keller, G. V. (1966). In *Handbook of physical constants* (ed. by Clark, S. P., Jr.). Geol. Soc. Am. Memoir 97, pp. 553–76. (G)

Keller, G. V. and Frischknecht, F. C. (1966). *Electrical methods in geophysical prospecting*, chapter 1. London, Pergamon. (G)

Parasnis, D. S. (1956). The electrical resistivity of some sulphide and oxide minerals and their ores. *Geophys. Prosp.* **4**, 249–79. (G)

Parasnis, D. S. (1966). *Mining geophysics*, chapter 6. Amsterdam, Elsevier. (G)

Parkhomenko, E. I. (1967). *Electrical properties of rocks* (trans. by G. V. Keller). New York, Plenum. (G)

6. Methods employing natural electrical sources

6.1 Self-potential method

6.1.1 *Origin of potentials*

Spontaneous ground potentials were discussed in §5.2.1. In mineral prospecting they may be classified as background potentials and mineralization potentials.

Background potentials are created by fluid streaming, bioelectric activity in vegetation, varying electrolytic concentrations in ground water and other geochemical action. Their amplitudes vary greatly but generally are less than 100 millivolts. On the average, over intervals of several thousand feet, the potentials usually add up to zero, since they are as likely to be positive as negative.

In addition there are several characteristic regional background potentials. One is a gradient of the order of 10 mV/1000 ft which sometimes extends over several miles and may be either positive or negative. It is probably due to gradual changes in diffusion and electrolytic potentials in ground water. Sometimes a more abrupt change will result in a baseline shift of background potential. Another regional gradient of similar magnitude seems to be associated with topography. It is usually negative going uphill and is probably caused by streaming potential. These background effects are not difficult to recognize.

Potentials arising from bioelectric activity of plants, trees, etc., sometimes are as large as several hundred millivolts. They have been observed as sharp negative anomalies – when passing from open ground into bush – which are quite similar to those appearing over sulphide zones.

Mineralization potentials are the main interest when prospecting with the self-potential method. They are associated with the sulphides of the metals, with graphite and sometimes with the metal oxides such as magnetite. The most common mineralization potential anomalies occur over pyrite, chalcopyrite, pyrrhotite, sphalerite, galena and graphite. Amplitudes range from a few millivolts to one volt; 200 mV would be considered a good SP anomaly. The potentials are almost always negative near the upper end of the body and are quite stable in time.

The mechanism of spontaneous polarization in mineral zones is not completely understood, although several theories have been developed to explain it. Field measurements indicate that some part of the mineral must be in a zone of oxidation in order that SP (self-potential) anomalies may appear at surface. The original explanation, based on this evidence, was that the body behaved like a galvanic cell with a potential difference being created between the oxidizing zone (generally the upper surface) and the remainder. The action of this cell is illustrated in fig. 6.1.

458

There are several weaknesses in this explanation. Graphite frequently is the source of large SP anomalies, although it does not oxidize appreciably. On the other hand, extensive oxidation, such as could occur in most metal sulphides, would leave the upper surface of the body with a net positive charge due to the loss of electrons. In fact the charge is negative.

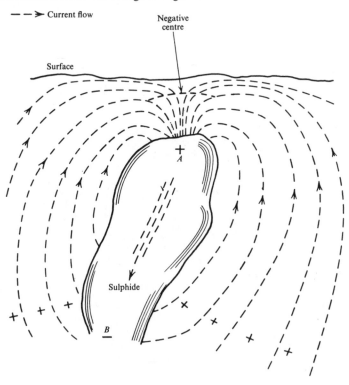

Fig. 6.1 Oxidizing sulphide zone as a galvanic cell.

Another hypothesis suggests that pH variations above and below the water table could provide the currents flowing around the source. There is considerable evidence that the solutions above sulphide bodies, and within the water table, are highly acidic (pH = 2 − 4), while those below the table are slightly basic (pH = 7 − 9). There is probably a close relation between pH and mineralization potentials, but a difference in pH alone is not sufficient to move electrons in and out of the mineral zone and maintain the flow of current.

The most reasonable and complete theory of mineralization potentials is that proposed by Sato and Mooney in 1960. They postulate two electrochemical half-cell reactions of opposite sign, one cathodic above the water table, the other anodic at depth. In the cathode half-cell there is chemical reduction of the substances in solution – that is, they gain electrons – while in the anode cell an oxidation reaction takes place and electrons are lost. The mineral zone itself functions only to transport electrons from anode to cathode. The magnitude of the overall SP effect is determined by the difference in oxidation potential (Eh) between the solutions at the two half-cells. This mechanism is illustrated in fig. 6.2, showing the flow of electrons and ions which leave the upper surface negatively charged, the lower positively.

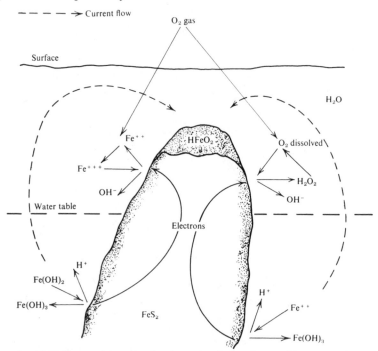

Fig. 6.2 Self-potential mechanism in pyrite. (From Sato and Mooney, 1960.)

This theory, although a considerable improvement on previous explanations, still fails to account for some observed anomalies. For instance, Sato and Mooney give maximum possible potentials for various sources, such as graphite (0·78 V), pyrite (0·73 V) and galena (0·33 V). For surface measurements this would imply a maximum no greater than these values, even when the body outcrops. Potentials as large as 1·5 volts, however, have been reported over graphite. A field study in which potentials were measured in drill holes penetrating a sulphide zone, as well as on the surface over the zone, gave surface anomalies of approximately the same size as those encountered in the sulphide itself, even though the latter was well below surface. These abnormally large SP results may be due to combinations of anomalies from adjacent mineralized zones, or enhancement by coincident background potentials. Other unusual field results are discussed in §6.1.4.

Most of the sulphides are good conductors, with the exception of sphalerite, cinnabar and stibnite. Self-potential anomalies have, however, been observed over sphalerite and in drill holes which passed through sphalerite bodies. The

Sato-Mooney theory assumes that the sulphide zone must be a good conductor to transport electrons from depth to the oxidizing zone near surface. Thus the case of sphalerite is puzzling, although it may behave as a semiconductor.

6.1.2 *Self-potential field equipment*

The SP method goes back to 1830, when Robert Fox used copper-plate electrodes and a string galvanometer as a detector in an attempt to find extensions of underground copper deposits in Cornwall. Since 1920 it has been employed in basemetal search, usually as a secondary method. The equipment required is extremely simple, consisting merely of a pair of electrodes connected by wire to a millivoltmeter. There are, however, two restrictions on the electrodes and detector which are most important.

If one were to use metal stakes driven into the ground as SP electrodes, the resultant electrochemical action at the ground contacts would create spurious potentials of the same size as those being measured. Furthermore these contact potentials are quite erratic in different ground and at different times, so that it would not be possible to make a fixed correction. Consequently nonpolarizing electrodes are essential. These consist of a metal immersed in a saturated solution of its own salt, such as Cu in $CuSO_4$, Zn in $ZnSO_4$, etc., and contained in a porous pot which allows the solution to leak slowly and make contact with the ground. A good electrode of this type is the walking stick arrangement shown in fig. 6.3.

The main requirement of the millivoltmeter is that its input impedance should be large enough that negligible current will be drawn from the ground during the measurement. This is achieved by using a potentiometer, or preferably a d.c. meter with an input impedance greater than 10^8 ohms. Such instruments, with ranges from 10 millivolts to 20 volts full scale, are readily available. It is sometimes necessary to enclose the meter in a shield can to prevent erratic readings caused by contact of the instrument case with the body or ground. Typical measuring instruments are illustrated in fig. 6.4.

6.1.3 *Field procedure*

The two porous pots should be filled from a uniform batch of salt solution. Otherwise the pots can be partly filled with salt crystals and water added; the first method is obviously preferable. When the loaded pots are standing side by side in a hole in the ground and the meter connected between them, the reading should be less than 2 mV. If not, the pots should be cleaned and recharged with fresh solution. Generally they can be used for a couple of days before running dry.

Where possible, traverses are carried out normal to the strike of suspected SP anomalies. Station intervals are usually not greater than 100 ft and may be as small as 10 ft. One of two electrode spreads may be employed: either one electrode is fixed at a base station while the other moves to successive stations along the line, or both electrodes are moved while maintaining a fixed interval between them.

The first arrangement requires a reel with a cable of several thousand feet. The meter may be at either electrode, but is usually at the base station. The advantages of this layout are that the potential is measured continually with respect to a fixed point – located, if possible, in a barren area; at the same time small zero errors between the electrodes do not accumulate. The only disadvantage is the long cable which inevitably slows down the measurements.

The fixed-electrode spread is maintained by moving the rear electrode up to the front pot hole and the forward electrode to a new station after each measurement. If the interval is small this is essentially a measurement of potential gradient, dV/ds, at a point midway between stations, where ds is the electrode spacing. Alternately the successive values of dV may be added algebraically to give the same potential profile as in the first method. This arrangement is faster than the other electrode layout, but has the disadvantage that the zero errors add up as the traverse progresses. To reduce this cumulative error the pots should be checked side by side in the same hole every 1000 ft or so.

An alternative procedure for the fixed-electrode spread is to leapfrog the electrodes in moving from station to station. If the rear pot is carried beyond the forward pot to the new station cumulative zero errors are eliminated. The relative

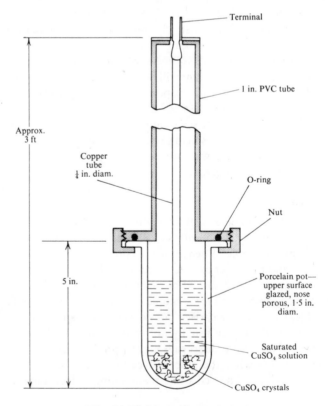

Fig. 6.3 Walking-stick SP electrode.

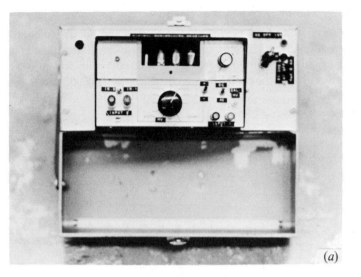

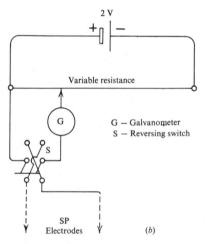

Fig. 6.4 Self-potential detectors. (*a*) Vacuum-tube millivoltmeter; (*b*) potentiometer.

polarities are reversed at successive stations in the process; this must be kept in mind to produce the potential profile.

Since the potential between the electrodes will be randomly positive and negative, the use of a centre-zero meter is a great advantage. At the same time it is essential to maintain a sign convention between the pots, i.e., forward pot with respect to rear, or moving pot with respect to base pot. On completion of a closed grid, the algebraic sum of potentials should be zero. This is obviously easy with a fixed base station, but may require some care in measuring with two moving electrodes. In all SP field work the individual traverse lines must be tied together by measuring the potentials between each line.

6.1.4 *Interpretation of self-potential data*

The end result of an SP survey is a set of profiles and possibly a contour map o
equipotentials. A typical profile and set of contours are illustrated in fig. 6.5. Note
that the negative maximum lies directly over the sulphide mass; where the
topography is steep, the centre may be displaced somewhat.

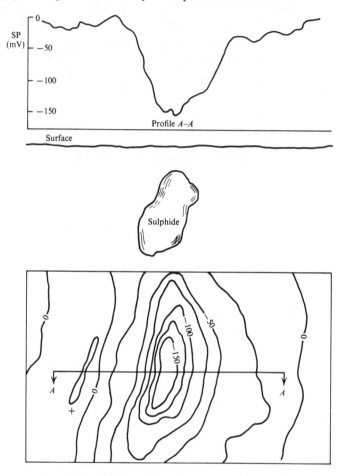

Fig. 6.5 Typical SP profile and contours over sulphide body.

It is possible to calculate the potential distributions around polarized bodies of
simple shape, such as the dipole, sphere and ellipsoid, by making some simplifications and assumptions concerning the potentials on the surfaces of the sources
themselves. For example, consider the polarized rod in fig. 6.6. The potential at a
point P on the ground above is given by

$$V = q(1/r_1 - 1/r_2), \tag{6.1}$$

where $\pm q$ is the charge at either end of the rod. Since $r_1 = \sqrt{(x^2 + z_1^2)}$, and $r_2 = \sqrt{\{(x-a)^2 + z_2^2\}}$, where $a = \ell \cos \alpha$, $\ell = $ length of rod, $\alpha = $ dip angle, this expression becomes

$$V = q[1/\sqrt{(x^2 + z_1^2)} - 1/\sqrt{\{(x-a)^2 + z_2^2\}}]. \tag{6.2}$$

Because of the air-ground interface, the rod has an electrical image above surface, which could make this potential twice as large (see §8.3.5). Typical profiles are plotted in fig. 6.6. It is apparent that the characteristic SP curve is fairly symmetric unless the dip angle is quite shallow.

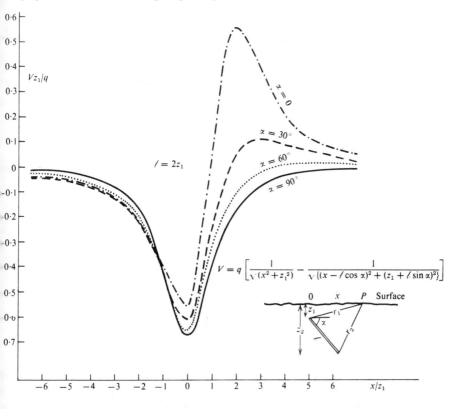

Fig. 6.6 SP profiles over buried polarized rod.

The polarized sphere is illustrated in fig. 6.7. To simplify the analysis, the sphere is assumed to be sliced in two with a constant potential $-V_0$ on the upper half and zero on the lower. To calculate the potentials off the vertical axis it is necessary to employ harmonic analysis (see §2.6.3e and §8.3.5).

Although these simple shapes give results which are similar to profiles obtained in the field, they are seldom used in SP interpretation, which is mainly qualitative. The shape of an anomaly, and its extent, are indicated by the contour map, or by a set of profiles normal to strike. An estimate of depth can be made from the shape of the profile. If $x_{\frac{1}{2}}$ is the total width of the profile at half the (negative) maximum, then the depth to the top of the body is of the order of half this distance. From the sphere in fig. 6.7, this is obviously a very crude rule, and judging from half a dozen random field examples, the estimate may be within $\pm 100\%$. In particular, if the anomalous profile is wide the source is also wide, rather than deep, because the depth of detection in SP is usually not greater than 200 ft.

$$V = 2V_0 \left[\frac{1}{2}\left(\frac{a}{r}\right) + \frac{3}{4}\left(\frac{a}{r}\right)^2 P_1(\cos\theta) - \frac{7}{16}\left(\frac{a}{r}\right)^4 P_3(\cos\theta) + \frac{11}{32}\left(\frac{a}{r}\right)^6 P_5(\cos\theta) - \ldots \right]$$

$P_n(\cos\theta) =$ Legendre polynomial

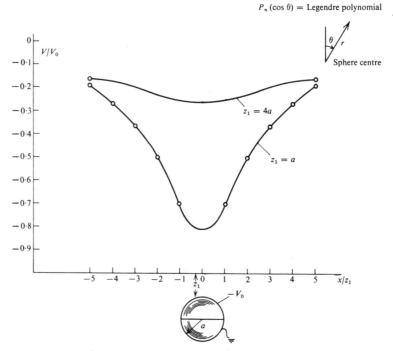

Fig. 6.7 SP-profiles over buried polarized sphere.

A rough idea of the attitude of the body may also be obtained from the lack of symmetry of the profile – that is, the steep slope and positive tail should be on the downdip side. It is often desirable to remove regional effects from the SP profiles in order to clarify the anomaly shapes. This can be done by inspection, to take out large-scale gradients, baseline shifts, effects of surface vegetation, known geologic structure and the like.

The type of overburden apparently has a pronounced effect on surface SP. Figures 6.8 and 6.9 show surface SP profiles over two sulphide bodies. Both have been drilled and the SP and core logs are also illustrated. In fig. 6.8, sulphides at 300 ft gave a good surface SP anomaly, while in fig. 6.9 the mineralization at 200 ft and 400 ft did not show up at all on surface. The topsoil in the vicinity of the first was sand, in the second clay. From other examples as well, the absence of surface SP seems to be associated with a clay cover.

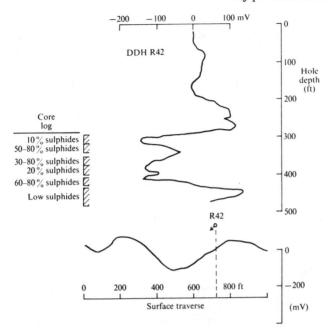

Fig. 6.8 Surface and drill-hole SP; sand overburden.

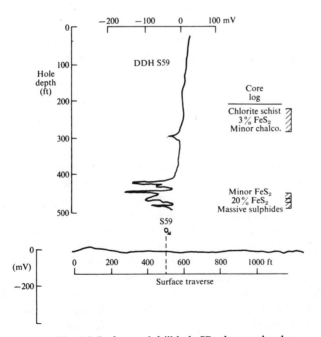

Fig. 6.9 Surface and drill-hole SP; clay overburden.

Figure 6.8 is also surprising for another reason. No sulphides were encountered above a depth of 300 ft; the surface SP anomaly is presumably due to mineraliza- tion between 300 and 500 ft. This is well below the water table in the area.

Self-potential has played a minor role in geophysical exploration. This is mainly due to the difficulty in making significant interpretation of the results – which are frequently quite erratic – and partly because of the shallow depth of penetration. It is, however, a simple, fast and cheap method which is useful in rapid ground reconnaissance for base metals when accompanied by magnetics, EM and geo- chemistry. Aside from the possibility of detecting sulphides, it would have some value in mapping shallow structures like faults, shear and fracture zones, contacts and so on.

6.2 Telluric and magnetotelluric methods

6.2.1 *Origin and characteristics of magnetotelluric fields and telluric currents*

In this section we shall consider techniques which employ certain large-scale (generally low-frequency) magnetic fields and the terrestrial current systems induced by these fields. The terms 'magnetotelluric' and 'telluric' are generally used to designate these fields and currents respectively. These topics are included in this chapter because they are natural-source electrical methods, like self-potential. Such magnetic fields, however, are identical with those discussed in §7.2 and §7.3.

The existence of natural large-scale earth currents was first established by Barlow in 1847 in the course of studies on the first British telegraph system. Long- term records of telluric currents were made at Greenwich, Paris and Berlin in the late nineteenth century; nowadays they are recorded at various observatories around the world.

The source of these currents has been fairly definitely located outside the earth. Periodic and transient fluctuations can be correlated with diurnal variations in the earth's magnetic field, caused by solar emission, aurora, etc. These activities have a direct influence on currents in the ionosphere; it is thought that the telluric currents are induced in the earth by ionospheric currents.

The inductive mechanism is an electromagnetic field propagated with slight attenuation over large distances in the space between the ionosphere and earth surface, somewhat in the manner of a guided wave between parallel conducting plates. That is to say, it proceeds by bouncing back and forth between these bound- aries and hence has a large vertical component. At large distances from the source this is a plane wave of variable frequency (from about 10^{-5} Hz up to the audio range at least). Obviously these magnetotelluric (MT) fields can penetrate the earth's surface to produce the telluric currents.

The pattern of these terrestrial current systems is shown in fig. 6.10. The huge whorls cover millions of square miles, are fixed with respect to the sun and rotate alternately clockwise and counterclockwise. In mid-latitudes there are two maxima and two minima per day, the average direction being mainly in the magnetic

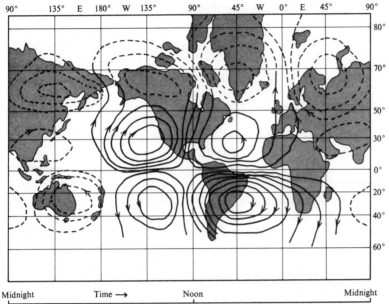

Fig. 6.10 World-wide telluric-current system on Mercator's projection. (After Dobrin, 1960.)

meridian. Near the equator (Peru, Madagascar), on the other hand, there appear to be only one maximum and one minimum per day, the amplitudes are considerably smaller and the average direction is east-west. The current intensity is larger on the daylight side of earth, while the electric field associated with the current is of the order of 10 mV/km. The frequencies are generally higher and more erratic during the day than at night.

Superimposed on this average telluric field are random fluctuations whose intensities vary with electrical disturbances in the ionosphere. These pulsations occur at frequencies as high as 100 kHz, although most are much lower. As a result the telluric record consists of erratic variations in potential, as illustrated in fig. 6.15 (§6.2.7a).

One source of the higher frequency current fluctuations is electric storms. Although their location is to some extent random, there are three major storm centres, all located in equatorial regions – Brazil, Central Africa and Malaya. Some of the thunderstorm energy is converted to electromagnetic fields which are propagated in the ionosphere-earth interspace. The weak currents induced by these fields in the subsurface may be useful in telluric and magnetotelluric prospecting, particularly because they have amplitude peaks at several distinct frequencies −8, 14, 760 Hz, etc. These same electromagnetic fields are also employed in the AFMAG method (see §7.5.4e).

6.2.2 *Elementary electromagnetic theory*

As mentioned above, an elementary development of electromagnetic theory can be employed to describe magnetotelluric wave propagation. To understand the propagation and attenuation of such waves it is necessary to use Maxwell's equations in a form relating the electric and magnetic field vectors:

$$\nabla \times \mathbf{E} = -\frac{\partial \mathbf{B}}{\partial t}, \tag{6.3}$$

$$\nabla \times \mathbf{H} = \mathbf{J} + \frac{\partial \mathbf{D}}{\partial t}, \tag{6.4}$$

where $\mathbf{J}$ = current density (A/m²), $\mathbf{E}$ = electric field intensity (V/m), $\mathbf{B}$ = magnetic flux density (webers/m²), $\mathbf{H}$ = magnetic field intensity (ampere-turns/m), $\mathbf{D}$ = electric displacement (C/m²).

Equation (6.3) is a mathematical statement of Faraday's law that an electric field exists in the region of a time-varying magnetic field, such that the induced emf is proportional to the negative rate of change of magnetic flux. Equation (6.4) is a mathematical statement of Ampere's law (taking into account Maxwell's displacement current, $\partial D/\partial t$), namely that a magnetic field is generated in space by current flow and that the field is proportional to the total current (conduction plus displacement) in the region (see fig. 6.11).

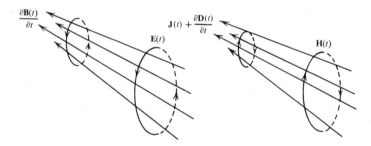

Fig. 6.11 Electric and magnetic fields of eqs. (6.3) and (6.4). (After Grant and West, 1965.)

Using the vector identity $\nabla \cdot \nabla \times \mathbf{A} = 0$ in these two equations, we get for time-varying fields

$$\left. \begin{aligned} \nabla \cdot \nabla \times \mathbf{E} = -\nabla \cdot \frac{\partial \mathbf{B}}{\partial t} = -\frac{\partial}{\partial t}(\nabla \cdot \mathbf{B}) = 0, \\[4pt] \text{that is,} \qquad \nabla \cdot \mathbf{B} = 0. \\[4pt] \text{Similarly,} \qquad \nabla \cdot \mathbf{J} + \nabla \cdot \frac{\partial \mathbf{D}}{\partial t} = \nabla \cdot \mathbf{J} + \frac{\partial}{\partial t}(\nabla \cdot \mathbf{D}) = 0. \end{aligned} \right\} \tag{6.5}$$

We also known that the divergence of current is equivalent to the rate of accumulation of charge, or

$$\nabla \cdot \mathbf{J} = -\frac{\partial q}{\partial t}.$$

In regions of finite conductivity, charge does not accumulate to any extent during current flow (omitting electrolytic conductors), hence $\partial q/\partial t = 0$ so that $\nabla \cdot \mathbf{J}$ also vanishes. Thus, from (6.5),

$$\nabla \cdot \mathbf{D} = \varepsilon\varepsilon_0 \nabla \cdot \mathbf{E} = 0. \tag{6.6}$$

Besides the relation between displacement and electric field we also have the following relation between $\mathbf{B}$ and $\mathbf{H}$ (see §3.2.1*h*):

$$\mathbf{B} = \mu\mu_0\mathbf{H}. \tag{6.7}$$

In eqs. (6.6) and (6.7), μ/μ_0 = relative magnetic permeability of the medium, $\varepsilon/\varepsilon_0$ = relative dielectric capacitivity, μ_0 = permeability of free space $= 4\pi \times 10^{-7}$ henrys/m, ε_0 = capacitivity of free space = $8\cdot85 \times 10^{-12}$ farads/m. Furthermore, in homogeneous isotropic media, we can express these relations, together with Ohm's law, as follows:

$$\mathbf{B} = \mu\mathbf{H}, \quad \mathbf{D} = \varepsilon\mathbf{E}, \quad \mathbf{J} = \sigma\mathbf{E}, \tag{6.8}$$

where σ = conductivity (mhos/m), and we can simplify eqs. (6.3) and (6.4) as follows:

$$\nabla \times \mathbf{E} = -\mu\partial\mathbf{H}/\partial t, \tag{6.9}$$

$$\nabla \times \mathbf{H} = \sigma\mathbf{E} + \varepsilon\partial\mathbf{E}/\partial t. \tag{6.10a}$$

If, as in some situations, there are independent current sources, $\mathbf{J}_0$, not related to the magnetic field (from SP, power devices, etc.), eq. (6.10*a*) becomes:

$$\nabla \times \mathbf{H} = \mathbf{J}_0 + \sigma\mathbf{E} + \varepsilon\partial\mathbf{E}/\partial t. \tag{6.10b}$$

Taking the curl of (6.9) and (6.10*a*) and using the following vector identity (which is valid in rectangular coordinates only),

$$\nabla \times \nabla \times \mathbf{A} = \nabla(\nabla \cdot \mathbf{A}) - \nabla \cdot \nabla\mathbf{A} = \nabla(\nabla \cdot \mathbf{A}) - \nabla^2\mathbf{A}, \tag{6.11}$$

we finally get

$$\nabla^2\mathbf{E} = \mu\frac{\partial}{\partial t}(\nabla \times \mathbf{H}) = \mu\sigma\frac{\partial\mathbf{E}}{\partial t} + \varepsilon\mu\frac{\partial^2\mathbf{E}}{\partial t^2}, \tag{6.12a}$$

$$\nabla^2\mathbf{H} = -\sigma(\nabla \times \mathbf{E}) - \varepsilon\frac{\partial}{\partial t}(\nabla \times \mathbf{E}) = \mu\sigma\frac{\partial\mathbf{H}}{\partial t} + \varepsilon\mu\frac{\partial^2\mathbf{H}}{\partial t^2}. \tag{6.12b}$$

If we choose time variations which are sinusoidal – which is generally done in MT work – we can put

$$E(t) = E_0\, e^{j\omega t}, \quad H(t) = H_0\, e^{j\omega t}, \quad \frac{\partial E}{\partial t} = j\omega E, \quad \frac{\partial H}{\partial t} = j\omega H,$$

where $\omega = 2\pi f = $ the angular frequency of the field. Thus eqs. (6.12a), (6.12b) are simplified to

$$\nabla^2 E = j\omega\mu\sigma E - \omega^2\varepsilon\mu E, \tag{6.13}$$

$$\nabla^2 H = j\omega\mu\sigma H - \omega^2\varepsilon\mu H, \tag{6.14}$$

the first and second terms on the right-hand side being related to the conduction and displacement currents respectively. These are the electromagnetic equations for propagation of electric and magnetic field vectors in an isotropic homogeneous medium having conductivity σ, relative permeability μ and dielectric capacitivity ε.

6.2.3 *Attenuation of EM fields*

The wave is attenuated in travelling through some media but not in free space. This can be shown as follows. Considering the relative magnitudes of the parameters ε, μ, ω and σ, we can say that the maximum normal value of ε occurs in water where $\varepsilon/\varepsilon_0 = 80$; for rocks, the ratio is generally less than 10. Similarly, $\mu/\mu_0 \leqslant 3$, even in ferromagnetic minerals; normally the value is unity. Thus we have

$$\varepsilon \approx 10\varepsilon_0 \approx 9 \times 10^{-11} f/\text{m}, \quad \mu \approx \mu_0 \approx 1.3 \times 10^{-6}\,\text{h/m}.$$

The periodic frequencies employed in MT work (and in EM methods as well, as we shall see in chapter 7) are usually less than 3000 Hz, hence $\omega \leqslant 2 \times 10^4$. The corresponding wavelengths, which are given by

$$\lambda = 2\pi c/\omega = 2\pi \times 3 \times 10^8/\omega \text{ meters,}$$

are greater than 90 km. Since the distances involved in field layouts are usually less than a mile, the phase variation resulting from propagation is negligible.

In the air $\sigma = 0$, $\varepsilon = \varepsilon_0$ and $\mu = \mu_0$. Thus the factor $\omega^2\varepsilon\mu$ in eqs. (6.13) and (6.14) is of the order of 5×10^{-9}, i.e., there is no attenuation of the wave in air.

The conductivity of rocks and minerals, however, varies enormously, as we have seen in §5.4.1. In rocks of low conductivity we might have $\varepsilon = 10\varepsilon_0$, $\mu = \mu_0$ and $\sigma \approx 10^{-3}$ mhos/m, so that, for $\omega = 2 \times 10^4$,

$$\nabla^2 E \approx (-4 \times 10^{-8} + 2.5 \times 10^{-5} j)E \approx 0.$$

However, in regions of high-conductivity – massive sulphides, graphite and the like – $\sigma \approx 10^3$ mhos/m and

$$\nabla^2 E \approx (-4 \times 10^{-8} + 25j)E \approx 25jE.$$

Comparison of eqs. (6.13) and (6.14) shows that identical relations hold for H also.

Thus in all cases the real part of the right-hand side of the equation (which corresponds to the displacement current) is negligible. As a result, in air and in poorly conducting rocks, we have

$$\nabla^2 E \approx 0, \quad \nabla^2 H \approx 0, \tag{6.15}$$

while in the presence of a good conductor the imaginary part of the expression may be significant, and eqs. (6.12a) and (6.12b) can then be written

$$\nabla^2 E \approx \mu\sigma \frac{\partial E}{\partial t} \approx j\omega\mu\sigma E, \quad \nabla^2 H \approx \mu\sigma \frac{\partial H}{\partial t} \approx j\omega\mu\sigma H. \tag{6.16}$$

This is the diffusion equation, which reduces to Laplace's equation (eq. (6.15)) in the air and in rocks of low conductivity. Equation (6.16) is generally difficult to solve; however, there is one important case in which a solution is readily obtained, that in which the wave is plane polarized. Assume the wave is propagating along the z-axis so that the xy-plane is the plane of polarization. We can then solve (6.16) by assuming the form

$$H = H_y(z, t) = H_0 e^{j\omega t + mz},$$

where H is the magnitude of **H**. Then,

$$\nabla^2 H = (\partial^2 H_y/\partial z^2) = m^2 H \quad \text{and} \quad (\partial H_y/\partial t) = j\omega H.$$

Substitution in (6.16) shows that we must have

$$m^2 = j\omega\mu\sigma, \quad \text{or} \quad m = \pm(1 + j)\sqrt{(\omega\mu\sigma/2)} = \pm(1 + j)a,$$

where $a = \sqrt{(\omega\mu\sigma/2)}$. Since H must be finite when $z = +\infty$, we discard the plus sign and obtain the solution:

$$H_y = H_0 e^{j\omega t - (1 + j)az} = H_0 e^{-az + j(\omega t - az)} \tag{6.17a}$$

Taking the real part as the required solution, we have:

$$H_y = H_0 e^{-az} \cos(\omega t - az). \tag{6.17b}$$

The second part of the expression represents simple harmonic motion with a phase shift while the exponential is the attenuation of the wave with propagation distance. This attenuation term may be written (taking $\mu = \mu_0$):

$$|H_y/H_0| \approx e^{-2 \times 10^{-3} z \sqrt{(f/\rho)}}.$$

Taking a few numerical examples, we get values as in tables 6.1 and 6.2 below.

A commonly used criterion for the penetration of electromagnetic waves is the *skin depth*, the distance in which the signal is reduced by $1/e$, that is, to 37%. This is given by

$$2 \times 10^{-3} z_s \sqrt{(f/\rho)} = 1, \quad \text{or} \quad z_s = 500\sqrt{(\rho/f)}.$$

Table 6.1. *Attenuation of EM waves*

f	ρ	$|H_y/H_0|$ for $z = 100$ ft	z for $|H_y/H_0| = 0.1$
1000 Hz	10^{-4} Ωm	0.00	0 ft
1000	10^{-2}	0.00	0
1000	1	0.15	120
1000	10^2	0.83	1200
1000	10^4	0.98	12,000
10	10	0.94	3800
10^2	10	0.83	1200
10^4	10	0.15	120
10^6	10	0.00	12

From these tables it is quite obvious that if the resistivity is low, or the frequency high, or both, the magnetic field will not penetrate the ground to any extent. As a crude rule of thumb we can say that if

$z\sqrt{(f/\rho)} > 10^3$, the attenuation will be large and vice versa.

For the same plane polarized wave in eq. (6.17b) we can also find the current, using eqs. (6.8) and (6.10a). Thus we have:

$$\nabla \times \mathbf{H} = \sigma \mathbf{E} = \mathbf{J}.$$

Hence

$$J_y = J_z = 0$$

Table 6.2. *Skin depth variation with frequency and resistivity*

	$\rho = 10^{-4}$ Ωm	10^{-2} Ωm	10^0 Ωm	10^2 Ωm	10^4 Ωm
f (Hz)	z_s (m)	z_s (m)	z_s (m)	z_s (m)	z_s (m)
10^{-3}	160	1600	1.6×10^4	1.6×10^5	1.6×10^6
10^{-2}	50	500	5000	5×10^4	5×10^5
10^{-1}	16	160	1600	1.6×10^4	1.6×10^5
1	5	50	500	5000	5×10^4
10	1.6	16	160	1600	1.6×10^4
10^2	0.5	5	50	500	5000
10^3	0.16	1.6	16	160	1600
10^4	0.05	0.5	5	50	500
10^6	0.005	0.05	0.5	5	50
10^8	—	0.005	0.05	0.5	5

and

$$J_x = -\frac{\partial H_y}{\partial z} = -\frac{\partial}{\partial z}\{H_0\,e^{-az}\cos(\omega t - az)\}$$

$$= aH_0\,e^{-az}\{\cos(\omega t - az) - \sin(\omega t - az)\}$$

$$= \sqrt{2}aH_0\,e^{-az}\cos(\omega t - az + \pi/4)$$

$$= \sqrt{(\omega\mu\sigma)}H_0\,e^{-z\sqrt{(\omega\mu\sigma/2)}}\cos\{\omega t - z\sqrt{(\omega\mu\sigma/2)} + \pi/4\}. \qquad (6.18)$$

This shows that the amplitude of the current is $\sqrt{(\omega\mu\sigma)}$ times that of the magnetic field at all points. Also, since J_x is proportional to H_y, the current flux exhibits the same skin effect as the magnetic field and in a good conductor, therefore, it is concentrated near the surface.

The physical result is that when $\sqrt{(\omega\mu\sigma/2)}$ is small, the magnetic field will propagate through the medium without much attenuation and in the process will fail to induce any appreciable current flow in it. Consequently there will be very little secondary magnetic field generated. On the other hand, when $\sqrt{(\omega\mu\sigma/2)}$ is large, the large surface current creates a large secondary magnetic field, out-of-phase with the original, which partially or completely cancels the primary field.

When the medium has intermediate conductivity there will be some secondary magnetic field developed. Even here the current density will not be uniform throughout the volume of the conductor but will be concentrated towards the outside. Because it is not generally possible to determine this current distribution, even for simple shapes, the analytical solution of most problems by electromagnetic theory is out of the question.

6.2.4 *Boundary conditions*

As in d.c. resistivity there are several boundary conditions for EM fields which must hold at interfaces where σ (and possibly μ) change abruptly. These can be derived from eqs. (6.3), (6.4), (6.6) and (6.8) as follows (see fig. 6.12):

$\mathbf{n} \times (\mathbf{E}_1 - \mathbf{E}_2) = 0$; electric field tangential to interface is continuous.
$\mathbf{n} \times (\mathbf{H}_1 - \mathbf{H}_2) = 0$; magnetic field tangential to interface is continuous.
$\mathbf{n}\cdot(\sigma_1\mathbf{E}_1 - \sigma_2\mathbf{E}_2) = 0$; current density normal to interface is continuous.
$\mathbf{n}\cdot(\mu_1\mathbf{H}_1 - \mu_2\mathbf{H}_2) = 0$; magnetic flux normal to interface is continuous.

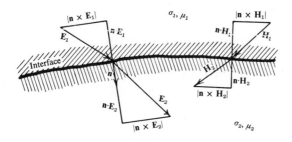

Fig. 6.12 Boundary conditions between two media.

All electromagnetic fields must satisfy the above boundary conditions at all interfaces.

6.2.5 *Magnetotelluric fields*

To adapt the wave equations to magnetotellurics, it is necessary to make certain simplifying assumptions. Certainly the frequencies are so low that displacement currents are negligible. Next, for plane waves of this type, it is clear that horizontal variations in **E** and **H** are small compared with vertical variations. Furthermore, we will consider only periodic frequency variations, since the fields are so erratic that it would be difficult to do otherwise. Taking the xy-plane as horizontal and z positive downward, these conditions can be expressed mathematically in the form

$$\frac{\partial D}{\partial t} = 0, \quad \frac{\partial}{\partial x} = 0 = \frac{\partial}{\partial y}, \quad E \propto e^{-j\omega t}, \quad H \propto e^{-j\omega t}, \quad \frac{\partial}{\partial t} = -j\omega.$$

If the wave is polarized in the xy-plane and travelling in the z-direction, we take the magnetic vector as H_0 at an angle θ to the x-axis so that the magnitudes of the magnetic components are $H_{z0} = H_0 \cos\theta$, $H_{y0} = H_0 \sin\theta$. We can write

$$H_z = (H_0 \cos\theta) \, e^{-az} \cos(\omega t - az), \quad H_y = (H_0 \sin\theta) \, e^{-az} \cos(\omega t - az) \quad (6.19a)$$

From (6.10a),

$$E_z = \frac{1}{\sigma}(x\text{-component of } \nabla \times H) = \frac{1}{\sigma}\left(-\frac{\partial H_y}{\partial z}\right)$$

$$= -\frac{1}{\sigma}(H_0 \sin\theta) \, e^{-az}\{-a \cos(\omega t - az) + a \sin(\omega t - az)\}$$

$$= \sqrt{2}\frac{a}{\sigma}(H_0 \sin\theta) \, e^{-az} \cos\left(\omega t - az + \frac{\pi}{4}\right). \quad (6.19b)$$

Similarly,

$$E_y = \sqrt{2}\frac{a}{\sigma}(H_0 \cos\theta) \, e^{-az} \cos\left(\omega t - az + \frac{\pi}{4}\right). \quad (6.19c)$$

Dividing (6.19b) and (6.19c) by (6.19a), the squares of the ratios of the amplitudes become

$$\left|\frac{E_y}{H_z}\right|^2 = \left|\frac{E_z}{H_y}\right|^2 = 2\left(\frac{a}{\sigma}\right)^2 = \omega\mu\rho. \quad (6.20)$$

If we assume σ to be the effective conductivity in a penetration depth Z, we can find approximate values of Z and σ by replacing $\partial/\partial z$ by $1/Z$ and ω by $2\pi/T$; eqs. (6.19) and (6.20) then give

$$Z \approx \frac{1}{\sigma}\left|\frac{H_y}{E_z}\right| \approx \frac{1}{\sigma\sqrt{(\omega\mu\rho)}} = \frac{\sqrt{(\omega\mu\rho)}}{\omega\mu} = \frac{T}{2\pi\mu}\left|\frac{E_z}{H_y}\right| \quad (6.21a)$$

and

$$\rho \approx \frac{T}{2\pi\mu} \left| \frac{E_x}{H_y} \right|^2. \tag{6.22a}$$

Clearly the x- and y-axes can be interchanged, hence these expressions can be stated in the following more general form:

$$Z \approx \frac{T}{2\pi\mu} \left| \frac{\mathscr{E}}{\mathscr{H}} \right| \tag{6.21b}$$

$$\rho \approx \frac{T}{2\pi\mu} \left| \frac{\mathscr{E}}{\mathscr{H}} \right|^2, \tag{6.22b}$$

where $\mathscr{E}/\mathscr{H}$ is equal to either E_x/H_y or E_y/H_x.

Setting $\mu = \mu_0 = 4 \times 10^{-7}$, substituting (6.22b) in (6.12b) and changing units to mV/km/ for $\mathscr{E}$, gammas for $\mathscr{H}$, km for Z, we get finally

$$Z \approx \frac{1}{2\pi} \sqrt{(5\rho T)} \text{ km}, \tag{6.21c}$$

$$\rho \approx 0 \cdot 2T \left| \frac{\mathscr{E}}{\mathscr{H}} \right|^2 \Omega\text{m}. \tag{6.22c}$$

The application of magnetotelluric theory to determine the electrical conductivity within the earth was originally described by Cagniard (1953). These relations are similar, except that the penetration depth is 70% of Cagniard's value.

By measuring the amplitudes of orthogonal horizontal components of the electric and magnetic fields at the surface, for various frequencies, one can determine the variation of resistivity with depth. This is an apparent resistivity (see §8.5.2).

The potential advantages in using these natural fields and currents are immediately obvious. With a relatively small electrode separation (few thousand feet) it should be possible to determine the depth and resistivity of horizontal beds. Furthermore the depth of penetration can be very great if low frequencies are selected. In practise, however, the possibilities are limited by the non-uniformity of the subsurface and by the fact that the signal is rarely sinusoidal so that measurements of E and H for a particular value of T are not easily achieved.

The magnetotelluric survey requires detection of both magnetic and electric field components. In the telluric method it is only necessary to measure the electric field associated with the earth currents. Thus the latter technique is simpler and requires less equipment. However, the amount of information derived from tellurics is considerably less than from MT work.

6.2.6 *Field equipment and operations*

(a) *Telluric current methods.* Since the currents cannot be measured directly, it is necessary to measure the potential gradients between electrodes planted on surface or possibly in drill holes.

As in *SP* work, non-polarizing electrodes should be used to reduce erratic potentials at the soil contact; lead plates, which are fairly inert chemically, are often employed instead. The electrodes are connected to an amplifier (bandwidth d.c. to 100 Hz, gain 2000) which will drive a strip chart or magnetic tape recorder. If specific frequencies are of interest, various bandpass and reject filters are incorporated in the amplifier section. Also, static potentials between the electrodes, which may be several hundred millivolts, must be balanced out; this is done with a potentiometer, or capacitance input, the latter having a time constant considerably longer than the maximum period to be recorded.

Because of large variations of signal amplitude with time, two electrode-spreads are necessary, one as a base station monitor, the other for the moving station. Since the signals also vary in direction with time, the base and field stations normally have two pairs of electrodes each, laid out perpendicular to each other, say N–S and E–W. Thus continuous records of two horizontal components are obtained at each station.

With this arrangement one can – at least in theory – compare the horizontal components of electric field variations between the base and field station, with regard to frequency, phase and amplitude. This is the method used in oil exploration applications.

A simpler field method has been used in initial attempts to apply tellurics to mineral exploration. A single pair of electrodes is set up at each station (one fixed base, the other moving), all electrodes being in a traverse line or perpendicular to it. The signals at the two stations are integrated over the same time intervals, using an integrating circuit and meter readout in place of the recorder, and the two compared for amplitude. In addition the amplifier has a narrow pass band, e.g. (8 ± 0.5) Hz.

Electrode spacings for oil exploration are usually 300–2000 ft. For mineral search the spread can be much smaller, 100 ft or less being reasonable. In both cases the field station is moved with respect to the base.

(b) *Magnetotelluric method.* The equipment for magnetotelluric work is quite complicated. The telluric set is similar to that used in telluric prospecting; two components are measured at each station, but no base station is required. A block diagram of an MT set is shown in fig. 6.13.

The magnetic field is usually measured with a coil, consisting of many turns on a large (3–4 ft) diameter frame. If the main interest is in very long periods, that is, large depth of penetration, it is possible to use a fluxgate magnetometer. With the coil, one is actually recording $\partial H/\partial t$ rather than variations of H. Again, two magnetic components are measured at each station.

Since the magnetic variations are in the milligamma range, the sensitivity requirements on the magnetic set are quite high. Very slight motions of the coil create noise voltages. This is particularly troublesome in wooded areas. A portable a.c. magnetometer would be an improvement over the coil; however, no suitable instrument is presently available.

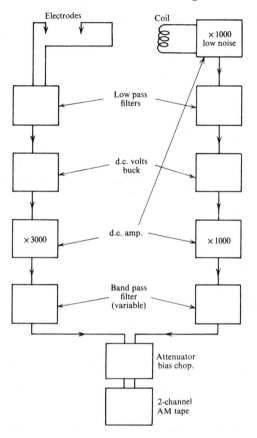

Fig. 6.13 Magnetotelluric field equipment (schematic).

Most field tests in magnetotellurics have been done on a large scale, using very low frequencies ($\ll 1$ Hz) in order to obtain deep penetration. However, there is no reason why the method cannot be used for shallow work by employing thunderstorm frequencies, as described in the section on tellurics.

6.2.7 *Interpretation*

(a) *Telluric currents*. If the ground were quite homogeneous at both base and field stations, the only difference in electric field signals between the two could be a slight phase shift, if they were far enough apart. However, any non-uniform geologic structure which distorts the current flow at one station will produce an anomaly in this field. An anticline or dome of high resistivity in the basement would cause such distortion by crowding the current lines over the apex, as shown in fig. 6.14.

The effect of certain simple geological structures on the electric field can be computed theoretically. These include two-dimensional structures such as the anticline, basement step and horizontal cylinder, and three-dimensional shapes like the sphere and ellipsoid. Examples of the first three, solved by using conformal transformations, are given in Keller and Frischknecht (1966).

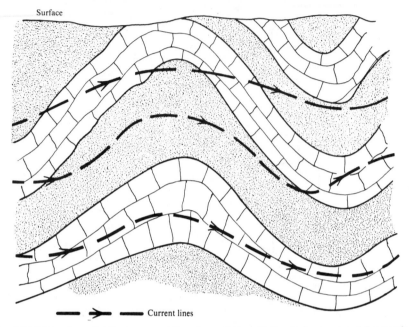

Fig. 6.14 Distortion of telluric current flow by anticline. (After Boissonnas and Leonardon, 1948.)

In these examples the current flow is assumed to be normal to the anomaly strike. In field situations, however, the currents tend to flow parallel to strike, that is, they flow in the good conductor rather than across it. Furthermore, since most rock structures are anisotropic, the flow will not usually be uniform in any case. The records may be further complicated by near-surface effects of overburden and erratic electrode contacts.

A typical set of records for two stations is illustrated in fig. 6.15. These can be analysed by the method of Boissonnas and Leonardon (1948) using selected intervals in which good data can be chosen from the record. The particular intervals chosen should show as good correlation as possible, while the interval length will depend on the depth of penetration desired and the character of the pulses.

Two assumptions are made: that the magnetic field is uniform over the survey area and that the particular frequencies selected from the record are low enough that the penetration depth is considerably greater than the actual depth of beds in which we are interested. Both are generally valid.

Figure 6.16 illustrates the instantaneous H-field vectors at two stations, one of which is movable. The other station is fixed, if possible, at a location where the

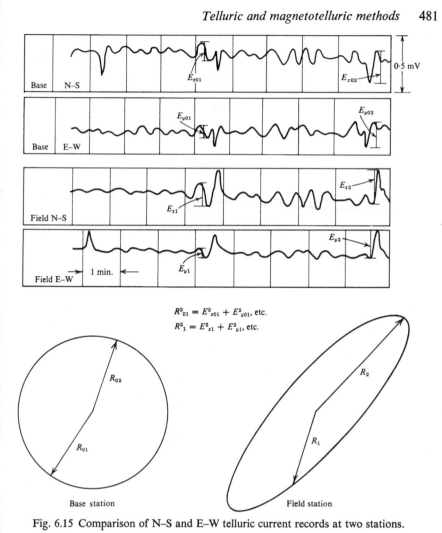

$$R^2_{01} = E^2_{x01} + E^2_{y01}, \text{ etc.}$$
$$R^2_1 = E^2_{x1} + E^2_{y1}, \text{ etc.}$$

Fig. 6.15 Comparison of N–S and E–W telluric current records at two stations.

subsurface is uniform, i.e., off the anomaly. Now if H_{x0} and H_{y0} are two orthogonal components of $\mathbf{H_0}$ at the base station, we have

$$H_{x0} = H_0 \cos \theta_0, \quad H_{y0} = H_0 \sin \theta_0.$$

At the base station we have assumed uniform ground, so that from eqs. (6.20) and (6.21a) we can write

$$\left| \frac{E_{x0}}{H_{y0}} \right| = \left| \frac{E_{y0}}{H_{x0}} \right| = \frac{1}{\sigma_0 Z_0} = c,$$

a constant. Expressing the components of $\mathbf{E_0}$ in terms of $\mathbf{H_0}$, we have

$$|E_{x0}| = c|H_0 \sin \theta_0|, \quad |E_{y0}| = c|H_0 \cos \theta_0|,$$

hence,

$$E_{x0}^2 + E_{y0}^2 = c^2 H_0^2 \tag{6.23}$$

which is the equation of a circle. That is, the field vector sweeps out a circle at the base station. If we make $cH_0 = 1$, this becomes a unit circle.

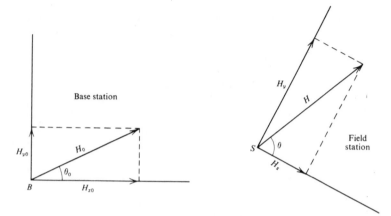

Fig. 6.16 Coordinate systems for telluric fields at base and field stations.

If the ground is not homogeneous at the moving or field station, then the conductivity will in general be different in two orthogonal directions and will be different from that at the base station. Also the depths will not be the same at the two stations. Then the ratio at the field station becomes

$$\left| \frac{E_x}{H_y} \right| = \left| \frac{E_x}{H \sin \theta} \right| = \frac{1}{\sigma_x Z} = c_x,$$

$$\left| \frac{E_y}{H_x} \right| = \left| \frac{E_y}{H \cos \theta} \right| = \frac{1}{\sigma_y Z} = c_y,$$

and, instead of a circle, we have an ellipse:

$$\frac{E_x^2}{(c_x H)^2} + \frac{E_y^2}{(c_y H)^2} = 1. \tag{6.24a}$$

The actual measurement in tellurics, of course, does not include H. Rewriting (6.24a) in terms of measured parameters, and assuming that $H \approx H_0 = 1/c$, we get

$$\frac{E_x^2}{(c_x/c)^2} + \frac{E_y^2}{(c_y/c)^2} = 1. \tag{6.24b}$$

The area of this ellipse is

$$\pi \left(\frac{c_x}{c} \right) \left(\frac{c_y}{c} \right) = \pi \left(\frac{\sigma_0^2 Z_0^2}{\sigma_x \sigma_y Z^2} \right),$$

while the area of the circle at the base station is π. Thus, if we compare the areas of the ellipse and circle, we have the quantity

$$\frac{\text{Area ellipse}}{\text{Area base circle}} = \frac{\sigma_0{}^2 Z_0{}^2}{\sigma_x \sigma_y Z^2}.$$ (6.25)

This ratio, while it does not determine the actual resistivities and depths at either station, compares their magnitudes. If we knew the conductivity and depth at any one station, by MT measurement or otherwise, we could find these quantities at all other stations. An example of the development of the circle and ellipse is shown in fig. 6.17, using vectors from the records of fig. 6.15.

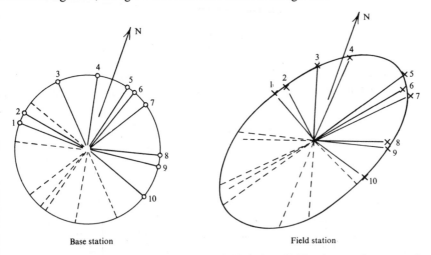

Base station   Field station

Fig. 6.17 Example of development of ellipse at field station. Field at base station assumed to produce circle, normalized radius = 1. (After Boissonnas and Leonardon, 1948.)

If the base station electric field vector varies considerably in magnitude during a cycle of rotation, then the original assumption of a circular pattern (i.e., homogeneous ground) is not valid, and will give an erroneous field pattern. If the local geology is known well enough, the base station can usually be set up in an area of uniform structure. Failing this, it may be necessary to select a base station from one of the field stations after the records have been studied.

The telluric method has not been used to any extent in petroleum prospecting in America, in spite of considerable application and some successes in Western Europe, North Africa and the USSR. Preliminary application of tellurics to mineral prospecting using the predominant 8-Hz frequency has given promising results. The first measurements were made over known faults and massive sulphides. Although continuous records have been made at both stations on occasion, the usual arrangement has been to integrate the two signals over a time interval of about 30 seconds. Signal ratios as large as 50 (between field station and background) over sulphides and 5 over a fault have been obtained in this way, when traversing perpendicular to the anomaly strike. In particular the field intensity is very low parallel to strike and correlation between base and field station signals is generally poor in the vicinity of a good conductor.

From eq. (6.22*a*) the apparent resistivity is proportional to $|E_x/H_y|^2$. Thus if we measure these quantities at two stations, we can write

$$\frac{E_{x1}}{E_{x2}} = \sqrt{\left(\frac{\rho_1}{\rho_2}\right)} \left|\frac{H_{y1}}{H_{y2}}\right|.$$

If H_y is chosen in the direction of strike for features which are mainly two-dimensional, it is easily shown from the boundary conditions that H_y does not vary along the ground surface. Then we have

$$\frac{\rho_1}{\rho_2} = \left|\frac{E_{x1}}{E_{x2}}\right|^2, \tag{6.26}$$

that is, we can obtain relative values of apparent resistivity directly from measurements of the telluric field alone, even though the absolute magnitude may not be known.

(b) *Magnetotellurics.* The end result of an MT survey is a paper and/or magnetic tape record of electric and magnetic field variations, as shown for the electric

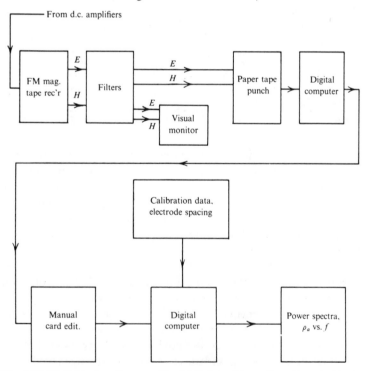

Fig. 6.18 Digital system for recording M-T data. (From Vozoff *et al.*, 1963.)

field only in fig. 6.15. The magnetic record is similar. To determine Z and ρ from eqs. (6.21a) and (6.22a), it is necessary to select groups of sinusoidal (or uniform) variations in E and H having the same periods. This might be done visually from simple clear records, but in general the curves are too complicated to be analysed so easily.

The usual method is to perform a power spectrum analysis on the records. This may be done by analog or digital computer; the former is faster but the latter is more commonly used. A typical digital system is shown in the block diagram in fig. 6.18. The magnetic tape is played back through various filters into an analog-digital converter to make punched paper tape records. Then the digital computer develops the power spectra by standard methods. Initially, because of gain instability of the amplifiers, a calibration of the systems at several frequencies is required. Information with regard to electrode spacing is also fed into the computer. The computer output gives the apparent resistivity as a function of frequency.

The magnetotelluric method clearly should give best results over uniform, isotropic, horizontal beds in which the resistivity varies only with depth. Cagniard (1953) has given an expression for the apparent resistivity in the case of two horizontal layers in the form

$$\frac{\rho_a}{\rho_1} = 1 + \frac{4 \cos 2zm}{n + 1/n - 2 \cos 2zm},\tag{6.27}$$

where $z =$ thickness of upper layer, $m = +\sqrt{(\omega\mu/\rho_1)} = \sqrt{(2\pi\mu/T\rho_1)}$, as in (6.18), $n = (\sqrt{\rho_2} + \sqrt{\rho_1})\,e^{2zm}/(\sqrt{\rho_2} - \sqrt{\rho_1})$.

A master curve for the two-layer earth is shown in fig. 6.19. With ρ_a/ρ_1 as ordinate and $\sqrt{T}$ as abscissa, both on log scales, a set of curves has been drawn for various values of ρ_2/ρ_1. Actually ρ_1 and z are considered to be 1 Ωm and 1 km respectively.

By plotting field results in the form of ρ_a versus $\sqrt{T}$ on the same scale log paper as the master curve, it is possible to solve for ρ_1, ρ_2 and z by superposition of the two curves. This is done by matching the field curve to one of the theoretical curves on the master; sometimes it may be necessary to interpolate between two of the latter. Of course the ordinate and abscissa scales must be maintained parallel.

When this superposition has been made, we find a particular value for ρ_a and $\sqrt{T}$ on the field curve corresponding to

$$\rho_a/\rho_1 = 1/zm = 1$$

on the master. Then this value of ρ_a on the field curve is ρ_1, the resistivity of the upper layer, while ρ_2/ρ_1 on the master yields the value of ρ_2 for the lower layer. Finally the depth of the upper layer can be found from the relation

$$z = \sqrt{(T\rho_1/2\pi\mu)},$$

where T is the abscissa value on the field curve, corresponding to $zm = 1$ on the master, and $\mu = \mu_0 = 4\pi \times 10^{-7}$.

Sets of master curves have also been computed for three horizontal layers; the

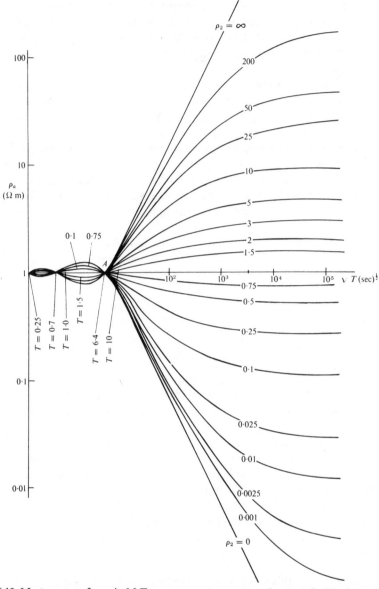

Fig. 6.19 Master curves for ρ_a in M-T measurements over a two-layer earth; T is the period for various resistivity contrasts, $\rho_1 = 1\ \Omega m$, $z_1 = 1$ km. (After Cagniard, 1953.)

procedure is similar. Other theoretical curves have been developed to show the effect of a fault and a dike on MT measurements when traversing normal to these features.

An example of magnetotelluric deep sounding in a relatively simple geologic area is given by Reddy and Rankin (1971) who carried out measurements in the period range 1–1000 seconds (1–0·001 Hz) at 16 stations in Alberta. Their results are summarized in fig. 6.20, which shows a geologic section for a SW–NE line through Edmonton extending from the Rocky Mountains for 350 miles to the Saskatchewan border. Sediments in the central Alberta plain vary in thickness from 4000 ft in the east to more than 15,000 ft where they abut the deep-rooted, deformed Precambrian mountain sediments in the southwest. The basement is composed of crystalline rocks of Precambrian age.

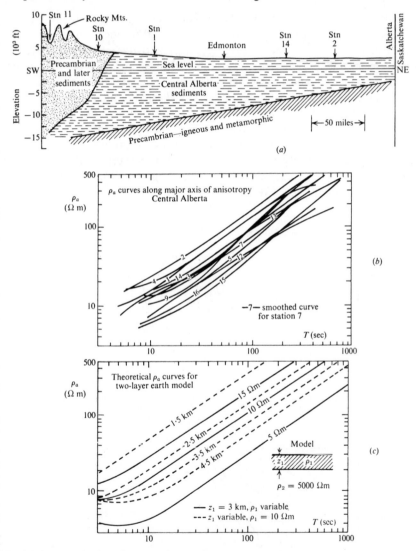

Fig. 6.20 Magnetotelluric deep soundings in Central Alberta. (After Reddy and Rankin, 1971.) (*a*) Geological section; (*b*) ρ_a-curves along major axis of anisotropy; (*c*) theoretical ρ_a-curves for two-layer earth.

The magnetotelluric fields are controlled by the strike of the contacts of these plains sediments with the other two sections, both of which have relatively high resistivities. As a result the currents at depth in the more conductive sedimentary basin are polarized NW–SE to produce a pronounced anisotropy in the apparent resistivities, i.e., the maximum values are generally parallel to strike for the longer period signals. The fact that there is no apparent anisotropy at short periods (10 seconds or so), on the other hand, would indicate that the basin sediments are reasonably homogeneous.

Figure 6.20 shows a set of smoothed curves of ρ_a versus period for the 16 stations, taken along the major axis of anisotropy, as well as theoretical curves for a two-layer earth model in which the thickness (z_1) and resistivity (ρ_1) of the top layer are varied. Clearly both parameters are variable in the field area. It is possible to produce more complex 3- and 4-layer sections for individual stations which match the measured results very well, although these models are not unique.

Although the results are satisfactory in the example of fig. 6.20, in general the agreement between field results and simple layer theory has not been particularly good. Discrepancies are often caused by anisotropic ground and lateral discontinuities in conductivity. The theory assumes vertical currents to be negligible. In fact large vertical current components extending over distances of more than 100 miles have recently been reported in the literature.

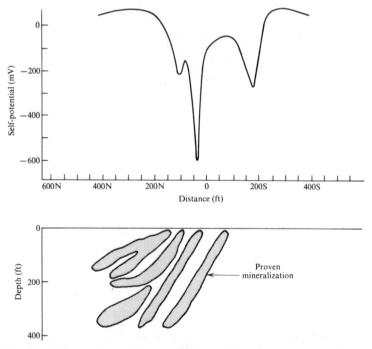

Fig. 6.21 Self-potential profile over massive sulphides, Senneterre Area, Quebec.

Since the equipment required is quite elaborate, it is also apparent that the method is not cheap. Among ground prospecting methods, only seismic is more expensive. As an alternative to seismic in oil exploration, it has been used with some success in Europe, particularly Russia, and North Africa, but has not made much impression in North America. In areas favourable for minerals, the more complex geology makes the method less attractive. However, since the telluric measurements seem promising in this environment, occasional spot measurements of magnetic as well as electric field would be useful to indicate resistivity.

6.3 Field examples

6.3.1 *Self-potential*

Figure 6.21 shows an SP profile obtained over a massive sulphide body in the Senneterre area of Quebec. The mineralization is pyrite and pyrrhotite, averaging

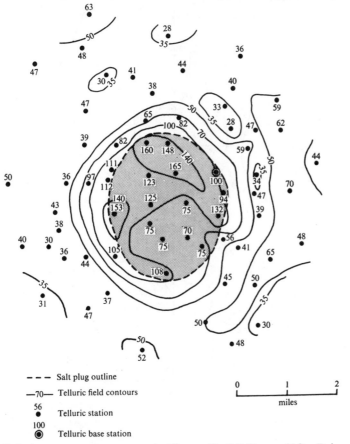

- – – Salt plug outline
- —70— Telluric field contours
- 56 ● Telluric station
- 100 ◉ Telluric base station

0   1   2
miles

Fig. 6.22 Telluric-current contours over the Haynesville Salt Dome. (After Boissonnas and Leonardon, 1948.)

about 30 % for the entire zone, 40–70 % in the more strongly mineralized sections. The host rocks are metasedimentary breccias and tuffs, interbedded with lava flows. Twenty-five diamond-drill holes outline the sulphides in some detail.

Although the maximum SP anomaly is very large (600 mV), it is also surprisingly narrow, appearing to reflect only one of the sulphide zones rather than a combined effect. North of this peak the two shallow sulphides produce very little surface potential, while the one to the south is not detected at all. The negative peak at 200S coinciding with a low swampy surface is not explained.

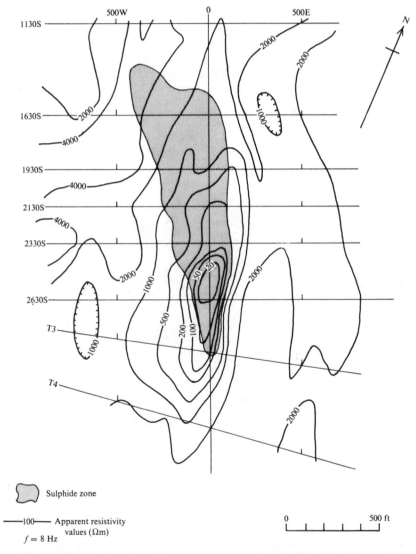

Sulphide zone

——100—— Apparent resistivity
values (Ωm)
$f = 8$ Hz

0   500 ft

Fig. 6.23 Telluric data from 8 Hz-survey over massive sulphides, New Brunswick.

6.3.2 *Tellurics*

1. An example of the analysis of Boissonnas and Leonardon (1948) is illustrated in fig. 6.22, which shows telluric contours over the Haynesville Salt Dome in Texas. The results fit the known geology very well. The salt plug is a poor conductor; consequently the current is concentrated in the layer above it. The numbers shown at the station locations correspond to ellipse areas, referred to the area of a unit circle at the base station, which is located in the east-central part of the map. These ellipse areas vary by a factor of five over the whole region, larger ellipses being over the salt dome.

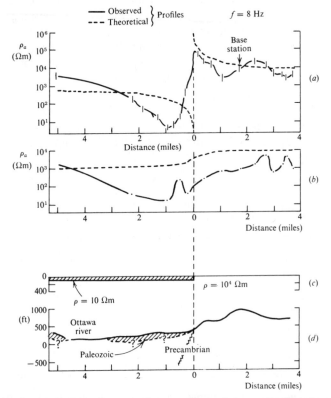

Fig. 6.24 Magnetotellurics over contact between Precambrian metasediments and Paleozoic sediments, north of the Ottawa River, Ontario. (*a*) Resistivity curves based on $E_\perp$; (*b*) resistivity curves based on $E_{||}$; (*c*) theoretical model; (*d*) geological section.

2. Results of an 8-Hz telluric survey made over a massive sulphide body in northern New Brunswick are shown in fig. 6.23. The zone, located in diorite-rhyolite host rock, has been well outlined by diamond drilling. It subcrops below the overburden to the south and plunges gently to a depth of 400 ft at the north end.

Contours are apparent resistivity values determined from relative telluric field strength measurements as in eq. (6.26). The sulphide zone is clearly indicated, with the lowest resistivities at the south end. The extension of the contours beyond the southern extremity of the sulphide outline is due to further shallow mineralization between lines T_3 and T_4. The telluric anomaly extends to line 1930S in the northern part of the mineralized zone, which is over 200 ft deep at this point.

6.3.3 Magnetotellurics

Figure 6.24 shows an 8-Hz magnetotelluric profile over a contact between Precambrian metasediments and Paleozoic shales, sandstones, limestones and dolomites. The area is about 20 miles west of Ottawa, north of the Ottawa river. The apparent resistivities in fig. 6.24a are obtained from the ratio $E_\perp/H_\parallel$, i.e., the telluric field was measured roughly normal to the contact, the magnetic field parallel to it. The agreement between the measured profile and a theoretical one, derived from a vertical contact of great depth with resistivity contrast of 100, is fair. The actual geologic profile is not so simple since the Paleozoic sediments (the low resistivity section) are only 600 ft thick. The MT profile, however, reflects the contact clearly.

Measurement of the telluric field at several azimuth angles other than $E_\perp$ showed that the apparent resistivity became progressively more anisotropic as the contact was approached from the north and that the minimum ρ_a was parallel to strike (see fig. 6.24b).

6.4 Problems

1. The following self-potential readings were obtained on a detailed traverse during a field school held in southern Quebec.

Station	SP	Station	SP	Station	SP
0 W	0 mV	6 W	−138 mV	12 W	−120 mV
1	−37	7	−210	13	−73
2	−100	8	−290	14	−40
3	−108	9	−335	15	−21
4	−158	10	−258	16	−17
5	−236	11	−170	17	−7

Stations are 10 ft apart and SP readings are given with reference to station 0. Make an interpretation of this anomaly based on the limited data available.

2. Figure 6.25 shows a set of SP contours for a prospect in northern New Brunswick. The area is mainly wooded with some open ground roughly parallel to, and slightly west of, the baseline between lines 10 N and 20 N. What interpretation can be made of this prospect – particularly the large SP negative centres – from the data? Is any additional work warranted and if so what would you recommend?

3. The SP profiles shown in fig. 6.26 were the result of a survey carried out on a large geochemical anomaly in eastern Nova Scotia. These profiles are part of a much larger

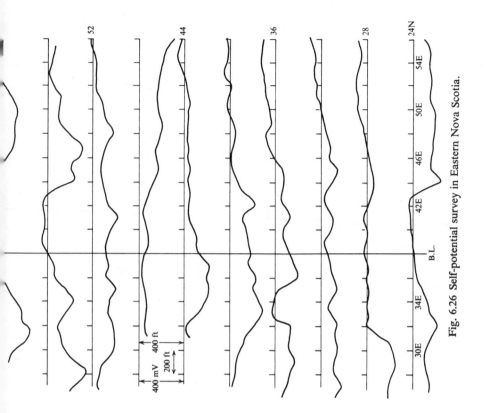

Fig. 6.26 Self-potential survey in Eastern Nova Scotia.

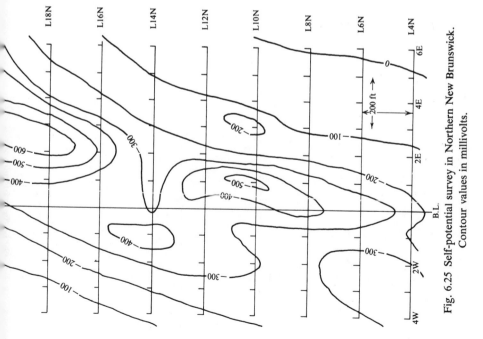

Fig. 6.25 Self-potential survey in Northern New Brunswick. Contour values in millivolts.

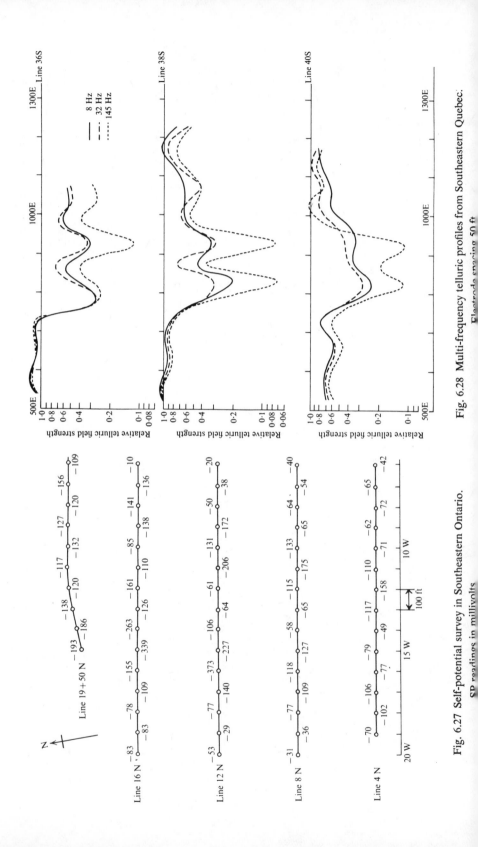

Fig. 6.28 Multi-frequency telluric profiles from Southeastern Quebec. Electrode spacing 50 ft.

Fig. 6.27 Self-potential survey in Southeastern Ontario. SP readings in millivolts.

SP survey, extending further in all directions except east, in which the general background SP is 140 mV more positive than the zero values shown. Convert the profiles into contours by adding −140 mV to all readings and make any interpretation you can of the result.

4. Readings from a self-potential survey made over suspected sulphide mineralization in southeastern Ontario are shown in fig. 6.27. By drawing profiles and contouring this data, make an interpretation for the area.

5. The following telluric field strength readings were obtained during a traverse over a prospect in eastern Nova Scotia. Four frequencies – 1 Hz, 8 Hz, 32 Hz and 145 Hz – are measured by integrating signals, obtained from 100 ft electrode spreads, for 30 seconds.

Stn	1 Hz	8 Hz	32 Hz	145 Hz	Stn	1 Hz	8 Hz	32 Hz	145 Hz
20 W	120	110	282	655	12 W	45	44	70	127
19	69	140	270	590	11	62	84	133	190
18	97	140	265	550	10	29	71	140	305
17	81	155	307	763	9	48	92	160	205
16	62	126	243	607	8	143	257	393	660
15	37	56	117	256	7	156	375	565	970
14	37	40	88	242	6	207	510	845	1700
13	46	39	81	242	5	223	445	715	1580

Normalize these readings for each channel, by taking the averages of all stations to obtain a common background of unity, plot on a log scale and interpret the result. (Note: presumably one can make use of the fact that depth of penetration varies inversely with frequency in some fashion.)

6. Figure 6.28 shows three profiles of multifrequency (8, 32 and 145 Hz) telluric field strength over a zone of sulphide mineralization in southeastern Quebec. As in Problem 5, the signals were integrated for 30 seconds, while the electrode separation was 50 ft. Make an interpretation of the results based on tellurics alone.

7. The profiles shown in fig. 6.29 were part of a multifrequency reconnaissance telluric survey. Only two of the four frequencies measured (1 Hz and 8 Hz) are reproduced here. Spacing between the two lines shown is 400 m. Relative telluric field readings have been converted to apparent resistivities, based on resistivity measurements made at several stations in the area. Make an interpretation of this limited data.

8. An experimental 8-Hz telluric survey was made over an area previously covered by IP in eastern Canada. The relative telluric field contours are shown in fig. 6.30. Interpret the broad features of this area with no other data available.

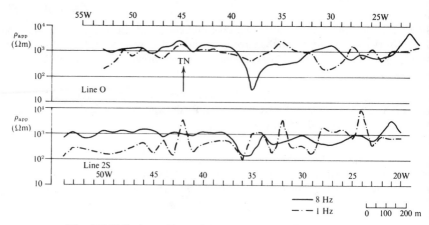

Fig. 6.29 Telluric profiles at 1 Hz and 8 Hz, Northeastern Brazil.

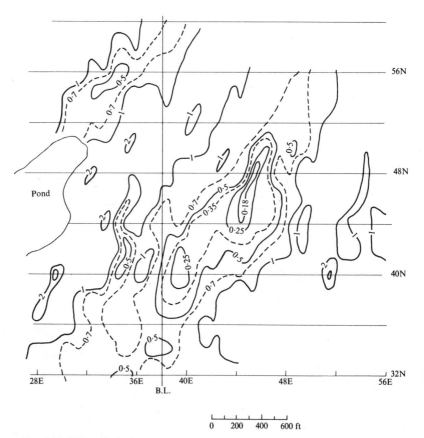

Fig. 6.30 8-Hz telluric data, normalized and contoured for large-scale interpretation.

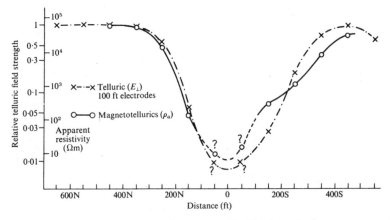

Fig. 6.31 8-Hz magnetotelluric profile, Northern Quebec.

9. The magnetotelluric profile illustrated in fig. 6.31 was obtained by measuring $E_\perp$ and $H_\parallel$ (electric field roughly normal, magnetic field parallel, to strike of the mineralization) at 8 Hz over a sulphide showing in northern Quebec. The telluric profile for $E_\perp$ is also shown. It indicates that the telluric measurement corresponds very closely to the magnetotelluric result, although the apparent resistivity values, of course, cannot be determined. What crude interpretation can be made from this profile?

References

F – field methods; G – general; I – instruments; T – theoretical

Self potential

Becker, A. and Telford, W. M. (1965). Spontaneous polarization studies. *Geophys. Prosp.* **13**, 173–88. (F)

De Witte, L. (1948). A new method of interpretation of self-potential field data. *Geophysics*, **13**, 600–8. (T)

Fox, R. W. (1830). On the electromagnetic properties of metalliferous veins in the mines of Cornwall. *Proc. Roy. Soc. London*, **2**, 411. (F)

Hansuld, J. A. (1966). Eh and pH in geochemical exploration. *Can. Inst. Mining & Met.* **69**, 77–84. (G)

Kelly, S. F. (1957). In *Methods and case histories in mining geophysics*, Canadian Inst. of Mining and Metallurgy (6th Commonwealth Mining and Metallurg. Congress), pp. 53–9: Montreal, Mercury. (F)

Petrovski, A. (1928). The problem of a hidden polarized sphere. *Phil. Mag.* **5**, 334–58. (T)

Sato, M. and Mooney, H. M. (1960). The electrochemical mechanism of sulphide self potentials. *Geophysics*, **25**, 226–49. (G)

Scott, B. I. H. (1962). Electricity in plants. *Sc. Amer.* **207**, 107–17. (G)

Schlumberger, C. (1930). *Étude de la prospection électrique du sous-sol* (2nd ed.). Paris, Gauthiers-Villars. (G)

Yungul, S. H. (1950). Interpretation of spontaneous polarization anomalies caused by spheroidal orebodies. *Geophysics*, **15**, 237–46. (T)

Tellurics

Berdichevskiy, M. N. (1965). Electrical exploration with telluric currents (trans. by Keller, G. V.). *Quart. Colo. Sch. Mines*, Jan. 1965. (F)

Boissonnas, E. and Leonardon, E. G. (1948). Geophysical exploration by telluric currents with special reference to a survey of the Haynesville Salt Dome, Wood, County, Texas. *Geophysics*, **13**, 387–403. (T)

Chapman, S. and Bartels, J. (1940). *Geomagnetism*, 417–48. Oxford, Clarendon. (G)

Dahlberg, R. S. Jr. (1945). An investigation of natural earth currents. *Geophysics*, **10**, 494–506. (F)

Jones, F. W. and Geldart, L. P. (1967). Vertical telluric currents. *Earth & Planetary Science Letters*, **2**, 69–74. (T)

Keller, G. V. and Frischknecht, F. C. (1966). *Electrical methods in geophysical prospecting*. Oxford, Pergamon. (G)

Mainguy, M. and Grepin, A. (1953). Some practical examples of interpretation of telluric methods in Languedoc (south-eastern France). *Geophs. Prosp.* **1**, 233–240. (F)

Rooney, W. J. (1939). Earth currents. In *Terrestrial magnetism and electricity* (ed. by J. A. Fleming), pp. 270–307, *Physics of the earth*, Series 8, N.R.C. New York, McGraw-Hill. (G)

Roy, A. (1963). New interpretation techniques for tellurics and some direct current fields. *Geophysics*, **28**, 250–61. (T)

Schlumberger, M. (1939). Application of telluric currents to surface prospecting. *Trans. Am. Geophys. Un.*, Pt. I, **20**, 271–7. (G)

Slankis, J. A. (1970). *Telluric and magnetotelluric surveys at 8 Hz*. Ph.D. thesis, McGill University, Montreal. (I)

Slankis, J. A., Telford, W. M. and Becker, A. (1972). 8 Hz telluric and magnetotelluric prospecting. *Geophysics*, **37**, 862–78. (F)

Tuman, V. S. (1951). The telluric method of prospecting and its limitations under certain geologic conditions. *Geophysics*, **16**, 102–14. (F)

Westcott, E. M. and Hessler, V. P. (1962). The effect of topography and geology on telluric currents. *Jour. Geophys. Res.* **67**, 4813–23. (T)

Y-Shu, L. (1963). Calculating the parameters used in telluric current prospecting. *Geophysics*, **28**, 482–5. (T)

Magnetotellurics

Blackman, R. B. and Tukey, J. W. (1958). *The measurement of power spectra.* New York, Dover. (G)

Cagniard, L. (1953). Basic theory of the magneto-telluric method of geophysical prospecting. *Geophysics*, **18**, 605–35. (T)

Cantwell, T. and Madden, T. R. (1960). Preliminary report on crustal magnetotelluric measurements. *Jour. Geophys. Res.* **65**, 4202–5. (F)

D'Erceville, I. and Kunetz, G. (1962). The effect of a fault on the earth's natural electromagnetic field. *Geophysics*, **27**, 651–65. (T)

Dobrin, M. (1960). *Geophysical prospecting*. New York, McGraw-Hill. (G)

Grant, F. S. and West, G. F. (1965). *Interpretation theory in applied geophysics*. New York, McGraw-Hill. (G)

Jain, S. (1966). A simple method of magnetotelluric interpretation. *Geophys. Prosp.* **14**, 143–8. (T)

Niblett, E. R. and Sayn-Wittgenstein, C. (1960). Variation of electrical conductivity with depth by the magnetotelluric method. *Geophysics*, **25**, 998–1008. (F)

Price, A. T. (1962). The theory of magnetotelluric methods when the source field is considered. *Jour. Geophys. Res.* **67**, 1907–18. (T)

Rankin, D. (1962). The magnetotelluric effect on a dike. *Geophysics*, **27**, 666–76. (T)

Reddy, I. K. and Rankin, D. (1971). Magnetotelluric measurements in central Alberta. *Geophysics*, **36**, 739–53. (F)

Srivastava, S. P., Douglass, J. L. and Ward, S. H. (1963). The applications of the magnetotelluric and telluric methods in central Alberta. *Geophysics*, **28**, 426–46. (F)

Srivastava, S. P. (1965). Method of interpretation of magnetotelluric data when the source field is considered. *Jour. Geophys. Res.* **70**, 945–54. (T)

Srivastava, S. P. (1967). Magnetotelluric two and three layer master curves. *Dom. Obs. (Ottawa, Canada) Pub.* **35**, No. 7. (T)

Vozoff, K., Hasegawa, H. and Ellis, R. M. (1963). Results and limitations of magnetotelluric surveys in simple geologic situations. *Geophysics*, **28**, 778–92. (T)

Vozoff, K. (1972). The magnetotelluric method in the exploration of sedimentary basins. *Geophysics*, **37**, 98–141. (G)

Yungul, S. H. (1961). Magneto-telluric sounding three-layer interpretation curves. *Geophysics*, **26**, 465–73. (T)

7. Electromagnetic methods

7.1 Introduction

With the exception of magnetics, the electromagnetic (EM) prospecting technique is the most commonly used in mineral exploration. It is not suitable for oil search, because it responds best to good electrical conductors at shallow depth. On the other hand, it has not been employed in civil engineering work either, although it is used occasionally to locate buried pipe and cable and for the detection of land mines.

As the name implies, the method involves the propagation of time-varying, low-frequency electromagnetic fields in and over the earth. There is a close analogy between the transmitter, receiver and buried conductor in the EM field situation and a trio of electrical circuits coupled by electromagnetic induction. In a few EM systems the source energy may be introduced into the ground by direct contact, although generally inductive coupling is used; invariably the detector receives its signal by induction.

Almost all EM field sets include a portable power source. However, limited use has also been made of radio transmission stations in the frequency range 100 kHz to 10 MHz and recently in the very low frequency range 5–25 kHz. One other field method which can reasonably be included with EM, known as AFMAG (audio-frequency magnetic fields), makes use of atmospheric energy resulting from world-wide thunderstorm activity.

A particular advantage of the inductive coupling is that it permits the use of EM systems in aircraft. Airborne EM, usually in combination with aeromagnetic equipment, has been widely applied in mineral exploration reconnaissance.

7.2 Electromagnetic theory

7.2.1 *Magnetic vector potential*

The propagation and attenuation of electromagnetic waves were discussed in §6.2.2 to §6.2.5 in connection with the magnetotelluric method. Although the frequencies employed in EM prospecting are somewhat higher than in MT work, the general theory, limiting assumptions (negligible displacement current and spatial phase shift) and boundary conditions are identical in the two methods.

In general potential theory it is usually easier to solve problems by starting with the potential and obtaining the field vectors by appropriate differentiation. The same rule applies in electromagnetics, where it is convenient to introduce a *vector magnetic potential* (one of several possible potentials) from which both

500

electric and magnetic field vectors may be derived. We define this potential in terms of the magnetic induction:

$$\nabla \times \mathbf{A} = \mathbf{B}. \tag{7.1a}$$

From eq. (6.3), §6.2.2, since $\nabla \times \mathbf{E} = -\partial \mathbf{B}/\partial t$, we have

$$\mathbf{E} = -\partial \mathbf{A}/\partial t. \tag{7.1b}$$

Again, from eq. (6.6), we have $\nabla \cdot \mathbf{E} = 0$, so that

$$\nabla \cdot \mathbf{A} = 0. \tag{7.1c}$$

Taking the curl of (7.1a), using the vector identity (6.11) plus eqs. (6.4), (6.8), (7.1b) and (7.1c), we get

$$\nabla \times \nabla \times \mathbf{A} = \nabla(\nabla \cdot \mathbf{A}) - \nabla^2 \mathbf{A} = -\nabla^2 \mathbf{A} = \nabla \times \mathbf{B} = \mu \nabla \times \mathbf{H},$$

and

$$\nabla^2 \mathbf{A} = -\mu \nabla \times \mathbf{H} = -\mu \sigma \mathbf{E} - \mu \varepsilon \frac{\partial \mathbf{E}}{\partial t} = \mu \sigma \frac{\partial \mathbf{A}}{\partial t} + \mu \varepsilon \frac{\partial^2 \mathbf{A}}{\partial t^2}, \tag{7.2}$$

where, as we have already noted in §6.2.3, the second term on the right is negligible. Hence the vector magnetic potential, like $\mathbf{E}$ and $\mathbf{H}$, satisfies Laplace's equation in poorly conducting media and the diffusion equation when σ is large enough. Also from (7.2) above and using eq. (6.8), we can write

$$\nabla^2 \mathbf{A} = -\mu \sigma \mathbf{E} = -\mu \mathbf{J}.$$

This relation resembles Poisson's equation in gravity and magnetics, the solution being

$$\mathbf{A} = \frac{\mu}{4\pi} \int_v \frac{\mathbf{J}\, dv}{r}, \tag{7.3a}$$

where $r^2 = x^2 + y^2 + z^2$.

The magnetic vector potential does not have the physical significance of the scalar potential in gravitation and electrostatics, and is solely a mathematical convenience in determining electromagnetic fields.

7.2.2 *Description of EM fields*

(a) *General.* The primary or source fields used in EM prospecting are normally generated by passing alternating current through long wires or coils. For simple geometric configurations the resultant fields can be calculated exactly for points in the surrounding region, although this is not generally possible. Furthermore, it is essential to know the primary field at the receiver, or at least to eliminate its effect, because it is always present, along with secondary fields due to currents induced in the subsurface. Consequently, one must measure the disturbing field in the presence of the original primary field.

(b) *Biot-Savart law*. Originally stated for static magnetic fields, this law is valid also for low frequency a.c., provided the linear distances involved are much less than the wavelength. We have already seen that this is so. The law relates the magnetic field at a point in space to a current flowing in a short straight element of conductor some distance away. The geometry is shown in fig. 7.1. This relation can be derived by using the vector potential expression in eq. (7.3*a*):

$$\mathbf{A} = \frac{\mu}{4\pi} \int_{\text{vol}} \frac{\mathbf{J}\, dv}{r} = \frac{\mu}{4\pi} \oint \frac{i\, d\mathbf{s}}{r},$$ (7.3*b*)

where the line integration is carried out over a closed path surrounding the wire element. Thus we have

$$d\mathbf{H} = \frac{d\mathbf{B}}{\mu} = \frac{\nabla \times d\mathbf{A}}{\mu} = \frac{i\, d\mathbf{s}}{4\pi} \left\{ \mathbf{r}_1 \times \nabla \left(\frac{1}{r} \right) \right\} = \frac{i\, d\mathbf{s} \times \mathbf{r}_1}{4\pi r^2},$$ (7.4)

where $d\mathbf{H}$ = magnetic field intensity at P, i = current, $d\mathbf{s}$ = conductor element carrying current, r = distance from $d\mathbf{s}$ to P, θ = angle between $\mathbf{r}$ and $d\mathbf{s}$, $\mathbf{r}_1$ = unit vector along $\mathbf{r}$.

The element of field is, of course, perpendicular to the plane containing $\mathbf{r}$ and $d\mathbf{s}$, while the direction is out of the paper in fig. 7.1.

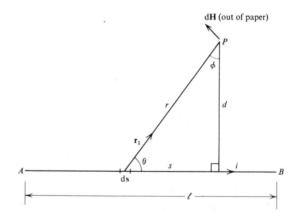

Fig. 7.1 Illustrating the Biot-Savart law.

(c) *Field due to straight wire*. Referring to fig. 7.1 again, we can calculate the total field at P for a length of straight wire between A and B, say ℓ. From the diagram,

$$r = d \sec \phi, \quad s = d \tan \phi, \quad ds = d \sec^2 \phi\, d\phi,$$

hence,

$$dH = \frac{id \sec^2 \phi \sin \theta\, d\phi}{4\pi d^2 \sec^2 \phi} = \frac{i}{4\pi d} \cos \phi\, d\phi.$$

Thus,

$$H = \frac{i}{4\pi d} \int_{\phi_1}^{\phi_2} \cos \phi \, d\phi = \frac{i}{4\pi d} (\sin \phi_2 - \sin \phi_1). \qquad (7.5a)$$

If the wire is very long, $\ell \gg d$, $\phi_2 = \pi/2$, $\phi_1 = -\pi/2$, and $H = i/2\pi d$. $\qquad (7.5b)$

(d) *Field of a rectangular loop.* By a simple extension of the above, we can obtain the field of a rectangular layout, which is illustrated in fig. 7.2.

Inside the rectangle, the field due to the segment AE is given by

$$H(AE) = \frac{i}{4\pi} \frac{\sin \phi_1}{EP} = \frac{i}{4\pi} \left(\frac{HP}{EP}\right)\left(\frac{1}{r_1}\right).$$

In the same way, for the segments EB, DG and GC, we obtain the following expressions for the fields:

$$\frac{i}{4\pi} \left(\frac{FP}{EP}\right)\left(\frac{1}{r_2}\right), \quad \frac{i}{4\pi} \left(\frac{HP}{GP}\right)\left(\frac{1}{r_4}\right), \quad \text{and} \quad \frac{i}{4\pi} \left(\frac{FP}{GP}\right)\left(\frac{1}{r_3}\right), \text{ respectively.}$$

Similar expressions hold for the fields due to the sides AD and BC. Adding to get the total field, we obtain the formula

$$H_p = \frac{i}{4\pi} \left(\frac{EP^2 + HP^2}{r_1 . HP . EP} + \frac{FP^2 + EP^2}{r_2 . EP . FP} + \frac{FP^2 + GP^2}{r_3 . GP . FP} + \frac{HP^2 + GP^2}{r_4 . GP . HP}\right),$$

$$= \frac{i}{4\pi} \left(\frac{r_1}{A_1} + \frac{r_2}{A_2} + \frac{r_3}{A_3} + \frac{r_4}{A_4}\right), \qquad (7.6a)$$

where A_1, A_2, A_3 and A_4 are the areas of the minor rectangles making up the loop $ABCD$.

Outside the rectangle, by the same analysis we obtain

$$H'_p = \frac{i}{4\pi} \left(\frac{r_1'}{A_1'} + \frac{r_2'}{A_2'} - \frac{r_3'}{A_3'} - \frac{r_4'}{A_4'}\right). \qquad (7.6b)$$

Within the loop the field varies about 40% over a rectangle concentric with $ABCD$ and one quarter the area, being a minimum at the centre. Outside the loop, the field will be mainly determined by the near side of the rectangle, provided the distance from it is considerably less than the distance to the far side.

It is important to note that the field in the neighbourhood of the long wire, or outside a large rectangle and relatively close to one side, falls off inversely as the distance, a relatively slow decrease in intensity.

Both the long wire with the ends grounded and the large rectangle or square have been extensively used for generating EM primary fields in systems such as the Bieler-Watson, compensator and Turam. The dimensions are generally 1000–5000 ft.

(e) *Field of small square loop.* If the rectangle is relatively small, or if the point P'

is at a great distance in fig. 7.2, the field is modified considerably. Consider the situation in fig. 7.3 in which a small square of side a has current flowing in it. The field at a distant point P, due to the near side, from eq. (7.5a) is

$$H_1 = \frac{i}{4\pi(y - a/2)}(\sin \phi_2 - \sin \phi_1) =$$

$$= \frac{i}{4\pi(y - a/2)} 2 \sin \left(\frac{\phi_2 - \phi_1}{2}\right) \cos \left(\frac{\phi_2 + \phi_1}{2}\right).$$

Thus,

$$H_1 \approx \frac{2i \sin \Delta\phi \cos \phi}{4\pi(y - a/2)},$$

where

$$\phi_2 - \phi_1 = 2\Delta\phi.$$

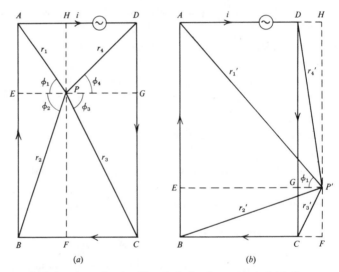

(a) $\qquad\qquad\qquad\qquad$ (b)

Fig. 7.2 Geometrical parameters involved in calculating the magnetic field of a rectangular loop. (a) Calculating the field at an internal point; (b) calculating the field at an external point.

We also know that

$$2 \sin \Delta\phi \approx 2\Delta\phi \approx \frac{2\delta}{\sqrt{\{x^2 + (y - a/2)^2\}}} \approx \frac{a \cos \phi}{\sqrt{\{x^2 + (y - a/2)^2\}}},$$

$$\cos \phi = \frac{y - a/2}{\sqrt{\{x^2 + (y - a/2)^2\}}}.$$

Therefore,

$$H_1 \approx \frac{ia(y - a/2)}{4\pi\{x^2 + (y - a/2)^2\}^{\frac{3}{2}}}.$$

Similarly, for the other three sides we have

$$H_2 = \frac{ia(x - a/2)}{4\pi\{(x - a/2)^2 + y^2\}^{\frac{3}{2}}},$$

$$H_3 = \frac{-ia(y + a/2)}{4\pi\{x^2 + (y + a/2)^2\}^{\frac{3}{2}}}, \quad H_4 = \frac{-ia(x + a/2)}{4\pi\{(x + a/2)^2 + y^2\}^{\frac{3}{2}}}.$$

Now if $r \gg a$, then $y \approx r \cos \phi$, $x \approx r \sin \phi$, $r^2 \approx x^2 + y^2$ and the total field is the sum of the four parts; thus, we have finally

$$H_p \approx \frac{ia}{4\pi r^3} \left\{ \frac{y - a/2}{\{1 - (a/r) \cos \phi + a^2/4r^2\}^{\frac{3}{2}}} - \frac{y + a/2}{\{1 + (a/r) \cos \phi + a^2/4r^2\}^{\frac{3}{2}}} + \right.$$

$$\left. + \frac{x - a/2}{\{1 - (a/r) \sin \phi + a^2/4r^2\}^{\frac{3}{2}}} - \frac{x + a/2}{\{1 + (a/r) \sin \phi + a^2/4r^2\}^{\frac{3}{2}}} \right\},$$

$$\approx -\frac{ia^2}{2\pi r^3}, \quad (7.7)$$

neglecting terms in a/r in comparison with unity.

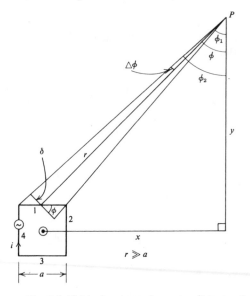

Fig. 7.3 Field of a square loop at a distant point.

(f) *Field of a small circular loop (a.c. magnetic dipole).* By using the vector potential we can calculate the field at a point in the neighbourhood of the loop, not necessarily in its plane or on the axis. The cylindrical coordinate system is shown in fig. 7.4. The cross-section of the winding on the loop is assumed to be very small with respect to its radius. Since the current is confined to a circular path there is only one component of $\mathbf{A}$, A_ϕ. Then $dv = a\,d\phi\,dr\,dz$, $J_\phi = i/dr\,dz$ and eq. (7.3) simplifies to

$$A_\phi = \frac{\mu}{4\pi} \int_{\text{vol}} \frac{\cos\phi\,J_\phi(r)\,dv}{R} = \frac{\mu i}{4\pi} \int_0^{2\pi} \frac{a\cos\phi\,d\phi}{\sqrt{(r^2 + z^2 + a^2 - 2ar\cos\phi)}}.$$

Now if we assume that $r^2 + z^2 \gg a^2$ and expand the denominator, we obtain

$$A_\phi = \frac{\mu i}{4\pi} \int_0^{2\pi} \frac{a\cos\phi}{\sqrt{(r^2 + z^2)}} \left\{ 1 - \frac{a^2}{2(r^2 + z^2)} + \frac{ar\cos\phi}{(r^2 + z^2)} \cdots \right\} d\phi,$$

$$\approx \frac{\mu i\,a^2 r}{4(r^2 + z^2)^{\frac{3}{2}}}. \quad (7.8)$$

Now we can get the magnetic field, since

$$\mathbf{H} = \frac{\nabla \times \mathbf{A}}{\mu} = \frac{ia^2}{4} \left\{ -\frac{\partial A_\phi}{\partial z}\mathbf{i}_r + \frac{1}{r}\frac{\partial(rA_\phi)}{\partial r}\mathbf{i}_z \right\}$$

$$= \frac{ia^2}{4(r^2 + z^2)^{\frac{5}{2}}} \{3rz\mathbf{i}_r + (2z^2 - r^2)\mathbf{i}_z\}, \quad (7.9a)$$

where $\mathbf{i}_r$, $\mathbf{i}_z$ are unit vectors along the r- and z-axes. Thus the magnetic field has two components, one in the r-, the other in the z-direction:

$$H_r = \frac{3ia^2 rz}{4(r^2 + z^2)^{\frac{5}{2}}} \quad \text{and} \quad H_z = \frac{ia^2(2z^2 - r^2)}{4(r^2 + z^2)^{\frac{5}{2}}}. \quad (7.9b)$$

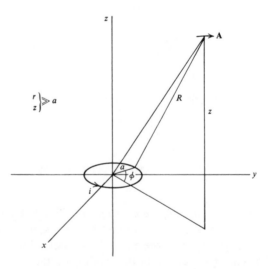

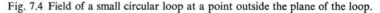

Fig. 7.4 Field of a small circular loop at a point outside the plane of the loop.

There are two important particular cases for these magnetic field components:

(i) $r = 0$ and we get for the field on the z-axis

$$H_z)_{r=0} = \frac{ia^2}{2z^3}, \qquad (7.9c)$$

($z \geqslant 7a$ for error $\leqslant 3\%$);

(ii) $z = 0$ and we get for the field outside the coil in the same plane

$$H_z)_{z=0} = -\frac{ia^2}{4r^3}, \qquad (7.9d)$$

($r \geqslant 7a$ for error $\leqslant 3\%$).

Both of these relations are approximations, since we have assumed that a^2 is small compared to $r^2 + z^2$. The exact expression for H_z when $r = 0$ is

$$H_z = \frac{ia^2}{2(z^2 + a^2)^{\frac{3}{2}}}.$$

When $z = 0$, this becomes

$$H_z)_{r=z=0} = \frac{i}{2a}. \qquad (7.9e)$$

Obviously this small coil is equivalent to an alternating magnetic dipole lying in the z-axis at the origin whose magnetic moment is given by

$$m = i\pi a^2. \qquad (7.10)$$

Formulae (7.9b), (7.9c) resemble the Gauss-*A* and Gauss-*B* positions (§3.2.2(a)). Equation (7.9d) is also quite similar to eq. (7.7), indicating that the shape of the current loop is not particularly significant for fields at a distance.

The electric field or current density could be derived from **A** by the relation in eq. (7.1b). The current i is, of course, sinusoidal and it is apparent that **J** and **E** have only one component, J_ϕ and E_ϕ. Thus the current flow is circular about the z-axis.

The dipole field intensity, then, falls off with the inverse cube of the distance; compared to the long-wire systems, the intensity decreases much faster. However, it should be kept in mind that both are approximations. The straight wire is assumed to be considerably longer than the perpendicular distance to the station at which the field is measured, while the closed-loop dimensions are considerably smaller than its distance from the field station.

In all the above magnetic field sources we have also assumed only one turn of wire. If there are n turns, the calculated fields are increased by a factor of n in all cases.

(g) *Field of a vertical straight wire (a.c. electric dipole).* One other artificial source of EM waves which has recently been employed in prospecting should be considered here. This is the high power VLF (very low frequency) transmission in the range 5–25 kHz, which is normally used for air and marine navigation. There are, of course, many RF stations available as well, but the frequencies are considerably higher (500 kHz and up) and the power lower than the VLF sources, so that the range and depth of penetration are limited.

The VLF antenna is effectively a grounded vertical wire, several hundred feet high. Consequently it is much shorter than a transmission wavelength which, for a frequency of 20 kHz, is 15 km. Whereas the small loop is equivalent to a magnetic dipole, the short wire behaves as an electric dipole.

There are several possible modes of radiation from this type of antenna, but in the low frequency range and at distances considerably greater than a wavelength, the propagation is a combination of ground wave and sky wave. The former travels over the earth's surface, while the sky wave is refracted and reflected by the ionized layers in the upper atmosphere (50 km and higher) to return to ground. At large distances from the antenna the VLF waves appear to be propagated in the space between the spherical reflecting shells formed by the earth surface and the lower ionosphere. The attenuation is comparatively small in both surfaces.

From fig. 7.5 it is apparent that the electric dipole is quite like the magnetic dipole if we interchange the E- and H-components of the wave. That is, there are two components of electric field, E_r and E_θ in spherical coordinates, and one magnetic component H_ϕ in the azimuth. The expressions for these components, which can be derived from the vector potential $\mathbf{A}$ in much the same manner as was done for the magnetic dipole, are given by the following:

$$H_\phi = \frac{i\delta\ell}{4\pi} e^{j\omega(t-r/c)} \left(\frac{j\omega}{rc} + \frac{1}{r^2}\right) \sin\theta, \tag{7.11a}$$

$$E_\theta = \frac{i\delta\ell}{4\pi} e^{j\omega(t-r/c)} \left(\frac{j\omega}{rc^2\varepsilon} + \frac{1}{r^2c\varepsilon} - \frac{j}{r^3\omega\varepsilon}\right) \sin\theta, \tag{7.11b}$$

$$E_r = \frac{i\delta\ell}{2\pi} e^{j\omega(t-r/c)} \left(\frac{1}{r^2c\varepsilon} - \frac{j}{r^3\omega\varepsilon}\right) \cos\theta, \tag{7.11c}$$

where $c = 1/\sqrt{(\mu\varepsilon)} = 3 \times 10^8$ m/sec.

At sufficiently large distances from the antenna only the so-called *radiation* or *far fields* are significant, since the terms in $1/r^3$ and $1/r^2$ are negligible. The expressions then simplify to

$$H_\phi = \frac{j\omega i\delta\ell}{4\pi cr} e^{j\omega(t-r/c)} \sin\theta, \tag{7.12a}$$

$$E_\theta = \frac{j\omega i\delta\ell}{4\pi c^2\varepsilon r} e^{j\omega(t-r/c)} \sin\theta, \tag{7.12b}$$

$$E_r \approx 0. \tag{7.12c}$$

The electric field component is practically vertical. The magnetic field lines are horizontal circles concentric about the antenna; at distances of several hundred miles the field is practically uniform over, say, a square mile of ground surface and is at right angles to the station direction.

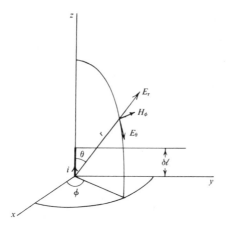

Fig. 7.5 Field produced by a short wire at a distant point.

The attenuation of the ground wave at VLF frequencies is small enough that, for distances greater than several hundred miles, it is reasonable to assume that the field varies as $1/r$. This fact, coupled with the large power output of the stations (500–1000 kW), makes it possible to use VLF transmitters as EM sources at distances of 2000–3000 miles.

Figure 7.6 compares the magnitude of primary fields from the various sources described in this section: VLF transmitter, long wire, square loop and circular coil. These were calculated from eqs. (7.5a), (7.7), (7.9d) and a modification of (7.12a), given below, which allows for attenuation:

$$H = \frac{i\delta\ell}{2\lambda r} e^{-1\cdot5 \times 10^{-3} r/\sqrt{\lambda}}, \qquad (7.12d)$$

where λ, r, $\delta\ell$ are in kilometres. The curves are drawn for the following parameters:

(1) VLF transmitter power $= 10^6$ W, frequency $= 20$ kHz, antenna current $= 5000$ A, height $= 300$ ft;
(2) long-wire power $= 1000$ W, current $= 3$ A, length $= 4000$ ft;
(3) square-loop power $= 300$ W, current $= 3$ A, section area $= 36$ ft^2, turns of wire $= 100$;
(4) circular-loop power $= 5$ W, current $= 100$ mA, diameter $= 3$ ft, turns of wire $= 1000$.

These values correspond roughly to (1) VLF, (2) Turam, (3) vertical-loop dip-angle, and (4) horizontal-loop systems used in EM field work.

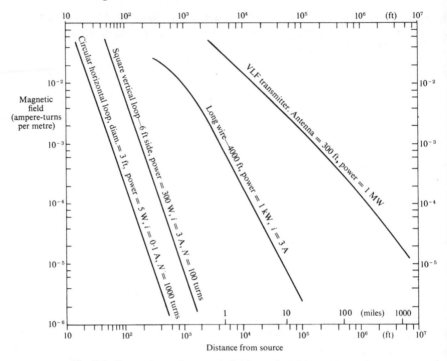

Fig. 7.6 Comparison of magnetic fields produced by various sources.

7.3 Combination of EM fields

7.3.1 *General*

So far we have described the propagation, attenuation and generation of alternating magnetic fields. We have seen that such fields can be initiated by various current configurations and attenuated more or less depending on their frequency and the conductivity (and permeability) of the medium through which they travel.

In §6.2.3 it was also noted that the EM field was shifted in phase on encountering a relatively good conductor. In fact this conductor becomes the source of a secondary field, which differs in phase from the primary field, while having the same frequency. Hence a suitable detector in the vicinity will be energized by both the primary and secondary fields simultaneously. The existence of the secondary field, indicating the presence of a subsurface conductor, may be established with respect to the primary field by a change of amplitude and/or phase in the normal detector signal. Some EM systems measure both quantities, some respond to one or the other.

7.3.2 *Amplitude and phase relations*

The character of the secondary magnetic field is best illustrated by a consideration of the coupling between a.c. circuits. We assume a trio of coils having inductance and resistance and negligible capacitance; the first is the primary source, the second is equivalent to the conductor, while the third is the detector (see fig. 7.7). The primary EM field at a point near the conductor (coil 2), resulting from a current i_p flowing in the first coil, is given by

$$H_p = Ki_p = KI_p \sin \omega t,$$

where K depends upon the geometry of the system, the area and number of turns of the primary coil and attenuation of the wave.

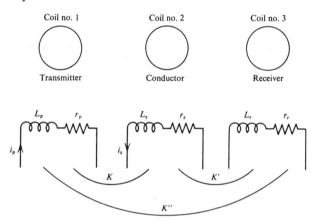

Fig. 7.7 Electric-circuit analogy for EM-system.

As a result of this field, coil 2 has an e.m.f. induced in it which lags behind the primary field by $\pi/2$. This is stated in the relation

$$e_s = - M\frac{di_p}{dt} = \frac{-j\omega M H_p}{K},$$

where M is the *mutual inductance* between coils 1 and 2. Then the current flowing in coil 2 will be

$$i_s = e_s/z_s = e_s/(r_s + j\omega L_s),$$

where $z_s = (r_s + j\omega L_s)$ is the effective impedance of the conductor of resistance r_s and inductance L_s. The secondary field near the detector (coil 3) as a result of this current will be

$$H_s = K'i_s = \frac{-K'j\omega M H_p}{K(r_s + j\omega L_s)} = \frac{-K'M H_p(j\omega r_s + \omega^2 L_s)}{K(r_s^2 + \omega^2 L_s^2)} = \frac{-K'M H_p(Q^2 + jQ)}{KL_s(1 + Q^2)},$$

where $Q = \omega L_s/r_s$ and K' is a constant similar to K.

The primary field at the detector coil will be

$$H_p' = \mathbf{K}''i_p = K''I_p \sin \omega t = K''H_p/K,$$

where K'' is similar to K and K'.

Thus the relative magnitude of the fields at the detector is

$$\left|\frac{H_s}{H_p'}\right| = \frac{K'i_s}{K''i_p} = \frac{K'M}{K''L_s}\left\{\frac{Q^4}{(1+Q^2)^2} + \frac{Q^2}{(1+Q^2)^2}\right\}^{\frac{1}{2}} = \frac{K'M}{K''L_s}\frac{1}{\sqrt{(1+1/Q^2)}}. \quad (7.13)$$

Since the ratio $K'M/K''L_s$ is generally very small, the ratio H_s/H_p' is small, regardless of the value of Q.

The phase difference between primary and secondary fields is

$$\theta_p - \theta_s = \left(\frac{\pi}{2} + \tan^{-1}\frac{\omega L_s}{r_s}\right) = (\pi/2 + \phi), \quad (7.14)$$

where $\tan \phi = \omega L_s/r_s$. The lag in phase of $\pi/2$ is due to the inductive coupling between coils 1 and 2, while the additional phase lag ϕ is determined by the properties of the conductor as an electrical circuit. That is,

$$H_s = K'I_s \sin\{\omega t - (\pi/2 + \phi)\} = -K'I_s \cos(\omega t - \phi). \quad (7.15)$$

The phase shift is most clearly illustrated by the vector diagram in fig. 7.8. (In this diagram the magnitude of H_s with respect to H_p is greatly exaggerated.) The resultant of H_p and H_s is H_r.

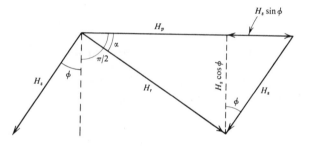

Fig. 7.8 Vector diagram showing phase shift between H_s and H_p.

From this diagram and equation (7.14) it can be seen that when we have a very good conductor, $Q = \omega L_s/r_s \to \infty$ and $\phi \to \pi/2$. In this case the phase of the secondary field is practically 180° (π) behind the primary field. For a very poor conductor $\omega L_s/r_s \to 0$ and $\phi \to 0$; the secondary field lags $\pi/2$ behind the primary. Generally H_s is somewhere between $\pi/2$ and π (90° and 180°) out-of-phase with H_p.

The component of H_s 180° out-of-phase with H_p is $H_s \sin \phi$, while the component 90° out-of-phase is $H_s \cos \phi$. In EM parlance, the 180° out-of-phase fraction of H_s is called the *real* or *in-phase* component. The 90° out-of-phase fraction of H_s is called the *imaginary*, *out-phase* or *quadrature* component. These terms originated in a.c. circuit theory and in fact there is nothing imaginary about the quadrature component.

7.3.3 *Elliptic polarization*

The detector in an EM field system, generally a small coil with many turns of fine wire, measures the secondary field produced by a subsurface conductor, in the presence of the primary field. It is not possible to eliminate the primary field at the time and place of detection as is done, for instance, in pulsed radar systems. Consequently when conductors are present, the detected signal is a combination of the primary and one or more secondary fields. In general the combination is a magnetic field which is elliptically polarized. From the previous section we can write

$$H_p = A \sin \omega t, \quad \text{and} \quad H_s = B \cos(\omega t - \phi),$$

where A, B are functions of the geometry of the transmitter, conductor and detector. Since

$$\cos(\omega t - \phi) = \cos \omega t \cos \phi + \sin \omega t \sin \phi$$
$$= \surd(1 - H_p^2/A^2) \cos \phi + H_p \sin \phi/A = H_s/B,$$

we get

$$\frac{H_p^2}{A^2} + \frac{H_s^2}{B^2} - \frac{2H_p H_s \sin \phi}{AB} = \cos^2 \phi,$$

that is,

$$\frac{H_p^2}{A^2 \cos^2 \phi} + \frac{H_s^2}{B^2 \cos^2 \phi} - \frac{2H_p H_s \sin \phi}{AB \cos^2 \phi} = 1. \tag{7.16}$$

This equation is of the form

$$Lz^2 - 2Mxz + Nx^2 = 1,$$

which is the equation of an ellipse.

We have made two simplifying assumptions in obtaining this equation. The first is that H_p and H_s are orthogonal in space, which is not generally true. However, if the angle between H_p and H_s is $\alpha \neq \pi/2$, these vectors may be resolved in two orthogonal components, say

$$H_z = H_p + H_s \cos \alpha \quad \text{and} \quad H_x = H_s \sin \alpha$$

in which case the expression for H_z and H_x is more complicated and has constant terms, but is still the equation of an ellipse.

The second assumption was that H_s is due to the current in only one conductor. This is not necessarily the case; however, a combination of vectors of different amplitudes, directions and phases, resulting from currents in several conductors, can be resolved into a single resultant for H_s.

Consequently the superposition of fields produces a single field which is elliptically polarized, the vector being finite at all times (although H_p and H_s become zero at $\omega t = n\pi$ and $(2n + 1)\pi/2$, respectively) and rotating in space with continuous amplitude change. Its extremity sweeps out an ellipse.

This ellipse may lie in any space plane, although the plane will normally be only slightly tilted off horizontal or vertical. This is because the major axis of the ellipse is determined by H_p – since it is usually much larger than H_s – and the primary field is normally either horizontal or vertical in EM systems.

There are two special cases of eq. (7.16) of considerable importance. Figure 7.8 is again helpful in visualizing these situations.

(i) $\phi = \pi/2$. Equation (7.16) then reduces to

$$\left(\frac{H_p}{A} - \frac{H_s}{B}\right)^2 = 0, \quad \text{or} \quad BH_p - AH_s = 0,$$

which is a straight line through the origin of the coordinates, having a slope $+B/A$. This case corresponds to a very good conductor, since

$$\phi = \tan^{-1}\omega L_s/r_s = \pi/2 \quad \text{or, since } \tan\phi = \infty, \quad r_s = 0.$$

The ellipse of polarization has collapsed into a straight line.

(ii) $\phi = 0$. The ellipse equation simplifies to

$$\frac{H_p^{\,2}}{A^2} + \frac{H_s^{\,2}}{B^2} = 1,$$

which signifies a poor conductor since $r_s \gg \omega L_s$ when $\phi = 0$. In the unlikely event that $A = B$ as well, the combination of H_p and H_s results in circular polarization.

Obviously a detector coil can always be oriented so that it lies in the plane of polarization, when a true null signal would be obtained. Some of the early EM methods were based on this fact; the dip and azimuth of the polarization ellipse and its major and minor axes were measured.

On the other hand, if the detector coil is rotated about its vertical or horizontal diameter it will not always be possible to find a perfect null position, because the plane of the coil will not, in general, coincide with the plane of the ellipse. There will, however, be a minimum signal at one coil orientation.

7.3.4 *Mutual inductance*

(a) *General theory.* It was noted in the discussion of phase shift that the inductive coupling between electrical circuits is proportional to the coefficient of mutual inductance M. This parameter can be used to some effect in determining the signal amplitude at the receiver due to both the transmitter and conductor. It has already been employed to estimate the current induced in the conductor as the result of the primary field.

If we can simulate the transmitter and receiver coils and the conductor by simple electric circuits, it may be possible to calculate the mutual inductances which couple them. Consider the coil system illustrated in fig. 7.9, in which the mutual inductances between transmitter and conductor, conductor and receiver and transmitter and receiver are respectively M_{TC}, M_{CR} and M_{TR}.

It was shown previously that the current i_s induced in the conductor by the transmitter is related to the transmitter current by the expression

$$i_s = -\frac{j\omega M_{TC} i_p}{r_s + j\omega L_s} = -\frac{M_{TC}}{L_s}\frac{(Q^2 + jQ)i_p}{1 + Q^2},$$

where $Q = \omega L_s/r_s$ was a figure of merit for the conductor circuit.

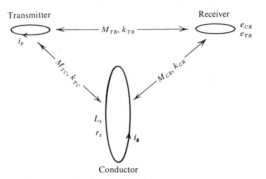

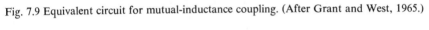

Fig. 7.9 Equivalent circuit for mutual-inductance coupling. (After Grant and West, 1965.)

Current i_s, in turn, will induce an e.m.f. in the receiver coil given by

$$e_{CR} = -j\omega M_{CR} \cdot i_s = \frac{j\omega M_{CR} M_{TC}}{L_s}\frac{(Q^2 + jQ)i_p}{1 + Q^2}.$$

At the same time the primary field induces an e.m.f. in the receiver

$$e_{TR} = -j\omega M_{TR} i_p.$$

Since the secondary or anomalous field is measured in the presence of the primary field, we have

$$\frac{e_{CR}}{e_{TR}} = -\frac{M_{CR} M_{TC}}{M_{TR} L_s}\left(\frac{Q^2 + jQ}{1 + Q^2}\right). \tag{7.17a}$$

Mutual inductance may also be written in terms of the coupled self-inductances:

$$M_{CR} = k_{CR}\sqrt{(L_s L_R)}, \quad M_{TC} = k_{TC}\sqrt{(L_s L_T)}, \quad M_{TR} = k_{TR}\sqrt{(L_T L_R)},$$

where k_{CR} = coupling coefficient between conductor and receiver, etc., L_T = inductance of transmitter loop, L_R = inductance of receiver loop. Then we can write (7.17a) in the form

$$\frac{e_{CR}}{e_{TR}} = -\frac{k_{CR} k_{TC}}{k_{TR}}\left(\frac{Q^2 + jQ}{1 + Q^2}\right). \tag{7.17b}$$

Although this eliminates L_s in the first part of the expression, it does not simplify the first ratio much because the k-values, like the M-values, involve complicated geometry of the system. Equation (7.17b) does indicate, however, that this ratio, sometimes called the *coupling parameter*, is usually a very small quantity since k_{TR} will tend to be much larger than the two coefficients in the numerator. That is to say, the transmitter and receiver are coupled through air, which means the attenuation is practically zero. In some EM field layouts, the source and detector coils are purposely oriented to reduce the direct coupling, while in others the decoupling is accomplished by electrical means.

In any case the mutual inductances between the various components within the range of the magnetic field are a controlling factor in EM systems. Consequently it would be useful to determine M for simple geometrical configurations which simulate field situations, as an aid in interpretation.

The mutual inductance M_{12} between two circuits 1 and 2 is defined as the total flux Φ_{12} through circuit 1, produced by unit current in circuit 2. Thus,

$$M_{12} = \frac{\Phi_{12}}{i_2} = \int_{S_1} \frac{\mathbf{B}_2 \cdot \mathrm{d}\mathbf{s}_1}{i_2} = \int_{S_1} \frac{\nabla \times \mathbf{A}_2 \cdot \mathrm{d}\mathbf{s}_1}{i_2}.$$

Now by Stokes' theorem,

$$\int_S \nabla \times \mathbf{A} \cdot \mathrm{d}\mathbf{s} = \oint \mathbf{A} \cdot \mathrm{d}\mathbf{s},$$

and from eq. (7.3b),

$$\mathbf{A} = \frac{\mu i}{4\pi} \oint \frac{\mathrm{d}\mathbf{s}}{r}.$$

This gives the general *Neumann formula* for mutual inductance:

$$M_{12} = \oint \mathbf{A}_2 \cdot \mathrm{d}\mathbf{s}_1 = \frac{\mu}{4\pi} \oint \oint \frac{\mathrm{d}\mathbf{s}_1 \cdot \mathrm{d}\mathbf{s}_2}{r} = \frac{\mu}{4\pi} \oint \oint \frac{\cos \theta}{r} \, \mathrm{d}s_1 \, \mathrm{d}s_2, \qquad (7.18a)$$

where $\mathrm{d}s_1$, $\mathrm{d}s_2$ = elements of length in circuits 1 and 2, r = distance between $\mathrm{d}s_1$ and $\mathrm{d}s_2$, θ = angle between $\mathrm{d}s_1$ and $\mathrm{d}s_2$.

In many cases it is necessary to integrate numerically to get a particular answer from this general formula. However, several simple circuits can be worked out exactly. These are illustrated in fig. 7.10. Units are microhenrys and meters.

(b) *Two parallel wires of equal length.* Let the length be ℓ, the distance apart s. Then,

$$M = 0 \cdot 2\ell[\log\{\ell/s + \sqrt{(1 + \ell^2/s^2)}\} - \sqrt{(1 + s^2/\ell^2)} + s/\ell]. \qquad (7.18b)$$

When $\ell/s > 1$, this can be modified to give

$$M = 0 \cdot 2\ell \left[\log\left(\frac{2\ell}{s}\right) - 1 + \frac{s}{\ell}\left\{1 - \left(\frac{s}{2\ell}\right)^2 + \frac{1}{2}\left(\frac{s}{2\ell}\right)^4 - \frac{2}{3}\left(\frac{s}{2\ell}\right)^6 \cdots \right\}\right].$$

There are various other expressions for straight conductors of unequal length, either parallel or making an angle with each other. These formulae are more complicated than the above. Several of them are given in Grover (1962).

(c) *Two coaxial circles.* Taking the radii as a and b, distance apart s, we can write

$$M = f(a, b, s), \tag{7.18c}$$

where the function f depends on the parameter k and

$$k^2 = \frac{(1 - b/a)^2 + s^2/a^2}{(1 + b/a)^2 + s^2/a^2}.$$

This function is usually tabulated or plotted in terms of k^2. When $s \gg a, b$ (10 times or more), the value M is approximately

$$M \approx \frac{0 \cdot 2\pi a^2 \pi b^2}{s^3} = \frac{0 \cdot 2AB}{s^3} \approx \frac{2a^2b^2}{s^3}, \tag{7.18d}$$

where A and B are the areas of the two circles.

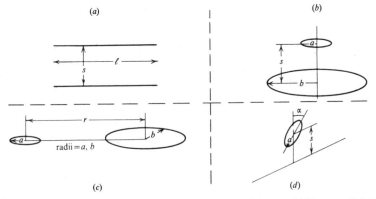

Fig. 7.10 Calculating mutual inductance of various geometrical figures. (*a*) Two parallel straight wires of equal length; (*b*) coaxial circles; (*c*) coplanar circles; (*d*) circle and long straight wire.

(d) *Two coplanar circles.* If the radii are a and b, distance between centres r, $r \gg a$, $r \gg b$, then

$$M \approx -\frac{0 \cdot 1\pi a^2 \pi b^2}{r^3} = -\frac{0 \cdot 1AB}{r^3} \approx -\frac{a^2b^2}{r^3}. \tag{7.18e}$$

It is also possible to obtain expressions for M where the circles are parallel but not coaxial, or when their axes are inclined to each other. However, these relations are quite cumbersome.

(e) *Circle and intersecting long straight wire.* Let a long straight wire intersect at right angles the extension of a diameter of the circle of radius a. If the plane of the circle makes an acute angle α with the plane through the diameter and the long wire, and if s is the distance between the wire and the centre of the circle, then

$$M = 4\pi \times 10^{-3}\{s \sec\alpha - \sqrt{(s^2 \sec^2\alpha - a^2)}\}, \tag{7.18f}$$

and when $s/a \geqslant 5$, this is simplified to

$$M \approx \frac{2\pi \times 10^{-3}a^2}{s \sec\alpha} = \frac{0.002A}{s \sec\alpha}. \tag{7.18g}$$

The self-inductances for the long straight wire and the circle are given by

$$L = 0.2\ell \left\{ \log\left(\frac{2\ell}{\rho}\right) - \frac{3}{4} \right\}, \tag{7.19a}$$

$$L \approx 4\pi \times 10^{-3}a \left\{ \log\left(\frac{8a}{\rho}\right) - 2 \right\}, \tag{7.19b}$$

respectively, where ρ is the radius of the wire, ℓ its length and a the radius of the circular loop. The self-inductance for a number of regular figures can be written

$$L = 0.2\ell_e \left\{ \log\left(\frac{4\ell_e}{\rho}\right) - \chi \right\}, \tag{7.19c}$$

where ℓ_e is the perimeter of the figure and χ is a constant which depends on the shape, being 2·45 for a circle, 2·85 for a square, etc.

Usually it is easy to determine the geometrical dimensions for regular figures like the loops used for transmitter and receiver. The problem is much more difficult when we attempt to simulate even the simplest shape of conductor by an equivalent electrical circuit. The main difficulty arises because we do not know the cross-section of current flow.

In all the above formulae, if circuit 1 has N_1 turns and circuit 2 has N_2 turns, the value of M is increased by N_1N_2; for a self inductance of N turns L increases by N^2.

(f) *Numerical example.* As an example of the magnitude of the coupling factor, $(M_{CR}M_{TC})/(M_{TR}L_s)$, in eq. (7.17a), consider the case of a horizontal-loop system (see §7.5.5d) straddling a long, thin vertical conducting sheet which outcrops at the surface. We assume that the transmitter and receiver coils are identical, the distance l between their centres being less than ℓ, the length of the sheet, and that the ratio $\ell/\rho \approx 2500$. The principal effect of the sheet is due to its upper edge; hence to a first approximation we can consider the sheet as a horizontal wire of length ℓ at the surface. Using eqs. (7.18e), (7.18g) and (7.19a), we get for the coupling factor in eq. (7.17a)

$$\frac{M_{TC}M_{CR}}{M_{TR}L_s} = \left(\frac{0.002A}{l/2}\right)\left(\frac{0.002A}{l/2}\right)\left(\frac{l^3}{0.1A^2}\right)\left(\frac{1}{0.2\ell\{\log(2\ell/\rho) - 3/4\}}\right)$$

$$\approx 10^{-4}(l/\ell) < 10^{-4}.$$

Thus the signal response has been reduced by the factor 10^4 or more.

7.3.5 *Conductor response*

Returning to eq. (7.17b), the second factor on the right-hand side depends only upon the conductor and the frequency (since $Q = \omega L_s / r_s$). In this situation Q is known as the *response parameter* of the conductor, while the complex ratio $(Q^2 + jQ)/(1 + Q^2)$ is called the *response function*.

Plotting the response function against Q, we get two curves resulting from the real and imaginary parts of the function:

$$A = Q^2/(1 + Q^2), \quad \text{and} \quad B = Q/(1 + Q^2),$$

where A and B are real. This plot is shown in fig. 7.11.

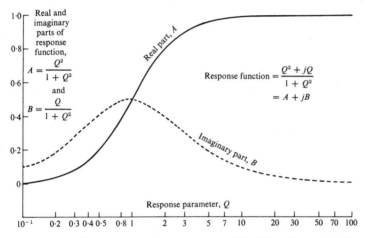

Fig. 7.11 Response function for a conductor in an a.c. field. (After Grant and West, 1965.)

When Q is very small, both real and imaginary parts of the function are very small. The ratio of secondary to primary response in the receiver will be

$$\frac{e_{CR}}{e_{TR}} \approx -\frac{jQk_{CR}k_{TC}}{k_{TR}}.$$

This is the case of a poor conductor. As Q increases, the imaginary part increases at a faster rate at first and its magnitude is larger than the real fraction until $Q = 1$ when they are both equal to 0·5. Beyond this point the imaginary part decreases until, at large values of Q, it is again zero. Meanwhile the value of A increases to an upper limit of unity when Q is large, or

$$\left| \frac{e_{CR}}{e_{TR}} \right| = \frac{k_{CR}k_{TC}}{k_{TR}},$$

which is the maximum value for a very good conductor.

When the value of Q is quite small the phase angle of this function is $\pi/2$. The angle increases to π for a very good conductor, that is, the secondary signal is opposed to the primary.

In the range $0 \leqslant Q \leqslant 1$ the imaginary or quadrature component is larger than the real component, while from $1 \leqslant Q \leqslant \infty$ the reverse is true. Thus the ratio of in-phase to quadrature components is somewhat diagnostic of the conductor. It is also clear that if one measures only the imaginary component in an EM system, a very good conductor will give a very poor response.

7.4 EM equipment

7.4.1 *General*

The measuring equipment for EM systems includes a local a.c. power source operating at one, or possibly two, frequencies, transmitter and receiver coils, receiver amplifier tuned to the transmitter frequency and an indicator such as headphones, meter, digital read-out or recorder. Some field sets require in addition an a.c. potentiometer (or phase and amplitude compensator) for comparison of primary and secondary field signals. There is very little difference in these components whether they are used for ground or airborne sets, except that the latter are more elaborate and bulky. Two field methods which properly come under the heading of EM, namely VLF and AFMAG, make use of remote power sources and consequently do not need a transmitter. One other, the INPUT or transient system, which employs a square-wave source, will also be considered.

7.4.2 *Power sources*

The power supply for EM transmitters is normally either a gas-driven alternator or a small, light battery-powered transistor oscillator with a power amplifier having a low impedance output. The choice between these depends on the type of field set, that is, whether the transmitter is only semi-portable or continually moving. In the long-wire, the large horizontal-loop and the vertical-loop fixed-transmitter systems the larger power source would be used, while for the various completely mobile transmitters such as horizontal-loop and vertical-loop broadside, the small unit is necessary.

Actual output from the semiportable power supplies varies between 250 and 2500 watts. The moving sources range from 1 to 10 watts. Weight of the equipment, of course, increases with the power output and may be anywhere from 5 to 250 lb.

The transmitter output, which is generally sinusoidal, is in the lower audio range, from 100 Hz to 5000 Hz. Usually the high power sets employ low frequencies, such as 400, 500, 870 and 1000 Hz. Some power sources are dual frequency, one low and one relatively high, for example 875 and 2200 Hz, 1000 and 5000 Hz, 400 and 2200 Hz. These are used in such systems as the Huntec horizontal-loop, McPhar dip-angle and INCO airborne sets. The advantage in using two frequencies is said to be a discrimination between shallow and deep conductors and/or an indication of the conductivity of the anomaly.

7.4.3 *Transmitter loops*

In order to generate the desired electromagnetic field, the output of the power source must be coupled to the ground by passing a current through some wire system. In the semiportable field sets this is done by coupling power into a long straight wire grounded at each end, a large (usually a single turn) rectangle or square laid out on the ground or a relatively large vertical loop supported on a tripod or hanging from a tree. In the first two arrangements the dimensions are of the order of a few thousand feet. The vertical loop, which may be single or multiple-turn winding, may be triangular, square or circular and of necessity has an area of only a few square metres. Some means must be provided for orienting this coil in any desired azimuth.

The completely mobile EM transmitters often employ multiple-turn coils (100 turns or more) wound on insulating frames of 1 metre diameter or less. Sometimes the coil may be merely a single turn of heavy conductor; in this case a matching transformer may be necessary in the source output. An alternative to this type of aircore coil, in which the wire is wound as a solenoid on a ferrite or other high permeability core, is now used in many EM systems. If the winding is distributed properly, it is possible to generate a field equivalent to a coil wound on a much larger frame, because the value of H increases with core permeability to compensate for the small area enclosed. Since the value of μ for certain ferrites is about 1000, the cross-section can be greatly reduced.

Many transmitter systems have a capacitance in series with the coil, the value being chosen to resonate approximately with the coil inductance at the source frequency. Because the coil has low resistance, this permits maximum current (within the limits of the power supply) to flow in the coil.

7.4.4 *Receiver coils*

These are always small enough to be entirely portable, that is, 3 ft diameter or less, and have many turns of fine wire. Coils with high-permeability cores are also used. The coil may be shunted by an appropriate capacitance to give parallel resonance at the source frequency. The resulting high impedance across the amplifier input acts as a bandpass filter to enhance the signal-to-noise ratio.

Receiver loops, and usually the portable transmitter loops as well, are electrostatically shielded to eliminate capacitive coupling between coil and ground and between coil and operator. The shielding is obtained by conductive paint or strips of conductive foil over the winding; the shield is then connected to the main circuit ground.

Generally the receiver coil must be oriented in a certain direction relative to the detected field. In some ground systems this merely means that the loop is maintained approximately horizontal or vertical. In others it is necessary to measure inclination and/or azimuth at each station. In airborne EM systems it is often a difficult problem to maintain a fixed orientation between transmitter and receiver coils.

7.4.5 *Receiver amplifiers*

Amplifiers are of fairly standard design, formerly using vacuum tubes but now transistorized to save weight, size and power. The overall voltage gain is usually between 10^4 and 10^5. One or more narrow bandpass filters, tuned to the source frequency or frequencies, are incorporated in the amplifier. A network of this type, called the *Twin-T*, which uses only resistors and capacitors, is illustrated in fig. 7.12. This circuit presents very high impedance at either end to a single frequency ω given by the formula in the diagram. When the Twin-T is connected across an amplifier stage, the feedback is practically zero at this frequency and increases rapidly either side of it. Sometimes band-reject filters, particularly for powerline frequencies, are included in the receiver system.

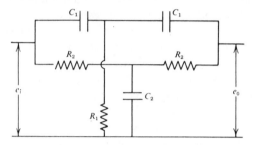

Fig. 7.12 Twin-T network for EM receivers. $\omega^2 = 1/(2C_1{}^2R_1R_2) = 2/(C_1C_2R_2{}^2)$. Generally $R_2 \approx 2R_1$, $C_2 \approx 2C_1$ as is usually the case for convenience in design.

7.4.6 *Indicators*

In many ground EM sets measurement is made by nulling or at least reducing the receiver signal to a minimum – sometimes by changing the orientation of the receiver loop, sometimes by manipulation of electrical components as in balancing an a.c. bridge. The indicator is normally a set of sensitive headphones on the output of the amplifier. Occasionally a suitable meter may be used, but in the audio range there is no particular advantage in the visual indicator. In any case, the significant parameters are noted when the minimum signal is obtained. These parameters might be the dip or azimuth angle of the receiver loop in degrees or per cent of a maximum value, or they could be the percentage change in amplitude and phase required to null the signal, this change being compared to a normal or background null.

Since the readings are continuous in airborne EM, a recorder of some type is required. These are generally of the strip chart type, although digital recording is now used in some up-to-date equipment.

7.4.7 *Compensating networks*

Unless the receiver loop is purposely oriented to minimize coupling of the primary wave, the secondary signal will be swamped by the primary, as we have seen from the numerical example in §7.3.4f. Since many EM systems, both ground and airborne, use transmitter-receiver loop geometry in which the coupling is maximum, it is necessary to reduce the primary signal by some other means. This is accomplished by introducing at the receiver input an artificial signal of the same frequency and amplitude but opposite in phase.

Compensation of this sort would be sufficient to permit measurement of an amplitude, that is, the real component, of the secondary field. However, from §7.3.2 and fig. 7.8 it is apparent that H_p and H_s generally differ in phase as well as amplitude and that, furthermore, this phase shift is diagnostic of the conductor. Thus some provision for changing phase should also be included in the compensator.

Referring again to fig. 7.8 we see that a vector equal in amplitude and opposite in direction to H_r will cancel the receiver signal. From the geometry of the vector triangle, when $H_s/H_p \ll 1$, it is easily shown that

$$H_r \approx H_p \left(1 - \frac{H_s}{H_p} \sin \phi + \frac{H_s^2}{2H_p^2} \right), \quad \text{and} \quad \sin \alpha \approx \frac{H_s}{H_p} \cos \phi.$$

Thus the required vector has approximately the same amplitude as H_p and is shifted in phase by $(\pi - \alpha)$. For a very good conductor,

$$H_r \approx (H_p - H_s), \quad \text{and} \quad \alpha \approx 0,$$

while for a very poor one,

$$H_r \approx H_p \left(1 + \frac{H_s^2}{2H_p^2} \right) \approx H_p, \quad \text{and} \quad \alpha \approx \frac{H_s}{H_p}.$$

The *compensator* (*ratiometer, a.c. potentiometer*) is an R-C or R-L-M network incorporated in the receiver section of the EM set. Three versions are illustrated in fig. 7.13. A reference voltage, usually taken from a small single-turn coil on or

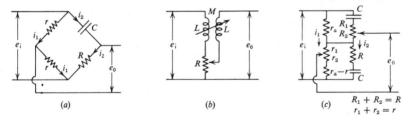

Fig. 7.13 Compensator networks. (a) R-C bridge; (b) Larsen compensator; (c) modified bridge.

near the transmitter loop, is changed in amplitude and phase and applied to the receiver input to cancel the main primary field and any other unwanted signals present. Having adjusted the compensator to give minimum signal in a barren or background region, we can measure relative amplitude and phase changes caused by secondary fields at other locations. This is done in airborne equipment by recording variations in separate amplitude and phase channels; with a ground set we note the change in amplitude (in-phase) and phase (quadrature, out-phase) controls necessary to restore the null.

The R-C bridge compensator shown in fig. 7.13a is easily analysed to show phase shift, since the input and output voltages are

$$e_i = i_1 2r = i_2(R + 1/j\omega C),$$
$$e_0 = i_1 r - i_2 R = -i_1 r + i_2/j\omega C,$$

and eliminating i_1, i_2 we get

$$\frac{e_0}{e_i} = \frac{1 - j\omega CR}{2(1 + j\omega CR)} = \frac{1 - \omega^2 C^2 R^2}{2(1 + \omega^2 C^2 R^2)} - \frac{2j\omega CR}{2(1 + \omega^2 C^2 R^2)}.$$

Thus, the phase angle between input and output is

$$\phi = \tan^{-1}\left(\frac{-2\omega CR}{1 - \omega^2 C^2 R^2}\right).$$

By varying R we can obtain practically $180°$ phase shift, since when $R = 0$, $\phi = \pi$; when $R = 1/\omega C$, $\phi = \pi/2$; and when $R \to \infty$, $\phi \to 0$.

Furthermore, the absolute value of e_0/e_i is always $\frac{1}{2}$, so that variations in R do not affect the relative amplitudes. Thus it is merely necessary to connect a potentiometer across the output to vary the amplitude.

In the second network in fig. 7.13b, known as the *Larsen compensator*, the phase shift is obtained by changing L in the variometer, the amplitude being controlled by varying the tap on R. The phase angle is approximately

$$\phi \approx \tan^{-1}\left(\frac{2\omega LR}{\omega^2 L^2 - R^2}\right),$$

where $M \approx L$, or $k \approx 1$.

Hence it is possible to vary ϕ by nearly $180°$ by changing L without affecting the amplitude.

The first of these networks has the advantage that it requires only resistance and capacitance, thereby saving space and weight. The compensator shown in fig. 7.13c is used in ground EM equipment. This is a refined version of the bridge in 7.13a.

7.5 EM field systems for ground surveys

7.5.1 *General*

A great variety of methods is available for EM fieldwork. These can be divided into ground and airborne systems and subdivided according to the actual measurement made, such as polarization ellipse, intensity and phase components and so on. There are in addition many techniques developed thirty or forty years ago which have gone out of style or have been superseded by improved versions; these will not be considered to any extent.

7.5.2 *Measurement of polarization ellipse*

As mentioned previously, a true null can always be obtained when the receiver loop lies in the plane of the polarization ellipse. To place the coil in this plane and keep it there requires a mounting which permits rotation in azimuth and dip, that is, about vertical and horizontal axes. The detector coil is rotated in azimuth until a minimum is found, then rotated through 90° from this azimuth and tilted until a null position is found. See fig. 7.14. However, not much information is obtained merely by locating this null plane. The only technique of this type which we will discuss is the Bieler-Watson method, shown schematically in fig. 7.15*a* (see also Bieler (1928), Watson (1931)).

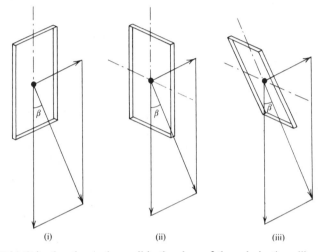

(i)   (ii)   (iii)

7.14 Orienting the receiver coil in the plane of the polarization ellipse.

The transmitter loop is a large rectangle, several thousand feet to the side. A long wire might also be used. The detector consists of two mutually perpendicular loops, one horizontal and one vertical, connected effectively in opposition through a transformer to amplifier and headphones. The vertical coil, with a fixed number of turns, is shunted by a capacitance to give parallel resonance at the source frequency and is connected through a reversing switch to the transformer primary. The horizontal coil has multiple taps for a variable inductance. The coils are about two feet in diameter and are mounted on a vertical post.

Although it would be possible to orient the vertical coil in the ellipse plane, this is troublesome and requires a more elaborate setup. Instead two readings are taken at each station, one with the vertical coil pointing N–S, the other E–W. Since the primary field is vertical, the vertical coil picks up no signal from it, while the horizontal coil is in maximum coupling position. With no secondary field present the null is obtained when there are zero turns switched into the horizontal coil and the reversing switch is in the proper sense to connect the vertical coil in series opposition to it. With a conductor in the vicinity, any horizontal field component is detected by the vertical coil. By varying the turns in the horizontal coil, this signal can be reduced to a minimum again.

The number of turns necessary to produce a minimum for each azimuth are then combined to give a vector whose amplitude and direction are an indication of the ellipse at the station. We have

$$\frac{H_h}{H_v} = kN_h,$$

where H_h, H_v are the horizontal and vertical components of the vector, k is a constant involving the impedances of the coils, N_h is the number of turns in the

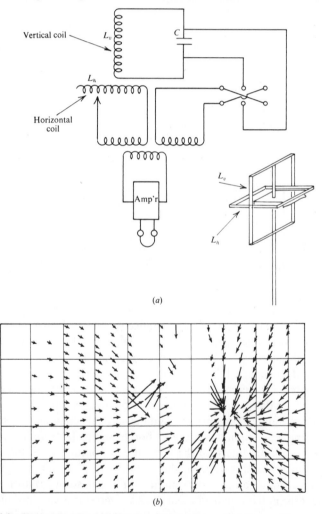

(a)

(b)

Fig. 7.15 Bieler-Watson method. (a) Schematic equipment diagram; (b) survey results – Aldermac Mine, Quebec.

horizontal coil. H_v can, of course, be calculated from the transmitter coil geometry. It is thus possible to determine the ratio of the in-phase field to the quadrature field.

In the presence of a good conductor, the minimum will be poor, since the polarization ellipse is not vertical. This is one of the disadvantages of the method, an inherent difficulty in this type of EM set.

Usually the resultant of H_h and H_v is plotted for the station reading, as illustrated in fig. 7.15b. Interpretation is mainly qualitative. The vectors point towards a conductor at stations near its edge, but their magnitudes are not particularly diagnostic of the relative conductivity. The Bieler-Watson technique is not used to any extent at present.

7.5.3 *Measurement of intensity*

Theoretically it is possible to measure the amplitude of any component of the EM field and compare it with the zero (background) value. The latter is determined by the station location and the orientation of the detector coil with respect to the transmitter. Practically this type of measurement is of little significance unless the background value is precisely known for each set-up; obviously the orientation of the detector coil would be extremely critical, as would small variations in the transmitter intensity in a dipole field, or errors in transmitter-receiver spacing.

With the long-wire or large horizontal-loop transmitter, however, this system is practical and formerly was quite popular, especially in Europe. It has now been replaced by more sophisticated techniques which provide more information. EM devices which measure only amplitude go under various names, such as the Swedish two-frame and long-wire systems.

In one of these, illustrated in fig. 7.16, traverses are carried out perpendicular to a long wire or normal to the longer side of a rectangular horizontal loop, maintaining the detector coil vertical with its axis parallel to the transmitter wire. Since the

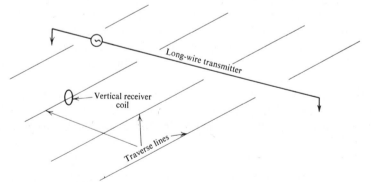

FIG. 7.16 Amplitude measurement with long-wire system. (Long wire may be replaced by a large rectangular loop.)

primary field is mainly vertical, the detector picks up only the horizontal component of the field parallel to the wire. Distortion in the normal pattern of this component indicates variation in subsurface conductivity.

Another version of this system is shown in fig. 7.17*a*, in which two receiver coils are maintained at a fixed spacing and moved on the same traverse line normal to the loop, one coil being closer to the transmitter than the other. The spacing may be 25–100 ft and the coils are initially horizontal and connected in series-opposition to the receiver amplifier. Normally the fields at the two coils are not equal and one of them is tilted to obtain a receiver minimum.

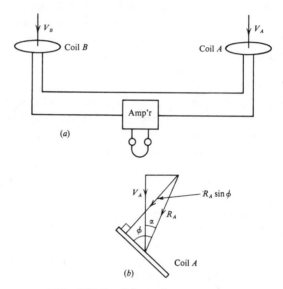

Fig. 7.17 Swedish two-frame system.

When the ground is uniform it should be possible to achieve this minimum merely by tilting the coil nearer to the transmitter wire about a horizontal axis normal to the traverse line. However, if there is a secondary field due to a conductor in the vicinity, it is preferable to find the polarization ellipse by orientation of the nearer coil A (see fig. 7.14), while coil B is disconnected from the receiver. If $\mathbf{R}_a$ is the total EM vector at A, making an angle α with the vertical, then the vertical field at A is $V_a = \mathbf{R}_a \cos \alpha$.

The two coils are now connected in opposition to the amplifier. The orientation of coil A will not generally give a minimum, since $\mathbf{R}_a$ is larger than V_b. Coil A is now rotated again about the same axis to find the minimum. At this position, as shown in fig. 7.17*b*,

$$V_b = \mathbf{R}_a \sin \phi = \frac{V_a \sin \phi}{\cos \alpha}, \quad \text{or} \quad \frac{V_a}{V_b} = \frac{\cos \alpha}{\sin \phi}.$$

In the single vertical-coil method, the variations in horizontal component, obtained with a meter detector, are plotted at the receiver station. For the two-coil system, the display of readings is more complicated. The ratio (V_a/V_b) is multiplied by $(x - \delta x/(x + \delta x)$, where x is the distance of the midpoint of the two coils from the transmitter wire and $2\delta x$ is the coil spacing. In homogeneous ground this would give a value of unity for each station, since the primary field falls off as $1/x$. Departures from unity indicate the presence of a conductor. This ratio is then plotted with the midpoint of the coils as the station location. A few profiles of this type are shown in fig. 7.45, §7.8.5e. They were obtained with model equipment, using a sheet conductor.

Both these field techniques have inherent drawbacks, the main one being that only limited information is obtained from the measurement. In the first set-up the effects of large-scale features, such as shear zones and faults, are accentuated at the expense of smaller conductors, which may in some cases be more attractive. Also the response for different shapes of conductor varies with distance from the transmitter coil.

The two-frame receiver is better than the vertical coil on the last two counts. However, the fieldwork is slower and more complicated and the minimum may be poorly defined in some cases.

Both of these systems have the advantage of large ground coverage and good sensitivity which is characteristic of the long-wire transmitter, but lack the additional information which is obtained by measuring the quadrature component as well. Detection is possible at stations as far away as one-half the long-wire length (perhaps out to 1000–2000 ft) and on traverses from one end to the other. With the horizontal loop, the detector may be moved either inside or outside the spread (see §7.2.2d), although the maximum area covered is less than with the long wire of the same dimension. On the other hand the long wire introduces conduction currents into the ground, which frequently complicate the interpretation, since the current flow at the ends of the wire always spreads laterally to some extent, as in resistivity work. The advantage in generating a field which decreases with the inverse first power of distance is partially offset by the decreased mobility of the transmitter.

7.5.4 Dip-angle measurements

(a) *General.* There are several field systems which measure, in effect, the direction of the combined primary and secondary fields at a receiver station. Whether they employ a natural or artificial source for the primary field and whether this source is fixed or movable, they all come under the heading of Dip-angle Measurements, because the tilt of the detector coil about a horizontal axis is recorded as the station reading. Furthermore, all the systems employ a primary field which is approximately horizontal. In one sense the two-frame method described in §7.5.3 is a dip-angle technique, but the primary field is vertical and the object is to compare two vertical fields rather than measure a resultant.

The dip-angle systems remain very popular in EM work, in spite of their limitations, primarily because the equipment is inexpensive, simple to operate and the technique is rapid and works quite well over steeply dipping sheet-like conductors, which are common geological features.

(b) *Fixed vertical-loop transmitter.*This is the oldest of the methods, developed in the twenties, and is still used quite widely. The transmitter coil, which may be square, triangular, or circular, usually has a few hundred turns with effective area of the order of 20 ft². The coil stands vertical and is free to rotate in the azimuth. The power source usually delivers several hundred watts.

The receiver coil, consisting of many turns of fine wire wound either on an open frame 1–2 ft in diameter or on a ferrite core, is connected to a tuned high-gain amplifier with headphones, or occasionally a meter, in the output. Provision is made for measuring the tilt angle of the coil.

Figure 7.18 shows the operating procedure. Traverses are made by moving the receiver along lines approximately normal to geologic strike. Station intervals are

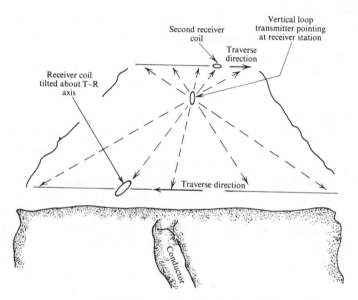

Fig. 7.18 Dip-angle system, fixed transmitter.

usually 50–200 ft. For each receiver setup, the transmitter coil is rotated to point at the receiver station (that is, its plane is in the transmitter-receiver line), either on a prearranged time schedule, or on receipt of a signal given by shouting or by walkie-talkie. The receiver is then tilted about the T-R axis for a minimum signal and the tilt angle recorded. Operations may be speeded up by having two receiver sets; in this event the second operator must occupy a station on the same axis at the same time as the first receiver (see fig. 7.18).

It is apparent that in the absence of conductors the minimum will be obtained with the receiver coil horizontal, since this represents zero-coupling geometry with respect to the transmitter coil. This would also be true when the receiver coil was directly over a rather narrow conductive zone, since the secondary field would have no vertical component. Characteristic profiles over this type of conductor are shown in figs. 7.32, 7.33, 7.42 and 7.43 (see §7.8.4b, §7.8.5a). The tilt angle either side of the conductor is such that the coil axis points towards the conductor until the receiver has moved a considerable distance away from it.

Range of receiver operation depends mainly on the size of the transmitter loop and power source. In practise the maximum T-R separation may be 600 to 1200 ft. There is also a minimum separation of about 200–400 ft; at smaller spacing it is difficult to obtain a minimum signal. Obviously there will be other situations as well when the minimum will be poorly defined, since there is no arrangement for balancing out the quadrature component.

The profiles in figs. 7.32, 7.33, etc., contain considerable information about the conductor. The *crossover point* (that is, the point where the dip angle changes sign) locates the top of the body, the slope near the crossover is an indication of its depth, as is the maximum dip angle, plus or minus. The symmetry of the profile is a clue to its dip, as can be seen in figs. 7.42 and 7.43.

For reconnaissance and ground follow-up of airborne EM, particularly where the strike is not known, the fixed-transmitter field procedure is modified. First the transmitter is set up roughly in the centre of the area of interest and pointed at successive receiver positions along the perimeter of the area. Dip angles are recorded at, say, 200 ft intervals in this fashion.

When a proper crossover (for discussion of this term see §7.8.5a and §7.8.6b) is found, the transmitter is moved to this station and dip angles measured on a traverse approximately across the centre of the area with the transmitter loop lined up on each station. If a second crossover is located the transmitter is moved to this station and the original crossover checked. Usually this point will be changed somewhat from the original perimeter crossover, unless the first transmitter location was fortuitously on top of the conductor.

When, as is often the case, more than one crossover has been found during reconnaissance, several interchanges of the transmitter and receiver may be necessary to establish the strike. Having clearly defined the latter, detailed dip-angle or other EM surveys may be carried out on suitable lines. If multiple conductors are present, usually the sharpest crossover (that is, the steepest slope or largest maximum tilt angle) will be obtained when the receiver is on the conductor nearest to, or directly below, the transmitter, since the coupling is then a maximum.

(c) *Broadside (parallel-line) method.* The transmitter is completely portable and is moved simultaneously with the receiver, the two moving along parallel lines. Readings are taken at intervals of 50–200 ft, with the transmitter coil pointed at the receiver for each station reading. The receiver coil, normally horizontal, is then rotated about the T-R axis to obtain a null. The transmitter-receiver line is maintained approximately parallel to geologic strike where possible, the T-R spacing being usually 400–600 ft.

This arrangement is shown in fig. 7.19, and typical profiles are given in figs. 7.34 and 7.35 (§7.8.4c). As in the fixed-transmitter method, two or more receivers can be used provided all are kept in line with the transmitter. The source power is inevitably lower than in the fixed-transmitter arrangement and is normally a battery-driven transistor or tube oscillator of 1–10 watts output.

Comparing the profiles of figs. 7.32 and 7.34, we see that the crossover is above the top of the conductor in both instances, while the slope of the curve near the crossover is somewhat steeper in fig. 7.34. The maximum dip angle is much more clearly defined by the parallel-line layout, since, as might be expected, the dip angle becomes zero again a relatively short distance off the conductor axis.

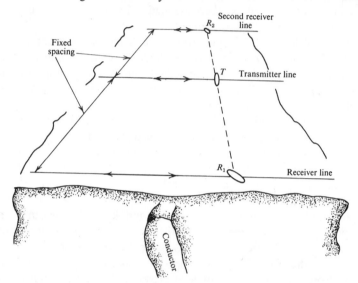

Fig. 7.19 Dip-angle system, parallel-line or broadside orientation.

A modification of this field method is often used in preliminary ground reconnaissance work. The transmitter and receiver are moved along the same traverse line with the transmitter coil pointing at the receiver station 200–400 ft away. Dip-angle measurements are recorded as described previously. The traverse lines are not usually perpendicular to strike, but rather at about 45°, since the purpose is merely to locate the conductor and perhaps get an idea of its extent. One of the standard dip-angle (or other) EM surveys is then carried out in detail.

(d) *Shoot-back method.* In very hilly terrain it is difficult to maintain correct alignment of the transmitter and receiver coils and as a result false dip angles are often obtained. The *shoot-back method* developed by Crone Geophysics, was intended to overcome this problem. The coil configuration resembles a modification of an early method of Mason (1927; see also Eve and Keys (1956, p. 176)) rather than the present dip-angle sets.

The field procedure is the same as that described above for reconnaissance with the portable transmitter. However, the shoot-back system requires a receiver and transmitter at each station; for this purpose the coils are convertible. The spacing

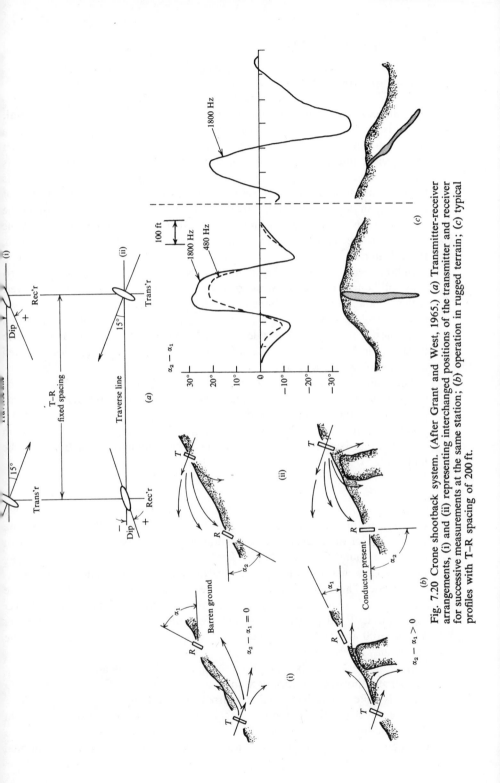

Fig. 7.20 Crone shootback system. (After Grant and West, 1965.) (a) Transmitter-receiver arrangements, (i) and (ii) representing interchanged positions of the transmitter and receiver for successive measurements at the same station; (b) operation in rugged terrain; (c) typical profiles with T–R spacing of 200 ft.

is usually 200 ft and the axis, rather than the plane, of the transmitter coil is pointed towards the receiver station.

Consider the situation shown schematically in fig. 7.20a. With unit 1 transmitting, the axis of coil 1 is directed towards coil 2 but dipping 15° below horizontal. Coil 2 is then rotated about a horizontal axis normal to the traverse line to get a minimum. Then a second reading is taken for the same station with transmitter and receiver interchanged. Coil 2, now the transmitter, is oriented so that its axis is directed at coil 1, but inclined 15° *above* the horizontal, while coil 1 is tilted for a minimum. In both set-ups, the possibility of misalignment and of obtaining an incorrect dip angle are eliminated by the relative orientation of the two coils; the axis of the transmitter coil, rather than its plane, determines the rotation of the receiver coil about an axis normal to, rather than coaxial with, the T-R line.

In homogeneous ground the difference between the two tilt angles will be zero. This will be true regardless of the relative elevations of the two coils. However, with a conductor present, the secondary field will affect the tilt angles at the two receiver positions in the opposite sense, as can be seen better from the distorted field lines in fig. 7.20b. A profile obtained over a sheet-like conductor is also illustrated in 7.20c. Normally the difference between the two dip angles is plotted at the midpoint of the two coils.

The equipment uses two frequencies, 480 and 1800 Hz. The latter is used alone for reconnaissance work. The profile in fig. 7.20c shows the resultant $(\alpha_2 - \alpha_1)$ positive over the conductor. A dipping sheet results in an asymmetric profile which is positive over the upper end and crosses zero to a negative maximum downdip. Flat-lying conductors produce a negative anomaly symmetric about the midpoint.

(e) *AFMAG method.* The initials denote 'audio-frequency magnetic fields'. This is a natural-source dip-angle method, introduced by Ward (1959). The main origin of the primary field is lightning discharge (sferics) associated with worldwide thunderstorm activity. There are other minor sources of energy such as corpuscular radiation interaction with the earth's magnetic field and man-made noise. This EM energy is propagated between the earth surface and the lower ionosphere as in a waveguide. The frequencies associated with AFMAG are in the ELF range, from 1 to 1000 Hz, with the best reception apparently between 100 and 500 Hz. Attenuation is quite small in this waveguide, so the energy may travel worldwide.

Since the sferic sources are random the signal is effectively noise with seasonal, diurnal and short-period variations in intensity. Over the ELF range an AFMAG record is quite similar to the telluric current record shown in fig. 6.15, §6.2.7a.

Generally the vertical component is small compared to the horizontal, except in the vicinity of a good conductor. Hence the AFMAG field may be detected by a tilt-angle technique. The receiver, however, is modified from the conventional dip-angle detector since the random variations in primary field intensity make it impossible to locate the minimum with a single coil.

Two mutually perpendicular coils, wound on an insulating frame or ferrite core, are mounted on a stand which allows rotation about vertical and horizontal axes. One of the coils is first connected to the receiver amplifier and rotated about a vertical axis to find the rough azimuth of the horizontal field (see fig. 7.21*a*). This azimuth, of course, is the direction of the horizontal component and is often quite fuzzy and erratic.

One of the orthogonal coils supplies a reference signal in the tilt-angle measurement. Using ferrite-core coils, as in the illustration of fig. 7.21, this reference coil is

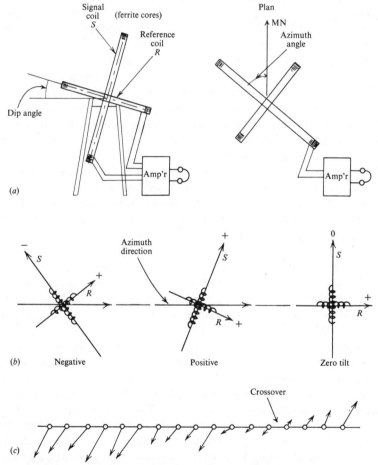

Fig. 7.21 AFMAG system. (*a*) AFMAG equipment (schematic) and measurement of azimuth; (*b*) measurement of tilt; (*c*) vector plot where azimuth is not in traverse line.

usually nearly horizontal and in the azimuth direction. The other (vertical) coil is now connected to the amplifier along with the reference coil and the pair are tilted about a horizontal axis normal to the main azimuth to get a minimum. Positive and negative tilt convention is illustrated.

Measurements are generally made at two frequencies, 140 and 510 Hz, with two narrow-band filters in the amplifier. The field procedure is otherwise quite similar to the fixed-transmitter method, with the transmitter considered to be at an infinite distance. Traverses are made at right angles to geologic strike where possible. If the AFMAG azimuth is not roughly along the traverse line, it is preferable to measure the tilt angles in the azimuth direction. The resultant crossover profiles may be plotted exactly as in the fixed-transmitter method, or they may be plotted as vectors (fig. 7.21c) if the minimum azimuth direction is distinctly different from that of the traverse line or if the azimuth varies appreciably over a short time interval. In the plot shown the length of each vector is proportional to the dip angle and the direction is that of the azimuth minimum with respect to the traverse. The crossover is then indicated by the reversal of the arrows.

AFMAG has several real and potential advantages over the artificial source methods. No transmitter is required. The frequency is comparatively low and hence the depth of penetration is probably greater than for a local source. Since the primary field is uniform, at least instantaneously, over the survey area, all the conductors are energized uniformly. At times this may be a disadvantage, however, since it may emphasize large-scale, relatively poor conductors at the expense of smaller concentrated bodies.

There are two specific disadvantages with AFMAG. The first is the effect of large random changes in the amplitude and direction of the inducing field, which produce corresponding variations in the signal strength. The second is that random fluctuations in direction may make it very difficult to locate the azimuth of the horizontal field, as mentioned earlier.

(f) *VLF method.* The use of VLF signals broadcast by certain marine and air navigation systems as sources for EM exploration has been mentioned in §7.2.2g. The main magnetic field component is horizontal like the AFMAG signal and theoretically is tangent to circles concentric about the antenna mast. Hence it is much less erratic in direction than AFMAG.

A worldwide network of high-power VLF stations is planned for marine navigation. The sites are arranged so that at least two stations can be detected anywhere over the earth's surface. At present suitable transmissions for EM prospecting in North America are located at Cutler, Maine; Annapolis, Md; Boulder, Colo.; Seattle, Wash. and Hawaii. The useful ranges are surprisingly large, since the Seattle station is easily detected on the east coast of Canada. The reception, at least in summer, is best in the morning, but is adequate all day.

The receiver for detecting VLF signals measures a tilt angle and a quadrature component by means of two mutually perpendicular coils wound on ferrite cores. The coil whose axis is normally vertical is first held in a horizontal position and rotated in azimuth to find a minimum. This direction is in line with the transmitter station and is usually well defined.

The same coil is next held vertically and tilted about a horizontal axis parallel to the direction of propagation. The second coil, which is rigidly mounted at right angles to the first and so is approximately horizontal initially, is similar to the reference coil in the AFMAG receiver. Its signal is shifted in phase by 90° and, connected in series with the vertical coil signal, is fed into the receiver. The amplitude of this signal is adjustable on the quadrature dial, which reads per cent plus or minus. A clinometer on the instrument allows tilt angle measurement. By tilt and quadrature adjustments, a good minimum is obtained.

The receiver amplifier incorporates two plug-in units tuned to frequencies of two VLF stations which can be detected in the survey area. It is useful to have extra units for other stations available, in case a particular station either is not operating, or its signal is weak, or the station direction is such that the azimuth minimum is not roughly normal to the direction of the traverse. The minimum signal indication is obtained on headphones. The receiver is illustrated schematically in fig. 7.22.

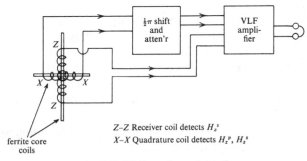

Fig. 7.22 VLF receiver schematic.

Clearly the VLF tilt-angle system is similar to AFMAG, with the advantage that the primary field direction is fixed and the signal level fairly uniform. The depth of penetration is not too well established, but seems to be somewhat less than that of dip-angle units using local power.

The field procedure and profile plotting are identical with AFMAG. The equipment is small, light, conveniently made and readings can be taken rapidly. The fact that the source is at infinity provides the same advantages and disadvantages in energizing the conductors in the survey area as described for AFMAG in the previous section. One drawback seems to be that it is not always possible to use a transmitting station which gives a primary field vector approximately across geologic strike in the survey area. The high frequency of the source is also an inherent weakness.

All the dip-angle methods have certain attractive features: simplicity, speed and relatively low price. They have one common disadvantage as well: the distinction between anomaly conductivity and depth is often difficult. Measurement at two frequencies should help in this regard. However, the higher frequency seems to enhance surface features (conductive overburden, groundwater concentration and the like) more than would be expected.

7.5.5 *Measurement of phase components*

(a) *General.* All the ground EM sets discussed so far record only a part of the available information. As we saw in §7.3.5, a measurement of both in-phase and quadrature components of the secondary field would provide us with some knowledge of the electrical properties of the conductor itself whereas the methods described so far are capable only of locating and outlining it. To record the additional data, it is necessary not only to cancel most of the primary field but to measure the phase as well as amplitude of the secondary field with respect to the primary. This is done with some form of compensating or ratiometer network. Three ground units of this type will be described.

(b) *Compensator (Sundberg) method.* This is a development of the two-frame equipment (which measured intensity only) and a forerunner of the later Turam system. The transmitter loop is a long grounded wire or large rectangle, that is, a semiportable unit. Traverses are made perpendicular to the long side with a single-coil detector connected through compensator and amplifier to headphones. The reference signal for the compensator unit is obtained from a small pickup coil lying near the transmitter cable and connected by wires to the receiver unit. Since the connecting cable varies in length as the receiver is moved, it is wound on a reel. Unless the ground is open it is necessary to reel in and start over when a new line is to be traversed. This layout is illustrated in fig. 7.23.

In a barren region the variation in phase from station to station would be zero, while the primary field, which is everywhere vertical, would fall off inversely as the distance of the receiver from the transmitter wire. Consequently the compensator dials are calibrated in per cent phase shift and per cent of the primary field amplitude, either plus or minus. The amplitude reading is based on the field at a particular station near the transmitter.

Usually the detector coil is held horizontally to measure the change in secondary vertical field, but it is also possible to determine all three components (real and imaginary parts for each component) of this field by orienting the coil in the vertical plane as well, in line with and across the traverse. These horizontal components would be zero in uniform ground, aside from background effects.

A typical profile over a long sheet conductor parallel to the transmitter cable is shown in fig. 7.24. The vertical component is quite like a dip-angle profile, while the horizontal is peaked over the conductor. In order to plot the real or in-phase vertical component, it is necessary to subtract the primary field at each station. This value can be obtained from the relation $H_p = i/2\pi l$, where i is the transmitter current and l the distance to the detector. In hilly ground a further correction for the difference in elevation between the transmitter and receiver coil will be necessary; the primary field is multiplied by $\cos \theta$ for the vertical component and $\sin \theta$ for the horizontal, where θ is the inclination of the transmitter-receiver line.

This method gives good results. However, the cable between the transmitter and compensator is a serious disadvantage. This cable can be eliminated by using a radio signal to carry the reference voltage, but such a scheme is quite elaborate. The field technique described next disposes of the cable by simpler means.

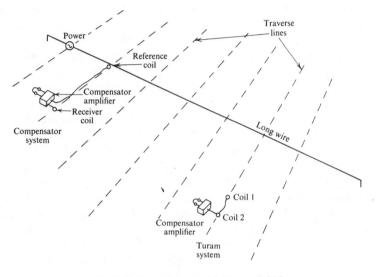

Fig. 7.23 Compensator and Turam field layouts.

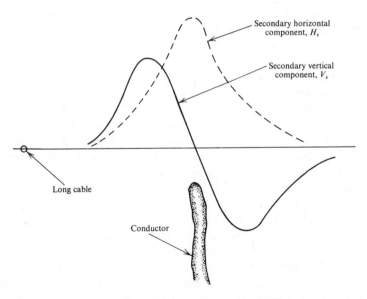

Fig. 7.24 Typical compensator-system profiles.

(c) *Turam method.* The transmitter system is identical to that used in the compensator method, while the receiver unit consists of two identical horizontal coils spaced 50–200 ft apart with a cable joining them. Traverses are made with fixed spacing on lines perpendicular to the transmitter wire. The field set-up is illustrated in fig. 7.23.

Measurements of amplitude ratio and phase difference between the two coils are made at each station, where the midpoint of the two coils is reckoned as the station location. As in the compensator system, with no conductor present the phase difference will be zero, while the amplitude ratio e_1/e_2 decreases with distance from the transmitter wire. Multiplying the amplitude ratio at each station by d_2/d_1, the ratio of the distances to the far and near coils, one obtains a constant amplitude of unity and a constant $\Delta\phi$ of zero for all stations under these conditions.

With a conductor near the receiver system, both parameters are changed— see figs. 7.40, 7.41, §7.8.4f and 7.45, §7.8.5e. The amplitude ratio is plotted in values greater or less than unity and the phase in degrees plus or minus. The profile is typical of a conductor of steep dip, its long axis roughly in line with the transmitter cable. Since the two receiver coils measure, in effect, the horizontal gradient of amplitude and phase of the vertical component of the secondary field, the profiles are horizontal derivatives of those obtained with the compensator equipment. Sensitivity is about 0·5% for the amplitude ratio, 0·2° in phase.

Thus Turam provides the extra information available when both real and imaginary components are measured, achieving a good null balance in the process. In addition the mechanical problem of the variable length reference signal cable, required in the compensator unit, is eliminated.

(d) *Moving source (horizontal-loop) method.* Known also as Slingram and Ronka EM, this system, like so many others, was developed in Sweden and has been popular in North America since about 1958. The field layout is illustrated in fig. 7.25.

Both transmitter and receiver are moved, a fixed spacing of 100–300 ft between them being maintained by a cable. The transmitter is low power (1–10 watts) and the transmitter coil is about the same size as the receiver. In some sets the coils are wound on insulating frames 2–3 ft in diameter, in others ferrite-core coils are employed. The coils are coplanar and almost always oriented to detect the vertical component, although this is not a necessary requirement. In at least one recent model, the connecting cable is replaced by a radio link; but the cable serves the additional purpose, on reasonably level ground, of maintaining correct spacing, which is quite critical.

A reference signal is obtained from a small pickup coil attached to the transmitter coil; its orientation can be adjusted for coarse amplitude control. This signal is carried by the cable to the compensator where its amplitude and phase may be adjusted. The compensator output is connected in series (opposing) with the receiver coil, both being across the input of the tuned amplifier. A null is obtained in the headphones connected to the amplifier output when the compensator is adjusted so that its output cancels the signal in the receiver coil. As in other systems of this type, the compensator unit is set to read zero on both dials (cancellation of the primary field in phase and amplitude) over barren ground where the signal is entirely due to H_p and its phase shift is $90°$ (see §7.3.2).

When readings are taken in the vicinity of a conductor, the amplitude and phase dials of the compensator then measure the in-phase and quadrature components of the secondary field, generally in per cent of the primary field intensity and phase at the receiver coil. (On the phase-shift dial 100% is equivalent to $90°$.) The sensitivity of both measurements is about $1–2\%$.

Traverses are made perpendicular to strike where possible and the readings plotted for the midpoint of the system. Typical profiles are shown in figs. 7.38, 7.39, 7.44 and 7.48 (§7.8.4e, §7.8.5d, §7.8.6b). As discussed later, the interpretation of anomalies with horizontal-loop system is generally simpler than with other EM field sets. However, the chief advantage is that the fixed relative positions of the receiver and transmitter maintain a constant mutual inductance M_{TR} (see eq. (7.18e), §7.3.4d). Thus the direction of traverse is immaterial, that is, we can interchange the receiver and transmitter and get the same reading at the same station.

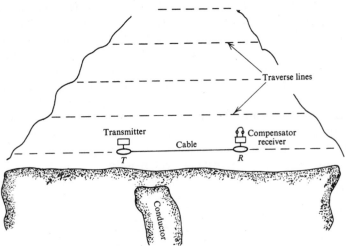

Fig. 7.25 Horizontal-loop system.

There are also some drawbacks to the horizontal-loop unit. The depth of penetration is inevitably limited by the portable low-power transmitter. Maximum depth for detecting a good conductor is often considered to be half the coil spacing; there would be no advantage, however, in increasing the separation much beyond 400 ft, because the depth would not increase correspondingly.

Both the spacing and orientation of the coils also are critical. Decreasing a 200-ft interval by 4 ft will produce an appreciable in-phase anomaly (6%), while a relative tilt of $10°$ between the coils will give a $1·5\%$ error or greater, depending on whether the measurement is made in the presence of a good conductor. Care must be taken in hilly ground to maintain both proper spacing and tilt. For instance, on a slope the coils must be parallel rather than horizontal, since their elevations are different. Correction may be necessary in this case since the total vertical component is not being measured. Again if the coils straddle a steep hilltop, the correct spacing may not be maintained.

Some horizontal-loop sets operate at two frequencies (e.g., 875 and 2200 Hz) with a view to distinguishing shallow anomalies which may mask better conductors at depth.

A recent development in EM ground equipment incorporates two methods in one field set: parallel-line dip-angle and moving-receiver horizontal-loop. Since both primary sources are portable and low power, the conversion from the dip-angle unit to the other system is quite simple, requiring merely the addition of a compensator and cable.

7.5.6 *Assessment of EM ground methods*

It would be attractive to compare the various EM ground sets in a sequence of increasing sophistication, so that we would arrive at the best possible equipment for field work, after consideration of simpler versions. This is not really possible, however, since there are inherent advantages and weaknesses in all the presently used systems. The criteria for judging the worth of a particular set include source power, reliability, speed and simplicity in field operation, information obtained and ease of interpreting the results. In a rough summary we can say that the depth of penetration increases with source power and hence the large transmitting loop systems have an advantage. By the same token they are less attractive for fast reconnaissance; in this area the numerous dip-angle techniques are particularly suitable. Finally the units which measure both in-phase and quadrature components provide more information about the anomalies, but at the same time are usually slower and require more competent operators.

7.6 Airborne EM systems

7.6.1 *General*

A prime attraction of EM prospecting, as mentioned before, is that reconnaissance exploration can be done from the air. Only magnetics and radioactivity, among the other methods, can be used in this way. A great variety of airborne EM systems has been developed to take advantage of this fact. They will be considered in the order in which they appeared, an order which corresponds roughly to the development of more complete and complex equipment.

The frequencies employed are the same as in ground EM, lower frequencies generally being used with higher power and greater altitude. The height of flight also determines the line spacing, which may vary from 500 ft to $\frac{1}{2}$ mile. Position of the aircraft is usually determined from continuous strip photography, occasionally by radio navigation, which, of course, adds considerably to the cost. Recording of data is continuous.

7.6.2 *Quadrature method*

The earliest airborne EM system used on any scale, this was developed jointly by International Nickel and McPhar Geophysics in Canada. It is essentially a high-power, large-aircraft unit. The transmitter coil is strung between wing tips and tail and is approximately horizontal. The receiver coil, nominally vertical, is towed in a bird, along with a preamplifier; the bird, like some airborne magnetometers, is roughly 400 ft behind and 200 ft below the plane. The plane usually flies at an altitude of 500 ft or more to prevent damage to the bird. The arrangement is shown in fig. 7.26.

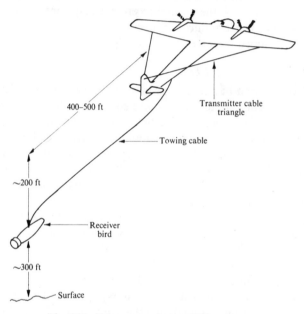

Fig. 7.26 Airborne quadrature EM-system.

Because the bird plunges, yaws and rotates in turbulent air, measurements of relative amplitude are impossible and it is practical to measure only the phase shift, which represents a time difference between primary and secondary fields. There is no equivalent ground EM of this type.

The phase of the receiver signal is compared with that of the transmitter output. Normally little could be gleaned from this one measurement, although such a system was used at 140 Hz in an early development. Generally two frequencies, 400 and 2300 Hz, are transmitted simultaneously. The ratio ϕ_{400}/ϕ_{2300} provides a rough criterion for conductivity of an anomaly, being greater than unity for good conductors and less than unity for bad. Sensitivity is about $0\cdot1°$ or $0\cdot2\%$.

Equipment which measures phase shift and nothing else suffers from the severe limitation that a very good conductor, producing 180° phase shift, will not be detected at all. Nevertheless, the quadrature system has had considerable application during the past 25 years.

7.6.3 *Long-wire system*

In a modification of the long-wire single-detector ground EM, the same ground transmitter is used, with a receiver coil in a light aircraft flying usually normal to the wire. There are two possible advantages to this arrangement. First, the ground transmitter gives better depth penetration than airborne transmitters. Second, the same transmitter may be used for immediate ground follow-up. At the same time it suffers from the usual limitations of the long-wire sets with the additional problem of variable sensitivity between transmitter and airborne receiver, which is more serious than with the ground receiver. Equipment of this type has been used in Russia and Western Europe for some time; a Canadian version, known as Turair, was developed recently.

7.6.4 *Airborne AFMAG, VLF*

The AFMAG method has been used in an airborne unit. Two identical orthogonally mounted coils are towed in a bird some 200 ft behind the aircraft. They are suspended with their axes in the line of flight and each 45° off horizontal. Signals from both coils are fed into an amplifier system tuned to two frequencies, 140 and 510 Hz, and compared in such a way that the variation is proportional to the tilt angle. This signal is recorded, together with altitude and a continuous record of the signal strength at each frequency (since the signal varies considerably and unpredictably with time).

Flight lines are roughly perpendicular to geologic strike. Zero or background readings may be checked in a barren area. Background noise results from vibration of the coils in the earth's magnetic field, thermal agitation in the coils and erratic movements of the bird.

A comparison of peak-signal amplitudes at the two frequencies provides some indication of conductivity, as in the quadrature system. AFMAG has the advantage of one-way transmission of signal, like the Turair type of set. Thus the method ought to have greater depth penetration than those using an airborne transmitter, which seems to be true, although this may apply particularly to large geologic features of minor economic interest. It has not been used to any great extent commercially.

All the arguments above would seem to apply to the VLF equipment converted to airborne use, although only limited information is available to date. Several airborne VLF sets are newly available or presently under development in Canada (Barringer Radiophase, McPhar KEM, Geonics EM18, Scintrex Delta-Air). A comparison of one of these with airborne AFMAG, over a very limited survey area, indicates that the 140-Hz AFMAG outlines structural features of contrasting conductivity more clearly than does the 510-Hz, while the VLF appears to emphasize near-surface conductors, such as water courses.

7.6.5 *Phase-component measurement*

There are several variations of this equipment, mounted usually on a light aircraft or helicopter. All are quite similar to the horizontal-loop moving-source ground set, except that the coils are vertical. The main requirement is that the receiver and transmitter coils be rigidly mounted to maintain fixed spacing and orientation. Because of the presence of the metal aircraft and the relatively large distance from the buried anomaly, the compensation must be more complete than in the ground EM circuit of the same type. Sensitivity of a few parts per million is necessary.

The block diagram for an airborne system of this type is illustrated in fig. 7.27.

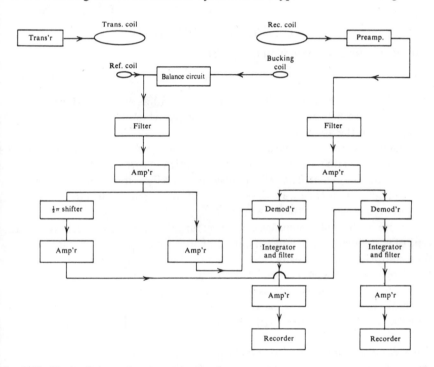

Fig. 7.27 Block diagram for the airborne phase-component system. (From Keller and Frischknecht, 1966.)

Obviously it is much more complex than the ground unit. The primary field and the fixed anomaly of the aircraft are cancelled at the receiver as completely as possible; any additional signals above noise level are recorded as anomalous conductors. In effect we are detecting H_T, which is made up of all the signals arriving at the receiver loop: the primary dipole field, the effect of the aircraft and whatever secondary field may also be present. That is,

$$H_T = H_S + H_P + H_A = H_S + H_P'.$$

Cancellation of the last term, H_P', by means of the bucking coil at the receiver, gives us H_S, which is measured in terms of H_P'. That is,

$$\frac{H_s}{H_P'} = \frac{H_T}{H_P'} - 1,$$

where the ratio is generally expressed in parts per million.

Figure 7.28 illustrates three coil arrangements used on aircraft and helicopters. The coils are either coplanar or coaxial with their axes in the line of flight. All three configurations are basically similar, since the primary field at the receiver is given by eqs. (7.9c) or (7.9d), §7.2.2f. That is, for either of the coaxial mounts on the helicopter,

$$H_P = Nia^2/2l^3,$$

and for the coplanar system on the wing tips,

$$H_P = -Nia^2/4l^3,$$

where N, i, a are the number of turns, current and radius of the transmitter coil and l is the distance between centres of the two coils, which varies from 20 to 80 ft, depending on the aircraft.

Frequencies used in the fixed separation systems range from 300 to 4000 Hz, the choice having to do with the type of aircraft which in turn determines the altitude of flying to some extent. The helicopter systems can fly as low as 50 ft and generally use the higher frequencies. Fixed-wing aircraft maintain an altitude of 300–600 ft and the frequencies are 1000 Hz or lower.

Flexure of the coil mountings caused by turbulent air is quite serious with this type of equipment. For a coil separation of 50 ft a change in the distance of a few thousandths of an inch is sufficient to change H_T by about 10 ppm, although only the in-phase component is affected. There are other sources of noise as well, some of which may be filtered out. Altitude changes also cause false anomalies. The depth of penetration is low, partly because of the inverse third-power fall-off of signal with a dipole source, partly due to the rather small coil separation.

7.6.6 *Rotary-field (two-aircraft) method*

This system, which was introduced in Sweden around 1958, has not as yet been used very much in North America. It has two novel features. Two light aircraft are generally required, one each for the transmitter and receiver, (although in one version the receiver coils have been towed in a bird from the transmitter plane). With two aircraft, the one containing the receiver flies in front, trailing the pickup coils and preamplifier about 100 ft behind. The transmitter is mounted in the rear aircraft, which maintains 800 ft separation between itself and the bird. Flying altitude is roughly 300 ft (height of the transmitter and bird). The tolerance on spacing, altitude and heading are ±20 ft, ± 20ft and ±4° respectively. The system is illustrated in fig. 7.29.

The second original feature is the rotating dipole field of the transmitter,

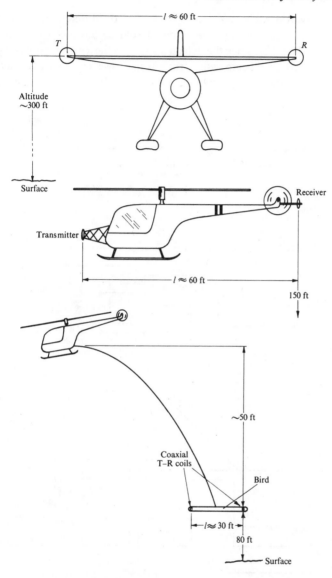

Fig. 7.28 Airborne double-dipole EM-systems. (After Grant and West, 1965.)

which is obtained by the use of two identical orthogonal coils supplied with current from a single source, one coil current being shifted in phase 90° with respect to the other. As a result of the circular polarization the primary field rotates in the xz-plane about the line of flight. The bird houses two orthogonal receiving coils nominally coplanar with the transmitter pair.

The two signals from these receiver coils, after amplification, are passed through phase shift networks such that, in a barren area, the differential output is zero. (Theoretically this means that one signal should be advanced or retarded $\pi/2$ with respect to the other, since the detected signal in the xy-coil is $\pi/2$ out of phase with that in the yz-coil to begin with; stray effects, however, require that additional phase adjustments be made.) The presence of a conductor results in a difference

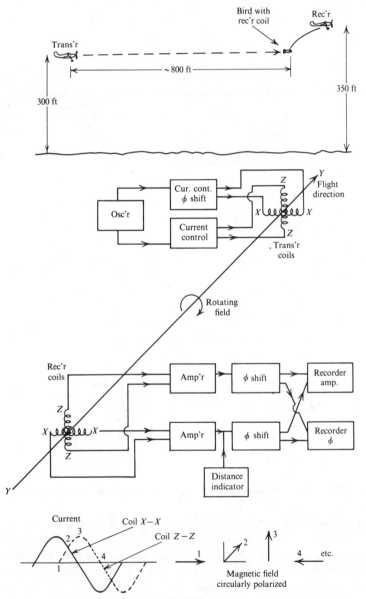

Fig. 7.29 Rotating-field two-airplane system. (After Ward, 1967.)

signal, which is continuously recorded on two channels, one for the in-phase, the other for the quadrature component.

Transmitter power is about 1500 watts and the frequency 880 Hz. Small changes in coil separation have much less effect than in the rigidly mounted systems described in the last section. Rotation of the receiver coils about the flight line is no problem because of the rotating field. However, rotation about the x- or z-axes, either of which is equivalent to changing the heading of one aircraft with respect to the other, will produce spurious signals. This appears to be the chief source of noise.

Among the airborne systems measuring both secondary field components, the two-airplane method has two specific advantages: low noise and greater depth of penetration, the latter due to the large coil separation. However, it is expensive and the interpretation may be complicated by the vertical coils in the transmitter and receiver.

7.6.7 *Transient (INPUT) method*

As we have seen from the previous descriptions, the fundamental problem in airborne EM is the isolation of very small secondary responses in the presence of a very large primary field. This difficulty, of course, applies to ground systems as well, but to a much lesser degree. One way of getting around the problem is to use a pulsed – rather than continuous – primary field and attempt to measure secondary response during transmitter off-time. This principle is the basis of the INPUT (Induced Pulse Transient) system developed by Barringer (1962). This is in effect a time-domain rather than frequency EM unit and the transmitting and receiving method has some resemblance to the induced polarization time-domain operation.

The transmitter loop, strung horizontally about a relatively large aircraft, is energized by half-sine-wave pulses of alternately opposite polarity. The on- and off-times are 1·5 and 2 msec respectively, in one model, resulting in a recurrence frequency of about 285 Hz. The receiver loop is vertical, with its axis in the line of flight. It is towed in a bird at the end of about 500 ft of cable. The voltage in the receiver coil will be proportional to dH_p/dt; the amplifier, however, is shut off (in electronic language, this is known as *gating*) during the transmitter-pulse interval so that there is no signal output. During the transmitter off-period the receiver amplifier is turned on. As a result the secondary magnetic field, arising from induced currents in the ground, which in this case decay exponentially because of the pulsating primary field, can be detected in the receiver during this 2-msec interval. The sequence of magnetic field and voltage pulses is shown in fig. 7.30.

The decay curve is sampled at several points (four in the first INPUT system, six in a later model) for intervals of 100 μsec or longer by electronic gating. These intervals are shown in fig. 7.30. The signals in each gate interval are integrated and reproduced on a four-channel strip chart. The signal amplitudes in the successive channels are diagnostic to some extent of the type of conductor causing the secondary field, since poor conductors and overburden (swamp, clay beds and the

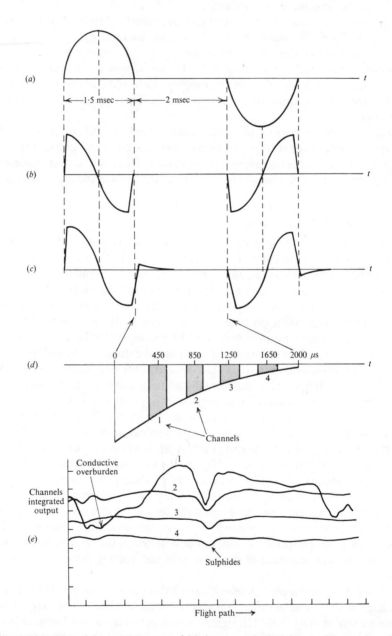

Fig. 7.30 INPUT airborne EM-system. (*a*) Primary magnetic field, H_p; (*b*) receiver-coil voltage due to H_p when $H_s = 0$; (*c*) receiver-coil voltage due to $H_p + H_s$; (*d*) decay signals in the different sampling channels; (*e*) four-channel records over Texas Gulf Sulphur, Timmins Ontario.

like) have shorter decay time constants than good conductors like massive sulphides. That is to say, the former are predominant only in the first, or first and second, channels, while the latter persist throughout all four. There is also some indication that the system responds to disseminated conductors in the sense of induced polarization, but this is not clearly established at present.

The chief source of noise is the spurious secondary field from the aircraft. The effect of this field is eliminated to a large extent by introducing a reference signal from the primary field voltage in the receiver coil, to cancel this signal in each channel of the receiver; relative motion of the bird is compensated at the same time. Other noise sources, such as atmospherics and power lines, are more difficult to suppress.

The INPUT method has certain attractive features. The depth of penetration appears to be larger (possibly as great as 400 ft) than most airborne EM, while the four-channel record reveals the character of conductors. This type of indicator, however, is a necessity as well as an inherent advantage, since in a pulsed system the coil detector, which records dH_s/dt, enhances the fast decay fields, whereas good conductors usually produces a secondary field which decays slowly. Some type of a.c. magnetometer capable of handling frequencies in the audio range would make a better detector than the coil, since the latter measures ΔH_s.

Because INPUT is relatively new, its possibilities and limitations have not yet been fully established. Like the two-airplane method, it is more expensive than the simpler airborne EM methods.

7.6.8 *Assessment of airborne EM*

It is apparent from the discussion that airborne EM is considerably different in detail from ground methods, although the fundamental principles are the same. The equipment is more elaborate, there is more of it, the requirements for background noise reduction are much more stringent, depth of penetration and discrimination between conductors are reduced and the interpretation possibilities restricted. The survey price, of course, is much increased, although not the cost per line-mile. In spite of all of these limitations, reconnaissance EM is invariably done from the air whenever the survey area is large enough and finances permit it. For fairly large-scale search, especially when accompanied by aeromagnetics, it becomes quite inexpensive per line-mile and is a powerful tool in mineral exploration.

7.7 EM field procedures

The standard field procedure is profiling along straight lines. Except in some ground reconnaissance, the surveying is done across geologic strike with the line and station spacing dictated by the amount of detail required. Procedures for the various ground and airborne systems have already been discussed in the previous sections.

EM may be used for vertical drilling in a manner similar to resistivity (see §8.5.4b). This can be accomplished either by increasing the transmitter-receiver separation while maintaining a constant frequency, or by varying the frequency with fixed spacing. The latter has the advantage that lateral changes in resistivity do not affect the readings. As in resistivity, however, if there are several horizontal layers of different conductivity, more than one spacing may be necessary even when the frequency is varied.

Several of the ground EM methods could be used for depth sounding. The long-wire transmitter systems have a greater depth potential than, say, the Ronka horizontal-loop set; however, the latter is more attractive for interpretation because of the symmetry.

Apart from the use of two distinct frequencies in some equipment, however, this drilling technique is seldom used. Magnetotellurics might be considered a form of EM depth sounding, but the analogy is rather artificial because magnetotellurics is a combination of very low frequency EM and resistivity. An EM set using variable frequency, say with a range of 100 to 5000 Hz, suitable for depth probing to perhaps 1500 ft would be much more elaborate and expensive than a Schlumberger resistivity layout. Furthermore the results obtained over multiple layers are more difficult to interpret than the corresponding resistivity data.

EM methods have been used to a limited extent in drill-hole logging; this application is discussed in §11.2.5.

7.8 Interpretation

7.8.1 *Introduction*

As in other geophysical methods, the interpretation of EM field results is done by comparison with the calculated and/or measured response of the same type of equipment to conductors of various simple shapes and conductivities.

Theoretical calculations for this purpose are limited to very elementary geometry. For instance, it is possible to solve the following configurations:

(i)   conducting sphere (cylinder) in uniform a.c. and dipole field,
(ii) conducting infinite horizontal thin sheet in uniform and dipole field,
(iii) conducting infinite half-space in uniform and dipole field,
(iv) conducting semi-infinite half-plane in dipole field.

This is a limited set of anomalies and the solutions are not at all simple. An alternative and more elementary theoretical approach is to assume the conductor to be a lumped circuit having resistance, self-inductance and mutual inductance with respect to transmitter and receiver. Here again the number of geometrical shapes is quite limited, being confined to circular and straight line elements – in effect the edges of thin conductors. The solutions, however, are comparatively simple.

For more complex geometry and variable conductivity it is necessary to resort to model measurements in order to match field results. This technique is more commonly used for EM interpretation than for other geophysical methods. It is necessary at this point to consider briefly the theory of scaling for model systems.

7.8.2 *Model systems*

From eq. (6.16), §6.2.3, we had the diffusion equation, which described the propagation of EM waves in the field in the form

$$\nabla^2 \mathbf{H} = \left(\frac{\partial^2}{\partial x^2} + \frac{\partial^2}{\partial y^2} + \frac{\partial^2}{\partial z^2} \right) \mathbf{H} = \mu\sigma \frac{\partial \mathbf{H}}{\partial t} = j\omega\mu\sigma\mathbf{H}.$$

In order that a model system may exactly simulate the field situation, both must satisfy this equation. Using subscripts m and f for model and field parameters, if l is a length and f a frequency and we scale distance and time linearly, then the following relations must hold:

$$l_m/l_f = x_m/x_f = y_m/y_f = z_m/z_f, \quad \text{and} \quad t_m/t_f = f_f/f_m.$$

Also,

$$\nabla^2 \mathbf{H}_m = 2\pi j f_m \sigma_m \mu_m \mathbf{H}_m, \quad \nabla^2 \mathbf{H}_f = 2\pi j f_f \sigma_f \mu_f \mathbf{H}_f,$$

hence

$$\nabla^2 \mathbf{H}_m = \left(\frac{l_f}{l_m} \right)^2 \nabla^2 \mathbf{H}_f = \left(\frac{f_m \mu_m \sigma_m}{f_f \mu_f \sigma_f} \right) \nabla^2 \mathbf{H}_f.$$

Therefore,

$$\frac{\rho_m}{\rho_f} = \frac{f_m \mu_m}{f_f \mu_f} \left(\frac{l_m}{l_f} \right)^2.$$

That is to say, the ratio $\rho/\mu f l^2$ has the same value in both systems. Now we generally measure EM effects in dimensionless form, for instance, the response ratio H_s/H_p or something similar to it; thus if the ratio $\rho/\mu f l^2$ has the same value in the two systems, the response ratio will be reproduced in going from one to the other. Practically, we may dispense with the permeability ratio, since magnetic permeabilities do not vary to any extent. Thus if we put $(l_m/l_f) = 1/n$, where $n \gg 1$, we get

$$\left(\frac{\rho_f}{\rho_m} \right)\left(\frac{f_m}{f_f} \right) = n^2. \tag{7.20}$$

This relation means that we can vary either ρ or f, or both, to satisfy the above requirements. As an example, suppose we want to model a massive sulphide conductor (say $\rho_f = 10^{-3}$ Ωm) on a dimensional scale of $1/500$. We can make $f_m/f_f = \rho_f/\rho_m = 500$.

If the field equipment has a frequency of 1000 Hz, the model parameters would be

$$6 \text{ in} = 250 \text{ ft}, \quad f_m = 500 \text{ kHz}, \quad \rho_m = 2 \times 10^{-6} \ \Omega\text{m}.$$

These values, however, are unsatisfactory for a practical model, partly because the frequency is too high, but mainly because it is difficult to vary the resistivities of suitable model conductor materials more than two or three orders of magnitude.

In fact it is preferable to choose the ρ_m value first and, if possible, to maintain the same frequency in the model as the field system, since the field-set receiver may be used for the model measurements. In this case, if the model conductor is aluminium ($\rho \approx 2 \cdot 8 \times 10^{-8}$ Ωm), we have

$$n = \sqrt{(10^{-3}/2 \cdot 8 \times 10^{-8})} = 190,$$

which makes the dimensional scale 6 in. = 105 ft.

As mentioned above, the choice of model conductors is limited. Resistivities range through aluminium, brass to stainless steel (7×10^{-7} Ωm). The only other solid material useful for models is commercial graphite (10^{-6} to 10^{-5} Ωm). Closed loops of copper or aluminium wire may be used as models for good conductors, as well as solid metal, since the induced currents do not penetrate the solid conductor to any extent. The wire loops have the additional advantage that the geometry may be changed easily.

If the host rock is a poor conductor it is permissible to measure the mod⸫ conductor in air. To simulate the effect of a conductive overburden, horizontal sheets of aluminium foil may be placed over the model orebody, possibly with an electrical connection between the two. Where the long-wire transmitter – which introduces currents directly into the ground – is being used, or where the host rock has appreciable conductivity, the EM model may be immersed in a water tank (see §8.6.2). The conductivity of the liquid is variable over a wide range by the addition of salt or acid. In the tank model it will usually be necessary to increase the frequency to maintain the scaling ratio.

Obviously the EM model equipment should be scaled the same as the conductor. Practically this may not be possible, since a 2-ft diameter receiver coil becomes 0·025 inches when scaled down 1000/1. But if the model coil is very large with respect to the station spacing, or receiver-transmitter spacing, it is preferable to adopt a smaller scale ratio. Coils may be conveniently wound on small diameter ($\frac{1}{8}$–$\frac{1}{4}$ in.) ferrite rods for model work; theoretically this limits the dimension ratio to about 200. However, considerable latitude is permissible in this respect.

7.8.3 *General interpretation procedure*

In discussing interpretation of ground EM we shall limit the types of field methods to dip-angle (fixed-transmitter, parallel-line and VLF) and phase-component techniques (Slingram and Turam), since these are most widely used at present.

7.8.4 *Semi-infinite vertical sheet*

(a) *General.* A very common configuration for conductive zones is a thin vertical sheet, long in one horizontal dimension and extending to great depth. If the conductivity is large, the induced currents flow mainly along the top edge, vertically at the ends and return at depth. Provided the larger dimensions of this sheet are several hundred feet, one can approximate the situation by current flowing in a long wire lying approximately along the upper edge of the sheet.

(b) *Dip-angle   measurement – fixed   transmitter.* By employing the mutual inductance formulae from §7.3.4e we can get some idea of the shape of the profiles obtained over this type of conductor. The geometry of the coils and conductor is shown in figs. 7.18 and 7.31. The transmitter coil, located, if possible, at a point directly over the conductor, is pointed at successive receiver stations and a minimum is obtained by tilting the receiver coil about the T–R axis.

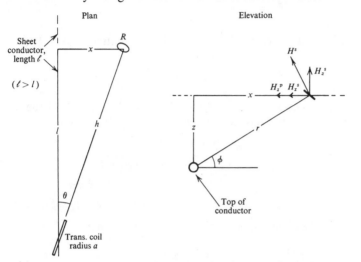

Fig. 7.31 Geometry for dip-angle fixed transmitter over steeply dipping sheet.

The measurement is a ratio of the vertical component to the total horizontal component at the detector. The former is entirely due to secondary field, while the horizontal is the sum of the secondary component and the direct field from the transmitter, which of course is horizontal.

From eq. (7.18g) the mutual inductance between the transmitter coil and horizontal edge of the conductor is given by

$$M = \frac{2\pi \times 10^{-3}a^2}{z \sec \theta} \mu H = \frac{2\pi \times 10^{-2}a^2 l}{z\sqrt{(x^2 + l^2)}} \text{ metres}$$

(since $1\mu H$ is equivalent to 10 m).

The voltage induced in the long conductor edge and the resultant current are

$$e_C = -j\omega M i_p, \quad i_C = \frac{e_C}{z_C} = -\frac{M i_p}{L}\left(\frac{Q^2 + jQ}{1 + Q^2}\right).$$

Since we are concerned only with the magnitude of the current, this is expressed as follows:

$$i_C = -\frac{M i_p}{L\sqrt{(1 + 1/Q^2)}} = -\frac{M i_p}{\beta L},$$

where $\beta = \sqrt{(1 + 1/Q^2)}$.

From fig. 7.31 it can be seen that the secondary components of the field at the receiver coil are given by

$$H_x^s = \frac{i_C z}{2\pi(x^2 + z^2)} = -\frac{Mi_p}{\beta L} \cdot \frac{z}{2\pi(x^2 + z^2)},$$

$$H_z^s = -\frac{Mi_p}{\beta L} \cdot \frac{x}{2\pi(x^2 + z^2)}.$$

In addition, the horizontal field from the transmitter is

$$H_x^p = -\frac{i_p a^2}{4(x^2 + l^2)^{\frac{3}{2}}}.$$

Now we are measuring $H_z^s/(H_x^s + H_x^p)$, the dip angle, δ, of the receiver coil for minimum signal. Thus, writing H_x^t for $(H_x^s + H_x^p)$, we have

$$\left| \frac{H_z^s}{H_x^t} \right| = \frac{-i_p M x/\{2\pi(x^2 + z^2)\beta L\}}{-i_p M z/\{2\pi(x^2 + z^2)\beta L\} - i_p a^2/\{4(x^2 + l^2)^{\frac{3}{2}}\}}$$

$$= \frac{x/z(x^2 + z^2)}{1/(x^2 + z^2) + 25\beta L/\{l(x^2 + l^2)\}}.$$

Introducing the dimensionless variable, $\alpha = x/z$, we have

$$\tan \delta = \left| \frac{H_z^s}{H_x^t} \right| = \frac{\alpha/(1 + \alpha^2)}{\left(\dfrac{1}{1 + \alpha^2} \right) + \dfrac{25\beta L}{l(l^2/z^2 + \alpha^2)}}$$

$$= \frac{\alpha}{1 + 25\beta(L/l)(z/l)^2 \dfrac{(1 + \alpha^2)}{(1 + \alpha^2 z^2/l^2)}}. \qquad (7.21)$$

The ratio L/l clearly affects the magnitude of the dip angle, along with z/l. Of course, β can vary enormously, since

$$\beta = \sqrt{(1 + R^2/\omega^2 L^2)} = \sqrt{\left[1 + \left(\frac{\rho\ell}{\pi r^2} \cdot \frac{10^6}{2\pi f \times 2\ell\{\log(2\ell/r) - (3/4)\}} \right)^2 \right]}.$$

That is,

$$\beta = \sqrt{\left\{ 1 + \left(\frac{50\rho}{f_k r^2 \gamma} \right)^2 \right\}},$$

where ℓ = effective conductor length (m), r = effective conductor radius (m), ρ = effective resistivity (Ωm), f_k = EM frequency (kHz), $\gamma = \log_e(2\ell/r) - (3/4)$. For example, if $f_k = 1$ kHz, $r = 1$ m, $\gamma \approx 5$, we have

$$\beta \approx 1 \text{ when } \rho \leqslant 0\cdot04 \ \Omega m, \quad \text{and} \quad \beta \approx 10^4 \text{ when } \rho \approx 1000 \ \Omega m.$$

About all that one can say is that $\beta \approx 1$ for a good conductor. The other term, L/l in dimensionless units, becomes

$$L/l = 0 \cdot 02 \ell \gamma / l,$$

and since the conductor is assumed to be very long and thin, this term is also of the order of unity.

Figure 7.32 shows two theoretical curves for $z/l = 1/5$ and 0, the second presumably meaning that the sheet outcrops, obtained from EM theoretical calculations, assuming infinite conductivity, or $\beta = 1$. The corresponding dotted curve for $z/l = \frac{1}{5}$ was obtained from eq. (7.21) with $\beta L/l = 1 \cdot 1$; the second made use of a modification of eq. (7.18g) in §7.3.4e,

$$M = 0 \cdot 004 \pi z \tan (\pi/4 - \alpha/2), \tag{7.18h}$$

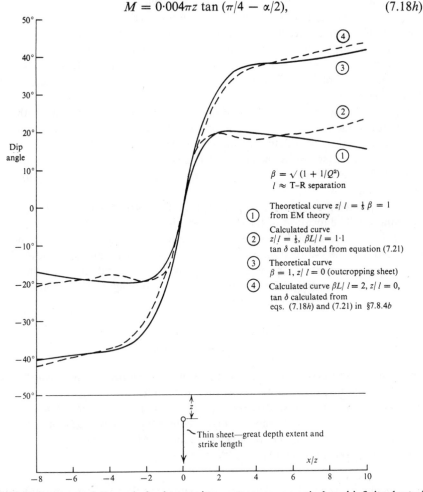

$\beta = \sqrt{(1 + 1/Q^2)}$
$l \approx$ T–R separation

① Theoretical curve $z/l = \frac{1}{5} \beta = 1$ from EM theory

② Calculated curve $z/l = \frac{1}{5}$, $\beta L/l = 1 \cdot 1$ tan δ calculated from equation (7.21)

③ Theoretical curve $\beta = 1$, $z/l = 0$ (outcropping sheet)

④ Calculated curve $\beta L/l = 2$, $z/l = 0$, tan δ calculated from eqs. (7.18h) and (7.21) in §7.8.4b

Thin sheet—great depth extent and strike length

x/z

Fig. 7.32 Response of dip-angle fixed-transmitter system over a vertical semi-infinite sheet of infinite conductivity.

which is used where $z \leqslant a$. In this case $z = 0$ and $\beta L/l = 2$. (It should be noted that in fig. 7.32 and several subsequent diagrams, various curves are referred to as theoretical, calculated and model profiles. The first type is calculated from EM theory, the second from electric circuit analogies such as were used in deriving eq. (7.21), while the third type is obtained from model measurements.)

The fit between these two pairs of curves is not bad, although at large values of x/z, eq. (7.21) shows that tan δ increases steadily with α. Practically this increase is limited by the fact that x and l have maximum values beyond which the various components cannot be detected at all.

Equation (7.21) may also be modified to allow for attenuation in a slightly conductive medium, by multiplying H_z^s and H_x^s by $e^{-z\sqrt{\omega\mu\sigma}/2}$ as given in eq. (6.17b), §6.2.3. On the other hand, H_x^p is not affected.

If the transmitter is offset from the position directly over the conductor, the characteristic profiles become asymmetrical, although the crossover is still directly over the top edge. The maximum dip angle is considerably reduced on the side where the transmitter is located, while the other half of the profile is not changed much. For example, an offset of $0 \cdot 4l$ decreases the maximum dip angle by more than a factor of two.

Figure 7.33 illustrates the effect of semi-infinite sheets of constant depth and different conductivities. The solid curves are obtained from model measurements in the case of finite conductivitity and from EM theory for infinite conductivity. The dotted curves were obtained from eq. (7.21) for $\beta = 1, 3, 10$, corresponding to $\rho = 0, 0 \cdot 06$ and $1 \cdot 4 \, \Omega m$ respectively (for the significance of the quantity $\sigma\mu_0\omega sl$, see §7.8.7c).

As mentioned previously, the dip-angle technique provides a limited amount of information. The location of the conductor – in this elementary case – is always under the crossover point, while its strike extent is indicated by the persistence of the profiles in the y-direction. Magnitude of the dip angles and sharpness of the slope near the crossover are somewhat diagnostic of the conductivity, although it is not possible to distinguish between conductivity and depth. For example, if we differentiate eq. (7.21) and put $x/z = 0$, we get

$$\frac{d(\tan \delta)}{d\alpha} = \frac{1}{1 + 25\beta(L/l)(z/l)^2},$$

which merely shows that both βL and z determine the slope. The same parameters, of course, will be present in the expression for tan δ_{max}.

By employing two transmitter frequencies, one high and one low, some idea of the depth can be obtained, since the high frequency profile will be enhanced by a shallow conductor more than the low. Thus these EM sets often use two frequencies, such as 5000 and 1000 Hz.

(c) *Dip angle – parallel line.* The analysis is similar to the fixed-transmitter layout, since the expressions for secondary-field components are the same and the trans-

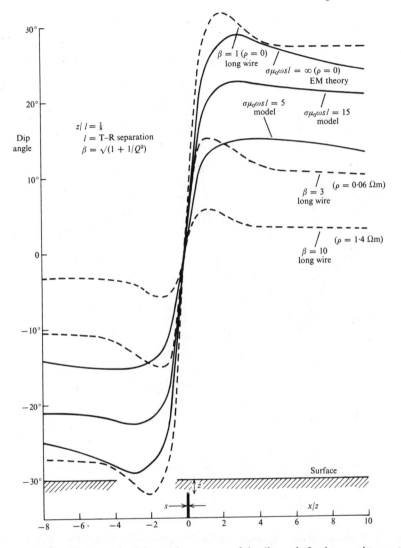

Fig. 7.33 Effect of finite conductivity on the response of the dip-angle fixed-transmitter system over a vertical semi-infinite sheet.

mitter-receiver coupling in this case is constant. The transmitter-conductor coupling, however, is different and the mutual inductance is given by

$$M \approx \frac{0 \cdot 02\pi a^2}{r} \cos \phi = \frac{0 \cdot 02\pi a^2 z}{x^2 + z^2},$$

where ϕ is the angle between the plane of the transmitter coil and the line r which

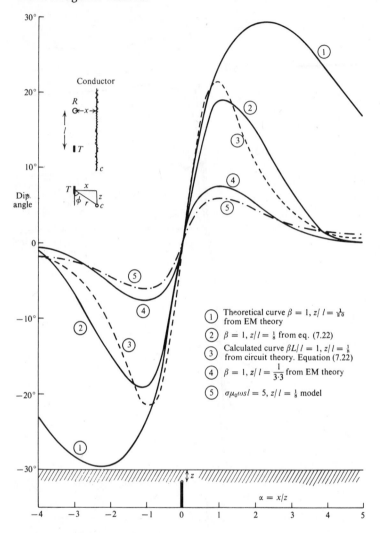

Fig. 7.34 Response of the dip-angle parallel-line (broadside) system over a semi-infinite vertical sheet.

joins its centre to the top edge of the conductor. See fig. 7.34. Then the field components are:

$$H_x^p = -\frac{i_p a^2}{4l^3}, \quad H_x^s = -\frac{Mi_p}{\beta L} \cdot \frac{z}{2\pi(x^2 + z^2)}, \quad H_z^s = -\frac{Mi_p}{\beta L} \cdot \frac{x}{2\pi(x^2 + z^2)},$$

and eq. (7.21) becomes

$$\tan \delta = \left| \frac{H_z^s}{H_x^t} \right| = \frac{\alpha}{1 + 25\beta\{(L/l)(z/l)\}^2(1 + \alpha^2)^2}. \tag{7.22}$$

Although this expression gives a profile of the proper general shape, that is, a crossover at $x = 0$, marked peak value and fairly rapid fall-off with large x, it does not fit the theoretical results obtained from EM theory or the model curves as well as eq. (7.21) does for fixed transmitter. This can be seen in fig. (7.34), which shows several profiles for a parallel-line, dip-angle method taken from EM theory and model results.

Equation (7.22) is a poor approximation since the slopes are too steep and the peaks too high. This is mainly due to the moving source; the value of M changes in some fashion as r changes, modifying the multiplier of the denominator term $(1 + \alpha^2)^2$.

The interpretation is again based on the peak dip angle and the slope near the origin. The ambiguity between conductivity and depth is perhaps better resolved than in the fixed-transmitter arrangement, since the maximum values of $\tan \delta$ are well defined and occur at $x \approx z$, unless the conductor is extremely shallow.

It is worth noting that the method of plotting profiles in fig. 7.32 and following figures makes all the slopes look much the same near the crossover. In fact they are very different when, as in field results, dip angle is plotted against x rather than x/z or x/l.

When the traverse is not carried out at 90° to strike, the profiles are considerably distorted, as might be expected. This is illustrated in fig. 7.35 which shows a 60° crossing. The distortion is more pronounced than for the fixed-transmitter layout.

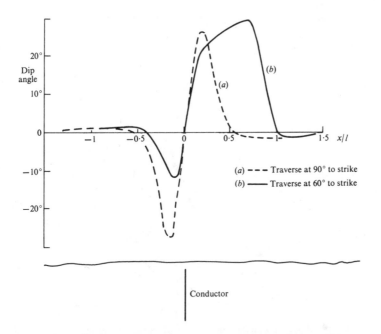

Fig. 7.35 Effect of skew traverse on the response of the dip-angle broadside array.

(d) *Dip angle – VLF and AFMAG.* The analysis in this case is simplified since the source is effectively at infinity and the field uniform over the survey area. The secondary field components are the same as before:

$$H_z{}^s = \frac{i_C x}{2\pi(x^2 + z^2)}, \quad \text{and} \quad H_x{}^s = \frac{i_C z}{2\pi(x^2 + z^2)},$$

while the primary horizontal component is given by eq. (7.12*d*), §7.2.2*g*,

$$H_x{}^p = \frac{i_p \delta l}{2\lambda r} \cdot e^{-1\cdot5 \times 10^{-3} r/\sqrt{\lambda}}.$$

Thus the dip angle is given by

$$\tan \delta = \left| \frac{H_z{}^s}{H_x{}^t} \right| = \frac{\alpha}{1 + K' z i_p (1 + \alpha^2)/i_C}, \tag{7.23a}$$

where $K' = (\pi \delta l/\lambda r) \cdot e^{-1\cdot5 \times 10^{-3} r/\sqrt{\lambda}}$, r = source distance (km), $\lambda = c/f$ = wavelength (km), δl = height antenna (km).

As a rough approximation we can say that both K' and i_C will vary inversely as r. Hence eq. (7.23*a*) may be simplified to

$$\tan \delta = \frac{\alpha}{1 + Kz(1 + \alpha^2)}, \tag{7.23b}$$

where $K = K'(i_p/i_C)$.

Figure 7.36 shows two typical profiles over model vertical sheets, obtained with

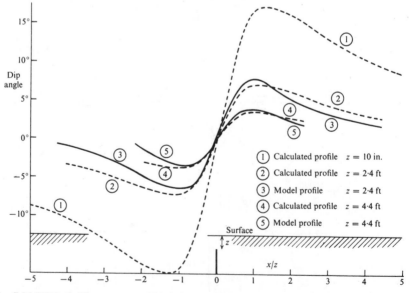

Fig. 7.36 VLF profiles over a semi-infinite vertical sheet. Model sheet 6 ft × 48 ft, $\beta = 1$, $\rho \approx 0$.

a VLF receiver, as well as three curves calculated from eq. (7.23b). The latter were solved for K by matching one of the model curves at one point. The fit between model and calculated curves is quite good, although the former have better defined maxima. This is because the models are of finite depth extent.

In general form they are all quite similar to the fixed-transmitter and parallel-line profiles. However, the slopes near the crossover are less steep, the maxima are better defined than in the fixed-transmitter measurement and the flanks of the curves fall off at a rate intermediate between the other two types. Consequently the depth may be estimated roughly, since $x \geqslant z$ at peak dip angle.

Equation (7.23b) does not involve the conductivity directly. However, the value of i_C varies with σ and hence tan δ will increase with σ, as in eqs. (7.21) and (7.22).

As mentioned previously, the VLF and AFMAG systems, which have little or no control over the primary field direction, frequently do not permit measurement of the dip angle in the direction of traverse. Although the profiles may be plotted as vectors, as shown in fig. 7.21, the results are inevitably distorted when the primary field is nearly parallel to conductor strike. This is a fundamental disadvantage of both methods.

VLF profiles are considerably distorted when the traverse line is not perpendicular to strike. Figure 7.37 shows two results, from the model used in

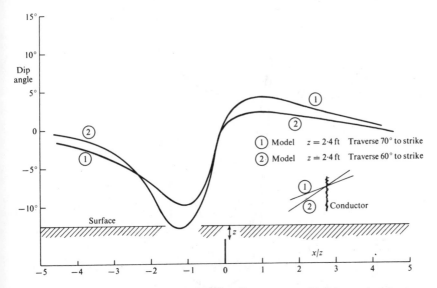

Fig. 7.37 Effect of skew traverse on VLF profiles over a semi-infinite vertical sheet.

fig. 7.36, for angles of 70° and 60°. The crossover point is not shifted, but the asymmetry increases as the angle between strike axis and profile decreases, which would give a false indication of a dipping sheet.

(e) *Phase-component systems – horizontal loop.* The circuit analogy for the long wire in this system is worked out easily enough. The geometry, shown in fig. 7.38, accounts for the mutual inductance

$$M = \frac{0 \cdot 02\pi a^2 \cos \theta_1}{r_1} = \frac{0 \cdot 02\pi a^2 (l/2 + x)}{z^2 + (l/2 + x)^2},$$

and the vertical component of the secondary field at the receiver becomes

$$H_z^{\,s} = -\frac{Mi_p}{\beta L} \cdot \frac{\cos \theta_2}{2\pi \sqrt{\{z^2 + (l/2 - x)^2\}}}$$

$$= \frac{-0 \cdot 01 a^2 i_p (l^2/4 - x^2)}{\beta L \{z^2 + (l/2 - x)^2\}\{z^2 + (l/2 + x)^2\}},$$

while the direct-coupled vertical component of the primary field is

$$H_z^{\,p} = -i_p a^2 / 4 l^3.$$

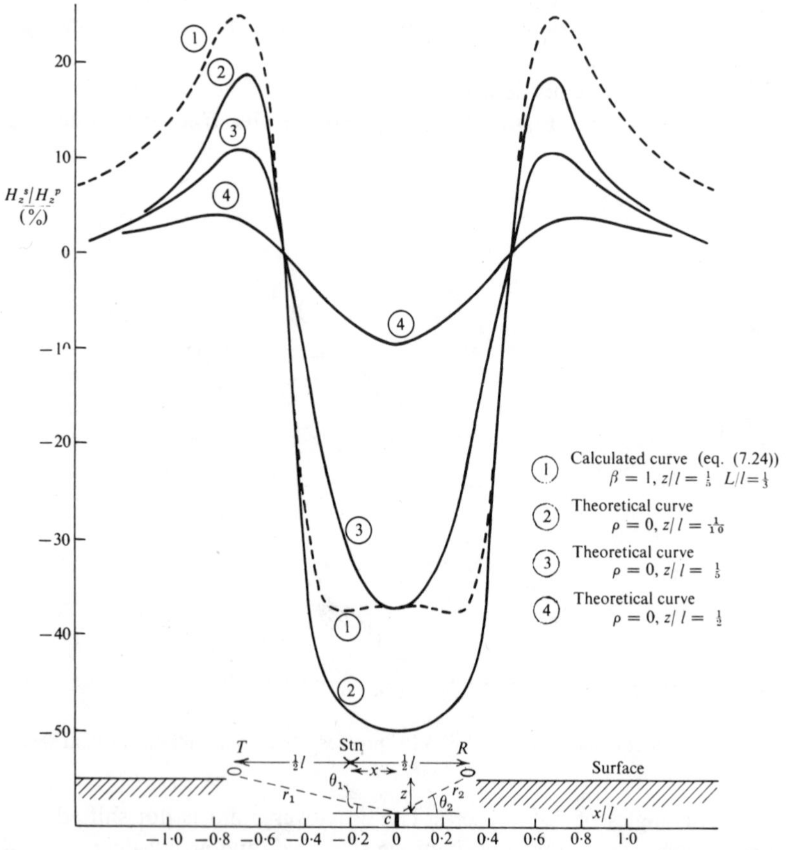

Fig. 7.38 Horizontal-loop profiles over a semi-infinite vertical sheet.

In this system we measure the ratio H_z^s/H_z^p. Putting $\alpha = x/l$ and simplifying the expression we get

$$\frac{H_z^s}{H_z^p} = \frac{4\alpha^2 - 1}{6\cdot25\beta(L/l)\{4z^2/l^2 + (1 - 2\alpha)^2\}\{4z^2/l^2 + (1 + 2\alpha)^2\}}. \quad (7.24)$$

A profile calculated from this expression for $z/l = \frac{1}{5}$, $\beta = 1$ and $L/l = \frac{1}{3}$ is plotted in fig. 7.38, which also shows three theoretical curves for $\rho = 0$, $z/l = \frac{1}{2}$, $\frac{1}{5}$ and $\frac{1}{10}$, derived from EM theory. Although the calculated profile has the right general shape, the positive maxima are too large, the slope near $x/l = \pm\frac{1}{2}$ is too steep and the negative maximum has a flat top. In fact, for still smaller values of z this flat top becomes a double peak either side of zero; it is obvious from eq. (7.24) that as $z/l \rightarrow 0$ there is a discontinuity at $x/l = \pm\frac{1}{2}$. This is a characteristic of tightly coupled electric circuits. Again, as in the parallel-line dip-angle system, the moving source makes the simple circuit analysis of little value. Clearly the inductive coupling between source and conductor, and hence the L and R values of the latter also, must vary with source position.

The profiles in fig. 7.38 have the following general characteristics. There is a single negative maximum when the receiver-transmitter spread straddles the conductor, two zeroes when either coil is directly over it (zero-coupling position) and two smaller positive peaks at approximately $0\cdot7l$, after which the flanks tail off to zero. The curves are entirely symmetrical, that is, the receiver and transmitter and/or the direction of traverse could be reversed without affecting the shape. The sizes of the maxima and hence the steepness of the curves are an indication of the depth of conductor. Maximum depth of conductor which can be detected is controlled by the coil separation, although the small power available in the portable transmitter is also a practical limitation.

A rule of thumb frequently used for the horizontal-loop method is that the maximum detectable depth is one half the coil separation. It is clear from fig. 7.38 that, even for a very good conductor, the maximum response will not be much above background if $z > l/2$.

There is zero quadrature response accompanying these profiles since we have assumed infinite conductivity. The effect of finite conductivity can be estimated from the equivalent circuit of the conductor by separating the real and imaginary parts of H_z^s, i.e., the complex function $1/\beta = (Q^2 + jQ)/(1 + Q^2)$. We then express the ratio H_z^s/H_z^p in two parts, one for the real, or in-phase, component, the other for the imaginary, or quadrature, portion. These are given by

$$\mathrm{Re}\left|\frac{H_z^s}{H_z^p}\right| = \frac{4\alpha^2 - 1}{6\cdot25(L/\ell)\{(2z/l)^2 + (1 - 2\alpha)^2\}\{(2z/l)^2 + (1 + 2\alpha)^2\}}\{Q^2/(1 + Q^2)\},$$

$$(7.25a)$$

$$\mathrm{Im}\left|\frac{H_z^s}{H_z^p}\right| = \frac{4\alpha^2 - 1}{6\cdot25(L/l)\{(2z/l)^2 + (1 - 2\alpha)^2\}\{(2z/l)^2 + (1 + 2\alpha)^2\}}\{Q/(1 + Q^2)\}.$$

$$(7.25b)$$

It is clear from these two expressions that the real component is Q times the imaginary and that the latter is zero when Q is large.

Figure 7.39 illustrates the effect of finite conductivity on the horizontal-loop profiles. Three pairs of curves for in-phase and quadrature response are plotted for $z/l = \frac{1}{5}$, of which two pairs were obtained by modelling and one pair from eqs. (7.25a) and (7.25b). The curve shapes are quite similar to those in fig. 7.38, but the maximum amplitudes increase with conductivity for the in-phase and decrease

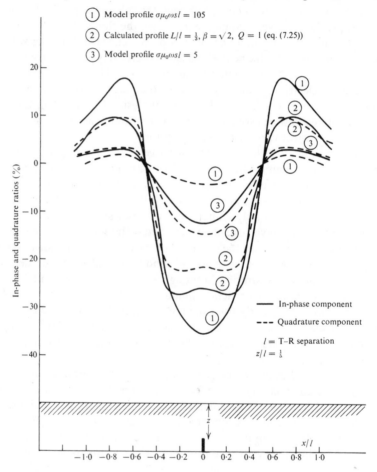

*Fig. 7.39 Effect of conductivity on horizontal-loop profiles over semi-infinite vertical sheet.

for the quadrature component. Thus the ratio of maximum in-phase to maximum quadrature response is quite diagnostic of conductivity. In the profiles of fig. 7.39 these ratios are roughly 9, 1·3 and 0·8.

The addition of the quadrature component clarifies, to a considerable extent, the ambiguity between conductivity and depth of burial, characteristic of dip-angle measurements. Some horizontal-loop field sets use two frequencies, one high and one low, as a further aid in estimating depth.

The effect of traverse direction is insignificant with the horizontal-loop system unless the angle with respect to conductor strike is less than 45°. Even then the profiles are quite symmetrical, although the in-phase has a smaller negative and larger positive maxima. This is a decided advantage of the method.

(f) *Phase-component systems – Turam.* The Turam system is particularly suitable for analysis with the semi-infinite sheet, since the source is fixed and the mutual inductance is effectively that between two long parallel wires, the top of the conductor and the near side of the transmitter loop. When the equivalent wire lengths are both equal to ℓ, where ℓ is greater than the separation between the wires, eq. (7.18b) of §7.3.4b takes the simplified form

$$M \approx 0.2\ell\{\log(2\ell/r) - \Delta\},$$

where $1 \geqslant \Delta \geqslant \frac{1}{2}$, if $0 \leqslant r/\ell \leqslant \frac{1}{2}$. See fig. 7.40 for the geometry.

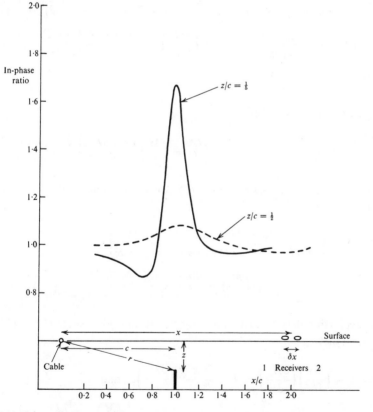

Fig. 7.40 Turam in-phase profiles over semi-infinite vertical sheet of infinite conductivity. Curves were obtained using eq. (7.26) taking $\delta x/c = 0.1$, $M/L = 0.2$, $\beta = 1$.

In Turam measurements the in-phase component is normally the vertical component in the form of a reduced ratio, since there are two coils (see §7.5.5c), and the quadrature is a difference in phase angle between the two coils. That is, we are measuring the total vertical fields at the two coils, $H_z{}^t(x - \delta x/2)$ and $H_z{}^t(x + \delta x/2)$, in the ratio

$$\left\{\frac{H_z{}^p(x - \delta x/2) + H_z{}^s(x - \delta x/2)}{H_z{}^p(x + \delta x/2) - H_z{}^s(x + \delta x/2)}\right\}\left\{\frac{H_z{}^p(x + \delta x/2)}{H_z{}^p(x - \delta x/2)}\right\}$$

$$= \left(\frac{H_1{}^p + H_1{}^s}{H_2{}^p + H_2{}^s}\right)\left(\frac{H_2{}^p}{H_1{}^p}\right) = \frac{1 + H_1{}^s/H_1{}^p}{1 + H_2{}^s/H_2{}^p} = \frac{R}{p},$$

and the difference in phase angles, $\Delta\phi = \phi_1 - \phi_2$. For the in-phase component the various fields are given by

$$H_1{}^p = \frac{i_p}{2\pi(x - \delta x/2)},$$

$$H_1{}^s = \frac{i_C(x - c - \delta x/2)}{2\pi\{(x - c - \delta x/2)^2 + z^2\}} = \left(\frac{-M}{\beta L}\right)\frac{i_p(x - c - \delta x/2)}{2\pi\{(x - c - \delta x/2)^2 + z^2\}},$$

$$H_2{}^p = \frac{i_p}{2\pi(x + \delta x/2)}, \quad H_2{}^s = \left(\frac{-M}{\beta L}\right)\frac{i_p(x - c + \delta x/2)}{2\pi\{(x - c + \delta x/2)^2 + z^2\}}.$$

Therefore,

$$\frac{H_1{}^s}{H_1{}^p} = \left(\frac{M}{\beta L}\right)\frac{(x - \delta x/2)(x - c - \delta x/2)}{z^2 + (x - c - \delta x/2)^2},$$

$$\frac{H_2{}^s}{H_2{}^p} = \left(\frac{M}{\beta L}\right)\frac{(x + \delta x/2)(x - c + \delta x/2)}{z^2 + (x - c + \delta x/2)^2}.$$

After rearranging, neglecting terms in $(\delta x/c)^2$ and putting x in the form $x/c = \alpha$ we get

$$\frac{R}{p} = \left\{1 + \left(\frac{M}{\beta L}\right)\frac{\alpha(\alpha - 1) - (\delta x/2c)(2\alpha - 1)}{(\alpha - 1)^2 - (\delta x/c)(\alpha - 1) + z^2/c^2}\right\} \bigg/$$

$$\left\{1 + \left(\frac{M}{\beta L}\right)\frac{\alpha(\alpha - 1) + (\delta x/2c)(2\alpha - 1)}{(\alpha - 1)^2 + (\delta x/c)(\alpha - 1) + z^2/c^2}\right\}. \quad (7.26a)$$

This expression may be written in a simpler form if we assume $H_z{}^s \ll H_z{}^p$ and $\delta x \ll x$, so that

$$\frac{R}{p} = 1 - \left(\frac{\delta x}{H_z{}^p}\right)\left(\frac{dH_z{}^s}{dx}\right) = 1 - \frac{2\pi x \delta x}{i_p}\left(\frac{dH_z{}^s}{dx}\right), \quad \text{and}$$

$$H_z{}^s = -\left(\frac{M}{\beta L}\right)\frac{i_p(x - c)}{2\pi\{z^2 + (x - c)^2\}}.$$

This gives for the ratio R/p the approximate expression

$$\frac{R}{p} = 1 + \frac{M}{\beta L}\left(\frac{\delta x}{c}\right)\frac{\alpha\{(z^2/c^2) - (\alpha - 1)^2\}}{\{(z^2/c^2) + (\alpha - 1)^2\}^2}. \tag{7.26b}$$

As in eqs. (7.25a) and (7.25b) we can separate the real and imaginary components in the practical situation when the conductivity is finite. The real component is obtained by multiplying eqs. (7.26a) and (7.26b) by $Q^2/(1 + Q^2)$. The Turam system, however, measures the phase difference between the two receiver coils, rather than the imaginary component. Thus the respective phase angles may be found from the ratio of the imaginary to real parts of H_z^t for each receiver coil, where

$$\tan \phi = IM|H^p + H^s|/RE|H^p + H^s|,$$

or,

$$\tan \phi_1 =$$
$$\frac{(M/L)Q(1 - \alpha - \delta x/2c)(\alpha + \delta x/2c)}{(1 + Q^2)\{(z/c)^2 + (1 - \alpha - \delta x/2c)^2\} + (M/L)Q^2(1 - \alpha - \delta x/2c)(\alpha + \delta x/2c)},$$

$$\tan \phi_2 =$$
$$\frac{(M/L)Q(1 - \alpha + \delta x/2c)(\alpha - \delta x/2c)}{(1 + Q^2)\{(z/c)^2 + (1 - \alpha + \delta x/2c)^2\} + (M/L)Q^2(1 - \alpha + \delta x/2c)(\alpha - \delta x/2c)},$$

from which

$$\Delta\phi = \phi_1 - \phi_2. \tag{7.27a}$$

The expression for phase angle is much simpler if we use the approximation in eq. (7.26b), when we have

$$\tan \Delta\phi \approx \Delta\phi \approx \frac{IM(R/p)}{1 + RE(R/p)} \approx IM(R/p).$$

Thus,

$$\Delta\phi = -\frac{M}{L}\left(\frac{\delta x}{c}\right)\frac{\alpha\{(z^2/c^2) - (\alpha - 1)^2\}}{\{(z^2/c^2) + (\alpha - 1)^2\}^2}\left(\frac{1}{Q}\right). \tag{7.27b}$$

Using eq. (7.26b), this can be written

$$\Delta\phi = \left(1 - \frac{R}{p}\right)\frac{1}{\sqrt{(1 + Q^2)}} \approx \left(1 - \frac{R}{p}\right)\frac{1}{Q}, \quad (Q \geqslant 3). \tag{7.27c}$$

Figure 7.40 shows the effect of depth of a semi-infinite sheet conductor on the Turam ratio for infinite conductivity, while fig. 7.41 has two profiles of ratio and phase difference for finite conductivity. From these illustrations and from model studies it appears that the depth to the top of the conductor is somewhat less than the width of either profile at half maximum.

As with other EM sets, Turam equipment is often operated at two frequencies

(220 and 660 Hz) to help discriminate between conductivity and depth of conductor and conductive overburden effects.

The typical Turam profile is slightly asymmetric because of the transmitter
cable location. It is possible to get an idea of the lateral extent of the conductor by
carrying out the same traverse twice, with the cable on opposite sides of the conductor and noting the difference in location between the positive peaks.

As mentioned previously, a long single wire, grounded at both ends, has
sometimes been used as a transmitter source instead of the large rectangular loop.
Such practise is not recommended because the ground is energized in a complicated
fashion, partly by induction, partly by galvanic currents as in the resistivity
method. This leads to high background noise and difficulties in interpretation.

As in the horizontal-loop system, the conductivity can be estimated from the
maximum values of ratios and phase differences. The usual method of plotting

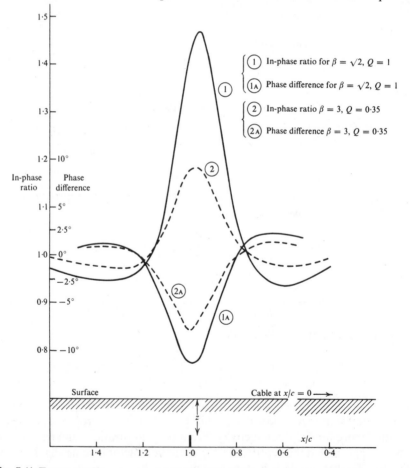

Fig. 7.41 Turam profiles over semi-infinite vertical sheet of finite conductivity; $z/c = 0.2$,
$\delta x/c = 0.1$, $M/L = 0.2$.

Turam results is not convenient for this, but it is relatively easy to determine the real and imaginary components of the resultant field from these values by converting the phase angle to a quadrature component. This procedure is explained with the aid of table 7.1, which includes a set of normal ratios H_1^p/H_2^p (which can be calculated from eq. (7.6*b*)) and sample data taken from a laboratory Turam survey on a model conductor.

Table 7.1. *Turam model data*

Line 4 + 00; transmitter loop – 1600 × 1200 ft; frequency 660 Hz

Station (ft)	Rec'r coils (ft)	Station readings: *FR*	Station readings: Phase ($\Delta\phi$)	*NR*	*RR*	$\Sigma\Delta\phi$ degrees	*V*	$R = V\cos\phi$	$I = V\sin\phi$
	200					0°	1·0	1·0	0
250		1·77	0·25°	1·76	1·01				
	300					0·25°	0·99	0·99	0·004
350		1·54	0·45	1·54	1·00				
	400					0·70	0·99	0·99	0·012
450		1·39	0·60	1·43	0·97				
	500					1·30	1·02	1·02	0·023
550		1·35	0·60	1·36	0·99				
	600					1·90	1·03	1·03	0·033
650		1·64	−0·25	1·31	1·25				
	700					1·65	0·83	0·83	0·023
750		1·39	1·45	1·27	1·10				
	800					3·10	0·75	0·75	0·041
850		1·26	1·50	1·25	1·01				
	900					4·60	0·74	0·74	0·059
950		1·24	1·65	1·22	1·02				
	1000					6·25	0·73	0·73	0·084
1050		1·23	1·80	1·21	1·02				
						8·05	0·70	0·70	0·098

The first column gives the location of the receiver station (midway between the two coils); *FR* is the '*field ratio*', that is, effectively R/p in eq. (7.26*a*) and elsewhere, while *NR*, *RR* and *V* refer to normal ratio, reduced ratio, total field respectively. Note that station readings, *NR* and *RR* values are referred to the station point, while in the remaining four columns the values are referred to the receiver coil nearer to the transmitter. This is because zero phase and unity total field are assumed to exist at the point of closest approach to the transmitter (presumably the latter has been located a reasonable distance from any conductors in the vicinity).

The reduced ratio is obtained by dividing the field ratio reading by the normal ratio for each station. Values of $\phi = \Sigma\Delta\phi$ are found by adding the successive phase readings and are entered opposite the location of the rear coil. Total field values, in the same row, are found from the ratio V/RR; for example, V_{300}

$= V_{200}/RR_{250} = 1.00/1.01 = 0.99$, $V_{400} = V_{300}/RR_{350}$, etc. Real and imaginary components are calculated from $R = V \cos \phi$ and $I = V \sin \phi$ respectively, where ϕ is the appropriate $\Sigma \Delta \phi$ value for each V. Note that R is always positive, while I may be either positive or negative.

By plotting the R- and I-curves we can get additional estimates of conductor depth – half the horizontal distance between maxima and minima on either curve. The current centre of the upper part of the conductor is located directly below the point of maximum slope of the R-curve. Also the maximum R- and I-values may be used with characteristic curves to determine the thickness-conductivity product.

7.8.5 *Dipping sheet of finite depth extent*

(a) *Dip-angle measurement – fixed transmitter.* This type of conductor is far more common and realistic than the sheet of infinite depth. Unfortunately the mathematical analysis is not possible in most cases and it is necessary to resort to modelling to simulate field profiles.

Figure 7.42 shows four profiles taken over model sheets of finite depth, dipping 30°, 45° and 60°. The effect of decreasing dip is to decrease the tilt angles on the

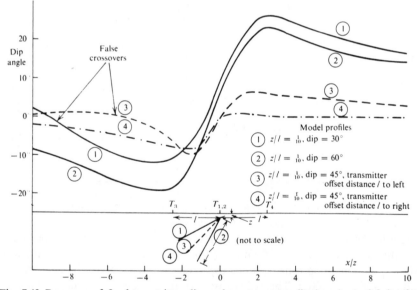

Fig. 7.42 Response of fixed-transmitter dip-angle system to a dipping sheet of finite depth extent.

downdip side and increase them on the updip side of the sheet. In addition, the zero crossover is shifted slightly toward the downdip side with decreasing dip angle of the sheet. As a result the profile asymmetry increases inversely with the dip angle.

The above characteristics apply to a set-up in which the transmitter is located directly over the top of the sheet. When the transmitter is offset on the updip or footwall side, the tilt angles on that side are practically zero, which is to be expected since the coupling is very weak. If the transmitter is located downdip or on the hanging wall side, the asymmetry is less, but the profile is sharply peaked on that side, again similar to the parallel-line EM system.

Profiles 1 and 3 exhibit reverse, or false, crossovers on the downdip side, so-called because the tilt angles on either side of the zero value point away from the crossover rather than towards it. False crossovers, caused either by beds of shallow dip or by multiple conductors, are common enough in field work with dip-angle EM.

The effect of increasing depth extent of the conducting sheet is to reduce the asymmetry of the profiles. The footwall side is not changed appreciably, but the reverse crossover on the downdip side is moved out and its slope decreased.

Figure 7.43 shows three profiles over a horizontal sheet conductor, corresponding to different positions of the transmitter. When the latter is roughly over the

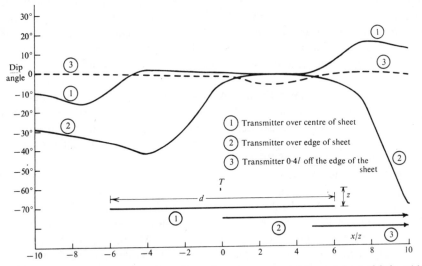

Fig. 7.43 Response of fixed-transmitter dip-angle system over a horizontal sheet of finite width and length; $d/l = 1\cdot2$, $z/l = 0\cdot1$, $\ell/l = 2\cdot8$.

middle of the sheet, the curve is symmetrical, with a proper crossover inside either edge and a very mild false crossover at the centre. If the transmitter is situated near one edge of the conductor there is not real crossover at all and large dip angles in the same sense appear near each edge. Finally, if the transmitter is located off the sheet, the dip angles are quite small.

None of these horizontal sheet profiles has the usual characteristics of the familiar crossover curves associated with dip-angle measurements. Although profile 1 has a certain symmetry it might be assumed, as we shall see later, to be caused by two distinct steeply dipping conductors. The other two would be difficult to interpret.

(b) *Parallel-line method*. Most of the remarks concerning the fixed transmitter system can be applied here as well. Asymmetry increases inversely with dip angle and inversely with depth extent of the conductor. The maximum dip angles occur on the updip side and the peaks, of course, are sharper than for the fixed transmitter system. The crossover is shifted slightly downdip with shallow dip angle. For a horizontal sheet the profile is similar to curve 1 in fig. 7.43, the peaks being much sharper, however, and the flanks falling off to zero more rapidly.

(c) *VLF and AFMAG systems*. Again the results are somewhat similar to the fixed transmitter, with one significant exception. Although the profile asymmetry decreases with dip, the maximum tilt angle is on the hanging wall or downdip side – just the opposite to the curves obtained with local power sources. In addition the profile asymmetry is not as marked for shallow dip angles as in the previous two methods, unless the conductor depth extent is relatively small.

The first difference may be explained on the basis of the circuit analogy for two long wires, representing the top and bottom (or current return) edges of the conductor, with current in the lower wire flowing in the opposite sense to that resulting from a local transmitter source. The fact that the dipping conductor of large depth extent has a fairly symmetric AFMAG or VLF profile indicates that the conductor appears to be a line source (see §7.8.4a and following).

(d) *Horizontal-loop method*. Figure 7.44 shows the response of a horizontal-loop EM over a dipping sheet conductor for dip angles of $60°$, $30°$ and $0°$. The first two profiles are for sheets of great depth extent, while the horizontal sheet has a width $d = 2l$ or twice the transmitter-receiver spread. The resistivities are in the medium range (order of 10^{-2} Ωm).

The dipping sheet results in an asymmetry of the profile, although this is not very pronounced for dip angles larger than $45°$. The in-phase peak negative response is increased as the dip angle decreases, the values being 57%, 30% and 25% for angles of $30°$, $60°$ and $90°$ respectively. (The latter curve is not shown here, but is roughly equivalent to profile 2 in fig. 7.39.) At the same time, as the dip decreases, the negative peak is displaced toward the hanging wall side, as are the zero crossover points, while the positive peak on the same side is enhanced and the positive peak on the footwall decreased. The result is an in-phase curve which has a steeper slope on the downdip side.

The quadrature response is also asymmetric. The peak negative increases slightly as the dip angle decreases, the zero crossover points are shifted downdip, but there is little significant change in the positive peaks. The negative peak, however, is shifted toward the footwall – just the opposite to the in-phase curve. Thus the steep slope on the quadrature profile is on the updip side. The direction and the approximate dip of the conductor can be estimated from the characteristics of the two curves.

The effect of a horizontal sheet, seen in curve 3, is to split the in-phase profile into two equal negative peaks, located in the interval $\frac{1}{2} \leqslant x/l \leqslant 1$, with a positive maximum over the centre of the sheet. The magnitudes of these peaks are greatly affected by the depth of the sheet as well as its width. For instance, when $z/l = \frac{1}{10}$, the negative peaks go to 110%, while the positive is 75% (but still negative); for $z/l = \frac{1}{2}$, the negatives are about 4% and the centre peak is 20% positive. For a narrower sheet (0·3l) the negative maxima are only 7%, while the centre is 5% positive and practically flat.

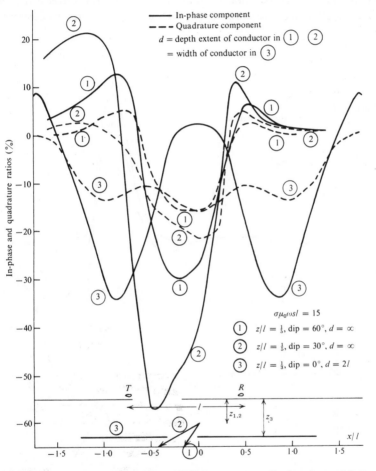

Fig. 7.44 Response of the horizontal-loop EM-system over a dipping sheet of finite depth extent. Traverse normal to strike.

The quadrature profile shows three minor negative maxima, the outside pair being closer to the edge of the sheet than the corresponding in-phase maxima, while the centre peak is also negative. Again, this profile is largely influenced by the depth of the conductor, while the width has less effect.

The horizontal sheet could easily be mistaken for a pair of vertical conductors, if the in-phase profile were considered alone. The quadrature curve, however, is generally different.

The depth extent of the conductor influences the horizontal-loop profiles only if it is small; if the depth extent is greater than $2l$, the profile is practically that of an infinite sheet.

(e) *Turam method.* Three Turam profiles for in-phase ratio over a model sheet dipping 60°, 30° and 0° are shown in fig. 7.45. Curve symmetry is not significantly

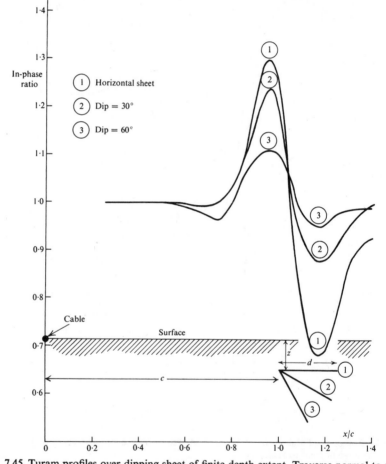

Fig. 7.45 Turam profiles over dipping sheet of finite depth extent. Traverse normal to strike; $z/d = \frac{1}{2}$; $z/c = \frac{1}{8}$.

affected until the dip angle is less than 45°, although the positive peak increases steadily with decreasing dip; a vertical sheet of the same size and depth produces a maximum of 1·09. The negative peak is on the downdip side, while the positive is slightly displaced toward the cable.

When the sheet dips toward the cable, rather than away from it, the maxima are reduced, although the general shape is similar. In fact there is a particularly geometry of thin dipping sheet and Turam cable for which the uniform primary field is parallel to the conductor face, when the resultant induced currents are practically zero, since they would have to flow across the narrow dimension. This situation can be demonstrated in model measurements and has also been found in the field. For dip angles even shallower than this (sheet still dipping towards the cable), the curves are inverted.

In such cases and also because the long, grounded, transmitter cable may provide galvanic as well as induction currents in the conductor (because of the ground contact), it is sometimes difficult to determine the direction of dip. By repeating the measurements with the cable on the other side of the conductor and/or by using two frequencies, the difficulty may be resolved.

7.8.6 *Sheet conductors – miscellaneous considerations*

(a) *Two conductors.* Using an extension of the coupled-circuit analysis (see §7.3.2 and §7.3.4a), Grant and West (1965, pp. 532–6) have shown that the effect of two conductors is not generally the sum of each one acting alone, but involves mutual coupling between them. This means that eq. (7.17b) becomes

$$\frac{e_s}{e_p} = - \left(\frac{k_{TC_1}k_{C_1R}}{k_{TR}}\right)\left(\frac{Q_1^{\,2} + jQ_1}{1 + Q_1^{\,2}}\right) - \left(\frac{k_{TC_2}k_{C_2R}}{k_{TR}}\right)\left(\frac{Q_2^{\,2} + jQ_2}{1 + Q_2^{\,2}}\right) -$$
$$- \frac{k_{C_1C_2}}{k_{TR}}\,(k_{TC_1}k_{C_2R} + k_{TC_2}k_{C_1R})\left\{\frac{Q_1Q_2(1 - Q_1Q_2) - jQ_1Q_2(Q_1 + Q_2)}{(1 + Q_1^{\,2})(1 + Q_2^{\,2})}\right\}, \quad (7.28a)$$

where $Q_1 = \omega L_1/R_1$ for conductor C_1, $Q_2 = \omega L_2/R_2$ for conductor C_2, and terms in $k^2{}_{C_1C_2}$ are neglected since the coupling between the conductors is assumed to be relatively weak.

Two limiting cases arise. If the conductors are very far apart, $k_{C_1C_2} = 0$ and

$$\frac{e_s}{e_p} = - \left(\frac{k_{TC_1}k_{C_1R}}{k_{TR}}\right)\left(\frac{Q_1^{\,2} + jQ_1}{1 + Q_1^{\,2}}\right) - \left(\frac{k_{TC_2}k_{C_2R}}{k_{TR}}\right)\left(\frac{Q_2^{\,2} + jQ_2}{1 + Q_2^{\,2}}\right), \quad (7.28b)$$

which is the sum of the two conductors separately, as one might expect. On the other hand, if the two are very close together, $k_{C_1C_2} \approx 1, k_{TC_1} = k_{TC_2}, k_{C_1R} = k_{C_2R}$, and

$$\frac{e_s}{e_p} = - \left(\frac{k_{TC}k_{CR}}{k_{TR}}\right)\left\{\frac{(Q_1 + Q_2)^2 + j(Q_1 + Q_2)}{1 + (Q_1 + Q_2)^2}\right\}, \quad (7.28c)$$

which is the response of a single loop having response parameter $Q_1 + Q_2$. The coupling between the two conductors may be positive or negative. Model studies have shown these expressions to be quite good approximations.

(b) *Multiple conductors.* The various EM ground methods show different responses to multiple conductors; clearly the differences are determined by the geometry of the particular system, as well as that of the conductors. The simplest dip-angle system to use as an illustration is VLF or AFMAG, in which the conductors are uniformly energized by the remote transmitter. Figures 7.46 and 7.47 show calculated profiles over several vertical sheet configurations. The plots were made from a modification of eq. (7.23*b*), assuming no coupling between the conductors. When the spacing between the sheets is less than 2*z*, these curves will not be exact, since the mutual coupling would make the maximum tilt angles somewhat larger.

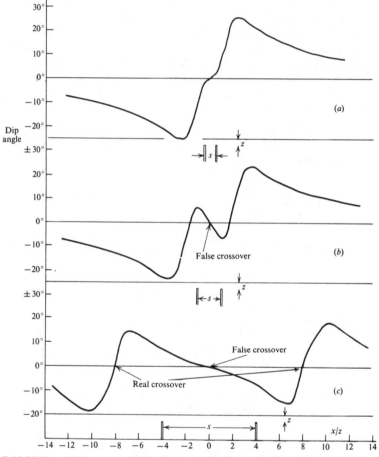

Fig. 7.46 VLF profiles over two identical vertical sheets of great depth extent. (*a*) $s = 2z$; (*b*) $s = 4z$; (*c*) $s = 16z$.

From fig. 7.46 it can be seen that the proper crossovers are displaced to the flanks of the curves, beyond the actual locations of the sheets, unless their separation is less than twice the depth, in which case the profile indicates a single thick conductor. The examples (*a*) and (*b*) of fig. 7.47 show that a conductor which is deeper or smaller (or of lower conductivity, although this situation is not illustrated), than its neighbour will give a relatively weak response.

Discrimination among multiple conductors is even more difficult with the fixed-transmitter dip-angle technique. Unless the transmitter is located fairly well between two identical sheets, the more remote will hardly respond at all. In this case the two are not energized equally. Consequently it is more difficult to resolve multiple conductors with this type of EM.

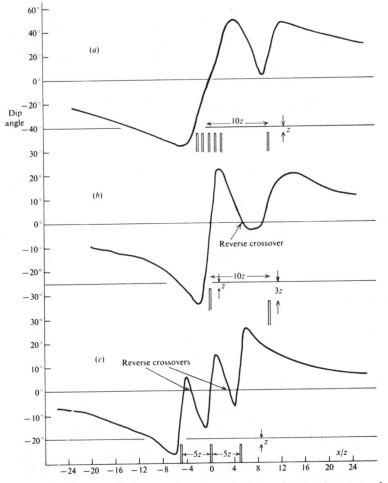

Fig. 7.47 VLF profiles over multiple vertical-sheet conductors. (*a*) Five sheets spaced *z* apart plus one sheet distant 10*z* from the mid-point of the five sheets, all at depth *z*; (*b*) two sheets at depths of *z* and 3*z*, spaced 10*z* apart; (*c*) three sheets at depth *z* and spaced 5*z* apart.

The parallel-line dip-angle resolves adjacent conductors better than the fixed-transmitter unit, as might be expected. Furthermore the crossovers are in the right place. Even so, when the separation is less than twice the depth, the broadside traverse will not distinguish between two distinct sheets, a single thick one, or a flat sheet.

The horizontal-loop EM is perhaps slightly better than the broadside array in resolving two conductors. The in-phase profile shows two peaks at the proper location when the separation is twice the depth or greater, while the quadrature component is almost the same. These curves are shown in fig. 7.48. Coupling between the sheets affects the quadrature response particularly, so that the conductivity of two separate sheets would appear to be lower than when they are close together.

The resolving power of Turam over multiple conductors is probably not as good as with the horizontal-loop EM, because of the fixed transmitter, which would generally be located to one side. However, the fact that two receivers are used with fairly close spacing may help the resolution somewhat.

(c) *Effect of conductor width.* With dip-angle EM equipment it is difficult to distinguish between a wide conductor and two thin sheets unless, as we have seen, the separation is appreciable.

One method of determining conductor width with the fixed-transmitter system is to offset the transmitter from the edges of the conductor 200 ft or less and make several traverses normal to the conductor. As the receiver-transmitter separation increases, the crossovers shift towards the edge of the conductor remote from the transmitter. By repeating this measurement with the transmitter on either side of the conductor, a fairly good outline of its lateral extent is achieved.

The broadside array is theoretically better than the other methods for thick conductors, since the crossover at the centre is reversed. Over separate thin sheets the centre crossover is in the normal sense. However, these crossovers are usually quite small and the difference between normal and reverse may be difficult to distinguish.

The effect of a dike-type of vertical conductor of considerable width on the horizontal-loop system is shown in fig. 7.48. Here the response is quite different from that obtained over two parallel thin sheets, as shown in the same figure, unless the width is somewhat greater than the receiver-transmitter separation (see also fig. 7.44).

It is possible to get a fair idea of conductor width with the horizontal-loop set. Since the zero crossovers occur at $x = \pm l/2$ for a thin sheet, any excess, greater than l, in the separation between these zeros is an estimate of conductor width. Here again the horizontal-loop EM has a distinct advantage over dip-angle EM.

As noted previously (see §7.8.4f), the near-surface width of conductor may be estimated with the Turam unit by taking the same profile twice with the transmitter cable on alternate sides.

(d) *Effect of conductive overburden.* In all preceding discussions the assumption has been that the conductors were immersed in a medium of very high resistivity. If the host rock and overlay are homogeneous and of relatively good conductivity, the attenuation will be appreciable, resulting in profiles of decreased amplitude. As mentioned before, this situation may be simulated in model work.

by making the measurements in a tank of suitable liquid; theoretically the homogeneous conducting medium can be allowed for as well.

However, when the overburden is a good conductor, lying as a horizontal slab over the anomaly source, the problem is more complicated. Field situations exist where swampy overburden of good conductivity (1–50 Ωm) has apparently masked out any response from metallic conductors lying no more than 50 ft below surface. In one location of this type in northern Quebec dip-angle, horizontal-loop

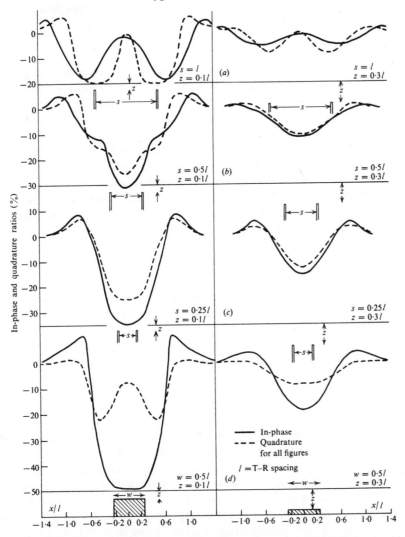

Fig. 7.48 Horizontal-loop EM profiles over various conductors of great depth extent. (a) Two thin sheets with spacing equal to the T–R spacing; (b) two thin sheets with spacing equal to one-half the T–R spacing; (c) two thin sheets with spacing equal to one-quarter the T–R spacing; (d) conducting dike with width equal to half the T–R spacing.

and Turam EM, as well as induced-polarization methods, all failed to give any significant response.

It is possible to test the effect of overburden in a model. The result of placing a conductive slab over the conductor of interest is to decrease the amplitude and shift the phase of the profiles, so that the anomaly appears to be deeper and of different conductivity than it actually is. Where two frequencies are employed, the ratio (that is, of dip angle, quadrature, amplitude, whatever the method) between them is decreased. Of course, if the overlying slab is of limited lateral extent it will also introduce an anomaly of its own, as for a horizontal sheet; such over-burden anomalies are, unfortunately, quite common.

(e) *Determining conductor strike length.* The strike of a relatively long sheet conductor is fairly well determined by joining crossovers on successive dip-angle profiles or the peaks on successive Slingram or Turam profiles. In the presence of broad or multiple conductors this is never as simple as it sounds. Additional knowledge of the geology is essential.

Since the fixed-transmitter system will give anomalous profiles on traverses beyond the ends of the conductor, it is necessary to relocate the transmitter on strike, but several hundred feet beyond the end of the conductor, to determine its strike length. The traverses near the transmitter will then be barren until the end of the conductor is crossed. With the other EM sets, strike length is quite well marked by the absence of the anomaly on traverses beyond the ends of the conductor.

7.8.7 *Sheet conductors – characteristic curves*

(a) *General.* The usual procedure in EM interpretation – comparison of field profiles with theoretical and model results – requires a good-sized library of the latter type of curves. A considerable advantage is achieved by combining the theoretical and model data into characteristic curves which emphasize certain features of the EM profiles, such as maximum tilt angle and crossover slope for dip-angle measurements, or in-phase and quadrature response for phase-component systems.

(b) *Dip-angle characteristic curves.* We have mentioned previously that the peak dip angle and the slope at the crossover are functions of both depth and conductivity. However, conductivity changes affect both parameters about the same, that is, the ratio of slope to peak response is approximately constant (see fig. 7.33). Thus a simple characteristic, which works fairly well for steeply dipping conductors, may be prepared by plotting this ratio against the known depth, or z/l ratio, taken from model results. An example is shown in fig. 7.49a; the slope near the zero crossover is plotted in the form of a ratio, degrees/(x/l), against z/l, where l is the transmitter-receiver spacing, in this case for the broadside array.

A more useful characteristic curve, which provides both dip and depth for the semi-infinite sheet conductor, has been developed by Grant and West (1965, pp. 559–60) and is illustrated in fig. 7.49*b*. Here the ratio of peak tilt-angle magnitudes on either side of the crossover is plotted against their total amplitudes. If the resultant point falls reasonably on this diagram, the dip angle may be interpolated and the depth calculated from the relation $z = kl \sin \alpha$. This plot is developed for the fixed-transmitter set and will give poor results if the sheet dimensions are less than the transmitter-receiver separation.

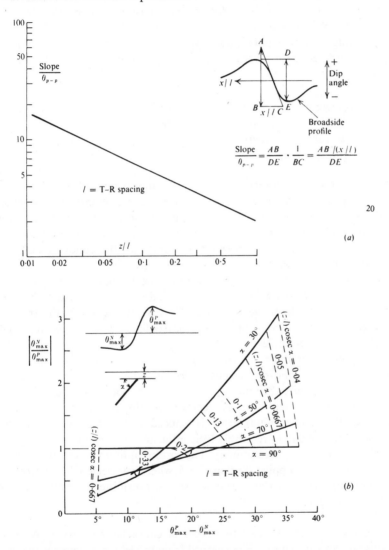

Fig. 7.49 Characteristic curves for semi-infinite dipping thin-sheet conductor. (*a*) Depth to the top of a steeply dipping thin-sheet conductor from the crossover slope and peak tilt. Broadside array. (From Ward, 1967.); (*b*) depth and dip using fixed-transmitter, dip-angle system. (From Grant and West, 1965.)

(c) *Parameters for characteristic curves.* The response parameter, discussed in §7.3.5 and illustrated in fig. 7.11 for the simple electric circuit analogy, is a significant quantity in EM characteristic curves. Recalling the discussion of scale models in §7.8.2, the ratio $\mu f l^2/\rho$ – or what is the same thing, $\mu\omega\sigma l^2$ – is dimensionless and invariant in passing from a field to a model system. Similarly the response parameter Q is dimensionless and has the same form, a product of conductivity, permeability, frequency and area (or linear dimension squared); this follows since $L \propto \mu l$, $R \propto \rho l/A$, hence $\omega L/R \propto \mu\omega\sigma A$ where A = area.

Employing a more sophisticated approach, electromagnetic theory gives the response parameters for several specific configurations. A few of these are listed below:

Sphere	$\mu\omega\sigma a^2$,	a = radius,
Disk	$\mu\omega\sigma s a$,	s = thickness,
Thin sheet	$\mu\omega\sigma s c$,	
Half space	$\mu\omega\sigma c^2$,	c = conductor-EM unit spacing.

Note that the linear dimensions a, s and c appearing in these expressions are particularly significant in determining the pattern and intensity of secondary current flow in the conductor. Frequently there are several dimensional parameters of this type, in which case we are free to choose any two of them. However, the effect of some of these parameters may vary with the conductor-EM system geometry; this is evidently true with moving source units.

In drawing up characteristic curves for thin sheet conductors, for example, it is customary to use as parameters the thickness and whichever fixed dimension of the EM system controls the response. Thus we would insert the transmitter-receiver spacing in artificial source dip-angle (see fig. 7.33), the height above ground in airborne double-dipole systems (see fig. 7.53b), or the horizontal distance between transmitter and conductor in Turam.

As mentioned in an earlier discussion of characteristic curves (see §2.6.5), we should try to minimize the number of variables in order to produce a reasonable number of curves. Since we are at liberty to employ one dimension as a scaling parameter, the number of unknowns is reduced by one. This scaling unit may be either a length which has relatively little influence on the conductor response, or one which can be measured directly. The former option was taken in the gravity example for the thin dipping sheet in §2.6.5, where we selected the depth extent; as an EM example of the latter, we could use the transmitter-receiver separation or the aircraft height.

(d) *Dip-angle curves.* At least two types of characteristic curves involving the response parameter have been developed from model measurements in dip-angle interpretation. In one of these, two source frequencies are required. The ratio of peak high-frequency response $(\theta^{neg}/\theta^{pos})_{HF}$ is plotted against low-frequency peak response $(\theta^{neg}/\theta^{pos})_{LF}$ for various depths and dip angles. A large set of curves is required for different dip angles, each set being drawn for constant depth and constant response parameter.

The second type, for single frequency, consists of plotting peak-to-peak ($\theta^{\mathrm{pos}} - \theta^{\mathrm{neg}}$) amplitude against response parameter for different depths, as illustrated in fig. 7.50, for fixed-transmitter and broadside array. The depth must be approximately known to use these curves. A complete catalogue again requires a characteristic curve for each dip angle. Presumably similar results could be developed for VLF.

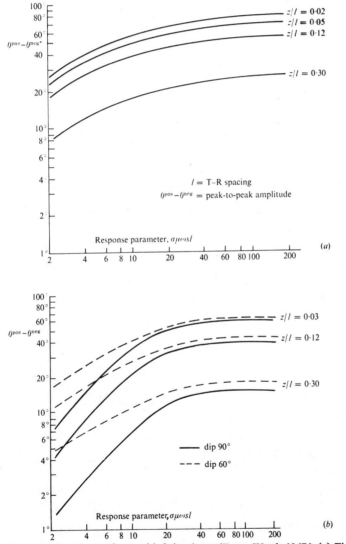

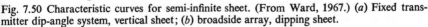

Fig. 7.50 Characteristic curves for semi-infinite sheet. (From Ward, 1967.) (*a*) Fixed transmitter dip-angle system, vertical sheet; (*b*) broadside array, dipping sheet.

From the plots illustrated in fig. 7.50 it is possible to find $\mu\omega\sigma sl$, if we can obtain the depth by other means. We know ω and l and can assume μ to be the free-space value in most cases. If a reasonable estimate of the width s can be made, the approximate conductivity may be determined.

(e) *Phase-component curves.* Standard characteristic curve sets for the thin sheet are illustrated in fig. 7.51 for the horizontal-loop EM. The ordinate and abscissa

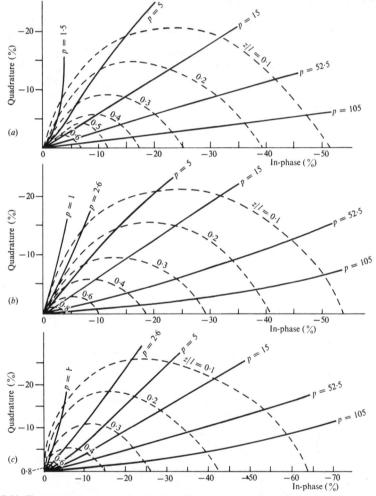

Fig. 7.51 Characteristic curves for horizontal-loop system over dipping sheet. $p = \sigma\mu\omega sl$. (From Strangway, 1966.) (*a*) Dip = 90°; (*b*) dip = 60°; (*c*) dip = 30°.

are, respectively, the quadrature and in-phase response taken midway between the zero crossovers. Three sets are shown, for dip angles of 90°, 60° and 30°. For the vertical sheet the responses are maximum values; this is not so for the dipping sheets, however, because of the asymmetry of the profiles. The depth and response parameter may be read off these curves. If the conductor width can be estimated by other means, it is then possible to get some idea of the conductivity.

There is some doubt, however, about the significance in separating the conductivity from the conductor width in the product σs. The chances of a conductor being homogeneous over its entire width are extremely small. As a result the values obtained for conductivity of massive sulphides are an average and generally fall in quite a narrow range – something like 0·1 to 10 mhos/m – which is a few orders of magnitude smaller than the conductivities usually assigned to the majority of sulphides in tables (see §5.4.1). Possibly the conductivity-thickness product is a more practical parameter and indeed it would give a rough estimate of volume or tonnage equally well.

Similar sets of curves are available for Turam, using peak quadrature and in-phase response. One of these, for a thin sheet dipping 60° and of 20 m depth, is shown in fig. 7.52. The response parameter $\lambda = 10^5/\sigma sf$, where σ is the conductivity in mhos/m, s is the sheet thickness in metres and f the transmitter frequency. This plot allows for a variable strike length of conductor, L; on the other hand the dimensions of the transmitter coil affect the curve somewhat.

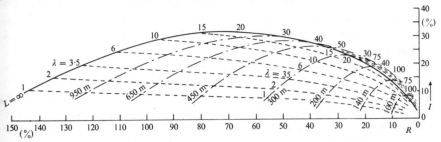

Fig. 7.52 Characteristic curves for Turam system over thin dipping-sheet conductor of variable length. Depth $h = 20$ m, dip $\phi = 60°$, trans. loop 600×1200 m, $L = $ conductor strike length, $\lambda = 10^5/\sigma sf$, $\mu \approx \mu_0$ (Mks units).

It should be mentioned that, although these universal plots are extremely useful for interpretation, they should not be used alone without comparison of the field profiles with a standard catalogue of theoretical and model curves as well. That is to say, the characteristic curves make use of only a couple of critical points on the standard profiles; the field plots should fit the latter reasonably well before the former may be used with confidence.

7.8.8 *Three-dimensional conductors – the sphere*

The theoretical response of a spherical conductor has been solved for two EM conditions, in a uniform a.c. field and in a dipole field. The solutions are in the form of Legendre polynomials, similar to the result for the sphere in resistivity (see §8.3.5), but considerably more complicated, particularly for the dipole field. The two solutions are the same when the dipole is far from the sphere.

As mentioned above, the response parameter for the sphere is $\mu\omega\sigma a^2$ and the response function a more complex form of $A + jB$ shown in fig. 7.11. When $\mu = \mu_0$ the uniform field response function is very similar to that for a circular loop. In fact reasonable profiles may be worked out from the circuit analogy on this basis, the mutual inductance being given by eqs. (7.18d) and (7.18e) for the dipole and eq. (7.18g) for long-wire coupling. The dipole coupling is suitable for analysis of airborne double-dipole results, where the source and detector, being relatively close together, can be considered common and the sphere is remote from both.

The lump conductor, approximated by a sphere, is not a very suitable target for dip-angle or long-wire transmitter methods. The response of a horizontal-loop system on a traverse directly over a sphere is similar to that for a dike; however, when the coupling is tight, that is, when the sphere is close to surface, a double-peak profile results.

7.8.9 *Interpretation of airborne EM results*

(a) *General.* Although fundamentally the same as ground work, the objective in airborne EM surveys is more modest. Generally the airborne results locate conductors, outline their approximate extent and perhaps provide information enough to estimate their characteristics. Ground follow-up is mandatory anyway, hence the airborne survey performs the function of rapid reconnaissance and elimination of barren ground.

As a rule the airborne operation produces an excess of anomalies, since it detects without discrimination swamps, shear zones, faults and similar large-scale features, as well as graphitic zones and metal conductors. It should be, and almost always is, carried out in conjunction with airborne magnetics. Obviously the combination of two or more geophysical methods generally will produce much more than the information obtained from each method separately; in airborne work the enhancement is even greater, because of the abundance of anomalies and because the geologic knowledge is frequently quite limited.

Scale modelling is very useful as an interpretation aid in airborne work. Three-dimensional (spherical) and sheet models are usually employed. The height of the aircraft, rather than transmitter-receiver separation, will normally be the significant control dimension. As a result, the response parameters for sphere and sheet may be $\mu \omega \sigma a h$ and $\mu \omega \sigma s h$, where h is vertical height.

(b) *Preliminary interpretation.* Since the classic asymmetric dip-angle curve is not recorded directly in airborne EM records (see §7.6.4), the interpretation in various systems involves a consideration of peak amplitude, width and y-axis extent of a basically symmetric response. Consequently the initial step is to classify the anomaly as three-dimensional, wide and flat-lying, or long, thin and probably steeply dipping. This rough classification may be made from the appearance of single profiles and correlation between adjacent flight lines.

Concurrently the interpreter notes the altitude and magnetic records to see if the anomaly may have been enhanced or created by changing height and whether there is any correlation between EM and magnetic results. This correlation is extremely useful, because with few exceptions, most massive sulphides contain some magnetite or pyrrhotite, so that coincident or adjacent magnetic anomalies are particularly attractive.

Having established this initial classification, any possible quantitative interpretation will depend on the type of EM system used and the simplicity of the anomaly pattern. We shall consider the dual-frequency quadrature and phase-component airborne sets in turn.

(c) *Quadrature system.* Considerable information on the Hunting Canso airborne EM has been provided by Paterson (1961). The records made with airborne sets of this type, which trail a bird, are somewhat asymmetrical – peak response is shifted opposite to the direction of flight. For the Hunting equipment, the height of the aircraft is 500 ft with the bird some 200 ft below; in this altitude range the signal response falls off at about the fourth power of h over a vertical half-plane. Practically a sheet of 1000 ft depth extent is equivalent to the half-plane. The response is halved if the flight direction is 45° to strike rather than 90°. At the same time the peak is shifted in the direction of flight.

For a dipping sheet the main peak shifts slightly, opposite to flight direction; a minor negative develops on the downdip side when the sheet is dipping in the flight direction. If the sheet dips towards the approaching aircraft, the main peak is shifted well downdip and a minor positive appears on the footwall side. The amplitude increases as the dip becomes shallower, although this is more pronounced when the traverse direction is downdip.

If the sheet is horizontal the peak response is about ten times that of the same sheet in a vertical position and this peak occurs over the edge nearest the approaching aircraft. A broad dike conductor gives a similar result. If the lateral extent is great enough, a secondary peak appears near the centre of the sheet.

Multiple conductors are not resolved until their separation is about 70 % of the altitude, after which the double peak resembles that for the horizontal sheet.

Under ideal conditions the maximum penetration for this type of equipment is perhaps 200 ft at 500 ft altitude and 350 ft at 300 ft altitude. In practise these depths may be reduced by a factor of two.

Typical model curves are used to estimate widths of simple shapes like the sheet and sphere. The profile widths are measured between inflection points (which may be difficult to find) and are, of course, related to altitude of flight and aircraft-bird separation as well as conductor width. For example, at a height of 500 ft the semi-infinite vertical sheet produces an anomaly width of about 300 ft. Any excess over this may be conductor width, although it could also be cause by a dipping sheet or a conductive overburden.

The ratio of peak responses at 400 and 2300 Hz is practically independent of aircraft height and provides some estimate of the conductivity-thickness product. A plot of these quantities is shown in fig. 7.53a.

(d) *Phase-component systems.* The interpretation problem is quite similar to the horizontal-loop EM ground set, although the coils are mounted vertically, as described in §7.6.5 (since the sensitivity to conductors of steep dip is greater than for flat-lying conductors when the ratio of h/l is appreciable). The system is symmetrical with respect to flight direction.

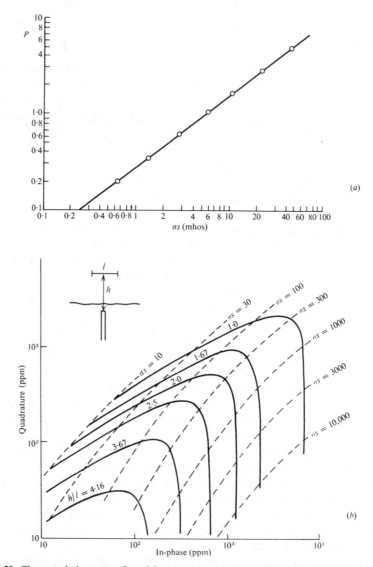

Fig. 7.53 Characteristic curves for airborne systems. (From Ward, 1967.) (*a*) Quadrature system over a half-plane. s = conductor thickness, P = (peak response at 400 Hz)/(peak response at 2300 Hz); (*b*) double-dipole system over a vertical sheet. σs in (mhos/metre) × feet;

The problem of the double-dipole system over a sphere or semi-infinite sheet lends itself to the lumped circuit analysis. For the sphere the coupling can be that of coplanar circular loops. Proceeding with the usual analysis, we get the following value for the in-phase ratio:

$$\frac{e_s}{e_p} = \frac{0 \cdot 8}{\beta} \left(\frac{R_C}{h}\right)^3 \left(\frac{l}{h}\right)^3 \frac{1}{(1 + \alpha^2)^3}. \tag{7.29a}$$

The phase angle or imaginary component, is given approximately by

$$IM \left|\frac{e_s}{e_p}\right| = \tan \phi \approx \frac{1}{Q} \left(\frac{e_s}{e_p}\right), \tag{7.29b}$$

where R_C = radius of sphere, h = distance between the centres of the dipole and the sphere, l = dipole separation, $\alpha = x/h$.

The resultant profiles are clearly symmetrical, fairly sharp (at $x = 2h$ the response is down by a factor of more than 100), and the maximum response falls off with the inverse third power. The depth may be roughly estimated from the value of α at half-maximum; knowing the aircraft altitude one can get some idea of the conductor dimensions. Clearly one flight line should pass almost directly over the conductor centre to get a reasonable response.

Characteristic curves may be prepared, using the in-phase and quadrature

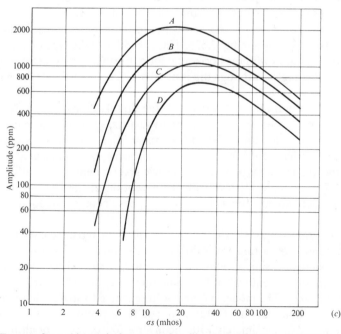

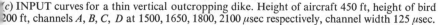

(c) INPUT curves for a thin vertical outcropping dike. Height of aircraft 450 ft, height of bird 200 ft, channels A, B, C, D at 1500, 1650, 1800, 2100 μsec respectively, channel width 125 μsec.

response, to estimate the depth and radius of the body and indicate the conductivity. These curves would be plotted for $\mu\omega\sigma R_C{}^2$ or $\mu\omega\sigma h^2$; they are compiled from model measurements.

Considerable data are available for the double dipole set over thin sheets. The profiles are similar to those for the sphere when the dip angle is large. Peak response decreases in proportion to $h^{-3\cdot3}$ from model measurements, where h in this case is the vertical distance to the top of the sheet. The total width of the profile (measuring from the points where the signal is lost in background noise) is about $8x/l$ when the sheet has steep dip; a long tail of low amplitude is present on the downdip side of the conductor. As the dip becomes shallower this tail and the peak amplitude both increase, although the main peak remains over the top of the sheet. Finally, for a horizontal sheet the response has three peaks, the largest being over the centre and the other two over the edges. The maximum response is about four times that for an equivalent vertical sheet.

Again a rough idea of the depth can be obtained from the width of the curve at half-maximum, unless the dip is very shallow. Like the equivalent ground system, this set is insensitive to direction of traverse. When the flight lines are at angles of 60° or more to the strike the profiles are not affected.

Multiple conductors are not resolved until the separation is about $0\cdot75h$, the value varying somewhat with the altitude. For larger separation the profile splits into two peaks. A similar curve results from a wide dike-like body; when the width is about $h/3$, two peaks appear. As a result it may not be possible to distinguish two separate conductors from a single wide one.

A characteristic curve for the vertical sheet conductor under a coaxial-coil airborne system is shown in fig. 7.53b. This is similar to the type of curve used in ground interpretation with the horizontal loop. Further curves might be required for different dip angles, although dip is usually difficult to estimate from airborne data.

The practical limit of penetration with the double-dipole system is probably about 200 ft. This value will vary with many factors, such as local geologic noise, separation of the coils, altitude, etc. In this regard the two-airplane system has an advantage of perhaps 50 % in depth. Other airborne systems, such as AFMAG, INPUT, the quadrature type and Turair have greater depth penetration (see, for instance, §7.9, example 11), but with the exception of the last, do not usually provide as much information as the airborne phase-component equipment.

(e) *INPUT system.* Figure 7.53c shows a set of characteristic curves for INPUT system response over a thin vertical dike which outcrops. These were obtained by scale modelling, using aluminium and stainless steel sheets 24 × 6 in. and thickness $\frac{1}{8}$ in. or less, equivalent to conducting dikes of 3000 ft strike length and 800 ft depth extent. The model system was insensitive to strike length and depth extent greater than about 1000 ft and 600 ft respectively, while the signal amplitude varied approximately with the inverse fourth power of altitude.

Curves of this type may be used to determine depth and conductivity-thickness product for dike conductors of steep dip, provided the channels and channel widths used in the field survey are similar to those in the diagram.

7.9 Field examples

Because of the great variety of EM field methods, the number of case histories and problems is necessarily large. We have tried to provide examples of all of the techniques commonly used at present.

It should be noted here that many of the diagrams contain abbreviated names or acronyms for different types of EM ground equipment. Thus, VLEM, VEM, and JEM refer to vertical-loop units of various kinds which employ a local transmitter (for the reader who is not familiar with the terms, it is necessary to specify in addition that the transmitter is fixed or movable). As mentioned earlier, VLF refers to 'very low frequency' sets (generally EM16) which use remote transmitters. The abbreviation HLEM stands for 'horizontal-loop (sometimes called Slingram or phase-component) ground EM equipment.

1. An illustration of the vertical-loop fixed-transmitter method is shown in fig. 7.54. This is the well-documented Mobrun sulphide body, 10 miles northeast of Noranda, Quebec. The deposit, whose section is shown in the diagram, contains

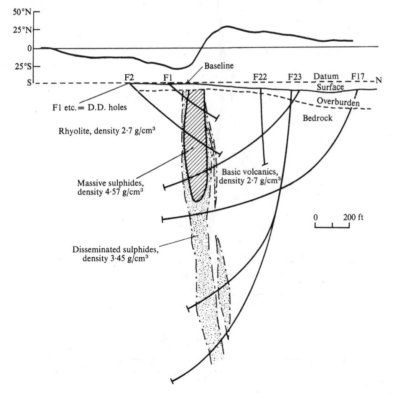

Fig. 7.54 Vertical-loop fixed-transmitter profiles over Mobrun orebody. (After Seigel, 1957.)

about 4 million tons of massive pyrite. This is a classical vertical-loop profile, obtained with the transmitter located 400 ft to the east, on the conductor strike axis. The large maximum dip angles on either side of the crossover are approximately equal, indicating a steeply dipping good conductor at shallow depth. Although the crossover appears to be displaced about 25 ft north of the geometrical centre of the section, this is not significant since the body is 100 ft wide. Depth estimates are not very satisfactory; using the curves of fig. 7.49b, the conductor appears to outcrop.

  2. Figure 7.55a is a good example of a vertical-loop broadside or parallel-line EM survey, carried out over the Uchi Lake sulphide body, some 50 miles east of the Red Lake gold area in northwestern Ontario. The mineralization, consisting of massive pyrite, chalcopyrite and sphalerite, occurs in a complex of rhyolitic and dacitic rocks, in the form of lenses striking and probably plunging northeast with nearly vertical dip.

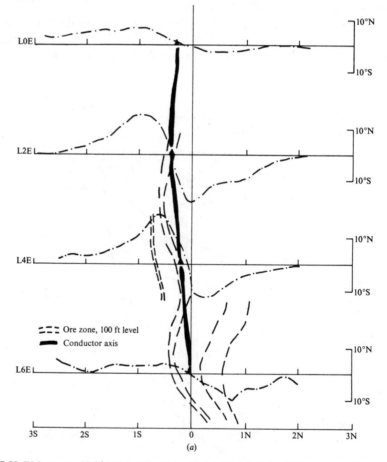

Fig. 7.55 EM survey, Uchi Lake, NW Ontario. (a) Dip-angle broadside profiles. Frequency

The transmitter and receiver, separated by 400 ft, were moved along parallel lines crossing the strike axis; readings were taken every 25 ft. The strong crossovers on lines 2E and 4E coincide with a sub-outcrop of one of these lenses. Strike length is indicated by the small response on lines 0 and 6E. A second ore lens, about 100 ft northeast, mainly rich in sphalerite, was not detected, probably because of poor conductivity.

A single traverse on line 4E with a vertical-loop fixed-transmitter unit is shown

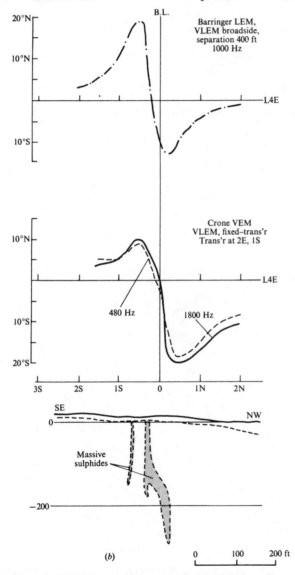

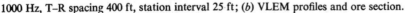

(b)

0   100   200 ft

1000 Hz, T–R spacing 400 ft, station interval 25 ft; (b) VLEM profiles and ore section.

596   Electromagnetic methods

in fig. 7.55*b*. For reference the broadside profile is included. The dual-frequency fixed transmitter (480 Hz, 1800 Hz) was at station 1*S* on line 2*E*. The similar response at both frequencies shows a highly conductive zone. The profile asymmetry, which is surprisingly large, is probably due to the limited depth extent of the lens, since it is known to be approximately vertical.

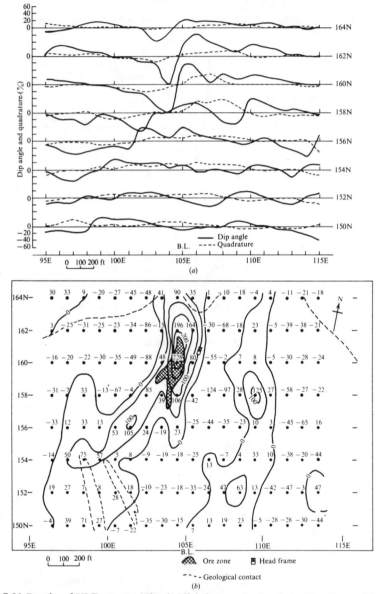

Fig. 7.56 Results of VLF survey, Atlantic Nickel property, Southern New Brunswick. Transmitter station: Cutler, Maine (17·8 kHz). (*a*) VLF profiles; (*b*) VLF contours.

3. The EM16 (VLF) profiles shown in fig. 7.56*a* are taken from a survey made on the Atlantic Nickel property near St Stephen in southern New Brunswick. The predominant geological feature in the area is a stock-like ultrabasic intrusive in metamorphosed sediments. The sulphide mineralization, consisting of pyrrhotite, chalcopyrite and pentlandite, occurs close to contacts between the intrusive and the altered sediments.

The VLF source was the transmitter at Cutler, Maine, located 70 miles to the south. Huge dip angles and steep crossovers on lines 160N and 162N are caused by a very good conductor which outcrops on line 160N. On lines 158N and 156N the profiles indicate that the zone is either becoming wider or splitting in two, one leg continuing south, the other southwest. On the northern lines the dip appears to be steep and to the west.

A different method of displaying VLF data is illustrated in fig. 7.56*b*. The profiles have been converted to contours in such a way that proper crossovers are transformed to positive peak readings, while reverse crossovers become negative values and are discarded; in the process there is also a smoothing effect which reduces the large geologic noise caused by the relatively high frequency of the transmitter (Cutler station operates at 17·8 kHz).

The procedure for conversion is very simple. If $\theta_1, \ldots, \theta_4$ are the dip angles obtained at stations $1, \ldots, 4$, the contour value C_{23} is

$$C_{23} = (\theta_3 + \theta_4) - (\theta_1 + \theta_2),$$

and is plotted midway between stations 2 and 3. The sign of C_{23} is positive in the vicinity of proper crossovers, negative for slopes in the opposite direction.

The contours in fig. 7.56*b* match the outline of the ore zone very well. The smaller anomaly in the vicinity of 109E, which is somewhat overshadowed on the profile display, is brought out more clearly in this plot.

Obviously this technique could be used on vertical-loop EM and AFMAG data as well. However, in the former method neither the anomaly amplitudes nor the source frequencies are large enough to warrant the application; for AFMAG, the erratic azimuth direction of the primary field could distort the contours.

4. The VLF method has been used for mapping shallow geologic structure. Although the significance of airborne results of this type appear slight, probably because the frequencies are too high, ground mapping with the EM16 VLF unit is more promising. Contacts between beds of contrasting resistivity produce profiles which are quite different from the crossovers associated with the dipping-sheet conductor. This, of course, is due to the fact that the profile over the contact corresponds to only half that over the sheet (see fig. 7.43, profile 1).

Using an extension of the analysis of §7.8.4d for the VLF response over a vertical sheet, it can be shown that the profiles for a vertical contact have the form shown in fig. 7.57*a*. Note that both profiles have steeper slopes over the lower resistivity bed and that the in-phase (dip-angle) response peaks at the contact, while the quadrature has a cusp going to zero. The effect of conductive overburden above the contact is to reduce the dip-angle peak and blunt the cusp.

An EM16 survey, made over the Gloucester fault southeast of Ottawa, Ontario, illustrates this type of VLF response. Profiles and a typical geologic section are shown in figs. 7.57*b* and 7.57*c*. The transmitter station was Cutler, Maine. This part of the Ottawa valley region is a sedimentary area well suited to such a tech-

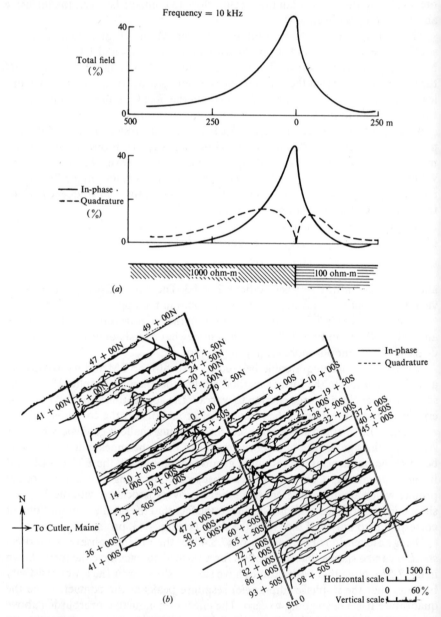

Fig. 7.57 VLF survey over the Gloucester Fault, Eastern Ontario. (*a*) Theoretical VLF profile

nique because of the contrast between the low-resistivity Carlsbad shales and the Nepean and March formations upthrust by Precambrian rocks to the west.

The Gloucester fault is clearly displayed near the middle of fig. 7.57*b*, with the steeper slope on the east, over the low-resistivity beds. A second anomaly, parallel to the Gloucester fault and about 2000 ft east of it, from 21 + 00S to 65 + 50S, indicates a contact with higher resistivity to the east. Still further east, from 32 + 00S to 60 + 50S, there appears to be a third feature which has higher resistivity on the west. At 32 + 00S it is about 1300 ft east of the second contact but, because of a more northerly strike, the separation is 800 ft at 60 + 50S, producing a crossover type of response on that line.

5. Figure 7.58*a* shows a pair of AFMAG profiles over the Mattagami Lake sulphide deposit at Watson Lake, about 90 miles north of Amos, Quebec. The ore zone, shown in rough outline, contains sphalerite, chalcopyrite and pyrite and is associated with basic rocks which have intruded the regional volcanics and sediments. A vertical section through line 4 + 00W is displayed in fig. 7.58*b*. Vectors drawn on the AFMAG profile give the azimuth direction of the field at various points of the traverse.

Both the high- and low-frequency anomalies are larger on line 1 + 00E than on 4 + 00W, because the zone plunges to the west; the depth of cover is about 35 ft on the east line and at least 300 ft on 4 + 00W. The 510-Hz response is greatly affected by the increased depth at 4 + 00W. It is difficult to get much quantitative information concerning the ore zone from these profiles. Their general shape indicates dip to the south, a widening of the zone to the west and a depth greater than the actual depth, particularly on line 1 + 00E.

6. Three horizontal-loop EM (HLEM) profiles, from a property near Woburn, Quebec, are shown in fig. 7.59. The mineralization consists of several dipping sulphide lenses located in a shear zone near a contact between volcanics and

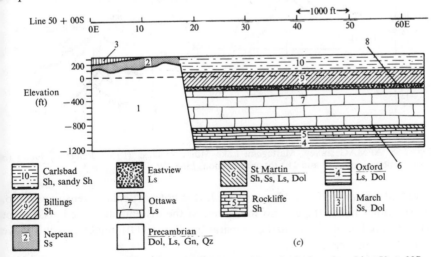

over a vertical contact; (*b*) plot of the field data; (*c*) geological section, Line 50 + 00S.

sediments. The sulphide content ranges from massive to disseminated, with considerable mineralization between the main zones.

Qualitatively, all three profiles indicate a single sheet-like conductor of shallow depth dipping west. In addition, it is probably a metallic conductor, since the ratio of peak in-phase to peak quadrature is about 4 in each case. Quantitatively we can use the peak values with the characteristic curves of fig. 7.51b, assuming the dip to be about 60°, to calculate some numerical results. For a transmitter receiver separation of $l = 200$ ft and frequency $f = 800$ Hz, we find the following

Lines	RE (max)	IM (max)	Depth	$\mu_0 \omega \sigma s l$	z	σs	
36S	−28%	−7%	0·27l	38	55ft	10	mhos
38S	−38%	−11%	0·17l	33	33	8·5	
40S	−40%	−9%	0·16l	38	30	10	

Obviously more information can be obtained from this type of EM survey than any we have discussed so far. Although the 60° dip angle is only an estimate, it is clear that there is a dip to the west and the results would not be much different

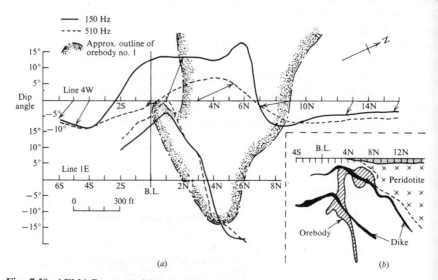

Fig. 7.58 AFMAG survey, Mattagami Lake Mines, Quebec. (From Paterson, 1966.) (*a*) Profiles along Lines 4W and 1E; (*b*) cross-section, Line 4W.

if we had assumed 90° or 45°, since HLEM is not particularly sensitive to dip. On the other hand, there is no indication of the disseminated sections, except perhaps on line 40S. A smaller separation ($l = 100$ ft) might possibly have detected these zones.

7. The Murray sulphide deposit, in Restigouche County, northern New

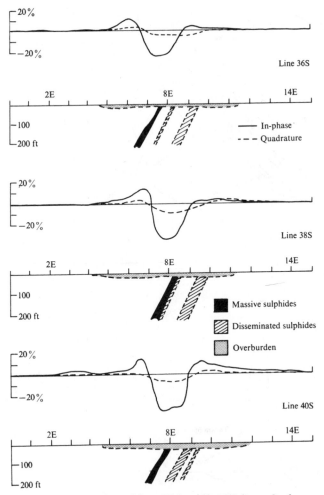

Fig. 7.59 Horizontal-loop EM profiles, Woburn, Quebec.

Brunswick, provides a good example of the Turam EM technique. This body, massive pyrite with sphalerite, galena and chalcopyrite, occurs as a replacement in a chlorite schist along the axis of a reverse drag fold in Ordovician sediments. Considerable leaching and weathering of the near-surface mineralization has produced a heavy gossan cover, varying in thickness from 50 to 200 ft, which masks detection of the main body by EM methods.

Turam field-strength ratios, shown in contour form in fig. 7.60a, correlate well with the outline of the sulphide at 200 ft depth. The phase-shift contours are said to be similar. This is only a small part of the whole Turam survey which covered an area about $1\frac{1}{4}$ miles E–W by $\frac{1}{2}$ mile N–S; extensive zones of graphitic schists also produced strong anomalies north and east of the sulphide body.

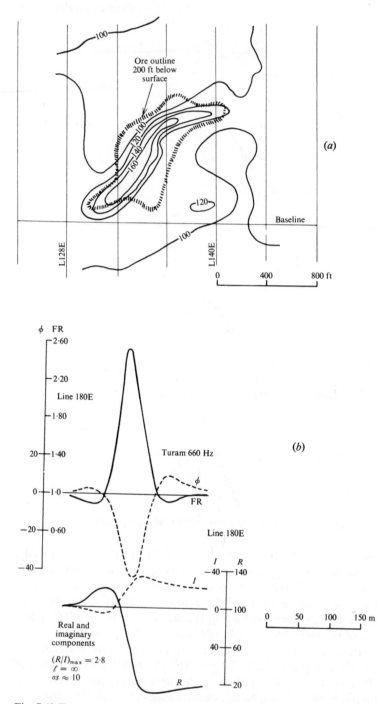

Fig. 7.60 Turam survey over the Murray sulphide deposit, Northern New Brunswick. (After

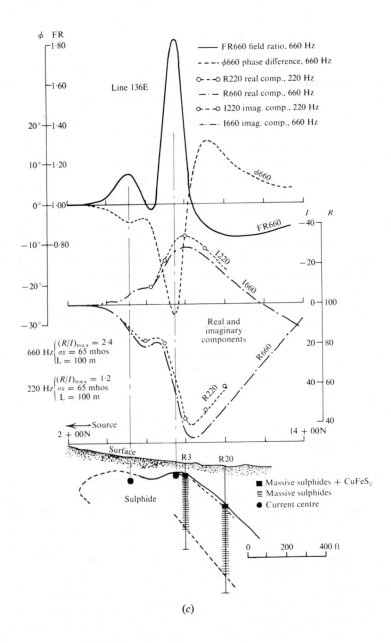

(c)

Fleming, 1961.) (a) Field-strength contours; (b) graphite response; (c) sulphide response.

Figures 7.60*b* and 7.60*c* show a profile across the sulphide area on line 136E, as well as a shorter one on line 180E over the main graphite zone – which is continuous for at least 1½ miles NW–SE. Both components in the first profile indicate a shallow dip to the north (see fig. 7.45) while the graphite zone appears to have a much steeper dip in the same direction. In both examples the response is strong. On line 136E there is, in addition, a minor anomaly 230 ft south of the main peak which corresponds to the isolated 120-contour in fig. 7.60*a*. Apparently this is caused by the undulation in the top surface of the sulphides.

Although the sulphide deposit is massive and appears homogeneous throughout the body, which is about 300 ft thick at 136E, the Turam anomaly resembles that of a thin conductor. Possibly this is explained by the core logs of drill holes R3 and R20, which show a high concentration of chalcopyrite in the first 15 ft of massive sulphides. The main bulk of the mineralization below this, mainly pyrite, appears to contribute little to the response. Massive pyrite is known to occur at times in a silicate matrix which isolates the sulphide particles from one another, reducing the conductivity greatly. Whether such a situation exists here is not known.

A very rough estimate of the depth of these conductors (the estimated value is invariably larger than the actual depth to the top) can be obtained from the width of the profile anomalies at half maximum. For the larger sulphide response we get 85 and 115 ft from the field-ratio and phase curves respectively, for the smaller peak about 140 ft. Since the actual depths are about 40 ft and 75 ft, it would appear that a better result would be obtained by using the half width rather than the full width. However, there is probably a widening of both parts of the profile because the anomalies are close together. In the case of the graphite zone the depth estimate is about 150 ft.

Real and imaginary components, derived from the Turam field curves are also displayed in figs. 7.60*b* and 7.60*c* for both zones. It is difficult to calculate the depth of the sulphide from these curves because they do not have well-defined minima, but for the graphite zone the estimates are about 125 ft and 90 ft from the real and imaginary components respectively. The σs values are 65 mhos for the sulphide and 10 mhos for the graphite.

8. Paterson (1967) has provided several excellent examples of the quadrature airborne EM system from surveys in the Canadian Shield. Typical profiles, together with ground follow-up by various methods, are illustrated in fig. 7.61. The ratio of low-frequency to high-frequency peaks is quite large in all three quadrature profiles (2·6, 3·7 and 4·5 respectively) indicating good conductivity. In fig 7.61*a* the anomaly is entirely due to graphitic shales. Figure 7.53 provides an estimate of the conductivity-thickness product from the quadrature peak-response ratio; it is about 20 mhos. Ratio of real to imaginary peaks from the horizontal-loop profiles is 1·5, which is not a particularly good conductor. Consulting the characteristic curves of fig. 7.51 we find the conductor depth to be about 25 ft and σs value about 30 mhos. The validity of using these characteristic curves might be questioned here, because the graphite zone appears to be about 200 ft wide.

The airborne system response is particularly large in fig. 7.61b, where the sulphides appear to be very close to surface. Figure 7.53a gives us a value of σs of about 35 mhos; here again we are probably not justified in taking this value too seriously, since the vertical-loop profile indicates a zone of about 400 ft width. The gravity anomaly is also broad. Significance of the magnetic profile is not apparent. The wide positive tails on the airborne EM profile are not explained.

The conductor in fig. 7.61c is more deeply buried, as indicated by the smaller quadrature responses and, of course, the 100 ft of overburden. The σs value is approximately 40. Both the ground magnetics and gravity profiles appear to reflect the very massive pyrite. This is rather surprising in the case of the magnetic anomaly because pyrite is only weakly magnetic and a section of 20 % magnetite at the bottom of the drill hole has no magnetic signature at all. The vertical loop is typical of a conductor with shallow dip to the left. This is also peculiar in view of the drill hole inclination, which would not normally be in the direction of dip. In fact the dip appears to be to the left in all three examples, although steeper in (a) and (b). There appears to be no indication of dip on the quadrature profiles, which is not surprising, since in general dip estimates from airborne EM data are difficult.

9. The Whistle Mine, located on the northeast rim of the Sudbury basin, northeastern Ontario, is a favourite test site for airborne electromagnetic equipment. The massive steeply dipping sulphides, mainly pyrrhotite, are both conductive and magnetic. Figure 7.62 shows an assortment of profiles from several airborne EM systems over this property. The upper three profiles in (a) include an airborne VLF (McPhar KEM), two-frequency quadrature (McPhar F-400) and aeromagnetics, for a single traverse. The next two in (b) show airborne AFMAG and a second airborne EM response, from a quadrature system, on the same flight line. The bottom set of three profiles in (c) is for a phase-component helicopter system (Aero-Newmont) with aeromagnetics on the same line. The relative locations of these three traverses are not known.

In (a) the ratio of peak low frequency (340 Hz) to peak high frequency (1070 Hz) response is about 0·78 giving a σs product of 4 from fig. 7.53. This is a moderately good conductor. The distinct crossover and field strength peak from the VLF trace correlates precisely with the quadrature peaks and indicates that the conductor is very shallow. There is in addition a large negative magnetic response to the south, whose north flank corresponds exactly to the electrical anomaly.

The AFMAG response in (b) shows a strong crossover with steep slope on both frequencies. From these traces we may infer that the conductor is near surface and has high conductivity, since the ratio of peak responses $(\phi_{150}/\phi_{510})_{max}$ is approximately unity.

The phase component system in (c) produces a very strong in-phase peak of nearly 1000 ppm. This is partly due to the high conductivity and shallow depth of the conductor, partly to the relatively low altitude which can be maintained by the helicopter. Using characteristic curves of the type shown in fig. 7.53b, the depth below the aircraft was found to be 115 ft and the σs product 140 mhos.

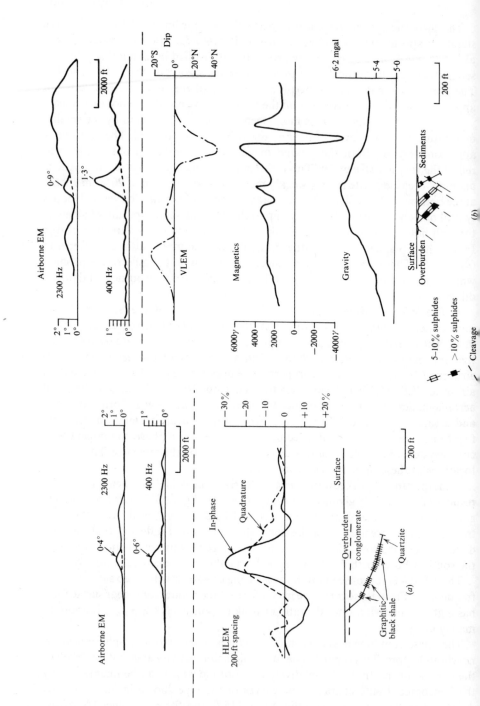

Airborne EM

2300 Hz

400 Hz

2000 ft

0·4°

0·6°

HLEM
200-ft spacing

In-phase

Quadrature

−30 %
−20
−10
0
+10
+20 %

200 ft

Surface

Overburden
conglomerate

Quartzite

Graphitic
black shale

(a)

Airborne EM

2300 Hz

400 Hz

2000 ft

0·9°

1·3°

2°
1°
0°

1°
0°

Dip

20°S
0°
20°N
40°N

VLEM

Magnetics

6000γ
4000
2000
0
−2000
−4000γ

Gravity

6·2 mgal

5·4

5·0

200 ft

Surface
Overburden

Sediments

5–10 % sulphides

> 10 % sulphides

Cleavage

(b)

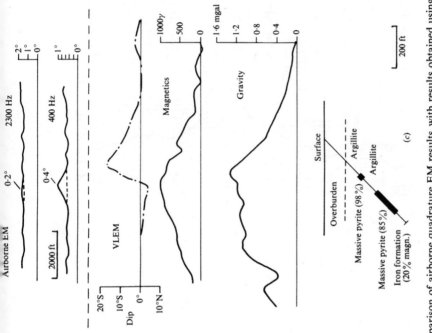

Fig. 7.61 Comparison of airborne quadrature EM results with results obtained using various ground methods. (After Paterson, 1967.) (a) Comparison with HLEM over graphite deposit; (b) comparison with several techniques over shallow sulphide deposit; (c) comparison with several techniques over deeper sulphide deposit.

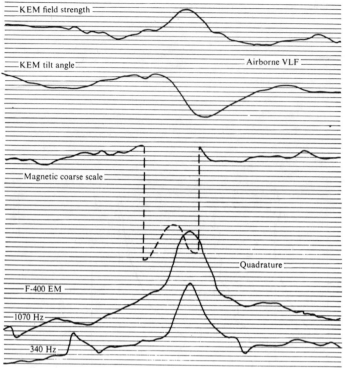

(a)

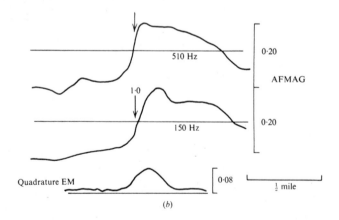

(b)

7.62 Comparison of various airborne EM systems and aeromagnetics, Whistle Mine,
ꜱury, Ontario. (a) Comparison of VLF, quadrature EM and aeromagnetics; (b) AFMAG
quadrature EM. (From Ward, 1959b.)

As in (*a*), a large magnetic anomaly occurs with the EM anomaly.

10. Airborne EM data obtained with the Aero-Newmont phase-component system are shown in figs. 7.63*a*, 7.63*b* and 7.63*c*. This equipment is mounted in a Sikorsky S-55 helicopter. The coaxial transmitter and receiver coils are mounted fore and aft 60 ft apart and the source frequency is 400 Hz.

In (*a*) a sharp peak, marked 160 (ppm), typical of a thin sheet or dike conductor is superimposed on a broader anomaly of smaller amplitude. The latter is due to a swamp; using characteristic curves for a horizontal sheet, we find that its σs product is roughly 0·8 mhos. If the swamp is about 25 ft thick its conductivity is 0·1 mhos/m. Whatever the thickness, it is not a particularly good conductor, as indicated by the ratio of peak in-phase to quadrature response. For the sharp peak, the ratio is 160/60; from fig. 7.53*b* we find $h/l \approx 3·7$, $\sigma s \approx 270$, or about 85 mhos. The altitude was 135 ft, so the sub-surface depth of the dike is about 85 ft. Drilling results indicated the zone to be 30 ft wide and only 45 ft deep.

The example in (*b*) is interesting because it shows a large response from a salt-water tidal flat. From the horizontal-sheet characteristic curve we find $\sigma s \approx 2·5$ mhos average. Assuming $\sigma = 5$ mhos/m for sea water, the depth is only about 1·5 ft. Although the anomaly is large, the ratio of peak responses is only 1·5.

Traces in fig. 7.63*c* show three dikes of different conductivities. The respective peak ratios are 260/0, 320/60 and 125/150 for helicopter altitudes of 140, 180 and 140 ft, giving σs values of about 2000, 500 and 90. Assuming conductor thicknesses of 50–100 ft in each case, the conductivities are 20–40, 5–10 and 1–2 mhos/m

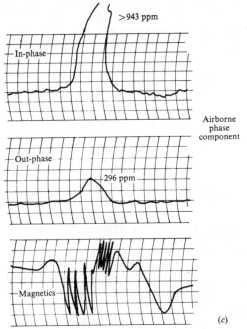

(*c*) phase-component EM and aeromagnetics. (From Ward, 1966.)

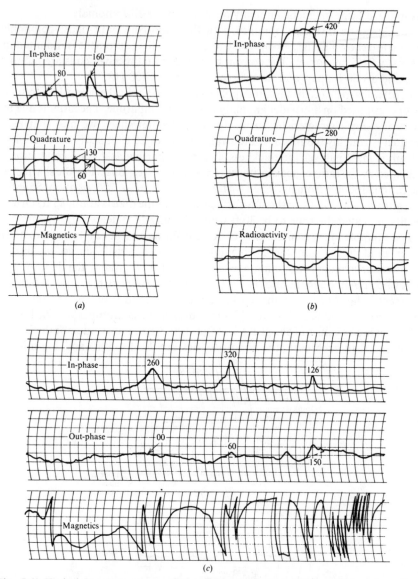

Fig. 7.63 Typical response curves of the airborne phase-component EM system (Aero-Newmont). (After Wiedewult, 1962.) (*a*) Swamp and buried dike; (*b*) salt-water tidal flat; (*c*) dikes of different conductivities.

respectively. The first zone is ore-grade nickel bearing sulphides, the others are massive pyrrohotite with graphite. Note that there are distinct magnetic anomalies associated with each zone.

11. A flight record of INPUT airborne EM response has been given in fig. 7.30*e*. Several additional examples, as well as one for Turair, are illustrated in figs.

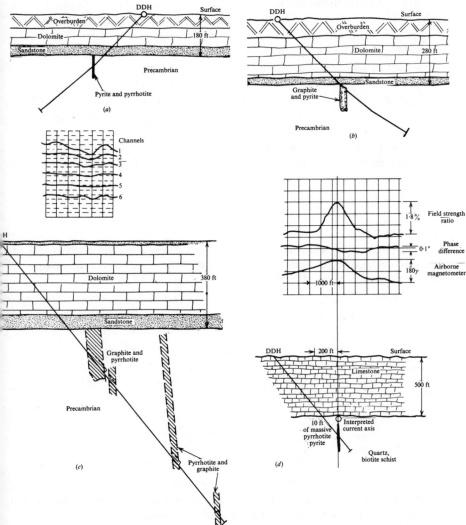

Fig. 7.64 INPUT and TURAIR responses over deep conductors, Manitoba Nickel Belt. (a) INPUT response, anomaly depth 180 ft; (b) INPUT response, anomaly depth 280 ft; (c) INPUT response, anomaly depth 380 ft; (d) TURAIR response, anomaly depth 500 ft.

7.64*a* to 7.64*d*. These surveys were made in the Manitoba nickel belt, northeast of Lake Winnipeg. This is an excellent test area, since the electrical conductors, occurring in Precambrian rocks, are quite deeply buried beneath Paleozoic sandstone and dolomite which are generally covered with unconsolidated overburden. Sulphide zones, with and without nickel, occur in ultrabasics and gneisses and frequently extend for miles. There is also considerable graphite with the sulphides. The mineralization appears to be associated with a huge fault structure.

The 6-channel INPUT data in (*a*), (*b*) and (*c*) are for progressively deeper anomalies at 180, 280 and 380 ft. The response is very strong in (*a*), showing on all 6 channels, still well defined in (*b*) and relatively weaker in (*c*). For some reason the peaks in (*b*) are displaced about 50 ft from the conductor location. Presumably the conductor width and depth extent have been defined by additional ground geophysics as well as by the drill logs.

The Turair profile in (*d*) shows a well-defined peak for a mineral zone 500 ft deep. As mentioned previously, this airborne EM system employs a Turam-type transmitter coil on the ground, which is a decided advantage in detecting deeply buried conductors.

Because of great strike length and large resistivity contrast, the examples illustrated here represent highly favourable targets for EM. Even so, the response for such depths is extremely good.

7.10 Problems

1. The following VLEM readings were obtained during a fixed-transmitter ground survey in northeast Brazil. Stations are 50 m apart, the lines are 400 m apart; the transmitter was located on line 1S, midway between stations 38 and 39.

	Line 0				Line 2S		
Station	ϕ (2400)	ϕ (600)	ϕ (c)	Station	ϕ (2400)	ϕ (600)	ϕ (c)
42W	$-28°$	$-22°$	$-28°$	43W	$-15°$	$-13°$	$-13°$
41	-35	-26	-36	42	-22	-15	-15
40	-35	-30	-35	41	-24	-20	-20
39	-30	-25	-30	40	-12	-10	-10
38	-2	-2	-2	39	13	12	12
37	20	15	15	38	20	17	18
36	25	20	25	37W	$18°$	$16°$	$17°$
35	27	18	27				
34W	$15°$	$10°$	$20°$				

ϕ(2400), ϕ(600) and ϕ(c) are dip angles measured respectively at 2400 Hz, 600 Hz and both frequencies simultaneously. Plot the profiles. Estimate the depth, dip, possible strike length and location of this anomaly, as well as its conductivity-width product. Do you see any advantages in the dip-angle measurement at two distinct frequencies? At two simultaneous frequencies?

2. The data tabled below are taken from a vertical-loop fixed-transmitter survey carried out in northern Quebec. The transmitter was located at 7 + 00S on line 4W. Station spacing was 100 ft and the two lines are 800 ft apart. Transmitter frequency was 1000 Hz.

	Line 8W				Line 0 + 00		
Station	Dip	Station	Dip	Station	Dip	Station	Dip
B.L.	3°	7S	10°	B.L.	12°	7S	10°
1S	3	7+50	13	1S	11	8	−11
2	2	8	9	2	14	9	−15
3	4	9	−8	3	13	10	−11
4	4	10	−22	4	12	11	−11
5	3	11	−24	5	15	12	−9
6S	5·5°	12	−26	6S	9°	13S	−8°
		13S	−25°				

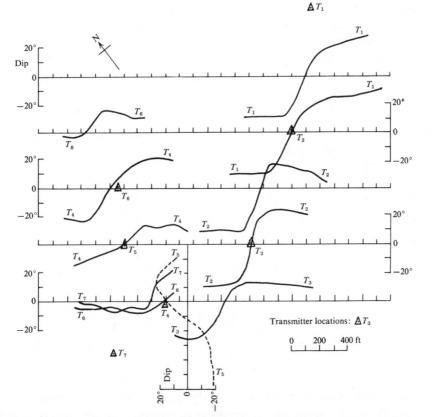

Fig. 7.65 Vertical-loop fixed-transmitter survey, Northern Manitoba. Plan view showing transmitter locations and lines of profiling, the data being plotted using the lines of profiling as horizontal axes.

614   *Electromagnetic methods*

Make an interpretation, similar to that in problem 1, of these results with the aid of the following additional information.'. . . Geological mapping in the vicinity of the conductor shows that it lies in a band of amphibolite intruded by gabbro sills . . . this band, about 2400 ft wide, is bounded north and south by andesitic volcanic rocks . . . outcrops of granite were located 2600 ft NW of the west end of the conducting zone . . . The axis of the anomaly is parallel to the strike of surrounding rocks . . . Within the band a weak dissemination of chalcopyrite and pyrite has been observed in one outcrop.'

3. VLEM profiles from a property in northern Manitoba are displayed in fig. 7.65. Transmitter locations are marked by the symbols ΔT_1, ΔT_2, etc., (the subscripts refer to the order of surveying) and corresponding curves are labelled T_1, T_2, etc. What is the purpose of the SW–NE traverse (bottom of diagram) made at right angles to the other lines? Outline the conductor indicated by these profiles, estimate its dip, depth and conductivity at various points along the axis. Assuming no other geophysical work had been done in the vicinity, what additional surveys would you recommend and why?

4. Dip angles obtained with broadside VLEM equipment are tabulated below. Four frequencies – 600, 1000, 2400 and 5000 Hz – were used in this survey. The receiver and transmitter were moved along parallel picket lines maintaining a fixed spacing of 400 ft and readings were taken every 100 ft.

	Line 12N				Line 8N				Line 4N			
Stn	600	1000	2400	5000	600	1000	2400	5000	600	1000	2400	5000
20W	—	2°	—	3°	—	1°	—	2°	0°	0°	0°	0°
19	—	0·5	—	2	0	2	2	3	7	5·5	8·5	9·5
18	—	3	—	4	7·5	2	1·5	3	14	20	22	24·5
17	6	7	8·5	8	2	5	2	11	−11·5	−19·5	−19·5	−23·5
16	−12	−16·5	−15	−18·5	2	−7	−5	−16	−12	−13	−18·5	−18
15	−14	−18	−18	−22	−1·5	−6	−10	−14	−4·5	−3	−9·5	−12·5
14	−1	−7·5	−14	−12	0	−1	−2·5	−2·5	−1·5	0	−5	2
13	0	0	−4	−2·5	1	1	−1	3	0	2	0	3
12	0	−1·5	−1	−3	0	1	1	3	0	0	−1·5	−2
11	0	−2	0	−4	0	1	1	1·5	0	−2	−1·5	−7
10	0	4	1	6	0	0	−1	−8	−2	−1	−4·5	−4·5
9	0	−2	−1·5	−12	0	−2	−3·5	−10	1	0	0	−1
8	0	−5	−8	−10	−1	−1	−2·5	−4	0	2	0	3
7	0	−3	−5	−10	1	1	0	3	1	1	0	1·5
6	1	−2	1	−4	3	0	2	0	0	1	0	1·5
5	0	1	1	2	2	0	3	1·5	0	2	0	3
4	0	1	1	1·5	1	3	2	5	0	1	1	2
3	0	1	1	2	2	2	3	3·5	0	0	0	2
2	0	0	1	−1	1	2	1·5	2	−1·5	1·5	−2	3·5
1	3·5	0	5	0	2	1	0	3	0	1	−1	2·5
B.L.	1	0	2·5	0	−2	2	0	3	0	0	1	0

Plot the profiles, preferably on two sheets by combining the 600–2400 Hz and the 1000–5000 Hz dip angles, since these are the dual frequencies on two different EM units. Locate any potential conductors and estimate their depth, dip and, if possible, the σs

product. Discuss the advantages of using two frequencies in this type of survey. Is there any point in employing four frequencies? Given a choice. how many or which of these would be preferable? Any other frequencies?

5. Figure 7.66 shows a set of VLF profiles, obtained with the EM16 unit, taken from a survey in eastern Nova Scotia. The numerical values of % dip angle are also shown for each station. The transmitter was Panama, roughly S30°W of the area. Real crossovers are in the sense of positive to negative going east, for example, near station 28E on line 28N.

At first glance there appears to be evidence both of steeply-dipping contacts and flat-lying conductors here. The overburden varies between 8 and 20 ft in thickness and is known to have a resistivity of about 100 Ωm in one zone to the southwest, although generally it is thought to have a higher resistivity than this over most of the area. Make what interpretation you can from the profiles. It is recommended that a contour plot, as described in §7.9, example 3, be made from the dip readings for clarification.

6. The following EM16 readings were taken during a survey in Nova Scotia, using the

Stn	Line 10N		Line 6N		Line 2N		Line 2S		Line 6S	
	Dip	Quad.	Dip	Quad.	Dip	Quad.	Dip	Quad.	Dip	Quad.
20W	—	—	14	−3	20	3	20	−2	10	−4
19	2	−7	20	−7	20	4	18	−2	8	2
18	−12	2	15	0	18	3	20	0	10	−2
17	13	−8	15	−1	15	3	17	−2	5	4
16	9	−3	17	0	17	−3	10	−3	−5	2
15	5	1	18	2	22	−2	5	−2	−5	0
14	7	1	18	5	14	0	5	−3	−10	2
13	5	2	18	4	15	2	−4	−3	−20	0
12	8	3	23	3	10	−2	−3	−3	−25	−1
11	10	3	25	2	10	−3	−11	3	−37	2
10	23	0	14	2	10	0	−25	3	−29	−2
9	15	3	6	0	5	0	−38	3	−32	0
8	1	−2	5	0	−3	−2	−30	−2	−33	4
7	−1	1	1	−3	−16	0	−27	−2	−13	7
6	2	−3	−8	1	−25	−1	−32	3	0	−2
5	−2	−3	−12	0	−25	3	−20	3	17	−1
4	−2	2	−20	5	−15	0	1	−3	15	−5
3	−6	3	−20	3	−3	−2	18	−3	21	−3
2	−2	0	0	0	−12	−2	24	−5	22	−5
1W	6	−2	6	−1	3	−3	13	3	15	−3
B.L.	8	−3	0	−6	8	−2	11	2	0	6
1E	5	−2	−10	1	11	5	5	2	3	0
2	7	0	3	0	19	−2	10	−1	0	0
3	12	0	15	−4	11	0	16	−5	−8	3
4	16	0	15	−2	9	0	5	−3	−2	−1
5	9	−1	11	2	7	1	0	−2	3	−1
6E	2	−2	11	4	12	−2	10	−2	13	−2

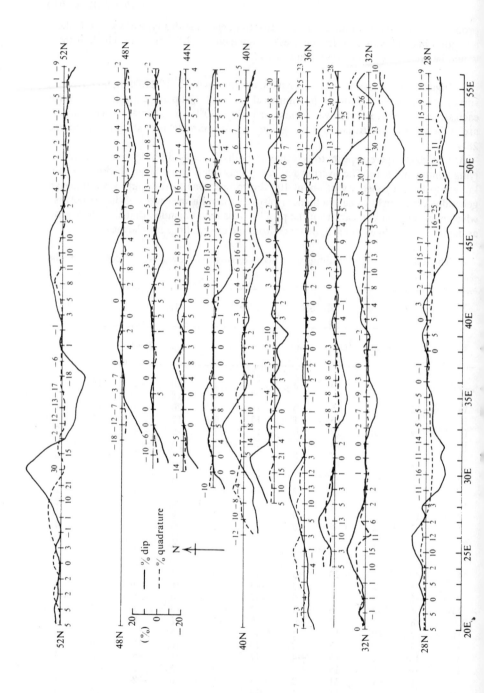

Panama transmitter. The topography here is quite rugged; there is an increase in elevation of 250 ft between line 10N just east of the baseline and the western portion of line 6S. This is the spine of a hill which has steep sides both east and west so that the terrain contours resemble the bowl of an inverted spoon. The hill is thought to be sandstone and there is a contact with limestone beds, which appear quite thin, on the east flank.

Stations are 100 ft apart, lines 400 ft. Real crossovers are positive to negative going east. Plot the profiles, contour the dip angles by the method of example 3, §7.9. In making an interpretation of the results, consider (a) overburden effects, (b) contacts between extensive beds of contrasting resistivity, (c) topographic effects, as well as the possibility of metallic conductors.

7. The AFMAG profiles shown in fig. 7.67 are taken from a large-scale survey. Geologically the area is Precambrian with numerous outcrops of ultrabasic rocks in the eastern portion, and gneiss, generally covered by thin overburden, to the west; the exposed rocks are extremely metamorphosed.

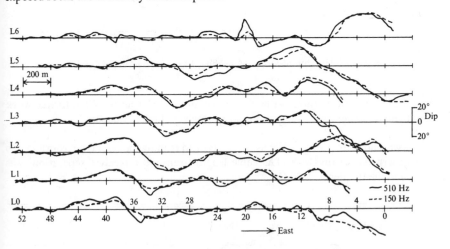

Fig. 7.67 AFMAG profiles, Northeastern Brazil.

Stations are 50 meters apart and the lines 200 meters; the AFMAG frequencies were 150 and 510 Hz. The predominant azimuth direction for these signals was about N60°W, although this was quite erratic and the tilt angle for minimum, particularly at the lower frequency, was often indeterminate over a considerable range. The effect of a high-voltage transmission line about 2 km to the east can be seen on the east end of several lines. Make an interpretation of these profiles with regard to location, depth and dip of potential conductors.

8. Figure 7.68 is a single horizontal-loop EM profile from a survey made in the Eastern Townships of Quebec. The main geological features consist of Ordovician and Cambrian slate, calcareous slate, greywacke, sandstone and quartzite conglomerate; the general geological strike is slightly east of north. The anomaly, shown here for line 81N, has a strike length of at least 1000 ft. Estimate its location, depth, dip, width, conductivity-thickness product and conductivity. Is there evidence for more than one conductor?

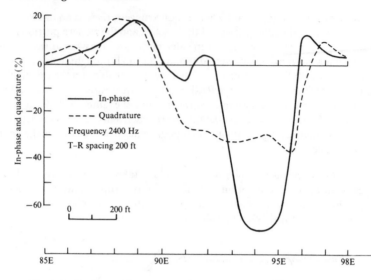

Fig. 7.68 Horizontal-loop EM profiles, Eastern Townships, Quebec.

With the limited information available, can you suggest the source? What further work would you do to verify the suggestion, possessing very limited finances?

9. As a contrast to problem 8, consider the HLEM readings below, taken from a survey in Northern New Brunswick. The cable length (transmitter-receiver separation) was 200 ft and the frequency 2400 Hz.

Stn	IN-PH.	Quad.	Stn.	IN-PH.	Quad.	Stn.	IN-PH.	Quad.
6S	2%	1·5%	3S	0	0	0	0	1·5%
5S	0	0	2S	−1·5%	0	1N	0	1·0
4S	0	1·0	1S	0	0			

This traverse crosses a conductive zone which has a strike length of 1100 ft. Is it possible to determine its properties from the horizontal-loop response? Would this response normally be considered significant? Would you recommend any further HLEM work or other geophysics on the basis of these results?

10. The profiles shown in fig. 7.69 were obtained during a HLEM survey in the Chibougamau area of Quebec. Geologically the region is Precambrian, typical of Northwestern Quebec, with metamorphosed, folded volcanics and sediments, intruded by acidic and basic rocks. Transmitter frequency was 880 Hz and cable length 200 ft. Make an interpretation of these profiles, as complete as possible.

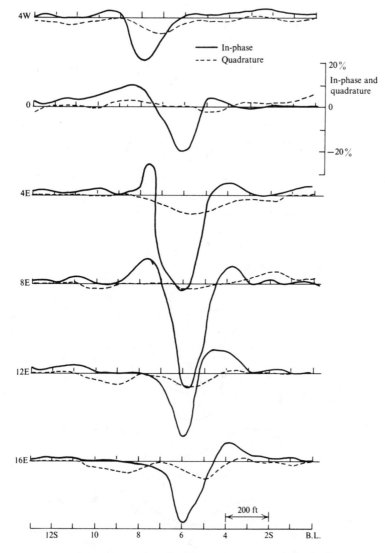

Fig. 7.69 Horizontal-loop EM profiles, Chibougamau Area, Quebec.

11. Determine all you can about the conductor associated with the HLEM profiles in fig. 7.70. The frequency was 600 Hz and transmitter-receiver separation 200 ft.

12. The Turam results tabulated below were obtained in a model survey over a sheet conductor in the laboratory. Figure 7.71 shows the 'field' layout. The rectangular transmitting loop is 2000 ft by 1000 ft and the field ratios and phase readings are for lines 4N and 16N, at stations 250W to 1050W in 100-ft steps. The transmitter frequency is 1000 Hz and the receiver coils are 100 ft apart.

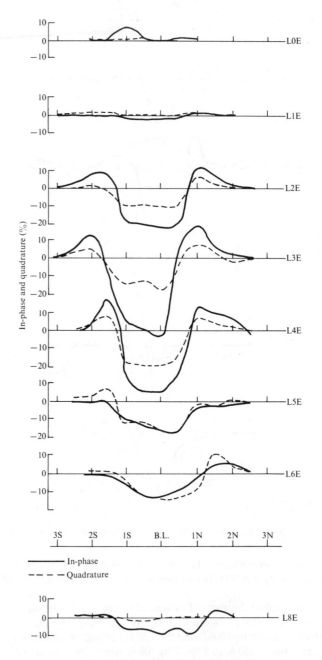

Fig. 7.70 Horizontal-loop EM profiles.

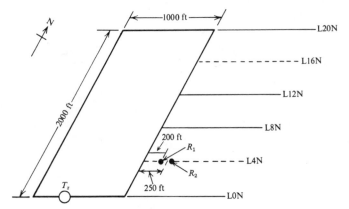

Fig. 7.71 Model Turam configuration.

Stn (ft)	First coil (ft)	Line 4N Station readings		Line 16N Station readings	
		FR	Phase ($\Delta\phi$)	FR	Phase ($\Delta\phi$)
	200				
250		1·77	0°	1·77	0·10°
	300				
350		1·54	0·20	1·54	0·20
	400				
450		1·39	0·35	1·43	0·50
	500				
550		1·35	0·35	1·35	0·80
	600				
650		1·64	−0·50	1·30	1·0
	700				
750		1·39	1·20	1·30	−1·0
	800				
850		1·26	1·25	1·33	−4·25
	900				
950		1·24	1·40	1·27	−0·95
	1000				
1050		1·23	1·55	1·22	1·15

Calculate the NR, RR, $\Sigma\Delta\phi$, V, R, I values from these data, using eq. (7.6b) and the routine outlined in §7.8.4f. Plot RR, $\phi(= \Sigma\Delta\phi)$, R and I and determine the location, approximate depth, dip and σs product for the conductor on each line.

13. Figure 7.72 shows HLEM and Turam profiles over a sulphide deposit in northern Quebec. The horizontal-loop survey was made first, while the Turam work was a follow-up in an attempt to increase the depth of investigation. The HLEM frequency was 3600 Hz, with cable separation 50 meters; the Turam frequency was 660 Hz. Interpret the results as completely as possible with this information. Why is the HLEM response so much larger than the Turam?

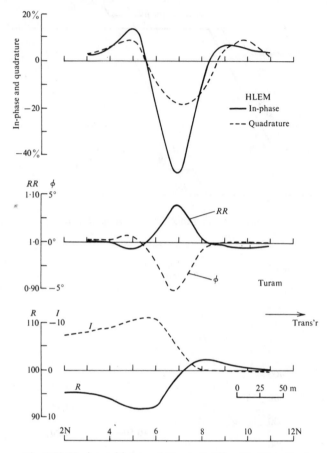

Fig. 7.72 Horizontal-loop and Turam profiles, Northern Quebec.

14. The readings tabled below are from a Turam survey over suspected sulphides on a property in Abitibi Ouest, Quebec. Because the overburden was thought to be thick and of low resistivity, it was hoped that Turam would have better possibilities of penetrating

Stn	FR	Phase ($\Delta\phi$)	NR	RR	$\Sigma\Delta\phi$	V	R	I
350S	1·81	−0·75	1·80					
250S	1·58	−1·50	1·59					
150S	1·51	−2·0	1·47					
050S	1·44	−2·0	1·40					
050N	1·37	−1·50	1·35					
150N	1·32	−1·0	1·31					
250N	1·28	−0·50	1·28					
350N	1·25	0·0	1·25					
450N	1·23	0·0	1·23					

to depth than other EM methods. The near leg of the transmitter rectangle, laid out E–W, was at station 6 + 00S and the dimensions of the rectangle were 1200 × 1000 ft. Transmitter frequency was 660 Hz and the receiver coils were 100 ft apart.

Complete the table for RR, $\Sigma\Delta\phi$, V, R, I columns, plot RR, $\phi(=\Sigma\Delta\phi)$, R and I profiles and interpret the results as completely as possible.

15. The Turam profiles in fig. 7.73 are taken from a survey in northern Manitoba. The geology is Precambrian, probably ultrabasics and gneisses, overlain by Paleozoic sediments. In addition to reduced ratio and phase profiles, the total field has been plotted,

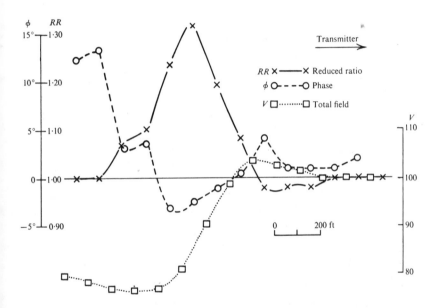

Fig. 7.73 Turam profile, Northern Manitoba.

with the points falling midway between the RR and ϕ values. Is it possible to derive the R and I values from these profiles? Make an interpretation of the source of this anomaly. The transmitter frequency was 660 Hz.

16. Figure 7.74 shows Crone-shootback, Turam and horizontal-loop EM profiles over a sulphide deposit in western Ontario. Interpretation of shootback EM results is mainly qualitative (see §7.5.4d); the ratio $(\phi_{480}/\phi_{1800})_{max}$ is somewhat diagnostic of the type of conductor as in AFMAG and the quadrature AEM system. Use all three profiles to assess the various parameters of this deposit.

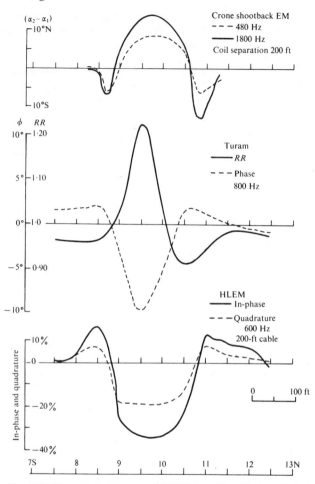

Fig. 7.74 Shootback, Turam and HLEM profiles, Western Ontario.

17. The readings tabled below were obtained from a survey over a copper prospect near Val d'Or, Quebec.

Stn (ft)	HLEM RE	HLEM IM	Shoot-back 2–1	VLEM	Stn (ft)	HLEM RE	HLEM IM	Shoot-back 2–1	VLEM
7N	0%	2·5%	—	0°	10+50	−12%	−10%	2·5°	7°
7+50	0	4	0°	−0·5	11N	−2·5	−1	0	6
8	−1	4	−1	−1	11+50	1·5	5	−2·5	4
8+50	−0	3	−1	−2	12N	1·0	4·5	−1·0	2
9	−6·5	−3	3	−3	12+50	0·0	2·5	−0·5	2
9+50	−13·0	−10	6·5	−2	13N	0·0	3·0	—	1
10N	−14·5	−12	5	2					

The respective EM set frequencies were HLEM -2000 Hz, shoot-back -1800 Hz, VLEM -1800 Hz and $T-R$ separations 200 ft, 300 ft, and 300 ft. (In the VLEM setup the transmitter was 300 ft east and on the conductor axis, which had been previously located by reconnaissance EM.) Plot these profiles and extract all the information you can from the data.

18. Quadrature airborne EM results over a test site in Ontario are illustrated in fig. 7.75. The location of VLEM and IP anomalies found in ground follow-up are illustrated for zones A and B in the upper diagram. This record is merely a small part of the airborne survey. It is obvious, as mentioned before, that airborne EM data do not suffer from a lack, but rather from a surplus, of anomalies; there are 17 marked in fig. 7.75a. These are loosely graded as 'definite', 'probable' and 'possible', based partly on the maximum response amplitude, partly on the ratio of low to high frequency maxima. These values are given for one 'probable', zone F on line $E_{\rm w}$.

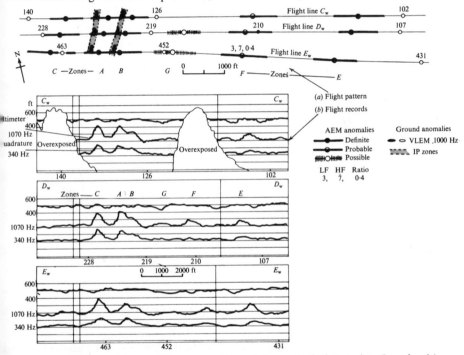

Fig. 7.75 Quadrature airborne EM survey, Cavendish geophysical test site, Ontario. (a) Flight pattern; (b) flight records.

From the flight records in fig. 7.75b, determine the low- and high-frequency response maxima and their ratios for the remaining 16 anomalies. Do you agree with the classifications in fig. 7.75a? Have they been influenced by response on adjacent flight lines, by altitude variations, by possible additional information – airborne magnetics, terrain, geology, etc., not available here? What is the reason for the relative displacement of airborne and ground anomalies (about 200 ft) in zones A and B? Can you estimate other parameters of the conductive zones – depth, dip, width, strike extent, σs product – from the records?

19. A small section of a quadrature EM survey, southwest of Chibougamau, Quebec, is shown in fig. 7.76. Flight line spacing is nominally ¼ mile. The anomalous zones are obvious enough; the numbers accompanying most of these refer to maximum low-frequency response and ratio $(\phi_{1f}/\phi_{hf})_{max}$ respectively. Do what you can with this limited data.

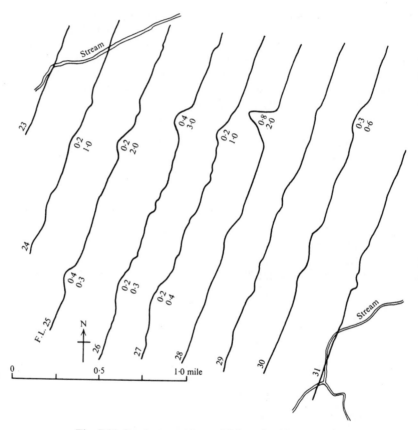

Fig. 7.76 Quadrature airborne EM results, Northern Quebec.

20. Three sets of airborne EM profiles – AFMAG, Rio-Mullard phase-component and the two plane rotary-field system – are displayed in fig. 7.77. These are taken from an integrated air-ground geophysical survey carried out in the Geita Goldfield area of Tanzania. The frequencies used in these airborne units were: AFMAG-340 and 90 Hz Rio-Mullard-320 Hz, rotary field-800 Hz. The Rio-Mullard aircraft flew at 250 ft while the coils in the rotary field system were maintained at 250 ft as well; the AFMAG equipment, mounted in a DC-3, trailed the coils in the bird at 350 ft altitude.

A partial geologic plan of this area, included in fig. 7.77, shows tuffs of the Nyanzian formation, one of the oldest in Africa, composed generally of metamorphosed sediments and volcanic rocks. The Nyanzian is commonly associated with granites in a complex geology.

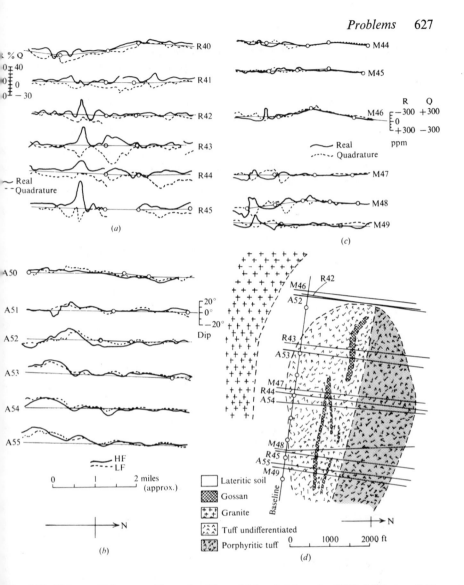

Fig. 7.77 Airborne EM survey, Tanzania. (From Makowiecki *et al.*, 1971.) (*a*) Rotary-field EM results; (*b*) AFMAG results; (*c*) phase-component EM results; (*d*) geological map with superimposed flight lines.

The interpretation of real- and quadrature-component anomalies shown in the rotary field profiles can be made, to a considerable extent, in the same way as horizontal-loop EM. However, the former is more sensitive to traverse direction than HLEM, since the peak quadrature shifts with respect to the in-phase response as the angle decreases from 90°.

Using all the information available in fig. 7.77, make an interpretation of these airborne survey results.

21. Three phase-component airborne EM profiles, each from a different area, are shown in fig. 7.78. Pertinent information with respect to each of these flight records is listed below.

(a) Transmitter frequency, 400 Hz; $T-R$ spacing, 60 ft; altitude, 140 ft.
(b) Transmitter frequency, 320 Hz; $T-R$ spacing, 62 ft; altitude, 140 ft.
(c) Transmitter frequency, 390 Hz; $T-R$ spacing, 60 ft; altitude, 140 ft.

One of these traverses was over a lake, one over a graphite zone and one crossed a copper deposit of economic grade. By making as complete an interpretation as possible from these profiles, can you locate and identify the three sources? Can you suggest any reason why the record is so noisy in (c)?

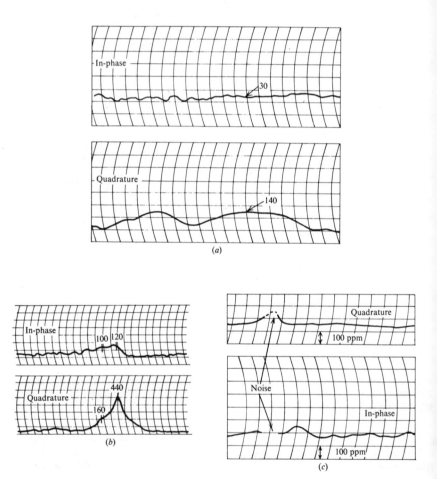

Fig. 7.78 Airborne phase-component EM profiles. (a) Helicopter survey, transmitter frequency 400 Hz; (b) helicopter survey, transmitter frequency 320 Hz; (c) fixed-wing aircraft, transmitter frequency 390 Hz.

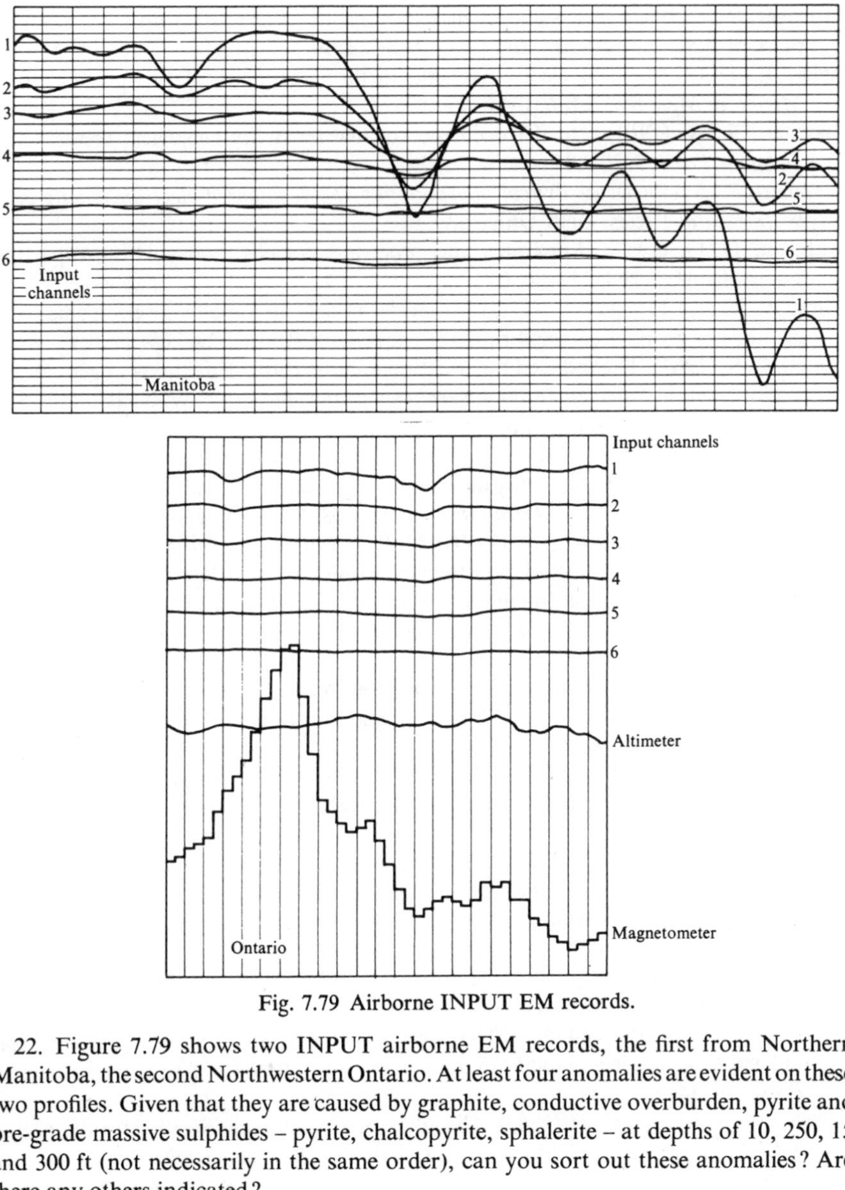

Fig. 7.79 Airborne INPUT EM records.

22. Figure 7.79 shows two INPUT airborne EM records, the first from Northern Manitoba, the second Northwestern Ontario. At least four anomalies are evident on these two profiles. Given that they are caused by graphite, conductive overburden, pyrite and ore-grade massive sulphides – pyrite, chalcopyrite, sphalerite – at depths of 10, 250, 15 and 300 ft (not necessarily in the same order), can you sort out these anomalies? Are there any others indicated?

References

F – field methods; G – general; I – instruments; T – theoretical

Auston, J. S. (1969). Uchi Lake – a geophysical case history. Paper presented at the SEG annual meeting, Calgary, Alberta, Sept. 1969. (F)

630   Electromagnetic methods

Barringer, A. R. (1962). A new approach to exploration – the INPUT airborne electrical pulse prospecting system. *Min. Cong. Jour.* **48**, 49–52. (I)

Bieler, E. S. (1928). The electromagnetic method, geophysical prospecting: *Can. Mining & Met. Bull.* **21**, 631–44. (F)

Bosschart, R. A. (1961). On the occurence of low resistivity geological conductors. *Geophys. Prosp.* **9**, 203–12. (F)

Bosschart, R. A. (1964). *Analytical interpretation of fixed source electromagnetic prospecting data.* Thesis, Delft University. (T)

Brubaker, D. G. (1957). Apparatus and procedure for electromagnetic prospecting. *Trans. A.I.M.E.* **208**, 777–80. (I)

Clark, A. R. and Mungal, A. G. (1951). Scale model experiments in electromagnetic methods of geophysical exploration. *Can. Jour. of Physics*, **29**, 285–93. (T)

Collett, L. S. (1964). The induced polarization and the INPUT methods in geophysical exploration. In *Some guides to mineral exploration*, (ed. E. R. W. Neale). Pub. S. 4, Brit. Comm. Geol. Surveys, pp. 47–53. (F)

Eve, A. S. and Keys, D. A. (1956). *Applied Geophysics.* London, Cambridge. (G)

Fleming, H. W. (1961). The Murray deposit, Restigouche County, N.B., a geochemical-geophysical discovery. *Bull. Can. Inst. Min. & Met.* **54**, 230–5. (F)

Fountain, D. K. and Bottos, F. B. *The McPhar 400 series dual frequency airborne electromagnetic system.* Pub. by McPhar Geophysics Ltd., Don Mills, Ont. (I)

Fraser, D. C. (1969). Contouring of VLF–EM data. *Geophysics*, **34**, 958–67. (T)

Grant, F. S. and West, G. F. (1965). *Interpretation theory in applied geophysics.* New York, McGraw-Hill.

Grover, F. W. (1962). *Inductance calculations.* New York, Dover. (G)

Hedstrom, E. H. (1940). Phase measurements in electrical prospecting. *Trans. A.I.M.M.E. Petr.* **138**, 456–72. (I)

Hedstrom, E. H. and Parasnis, D. S. (1958). Some model experiments relating to. electromagnetic prospecting with special reference to airborne work. *Geophys. Prosp.* **6**, 322–41. (T)

Hughes, W. J. and Wait, J. R. (1975). Electromagnetic induction over a two-layer earth with a sinusoidal overburden. *Pure & App. Geophysics,* **113**, 591–9. (T)

Keller, G. V. and Frischknecht, F. C. (1966). *Electrical methods in geophysical prospecting.* London, Pergamon.

King, W. F. (1971). *Studies of geologic structure with the VLF method.* M.Sc. thesis, McGill University, Montreal. (F)

Koefoed, O. and Kegge, G. (1968). The electrical current pattern induced by an oscillating magnetic dipole in a semi-infinite thin plate of infinitesimal resistivity. *Geophys. Prosp.* **16**, 144–58. (T)

Lowrie, W. and West, G. F. (1965). The effect of a conducting overburden on electromagnetic prospecting measurements. *Geophysics*, **30**, 624–32. (T)

Makowiecki, L. Z., King, A. J. and Cratchley, C. R. (1971). *A comparison of selected geophysical methods in mineral exploration.* Inst. Geol. Science, Geophys. Paper No. 3, London. (I)

Mason, Max. (1927). Geophysical exploration for ores. *Tech. Pub., Amer. Inst. Min. Eng.* no. 45.

Methods and case histories in mining geophysics (1957). 6th Commonwealth Mining and Met. Congress. Montreal, Mercury Press. (G)

Mining geophysics, vol. 1, chapter 2, parts A, C, 1966. Tulsa, Soc. of Explor. Geophys. (G)

Norton, K. A. (1941). The calculation of ground wave field intensity over a finitely conducting spherical earth. *Proc. Inst. Radio Eng.* **29**, 623–39. (T)

Paterson, N. R. (1961). Experimental and field data for the dual-frequency phase-shift method of airborne electromagnetic prospecting. *Geophysics*, **26**, 601–17. (F)

Paterson, N. R. (1966). In *Mining Geophysics*, vol. 1, pp. 185–96. Tulsa, Soc. of Explor. Geophys. (F)

Paterson, N. R. (1967). In *Mining and groundwater geophysics* (ed. by L. W. Morley). Econ. Geol. Report No. 26, Geol. Survey of Canada, pp. 275–89.

Pemberton, R. H. (1962). Airborne EM in review. *Geophysics*, **27**, 691–713. (I)

Seigel, H. O. (1957). In *Methods and case histories in mining geophysics*. 6th Commonwealth Mining and Met. Congress. Montreal, Mercury Press. (F)

Sinclair, G. (1948). Theory of models of electromagnetic systems. *Proc. Inst. Radio Eng.* **36**, 1364–70. (T)

Smythe, W. R. (1950). *Static and dynamic electricity*. New York, McGraw-Hill. (G)

Strangway, D. W. (1966a). Electromagnetic scale modelling. In *Methods and techniques in geophysics* (ed. S. K. Runcorn). Interscience. (T)

Strangway, D. W. (1966b). Electromagnetic parameters of some sulphide ore bodies. In *Mining geophysics*, vol. 1, pp. 227–42. Tulsa, Soc. of Explor. Geophys. (T)

Tikkanen, G. D. Deeply penetrating surveys in Northern Manitoba. Personal communication. (F)

Tornqvist, G. (1958). Some practical results of airborne electromagnetic prospecting in Sweden. *Geophys. Prosp.* **6**, 112–26. (F)

Wait, J. R. (1951). A conducting sphere in a time varying magnetic field. *Geophysics*, **16**, 666–72. (T)

Wait, J. R. (1954). Mutual coupling of loops lying on the ground. *Geophysics*, **19**, 290–6. (T)

Wait, J. R. (1959). On the electromagnetic response of an imperfectly conducting thin dyke. *Geophysics*, **24**, 167–71. (T)

Ward, S. H., Carter, W. O., Harvey, H. A., McLaughlin, G. H. and Robinson, W. A. (1958). Prospecting by use of natural alternating magnetic fields of audio and sub-audio frequencies. *Can. Inst. Min. & Met. Bull.* **51**, 487–94. (F)

Ward, S. H. (1959a). Unique determination of conductivity, susceptibility, size and depth in multifrequency electromagnetic exploration. *Geophysics*, **24**, 531–46. (T)

Ward, S. H. (1959b). AFMAG – airborne and ground. *Geophysics*, **24**, 761–89. (F)

Ward, S. H. (1966). In *Mining geophysics*, vol. 1, pp. 117–29. Tulsa, Soc. of Explor. Geophys. (F)

Ward, S. H. (1967). In *Mining geophysics*, vol. 2, pp. 10–196, 224–372. Tulsa, Soc. of Explor. Geophys. (F)

Watson, H. G. I. (1931). The Bieler-Watson method. *Mem. Geol. Surv. Can.* no. 165, pp. 144–60. (F)

Vieduwilt, W. G. (1962). Interpretation techniques for a single frequency airborne electromagnetic device. *Geophysics*, **27**, 493–506. (T)

8. Resistivity methods

8.1 Introduction

Among the resistivity prospecting methods, of which there is a considerable variety, we include also equipotential line and point and potential ratio measurements. The latter are not much used at present. All resistivity methods employ an artificial source of current which is introduced into the ground through point electrodes or long line contacts. The procedure then is to measure potentials at other electrodes in the vicinity of the current flow. In most cases the current is noted as well; it is then possible to determine an effective or apparent resistivity of the subsurface.

In this regard the resistivity technique is superior, theoretically at least, to all the other electrical methods, since quantitative results are obtained by using a controlled source of specific dimensions. Practically – as in the other geophysical methods – the maximum potentialities of resistivity are not usually realized. The chief drawback is its large sensitivity to minor variations in conductivity near surface; in electronic parlance the noise level is high. An analogous situation would exist in magnetics if one were to employ a magnetometer with sensitivity in the milligamma range.

This limitation, added to the practical difficulty involved in dragging several electrodes and long wires over rough wooded terrain, has made the electromagnetic method more popular than resistivity in mineral exploration. Nor is resistivity particularly suitable for oil prospecting.

8.2 Elementary theory

8.2.1 *Potentials in homogeneous media*

Consider a continuous current flowing in an isotropic homogeneous medium. (This analysis will also apply to a.c. if the frequency is low enough that displacement currents are insignificant.) If δA is an element of surface and $\mathbf{J}$ the current density in amperes/metre^2, then the current passing through δA is $\mathbf{J} \cdot \delta A$. The current density $\mathbf{J}$ and the electric field $\mathbf{E}$ are related through Ohm's law:

$$\mathbf{J} = \sigma \mathbf{E}, \tag{8.1}$$

where $\mathbf{E}$ is in volts/metre and σ is the conductivity of the medium in mhos/metre.

The electric field is the gradient of a scalar potential,

$$\mathbf{E} = -\nabla V, \tag{8.2}$$

Elementary theory   633

where V is in volts. Thus we have

$$\mathbf{J} = -\sigma \nabla V. \tag{8.3}$$

If charge is conserved (no current sources or sinks) within a volume enclosed by a surface A, we can write

$$\int_A \mathbf{J} \cdot d\mathbf{A} = 0. \tag{8.4}$$

Gauss' Theorem states that the volume integral of the divergence of current throughout a given region is equal to the total charge enclosed, so that in this case,

$$\int_V \nabla \cdot \mathbf{J} dV = 0.$$

Taking V as an infinitesimal volume enclosing a given point, we get for this point

$$\nabla \cdot \mathbf{J} = -\nabla \cdot \nabla(\sigma V) = 0,$$

hence,

$$\nabla \sigma \cdot \nabla V + \sigma \nabla^2 V = 0. \tag{8.5}$$

If σ is constant throughout, the first term vanishes and we have Laplace's Equation, i.e., the potential is harmonic:

$$\nabla^2 V = 0. \tag{8.6}$$

There are two boundary conditions which must hold at any contact between two regions of different conductivity. Firstly, the potential must be continuous across the boundary; secondly, the normal component of $\mathbf{J}$ must also be continuous. The continuity of V means that $\partial V/\partial x$, x being parallel to the boundary, is also continuous. Thus, we have

$$V^{(1)} = V^{(2)}; \quad \left(\frac{\partial V}{\partial x}\right)^{(1)} = \left(\frac{\partial V}{\partial x}\right)^{(2)}; \quad J_n^{(1)} = J_n^{(2)}, \tag{8.7a}$$

where n and t (see below) denote the normal and tangential components. In terms of electric fields these conditions become

$$E_t^{(1)} = E_t^{(2)}; \quad \sigma_1 E_n^{(1)} = \sigma_2 E_n^{(2)}. \tag{8.7b}$$

There are several field configurations used in resistivity which we will consider in turn.

8.2.2 Single current electrode at depth

We have an electrode of small dimensions buried in a homogeneous isotropic medium. This corresponds to the *mise-à-la-masse* method (see §8.5.4d) where the single electrode is down a drill hole or otherwise under the ground. The current circuit is completed through another electrode, usually at surface, but in any case far enough away that its influence is negligible.

From the symmetry of the system, the potential will be a function of r only, where r is the distance from the first electrode. Under these conditions Laplace's equation, in spherical coordinates, simplifies to

$$\nabla^2 V = \mathrm{d}^2 V/\mathrm{d}r^2 + (2/r)\,\mathrm{d}V/\mathrm{d}r = 0. \tag{8.8}$$

Multiplying by r^2 and integrating, we get

$$\frac{\mathrm{d}V}{\mathrm{d}r} = \frac{A}{r^2}; \tag{8.9}$$

integrating again, we have

$$V = -A/r + B, \tag{8.10}$$

where A and B are constants. Since $V = 0$ when $r \to \infty$, we get $B = 0$. In addition, the current flows radially outward in all directions from the point electrode. Thus the total current crossing a spherical surface is given by

$$I = 4\pi r^2 J = -4\pi r^2 \sigma \frac{\mathrm{d}V}{\mathrm{d}r} = -4\pi\sigma A,$$

from (8.3) and (8.9), so that

$$A = -\frac{I\rho}{4\pi},$$

hence,

$$V = \left(\frac{I\rho}{4\pi}\right)\frac{1}{r} \quad \text{or} \quad \rho = 4\pi r V/I. \tag{8.11}$$

The equipotentials, which are everywhere orthogonal to the current flow lines, will be spherical surfaces given by $r = $ constant. These are illustrated in fig. 8.1.

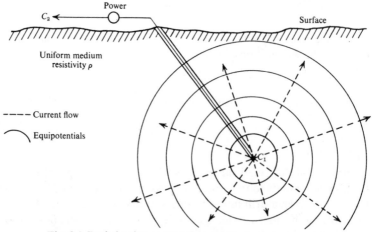

Fig. 8.1 Buried point source of current in homogeneous ground.

8.2.3 *Single current electrode at surface*

If the point electrode delivering I amperes is located at the surface of the homogeneous isotropic medium and if the air above has zero conductivity, then we have the single probe or three-point system used in surface resistivity layouts. Again the return current electrode is at a great distance.

The boundary conditions are somewhat different from the preceding case, although $B = 0$ as before since $V = 0$ as $r \to \infty$. In addition $dV/dz = 0$ at $z = 0$ (since $\sigma_{\text{air}} = 0$). This is already fulfilled since

$$\frac{\partial V}{\partial z} = \frac{\partial}{\partial z}\left(-\frac{A}{r}\right) = -\frac{\partial}{\partial r}\left(\frac{A}{r}\right)\frac{\partial r}{\partial z} = \frac{Az}{r^3} = 0 \text{ at } z = 0$$

(recalling that $r^2 = x^2 + y^2 + z^2$).

In addition all the current now flows through a hemispherical surface in the lower medium, or

$$A = -\frac{I\rho}{2\pi},$$

so that in this case

$$V = \left(\frac{I\rho}{2\pi}\right)\frac{1}{r}, \quad \text{or} \quad \rho = 2\pi r V/I. \tag{8.12}$$

Here the equipotentials are hemispherical surfaces below ground as shown in fig. 8.2.

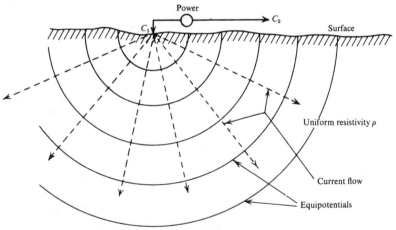

Fig. 8.2 Point source of current at the surface of a homogeneous medium.

8.2.4 *Two current electrodes at surface*

When the distance between the two current electrodes is finite (see fig. 8.3), the potential at any nearby surface point will be affected by both current electrodes.

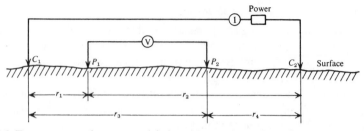

Fig. 8.3 Two current and two potential electrodes on the surface of homogeneous isotropic ground of resistivity ρ.

As before, the potential due to C_1 at P_1 is

$$V_1 = -\frac{A_1}{r_1}, \quad \text{where } A_1 = -\frac{I\rho}{2\pi}.$$

Similarly the potential due to C_2 at P_1 is

$$V_2 = -\frac{A_2}{r_2}, \quad \text{where } A_2 = \frac{I\rho}{2\pi} = -A_1$$

(since the currents at the two electrodes are equal and opposite in direction). Thus, we have

$$V_1 + V_2 = \frac{I\rho}{2\pi}\left(\frac{1}{r_1} - \frac{1}{r_2}\right).$$

Finally, by introducing a second potential electrode at P_2 we can measure the difference in potential between P_1 and P_2, which will be

$$\Delta V = \frac{I\rho}{2\pi}\left\{\left(\frac{1}{r_1} - \frac{1}{r_2}\right) - \left(\frac{1}{r_3} - \frac{1}{r_4}\right)\right\}. \tag{8.13}$$

Such an arrangement corresponds to the many-four electrode spreads normally used in resistivity field work. In this configuration the current-flow lines and equipotentials are distorted by the proximity of the second current electrode C_2. The equipotentials, obtained by plotting the relations

$$\frac{1}{R_1} - \frac{1}{R_2} = \text{constant}, \quad R_1{}^2 + R_2{}^2 - 2R_1R_2\cos\theta = 4L^2$$

are shown in fig. 8.4, along with the orthogonal current lines. The distortion from spherical equipotentials is most evident in the region between the current electrodes.

8.2.5 *Line electrodes at surface*

Consider now one of the line electrodes of length ℓ in fig. 8.5, with current I flowing out of it. The electrode makes good contact with the ground throughout its length, so the current flow is normal to the line into the ground and very little

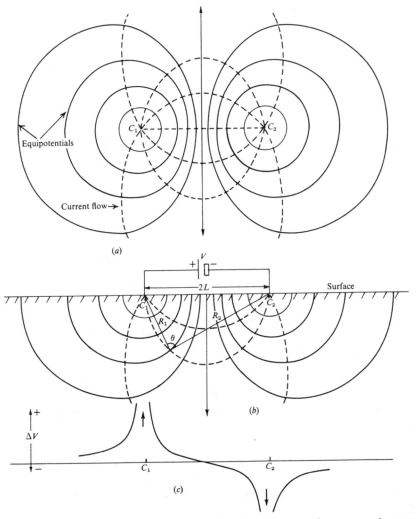

Fig. 8.4 Distortion of equipotentials and current flow-lines for two point sources of current. (After Dobrin, 1960.) (*a*) Plan view; (*b*) vertical section; (*c*) plot of potential variation at the surface along a straight line passing through the point sources.

goes out the ends. The equipotential surfaces will be half-cylinders of length ℓ and radius r as shown in section in fig. 8.6. Laplace's equation, with cylindrical symmetry, is simplified to

$$\nabla^2 V = \frac{1}{r}\frac{d}{dr}\left(\frac{r\,dV}{dr}\right) = 0, \tag{8.14}$$

that is,

$$\frac{dV}{dr} = \frac{A}{r}.$$

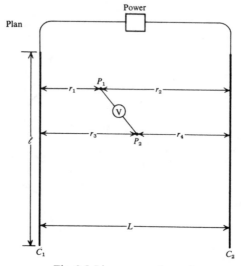

Fig. 8.5 Line current-electrode system.

Since the current is everywhere perpendicular to the cylindrical equipotential surfaces, the current density, j_r, is equal to the total current per unit length of electrode, I/ℓ, divided by the area of the cylindrical strip through which it flows. Thus,

$$j_r = \left(\frac{I}{\ell}\right)\Big/ \pi r = -\sigma\frac{dV}{dr} = -\sigma\frac{A}{r},$$

using eq. (8.3), and recalling that the current flows through only half of the cylindrical surface. Thus, we have $A = -I\rho/\pi\ell$ and the potential at P_1 becomes

$$V_1 = -\frac{I\rho}{\pi\ell}\int_{r_1}^{r_2}\frac{dr}{r} = \frac{I\rho}{\pi\ell}\log\left(\frac{r_1}{r_2}\right).$$

Finally, the potential difference between P_1 and a second potential electrode at P_2 is

$$\Delta V = V_1 - V_2 = \frac{I\rho}{\pi\ell}\left\{\log\left(\frac{r_1}{r_2}\right) - \log\left(\frac{r_3}{r_4}\right)\right\} = \frac{I\rho}{\pi\ell}\log\left(\frac{r_1 r_4}{r_2 r_3}\right). \tag{8.15}$$

Clearly, in homogeneous ground, the potential is constant along any line parallel to the electrodes, i.e., when $r_1 = r_3, r_2 = r_4$. This is the arrangement used in equipotential line measurements. Figure 8.6 illustrates the equipotentials and current flow for this case. If the line electrode C_2 is removed to a great distance, the equipotentials are the lower halves of circular cylindrical surfaces. When C_2 is not at infinity, the surfaces are elliptical with foci at the two electrodes; the equipotentials may be plotted in this case from the equations

$$R_1/R_2 = \text{constant}; \quad R_1{}^2 + R_2{}^2 - 2R_1 R_2 \cos\theta = L^2$$

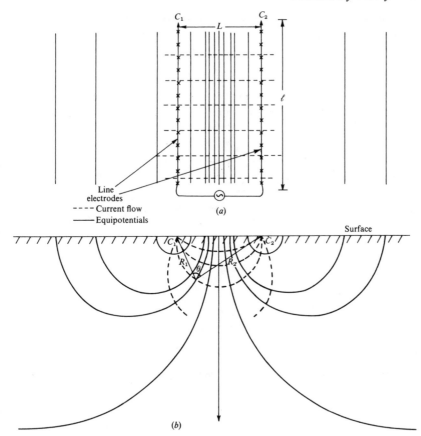

Fig. 8.6 Distortion of equipotentials and current flow-lines in vicinity of line electrodes. (*a*) Plan view; (*b*) vertical section.

(see fig. 8.6). As C_2 approaches closer and closer to C_1, the surfaces become more and more distorted from the circular shape which they would have for infinite electrode separation.

8.2.6 *Current distribution*

Figures 8.1, 8.2, 8.4 and 8.6 illustrate in a general way the flow of current in homogeneous ground. They do not, however, show the quantitative distribution in depth. Obviously it is necessary to increase the electrode separation to increase the penetration. Consider the current flow in a homogeneous medium between two point electrodes, C_1 and C_2, in fig. 8.7. The horizontal current density at point P is

$$j_x = -\frac{1}{\rho}\frac{\partial V}{\partial x} = -\frac{I}{2\pi}\frac{\partial}{\partial x}\left(\frac{1}{r_1} - \frac{1}{r_2}\right) = \frac{I}{2\pi}\left(\frac{x}{r_1^3} - \frac{x-L}{r_2^3}\right),$$

and if this point is on the vertical plane midway between C_1 and C_2, we have $r_1 = r_2 = r$, and

$$j_x = \frac{I}{2\pi} \frac{L}{(z^2 + L^2/4)^{\frac{3}{2}}}. \tag{8.16}$$

Figure 8.8 shows the variation in current density with depth across this plane when the electrode separation is maintained constant. If, on the other hand, the electrode spacing is varied, it is found that j_x is a maximum when $L = \sqrt{2}z$.

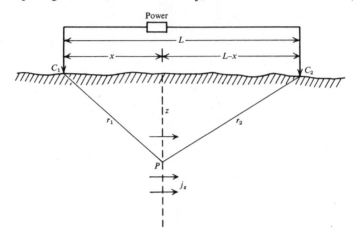

Fig. 8.7 Determining the current density in uniform ground below two surface electrodes.

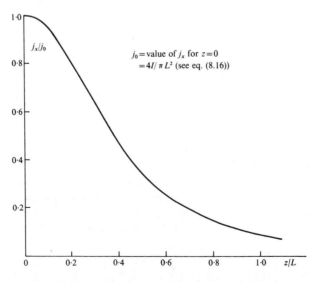

Fig. 8.8 Current density versus depth, fixed electrode spacing.

We can also calculate the fraction of current flowing through any particular area of this vertical plane, say between depths z_1 and z_2. Since the current through an element $dy\,dz$ in the plane is

$$\delta I_x = j_x\,dy\,dz = \frac{I}{2\pi}\,\frac{L}{\{(L/2)^2 + y^2 + z^2\}^{\frac{3}{2}}}\,dy\,dz,$$

the fraction of total current through a long strip $(z_2 - z_1)$ wide will be

$$\frac{I_x}{I} = \frac{L}{2\pi}\int_{z_1}^{z_2} dz \int_{-\infty}^{\infty} \frac{dy}{\{(L/2)^2 + y^2 + z^2\}^{\frac{3}{2}}} = \frac{2}{\pi}\left(\tan^{-1}\frac{2z_2}{L} - \tan^{-1}\frac{2z_1}{L}\right).$$

(8.17a)

This fraction has a minor maximum when $L = 2\sqrt{(z_1 z_2)}$. Taking a numerical example, if $z_1 = 500$ ft, $z_2 = 1000$ ft, the electrode spacing should be 1400 ft to get the maximum horizontal current density in the slab. The concentration, however, is not very significant.

Otherwise, if $z_2 \to \infty$, eq. (8.17a) becomes

$$\frac{I_x}{I} = 1 - \frac{2}{\pi}\tan^{-1}\frac{2z_1}{L}.$$

(8.17b)

Figure 8.9 shows the electrode spacing necessary to force a large fraction of the current into the ground below a depth z_1. From this plot we see that, when $L = 2z_1$, half the current flows in the top layer, half below it.

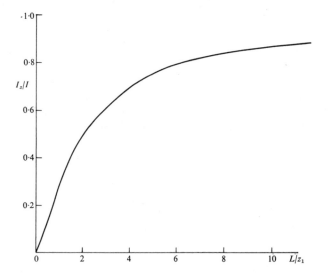

Fig. 8.9 Fraction of current flowing below a depth z_1 for an electrode spacing L. (See eq. (8.17b).)

642   *Resistivity methods*

Equation (8.16) is slightly changed for two line electrodes spaced L apart. The current density is given by

$$j_x = \frac{IL}{\pi \ell \{z^2 + (L/2)^2\}}.$$ (8.18)

In this case the current density is a maximum above z when $L = 2z$, and the fraction of total current between depths z_1 and z_2 is the same as in eq. (8.17a).

Since the variations in potential, measured at surface, depend on the current flow below, it is desirable to get as much current into the ground as possible. The above relations indicate the proper current electrode spread required to energize the ground at a particular depth with a limited power source; i.e., for a good penetration to 500 feet, the electrodes must be at least 500 feet apart and preferably more than this. Compared to magnetotellurics, for instance, this places an inherent limitation on the resistivity method. However the controlled power source provides certain advantages.

8.3 Effect of inhomogeneous ground

8.3.1 *Introduction*

So far we have considered current flow and potential in and over homogeneous ground, a situation which is extremely rare in the field and which would be of no practical significance anyway. What we want to detect is the presence of anomalous conductivity in various forms, such as lumped (three-dimensional) bodies, dikes, faults and vertical or horizontal contacts between beds. The resistivity method is most suitable for outlining horizontal beds and vertical contacts, less useful on bodies of irregular shape.

8.3.2 *Distortion of current flow at a plane interface*

Consider two homogeneous media of resistivities ρ_1 and ρ_2 separated by a plane boundary as in fig. 8.10. Suppose that a current of density $\mathbf{j}_1$ is flowing in medium (1) in such a direction as to meet the boundary at an angle θ_1 to the normal. To determine the direction of this current in medium (2) we recall the conditions given

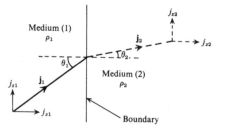

Fig. 8.10 Distortion of current flow at a plane boundary.

in eq. (8.7b), §8.2.1; using Ohm's law to express these results in terms of the current density, we obtain

$$j_{x1}\rho_1 = j_{x2}\rho_2, \quad \text{and} \quad j_{z1} = j_{z2}.$$

Dividing these expressions, we have

$$\rho_1(j_{x1}/j_{z1}) = \rho_2(j_{x2}/j_{z2}), \quad \text{or} \quad \rho_1 \tan \theta_1 = \rho_2 \tan \theta_2,$$

so that

$$\tan \theta_2/\tan \theta_1 = \rho_1/\rho_2. \tag{8.19}$$

Thus the current lines are bent in crossing the boundary. If $\rho_1 < \rho_2$, they will be bent towards the normal and vice versa.

8.3.3 *Distortion of potential at a plane interface*

Clearly if the current flow is distorted in passing from a medium of one resistivity into another, the equipotentials also will be distorted. It is possible to determine the potential field mathematically by solving Laplace's equation for the appropriate boundary conditions or by integrating it directly. Both these approaches require considerable mathematics. A much simpler approach employs electrical images, in analogy with geometrical optics. The use of images is valid in solving only a limited number of potential problems, including the plane boundary and the sphere.

The analogy between the electrical situation and optics is based on the fact that current density, like light ray intensity, decreases with the inverse square of distance from a point source. The problem is to determine the potential distribution resulting from a point source in a medium of resistivity ρ_1, separated from an adjacent medium ρ_2 by a plane boundary.

In optics the analogous case would be a point source of light in one medium separated from another by a semi-transparent mirror, having reflection and transmission coefficients k and $1 - k$. Then the light intensity at a point in the first medium is partly due to the point source and partly to its image in the second medium, the latter effect diminished by reflection from the mirror. On the other hand, the intensity at a point in the second medium is due only to the source in the first, diminished by transmission through the mirror (see fig. 8.11a).

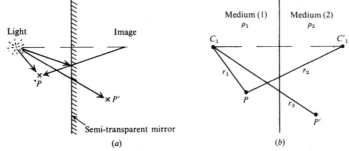

Fig. 8.11 Analogy between optical and electrical images. (a) Optical image; (b) electrical image.

If we replace the point source of light by a point source of current and the light intensity at a point by potential, the problem is now in the electrical domain. From fig. 8.11b we see that the potential at P in the first medium is

$$V = \frac{I\rho_1}{4\pi} \left(\frac{1}{r_1} + \frac{k}{r_2} \right), \tag{8.20}$$

and in the second medium at P' it is

$$V' = \frac{I\rho_2}{4\pi} \left(\frac{1 - k}{r_3} \right). \tag{8.21}$$

Applying the boundary conditions, these potentials must be equal at the interface, when $r_1 = r_2 = r_3$. Thus we have

$$\frac{\rho_1}{\rho_2} = \frac{1 - k}{1 + k}, \quad \text{or} \quad k = \frac{\rho_2 - \rho_1}{\rho_2 + \rho_1}. \tag{8.22}$$

In this expression k is a *reflection coefficient* whose value lies between ± 1, depending on the relative resistivities in the two media.

Figure 8.12 shows the traces of equipotential surfaces plotted from the relations in eq. (8.20) for $k = \pm\frac{1}{2}$. A few current flow lines are also drawn. This situation corresponds to the practical case of resistivity logging with respect to a plane boundary underground or the measurement of surface potentials across a vertical contact.

8.3.4 *Surface potential due to horizontal beds*

If the current source and potential point are located on surface, above a horizontal boundary separating two media, the upper having resistivity ρ_1, the lower ρ_2, the analysis is more complicated. In fact because of the ground surface there are now three media, separated by two interfaces. As a result there is an infinite set of images above and below the current electrode, as illustrated in fig. 8.13. The original image C_1', at depth $2z$ below surface, is reflected in the surface boundary to give an image C_1'' a distance $2z$ above C_1. This second image produces a third C_1''' at a depth $4z$, reflected in the lower boundary, and so on.

The effect of each successive image on the potential at P is reduced by the reflection coefficient between the boundaries. For the current source and its first image below ground, the potential, is, as in eq. (8.20),

$$V' = \frac{I\rho_1}{2\pi} \left(\frac{1}{r} + \frac{k}{r_1} \right).$$

The effect of the second image at $2z$ above ground is

$$V'' = \frac{I\rho_1}{2\pi} \left(\frac{k \times k_a}{r_1} \right),$$

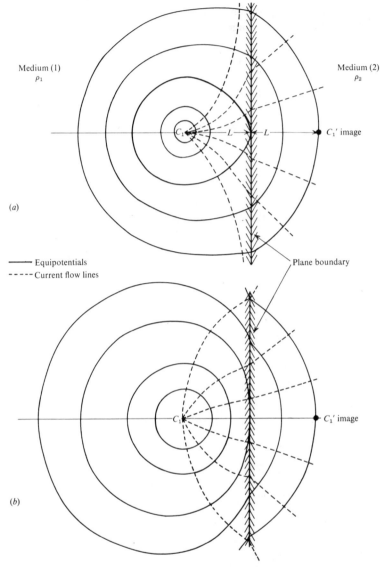

Fig. 8.12 Distortion of equipotentials and current flow-lines at a boundary between two media of different resistivities. (a) $\rho_2/\rho_1 = 3$, $k = 0\cdot5$; (b) $\rho_2/\rho_1 = \frac{1}{3}$, $k = -0\cdot5$.

where k_a is the reflection coefficient at the surface boundary. Since ρ_a is essentially infinite this coefficient is unity, and

$$V' + V'' = \frac{I\rho_1}{2\pi}\left(\frac{1}{r} + \frac{2k}{r_1}\right).$$

The potential due to the third image C_1''', $4z$ below ground, will be further reduced, as will that of its image $4z$ above ground, hence

$$V''' + V^{IV} = \frac{I\rho_1}{2\pi} \left(\frac{k \times k}{r_2} + \frac{k \times k \times k_a}{r_2} \right) = \frac{I\rho_1}{2\pi} \left(\frac{2k^2}{r_2} \right).$$

The resultant total potential at P can thus be expressed as an infinite series of the form

$$V = \frac{I\rho_1}{2\pi} \left\{ \frac{1}{r} + \frac{2k}{r_1} + \frac{2k^2}{r_2} + \ldots + \frac{2k^m}{r_m} + \ldots \right\}, \qquad (8.23)$$

where

$$r_1 = \sqrt{\{r^2 + (2z)^2\}}, r_2 = \sqrt{\{r^2 + (4z)^2\}}, \ldots r_m = \sqrt{\{r^2 + (2mz)^2\}}.$$

This series can be written in the compact form

$$V = \frac{I\rho_1}{2\pi} \left[\frac{1}{r} + 2 \sum_{m=1}^{\infty} \frac{k^m}{\sqrt{\{r^2 + (2mz)^2\}}} \right] = \frac{I\rho_1}{2\pi r} \left[1 + 2 \sum_{m=1}^{\infty} \frac{k^m}{\sqrt{\{1 + (2mz/r)^2\}}} \right].$$

$$(8.24)$$

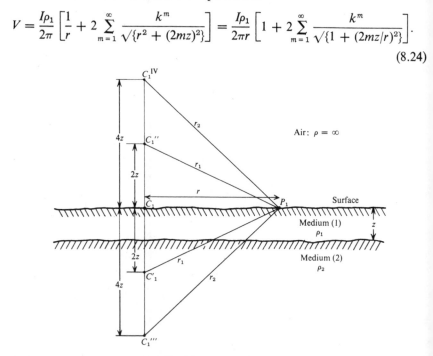

Fig. 8.13 Images resulting from two horizontal beds.

This series is convergent, since $|k| < 1$, while the denominator increases indefinitely. The number of terms necessary to get a reasonable answer depends mainly on the value of k and partly on the ratio z/r. For a fixed value of r, the potential differs from that measured over uniform ground. The latter is given by the first term in the bracket of eq. (8.24) and is called the *normal potential*. The portion expressed by the infinite series is the *disturbing potential*. When k is positive and approximately unity, the total potential at P may be increased by a factor of two or more.

8.3.5 *Potential due to buried sphere*

A three-dimensional body for which the external potential may be developed is the sphere. Figure 8.14 illustrates this case, in which we use spherical coordinates with the sphere centre as origin and the polar axis parallel to the x-axis. The problem is to find solutions of Laplace's equation for particular boundary conditions; for simplificity we assume the sphere to be in a uniform field E_0 which is parallel to the x-axis. This is equivalent to having the current electrode at some distance from the sphere.

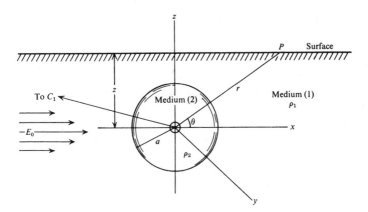

Fig. 8.14 Buried conductive sphere in uniform field.

It is now necessary to solve Laplace's equation in the form

$$\frac{1}{r^2}\frac{\partial}{\partial r}\left(r^2\frac{\partial V}{\partial r}\right) + \frac{1}{r^2\sin\theta}\frac{\partial}{\partial \theta}\left(\sin\theta\frac{\partial V}{\partial \theta}\right) = 0, \qquad (8.25)$$

where the system is independent of ϕ, the longitude, and subject to the usual boundary conditions given by (8.7), namely,

$$\left.\begin{array}{l} V_1 = V_2, \\[2mm] \dfrac{1}{\rho_1}\dfrac{\partial V_1}{\partial r} = \dfrac{1}{\rho_2}\dfrac{\partial V_2}{\partial r}, \end{array}\right\} \text{ at } r = a; \qquad (8.26)$$

also V_1 must approach the value of the undisturbed potential remote from the sphere, that is, for large x or r,

$$V_1 = -E_0 x = -E_0 r \cos\theta.$$

The solution is in the form of a series of Legendre polynomials satisfying the following conditions inside and outside the sphere (see §2.6.3e, §6.1.4 and fig. 6.7)

$$
\left.
\begin{aligned}
V_1 &= -E_0 r \cos\theta + \sum_{n=0}^{\infty} \frac{b_n P_n(\mu)}{r^{n+1}}, \quad r > a \\
V_2 &= \sum_{n=0}^{\infty} a_n r^n P_n(\mu), \quad r < a,
\end{aligned}
\right\}
\tag{8.27}
$$

where a_n, b_n are constants, $\mu = \cos\theta$, and $P_n(\mu)$ is the Legendre polynomial. Expanding the series in (8.27) and noting that $P_1(\mu) = \mu = \cos\theta$, we get

$$
V_1 = -E_0 r P_1 + \frac{b_0}{r} + \frac{b_1}{r^2} P_1 + \frac{b_2}{r^3} P_2 + \ldots,
\tag{8.28a}
$$

$$
V_2 = a_0 + a_1 r P_1 + a_2 r^2 P_2 + \ldots,
\tag{8.28b}
$$

$$
\frac{1}{\rho_1} \frac{\partial V_1}{\partial r} = \left(-E_0 P_1 - \frac{b_0}{r^2} - \frac{2 b_1 P_1}{r^3} - \frac{3 b_2 P_2}{r^4} + \ldots \right) \frac{1}{\rho_1},
\tag{8.28c}
$$

$$
\frac{1}{\rho_2} \frac{\partial V_2}{\partial r} = (a_1 P_1 + 2 a_2 r P_2 + \ldots) \frac{1}{\rho_2}.
\tag{8.28d}
$$

If the coefficients are correct for the space inside and outside the sphere, they must be correct on its surface as well. Then we can equate coefficients of like orders of $P_n(\mu)$ at $r = a$ in eqs. (8.28a) and (8.28b), also in (8.28c) and (8.28d), and obtain

$$
a_0 = b_0/a \quad \text{and} \quad b_0 = 0; \quad \text{so} \quad a_0 = 0,
$$

$$
a_1 a = -E_0 a + \frac{b_1}{a^2} \quad \text{and} \quad -\frac{E_0}{\rho_1} - \frac{2 b_1}{\rho_1 a^3} = \frac{a_1}{\rho_2},
$$

whence

$$
b_1 = \frac{E_0 (\rho_1 - \rho_2) a^3}{(\rho_1 + 2\rho_2)} \quad \text{and} \quad a_1 = -E_0 \frac{3 \rho_2}{(\rho_1 + 2\rho_2)},
$$

while all subsequent coefficients are zero. Thus the solution outside the sphere is

$$
V_1 = -E_0 r \cos\theta + \frac{E_0 (\rho_1 - \rho_2) a^3 \cos\theta}{r^2 (\rho_1 + 2\rho_2)},
$$

or,

$$
V_1 = -E_0 r \cos\theta \left\{ 1 - \frac{(\rho_1 - \rho_2)}{(\rho_1 + 2\rho_2)} \left(\frac{a}{r} \right)^3 \right\}.
\tag{8.29}
$$

If the potential is measured at the ground surface, the sphere will have an image which will double the second term. In addition, if we consider the field to be generated by a current source I at a distance R from the origin, we can write

$$
V_1 = -\frac{I \rho_1}{2\pi R^2} \left\{ 1 - 2 \frac{(\rho_1 - \rho_2)}{(\rho_1 + 2\rho_2)} \left(\frac{a}{r} \right)^3 \right\} r \cos\theta.
\tag{8.30}
$$

As in eq. (8.24) we have two terms, the first being the normal potential, the second the disturbing potential caused by the sphere. Equipotential and current flow lines are illustrated in the section shown in fig. 8.15.

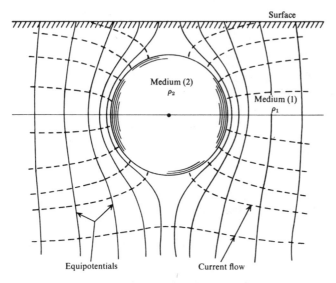

Fig. 8.15 Equipotentials and current flow-lines for buried conductive sphere.

Note that we have made two assumptions here which are not necessarily valid, first that the external or normal field is uniform and second that there is no interaction between the sphere and its image. Both are strictly true only when the sphere is a great distance from both the current source and surface, in which case the anomaly could not be detected anyway. However if the distance between the sphere's centre and the surface is not less than 1·3 times the radius, the approximation is reasonably good.

8.3.6 *Effect of anisotropic ground*

Most rock masses are anything but homogeneous and isotropic in the electrical sense because they are full of fractures. In particular, shales, slates, and frequently limestones and schists have a definite anisotropic character, especially with respect to the bedding planes.

As an example of this type of anisotropy, consider a semi-infinite medium in which the resistivity is uniform in the horizontal direction and has the value ρ_t; in the vertical direction it is also constant and has a different magnitude ρ_v, ρ_v almost invariably being larger than ρ_t (see §5.2.2c). The current-density components will be

$$j_x = -\frac{1}{\rho_t}\frac{\partial V}{\partial x}, \quad j_y = -\frac{1}{\rho_t}\frac{\partial V}{\partial y}, \quad \text{and} \quad j_z = -\frac{1}{\rho_v}\frac{\partial V}{\partial z}. \tag{8.31}$$

Otherwise conditions are identical with homogeneous isotropic ground and the divergence of current is zero except at the source. Thus,

$$\nabla \cdot \mathbf{J} = \frac{1}{\rho_t} \left(\frac{\partial^2 V}{\partial x^2} + \frac{\partial^2 V}{\partial y^2} \right) + \frac{1}{\rho_v} \frac{\partial^2 V}{\partial z^2} = 0.$$

By the following change in variables,

$$\alpha = \sqrt{(\rho_t)}x, \ \beta = \sqrt{(\rho_t)}y, \ \gamma = \sqrt{(\rho_v)}z, \ \text{and} \ \tau^2 = \alpha^2 + \beta^2 + \gamma^2, \ (8.32)$$

we can transform this relation into Laplace's equation in polar coordinates. The equation becomes, as in §8.2.2,

$$\nabla^2 V = \frac{1}{\tau^2} \frac{d}{d\tau} \left(\tau^2 \frac{dV}{d\tau} \right) = 0.$$

Thus the solution is

$$V = -A/\tau + B.$$

The constant B is zero if we assume, as usual, that $V = 0$ at $\tau = \infty$. Now the relations in (8.31) may be used to determine A:

$$j_x = -\frac{Ax}{\tau^3} = \frac{-Ax}{\{\rho_t(x^2 + y^2 + \lambda^2 z^2)\}^{\frac{3}{2}}} \tag{8.33a}$$

$$j_y = -\frac{Ay}{\tau^3} = \frac{-Ay}{\{\rho_t(x^2 + y^2 + \lambda^2 z^2)\}^{\frac{3}{2}}}, \tag{8.33b}$$

$$j_z = -\frac{Az}{\tau^3} = \frac{-Az}{\{\rho_t(x^2 + y^2 + \lambda^2 z^2)\}^{\frac{3}{2}}} \tag{8.33c}$$

where $\lambda = \sqrt{(\rho_v/\rho_t)}$ is the *coefficient of anisotropy*.

Then the current density is

$$J = \sqrt{(j_x^2 + j_y^2 + j_z^2)} = \frac{A(x^2 + y^2 + z^2)^{\frac{1}{2}}}{\rho_t^{\frac{3}{2}}(x^2 + y^2 + \lambda^2 z^2)^{\frac{3}{2}}}.$$

In polar coordinates, since

$$x^2 + y^2 = r^2 \sin^2 \theta, \ \text{and} \ z^2 = r^2 \cos^2 \theta,$$

$$J = \frac{A}{\rho_t^{\frac{3}{2}} r^2 \{1 + (\lambda^2 - 1) \cos^2 \theta\}^{\frac{3}{2}}}. \tag{8.34}$$

Near a source electrode, all the current will flow through a bowl-like surface which, because of the uniform resistivity in the horizontal direction, will be symmetrical about the z-axis, as in fig. 8.16. If we assume that this surface is spherical rather than spheroidal (an approximation which is in error by about the same amount as the degree of anisotropy, i.e., if $\lambda^2 = 1.2$, the error is 19 %), then a strip of this surface of slant height dt will have an area

$$dS \approx (2\pi r \sin \theta)(r \, d\theta) = 2\pi r^2 \sin \theta \, d\theta$$

(since the surface is approximately spherical).

Then,

$$I = \int_s j_n \, \mathrm{d}S = \int_0^{\pi/2} \frac{2\pi A \sin\theta \, \mathrm{d}\theta}{[\rho_t\{1 + (\lambda^2 - 1)\cos^2\theta\}]^{\frac{3}{2}}}.$$

Setting $\cos\theta$ equal to x, the above can be integrated to give

$$I = \frac{2\pi A}{\rho_t^{\frac{3}{2}}\lambda} \tag{8.35}$$

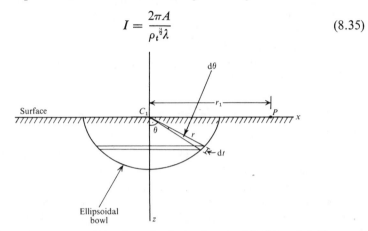

Fig. 8.16 Point current source at the surface of anisotropic ground having resistivities ρ_t and ρ_v in the horizontal and vertical directions respectively.

Then the potential is

$$V = -\frac{I\rho_t^{\frac{3}{2}}\lambda}{2\pi r} = -\frac{I\rho_t\lambda}{2\pi\sqrt{(x^2 + y^2 + \lambda^2 z^2)}} \tag{8.36}$$

The equipotential surfaces are ellipsoidal and symmetrical about z with the equation

$$x^2 + y^2 + \lambda^2 z^2 = \text{constant}.$$

The potential at a surface point P, a distance r_1 from the current electrode C_1, will be

$$V_p = \frac{-I\rho_t\lambda}{2\pi r_1} = \frac{-I\sqrt{(\rho_t\rho_v)}}{2\pi r_1}, \tag{8.37}$$

that is, the potential is equivalent to that for an isotropic medium of resistivity $\sqrt{(\rho_t\rho_v)}$. Thus it is not possible to detect this type of anisotropy from field measurements.

From eq. (8.37) and fig. 8.16 it is obvious that the resistivity measured over horizontal beds is larger than the actual horizontal resistivity in the beds, but smaller than the vertical resistivity. On the other hand if the beds have a steep dip and the measurement is made with a spread perpendicular to strike, the apparent resistivity will be smaller than the true resistivity normal to the bedding, just the opposite to the result over horizontal layers; if the array is parallel to the strike of the dipping beds, the apparent resistivity will be too large.

8.3.7 *Effect of topography*

As mentioned earlier, resistivity measurements are strongly influenced by local variations in surface conductivity, caused by weathering and moisture content. Rugged topography will have a similar effect, since the current flow tends to follow the surface, particularly if the shallow layers are somewhat conductive. The equipotential surfaces are distorted as a result, producing false anomalies due to the topography alone. This effect may distort or mask a real anomaly.

As in the case of anisotropic ground there is no method of isolating the topography effect. In simple cases such as sharp depressions and knolls, the potential distortion might be roughly corrected by graphical smoothing.

8.4 Equipment for resistivity field work

8.4.1 *Power sources*

The necessary components for making resistivity measurements include a power source, meters for measuring current and voltage (which may be combined in one meter to read resistance), electrodes, cable and reels. The power may be either d.c. or low frequency a.c., preferably less than 60 Hz. If d.c. is used, a set of B-batteries (45 to 90 volts) may be connected in series to give several hundred volts total. Because of the limited current capacity and short life, battery sources have little advantage except portability. For large-scale work it is preferable to use a motor-generator having a capacity of several hundred watts. Equipment of this type, because of its bulk and weight, is only semiportable; it would not be moved each time the electrodes were shifted.

To avoid the effects of electrolytic polarization caused by unidirectional current, the d.c. polarity should be reversed periodically, either by hand with a reversing switch, or by a mechanical commutator, relay system or vibrator. The rate of commutation may range from three or four times a minute to 100 times per second.

Alternating current is also employed in place of commutated (effectively square-wave) d.c. A low-frequency sine-wave transistor oscillator with transformer output of a few watts makes a convenient portable source. Larger power can be obtained from a motor-driven alternator.

Each of these devices obviously has particular advantages and limitations. The d.c. source permits measurement of d.c. resistivity – which is desirable – but it also measures spontaneous potentials. This requires that porous pots be used as potential electrodes; then, the SP effect must be noted before the source is turned on, and then subtracted, either directly or by means of a compensating voltage, from the potential measured when current is flowing.

The use of a.c. or rapidly interrupted d.c. eliminates the SP effect. In addition, narrow-band amplifiers tuned to the source frequency can be employed to increase the signal-to-noise ratio. However, the resistivity measured will generally be lower than the true d.c. value. More serious, inductive coupling between long current and adjacent potential leads, as well as leakage currents, particularly on wet ground, may give erratic readings. All these effects increase with the frequency.

8.4.2 *Meters*

With d.c. or long-period commutated d.c. sources, the current is measured with a d.c. milliammeter, whose range should be from about 5 to 500 mA, depending on the electrode spread, type of ground and power used. Potential is normally measured with a d.c. voltmeter of high input impedance (1 megohm or greater) and range 10 mV to perhaps 20 V. When a.c. sources are used a.c. meters are of course necessary.

A typical resistivity set with voltage and current meters is illustrated schematically in fig. 8.17. In some resistivity equipment the current is maintained constant with a regulator, which eliminates the current measurement.

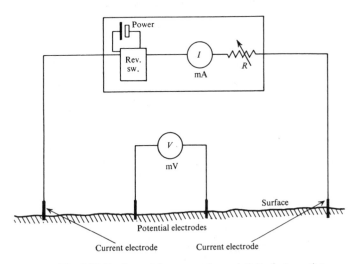

Fig. 8.17 Equipment for measuring resistivity (schematic).

An instrument which measures the ratio of potential to current (that is, resistance) – usually associated with the trade name *Megger* – has been frequently employed for resistivity work. Originally developed for testing cable insulation, this device was easily modified to measure ground resistance. Power is developed by hand cranking a d.c. generator or magneto; the output is about 100 volts and a d.c. current coil is connected in series with one side. The output then is commutated on the generator shaft and applied to the current electrodes, the rate of reversal, 10 to 50 times per second, being regulated by a governor. The potential electrodes are connected to a second commutator, synchronized with the other, which rectifies the a.c. potential and applies it to the potential coil. The latter is mounted with the current coil in such a way as to make the needle deflection proportional to V/I. This instrument is shown schematically in fig. 8.18.

Several other all-in-one resistivity instruments are also available, employing a vibrator powered by dry cells or low frequency transistor oscillator. Such devices, like the Megger, necessarily have low power output. Furthermore, with some electrode spreads, the combination of power source and both meters in one box may be a definite disadvantage. However, such instruments are compact and completely portable.

8.4.3 *Electrodes and wire*

With a.c. power sources all the electrodes may be steel, aluminium or brass; stainless steel is probably best for combined strength and resistance to corrosion. If d.c. power is used, the potential electrodes, however should be porous pots as in SP work. Metal electrodes should be at least one foot long so that they can be driven into the ground several inches for good electrical contact. In very dry ground this contact may be improved by watering the electrodes.

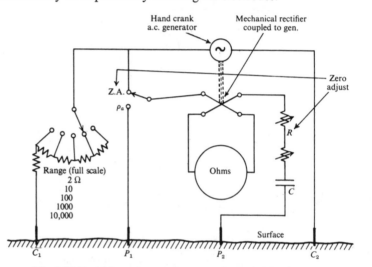

Fig. 8.18 Portable equipment for measuring resistivity (schematic).

Wire, usually wound on portable reels, obviously must be insulated and should be as light as possible. Plastic insulation is more durable than rubber against abrasion and moisture. However, some plastics deteriorate in cold weather and animals seem to find them very tasty in any season.

8.5 Electrode layouts and field procedure

8.5.1 *General*

An enormous number of electrode spreads have been used in resistivity at various times; not more than six have survived to any extent. In principle it is not necessary to use a collinear array. Practically, however, the electrodes are almost always in line, since interpretation of results is otherwise difficult and the field work is complicated.

One drawback in resistivity work is the practical difficulty of moving stakes with great lengths of wire attached, a slow and expensive task in relation to magnetics, EM and some other electrical survey methods. Thus it is an advantage to use the electrode spreads which may require only one or two electrodes to be moved, and these at close spacing where possible.

8.5.2 *Apparent resistivity*

Before discussing the various electrode spreads it is necessary to consider what is actually measured by an array of current and potential electrodes. We can rearrange the terms in eq. (8.13), §8.2.4, to obtain

$$\rho = \frac{2\pi\Delta V}{I} \frac{1}{\{(1/r_1 - 1/r_2) - (1/r_3 - 1/r_4)\}} = \left(\frac{2\pi\Delta V}{I}\right)p, \qquad (8.38)$$

where the parameter p has to do with the electrode geometry. By measuring ΔV and I and knowing the electrode configuration, we obtain a resistivity ρ. Over homogeneous isotropic ground this resistivity will be constant for any current and electrode arrangement. That is, if the current is maintained constant and the electrodes are moved around, the potential ΔV will adjust at each configuration to keep the ratio $(\Delta Vp/I)$ constant.

If the ground is inhomogeneous, however, and the electrode spacing is varied, or the spacing remains fixed while the whole array is moved, then the ratio will, in general, change. This results in a different value of ρ for each measurement. Obviously the magnitude is intimately involved with the arrangement of electrodes.

This measured quantity is known as the *apparent resistivity*, ρ_a. Although it is diagnostic, to some extent, of the actual resistivity of a zone in the vicinity of the electrode array, this apparent resistivity is definitely not an average value. Only in the case of homogeneous ground is the apparent value equivalent to the actual resistivity.

Another term which is frequently found in the literature is the so-called *surface resistivity*. This is the value of ρ_a obtained with small electrode spacing. Obviously it is equal to the true surface resistivity only when the ground is uniform over a volume roughly of the dimensions of the electrode separation.

8.5.3 *Electrode spreads*

(a) *Wenner spread*. The most commonly used point-electrode systems are illustrated in fig. 8.19. The first two examples, the *Wenner* and *Schlumberger arrays*, are presently most popular.

In the Wenner spread the electrodes are uniformly spaced in a line. From eq. (8.38), $r_1 = r_4 = a$ and $r_2 = r_3 = 2a$. Thus the apparent resistivity is

$$\rho_a = 2\pi a \Delta V/I. \qquad (8.39)$$

In spite of the simple geometry, this arrangement is often quite inconvenient for field work, and has some disadvantages from a theoretical point of view as well. For depth exploration using the Wenner spread, the electrodes are expanded about a fixed centre, increasing the spacing a in steps. For lateral exploration or mapping, the spacing remains constant and all four electrodes are moved along the line, then along another line, and so on. In mapping, the apparent resistivity for each array position is plotted against the centre of the spread.

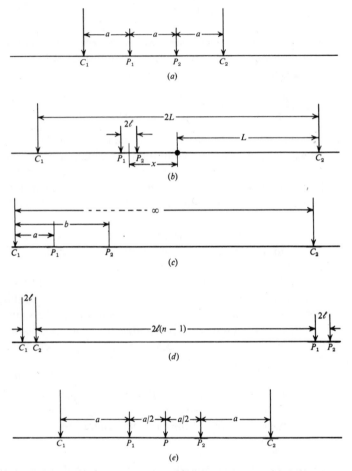

Fig. 8.19 Electrode spreads in common use. (*a*) Wenner spread; (*b*) Schlumberger spread; (*c*) three-point spread; (*d*) double-dipole spread; (*e*) Lee-partition spread.

(b) *Schlumberger spread.* For the Schlumberger array the current electrodes are spaced much further apart than the potential electrodes. From figs. 8.3 and 8.19*b*, we find that

$$r_1 = (L - x) - \ell, \quad r_2 = (L + x) + \ell, \quad r_3 = (L - x) + \ell$$

and

$$r_4 = (L + x) - \ell.$$

If the smallest current-potential electrode distance is always considerably greater than the distance between the two potential electrodes (factor of 10 or more), then eq. (8.38) gives

$$\rho_a = \frac{2\pi \Delta V}{I} \frac{1}{\left[\left\{\dfrac{1}{(L-x)-\ell} - \dfrac{1}{(L+x)+\ell}\right\} - \left\{\dfrac{1}{(L-x)+\ell} - \dfrac{1}{(L+x)-\ell}\right\}\right]}$$

Simplifying, since $(L - x) \gg \ell$, we get

$$\rho_a = \frac{\pi}{2\ell} \frac{(L^2 - x^2)^2}{(L^2 + x^2)} \left(\frac{\Delta V}{I}\right). \tag{8.40}$$

This array is often used symmetrically, i.e., $x = 0$, in which case;

$$\rho_a = \frac{\pi L^2}{2\ell} \left(\frac{\Delta V}{I}\right). \tag{8.41}$$

In depth probing the potential electrodes remain fixed while the current-electrode spacing is expanded symmetrically about the centre of the spread. For large values of L it may be necessary to increase ℓ also in order to maintain a measurable potential. Equation (8.41) applies in this case. This procedure is more convenient than the Wenner expanding spread because only two electrodes need move. In addition the effect of shallow resistivity variations is constant with fixed potential electrodes.

Lateral exploration may be done in two ways. With a very large fixed separation of the current electrodes (1000 feet or more), the potential pair is moved between them, also with fixed spacing, subject to the limitation $(L - x) \gg \ell$, (see eq. (8.40)). Apparent resistivity is plotted against the midpoint of the potential electrodes.

The other layout is similar to the Wenner, in that the electrode spacing remains fixed ($L \gg \ell$) and the whole array is moved along the line in suitable steps. This arrangement is less convenient than the first since it requires that all four electrodes be moved for each station.

In lateral exploration with the Schlumberger array, it is permissible to measure potential somewhat off the line between fixed current electrodes, that is, to map the surface in two dimensions.

(c) *Three-point (gradient) system.* One of the current electrodes is fixed at a great distance from the other three, all of which can have various spacings. The values in eq. (8.38) are now

$$r_1 = a, \quad r_3 = b, \quad r_2 = r_4 = \infty,$$

so that

$$\rho_a = \frac{2\pi a b}{b - a} \left(\frac{\Delta V}{I}\right). \tag{8.42a}$$

When $b = 2a$ this becomes

$$\rho_a = 4\pi a \left(\frac{\Delta V}{I}\right), \tag{8.42b}$$

or double the ratio in the Wenner array. When the potential spacing is very small compared to the distance of either potential electrode from C_1, (C_2 still at ∞), we write: $r_1 = a - \delta a/2$, $r_3 = a + \delta a/2$, and the apparent resistivity becomes

$$\rho_a = \frac{2\pi a^2}{I} \left(\frac{\partial V}{\partial a}\right). \tag{8.42c}$$

This arrangement is equivalent to a half-Schlumberger spread.

Since the electrode C_2 is remote, it is not necessary to have it in line with the other three. This permits lateral exploration on radial lines from a fixed position of C_1, by moving one or both potential electrodes, a particularly convenient method for resistivity mapping in the vicinity of a conductor of limited extent.

This electrode arrangement is effectively the same as the *lateral spread* used in well logging, described in §11.2.2e. It is also similar to the *mise-à-la-masse* method (see §8.5.4d)) in which the electrode C_1 is in contact with the conducting zone.

A further variation on the three-point system is obtained by moving one of the potential electrodes, say P_2, to a distant point, which is also remote from C_2. In this case $r_3 = b = \infty$ as well, and the expression (8.42a) is the same as for the Wenner spread, hence this array is known as the *Half-Wenner array*. Although it is only necessary to move one potential electrode, the long connecting wire to the other is a disadvantage.

In field work the location of an electrode at infinity requires that it have very little influence on the rest of the array. For instance, when using a Wenner spread, the remote electrode or electrodes must be at least ten times the spacing to reduce the effect to 10% or less. With the Schlumberger system, since the potential electrodes are close together, the far current electrode need only be about three times as far away as the one nearby to get the same result.

However, since the subsurface resistivity may vary laterally, these spacing estimates can be much too low and may have to be increased by a factor of ten or more, depending on the resistivity contrast.

(d) *Double-dipole system.* The potential electrodes are closely spaced and remote from the current electrodes, which are also close together. In this case, from fig. 8.19d and eq. (8.38) we get

$$r_1 = r_4 = 2n\ell, \ r_2 = 2\ell (n - 1), \ r_3 = 2\ell (n + 1), \text{ where } n \gg 1.$$

Then

$$\rho_a = -2\pi n^3 \ell \Delta V/I. \tag{8.43}$$

If the dipoles are broadside instead of end on,

$$r_1 = r_4 = 2n\ell, \quad r_2 = r_3 = 2\sqrt{\{(n\ell)^2 + \ell^2\}} \approx 2n\ell(1 + 1/2n^2).$$

Then

$$\rho_a \approx -4\pi n^3 \ell \Delta V/I. \tag{8.44}$$

If the spacing between the two inner electrodes is not considerably greater than the dipole lengths, (8.43) becomes, in exact form,

$$\rho_a = -2\pi(n-1)n(n+1)\ell\Delta V/I. \tag{8.45}$$

This last spread is more commonly employed in IP work than resistivity; there the separation between C_2 and P_1 is usually from 1 to 5 times the dipole lengths themselves (see §9.4.3 and fig. 9.8). Inductive coupling between potential and current cables is reduced with this arrangement.

(e) *Lee partition method.* This layout, which is now rarely used, is identical to the Wenner, except that a third potential electrode is located at the centre P of the spread. Two measurements of potential are made, first between P_1 and P, then between P and P_2. The resultant apparent resistivities are

$$\rho_{a1} = \frac{4\pi a \Delta V_1}{I}, \quad \rho_{a2} = \frac{4\pi a \Delta V_2}{I}. \tag{8.46}$$

In all the above electrode layouts the potential and current electrodes may be interchanged. By the principle of reciprocity, the apparent resistivity should be the same in either case. The switching of current and potential electrodes could be desirable, for instance, in using high voltages with large spreads in Schlumberger and, possibly, Wenner layouts.

(f) *Line-electrode spread.* Long-line current-electrode layouts, which were at one time quite fashionable, are not much used now. Although bare copper wire lying in water or mud would make a uniform contact throughout its length, under normal field conditions it is necessary to peg the wire to ground with point electrodes every 20 feet or so. Even then the contact may not be the same at all points of the line.

The potential difference between two point electrodes located inside the lines was given in eqs. (8.15), §8.2.5. If the separation between potential electrodes is small and we put

$$r_1 = L - x - \ell, \quad r_2 = L + x + \ell, \quad r_3 = L - x + \ell,$$
$$\text{and} \quad r_4 = L + x - \ell$$

as in the Schlumberger array, this potential can be written

$$V = -\frac{I_u\rho_a}{\pi}\left\{\log\left(\frac{L+x+\ell}{L+x-\ell}\right) + \log\left(\frac{L-x+\ell}{L-x-\ell}\right)\right\}$$

$$= -\frac{I_u\rho_a}{\pi}\left[\log\left\{\frac{1+\ell/(L+x)}{1-\ell/(L+x)} \times \frac{1+\ell/(L-x)}{1-\ell/(L-x)}\right\}\right],$$

where I_u is the current per unit length of the line.
Since

$$\log\left(\frac{1+a}{1-a}\right) = 2\left(a + \frac{a^3}{3} + \frac{a^5}{5} + \ldots\right) \quad \text{for} \quad a^2 < 1,$$

if we maintain ℓ small with respect to $L - x$, we have

$$V \approx -\frac{I_u\rho_a}{\pi}\left(\frac{2\ell}{L+x} + \frac{2\ell}{L-x}\right) \approx \left(\frac{4\ell L}{L^2-x^2}\right)\left(\frac{I_u\rho_a}{\pi}\right).$$

Thus the apparent resistivity is

$$\rho_a \approx \frac{\pi(L^2-x^2)}{4\ell L}\left(\frac{\Delta V}{I_u}\right). \tag{8.47}$$

One advantage of line electrodes over point current sources is that the electric field is uniform within the area enclosed, so that potential measurements can be carried out over the whole rectangle $ABCD$ (see fig. 8.20). It is also unnecessary to measure current repeatedly.

The presence of a good conductor between the current lines distorts the equipotentials, as shown in fig. 8.20, by bending them away from the conductor. If the

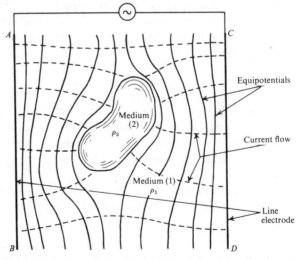

Fig. 8.20 Distortion of equipotentials of a line-electrode system due to the presence of a conductor.

body had a high resistivity with respect to the surroundings the equipotentials would be bent in the opposite sense. Of course this is also true with point current electrodes (see fig. 8.15) but the effect is more obvious in the plan view.

8.5.4 *Resistivity method field procedures*

(a) *Introduction.* Regardless of the specific electrode spread employed, there are really only two basic procedures in resistivity work. The particular procedure to be used depends on whether one is interested in resistivity variations with depth or with lateral extent. The first is called *electric drilling*, the second *mapping* or *trenching*.

(b) *Electric drilling.* Since the fraction of total current which flows at depth varies with the current-electrode separation, as described in §8.2.6, the field procedure is to use a fixed centre with an expanding spread. The Wenner and Schlumberger layouts are particularly suited to this technique, the Schlumberger having certain advantages as outlined in §8.5.3b. The presence of horizontal or gently dipping beds of different resistivities is best detected by the expanding spread. Hence the method is useful in determining depth of overburden, depth, structure and resistivity of flat-lying sedimentary beds and possibly of the basement also if it is not too deep.

It is frequently necessary to carry out this drilling procedure at several locations in an area, even when the main interest may be in lateral exploration, in order to establish proper electrode spacings for the lateral search.

(c) *Electric mapping.* This method is particularly useful in mineral exploration, where the detection of isolated bodies of anomalous resistivity is required. Any of the electrode arrangements described in §8.5.3 may be used, the selection depending mainly on the field situation. Again the two most commonly used layouts are the Wenner and Schlumberger. In all cases the apparent resistivity is plotted at the midpoint of the potential electrodes, except where one of these is effectively at infinity, as in the modified three-probe system, when the station is reckoned at the near potential electrode.

When the potential electrodes are closely spaced with respect to the current spread, as in the Schlumberger, dipole and possibly the three-point system, the measurement is effectively of potential gradient at the midpoint. This can be seen from eq. (8.41) where, putting $2\ell = \Delta r$, we can write

$$\rho_a = \frac{\pi L^2}{I} \left(\frac{\Delta V}{\Delta r} \right). \tag{8.48}$$

If in addition, the current electrodes are close together and remote from the potential pair, the measurement is essentially that of the curvature of the field or the second derivative. For the double-dipole spread in fig. 8.19d, the potential gradient at the midpoint of $P_1 P_2$ due to C_1 only is $\Delta V_1/\Delta r$, where Δr is the spacing of the potential electrodes. Similarly the potential due to C_2 only is $\Delta V_2/\Delta r$. Then the measured potential becomes in the limit as $\Delta r \to 0$,

$$\frac{\Delta V}{\Delta r} = \frac{\Delta V_1 - \Delta V_2}{\Delta r} \to \left(\frac{\partial V}{\partial r}\right)_{C_1} - \left(\frac{\partial V}{\partial r}\right)_{C_2} = \Delta r \left(\frac{\partial^2 V}{\partial r^2}\right),$$

or,

$$\Delta V = (\Delta r)^2 \frac{\partial^2 V}{\partial r^2}. \tag{8.49}$$

Also, from the usual expression for potential, with $r_1 = r_4 = r$, $r_2 = r - \Delta r$ and $r_3 = r + \Delta r$, we obtain

$$\Delta V = \frac{I\rho_a}{2\pi}\left(\frac{1}{r} - \frac{1}{r - \Delta r} - \frac{1}{r + \Delta r} + \frac{1}{r}\right) \approx -\frac{I\rho_a}{\pi}\frac{(\Delta r^2)}{r^3}.$$

Thus we have

$$\rho_a \approx -\frac{\pi r^3}{(\Delta r)^2}\left(\frac{\Delta V}{I}\right) \approx -\frac{\pi r^3}{I}\left(\frac{\partial^2 V}{\partial r^2}\right). \tag{8.50}$$

Lateral exploration by resistivity measurements is best suited to detection of steeply dipping contacts and dykes of contrasting resistivity, that is, two-dimensional anomalies, and to a lesser extent for location of anomalous three-dimensional conductors such as could be roughly simulated by the sphere.

(d) *Mise-à-la-masse.* This is a variation on the three-point electrode system, used where some part of the conductive zone is already located and exposed, either as outcrop or in a drill hole. The near current electrode is embedded in the zone itself, the other being a large distance away on surface. The potential electrodes are moved about, either on surface or in drill holes. The extent, dip, strike and continuity of the zone will be better indicated by introducing the current directly into it than by the usual mapping techniques.

The effect of a dipping mineralized zone on the equipotentials is shown in fig. 8.21. Since the second current electrode is at infinity, it is possible to map the potentials in all directions around the zone without shifting the current stakes.

(e) *Potential ratio; Racom.* These include several methods, modifications of standard resistivity equipment, which were popular about forty years ago. They all made use of three potential electrodes, as in the Lee partition spread, and some employed relatively high frequencies (up to 1000 Hz) and bridge-type detectors, as illustrated in fig. 8.22a.

With the arrangement shown in fig. 8.22a, it is possible to measure both the potential ratio, $P_1 P$ to $P_2 P$, and the phase shift in the two arms. Assuming that

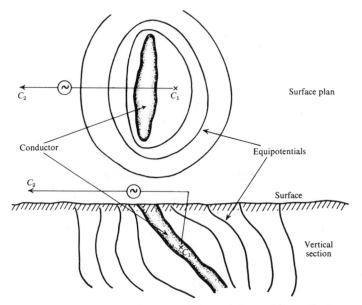

Surface plan

Conductor

C_2

Equipotentials

Surface

Vertical section

Fig. 8.21 Distortion of the equipotentials around the near electrode by a dipping conductor when using the mise-à-la-masse method.

the subsurface is effectively a pair of resistances shunted by capacitances, we have a bridge network. The balance condition is

$$\left.\begin{aligned}
\frac{V_2}{V_1} &= \frac{Z_2}{Z_1} = \sqrt{\left(\frac{R_2{}^2 + 1/\omega^2 C_2{}^2}{R_1{}^2 + 1/\omega^2 C_1{}^2}\right)}, \\
\Delta\theta &= \theta_2 - \theta_1 = \tan^{-1}\omega C_2 R_2 - \tan^{-1}\omega C_1 R_1.
\end{aligned}\right\} \tag{8.51}$$

In order to reduce the effects of stake resistance the values of R_1 and R_2 are large (50 kΩ or so). A modification of the same scheme used a double balance with an extra set of ratio arms in the bridge to eliminate the stake resistance entirely.

Another layout had one of the current electrodes at infinity. In this case a correction must be applied to obtain the normal potential ratio, since V_1 and V_2 are not equal in homogeneous ground. The corrected ratio is given by

$$\frac{V_2}{V_1} = \frac{V_2}{V_1}\left(\frac{r + \ell}{r - \ell}\right) \tag{8.52}$$

where the spacings are as shown in fig. 8.22a with C_2 removed to infinity and $C_1 P = r$.

The example of the potential-ratio method in fig. 8.22b shows a traverse across a graphite zone in South Australia. Line electrodes were employed for the current, located at stations 0 and 4000W, while the three potential electrodes were spaced 50 ft apart. The profile of potential gradient drops to a minimum over the conductive graphite zone; minor peaks on each side indicate an increased density of equipotential lines.

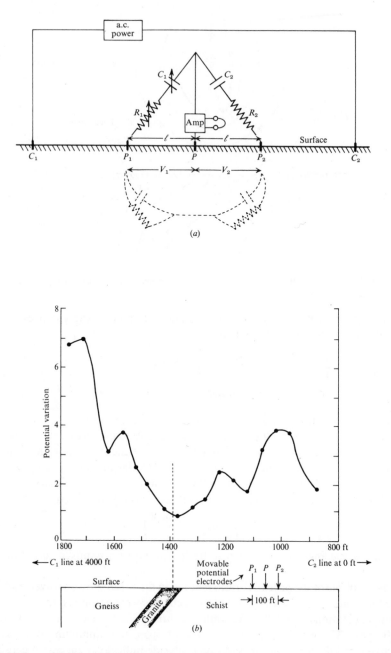

(a)

(b)

Fig. 8.22 Potential-ratio (Racom) method. (*a*) Field instrument (schematic). (*b*) profile over graphite zone.

8.6 Interpretation

8.6.1 *Introduction*

The mathematical analysis for quantitative interpretation of resistivity results is most highly developed for the depth probing or drilling technique, reasonably so for lateral exploration over large-scale contacts of steep dip and least developed for the detection of three-dimensional anomalies. As in other geophysical methods where quantitative interpretation is possible, the assessment of results should progress from rough preliminary estimates made in the field towards more sophisticated methods of interpretation, eventually based on the complete survey. Such a procedure keeps the field work up-to-date, controls the day-by-day programme and indicates where more intensive work is warranted, both in the field survey and its interpretation.

8.6.2 *Resistivity models*

The use of models, although not as common in resistivity as in EM, can be a useful aid in interpretation. Resistivity modelling is generally done in a water tank, the resistivity of the liquid being varied by the addition of salts such as NaCl or acids such as H_2SO_4. Sand may be used instead of liquid, provided reproducible contacts between the model electrodes and the surface can be made.

Various conducting or insulating sheets, cylinders and blocks are immersed in the tank to simulate the field situation. The electrode spread is moved about laterally in a convenient jig mount on the surface of the tank; alternatively, in a liquid medium, the model anomaly may be moved past the electrode spread.

Scaling is not a problem in resistivity model work (see, for example, §7.8.2). The usual relation in resistivity is: $\Delta V/I \propto \rho/\ell$, where ℓ is scaled linearly. Rather than vary the resistivity ρ, it is simpler to change ΔV or I. However, higher frequencies would normally be used in the model than in the field.

8.6.3 *Electric drilling, two horizontal beds*

(a) *Basic formula.* Equation (8.24) relates the potential of a single electrode to the resistivity of the upper layer in terms of the electrode spacing, the depth to the interface and the resistivity contrast between the two beds. We want this expression in the form of an apparent resistivity which would be measured by a four-electrode system. Using the symbols of fig. 8.3 and eqs. (8.13), (8.22), (8.24) and (8.38), the potential of electrode P_1 with respect to current electrodes C_1 and C_2 becomes

$$V_1 = \frac{I\rho_1}{2\pi}\left[\left\{\frac{1}{r_1} + 2\sum_{m=1}^{\infty}\frac{k^m}{\sqrt{(r_1{}^2 + 4m^2z^2)}}\right\} - \left\{\frac{1}{r_2} + 2\sum_{m=1}^{\infty}\frac{k^m}{\sqrt{(r_2{}^2 + 4m^2z^2)}}\right\}\right],$$

while for P_2 we have

$$V_2 = \frac{I\rho_1}{2\pi}\left[\left\{\frac{1}{r_3} + 2\sum_{m=1}^{\infty}\frac{k^m}{\sqrt{(r_3{}^2 + 4m^2z^2)}}\right\} - \left\{\frac{1}{r_4} + 2\sum_{m=1}^{\infty}\frac{k^m}{\sqrt{(r_4{}^2 + 4m^2z^2)}}\right\}\right].$$

Thus, the measured potential difference between P_1 and P_2 will be

$$\Delta V = V_1 - V_2 = \frac{I\rho_1}{2\pi}\left[\left(\frac{1}{r_1} - \frac{1}{r_2}\right) - \left(\frac{1}{r_3} - \frac{1}{r_4}\right) + 2\sum_{m=1}^{\infty} k^m \left\{\frac{1}{\sqrt{(r_1^2 + 4m^2z^2)}} - \right.\right.$$

$$\left.\left. - \frac{1}{\sqrt{(r_2^2 + 4m^2z^2)}} - \frac{1}{\sqrt{(r_3^2 + 4m^2z^2)}} + \frac{1}{\sqrt{(r_4^2 + 4m^2z^2)}}\right\}\right]. \quad (8.53)$$

(b) *Wenner spread.* Since $r_1 = r_4 = a$, $r_2 = r_3 = 2a$ (see fig. 8.19a), eq. (8.53) is simplified to give

$$\Delta V = \frac{I\rho_1}{2\pi a}\left[1 + \sum_{m=1}^{\infty} \frac{4k^m}{\sqrt{\{1 + (2mz/a)^2\}}} - \sum_{m=1}^{\infty} \frac{4k^m}{\sqrt{\{4 + (2mz/a)^2\}}}\right]$$

$$= \frac{I\rho_1}{2\pi a}(1 + 4D_w),$$

so that the apparent resistivity is

$$\rho_a = \rho_1\left[1 + \sum_{m=1}^{\infty} \frac{4k^m}{\sqrt{\{1 + (2mz/a)^2\}}} - \sum_{m=1}^{\infty} \frac{4k^m}{\sqrt{\{4 + (2mz/a)^2\}}}\right] = \rho_1(1 + 4D_w),$$

$$(8.54)$$

where

$$D_w = \sum_{m=1}^{\infty} k^m \left[\frac{1}{\sqrt{\{1 + (2mz/a)^2\}}} - \frac{1}{\sqrt{\{4 + (2mz/a)^2\}}}\right].$$

(c) *Schlumberger spread.* In this case $r_1 = r_4 = L - \ell$, $r_2 = r_3 = L + \ell$ (see fig. 8.19b), and the potential is

$$\Delta V = \frac{I\rho_1}{2\pi}\left[\left(\frac{2}{L - \ell} - \frac{2}{L + \ell}\right) + 4\sum_{m=1}^{\infty} k^m \left\langle\frac{1}{(L - \ell)\{1 + (2mz)^2/(L - \ell)^2\}^{\frac{1}{2}}} - \right.\right.$$

$$\left.\left. - \frac{1}{(L + \ell)\{1 + (2mz)^2/(L + \ell)^2\}^{\frac{1}{2}}}\right\rangle\right],$$

$$= \frac{I\rho_1 2\ell}{\pi(L^2 - \ell^2)}\left[1 + \left(\frac{L + \ell}{\ell}\right)\sum_{m=1}^{\infty} \frac{k^m}{\{1 + (2mz)^2/(L - \ell)^2\}^{\frac{1}{2}}} - \right.$$

$$\left. - \left(\frac{L - \ell}{\ell}\right)\sum_{m=1}^{\infty} \frac{k^m}{\{1 + (2mz)^2/(L + \ell)^2\}^{\frac{1}{2}}}\right].$$

When $L \gg \ell$, the terms inside the square brackets can be simplified by using the relation

$$\frac{1}{\{(L - \ell)^2 + (2mz)^2\}^{\frac{1}{2}}} - \frac{1}{\{(L + \ell)^2 + (2mz)^2\}^{\frac{1}{2}}} \approx \frac{2\ell}{L^2\{1 + (2mz/L)^2\}^{\frac{3}{2}}}.$$

Then the potential becomes

$$\Delta V \approx \frac{I\rho_1 2\ell}{\pi L^2} \left[1 + 2 \sum_{m=1}^{\infty} \frac{k^m}{\{1 + (2mz/L)^2\}^{\frac{3}{2}}} \right] \approx \frac{I\rho_1 2\ell}{\pi L^2} (1 + 2D_s')$$

(see below for the definition of D_s').

The exact expression for apparent resistivity is

$$\rho_a = \rho_1 \left[1 + \left(\frac{L+\ell}{\ell}\right) \sum_{m=1}^{\infty} \frac{k^m}{\{1 + (2mz)^2/(L-\ell)^2\}^{\frac{1}{2}}} - \right.$$

$$\left. - \left(\frac{L-\ell}{\ell}\right) \sum_{m=1}^{\infty} \frac{k^m}{\{1 + (2mz)^2/(L+\ell)^2\}^{\frac{1}{2}}} \right] = \rho_1(1 + D_s), \quad (8.55a)$$

D_s being defined below.

Approximately, we have

$$\rho_a \approx \rho_1 \left[1 + 2 \sum_{m=1}^{\infty} \frac{k^m}{\{1 + (2mz/L)^2\}^{\frac{3}{2}}} \right] = \rho_1(1 + 2D_s'), \quad (8.55b)$$

where

$$D_s = \left(\frac{L+\ell}{\ell}\right) \sum_{m=1}^{\infty} \frac{k^m}{\{1 + (2mz)^2/(L-\ell)^2\}^{\frac{1}{2}}} -$$

$$- \left(\frac{L-\ell}{\ell}\right) \sum_{m=1}^{\infty} \frac{k^m}{\{1 + (2mz)^2/(L+\ell)^2\}^{\frac{1}{2}}},$$

$$D_s' = \sum_{m=1}^{\infty} \frac{k^m}{\{1 + (2mz/L)^2\}^{\frac{3}{2}}}.$$

This result can also be obtained by differentiating equation (8.24) with respect to r, multiplying the result by 2 (since there are two current electrodes) and applying eq. (8.48) to get ρ_a.

(d) *Double-dipole spread.* Since $r_1 = r_4 = 2n\ell$, $r_2 = 2(n-1)\ell$, $r_3 = 2(n+1)\ell$ (see fig. 8.19d) the exact expression for the potential is

$$\Delta V = - \frac{I\rho_1}{2\pi(n-1)n(n+1)\ell} \left[1 + n(n+1) \sum_{m=1}^{\infty} \frac{k^m}{[1 + (2mz)^2/\{2(n-1)\ell\}^2]^{\frac{1}{2}}} + \right.$$

$$+ n(n-1) \sum_{m=1}^{\infty} \frac{k^m}{[1 + (2mz)^2/\{2(n+1)\ell\}^2]^{\frac{1}{2}}} -$$

$$\left. - 2(n-1)(n+1) \sum_{m=1}^{\infty} \frac{k^m}{\{1 + (2mz/2n\ell)^2\}^{\frac{1}{2}}} \right],$$

The apparent resistivity is given by

$$\rho_a = \rho_1(1 + D_d), \quad (8.56a)$$

where $(1 + D_d)$ is the expression inside the square brackets above.

If we make $n \gg 1$, the above result is simplified and we can make use of eq. (8.24). Differentiating twice,

$$\frac{\partial^2 V}{\partial r^2} = \frac{I\rho_1}{\pi r^3} \left[1 - \sum_{m=1}^{\infty} \frac{k^m}{\{1 + (2mz/r)^2\}^{\frac{3}{2}}} + 3 \sum_{m=1}^{\infty} \frac{k^m}{\{1 + (2mz/r)^2\}^{\frac{5}{2}}} \right],$$

and using eq. (8.50),

$$\rho_a = \rho_1 \left[1 - \sum_{m=1}^{\infty} \frac{k^m}{\{1 + (2mz/r)^2\}^{\frac{3}{2}}} + 3 \sum_{m=1}^{\infty} \frac{k^m}{\{1 + (2mz/r)^2\}^{\frac{5}{2}}} \right] = \rho_1(1 + D_a').$$

(8.56b)

(e) *Discussion of theoretical results.* Quantitatively we can see how the apparent resistivity varies from eq. (8.54) through (8.56) for the different electrode spreads. When the electrode spacing is very small, that is, $r \ll z$, the series terms in all cases tend to zero, so that we measure the resistivity in the upper formation. This is the surface resistivity defined in §8.5.2.

Since the reflection coefficient is less than unity, when the C − P electrode spacing is very large compared to z, the depth of the bed, the series expansions in all of the equations become the same:

$$\rho_a \approx \rho_1(1 + 2 \sum_{m=1}^{\infty} k^m) \tag{8.57}$$

(because the denominators are approximately unity). Since $k^2 < 1$, we can write the summation term in the form

$$\sum_{m=1}^{\infty} k^m = 1/(1 - k) - 1;$$

substituting $k = (\rho_2 - \rho_1)/(\rho_2 + \rho_1)$, we get $\rho_a = \rho_2$.

That is to say, at very large spacing, the apparent resistivity is practically equal to the resistivity in the lower formation. If, however, the lower bed is an insulator, $\rho_2 = \infty$ and $k = 1$. Then the apparent resistivity increases indefinitely with electrode spacing, as is obvious from equation (8.57). Since all the current will flow in the upper bed, it is possible to determine the value of ρ_a in the limiting case by calculating the electric field at the mid-point of the current electrodes. Since their separation is much larger than the thickness of the upper bed, it is reasonable to assume a uniform current density from top to bottom. Then the current from either electrode is found by integrating over a cylindrical equipotential surface of radius r and height z. Thus,

$$I = \int_0^{2\pi} \int_0^z Jr \, d\theta \, dz = 2\pi rzJ.$$

From eq. (8.1) we have in this case (noting that the current is doubled because there are two current electrodes)

$$E = 2\rho_1 J = \rho_1 I/\pi rz.$$

For the Wenner array, we get an apparent resistivity

$$\rho_a = \frac{2\pi a \Delta V}{I} = \frac{2\pi a}{I} \int_a^{2a} E\,dr = \left(\frac{2a\rho_1}{z}\right)\log 2 = 1\cdot38\left(\frac{a\rho_1}{z}\right). \qquad (8.58a)$$

For the Schlumberger layout,

$$\rho_a = \frac{\pi L^2}{I}\frac{\partial V}{\partial l} = \frac{L\rho_1}{z}, \qquad (8.58b)$$

and for the double-dipole system,

$$\rho_a = -\left(\frac{\pi r^3}{I}\right)\frac{\partial^2 V}{\partial r^2} = \frac{r\rho_1}{2z}, \qquad (8.58c)$$

where r is the distance between centres of the current and potential dipoles.

In all three spreads the ratio of apparent resistivity to electrode spacing in eqs. (8.58) equals $c(\rho_1/z)$ where the constant c varies with spread type; thus,

$$\rho_a/\rho_1 = c(\text{Electrode spacing/depth to interface}). \qquad (8.58d)$$

Thus if we plot ρ_a/ρ_1 versus a/z under these conditions the curve is a straight line.

On the other hand, if the lower bed is a very good conductor, $\rho_2 \approx 0$ and $k \approx -1$. In this case $\rho_a \approx \rho_2 \approx 0$ for large spacing.

(f) *Crude interpretation.* Before applying the more complicated methods of interpretation it is useful to consider a few rough ideas. Figure 8.23 shows a pair

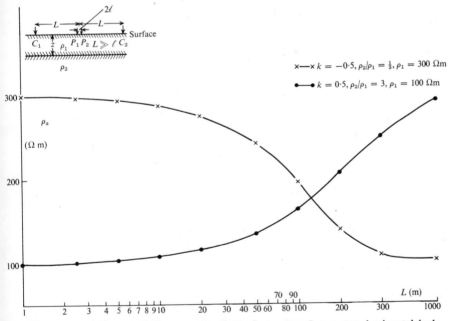

Fig. 8.23 Schlumberger expanding spread – plots of ρ_a versus L over two horizontal beds. Curves calculated from eq. (8.55b), $z = 100$ m.

of resistivity curves for two layers with contrasts of 3 and $\frac{1}{3}$. The upper bed resistivity is 100 Ωm and 300 Ωm for the two cases (to put the two curves on the same ordinate scale), while the thickness is 100 m.

The curve for $\rho_2/\rho_1 = \frac{1}{3}$ is clearly asymptotic to ρ_1 and ρ_2 at the limits of small and large spacing, while its point of maximum slope is approximately at 100 m. Thus we can estimate the depth and the resistivities of the two beds for this simple example.

The other curve gives the upper layer resistivity at small spacing, but it is not so clear what the value of ρ_2 may be. Had the spacing been increased to several thousand metres, it would also be asymptotic to 300 Ωm. The point of inflexion is not at 100 m but at a rather larger spacing.

Approximately then, we can get some idea of the unknown parameters ρ_1 and ρ_2

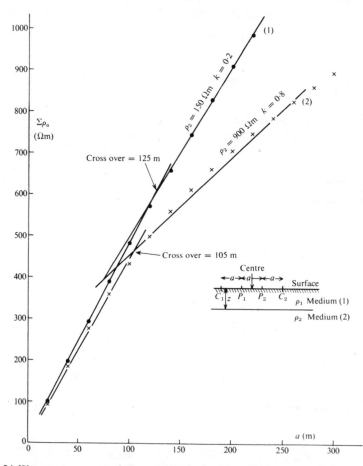

Fig. 8.24 Wenner array – cumulative ρ_a-plot over horizontal beds; $\rho_1 = 100$ Ωm, $z = 100$ m. (1) $\rho_1/\rho_2 = 0.67$, $k = 0.2$; (2) $\rho_1/\rho_2 = 0.11$, $k = 0.8$.

and z from the field curve, provided the resistivity contrast is not too great and particularly if the lower bed is the more conductive of the two.

A better estimate of the depth to an interface can sometimes be made from the cumulative $\Sigma\rho$ plot which is illustrated in fig. 8.24. The ordinates are obtained by summing all measured resistivity values up to and including the spacing against which each is plotted. That is, if the successive ρ_a values were 100, 200, 300 Ωm, for spacings of 10, 20 and 30 m, one would plot 100, 300, 600 Ωm versus 10, 20 and 30 metres. An attempt is then made to draw segments of straight lines through as many points as possible; the breaks in these segments indicate depths to interfaces. Alternately the depth may be estimated by the point where the curve departs from a straight line.

The cumulative $\Sigma\rho$ plot gives reasonable estimates for shallow depths and will show the presence of several contrasting beds. It does not give good results on thick beds. The method is of no use unless the increments in electrode spacing are constant and small compared to the thickness of the beds.

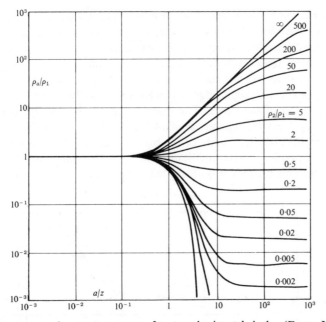

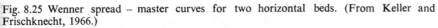

Fig. 8.25 Wenner spread – master curves for two horizontal beds. (From Keller and Frischknecht, 1966.)

(g) *Curve matching.* A much more accurate and dependable method of interpretation in electric drilling involves the comparison of field profiles with characteristic curves. It is quite similar to the interpretation from master curves in magnetotellurics, described in §6.2.7b.

The master curves are prepared with dimensionless coordinates. All of the eqs. (8.54) to (8.56) can be put in this form by dividing ρ_a by ρ_1. The ratios ρ_a/ρ_1 are then plotted against $a/z, L/z, r/z$, that is, the electrode spacing divided by the depths of the upper bed for whatever electrode system is used. The curves are on logarithmic paper, usually six decades each way to provide a large range of both ratios on one sheet. Thus we are plotting ($\log \rho_a - \log \rho_1$) against ($\log a - \log z$). If we make $\rho_1 = 1\ \Omega m$ and $z = 1$ km, all the characteristic curves are preserved in shape, no matter what the multiplier is for the coordinates (ρ_a, a). The sets of curves are constructed either for various values of k between ± 1 or for various ratios of ρ_2/ρ_1 between $\pm\infty$. A typical set of curves is shown in fig. 8.25.

The characteristic curves are generally drawn on a transparency. To match a field result it is only necessary to slide the master sheet around on the field profile until the latter coincides more or less with one of the master curves (or can be interpolated between adjacent master curves). The respective coordinate axes must be kept parallel. The point where $\rho_a/\rho_1 = a/z = 1$ on the master sheet then determines the values of ρ_1 and z on the field curve axes, while the actual curve fit gives the value of k and hence ρ_2.

(h) *Interpretation by asymptotes.* In the event that the lower bed has very large resistivity we have seen that the characteristic two-layer curve becomes a straight line for large electrode spacing. In the logarithmic plots this line has a slope of 45° for all of the arrays considered, since we have made ρ_1 and z unity.

The master curves are not necessary in this case. After plotting the field profile on log-log paper, a straight edge is placed horizontally as a best fit along the left-hand portion of the curve. The intersection of this straight edge with the ρ_a-axis gives ρ_1. Next the hypotenuse of a 45°-triangle is fitted to the sloping part of the curve on the right-hand side of the profile. The interface depth can then be found on the horizontal axis from the intersection of the triangle and the horizontal straight edge. This procedure is illustrated in fig. 8.26.

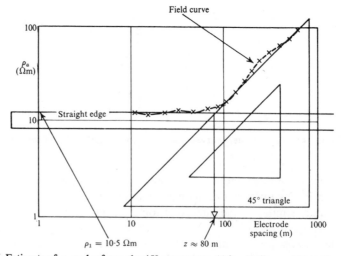

Fig. 8.26 Estimate of ρ_1 and z from the 45°-asymptote. (After Keller and Frischknecht, 1966.)

The asymptote method may also be used even when the maximum spacing has not been large enough to establish that the bottom layer has a very high resistivity. In this case the 45°-triangle is placed to intersect the point of maximum spacing as, for example, in fig. 8.27. In this case the depth estimate can only be a minimum.

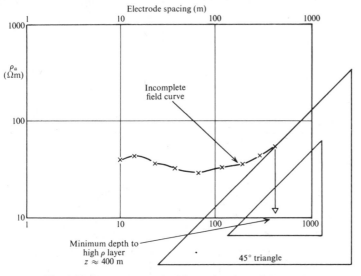

Fig. 8.27 Asymptote method for estimating minimum depth.

8.6.4 *Electric drilling, multiple horizontal beds*

(a) *Introduction.* When there are more than two horizontal beds present, as is usually the case, the single overburden analysis discussed above is first used for relatively small electrode spacing. This gives the depth and resistivity of the upper layer.

Next it is possible to estimate the minimum conductance of all layers above the bottom by drawing the 45°-line through the point given for maximum electrode separation, as shown in fig. 8.27. The ratio of spacing to ρ_a for any point on this line will be a conductance representing all the rocks above an insulating layer; in fig. 8.27, for example, it is about 9 mhos. If the right hand extreme of the field profile is itself a 45°-line on the log-log plot, the bottom layer is highly resistive. In this case the actual, rather than minimum, conductance is determined.

(b) *Crude interpretation.* The overall shape of the middle portion of the profile will give us some idea of the character of the beds between surface and basement. Several shapes are illustrated in fig. 8.28. Types H and K have a definite minimum and maximum, indicating a bed, or beds, of anomalously low or high resistivity respectively, at intermediate depth. Types A and Q show fairly uniform change in resistivity, the first increasing, the second decreasing with depth. Obviously these curves also may be combined. It is generally possible to tell from the shape of the adjacent parts of the profile which layer corresponds to the maximum or minimum on the first two curve types.

(c) *Use of maximum and minimum points.* The coordinates of the extreme points in curves of type H and K, fig. 8.28 (i.e., maximum or minimum ρ_a and electrode

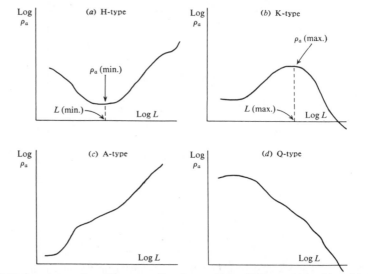

Fig. 8.28 Various types of multilayer profiles for ρ_a. (After Keller and Frischknecht, 1966.)

separation) may be used with certain characteristic curves for three layers employing a particular electrode spread. Figure 8.29a and 8.29b show a set for the Schlumberger array in which: (a) $\rho_a(\text{max.})/\rho_1$ is plotted against ρ_2/ρ_1 for various ratios of z_2/z_1, and (b) the ratio $L(\text{max.})/z_1$ is plotted against z_2/z_1 for various ratios of ρ_2/ρ_1, $L(\text{max.})$ being the electrode spacing at which ρ_a is a maximum or minimum.

Since we know the value of $\rho_a(\text{max.})/\rho_1$ and $L(\text{max.})/z_1$ (presumably ρ_1 and z_1 can be found from a two-layer curve match on the left of the profile), a horizontal line drawn across the characteristics in figs. 8.29a and 8.29b gives a set of possible values of ρ_2/ρ_1 and z_2/z_1, corresponding to the intersection. If we now plot these values of z_2/z_1 versus ρ_2/ρ_1, we get two curves which intersect at one point. This point represents the correct values of z_2 and ρ_2 for the layer in question as shown in fig. 8.30.

(d) *Partial curve matching.* This technique requires matching of small segments of the field profile with theoretical curves for two horizontal layers. Generally one would start from the left-hand (small spacing) side of the profile and match successive segments towards the right (large spacing). When a portion of the field curve is reasonably matched in this way, all the layers in this segment are lumped together and assumed to have an effective resistivity ρ_e and depth z_e. This lumped

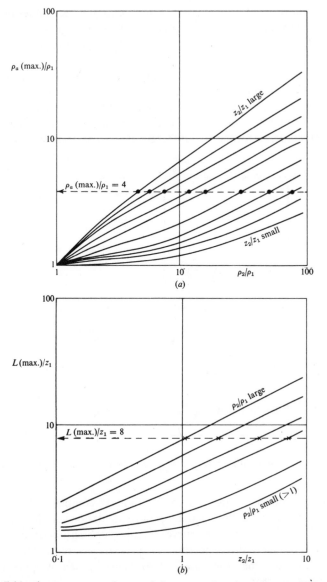

Fig. 8.29 Schlumberger array – characteristic curves for three layers. (After Keller and Frischknecht, 1966.) (*a*) Plots of ρ_a (max.)/ρ_1 (see type *B* curves in fig. 8.28) versus ρ_2/ρ_1 for various z_2/z_1 ratios; (*b*) plots of $L(\text{max.})/z_1$ (type *B* curve in fig. 8.28) versus z_2/z_1 for various ρ_2/ρ_1 ratios.

layer is used as a surface layer and the next portion of the field curve is interpreted in a similar way.

Generally it is not possible to determine ρ_e and z_e uniquely. By assuming a number of values of z_2 we obtain a corresponding set for ρ_e and z_e and match these to the field profile to get the best fit and hence the best values of ρ_e, z_e and eventually ρ_3. Auxiliary characteristic curves may also be drawn up for the four standard profiles in fig. 8.28, using a range of ratios ρ_2/ρ_1 and z_2/z_1.

(e) *Complete curve matching.* Master curves have been drawn up by Mooney and Wetzel (1956) for two and three layers over a basement of infinite depth, also by La Compagnie Générale de Géophysique (1955) for two layers over an infinite basement. The first set, about 2300 curves, is for the Wenner array, the second, which has 480 curves, is for the Schlumberger array. The procedure is exactly as in the two-layer case.

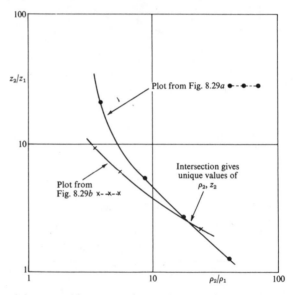

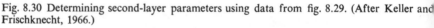

Fig. 8.30 Determining second-layer parameters using data from fig. 8.29. (After Keller and Frischknecht, 1966.)

The enormous number of ratios necessary to make complete albums of master curves prohibits their development by hand calculation. A digital computer, however, allows one to make exact curves for any combination of resistivities and layer thicknesses. Where this is possible, the procedure is to match the field data approximately to, say, the Mooney and Wetzel curves, then compute more exact curves by trial and error.

(f) *Drilling interpretation errors.* Clearly the interpretation of horizontal layers is quite highly developed mathematically. A very good treatment is given in Keller and Frischknecht (1966, chapter 3). Practically, however, there are often erratic variations in the measured results for ρ_a, due mainly to local changes in resistivity near surface and sometimes to poor electrode contacts, both of which affect the interpretation.

Errors in interpretation arise when it is possible to match field curves with more than one master curve. This may occur because interpolation between the curves is rather coarse. Generally the error in estimating ρ in this case is about the same as for z.

As mentioned previously, anisotropic ground will lead to errors in estimating ρ and z, as will the effect of topography. An expanding spread, in homogeneous ground, parallel and adjacent to a vertical contact will give a profile which is somewhat similar to that obtained over horizontal layers, particularly if the bed on the other side of the contact has low resistivity (see fig. 8.33e and §8.6.5d). In all depth-probing operations the expanding spread should be carried out in at least two azimuths, a sound preliminary procedure in any case.

An example of pronounced topography effect is shown in fig. 8.31. Apparent resistivities were obtained from an expanding Wenner system over relatively homogeneous dolomite and limestone. The spread, however, is parallel to a 100 ft cliff, which produces a linearly increasing curve for ρ_a versus electrode separation. The bumps in curve A are probably the result of local variations in surface resistivity near the cliff edge. Obviously this is an extreme case, but the curves would be similar if the cliff face were a contact and the void filled with high resistivity rock. In the latter case the result could be erroneously interpreted as a subsurface layer of high resistivity.

The effect of dipping beds is not serious in drilling investigations. With the Schlumberger and Wenner spreads it is possible to get reasonable results over beds dipping as much as 45°. The double-dipole array appears to be much more sensitive to dip angles, the standard curves apparently being unsuitable for a dip greater than 10°. Again it is necessary to measure in two horizontal directions to determine the dip.

8.6.5 *Lateral exploration-vertical contact*

(a) *General equations.* Also called electric mapping, profiling or trenching, this technique is of considerable importance in mineral prospecting. It is also used for the measurement of overburden depth in civil engineering work. The mineral exploration interpretation includes location of vertical contacts – faults, dikes, shear zones and steeply dipping veins – and three-dimensional bodies such as massive sulphides of anomalous conductivity.

In §8.3.3 the variation in potential crossing a plane interface was established for a single current and potential electrode system. Equation (8.20) gave the potential with both electrodes in the same medium, eq. (8.21) when they were on opposite sides of the contact.

Any of the electrode arrays may be used for this type of mapping, but the profiles differ considerably from one to another. There is also a practical consideration: the traverse can be made faster and more easily if it is not necessary to move all the electrodes for each station measurement. It turns out also that the profiles are usually easier to understand as well in this case.

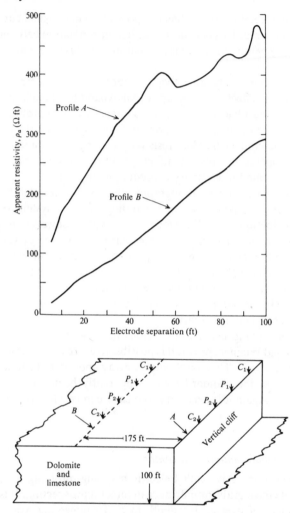

Fig. 8.31 Effect of topography on expanding Wenner spread.

For the general case, with spacing r_1, r_2, r_3 and r_4 we have five possible situations, depending on electrode positions with respect to the contact (see fig. 8.32):

(i) All electrodes on LHS:

$$V_1 = \frac{I\rho_1}{2\pi} \left\{ \left(\frac{1}{r_1} + \frac{k}{2s - r_1} \right) - \left(\frac{1}{r_2} + \frac{k}{2s - 2r_1 - r_2} \right) \right\},$$

$$V_2 = \frac{I\rho_1}{2\pi} \left\{ \left(\frac{1}{r_3} + \frac{k}{2s - r_3} \right) - \left(\frac{1}{r_4} + \frac{k}{2s - 2r_3 - r_4} \right) \right\},$$

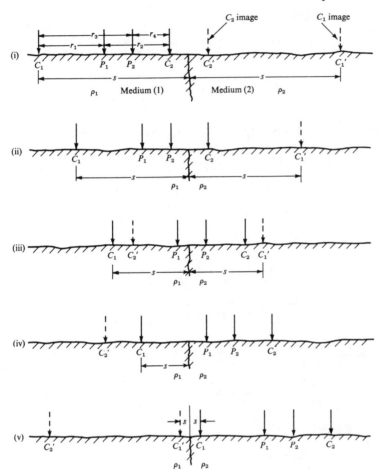

Fig. 8.32 General resistivity spread over a vertical contact.

$$\Delta V = \frac{I\rho_1}{2\pi} \left[\left(\frac{1}{r_1} - \frac{1}{r_2} \right) - \left(\frac{1}{r_3} - \frac{1}{r_4} \right) + k \left\{ \left(\frac{1}{2s - r_1} - \frac{1}{2s - 2r_1 - r_2} \right) - \right. \right.$$
$$\left. \left. - \left(\frac{1}{2s - r_3} - \frac{1}{2s - 2r_3 - r_4} \right) \right\} \right]. \quad (8.59a)$$

(ii) C_2 on RHS (note that k changes sign for a current electrode to the right of the vertical contact):

$$\Delta V = \frac{I\rho_1}{2\pi} \left[\left(\frac{1}{r_1} - \frac{1}{r_2} \right) - \left(\frac{1}{r_3} - \frac{1}{r_4} \right) + k \left\{ \left(\frac{1}{2s - r_1} - \frac{1}{r_2} \right) - \right. \right.$$
$$\left. \left. - \left(\frac{1}{2s - r_3} - \frac{1}{r_4} \right) \right\} \right]. \quad (8.59b)$$

(iii) C_2, P_2 on RHS:

$$V_1 = \frac{I\rho_1}{2\pi}\left\{\left(\frac{1}{r_1} + \frac{k}{2s - r_1}\right) - \frac{1 + k}{r_2}\right\},$$

$$V_2 = \frac{I\rho_2}{2\pi}\left\{\frac{1 - k}{r_3} - \left(\frac{1}{r_4} - \frac{k}{2r_3 + r_4 - 2s}\right)\right\}.$$

Using eq. (8.22), §8.3.3, to express ρ_2 in terms of ρ_1, we obtain

$$\Delta V = \frac{I\rho_1}{2\pi}\left[\left(\frac{1}{r_1} - \frac{1}{r_2}\right) - \left(\frac{1 + k}{1 - k}\right)\left(\frac{1}{r_3} - \frac{1}{r_4}\right) + k\left\{\left(\frac{1}{2s - r_1} - \frac{1}{r_2}\right) + \right.\right.$$
$$\left.\left. + \left(\frac{1 + k}{1 - k}\right)\left(\frac{1}{r_3} - \frac{1}{2r_3 + r_4 - 2s}\right)\right\}\right]. \quad (8.59c)$$

(iv) C_2, P_2, P_1 on RHS:

$$V_1 = \frac{I\rho_2}{2\pi}\left\{\left(\frac{1}{r_1} - \frac{1}{r_2}\right) - k\left(\frac{1}{r_1} - \frac{1}{2r_1 + r_2 - 2s}\right)\right\},$$

$$V_2 = \frac{I\rho_2}{2\pi}\left\{\left(\frac{1}{r_3} - \frac{1}{r_4}\right) - k\left(\frac{1}{r_3} - \frac{1}{2r_3 + r_4 - 2s}\right)\right\},$$

$$\therefore\ \Delta V = \frac{I\rho_1}{2\pi}\left[\frac{(1 + k)}{(1 - k)}\left\langle\left(\frac{1}{r_1} - \frac{1}{r_2}\right) - \left(\frac{1}{r_3} - \frac{1}{r_4}\right) - \right.\right.$$
$$\left.\left. - k\left\{\left(\frac{1}{r_1} - \frac{1}{2r_1 + r_2 - 2s}\right) - \left(\frac{1}{r_3} - \frac{1}{2r_3 + r_4 - 2s}\right)\right\}\right\rangle\right]. \quad (8.59d)$$

(v) All electrodes on RHS:

$$V_1 = \frac{I\rho_2}{2\pi}\left\{\left(\frac{1}{r_1} - \frac{k}{2s + r_1}\right) - \left(\frac{1}{r_2} - \frac{k}{2s + 2r_1 + r_2}\right)\right\},$$

$$V_2 = \frac{I\rho_2}{2\pi}\left\{\left(\frac{1}{r_3} - \frac{k}{2s + r_3}\right) - \left(\frac{1}{r_4} - \frac{k}{2s + 2r_3 + r_4}\right)\right\},$$

$$\therefore\ \Delta V = \frac{I\rho_1}{2\pi}\left[\frac{(1 + k)}{(1 - k)}\left\langle\left(\frac{1}{r_1} - \frac{1}{r_2}\right) - \left(\frac{1}{r_3} - \frac{1}{r_4}\right) - \right.\right.$$
$$k\left\{\left(\frac{1}{2s + r_1} - \frac{1}{2s + 2r_1 + r_2}\right) - \left(\frac{1}{2s + r_3} - \frac{1}{2s + 2r_3 + r_4}\right)\right\}\right\rangle\right]. \quad (8.59e)$$

If we ignore the existence of the contact, we can substitute ΔV and I in eq. (8.38) and get ρ_a. If we divide eq. (8.38) by eqs. (8.59), we get

$$\rho_a/\rho_1 = p/p',$$

where $1/p'$ is the quantity in square brackets in eqs. (8.59) and p is the quantity in eq. (8.38) evaluated for the particular electrode spacing which holds for (8.59).

The expressions for ρ_a for the Wenner and Schlumberger arrays are cumbersome in general. We shall consider certain special cases in the following section.

(b) *Half-Wenner array.* $r_1 = a, r_2 = r_3 = r_4 = \infty$. For the above configurations, the results are:

(i) $\rho_a/\rho_1 = 1 + \{ka/(2s - a)\}$, (8.60a)

(ii), (iii) Same as in (i), (8.60b, c)

(iv) $\rho_a/\rho_1 = (1 + k)$, (8.60d)

(v) $\rho_a/\rho_1 = \left(\dfrac{1 + k}{1 - k}\right)\left(1 - \dfrac{ka}{2s + a}\right)$. (8.60e)

(c) *Half-Schlumberger array.* $r_1 = L - \ell, r_3 = L + \ell, r_2 = r_4 = \infty, L \gg \ell$.

(i) $\dfrac{\rho_a}{\rho_1} = 1 - \dfrac{kL^2}{(2s - L)^2}$ and for (ii) as well, (8.61a)

(iii) $\dfrac{\rho_a}{\rho_1} = 1 + \dfrac{k(L - s)(L - \ell)}{\ell(2s - L + \ell)}$, (8.61c)

(iv) $\dfrac{\rho_a}{\rho_1} = 1 + k$, (8.61d)

(v) $\dfrac{\rho_a}{\rho_1} = \dfrac{1 + k}{1 - k}\left\{1 - \dfrac{kL^2}{(2s + L)^2}\right\}$. (8.61e)

(d) *Double-dipole array.* (Potential electrodes to right of current pair). $r_1 = r_4 = r$, $r_2 = r - \ell, r_3 = r + \ell, r \gg \ell$.

(i) All four electrodes on LHS:

$$\frac{\rho_a}{\rho_1} = 1 - \frac{kr^3}{(2s - r)^3},$$ (8.62a)

(ii) Dipole straddles contact:

$$\frac{\rho_a}{\rho_1} = 1 + k,$$ (8.62b)

(iii) All four on RHS:

$$\frac{\rho_a}{\rho_1} = \frac{1 + k}{1 - k}\left\{1 + \frac{kr^3}{(2s + r)^3}\right\}.$$ (8.62c)

Profiles for these three electrode systems, plus some other possible arrangements, are shown in fig. 8.33. The profiles are characteristic of the array used. Except for fig. 8.33e, they all have discontinuities in the vicinity of the contact, related to the electrode spacing. In fig. 8.33*e* the electrodes are aligned parallel to the contact and are moved broadside, so that they all cross it simultaneously. The value of ρ_a/ρ_1 can be obtained in this case by substitution in eqs. (8.59*a*) and (8.59*e*). With the usual approximations ($L \gg \ell$), we have

$$\rho_a/\rho_1 = 1 + k\left\{1 + \left(\frac{2s}{L}\right)^2\right\}^{-\frac{3}{2}} \qquad \text{electrodes in medium (1)}$$

$$= \left[1 - k\left\{1 + \left(\frac{2s}{L}\right)^2\right\}^{-\frac{3}{2}}\right]\left(\frac{1+k}{1-k}\right) \qquad \text{electrodes in medium (2)}$$

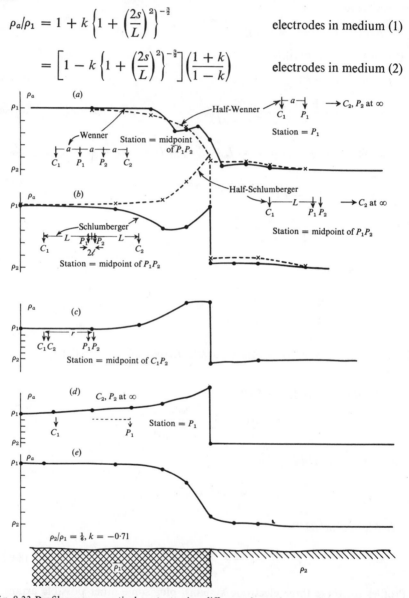

Fig. 8.33 Profiles over a vertical contact using different electrode spreads. (*a*) Wenner spread; (*b*) Schlumberger spread; (*c*) double-dipole spread; (*d*) spread with fixed C_1, movable P_1; (*e*) Schlumberger broadside spread.

This profile is by far the best for interpretation but is not very practical in the field. In some of the spreads the profile shape varies with the direction of traversing, that is, it depends upon which electrode crosses the contact first.

8.6.6 *The vertical dike*

When a dike of anomalous resistivity and finite width is traversed, the profiles are even more affected by the electrode spacing than in the case of the vertical contact. The expressions for potential are as in §8.6.5 plus the effect of one or two sets of images caused by reflections in the two boundaries. The development is not fundamentally difficult, but is tedious. We will describe only one electrode configuration in detail, that in which we have one current electrode in medium (1) (see fig. 8.34).

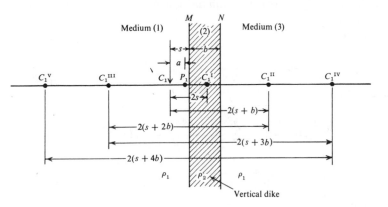

Fig. 8.34 Locations of current images near a vertical dike.

The current source I at C_1 has an image in the dike at $C_1{}^I$ (provided $s < b$), caused by reflection in boundary M. Its strength is kI and it is located $(2s - a)$ from P_1, the potential electrode.

There is a second image of C_1 reflected in boundary N at $C_1{}^{II}$, which is located in medium (3) at a distance $(2b + 2s - a)$ from P_1. This in turn causes an image in medium (1) at $C_1{}^{III}$, $(2b + a)$ from P_1. It is reflected in N to produce image $C_1{}^{IV}$, at $(4b + 2s - a)$ from P_1. These reflections are repeated indefinitely in M and N to give two infinite series, of which only the set of images in medium (3) have any effect on the potential in medium (1).

When the potential electrode is in medium (2), the image at $C_1{}^I$ does nothing, but both the series in medium (1) and (3) influence the potential. Finally when P_1 is in medium (3), it is perturbed by $C_1{}^I$ and all the images in medium (1). As a result we obtain the following potential expressions, for the different locations of P_1:
Medium (1):

$$V_1 = \frac{I\rho_1}{2\pi}\left\{\frac{1}{a} + \frac{k}{2s - a} + A\sum_{m=0}^{\infty}\frac{(k^2)^m}{2(m+1)b + 2s - a}\right\}, \qquad (8.63a)$$

Medium (2):

$$V_2 = \frac{I\rho_1}{2\pi}\left\{ B \sum_{m=0}^{\infty} \frac{(k^2)^m}{2(m+1)b + 2s - a} + C \sum_{m=0}^{\infty} \frac{(k^2)^m}{2mb + a} \right\}, \quad (8.63b)$$

Medium (3):

$$V_3 = \frac{I\rho_1}{2\pi}\left\{ D \sum_{m=0}^{\infty} \frac{(k^2)^m}{2mb + a} \right\}. \quad (8.63c)$$

A, B, C and D are constants which are evaluated by applying the boundary conditions of eq. (8.7) at the boundaries M and N. They turn out to be:

$$A = -k(1 - k^2), \quad B = -k(1 + k), \quad C = 1 + k, \quad D = 1 - k^2.$$

By a similar analysis, we can obtain the potentials when the current electrode is in the dike.

Medium (1):

$$V_1' = \frac{I\rho_1}{2\pi}(1 + k)\left\{ \sum_{m=0}^{\infty} \frac{k^{2m}}{2mb + a} - k \sum_{m=0}^{\infty} \frac{k^{2m}}{2(m+1)b + 2s + a} \right\}, \quad (8.64a)$$

Medium (2):

$$V_2' = \frac{I\rho_1(1 + k)}{2\pi(1 - k)}\left[\frac{1}{a} + k^2 \left\{ \sum_{m=0}^{\infty} \frac{k^{2m}}{2(m+1)b + a} + \sum_{m=0}^{\infty} \frac{k^{2m}}{2(m+1)b - a} \right\} - \right.$$
$$\left. - k \left\{ \sum_{m=0}^{\infty} \frac{k^{2m}}{2(m+1)b + 2s \pm a} + \sum_{m=0}^{\infty} \frac{k^{2m}}{2mb - (2s \pm a)} \right\} \right], \quad (8.64b)$$

Medium (3):

$$V_3' = \frac{I\rho_1}{2\pi}(1 + k)\left\{ \sum_{m=0}^{\infty} \frac{k^{2m}}{2mb + a} - k \sum_{m=0}^{\infty} \frac{k^{2m}}{2mb - (2s - a)} \right\}. \quad (8.64c)$$

In these equations the relative positions of the potential and current electrodes must be specified. In (8.64a) the potential electrode is always to the left of the current electrode, while in (8.64c) it is always to the right. In eq. (8.64b), however, it may be on either side of the current electrode. When it is on the left, one uses the upper signs for a in the denominators of the last bracket; when on the right, the lower sign.

Finally, when the current electrode is on the RHS in medium (3), and P_1 is to the left of C_1, in all cases the potentials are:

Medium (1):

$$V_1'' = \frac{I\rho_1}{2\pi}(1 - k^2) \sum_{m=0}^{\infty} \left(\frac{k^{2m}}{2mb + a} \right), \quad (8.65a)$$

Medium (2):

$$V_2'' = \frac{I\rho_1}{2\pi}(1 + k)\left\{ \sum_{m=0}^{\infty} \left(\frac{k^{2m}}{2mb + a} \right) - k \sum_{m=0}^{\infty} \frac{k^{2m}}{2mb - (2s + a)} \right\}, \quad (8.65b)$$

Medium (3):

$$V_3'' = \frac{I\rho_1}{2\pi} \left\{ \frac{1}{a} + k \sum_{m=0}^{\infty} \frac{k^{2m}}{2(m-1)b - (2s+a)} - k \sum_{m=0}^{\infty} \frac{k^{2m}}{2mb - (2s+a)} \right\}.$$

$$(8.65c)$$

From these relations one can obtain the value of ρ_a in terms of ρ_1, in the usual way, for complete profiles across the dike. In addition, the expressions can be made more general by assuming a resistivity ρ_3 in medium (3).

The formulae are modified in all cases by differentiating the potentials for a half-Schlumberger array and by using the second derivative for the double dipole.

Profiles obtained with different spreads in traversing a thin dike are shown in fig. 8.35. On the whole the Schlumberger curves reproduce the shape of the dike

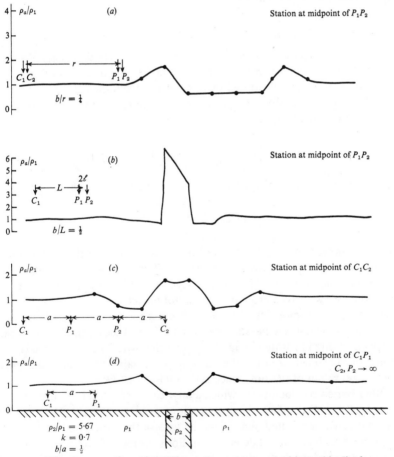

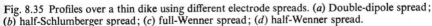

Fig. 8.35 Profiles over a thin dike using different electrode spreads. (*a*) Double-dipole spread; (*b*) half-Schlumberger spread; (*c*) full-Wenner spread; (*d*) half-Wenner spread.

best, particularly for thin sections. The corresponding dipole profile has two peaks, the gap between being equal to the dipole spacing. This double anomaly could be quite misleading. On the other hand, the half-Wenner spread over a thin dike of high resistivity shows a 'conductor' of width greater than the actual dike. The full-Wenner system, however, gives better results, although there are discontinuities near the edges of the dike, as illustrated in fig. 8.35c.

As in the case of the single vertical contact, better profiles would be obtained by moving the array broadside to the structure. In fact, the profiles are considerably better over thin dikes when the traverse is made at an oblique angle, although the anomalies will be wider than the actual dike.

Lateral exploration may also be applied to channels and filled sinks of anomalous resistivity when such features outcrop or lie very close to surface. The profiles are similar to the dike, although the latter was assumed to have infinite depth in the previous discussions.

8.6.7 *Mapping three-dimensional anomalies*

The resistivity method is not particularly sensitive to three-dimensional anomalies for the same reason that it is ineffective over buried structures of finite extent. This limitation is well illustrated by reference to the buried sphere considered in §8.3.5. Using a Schlumberger spread, the apparent resistivity can be calculated by differentiating eq. (8.29). The approximate value is

$$\rho_a \approx \rho_1 \left\{ 1 + \left(\frac{\rho_1 - \rho_2}{\rho_1 + 2\rho_2} \right) \left(\frac{a}{z} \right)^3 \frac{(2x^2/z^2) - 1}{(1 + x^2/z^2)^{\frac{5}{2}}} \right\},$$

and assuming that the sphere is a very good conductor, so that $\rho_2 \approx 0$,

$$\frac{\rho_a}{\rho_1} \approx 1 + \left(\frac{a}{z} \right)^3 \frac{(2x^2/z^2) - 1}{(1 + x^2/z^2)^{\frac{5}{2}}}, \tag{8.66}$$

where x is the distance of the potential electrode from the surface point above the origin and z the depth to the sphere centre. When $z = 2a$ the maximum contrast between ρ_a and ρ_1 is only 12%. Thus a sphere 100 ft in diameter whose top lies only 50 ft below surface probably would not be detected.

A similar limitation exists when the body outcrops, for instance, a hemispherical sink. Unless the traverse passes very close to the rim the anomaly will be missed. These effects are illustrated in fig. 8.36. Note that when the survey line is over the centre of the bowl ($d = 0$), the ratio ρ_a/ρ_1 remains zero until the potential electrodes are out of the sink, since $\Delta V = 0$ for $a \leqslant 2R$.

Other buried conductors for which analytical or graphical solutions have been successfully applied include the dike, dome and oblate spheroid. The mathematics are relatively complicated and unless the tops of the structures are near surface, resistivity profiling will not detect them. For solutions of bodies having irregular cross-sections, dot charts, similar to those in gravity and magnetic problems, have been employed.

8.6.8 *Measuring overburden depth and resistivity*

Obviously the depth of overburden can be found using an expanding spread. However, if the bedrock surface is irregular, a great many stations must be occupied, entailing considerable time and expense. Where the overburden has much lower resistivity than bedrock, which is the usual case, the measurements may be made with two traverses, employing different electrode separations.

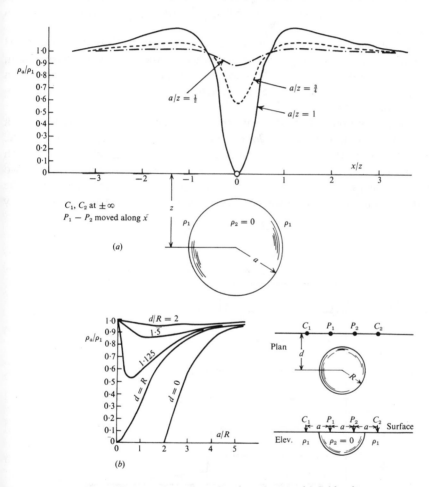

Fig. 8.36 Resistivity profiles over three-dimensional conductors. (*a*) Schlumberger array over buried sphere; (*b*) expanding Wenner array over outcropping hemispherical conductor.

If two spreads of constant spacing, one smaller and the other somewhat larger than the minimum and maximum overburden thicknesses, are moved laterally, we can estimate the depth and resistivity of the soil fairly accurately. Since the larger spread will give an apparent resistivity which falls on the 45° portion of the profile, while the short spread measures ρ_1, we have, from eq. (8.58d),

$$z = c(\rho_1/\rho_a)/(\text{electrode spacing}),$$

and all the quantities on the right-hand side are measured.

If the bedrock resistivity is considerably lower than that of the overburden, it is necessary to use expanding spreads to get quantitative values for the depth. There is no equivalent relation between ρ_a and z when $\rho_2 < \rho_1$, as can be seen from fig. 8.25.

The cumulative ρ-plot has frequently been used for depth of overburden and resistivities, since it works quite well on thin layers. Another type of interpretation, known as the Barnes layer method, has been applied with the Wenner spread, in which the spacing is increased in approximately equal increments. A brief description of the method follows.

The assumption is made that with spacing a, one measures the resistivity in a layer of depth a, with spacing $2a$, one measures the resistivity of depth $2a$, etc. Then the layers are considered equivalent to resistors connected in parallel. With spread a the resistance is that of one resistor, when the spread is $2a$ it is two resistors in parallel, of which the second resistor is unknown, although the parallel combination is known. Mathematically for the nth layer, this is given by

$$1/R_n = 1/\bar{R}_n - 1/\bar{R}_{n-1},$$

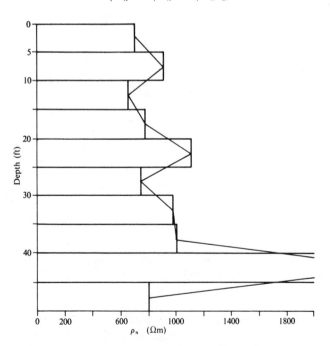

Fig. 8.37 Resistivity versus depth from the Barnes layer method.

where R_n = resistance of nth layer, $\bar{R}_n$ = resistance between surface and bottom of the nth layer.

The resulting resistivity of the nth layer is

$$\rho_n = 2\pi a_n R_n. \tag{8.67}$$

The data are then plotted as a bar graph with the ρ_n values taken at the midpoint of each increment of electrode spacing, the latter being converted to vertical depth. An example is shown in fig. 8.37.

Still another method for determining depth of overburden is that of Tagg (1934), which is essentially an earlier form of two-layer master curves of fig. 8 25. Tagg's

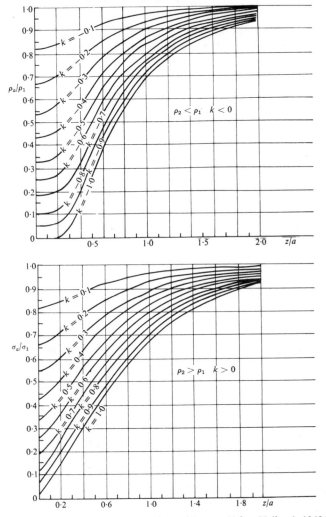

Fig. 8.38 Tagg's curves for two horizontal layers. (After Heiland, 1940.)

curves were developed for the expanding Wenner spread over two horizontal layers (although they can be modified for other electrode arrangements as well). There are two sets of characteristic curves, in which either ρ_a/ρ_1 or σ_a/σ_1 is plotted against z/a for various values of k. They are illustrated in fig. 8.38. The appropriate set of curves is selected on the basis of the relative values of ρ_1 and ρ_2. If $\rho_2 > \rho_1$, the σ_a/σ_1 curves are used ($k > 0$), while if $\rho_2 < \rho_1$, we use the ρ_a/ρ_1 set ($k < 0$). It is normally obvious, from the shape of the field curve for an expanding spread, which situation prevails.

The following table will make the procedure clear. Knowing the value of ρ_1 from a measurement at small spacing and noting that $\rho_a/\rho_1 > 1$ for larger spacings, we select the curves for σ_a/σ_1. Writing the values for a and σ_a/σ_1 along the top rows of table 8.1 and various values of k in the first column, we enter the master curve from the appropriate value of σ_a/σ_1 to find where it is intersected by one of the k curves. The corresponding value of z/a can then be found from the abscissa, to be entered in the column for z/a under the correct separation a. The correct k value will be the one which makes z constant for different electrode separations. When k is selected too large, the z-values increase with increasing a, when too small, they decrease with increasing a.

Table 8.1. *Application of Tagg's curves*

a (ft)	ρ_a (Ωm)	σ_a/σ_1	a (ft)	ρ_a (Ωm)	σ_a/σ_1	a (ft)	ρ_a (Ωm)	σ_a/σ_1
150	35·3	0·748	250	48·6	0·544	350	60·0	0·441
200	42·3	0·625	300	54·5	0·483	400	64·8	0·407

$$\rho_1 = 26\cdot4 \ \Omega\text{m}$$

$a \to$ 150 ft		200 ft		250 ft		300 ft		350 ft		400 ft	
$\sigma_a/\sigma_1 \to$ 0·748		0·625		0·544		0·483		0·441		0·407	
$k\downarrow$ z/a	z	z/a	z	z/a	z	z/a	z	z/a	z	z/a	z
1·0 1·19	179	0·91	183	0·77	193	0·67	202	0·61	214	0·56	224
0·8 1·04	157	0·77	155	0·64	160	0·54	163	0·48	170	0·43	174
0·7 0·96	144	0·70	140	0·56	141	0·48	143	0·41	144	0·36	144
0·6 0·87	130	0·62	124	0·48	121	0·39	117	0·32	114	0·28	112
0·4 0·66	99	0·42	84	0·27	67	0·16	40	0·06	21	—	—
0·2 0·31	47	—	—	—	—	—	—	—	—	—	—

The best value of k here is 0·7. This gives a mean value of 143 ft for z and from the relation $\rho_2/\rho_1 = (1 + k)/(1 - k)$ we obtain $\rho_2 = 150\ \Omega m$. This method has been extended to the three-layer problem as well.

8.7 Field examples

In recent years most resistivity data related to mineral exploration are included in the results of IP surveys; the resistivity method is not much used as an independent technique in this application. It has, however, been employed to a considerable extent in ground water search and for engineering geology – preparation of damsites, highway routes, building foundations, etc. Consequently the case histories and problems in this section include several examples not directly related to conventional prospecting. Further examples may be found in chapter 9 of surveys where resistivity data were obtained in conjunction with IP data.

1. Figure 8.39 shows the results of an expanding Lee spread centred approximately over a shallow filled sinkhole, hemispheroidal in shape, in Cherokee County, Kansas. The east and west profiles refer to measurements between

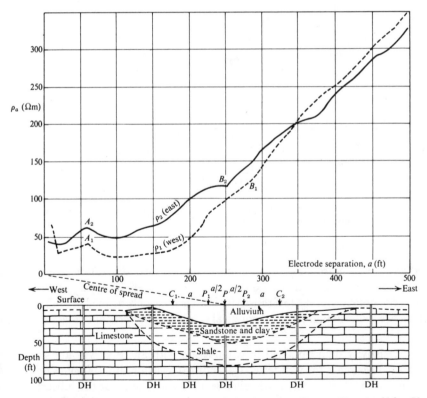

Fig. 8.39 Expanding Lee spread over filled sinkhole, Cherokee County, Kansas. (After Van Nostrand and Cook (1966).)

$P_2 - P$ and $P_1 - P$ respectively (see fig. 8.19). The peaks A_2 and A_1 at $a = 60$ ft occur when the current electrodes pass the edge of the alluvium bowl. The discontinuity in slope, B_2, at $a = 250$ ft corresponds to P_2 arriving at the east edge of the shale sink, while B_1 at $a = 280$ ft indicates that P_1 has crossed the west limestone-shale contact. The latter, however, is not well defined. For separations larger than 340 ft the two curves are practically parallel, reflecting high resistivity limestone.

2. Apparent resistivity profiles obtained with a half-Schlumberger, or gradient, array traversing vertical contacts are shown in fig. 8.40. The two profiles correspond to different fixed locations of the current electrode C_1. The potential

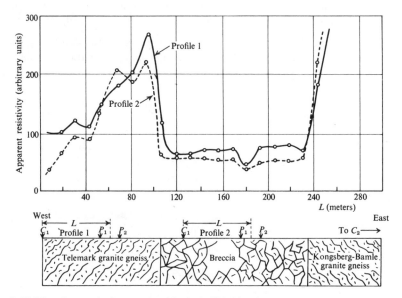

Fig. 8.40 Mapping vertical contacts with the half-Schlumberger (gradient) array, Kongsberg, Norway. (After Van Nostrand and Cook, 1966.)

electrodes, which are close together compared to their separation from C_1, are moved together, while the second current electrode is effectively at infinity to the east. The contact with the low resistivity breccia is sharply defined in both profiles, which are quite similar to the theoretical result for the vertical contact shown in fig. 8.33*b*.

3. Resistivity data obtained in conjunction with an IP survey are shown in fig. 8.41. The area is the well-tested Cavendish Township site, 100 miles northeast of Toronto. Unfortunately no detailed section is available for the subsurface.

The rocks are mainly metasedimentary with small areas of acidic and basic igneous types. The trend is northeast. Sulphides are present throughout the area, at least in trace amounts, and graphite occurs in the western band of calcareous-siliceous rocks.

Figure 8.41*a* shows apparent resistivities plotted in profile for four separations of the double-dipole electrode system ($x = 200$ ft, $n = 1, 2, 3, 4$) on line *B*; fig. 8.41*b* shows the usual pseudo-vertical sections employed in IP work. (For the method of plotting these sections, see §9.5.1.) Clearly there is a low resistivity zone, continuous at depth from 4W to 18W, which is capped by a higher resistivity bowl near surface, located between 10W and 14W. A variety of EM surveys made on the Cavendish test site agree with the shallow resistivity profiles, since they outline two distinct zones trending NE, located at 8W and 15W.

8.8 Problems

1. In an investigation to determine the depth of a conducting layer of brine at Malagash, Nova Scotia, the following readings were taken with a Megger using an expanding Wenner spread. The surface layer was found to have a resistivity of 29 Ωm. Determine the depth and resistivity of the brine layer.

Separation (ft)	ρ_a (Ωm)	Separation (ft)	ρ_a (Ωm)	Separation (ft)	ρ_a (Ωm)
40	28·5	160	18·0	280	8·7
60	27·1	180	16·3	300	7·8
80	25·3	200	14·5	320	7·1
100	23·5	220	12·9	340	6·7
120	21·7	240	11·3	360	6·5
140	19·8	260	9·9	380	6·4

2. The following readings were made with an expanding spread (Wenner configuration) to find the depth of glacial drift above bed rock.

Separation (ft)	ρ_a (Ωm)	Separation (ft)	ρ_a (Ωm)	Separation (ft)	ρ_a (Ωm)
50	28·3	250	19·8	450	14·6
100	26·0	300	18·1	500	14·6
150	23·8	350	16·7	550	15·0
200	21·9	400	15·4	600	16·2

Find the depth and resistivity of bed rock by three methods – cumulative ρ-plot, Tagg's curves and matching with the master two-layer curve. Discuss the agreement between the results and the accuracy of the different techniques.

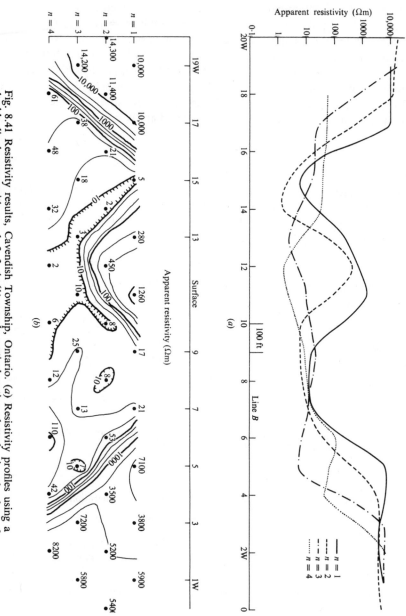

Fig. 8.41 Resistivity results, Cavendish Township, Ontario. (a) Resistivity profiles using a double-dipole array with n = 1, 2, 3, 4; (b) pseudo-vertical section plotted using the data of fig. 8.41a.

3. In a resistivity survey performed for highway construction, the following readings were obtained with an expanding Wenner spread.

Electrode separation (ft)	Resistivity (Ωm)	Electrode separation (ft)	Resistivity (Ωm)
5	78·1	30	51·2
10	56·0	35	59·8
15	49·8	40	76·0
20	47·1	45	79·8
25	46·0	50	72·2

Plot ρ_a versus a. How many layers are indicated by this curve? Is it possible to get a reasonable result by using Tagg's curves for a two-layer section? Can you use the method of partial curve matching for multiple layers and if so, do the results agree with those obtained by the cumulative ρ-plot? Find the depth of overburden by any or all of these methods.

4. Figure 8.42 shows a profile taken with a Wenner spread having a fixed spacing of 100 ft. Station intervals are 100 ft. The geologic section includes sandstone and limestone beds with practically vertical contacts. Locate these beds and speculate on the source of the small positive anomaly at 3100 ft.

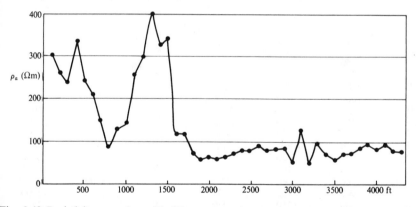

Fig. 8.42 Resistivity mapping with Wenner spread over limestone and sandstone section separated by vertical contacts. Station interval = 100 ft, a = 100 ft. (After Van Nostrand and Cook, 1966.)

5. The profile of fig. 8.43 was obtained in exactly the same manner as that of Problem 4; this is an area of karst topography in Hardin County, Illinois. The limestone contains numerous sinkholes and channels, most of which are filled with clay. There are occasional empty caverns as well. Make a rough interpretation of the near-surface section from the resistivity profile by locating the clay-filled sinks and/or caverns in the limestone host rock.

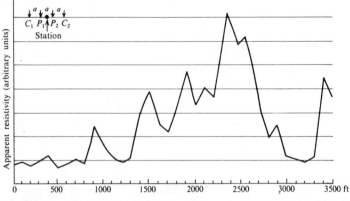

Fig. 8.43 Resistivity mapping with Wenner spread over karst topography, Hardin County, Illinois. Station interval = 100 ft, a = 100 ft. (After Van Nostrand and Cook, 1966.)

6. Apparent resistivities in Ωm are given below for portions of three lines from an area in eastern Nova Scotia where an IP survey was carried out. The topography is generally flat except for the west portion (line 0 + 00 has an elevation change of +250 ft between stations 13 and 20, line 10 + 00N has a change of +100 ft between stations 15 and 20,

Station	Line 10 + 00S		Line 0 + 00		Line 10 + 00N	
	$\rho_a(n = 1)$	$\rho_a(n = 4)$	$\rho_a(n = 1)$	$\rho_a(n = 4)$	$\rho_a(n = 1)$	$\rho_a(n = 4)$
26W	2750					
25	2000					
24	1700					
23	1850					
22	2250	1450				
21	2000	700				
20	1000	450	1700		500	1500
19	400	250	3000		150	1900
18	250	200	2150	500	400	2000
17	200	150	850	300	850	1600
16	150	200	350	500	1500	800
15	250	150	250	500	1700	2400
14	400	100	250	450	1200	5300
13	600	20	200	50	3800	10000
12	900	150	150	100	8900	12700
11	1000	350	250	450	9300	13500
10	150	200	450	850	9200	12700
9	550	50	650	1100	6650	12000
8	4100	20	1750	550	2460	10900
7	3600	25	2350	800	5750	7800
6	2800	100	3800	1550	6600	5700
5	1000	350	1000	850	4000	5500
4		700	950	1250	2250	5200
3			1000	850	4000	5500
2			1100	2000	7400	4300
1W			1600	2850	7000	6500
B.L.			3200	2350	6600	6800

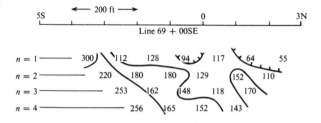

Fig. 8.44 Pseudo-depth section showing apparent resistivity ($\rho_a/2\pi$ Ωft) for a double-dipole array with separation 100 ft, Abitibi area, Quebec.

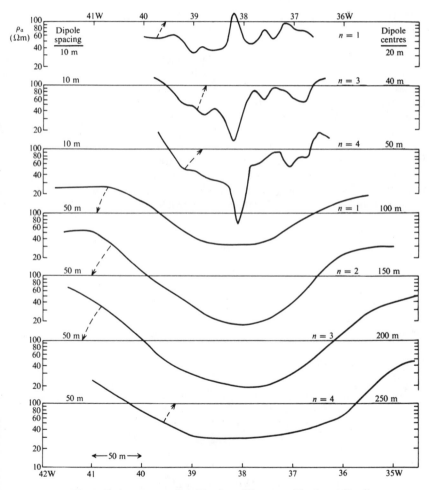

Fig. 8.45 Apparent resistivities from IP survey, Northeast Brazil.

line 10 + 00S has a change of +75 ft between stations 21 and 26). The rocks are known to be sedimentary in the valley while the hills in the vicinity are mainly granitic. Lines are 1000 ft apart and stations 100 ft. The double-dipole spread was used with dipole spacings of $\ell = 100$ ft; resistivities are given for $n = 1$ and $n = 4$ only, i.e., the distances between the inner electrodes were 100 ft and 400 ft respectively. Plot these profiles and interpret the results.

7. The pseudo-depth section of fig. 8.44 shows apparent resistivities $(\rho_a/2\pi)$ in ohm-feet in the usual method of displaying resistivity results obtained in IP surveys. The work was done as a follow-up on a strong magnetic anomaly in the course of sulphide exploration in the Abitibi area of Quebec. The double-dipole electrode separation was $\ell = 100$ ft.

What is the equivalent resistivity in ohm-metres for the $\rho_a/2\pi$ value at station 3S, $n = 2$? Are there any resistivity contrasts which might indicate sulphide zones? Would there be any advantage in plotting these results in profile form? Would you recommend any other type of geophysical work to aid the interpretation? Are the electrode configurations used here suitable for determining the resistivity of the overburden?

8. The resistivity profiles shown in fig. 8.45 are taken from an IP survey. The double-dipole spread was used with two dipole spacings: $\ell = 10$ metres for the top three profiles, 50 metres for the remaining four. Distances between the dipole centres are noted on the right-hand column beside the profiles, corresponding to $n = 1, 3$ and 4 for $\ell = 10$ m, and $n = 1, 2, 3, 4$ for $\ell = 50$ m. Apparent resistivities are plotted on a log scale and vary from a maximum of about 700 Ωm (west end of profile for $\ell = 50$ m, $n = 3$) to a minimum of 7 Ωm (at 38W on the profile for $\ell = 10$ m, n = 4). The profiles represent successively larger depths of penetration from top to bottom of the figure. The overburden is considerably oxidized but is known to be thin, about 1–2 metres. What interpretation can be made from these profiles? Would there be any advantage in plotting expanding spreads, i.e., depth sounding profiles, for fixed station locations?

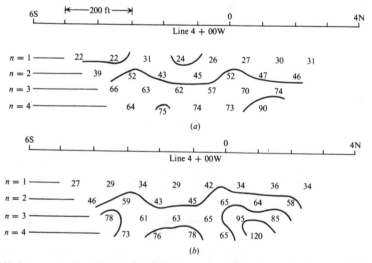

Fig. 8.46 Apparent resistivities $(\rho_a/2\pi$ Ωft) from two IP surveys using a double-dipole array with separation 100 ft. (*a*) Data from frequency-domain IP; (*b*) data from time-domain IP.

9. Two depth sections of apparent resistivity are shown in fig. 8.46 for an area in northern Quebec. Both employed the same double-dipole electrode system with $\ell = 100$ ft and $n = 1, 2, 3, 4$. One was done with a time-domain, the other with a frequency-domain IP set. The traverse is in the vicinity of an old mining operation in which zinc, copper, lead and some silver were recovered. Compare the results obtained by the two methods by plotting the profiles for $n = 1$ to 4. Can you explain the differences? Are there any obvious interesting features in these plots? The survey was done during winter because of swampy terrain.

10. Figure 8.47 shows apparent resistivity contours obtained from an IP survey in eastern Nova Scotia. The electrode arrangement was double-dipole with $\ell = 200$ ft and

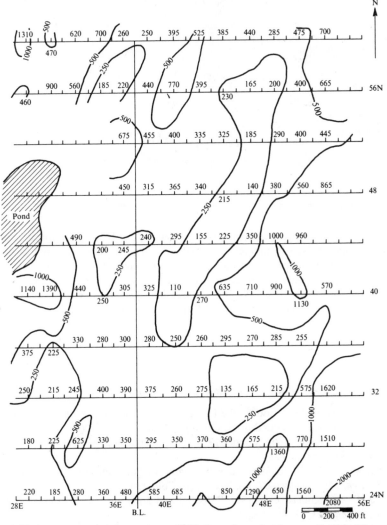

Fig. 8.47 Apparent resistivity contours (Ωft) for a base-metal zone, Eastern Nova Scotia.

$n = 1$. The rocks in the area are generally volcanics although in the section shown there are no outcrops; the overburden is not expected to be anywhere more than 25 ft deep and usually is less than 15 ft. There is a large-scale geochemical anomaly (Cu, Pb, Zn) associated with the area. Drainage is to the south while the glaciation direction is approximately northeast. With these data make an interpretation of the zone.

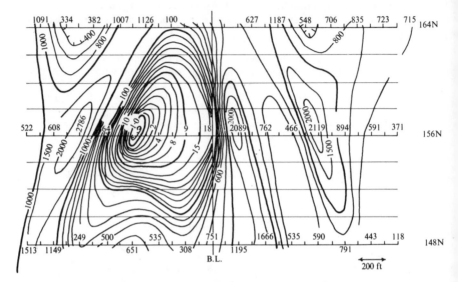

Fig. 8.48 Apparent resistivity contours (Ωm), Southern New Brunswick.

11. The contours of apparent resistivity illustrated in fig. 8.48 were developed from an IP survey in southern New Brunswick. The predominant geological feature in this area is a stock-like basic intrusive of gabbro-norite in an anticlinal structure of metasediments – argillite, slate, quartzitic mica schist and gneiss. Note that the only lines actually surveyed are 148, 156 and 164N. Would you consider this coverage sufficient to interpolate contours of this type? Take off profiles from lines 156 and 160N and make an interpretation.

References

F – field method; G – general; I – instruments; T – theoretical.

Alfano, L. (1959). Introduction to the interpretation of resistivity measurements for complicated structural conditions. *Geophys. Prosp.* **7**, 311–66. (T)

Compagnie Générale de Géophysique (1955). Abaque de sondage électrique. *Geophys. Prosp.* **3**, Supp. no. 3. (T)

Dakhnov, V. N. (1962). *Geophysical well logging* (tr. by Keller, G. V.), chapter 3, *Colo. Sch. Mines Quart.* **57**, no. 2, Golden, Colo. (G)

Degery, J. C. and Kunetz, G. (1956). Potential and apparent resistivity over dipping beds. *Geophysics*, **21**, 780–93. (T)

Dobrin, M. (1960). *Geophysical prospecting*. New York, McGraw-Hill. (G)

Edge, A. B. and Laby, T. H. (1931). *Principles and practice of geophysical prospecting.* London, Cambridge. (G)

Geological Survey of Canada (1967). *Mining and groundwater geophysics/1967* (ed. by Morley, L. W.), Econ. geol. rep. no. 26. Ottawa, Queen's Printer. (G)

Grant, F. S. and West, G. F. (1965). *Interpretation theory in applied geophysics*, chapter 14. New York, McGraw-Hill. (G)

Heiland, C. A. (1940). *Geophysical exploration.* New York, Prentice-Hall. (G)

Keller, G. V. and Frischknecht, F. C. (1966). *Electrical methods in geophysical prospecting*, chapter 3. London, Pergamon. (G)

Maeda, K. (1955). Apparent resistivity for dipping beds. *Geophysics*, **20**, 123–39. (T)

Mooney, H. M. and Wetzel, W. W. (1956). *The potentials about a point electrode and apparent resistivity curves for a two-, three- and four-layer earth.* Minneapolis, Univ. Minn. Press. (T)

Moore, R. W. (1945). An empirical method of interpretation of earth resistivity methods. *Trans. A.I.M.M.E.* **164**, 197–223. (T)

Muskat, M. and Evinger, H. H. (1941). Current penetration in direct-current prospecting. *Geophysics*, **6**, 397–427. (F)

Parasnis, D. S. (1966). *Mining geophysics*, chapter 6. Amsterdam, Elsevier. (G)

Seigel, H. O. (1952). Ore body size determination in electrical prospecting. *Geophysics*, **17**, 907–14. (T)

Stratton, J. A. (1941). *Electromagnetic theory.* New York, McGraw-Hill. (G)

Tagg, G. F. (1934). Interpretation of resistivity measurements. *Trans. A.I.M.M.E.* **110**, 135–47. (T)

Van Nostrand, R. G. and Cook, K. L. (1955). Apparent resistivity for dipping beds – a discussion. *Geophysics*, **20**, 140–7. (T)

Van Nostrand, R. G. and Cook, K. L. (1966). Interpretation of resistivity data (complete bibliography). U.S.G.S. Prof. paper no. 499 (Washington, D.C.). (T)

Vozoff, K. (1960). Numerical resistivity interpretation: general inhomogeneity. *Geophysics*, **25**, 1184–94. (T)

9. Induced polarization

9.1 Introduction

Induced polarization (IP) is a relatively new technique in geophysics, and has been employed mainly in base-metal exploration and to a minor extent in ground-water search. Although the Schlumberger brothers, the great pioneers in geophysical exploration, had recognized the phenomenon of induced polarization some 50 years ago, during their original work in self potential, its popularity as a geophysical tool dates from the mid-fifties, following further development work from 1948 to 1953. One form of polarization, the overvoltage effect, has been familiar in the field of physical chemistry for an even longer time.

An illustration of induced polarization can be obtained with a standard four-electrode d.c. resistivity spread by interrupting the current abruptly. The voltage across the potential electrodes generally does not drop to zero instantaneously, but decays rather slowly, after an initial large decrease from the original steady state value. This decay time is of the order of seconds or even minutes. If the current is switched on again the potential, after a sudden initial increase, builds up over a similar time interval to the original d.c. amplitude.

In one type of IP detector the decay voltage is measured as a function of time in various ways; this method is known as *time-domain IP*. Since the build-up time is also finite, it is clear that the apparent resistivity (actually a complex impedance) must vary with frequency, decreasing as the latter increases. Thus the measurement of ρ_a at two or more a.c. frequencies, generally below 10 Hz, constitutes another method of detection. This is known as *frequency-domain IP*.

Superficially this decay and build-up time resembles the discharge or charge time of a condenser through a finite resistance. But the decay curve is not exponential, as in the R-C circuit, nor does it commence at the static potential maximum. The difference between IP and R-C transients is illustrated in fig. 9.1.

Since the equipment employed, although more elaborate, is quite similar to resistivity, IP and resistivity field sets are much the same and it is customary to measure apparent resistivity, in addition to the IP effect, at each station. However, induced polarization, being mainly electrochemical in origin, has more in common with spontaneous polarization than bulk resistivity. It is necessary to consider these origins briefly in order to understand IP.

9.2 Sources of the induced polarization effects

9.2.1 *General*

The decay curve shown in fig. 9.1 represents a return to the original state following

702

the disturbance due to applied current. During the time of the original current flow, presumably some energy storage took place in the material. Although this stored energy theoretically could – and probably does – exist in several forms, for example, mechanical, electrical and chemical, laboratory studies of polarization in various rock types have established that the chemical energy is by far the most important.

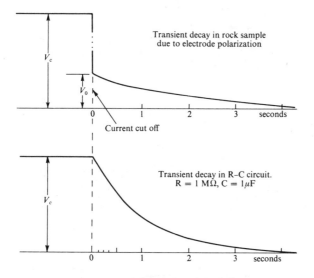

Fig. 9.1 Comparison of IP and R–C decay curves.

This chemical energy storage is the result of: (a) variations in the mobility of ions in fluids throughout the rock structure, (b) variations between ionic and electronic conductivity where metallic minerals are present. The first of these effects is known as *membrane* or *electrolytic polarization* and constitutes the background or so-called *normal IP effect*. It may occur in rocks which do not contain metallic minerals. The second is known as *electrode polarization* or *overvoltage*. It is generally larger in magnitude than the background IP and depends on the presence of metallic minerals in the rock. The two effects are indistinguishable by IP measurement. Furthermore, they appear to be independent of the atomic or molecular structure in rocks and minerals, that is, IP is a bulk effect.

9.2.2 *Membrane polarization*

Electrolytic conduction is the predominating factor in most rocks (see §5.2.2 and §5.2.4), being the only form of conduction when no minerals are present and the frequency is low. Thus a rock structure must be somewhat porous to permit current flow when metallic minerals are absent. Most rock minerals have a net negative charge at the interface between the rock surface and pore fluid. Consequently positive ions are attracted towards, negative repelled from, this interface; this positive ion concentration may extend into the fluid zone to a depth of about 10^{-6} cm. If this is the order of width of the pore itself, negative ions will accumulate at one end of the zone and leave the other when a d.c. potential is applied across it. As a result of this polarized distribution, current flow is impeded. At a later time when the current is switched off the ions return to their original positions, taking a finite time to do so. This situation is illustrated in fig. 9.2.

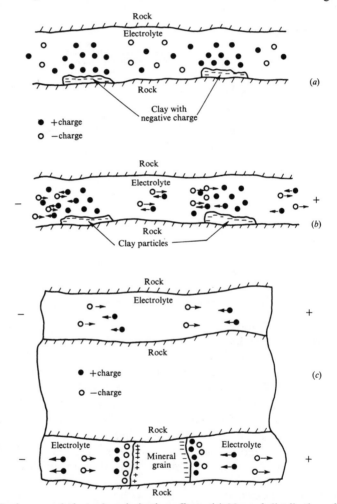

Fig. 9.2 Membrane and electrode polarization effects. (*a*) Normal distribution of ions in a porous sandstone; (*b*) membrane polarization in a porous sandstone due to an applied d.c. voltage; (*c*) electrolytic flow in upper pore, electrode polarization in lower pore.

The membrane IP effect is most pronounced in the presence of clay minerals, in which the pores are particularly small. The magnitude of polarization, however, does not increase steadily with the clay mineral concentration, but reaches a maximum and then decreases again. This is because there must be an alternate passage of larger cross-section in the material where ion accumulation does not take place; otherwise both total current flow and polarization are reduced. Optimum concentration varies in different types of clay, being low in montmorillonite and higher in kaolinite. Shales, with a high percentage of clay minerals, have a relatively low polarization. The membrane effect also decreases with the salinity of the pore fluid.

As a result of these factors membrane polarization is generally a maximum in a rock containing clay minerals scattered through the matrix in rather small ($\leqslant 10\%$) concentration and in which the electrolyte has low salinity.

Other sources of background polarization include normal dielectric and electrokinetic effects, presence of conducting minerals in very small amounts and possibly surface conduction on normally non-conducting material. Of these, the electrokinetic response due to variations in pore cross-section affecting fluid flow is probably more significant than the others. None of these sources, however, is comparable in magnitude to membrane polarization.

The overall background polarization is about what one would expect from a rock containing one to two per cent conducting minerals, but may vary from one tenth to ten times this value. Since it cannot be distinguished from electrode polarization, the background provides a level of geologic noise varying from place to place.

9.2.3 *Electrode polarization*

This type, similar in principle to membrane polarization, exists when metallic material is present in the rock and the current flow is partly electronic, partly electrolytic. A chemical reaction occurs at the interface between the mineral and solution.

Consider the two pore passages shown in the rock section in fig. 9.2c. In the upper one the current flow is entirely electrolytic. In the lower, the presence of a metallic mineral, having net surface charges of opposite sign on either face, results in an accumulation of ions in the electrolyte adjacent to each. The action is that of electrolysis, when current flows and an electron exchange takes place between the metal and the solution ions at the interface; in physical chemistry this effect is known as *overvoltage*.

Since the velocity of current flow in the electrolyte is much slower than in the metal, the pileup of ions is maintained by the external voltage. When the current is interrupted, the residual voltage decays as the ions diffuse back to their original equilibrium state.

Minerals which are electronic conductors exhibit electrode polarization. These include almost all the sulphides (excepting sphalerite and possibly cinnabar and stibnite), some oxides such as magnetite, ilmenite, pyrolusite and cassiterite and – unfortunately – graphite.

The magnitude of this electrode polarization depends, of course, on the external current source and also on a number of characteristics of the medium. It varies with the mineral concentration, but since it is a surface phenomenon, it should be larger when the mineral is disseminated than when it is massive. Actually the situation is not as simple as this. The optimum particle size varies to some extent with the porosity of the host rock and its resistivity. The fact that disseminated mineralization gives good IP response is a most attractive feature, since other electrical methods do not work very well in these circumstances.

For a particular concentration the polarization decreases with the rock porosity, since there is an increasing number of alternate paths for electrolytic conduction. Thus one would expect a larger IP effect in a disseminated sulphide occurring in dense igneous rock than in a porous host rock. Polarization also varies with the fluid content of the rock; from sample experiments it has been shown that a maximum occurs when approximately 75% of the pore space is filled with water.

The overvoltage varies inversely with the current density to some extent; over a wide range of current densities, the IP effect decreases by a factor of 2 as the current density increases tenfold. Finally the IP effect decreases with increasing source frequency (using a steady state a.c. of variable frequency rather than interrupted d.c.). This is true of membrane as well as electrode polarization, but in the latter the decrease is perhaps two orders of magnitude larger than in the former.

9.2.4 *Equivalent electrical circuits*

It is attractive to replace the porous rock structure, with or without mineral and membrane zones, by an equivalent electrical circuit. From fig. 9.1 we have already seen that a simple capacitance-resistance network will not explain the current flow and consequent IP effect. The circuits illustrated in fig. 9.3 provide a better analogue.

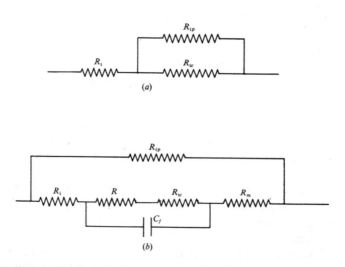

Fig. 9.3 Equivalent electrical circuits to simulate the IP effect. (*a*) Equivalent circuit for membrane polarization; (*b*) equivalent circuit for electrode polarization.

Figure 9.3*a* shows an equivalent circuit for membrane polarization. The effective pore fluid resistance is shown as R_i and R_{ip}, the series section representing normal unblocked passages while the parallel section R_{ip} simulates the leakage of ions past the membrane zone. The latter is in effect a complex impedance, R_w, known as the *Warburg impedance*. Its magnitude varies inversely as the square root of frequency. Typical values for a clay-bearing sandstone are:

$$R_i = 1000 \ \Omega, \quad R_{ip} = 200 \ \Omega, \quad R_w = 150(1 + j)/\sqrt{f} \ \ \Omega/\text{cm}^2.$$

The equivalent circuit for electrode polarization, shown in fig. 9.3*b*, is more complicated. Here we have an additional component, C_f, to simulate an effective capacitance at the double layer of positive and negative ions adjoining the interface, as well as R_m, a resistance for the mineral zone, and a reaction resistance, R, representing the electrochemical reaction. In this circuit the effect of frequency is more complex. For instance, at high frequencies ($\geqslant 1000$ Hz) the reactance of C_f is very small and the circuit impedance is mainly due to R_i and R_m in series. Hence the system becomes independent of frequency. This is true also at very low frequencies ($\leqslant 0.01$ Hz), when both C_f and R_w are practically open circuits and the current must flow by some other path entirely, presumably electrolytic. In an intermediate range (0·1 to 100 Hz) the impedance will vary with frequency.

The Warburg impedance for electrode polarization also varies as $f^{-\frac{1}{2}}$. Marshall and Madden (1959) found that it is proportional to the concentration and that for several types of electrodes (pyrite, galena, magnetite, graphite, copper, etc.) it could be approximated in the range from 100 to 0·01 Hz by the expression

$$R_w = 1.5 \times 10^3/\sqrt{f} \ \ \Omega/\text{cm}^2.$$

Needless to say, these circuit analogues are oversimplified. They do provide, however, a fairly good insight into the polarization mechanism.

9.3 Induced polarization measurements

9.3.1 *General*

As mentioned in §9.1, measurements of IP may be made either in the time or the frequency domain. The former are known as *pulse transient measurements*, the latter as *frequency variations*. Several terms of measurement which have been used in the two methods are defined in the following.

9.3.2 *Time-domain measurements*

(a) *Millivolts per volt* (*IP per cent*). The simplest way to measure IP effect with time-domain equipment is to compare the residual voltage $V(t)$ existing at a time t after the current is cut off with the steady voltage V_c during the current-flow interval (see fig. 9.4*a*). It is not possible to measure potential at the instant of cut-off because of large transients caused by breaking the current circuit. On the other hand $V(t)$ must be measured before the residual has decayed to noise level.

Since $V(t)$ is much smaller than V_c, the ratio $V(t)/V_c$ is expressed as millivolts/volt, or as a per cent, $100\,V(t)/V_c$, where both are in millivolts. The time interval t may vary between 0·1 and 10 seconds.

(b) *Decay-time integral.* Commercial IP sets generally measure potential integrated over a definite time interval of the transient decay, as shown in fig. 9.4*b*. If this integration time is very short and if the decay curve is sampled at several points, the values of the integral are effectively a measure of the potential existing at different times, i.e., $V(t_1)$, $V(t_2)$, . . ., $V(t_n)$. This is an extension of the measurement in (*a*) from which one also obtains the decay curve shape.

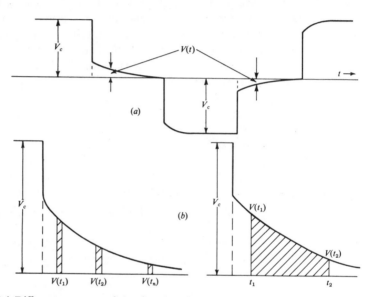

Fig. 9.4 Different measures of the time-domain IP effect. (*a*) Comparison of $V(t)$ with V_c (*b*) integral of V(*t*) over a time interval.

(c) *Chargeability.* This is defined as

$$M = \frac{1}{V_c} \int_{t_1}^{t_2} V(t)\,\mathrm{d}t, \tag{9.1}$$

and is the most commonly used quantity in time-domain IP measurement. When $V(t)$ and V_c have the same units, the chargeability M is in milliseconds.

9.3.3 *Frequency-domain measurements*

(a) *Frequency effect.* In frequency-domain IP, one measures the apparent resistivity at two or more frequencies. The *frequency effect* is usually defined as

$$fe = (\rho_{dc} - \rho_{ac})/\rho_{ac} = (\rho_{dc}/\rho_{ac}) - 1, \tag{9.2a}$$

while the *per cent frequency effect* is given by

$$\text{PFE} = 100(\rho_{dc} - \rho_{ac})/\rho_{ac}, \tag{9.2b}$$

where ρ_{dc}, ρ_{ac} are apparent resistivities measured at d.c. and very high frequency. As we have seen in §9.2.4 and fig. 9.3b, ρ_{dc} is determined by the alternate path R_{ip} only, while ρ_{ac} depends on R_{ip} shunted by R_i and R_m. Hence $\rho_{ac} < \rho_{dc}$. In practice measurements are made at two or more frequencies in the range 0·1 to 10Hz, ρ_{dc} being taken as the value obtained at the lowest frequency.

(b) *Metal factor.* We have mentioned that the IP effect varies with effective resistivity of the host rock, that is, the type of electrolyte, temperature, pore size, etc. The *metal-factor parameter*, originally suggested by Marshall and Madden (1959), corrects to some extent for this variable. It is a modification of the expression in eq. (9.2a):

$$\text{MF} = 2\pi \times 10^5(\rho_{dc} - \rho_{ac})/\rho_{dc}\rho_{ac} = 2\pi \times 10^5 fe/\rho_{dc}. \tag{9.3a}$$

Because apparent resistivities are frequently given in ohm-feet (actually in the form $\rho_a/2\pi$ ohm-ft) on frequency domain IP equipment, metal-factor values may have units of mhos/ft, rather than mhos/m. Thus, a more conveninent form of eq. (9.3a) is

$$\text{MF} = 10^5 fe/(\rho_{dc}/2\pi) = 10^3 \text{PFE}/(\rho_{dc}/2\pi). \tag{9.3b}$$

9.3.4 *Relation between time- and frequency-domain IP measurements*

In theory, since both IP methods measure the same phenomenon, their results ought to be the same; practically the conversion of time domain to frequency domain and vice versa is quite difficult. The square wave used in time-domain IP contains all frequencies, assuming that the fronts are infinitely steep.

Seigel (in Wait, 1959) defines the chargeability as

$$M = \{\operatorname*{Lim}_{t \to \infty} V(t) - \operatorname*{Lim}_{t \to 0} V(t)\}/\operatorname*{Lim}_{t \to \infty} V(t).$$

By Laplace transform theory, it can be shown that

$$\operatorname*{Lim}_{t \to \infty} V(t) = J\rho_{dc} \quad \text{and} \quad \operatorname*{Lim}_{t \to 0} V(t) = J\rho_\infty,$$

where ρ_∞ is the apparent resistivity at very high frequency and J is the current density. Consequently, using eq. (9.2a) and assuming that $\rho_{ac} = \rho_\infty$, we can write for the chargeability

$$M = \frac{\rho_{dc} - \rho_\infty}{\rho_{dc}} = 1 - \frac{\rho_{ac}}{\rho_{dc}} = 1 - \frac{1}{1 + fe} = \frac{fe}{1 + fe} \approx fe \tag{9.4}$$

when $fe \ll 1$.

In practical situations this simple relation is not valid, partly because an exact theoretical analysis of the IP effect is not available (that is, the basic premises of the two systems of measurement are only approximately valid), partly because the measurements are not made at d.c. and VHF in either IP system. Thus, in general it is not possible to convert one result to the other. There has been considerable controversy between the manufacturers and users of the two types of equipment as to their respective merits; from a theoretical point of view these arguments appear to be superfluous.

9.3.5 *Laboratory investigations*

Considerable laboratory work has been carried out over the past twenty years on the polarization of rocks and minerals, the effects of electrolytic concentration, temperature and pressure, variations in mineral grain sizes and concentrations, current density and, of course, type of mineral.

Block diagrams of the pulse transient and variable frequency test equipment are shown in fig. 9.5. The current, pulse length and decay interval are usually continuously variable. In the variable frequency method the oscillator sometimes operates at fixed frequencies in the range 0·1–10 Hz in multiples of three; otherwise it may be continuously variable as well.

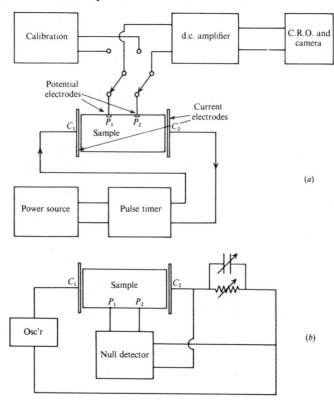

Fig. 9.5 Laboratory equipment for IP measurements. (*a*) Pulse transient test equipment (After Collett, in Wait, 1959.); (*b*) variable frequency test equipment.

It is important that the input impedance of the detector connected across the potential electrodes be very high ($\geqslant$ 10 MΩ) compared to the impedance seen across $P_1 - P_2$. Also the impedance $C_1 - C_2$ should not be unreasonably high ($\leqslant$ 500 $k\Omega$).

Either artificial or field samples, cut to size, may be measured. The chief problem with the samples is the electrode connections. The current electrode discs are sometimes faced with graphite, while the potential contacts are made with rings of copper wire, sometimes immersed in a mixture of gelatin and $CuSO_4$ to maintain stability.

Prepared samples can be made up by combining the required concentration and size of mineral particles in a matrix of non-mineralized rock or cement and immersing the whole in a lucite tube containing electrolyte of the desired type and strength (NaCl, $CuSO_4$, etc.). Solid rock samples generally have to be filled with electrolyte under pressure and kept under water until tested.

Figure 9.6 illustrates typical decay curves and conductivity-frequency variations for several mineral samples, all of the same concentration and grain size in the same electrolyte. The most obvious feature of the decay curves is the wide variation for different minerals. The curves, in some cases, are roughly linear with log t. It is also evident that graphite has a large IP response. The frequency variation curves are normalized to the conductivity at 10 Hz.

Further laboratory measurements in which other parameters were varied systematically lead to the following conclusions:

(i) in non-mineralized specimens of andesite, as the amount of electrolyte is increased, the IP response increases to a maximum, then falls off again. As the concentration of NaCl in a 5% electrolyte is increased, the response increases. Similarly the response is roughly proportional to temperature.

(ii) in mineralized samples of andesite (2% pyrite) the response decreases both with the amount and concentration of the electrolyte and also with increasing temperature. When the mineral particle size is varied, keeping the relative volume of pyrite constant, the maximum response generally falls off as the particles get bigger, for initial decay times of 0·003 seconds or less. As a result the response for a given particle size reaches a maximum at some fixed time on the decay curve. Finally the response increases with mineral concentration, at least in the range 1 to 30% of total volume.

Laboratory measurements, with few exceptions, have been carried out at current densities much greater than those employed in field work. Since the IP effect decreases with increasing current density (see §9.2.3), it is not possible to extrapolate from the laboratory measurements to predict field results.

9.3.6 *IP response examples*

Although the type and grade of mineralization is not apparent from the character of the decay curves, the following tables may be of some use in crude assessment of field results. Table 9.1 lists a variety of minerals at 1 % volume concentration and

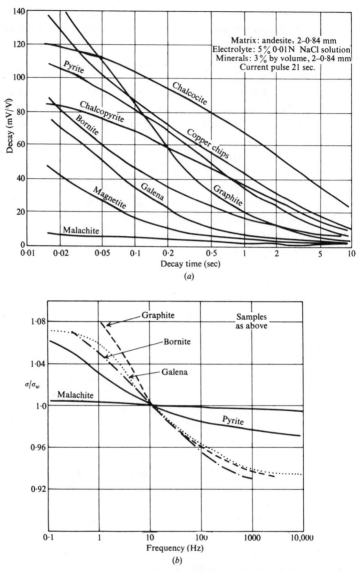

Fig. 9.6 IP measurements on laboratory samples. (After Collett, in Wait, 1959.) (*a*) Transient decay curves for various minerals; (*b*) frequency characteristics of various minerals.

their IP response. The duration of the current square wave was 3 seconds and the decay was integrated over 1 second.

Table 9.2 shows the response of a variety of mineralized and barren rocks. Here the charging time is long (about one minute) and the decay curve is integrated over its entire duration (excluding the initial transient and final noise).

These values appear high with respect to usual field measurements and in fact they are, since it is not customary to employ such a long timing cycle or to integrate the complete decay curve. However, they do illustrate the variation between different IP sources.

Table 9.1. *IP chargeability of minerals*

Mineral	Chargeability (msec)	Mineral	Chargeability (msec)
Pyrite	13·4	Bornite	6·3
Chalcocite	13·2	Galena	3·7
Copper	12·3	Magnetite	2·2
Graphite	11·2	Malachite	0·2
Chalcopyrite	9·4	Hematite	0·0

Table 9.2. *IP chargeability of various minerals and rocks*

Material	Chargeability (msec)	Material	Chargeability (msec)
20% sulphides	2000–3000	Dense volcanic rocks	100–500
8–20% sulphides	1000–2000	Shale	50–100
2–8% sulphides	500–1000	Granite, Granodiorite	10–50
Volcanic tuffs	300–800	Limestone, Dolomite	10–20
Sandstone, siltstone	100–500		

Table 9.3 shows further values of chargeability for various materials. The charging time was 3 seconds and the integration time from 0·02 to 1 second of the decay curve.

Table 9.3. *IP chargeability of various materials*

Material	Chargeability (msec)	Material	Chargeability (msec)
Ground water	0	Schists	5–20
Alluvium	1–4	Sandstones	3–12
Gravels	3–9	Argillites	3–10
Precambrian volcanics	8–20	Quartzites	5–12
Precambrian gneisses	6–30		

Table 9.4 lists typical metal factors for a variety of igneous and metamorphic rocks.

Table 9.4. *IP metal factor of various materials*

Material	Metal factor (mhos/cm)	Material	Metal factor (mhos/cm)
Massive sulphides	10,000	Graphitic sandstone	
Fracture-filling		and limestone	4–60
sulphides	1000–10,000	Gravels	0–200
Massive magnetite	3–3000	Alluvium	0–200
Porphyry copper	30–1500	Precambrian gneisses	10–100
Dissem. sulphides	100–1000	Granites, monzonites,	
Shale-sulphides	3–300	diorites	0–60
Clays	1–300	Various volcanics	0–80
Sandstone-1–2%		Schists	10–60
sulphides	2–200	Basic rocks (barren)	1–10
Finely dissem.		Granites (barren)	1
sulphides	10–100	Groundwater	0
Tuffs	1–100		

Obviously because of the considerable overlap in values, it is not possible to distinguish between poorly mineralized rocks and several barren types, such as tuffs and clays.

9.4 IP field operations

9.4.1 *General*

As mentioned earlier, the equipment and field procedure for induced polarization surveys are similar to that used in resistivity exploration. This usually results in a combined resistivity-IP survey; sometimes SP may be measured as well. The equipment is relatively elaborate and bulky. Of the commonly used ground-exploration methods (excluding seismic), it is the most expensive, being roughly comparable to gravity in cost per month. The field work also is slow compared to magnetics, EM and SP.

9.4.2 *Field equipment*

(a) *Time-domain equipment.* A block diagram of a standard IP set of this type is illustrated in fig. 9.7a. The timing cycle is variable, but the current flows for a period of at least one second, while the off-time is usually one half of this, or less. A typical cycle in use is 1·5 seconds on, 0·5 seconds off. Much longer cycles, up to one minute, are sometimes used. Vacuum relays are necessary to handle the large switching currents. A dummy load to take the output during off-time, capable of handling the maximum power from the transmitter, is included in the H.T. set.

The prime mover is a gasoline engine driving a generator, usually 400 Hz at

110 or 208 V, single or three phase, with outputs varying from 1 kVA minimum to 10 kVA. (At least one low-power, easily portable commercial set, operating from batteries, is also available, but this has limited range.) The d.c. power supply consists of a rectifier and filter with output from about 300 V to 5 kV, maximum current about 10 A on the larger units. The current regulation is generally maintained to $\pm 2\%$ or better. In addition, cutouts are provided to protect the supply when excessive variations of input or load occur. The large power requirements call for heavy equipment; the weight of the combined generator and H.T. set varies between 150 and 750 lb for different models.

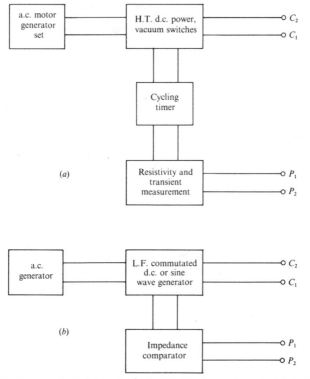

Fig. 9.7 Block diagram of IP field sets. (a) Time-domain equipment; (b) frequency-domain equipment.

The timing circuit controls the on-off cycle at a low power level and also switches on the receiver to read both the steady voltage V_c (see eq. (9.1)) and the transient voltage over a selected period of the decay. Integration of the decay voltage $V(t)$, performed electronically, is carried out over a desired interval of the decay.

In most IP sets of this type, measurements are made by nulling a meter with a calibrated compensator circuit. The readings are then obtained, one for V_c and the other for integrated $V(t)$. On a more recent model, V_c is balanced out and a direct reading of the chargeability obtained.

It is necessary to buck out any self-potential existing across the potential electrodes. This was formerly done manually, but is semi-automatic on the latest

models. Telluric currents are more troublesome since they produce random
potentials. In time-domain receivers the telluric effect is reduced by averaging
readings over several decay cycles.

(b) *Frequency-domain equipment.* Figure (9.7b) shows a block diagram of the
variable frequency IP set. From the diagram it appears to be considerably simpler
than the pulse transient equipment, although the price is about the same. Power
output from the generator varies in different sets from 800 W to 5 kW.

At least two, and sometimes several, discrete frequencies are available from the
transmitter, in the range 0·1 to 10 Hz. A typical field set operates at 0·3 and 5 Hz
as well as d.c. For such low frequency a.c. it is more efficient to commutate a
d.c. voltage with mechanical switching, although occasionally sine wave output
are used. The output current is regulated to 0·5% or better.

Usually the apparent resistivity is measured directly on a stable bridge or
potentiometer in the receiver. Sharp band-pass filters are required at the receiver
when square-wave a.c. is employed in the transmitter. Telluric noise can be reduced quite successfully by a high-pass filter, particularly if the lower transmitter
frequency is not less than 0·3 Hz.

It might appear attractive to make an extremely cheap and simple variable
frequency IP unit by using d.c. and 60 Hz a.c., since the only requirement would be
a standard 60-Hz generator and a rectifier. Unfortunately, electromagnetic
coupling between potential and current circuits (especially with large electrode
spacings which result in long-wire layouts) may introduce spurious variations in
apparent resistivity at the 60-Hz frequency. The EM coupling effect will be
discussed in §9.4.4c.

(c) *Electrodes and cables.* Current electrodes are usually metal stakes as in
resistivity work. Sometimes it is necessary to use aluminum foil in shallow holes. I
may also be necessary to wet the electrodes with salt water to provide sufficiently
good contact for the relatively high currents desired. Porous pots are often used fo
the potential electrodes because of the very low frequencies. The current wires mus
be capable of withstanding high voltages of the order of 5–10 kV.

9.4.3 *Field procedures*

Since the IP electrode system is identical to resistivity, theoretically one can use
any of the field spreads described in §8.5.3. In practice the Schlumberger or gradien
array, the pole-dipole in which one current electrode is removed a great distance
and the double-dipole, with a rather small value of n, are the three commonly use
IP spreads.

The latter two configurations are illustrated in fig. 9.8. Using the dimensions a
shown and eq. (8.13), §8.2.4, the apparent resistivities for these two spreads, ove
homogeneous ground, are:

Double dipole: $\rho_a = \pi n(n + 1)(n + 2) \, x\Delta V/I,$ (9.5

Pole dipole: $\rho_a = 2\pi n(n + 1) \, x\Delta V/I.$ (9.6

Values of n range from one to ten, although six is usually the upper limit. The electrode spacing may be as small as 10 ft and as large as 1000 ft. To reduce the work of moving the current electrodes and particularly the heavy transmitter unit, several pairs of current electrodes are often placed in suitable locations and wired to a fixed transmitter; the latter is then switched from one to the other.

Results are usually plotted at the midpoint of the spread (or in pole-dipole, the midpoint of C_1P_1), although occasionally the midpoint of either current or potential pair is considered as the station location.

The larger electrode spacings are mainly for reconnaissance although, as in resistivity, the depth of penetration is controlled in part by the spacing. Frequently the same line is traversed several times with different spacings, for example, $x = 100$ or 200 ft and $n = 1, 2, 3, 4$, etc.; by so doing, one obtains a combination of lateral profiling and depth probing.

As mentioned previously, apparent resistivities are also obtained at each station. Self-potential may also be recorded by noting the bucking potential required before current is switched on.

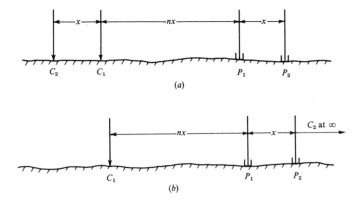

(a)

(b)

Fig. 9.8 Typical IP spreads. (a) Double-dipole spread; (b) pole-dipole spread.

9.4.4 *Noise sources*

(a) *General.* Besides SP, which is easily compensated, other sources of background noise are telluric currents, capacitive and electromagnetic coupling and the general IP effect from barren rocks (see §9.2.2). The reduction of telluric noise has already been mentioned.

(b) *Capacitive coupling.* This may occur between current electrodes and potential wires, or vice versa, as a result of leakage currents, or by leakage between current and potential wires. The capacitive effect is usually small enough to be negligible, unless the insulation of the wires is defective or the wires lie very close to electrodes other than their own. In IP well logging, where the cables are side by side, it is necessary to use shielded wire.

(c) *Electromagnetic coupling.* This effect is extremely troublesome. It results from mutual inductance between current and potential wires, both directly and through the ground in their vicinity. The EM effect can become quite large when long wire layouts or higher frequencies are used. Double-dipole and pole-dipole spreads are employed to reduce coupling due to long wires while the frequencies are usually kept below 10 Hz.

It is possible to calculate approximately the coupling between two wires in the presence of homogeneous ground. Resistivity variations in the vertical plane also influence the EM effect considerably. Madden and Cantwell give a rule-of-thumb for limiting either the frequency or electrode spacing for a particular array, in order to keep the EM coupling effect within background. For double-dipole electrode spreads the expression is

$$nx\sqrt{(f/\rho)} < 200 \qquad (9.7a)$$

in variable frequency equipment, where x is in metres, ρ in ohm-metres. For time-domain measurements the limit is

$$t < 2\pi/f_c, \qquad (9.7b)$$

where

$$f_c \approx 10^4\rho/(nx)^2.$$

Table 9.5 shows the maximum spreads permissible in frequency-domain measurement for double dipole, at various frequencies and ground resistivities. When pole-dipole spreads are used the situation is somewhat better (longer spreads can be used), while for the Schlumberger or gradient array, the maximum nx is reduced by two.

Table 9.5. *Maximum spreads for various frequencies and ground resistivities*

f (Hz)	ρ (Ωm)	nx (max) (ft)	f (Hz)	ρ (Ωm)	nx (max) (ft)	f (Hz)	ρ (Ωm)	nx (max) (ft)
50	1000	3000	10	1000	6500	3	1000	12,000
	100	1000		100	2000		100	3700
	10	300		10	650		10	1200
	1	100		1	200		1	370

9.5 Interpretation

9.5.1 *Plotting methods*

The display of IP results is often made in simple profiles, in which chargeability (milliseconds), metal factor or per cent frequency effect is plotted as ordinate against station location on the horizontal axis. Such plots are illustrated in fig. 9.9.

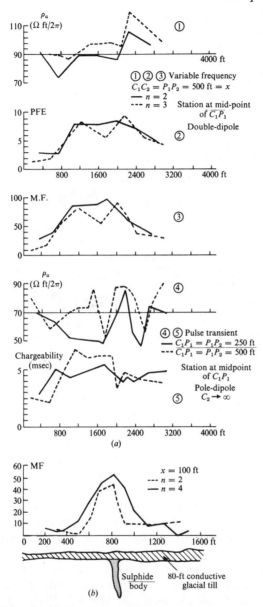

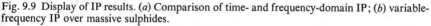

Fig. 9.9 Display of IP results. (*a*) Comparison of time- and frequency-domain IP; (*b*) variable-frequency IP over massive sulphides.

The profiles in fig. 9.9*a* show the same anomaly traversed with both time- and frequency-domain IP. There is little difference between the frequency-effect and metal-factor plots, while the chargeability profile is somewhat similar. However, the resistivity profiles are quite different for the two methods. This is probably due to the fact that the variable frequency IP used a double-dipole spread, while the pulse system employed pole-dipole. These profiles are taken from line 29 + 00 on the contour plot of fig. 9.10 which is a form of display occasionally used. From this illustration the two methods appear to give similar results.

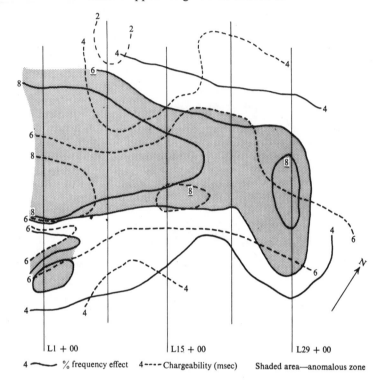

4 ⌒ % frequency effect   4‑‑‑‑ Chargeability (msec)   Shaded area—anomalous zone

Fig. 9.10 Contours for time- and frequency-domain IP.

Figure 9.9*b* shows a variable-frequency profile over a massive sulphide covered by some 80 ft of overburden (glacial till), which was a relatively good conductor. In the absence of this cover, the response would presumably be very much larger. It is also worth noting that the larger dipole separation gave slightly better response.

An alternative display method, which has been used in plotting IP to illustrate the effects of variable electrode spacing, was originally developed by Madden, Cantwell and Hallof (see Marshall and Madden (1959)). It is illustrated in fig. 9.11, for the sulphide deposit shown in fig. 9.9*b*. Values of frequency effect and apparent resistivity for each station are plotted on a vertical section at the points of intersection of 45° lines drawn from the base line or surface, starting at the mid-points of the current and potential electrodes (double-dipole array). In this way the PFE values appear at points directly below the centre of the electrode spread, at a vertical distance which increases with the *n* value for the spread. Similarly the

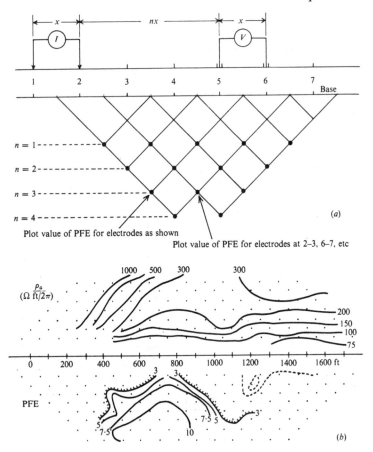

Fig. 9.11 Variable-frequency IP pseudo-depth sections. (After Marshall and Madden, 1959.) (*a*) Graphical construction for locating data points; (*b*) pseudo-depth plot for the data of fig. 9.9*b*.

ρ_a values are located at mirror image points above the centre line. Finally contours of equal PFE and apparent resistivity are drawn on these vertical sections; the result is a form of two-dimensional plot in vertical section.

The attractive feature of this display is that it gives some idea of the relative depths of anomalous conducting zones. The justification for such a plot is that as the dipole separation is increased, the measured values are influenced by increasingly deeper zones. The resultant contours may be misleading, however, because they appear to provide a vertical section of the ground conductivity. As pointed out in §8.5.2, the apparent resistivity is not in fact the actual resistivity in a volume of ground below the electrode array, but depends on the geometry of the electrodes as well as the surface resistivities. Consequently it should not be assumed that this type of plot is a representation of the actual subsurface.

9.5.2 *Theoretical and model work*

IP response has been worked out analytically for a few simple shapes like the sphere, ellipsoid and two-dimensional features, such as a vertical contact, dike and two horizontal beds. The development is similar to resistivity. Figure 9.12 illustrates the theoretical IP and resistivity response over a sphere and a triaxial ellipsoid, the latter having its long axis vertical. In both cases the centre of the body is two units below surface and the resistivity contrast is $k = (\rho_2 - \rho_1)/(\rho_2 + \rho_1)$

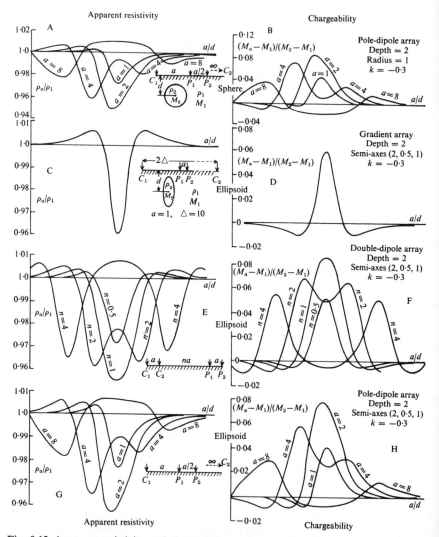

Fig. 9.12 Apparent resistivity and chargeability curves for three-dimensional bodies. (After Dieter *et al.* 1969.) Curves A and B: pole-dipole array over sphere; curves C and D: gradient array over triaxial ellipsoid; curves E and F: double-dipole array over triaxial ellipsoid; curves G and H: pole-dipole array over triaxial ellipsoid.

$= -0.3$. Resistivity response is given in the usual form ρ_a/ρ_1, while the IP parameter is plotted as $(M_a - M_1)/(M_2 - M_1) = (\rho_2/\rho_a)/(\partial\rho_a/\partial\rho_2)$. For interpretation, one attempts to match the field profiles of ρ_a and M_a to a particular pair of theoretical curves. Larger resistivity contrasts do not increase the maximum IP response appreciably, while the effect of dip, in the case of the ellipsoid, is insignificant. Note also that in profiles C to H the peak responses are about the same over the ellipsoid, regardless of the electrode spread, although the profile shapes are quite different.

Some model work has been done. Apparent resistivities have been measured at different frequencies for various shapes and resistivities of a $CuSO_4$-gelatin model immersed in a tank of water. Typical results are shown in fig. 9.13. Clearly the resistivity and frequency variation curves match up quite closely and are similar to the field examples in fig. 9.11, where the sulphide has a comparable cross-section.

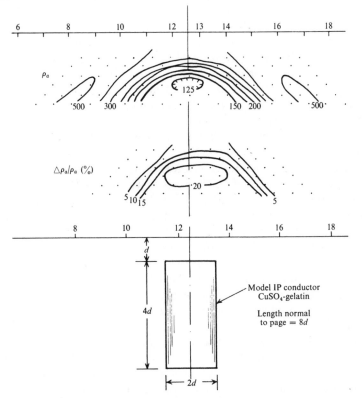

Fig. 9.13 Model measurements of ρ_a and IP effects. (After Madden and Cantwell, 1967.)

9.5.3 *General interpretation*

To date, IP interpretation has tended to be mainly qualitative. Location, lateral extent and depth of anomalies can be estimated from the characteristics of the profiles, such as sharpness, symmetry and amplitude above background. By traversing with several different electrode separations it is often possible to resolve multiple anomalies and gain some idea of depth extent. The inherent advantages and weaknesses of resistivity apply to IP as well. Among the former are fairly good depth estimate and the possibility of considerable depth of penetration, while the latter include ambiguity as to location, undesirable effects of near-surface variations and, in particular, slow field operations.

Certain claims as to unique features of IP are probably valid only in specific situations. For instance, highly conductive overburden overlying mineral conductors will generally mask out the detection of the latter by IP as well as by EM and resistivity, although in particular instances it may be possible to distinguish between the two IP responses. Similarly, water-filled shear zones are generally indistinguishable from mineral zones; however, in special circumstances, for example, if the electrolytic effect is not as pronounced as the electrode polarization, it may be possible to distinguish between the two with IP.

At one time it was thought that massive sulphides should have a lower IP response than disseminated mineralization; this is theoretically reasonable, as discussed in §9.2.3. However, it is probable that the opposite is true. This may be due to the halo of disseminated mineralization which usually surrounds a massive zone. Another explanation is that truly homogeneous massive sulphide deposits do not exist; rather they are broken up into a great number of smaller conducting zones within a non-conducting, or poorly conducting matrix. Self-potential well logs generally indicate this internal subdivision for sections designated massive in the descriptive log.

The steeply dipping thin-sheet conductor, commonly used in EM modelling, is not a particularly good target for IP or resistivity surveys. The principal reason for this is that the electrode spacings are normally too large to respond strongly to such a structure. Although a disadvantage, this is hardly a fundamental weakness of IP, since the technique would not usually be employed (and should not be necessary) to detect conductors of this nature. However, it does account for the lack of response directly over some of these structures and in certain cases, an apparently displaced IP anomaly on the flanks, the latter probably caused by the disseminated halo. In fact, an IP traverse made with dipole separations of 25 and 50 ft in one area produced a strong response directly above the sheet-like conductor.

The induced polarization method has become very popular in base-metal exploration. This is certainly not because it is cheap or fast. Average monthly coverage varies enormously, depending on terrain and other factors such as surface conductivity, but 10–40 line miles per month is common. The price per line mile is thus about $200–$1000, which is considerably higher than magnetics or EM.

This popularity is based on definite base-metal discoveries, particularly of large low-grade bodies, made with the aid of IP. A study of various field results indicates that the IP and resistivity anomalies (generally IP highs and resistivity lows) very often occur together. One might argue, therefore, that the expense of the IP survey was not warranted. It is quite unlikely, however, that resistivity alone would provide enough information to justify itself. There are also numerous case histories of IP successes in areas of disseminated mineralization, such as porphyry coppers, where the resistivity anomaly could be considered non-existent.

9.6 Field examples

Several examples of IP field results have already been given in figs. 9.9, 9.10 and 9.11. Three further illustrations are described below.

1. Figure 9.14 is a profile of apparent resistivity and chargeability obtained during a time-domain IP survey on the Gortdrum copper-silver orebody in

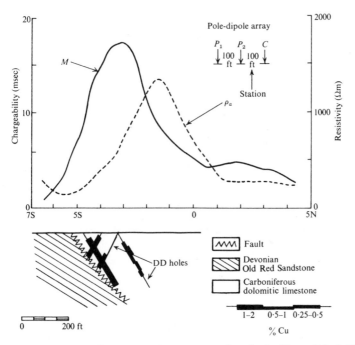

Fig. 9.14 Time-domain IP over Gortdrum copper-silver body. (From Seigel, 1967.)

Ireland. This is a low-grade deposit, averaging only $1\cdot2\%$ by volume of copper and $0\cdot75$ oz of silver per ton, that is, less than 2% metallic conducting minerals. Neither pyrite nor pyrrhotite is present to enhance the bulk conductivity, as is often the case with such mineralization. However, the chargeability anomaly is very strong and well located. The ρ_a-profile shows a large resistivity contrast between the dolomitic limestone and sandstone with a minimum directly over the fault; there is no indication of the sulphide zones containing chalcocite, bornite and chalcopyrite. The three-electrode or pole-dipole spread was used in this work, with spacing as shown in the diagram.

2. Pseudo-depth sections plotted from the results of a double-dipole traverse using frequency-domain IP are shown in fig. 9.15. This is in the Timmins area of northern Ontario where the glacial overburden is frequently 100–200 ft thick and, being of low resistivity, effectively masks the response of conductors lying beneath it. Using 200 ft dipole spacing and separations of 200, 400, 600 and 800 ft, a good IP response was obtained. The shape of the metal-factor contours indicates a source at depth. The resistivity section shows low resistivity continuing to depth with a westward dip, as well as the effect of the conductive overburden. Subsequent drilling intersected massive sulphide mineralization over 100 ft wide at a depth of 240 ft. It is not surprising that EM methods failed to detect this zone.

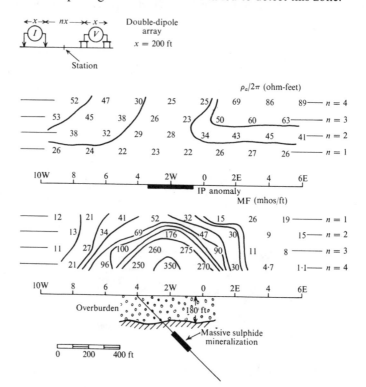

Fig. 9.15 Frequency-domain IP results for massive sulphides overlain by thick conductive overburden.

3. Chargeability and apparent resistivity profiles, using three separations of the pole-dipole array in traversing the Lornex porphyry copper deposit in British Columbia, are displayed in fig. 9.16. This is a type of mineralization for which the IP technique is particularly effective, since no other electrical method would be capable of detecting the main body, although there might be minor indications on the flanks. Moreover, it is unlikely that gravity would produce any response.

The resistivity profiles for 400-ft and 800-ft electrode separations might be

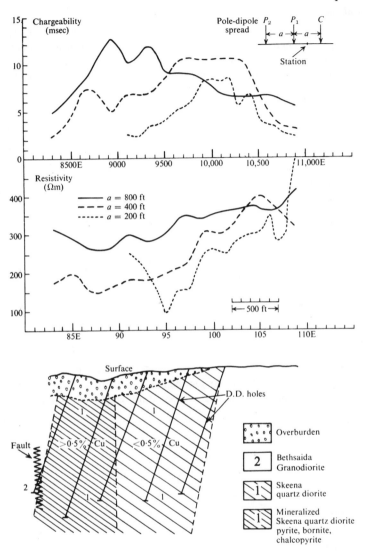

Fig. 9.16 Time-domain IP results over porphyry-copper deposit. (After Seigel, 1967.)

interpreted as showing a mild reflection of the mineralization, were it not for the fact that the apparent resistivity increases with depth. This tells us that the overburden, which is 200 ft thick on the east, has a higher conductivity than the ore zone below it. On the other hand, the chargeability response increases with electrode separation and determines the lateral extent and depth of the zone quite well.

9.7 Problems

1. The following results were obtained using frequency-domain IP in a survey over suspected sulphide mineralization in northern New Brunswick. The double-dipole array was used with dipole separations of 100 ft and $n = 1$, 2, 3. Resistivity values are in the form $\rho_a/2\pi$ ohm-feet. The grid line is roughly N–S with pickets every 100 ft. In all cases the potential dipole was south of the current dipole.

Potential dipole	$n = 1$		$n = 2$		$n = 3$	
	$\rho_a/2\pi$	MF	$\rho_a/2\pi$	MF	$\rho_a/2\pi$	MF
10S–9S	—	—	—	—	280	27
9S–8S	180	28	190	24	270	33
8S–7S	210	31	275	36	290	60
7S–6S	270	42	280	35	72	219
6S–5S	315	39	80	172	70	175
5S–4S	480	40	220	17	675	99
4S–3S	330	88	1120	41	1751	61
3S–2S	1091	46	1130	29	1830	31
2S–1S	1200	31	1510	27	1710	28

Plot the values in pseudo-depth sections for $\rho_a/2\pi$ and MF; draw contours and interpret the results.

2. A time-domain IP profile of chargeability and apparent resistivity is shown in fig. 9.17. This is from the Pine Point sedimentary area of the Canadian Northwest Territories, where IP methods have been successfully employed to locate large lead-

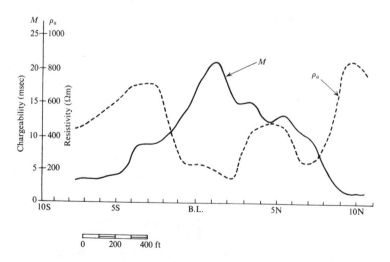

Fig. 9.17 IP chargeability and apparent resistivity, Pine Point Area, Northwest Territories, Canada. (After Seigel, 1967.)

zinc deposits. The host rocks are carbonates and the background IP is generally low and uniform. With no additional information, try to answer the following questions.

(a) What type of electrode array was used?
(b) Was the electrode separation relatively large or small?
(c) Is the anomaly caused by electrode or membrane polarization?
(d) Is the anomalous source deep, shallow, wide, of great depth extent?
(e) Would you recommend further geophysical work, and if so, what?
(f) Would you drill this anomaly, and if so, where?

3. In the course of sulphide exploration in northwestern Quebec, both frequency- and time-domain IP techniques were employed. Figure 9.18 shows pseudo-depth sections for PFE, metal factor and chargeability from a particular line traverse; as noted, the double-dipole array was used in both cases, with 100 ft separation. Compare the results obtained with the two methods and make whatever interpretation you can from all the data. What is the significance of the negative chargeability values?

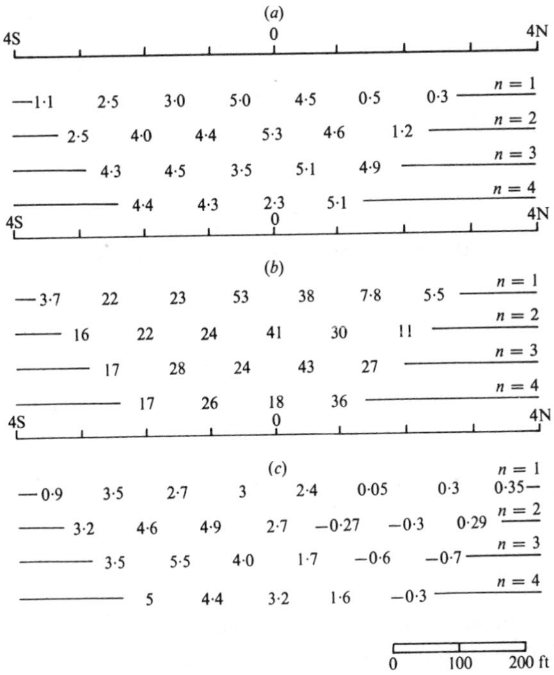

Fig. 9.18 Time- and frequency-domain IP pseudo-depth plots. Double-dipole array, $x = 100$ ft. (a) Per cent frequency effect; (b) metal factor (mhos/ft); (c) chargeability (msec).

4. Figure 9.19 shows chargeability contours from a time-domain IP survey carried out on a base-metal property in southern New Brunswick. From previous drilling, massive

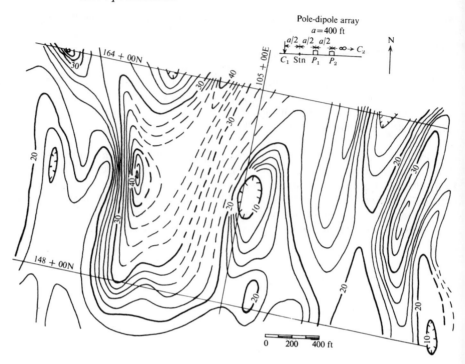

Fig. 9.19 Time-domain IP survey in Southern New Brunswick. Contour interval: 2 msec.

sulphide mineralization, striking N–S, had been found in the vicinity of line 105E, about the middle of the map; the zone was not very wide. Take off an E–W profile across the sheet around 156N. From this profile and the contours, make whatever interpretation you can of the data. Can you explain why the known mineral zone was not detected by IP?

5. Data for the metal-factor contours in fig. 9.20 were obtained from a survey in Nova Scotia, using the double-dipole array with $x = 100$ ft and $n = 1$. Make an interpretation of the area based on these results. Can you match this map with the one from Problem 10 in chapter 8 and if so is the additional information an aid to the interpretation?

6. A frequency-domain survey, similar to that in Problem 1, carried out over two lines on a property in Brazil, produced the results tabled below. The dipole separation was 50 metres with $n = 1, 2, 3$ 4. Lines are E–W and separated by 400 metres, with stations 50 metres apart; the current dipole was to the west in all cases. Resistivities are in ohm-metres.

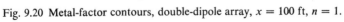

Fig. 9.20 Metal-factor contours, double-dipole array, $x = 100$ ft, $n = 1$.

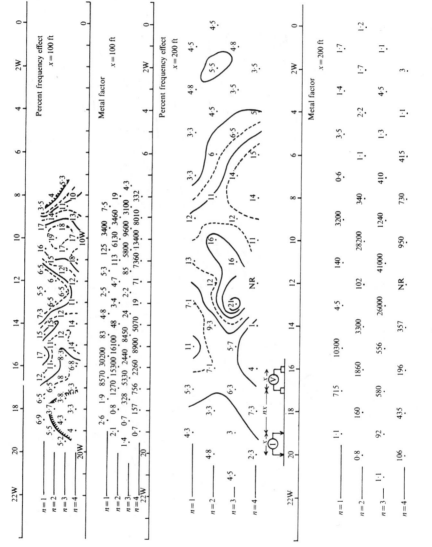

Fig. 9.21 Pseudo-depth plots for frequency-domain IP, Western Ontario.

Current dipole	$n = 1$			$n = 2$			$n = 3$			$n = 4$		
	ρ_a	PFE	MF	ρ_a	PFE	MF	ρ_a	PFE	MF	ρ_a	PFE	MF
Line 0												
44W–43W	228	1·6	13	390	1·8	9	637	1·6	5	250	4·8	37
43W–42W	248	1·7	13	520	1·5	5	217	4·5	40	71	14·5	390
42W–41W	220	1·2	10	128	3·0	45	44	14·5	630	29	11·5	740
41W–40W	76	2·5	62	34	13·5	750	22	11·5	1000	29	9·5	630
40W–39W	30	11	760	17	11·5	1330	19	9·8	980	34	8·4	470
39W–38W	36	3·7	195	38	3·5	175	59	4·0	130	65	4·2	124
38W–37W	114	1·5	25	217	1·5	13	275	2·2	15	340	1·5	8
37W–36W	190	0·2	2	305	0·7	4	410	1·0	5	650	1·2	3·5
Line 2S												
40W–39W	150	1·7	24	120	4·5	72	59	9·0	290	105	8·0	150
39W–38W	86	4·5	100	52	8·5	315	88	8·7	190	80	8·5	200
38W–37W	36	7·4	390	71	8·0	215	61	8·5	265	69	8·3	230
37W–36W	260	0·2	5	305	1·3	8	380	1·4	7	450	1·3	5·5
36W–35W	240	1·3	10	355	0·5	2·5	460	0·5	2	670	0·5	1

Plot these results in pseudo-depth contours in ρ_a, PFE and MF, and interpret the results.

7. Pseudo-depth plots for frequency effect and metal factor, on a base-metal prospect in Western Ontario are shown in fig. 9.21. Two spacings of double-dipole array were employed – 100 ft and 200 ft – as noted on the diagram. Contour the metal-factor data and compare the results with the PFE contours. Can you see any particular advantages in using two spreads? Is one more suitable for this particular job than the other? Interpret the data.

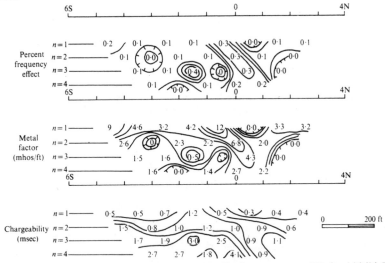

Fig. 9.22 Time- and frequency-domain IP, double-dipole array, $x = 100$ ft, Abitibi West, Quebec.

8. Figure 9.22 shows frequency- and time-domain contours in psuedo-depth for an area in the Abitibi West region of Quebec. As noted in the diagram, the double-dipole array had a separation of 100 ft with $n = 1, 2, 3, 4$. The IP results are obviously not very promising, particularly in the frequency domain. There is, however, a base-metal orebody here of economic grade. Can you make any estimate of its location, depth and width from the IP survey? Can you explain the poor response?

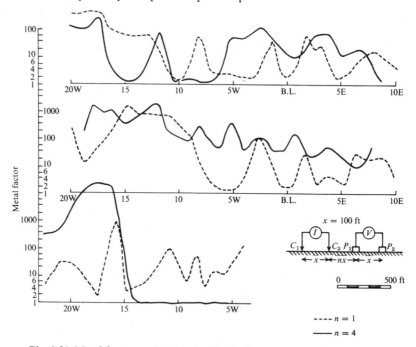

Fig. 9.23 Metal-factor profiles for double-dipole array, Eastern Nova Scotia.

9. Metal-factor profiles from part of an IP survey in Nova Scotia are displayed in fig. 9.23. The rather unusual representation is due to the fact that data for $n = 2, 3$ are not available; because of the large range in metal-factor values, it was necessary to use a log scale, which is somewhat deceptive because of the generally high background. Make what interpretation you can from the limited data.

References

F – field method; G – general; I – instruments; T – theoretical

Bleil, D. F. (1953). Induced polarization: a method of geophysical prospecting. *Geophysics*, **18**, 636–61. (I)

Dieter, K., Paterson, N. R. and Grant, F. S. (1969). IP and resistivity type curves for three-dimensional bodies. *Geophysics*, **34**, 615–32. (T)

Frische, R. H. and von Buttlar, H. (1957). A theoretical study of induced electrical polarization. *Geophysics*, **22**, 688–706. (T)

Hallof, P. G. (1960). The uses of induced polarization in mining exploration. *Trans. A.I.M.M.E.* (*Mining*), **217**, 319–27. (F)

Hallof, P. G. (1964). A comparison of the various parameters employed in the variable-frequency induced-polarization method. *Geophysics*, **29**, 425–33. (T)

Keevil, N. B. and Ward, S. H. (1962). Electrolyte activity: its effect on induced polarization. *Geophysics*, **27**, 677–90. (G)

Madden, T. R. and Cantwell, T. (1967). In *Mining Geophysics*, II, 373–400: Tulsa, Soc. of Explor. Geophys. (T)

Marshall, D. J. and Madden, T. R. (1959). Induced polarization: a study of its causes. *Geophysics*, **24**, 790–816. (T)

Mining geophysics, vol. I, chap. 3–4, vol. II, chap. 11 (parts B and D); 1966, 1967. Tulsa, Soc. of Explor. Geophysicists. (G)

Scott, W. J. (1965). M.Sc. thesis, Univ. of Toronto, Toronto. (I)

Seigel, H. O. (1959). Mathematical formulation and type curves for induced polarization. *Geophysics*, **24**, 547–65. (T)

Seigel, H. O. (1962). Induced polarization and its role in mineral exploration. *Can. Inst. Min. Met. Bull.*, **55**, 242–9. (F)

Seigel, H. O. (1967). The induced polarization method. In *Mining and Groundwater Geophysics* (ed. L. W. Morley). Econ. Geol. Report No. 26, Geol. Survey of Canada, 123–37.

Sumi, F. (1965). Prospecting for non-metallic minerals by induced polarization. *Geophys. Prosp.* **13**, 603–16. (T)

Vacquier, V., Holmes, C. R., Kintzinger, P. P. and Lavergne, M. (1957). Prospecting for ground water by induced electrical polarization. *Geophysics*, **22**, 660–87. (F)

Wait, J. R. (ed.) (1959). *Overvoltage research and geophysical applications*. London, Pergamon. (G)

10. Radioactivity method

10.1 Introduction

The geophysical techniques described in previous chapters have depended upon variations in the mechanical, chemical, electrical or magnetic properties of rocks and minerals. Within the last twenty-five years another property of certain elements has become of great economic importance. This property is known as *radioactivity*.

The original discovery was made by Becquerel in 1896, shortly after Röentgen had announced in 1895 the discovery of X-rays. Becquerel found that minerals containing uranium, as well as salts of uranium, emitted radiations which passed through material opaque to ordinary light and which affected photographic emulsions in a manner similar to X-rays. He further found that the radiation was characteristic of the particular uranium compound and that it would ionize a gas.

The discovery of other radioactive elements soon followed. Mme Curie, investigating minerals of uranium, extracted two new elements, polonium and radium, which were much more active than uranium. About the same time Schmidt discovered that thorium was radioactive and Debierne found the new radioactive element actinium.

Although at least twenty naturally occurring elements are now known to be radioactive, only uranium (U), thorium (Th) and an isotope of potassium (K) are of importance in exploration. One other, rubidium, is useful in determining ages of rocks, but the rest are either so rare or so weakly radioactive, or both, as to be of no significance in applied geophysics. A complete list, with characteristic radiations and other pertinent data, is given in table 10.1.

The two elements, uranium and thorium, are important today as a source of fuel for the generation of heat and power in nuclear reactors. Thousands of square miles have been surveyed from the air and on the ground in all parts of the world in the search for uranium, using special detectors which will be discussed later.

The radioactive method is relatively unimportant in comparison with other geophysical techniques. It was first used in the late thirties for stratigraphic correlation in oil well logging (see §11.3.2). Radioactivity prospecting became quite popular in the period 1945–57, fell off with the decrease in demand for uranium, and was revived again in the late sixties.

10.2 Principles of radioactivity

10.2.1 *Constituents of the nucleus*

(a) *Introduction.* Although much of the original work on emanations from radioactive substances was done by Rutherford and others nearly 70 years ago, their source – the nucleus of the atom – was not well understood at the time. We shall now consider this source and its elementary parts.

(b) *Atoms.* The *atom*, which is the fundamental part of all the elements, consists of a dense, small ($\sim 10^{-13}$ cm in radius), positively charged nucleus surrounded by negatively charged electrons, in number equal to the nuclear charge. The arrangement is quite analogous to a solar system, with planets moving about a central sun. Since there are never more than 92 electrons and since the atomic radii are of the order 10^{-8} cm, most of the atom, like the solar system, is empty.

(c) *Protons.* The nucleus is composed of tightly packed *protons* and *neutrons*. The proton, carrying unit positive charge, has a mass $1 \cdot 00812$ on the physical scale ($0 = 16 \cdot 0000 \ldots$), the actual mass being $1 \cdot 7 \times 10^{-24}$ g. The number of protons in a nucleus determines the element itself. For example, the first element in the periodic table, hydogen, has one proton, oxygen has 8, cadmium 48 and so on up to uranium, with 92 protons.

(d) *Neutrons.* The other nuclear particle, the neutron, has zero charge and a slightly greater mass than the proton ($1 \cdot 00893$). The only element lacking neutrons is common hydrogen. As we proceed through the periodic table, the ratio of the number of neutrons to the number of protons increases from 1 to about $1 \cdot 5$. Thus helium has two neutrons and two protons, while uranium contains 146 neutrons and 92 protons.

(e) *Isotopes.* Most elements are composed of a mixture of nuclei having different numbers of neutrons, the number of protons, of course, being the same. These are called *isotopes*, that is, forms of the same element having different atomic weights. (Practically all the mass of an element is contained in the nucleus, hence is determined by the number of protons and neutrons.) For instance, hydrogen is a mixture of two isotopes: $_1H^1$, which is a single proton ($99 \cdot 985 \%$ abundance), and $_1H^2$, 1 proton and 1 neutron, familiarly known as deuterium ($0 \cdot 015 \%$ abundance). Titanium has 5 isotopes, tin has 10, tungsten 5, lead 4 and so on.

(f) *Alpha particles.* Actually these are the equivalent of a helium nucleus, $2p + 2n$; the name was attached in the pioneer days of radioactivity, before the nature of the particle was understood. It has a charge $+2$, mass $4 \cdot 00389$ and is frequently a tightly bound entity within nuclei heavier than helium. It may be ejected from the nucleus during a disintegration.

(g) *Electrons.* The outer atomic constituent, the electron, has a charge -1 and mass about $1/1840$ of the proton. Although the electron does not exist as a separate entity in the nucleus, it is ejected in certain nuclear disintegrations when a neutron splits into a proton and an electron, the proton remaining in the nucleus. This transmutation results in a gain of $+1$ unit of charge and practically zero mass change, that is to say, the element moves up one place in the periodic table. Electrons ejected from the nucleus were originally called *beta* (β) particles or rays.

(h) *Gamma radiation.* During nuclear disintegrations, pure electromagnetic radiation, representing excess energy, is frequently emitted from the excited nucleus. The early name assigned, *gamma-ray* (γ-ray) is quite appropriate in this case (α- and β-rays are really discrete particles). They differ from X-rays only in name, although usually the latter term is used for radiation of lower energy. The relative location of γ-rays in the electromagnetic spectrum is illustrated in fig. 10.1.

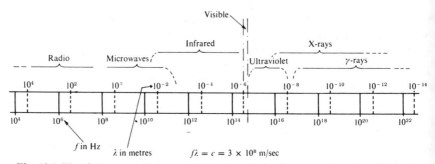

Fig. 10.1 The electromagnetic spectrum showing relative frequency (wavelength) bands.

10.2.2 *Nuclear disintegrations*

While carrying on pioneer work in nuclear physics, Sir Ernest Rutherford investigated the radiations from naturally occurring radioactive elements and showed that they consisted of the three distinct types mentioned above, α-, β-, and γ-rays. Each of these rays produces three different effects in varying degrees, namely:

(i) they affect photographic emulsions in much the same way as light and X-rays,
(ii) they ionize gas, making it electrically conducting,
(iii) they produce scintillations or phosphorescence in certain minerals and chemical compounds.

All three effects are used in geophysical prospecting by the radioactivity method.

The three 'rays' characteristic of natural nuclear disintegrations have very different penetrating powers. Thus, α-rays are easily stopped by a sheet of paper, β-rays by a few millimetres of aluminium, whereas γ-radiation requires several inches of lead. Their equivalent range in overburden or rock is thus practically zero for the first two and not more than a foot or two of rock for γ-rays.

In fact, this range is a complicated function of the energy and character of the particles or radiation and of the density or atomic number of the medium through which they pass. It is clear that the range varies with initial energy and the rate of dissipation of energy. The latter is a complex process of scattering, collision and absorption involving the atoms of the host material and resulting in ionization along the path. Charged particles (α, β) ionize strongly, uncharged electromagnetic radiations (γ-rays, X-rays) do not.

Maximum energy in natural nuclear disintegrations is generally less than 3 MeV (1 MeV = 10^6 electron-volts, the energy acquired by a particle of unit charge, falling through a potential of 10^6 volts). Even in air, the range of 3 MeV α- and β-particles is only a few centimetres and feet respectively. On the other hand, γ-rays of this energy will travel several hundred feet in air.

Table 10.1. *Naturally occurring radioactive isotopes*

Element	Isotope	Abundance %	Half-life (years)	Type of radiation	Energy (MeV)
Potassium	$_{19}K^{40}$	0·012	$1·3 \times 10^9$	β,K — cap $+ \gamma^*$	1·46
Calcium	$_{20}Ca^{48}$	0·18	$> 2 \times 10^{16}$	β	0·12
Vanadium	$_{23}V^{50}$	0·24	6×10^{15}	β, K — cap $+ \gamma^*$	0·71, 1·59
Rubidium	$_{37}Rb^{87}$	27·8	$4·7 \times 10^{10}$	β	0·27
Indium	$_{49}In^{115}$	95·72	6×10^{14}	β	0·60
Lanthanum	$_{57}La^{138}$	0·089	$1·1 \times 10^{11}$	β, K — cap $+ \gamma^*$	0·54, 0·81, 1·43
Cerium	$_{58}Ce^{142}$	11·1		α	1·5
Neodymium	$_{60}Nd^{144}$	23·8	5×10^{15}	α	1·8
Samarium	$_{62}Sm^{147}$	14·97	10^{11}	α	2·32
Samarium	$_{62}Sm^{148}$	11·2	$1·2 \times 10^{13}$	α	2·14
Samarium	$_{62}Sm^{149}$	13·8	$\sim 4 \times 10^{14}$	α	1·84
Gadolinium	$_{64}Gd^{152}$	0·2	$1·1 \times 10^{14}$	α	2·24
Lutecium	$_{71}Lu^{176}$	2·6	3×10^{10}	β, γ	0·088, 0·20, 0·31
Hafnium	$_{72}Hf^{174}$	0·16	2×10^{15}	α	2·5
Rhenium	$_{75}Re^{187}$	62·9	7×10^{10}	β	$\leqslant 0·008$
Platinum	$_{78}Pt^{190}$	0·013	6×10^{11}	α	3·11
Platinum	$_{78}Pt^{192}$	0·78	$\sim 10^{15}$	α	2·6
Lead	$_{82}Pb^{204}$	1·48		α	
Thorium**	$_{90}Th^{232}$	100	$1·39 \times 10^{10}$	α, β, γ	0·03–2·62
Uranium**	$_{92}U^{235}$	0·72	$7·1 \times 10^8$	α, β, γ	0·02–0·9
Uranium**	$_{92}U^{238}$	99·3	$4·5 \times 10^9$	α, β, γ	0·4–2·5

* K-electron capture followed by γ-ray emission.
** Each of these undergoes a long series of disintegrations yielding lead isotopes 208, 207, 206, respectively. During these disintegrations numerous γ-rays are emitted, in addition to the α- and β-particles.

In addition to α-, β- and γ-emissions, there is one other type of nuclear transmutation, called *K-capture* (see table 10.1 and fig. 10.2), which occurs in several of the natural radioelements. In this process, an electron from the innermost *K*-orbit enters the nucleus, which then emits γ-rays; as a result of the electron capture the atomic number decreases by one and a different element is created.

The equations representing transitions of element X to Y by α- and β-ray emission and electron capture are given below:

$$_p X^{p+n} \rightarrow {_{p-2}} Y^{p+n-4} + {_2}He^4 \quad (\alpha\text{-emission})$$
$$_p X^{p+n} \rightarrow {_{p+1}} Y^{p+n} + e^- \quad (\beta\text{-emission})$$
$$_p X^{p+n} + e^- \rightarrow {_{p-1}} Y^{p+n} \quad (K\text{-capture})$$

10.2.3 *Radioactive decay processes*

In 1902 Rutherford and Soddy announced the theory of radioactive transformation, in which they stated that when an element emitted α- or β-rays, it was transmuted into a new element, the rate of disintegration being a characteristic of each radioactive nucleus. They showed that the rate of change was proportional

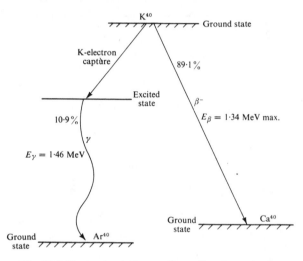

Fig. 10.2 Energy-level diagram for radioactive potassium.

to the number of atoms present and was not affected by physical or chemical processes in the surroundings.

Thus, for any type of radioactive atom, we have the relation

$$dN/dt = -\lambda N,$$

where N is the number of atoms present at time t and λ a decay constant which is characteristic of each element. Therefore,

$$\int_{N_0}^{N} \frac{dN}{N} = -\lambda \int_{0}^{t} dt, \quad \text{or} \quad N = N_0 e^{-\lambda t}, \tag{10.1}$$

where N_0 is the number of atoms at an arbitrary time $t = 0$. If $T_{\frac{1}{2}}$ is the time required for half of the nuclei to disintegrate, we have

$$N/N_0 = \tfrac{1}{2} = e^{-\lambda T_{\frac{1}{2}}}, \quad \text{or} \quad \lambda = (\log 2)/T_{\frac{1}{2}} = 0.693/T_{\frac{1}{2}}. \tag{10.2}$$

Half-life values of radioactive nuclei vary enormously, from $_{84}Po^{212} \approx 10^{-7}$ sec to $_{82}Pb^{204} \approx 10^{19}$ years. Obviously a short half-life goes with a vigorous rate of disintegration, while the lead isotope 204 above is, for all practical purposes, stable – the disintegration rate would be three or four nuclei per week per gram.

As mentioned earlier, only three radioactive elements, U, Th, and K are of practical significance in prospecting. The potassium is mainly a nuisance when searching for the other two; although the K^{40} isotope is, apparently, no more plentiful than U or Th, the widespread occurrence of potassium-rich rocks and particularly the association of these with U and Th, e.g. in pegmatites, creates a problem somewhat analogous to that of graphite versus metal sulphides in electrical prospecting.

As noted in tables 10.1 and 10.2, there are three radioactive series for uranium and thorium, starting with $_{90}Th^{232}$, $_{92}U^{235}$ (the so-called actinium series) and $_{92}U^{238}$. All decay eventually to stable isotopes of lead, with 10, 15 and 15 intermediate radioactive stages respectively.

It is useful to calculate the number of daughter atoms present at any time, given the number N_0 of the parent at time $t = 0$. Then the number of parent atoms left at a later time t will be: $N_1 = N_0 e^{-\lambda_1 t}$, where λ_1 is its decay constant. But the rate of decay of the parent atoms, $dN_1/dt = -\lambda_1 N_1$, is just the rate of production of the daughter. At the same time the daughter atoms are disintegrating at a rate $\lambda_2 N_2$, where N_2 is the number present at time t and λ_2 its decay constant. Hence the rate of accumulation of the daughter atoms is the difference between production and decay, or

$$dN_2/dt = \lambda_1 N_1 - \lambda_2 N_2 = \lambda_1 N_0 e^{-\lambda_1 t} - \lambda_2 N_2. \qquad (10.3)$$

We can solve this equation by assuming $N_2 = A e^{-\lambda_1 t} + B e^{-\lambda_2 t}$, with the condition that, when $t = 0$, $N_2 = 0$. The result is

$$N_2 = \frac{\lambda_1 N_0}{\lambda_2 - \lambda_1}(e^{-\lambda_1 t} - e^{-\lambda_2 t}). \qquad (10.4)$$

This calculation can be carried on for successive members of the series. The number of atoms of the nth product produced after time t is given by

$$N_n = C_1 e^{-\lambda_1 t} + C_2 e^{-\lambda_2 t} + \ldots C_n e^{-\lambda_n t},$$

where

$$\left.\begin{aligned}
C_1 &= \lambda_1 \lambda_2 \ldots \lambda_{n-1} N_0/(\lambda_2 - \lambda_1)(\lambda_3 - \lambda_1) \ldots (\lambda_n - \lambda_1), \\
C_2 &= \lambda_1 \lambda_2 \ldots \lambda_{n-1} N_0/(\lambda_1 - \lambda_2)(\lambda_3 - \lambda_2) \ldots (\lambda_n - \lambda_2), \\
&\ldots \\
C_n &= \lambda_1 \lambda_2 \ldots \lambda_{n-1} N_0/(\lambda_1 - \lambda_n)(\lambda_2 - \lambda_n) \ldots (\lambda_{n-1} - \lambda_n).
\end{aligned}\right\} \qquad (10.4a)$$

10.2.4 *Radioactive equilibrium*

From eq. (10.3) it follows that when a radioactive series is in equilibrium, we have

$$\lambda_1 N_1 = \lambda_2 N_2 = \lambda_3 N_3 = \ldots \lambda_n N_n. \qquad (10.5)$$

That is to say, at equilibrium the number of daughter atoms disintegrating per second is the same as the number being created by disintegrations of the parent. This state of radioactive equilibrium merits further explanation. Consider radium and radon, the successive intermediate products in the U^{235} series (see $_{88}Ra^{223}$ and $_{86}Rn^{219}$, table 10.2, and fig. 10.3). Here the daughter product decays about 10^5 times faster than its parent. If we start with a sample of pure radium, we find that its decay rate is practically constant for the first day or two, since the half-life is about 12 days. At the same time the supply of radon atoms is building up at the same rate, although the radon is decaying considerably faster than the radium. From eqs. (10.1) and (10.4) we can get the ratio of the number of atoms of parent to daughter at any time

$$\frac{N_2}{N_1} = \frac{\lambda_1}{\lambda_2 - \lambda_1} \{1 - e^{(\lambda_1 - \lambda_2)t}\}.$$

When equilibrium has been reached, $N_2/N_1 = \lambda_1/\lambda_2$. Thus we have

$$\frac{\lambda_1}{\lambda_2} = \frac{\lambda_1}{\lambda_2 - \lambda_1} \{1 - e^{(\lambda_1 - \lambda_2)t_{eq}}\}, \quad \text{or} \quad e^{(\lambda_1 - \lambda_2)t_{eq}} = \lambda_1/\lambda_2,$$

hence,

$$t_{eq} = \left(\frac{1}{\lambda_1 - \lambda_2}\right) \log\left(\frac{\lambda_1}{\lambda_2}\right) \tag{10.6a}$$

For this example the value of t_{eq} is about one minute, after which the two will be in equilibrium, as long as the radium holds out.

In the case of a series with n products, the time to reach equilibrium can be found from equation (10.4a). It is

$$\frac{N_n}{N_1} = K\left\{\frac{1}{\delta_1} + \frac{e^{(\lambda_1 - \lambda_2)t}}{\delta_2} + \frac{e^{(\lambda_1 - \lambda_3)t}}{\delta_3} + \ldots \frac{e^{(\lambda_1 - \lambda_n)t}}{\delta_n}\right\} = \frac{\lambda_1}{\lambda_n}, \tag{10.6b}$$

where

$$K = \lambda_1\lambda_2\lambda_3\ldots\lambda_{n-1},$$
$$\delta_1 = (\lambda_2 - \lambda_1)(\lambda_3 - \lambda_1)\ldots(\lambda_n - \lambda_1),$$
$$\delta_2 = (\lambda_1 - \lambda_2)(\lambda_3 - \lambda_2)\ldots(\lambda_n - \lambda_2),$$
$$\ldots$$
$$\delta_n = (\lambda_1 - \lambda_n)(\lambda_2 - \lambda_n)\ldots(\lambda_{n-1} - \lambda_n).$$

For thorium this time interval is less than 100 years and for the two uranium chains, of the order 10^6 years. Measurement of series products, in which the equilibrium situation is significant, will be discussed later; under these conditions it is possible to determine the amount of a parent product in a sample by measuring the amount of one of the succeeding members.

Table 10.2 and fig. 10.3 show the three radioactive series in detail, with the principal radiation acccompanying each disintegration. There is in addition a wide spectrum of γ-rays accompanying both α - and β-emission, some of which are

Table 10.2. *Natural radioactive series of thorium and uranium*

Element	Isotope	Half-life	Decay constant sec^{-1}	Radiation	γ-ray energies (MeV)	No. of γ-rays
Thorium	$_{90}Th^{232}$	1·4 × 10^{10} yr	1·58 × 10^{-18}	α, SF*, γ	0·059	
Radium	$_{88}Ra^{228}$	6·7 yr	3·3 × 10^{-9}	β, γ	0·03	
Actinium	$_{89}Ac^{228}$	6·1 hr	3·1 × 10^{-4}	β, γ	0·06–0·97	>10
Thorium	$_{90}Th^{228}$	1·91 yr	1·15 × 10^{-8}	α, γ	0·085–0·214	5
Radium	$_{88}Ra^{224}$	3·64 day	2·2 × 10^{-6}	α, γ	0·24, 0·29	
Radon	$_{86}Rn^{220}$	51 sec	1·3 × 10^{-2}	α, γ	0·54	
Polonium	$_{84}Po^{216}$	0·16 sec	4·3	α		
Lead	$_{82}Pb^{212}$	10·6 hr	1·8 × 10^{-5}	β, γ	0·11–0·41	5
Bismuth	$_{83}Bi^{212}$	60·6 min	1·9 × 10^{-4}	β, α, γ	0·04–2·2	>10
Polonium	$_{84}Po^{212}$	0·3 × 10^{-6} sec	2·3 × 10^{6}	α		
Thallium	$_{81}Tl^{208}$	3·1 min	3·7 × 10^{-3}	β, γ	0·28–2·62	5
Lead	$_{82}Pb^{208}$	Stable				
Uranium	$_{92}U^{235}$	7·1 × 10^{8} yr	3·1 × 10^{-17}	α, SF*, γ	0·07–0·38	10
Thorium	$_{90}Th^{231}$	25·6 hr	7·4 × 10^{-6}	β, γ	0·08–0·31	>10
Protact.	$_{91}Pa^{231}$	3·4 × 10^{4} yr	6·5 × 10^{-13}	α, γ	0·29–0·36	>10
Actinium	$_{89}Ac^{227}$	21·6 yr	10^{-9}	β, α, γ	0·09–0·19	9
Thorium	$_{90}Th^{227}$	18·2 day	4·35 × 10^{-7}	α, γ	0·05–0·33	>10
Francium	$_{87}Fr^{223}$	22 min	5·2 × 10^{-4}	β, α, γ	0·05–0·31	4
Radium	$_{88}Ra^{223}$	11·7 day	6·76 × 10^{-7}	α, γ	0·03–0·45	>10
Radon	$_{86}Rn^{219}$	4 sec	0·17	α, γ	0·27, 0·4	
Astatine	$_{85}At^{219}$	54 sec	1·28 × 10^{-2}	α, β		
Polonium	$_{84}Po^{215}$	1·8 × 10^{-3} sec	3·8 × 10^{2}	α, β		
Astatine	$_{85}At^{215}$	10^{-4} sec	6·9 × 10^{3}	α		
Bismuth	$_{83}Bi^{215}$	8 min	1·44 × 10^{-3}	β		
Bismuth	$_{83}Bi^{211}$	2·15 min	5·35 × 10^{-3}	α, β, γ	0·35	
Polonium	$_{84}Po^{211}$	0·52 sec	1·32	α, γ	0·56, 0·88	
Lead	$_{82}Pb^{211}$	36 min	3·2 × 10^{-4}	β, γ	0·065–0·83	4
Thallium	$_{81}Tl^{207}$	4·8 min	2·4 × 10^{-3}	β, γ	0·89	
Lead	$_{82}Pb^{207}$	Stable				
Uranium	$_{92}U^{238}$	4·51 × 10^{9} yr	4·9 × 10^{-18}	α, SF*, γ	0·048	
Thorium	$_{90}Th^{234}$	24·1 day	3·3 × 10^{-7}	β, γ	0·03–0·09	3
Protoact.	$_{91}Pa^{234}$	6·7 hr	2·84 × 10^{-5}	β, γ	0·044–1·85	>10
Uranium	$_{92}U^{234}$	2·48 × 10^{5} yr	8·9 × 10^{-14}	α, SF*, γ	0·053, 0·118	
Thorium	$_{90}Th^{230}$	8 × 10^{4} yr	2·75 × 10^{-10}	α, γ	0·068–0·25	7
Radium	$_{88}Ra^{226}$	1622 yr	1·35 × 10^{-11}	α, γ	0·19–0·64	4
Radon	$_{86}Rn^{222}$	3·82 day	2·07 × 10^{-6}	α, γ	0·51	
Polonium	$_{84}Po^{218}$	3·05 min	3·8 × 10^{-3}	α, β		
Astatine	$_{85}At^{218}$	1·35 sec	0·51	α		
Radon	$_{86}Rn^{218}$	0·03 sec		α	0·61	
Bismuth	$_{83}Bi^{214}$	19·7 min	5·85 × 10^{-4}	β, α, γ	0·45–2·43	>10
Polonium	$_{84}Po^{214}$	1·64 × 10^{-4} sec	4·2 × 10^{3}	α		
Lead	$_{82}Pb^{214}$	26·8 min	4·3 × 10^{-4}	β, γ	0·05–0·35	>10
Lead	$_{82}Pb^{210}$	21 yr	1·05 × 10^{-9}	β, γ	0·047	
Bismuth	$_{83}Bi^{210}$	5 day	1·58 × 10^{-6}	β		
Polonium	$_{84}Po^{210}$	138·4 day	5·7 × 10^{-8}	α, γ	0·79	
Thallium	$_{81}Tl^{210}$	1·3 min	8·85 × 10^{-3}	β, γ	0·3, 0·78, 1·1	
Thallium	$_{81}Tl^{206}$	4·2 min		β		
Lead	$_{82}Pb^{206}$	Stable				

SF* = spontaneous fission.

included in the table (see also table 11.1, §11.3.1*b*). The thorium series has an isolated γ-ray from Tl^{208} at 2·62 MeV; the uranium series do not have such well isolated β-rays, although the 1·76 MeV β-ray from $_{83}Bi^{214}$ is reasonably distinctive.

10.2.5 *Units*

The unit used for measuring the activity of a radioactive specimen is the *curie*, named for the discoverer of radium, Mme Curie. It is the activity which results in 3·7 × 10^{10} disintegrations per second, this being the number of α-particles emitted by one gram of pure radium, $_{88}Ra^{226}$, in one second. Multiple and submultiple units are also used, for example, kilocurie (1000), millicurie (10^{-3}) and microcurie (10^{-6}).

Since γ-rays are similar in nature to X-rays, the strength or intensity of gamma radiation (as well as α- and β-particles) is also measured in the X-ray unit, called the *röentgen*. This is the quantity of radiation which will produce one electrostatic unit of charge (2·08 × 10^9 ion pairs) per cm^3 in air at 0°C and 760 mm Hg (NTP).

Table 10.3. *Radioactive minerals*

Potassium	Mineral	(i)	Orthoclase and microline feldspars $[KAlSi_3O_8]$
		(ii)	Muscovite $[H_2KAl(SiO_4)_3]$
		(iii)	Alunite $[K_2Al_6(OH)_{12}SO_4]$
		(iv)	Sylvite, carnallite $[KCl, MgCl_2 . 6H_2O]$
	Occurrence	(i)	Main constituents in acid igneous rocks and pegmatites
		(ii)	Same
		(iii)	Alteration in acid volcanics
		(iv)	Saline deposits in sediments
Thorium	Mineral	(i)	Monazite $[ThO_2 +$ Rare earth phosphate]
		(ii)	Thorianite $[(Th, U)O_2]$
		(iii)	Thorite, uranothorite $[ThSiO_4 + U]$
	Occurrence	(i)	Granites, pegmatites, gneiss
		(ii), (iii)	Granites, pegmatites, placers
Uranium	Mineral	(i)	Uraninite [Oxide of U, Pb, Ra + Th, Rare earths]
		(ii)	Carnotite $[K_2O . 2UO_3 . V_2O_5 . 2H_2O]$
		(iii)	Gummite [Uraninite alteration]
	Occurrence	(i)	Granites, pegmatites and with vein deposits of Ag, Pb, Cu, etc.
		(ii)	Sandstones
		(iii)	Associated with uraninite

Subunits are the milliröentgen (mr) and microröentgen (μr). This is the unit used in defining maximum dosage permissible to humans exposed to radioactivity, about 300 mr/week.

Some field instruments indicate radioactivity as counts per minute, generally marked on the scale of a microammeter in an integrating circuit which adds up pulses to measure total intensity. None of these units takes into account the energy of the radiation.

10.2.6 *Radioactivity of rocks and minerals*

Some of the common radioactive minerals of Th and U are listed in table 10.3. The potassium minerals, as mentioned previously, are very widespread. Large deposits of monazite are found in Brazil, India and South Africa. Thorite and uraninite (pitchblende) occur particularly in Canada (Great Bear Lake, northern Saskatchewan; Blind River, Ontario), in Zaïre, Central Europe (Saxony and Czechoslovakia), Malagasy, etc.

Trace quantities of radioactive material are found in all rocks. Along with minute amounts of cosmic radiation always present in the air, these trace amounts produce a continuous background reading, which may vary from place to place by as much as a factor of five. The following table gives the activity and/or trace amounts of radioactivity of a number of typical rocks, as well as the amount of radium in waters.

10.4. *Background radioactivity in rocks and waters*

Rock	Curies/gm ($\times 10^{-12}$)	K (ppm)	Th (ppm)	U (ppm)	Water (radium)	Curies/gm ($\times 10^{-12}$)
Hornblende	1·2				Saratoga, N.Y.	0·01–0·1
Granite	0·7–4·8	35,000	15	4	Bath, England	0·14
Basalts	0·5	9000	2	0·6	Carlsbad, Czech.	0·04–0·1
Olivine	0·33				St Lawrence River	0·00025
Ultramafics		10	0·2	0·05	Valdemorillo, Spain	0·02
Marble	1·9				Aix-les-bains, France	0·002
Quartzite	5·0				Manitou, Colorado	0·003
Sandstone	2–4				Hot Springs, Ark.	0·0009
Slates	3–8				Atlantic Ocean	0·014–0·034
Dolomites	8				Indian Ocean	0·007
Chalk	0·4					
Chondrites		850	0·08	0·02		
Iron meteor.			0·015	0·04		

In general the activity in sedimentary rocks and metamorphosed sediments is higher than that in igneous and other metamorphic types, with the exception of potassium-rich granites.

10.3 Instruments

10.3.1 *Introduction*

Various devices have been used for the detection of radioactivity. One of the earliest was the *ionization chamber*. At present there are two principal instruments, the *Geiger counter* and the *scintillation meter*, plus the *pulse-height analyser* or

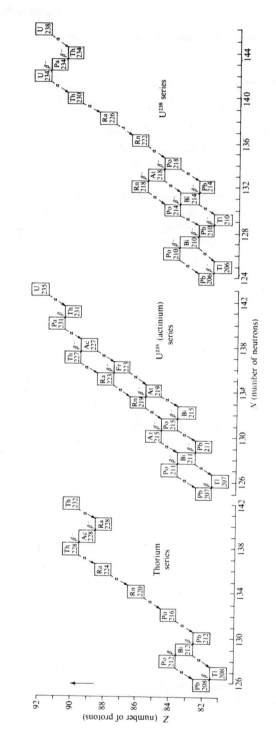

Fig. 10.3 Uranium and thorium radioactive series.

γ-ray spectrometer, which is an extension of the scintillation meter. Both these detectors also were used very early in the course of radioactivity study. In their present prospecting form they are adaptations of laboratory instruments developed in the period 1944–50.

The heart of the radioactive detector is a device which will respond efficiently to β- and γ-radiation (α-particles have such short ranges that they need not be considered generally). As noted previously, natural β-rays also have a short range, even in air; hence a β-detector is effective only within a few feet of the source. Consequently gamma-radiation detection is most desirable.

10.3.2 *Ionization chamber*

Used very seldom now, except in some well-logging devices for neutron detection, the ionization chamber is similar in construction to the Geiger counter, consisting of a cylindrical thin-walled metal case surrounding an axial metal rod. The latter is usually maintained a few hundred volts negative with respect to the case and acts as a collector of positive ions. The sensitivity can be made very large for particular radiation by coating the chamber walls with cadmium or boron, or by using a gas such as BF_3, for detection.

10.3.3 *Geiger-Müller counter*

This is a very simple device which responds primarily to β-radiation. Consequently it can be used only in ground traversing. A diagram showing the essential parts is given in fig. 10.4*a*. Like the ionization chamber, the detector is a thin-walled cylindrical tube, often with a very thin ($\leqslant 0.001$ in.) mica window in the end, to permit the passage of β-particles.

The tube contains an axial anode wire with a coaxial cathode cylinder and is filled to a pressure of about 0·1 atmosphere with an inert gas, such as argon, plus a trace of alcohol, methane, water vapour, or a combination of these. This gas mixture produces a quenching action. A dry cell or other low-current source supplies several hundred volts across the diode, as shown in the diagram.

Radiation entering the tube ionizes gas atoms and the positive ions and electrons are accelerated by the high voltage to the cathode and anode respectively. These charges also ionize other gas atoms en route. The ionization is cumulative and the original ray produces a discharge pulse across the anode resistor, which is amplified in the transistor stage to produce a click in the headphones. Figure 10.4*a* also shows a simple integrating circuit in series with the headset. Successive pulses charge up the condenser which then leaks off slowly through the high resistance R, in series with the microammeter A. The meter registers a current proportional to the integral of the charge entering the condenser.

The purpose of the quenching agent is to suppress secondary electron emission from the cathode, caused by positive ion bombardment. This effect tends to prolong the discharge. Fast quenching of the discharge allows the tube to return quickly to the non-conducting state and hence respond to succeeding rays entering the chamber shortly after the first. Although even in a good tube the clean-up time is still appreciable, it can be reduced to less than 100 μsec with a suitable quenching gas mixture.

The electronic section of fig. 10.4*a* is oversimplified. Normally the Geiger tube pulse drives a multivibrator circuit which shapes the pulses as well as amplifying them.

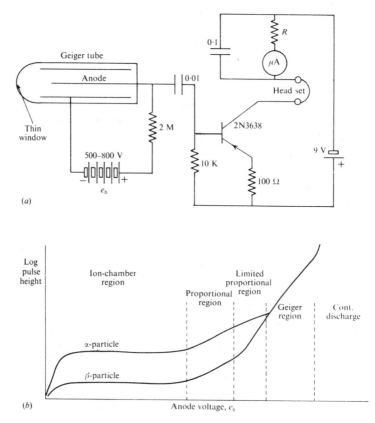

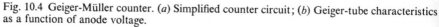

Fig. 10.4 Geiger-Müller counter. (*a*) Simplified counter circuit; (*b*) Geiger-tube characteristics as a function of anode voltage.

Figure 10.4*b* shows the characteristics of the Geiger tube as a function of anode voltage. Normally the voltage is kept within the plateau (ion-chamber) region of the curve, where pulse height is independent of the voltage, thus reducing the effect of variations in battery voltage.

The prospecting Geiger counter has the virtue of being simple and cheap (about $100). However, it has little else to recommend it. It must be held close to the outcrop to detect β-rays (since it is an extremely inefficient detector of γ-rays, which are weak ionizers and tend to pass right through the tube without being registered). Lead fins have been mounted on the outside of the tube to degrade and convert the γ-rays to β-rays; however, this has not improved the efficiency enough to make the instrument competitive with the scintillation meter. Thus the Geiger counter remains a tool of limited application.

10.3.4 *Scintillation meter*

(a) *General.* The counting of scintillations produced by radiation bombardment of a zinc sulphide screen was one of the earliest methods of detection. Other materials which have been used for this purpose include anthracene, stilbene and scheelite. One of the best scintillation detectors is made by growing natural crystals of sodium iodide (NaI), treated with thallium (Tl). The NaI is, of course, transparent to its own fluorescent emission and all faces but one are coated with light reflecting material. If the crystal is large enough, its conversion efficiency for γ-rays is practically 100%. A portable device of this type became possible following the development of the photomultiplier tube.

(b) *Gamma-ray interactions.* As mentioned briefly in §10.2.2, the dissipation of energy as radiation passes through matter is a complex process. In order to explain the operation of the scintillation meter, it is necessary to discuss the sequence of events in which the radiation is absorbed. The interaction of γ-radiation with matter takes place by the following processes (see also §11.3.1c).

(i) The *photoelectric effect*, in which the γ-ray loses all of its energy to a bound atomic electron, part of the energy being used to overcome its binding to the atom, while the remainder appears as kinetic energy of the electron. This effect predominates at low energy ($\leqslant 200$ keV) although it also varies greatly with the atomic number of absorbing material.

(ii) *Compton scattering* by atomic electrons, in which the γ-ray is deflected in its path. When the γ-ray energy is much larger than the electron binding energy (which varies from $\sim 10^5$ eV for innermost K electrons of heavy elements to a few electron volts in light elements) the scattering takes place as though the electrons were unbound and at rest. This is the dominant interaction at intermediate energies (100 keV–2 MeV) and the effect of atomic number is not so pronounced.

(iii) *Pair production*, in which the γ-ray is annihilated near a nucleus or electron while creating an electron-positron (positive electron) pair. The energy required for this process must be greater than the rest energy (energy equivalent to the mass) of the pair; any excess appears as kinetic energy of the electron and positron. Since the electron rest energy is $0\cdot51$ MeV, pair production cannot take place unless the γ-ray energy originally was larger than $1\cdot02$ MeV. Hence it is essentially a high energy phenomenon.

Of these three modes of interaction, the first is most desirable for γ-ray spectroscopy (see next section), since the original radiation is converted to a light photon, giving up all its energy in the process. For the ordinary scintillation meter, however, the only requirement is that the input γ-rays be eventually converted to light, regardless of the mode of absorption.

(c) *Description of the scintillation meter.* A schematic of the scintillation meter is shown in fig. 10.5. Light generated in the NaI crystal by γ-conversion falls on the semi-transparent photocathode of the photomultiplier tube, causing electron emission. The crystal and multiplier tube are mounted as a single unit in a light-tight cylindrical can, the crystal face being in contact with the photocathode end.

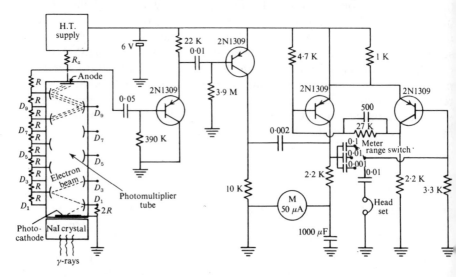

Fig. 10.5 Scintillation-meter schematic.

These electrons emitted from the photocathode are accelerated towards the first electrode, D_1, operating at about 150 volts positive with respect to the grounded cathode. The intermediate electrodes, D_1 to D_{10}, called dynodes and usually about ten in number, provide electron multiplication by secondary emission from surfaces coated with low work-function material, e.g. Cs_3Sb, the chain being so mounted that the electrons must proceed from D_1 to D_2, etc., then finally to the anode. With a gain factor of about four per stage, the total current amplification is roughly 10^6. This produces a current pulse of about 0·5 μA through the anode resistor, R_a, i.e. a 20-mV pulse. From this point on, the corresponding voltage pulse across R_a is amplified and integrated as in the Geiger counter circuit.

The great advantage of this instrument is in the efficiency of γ-ray detection. It will also detect β-rays. The price is about ten times that of a Geiger counter and the size and weight are somewhat greater than the Geiger. It can be used as a completely portable unit (frequently with detachable crystal-multiplier head for entering a confined space), or as a semi-portable unit in a car or aircraft. The airborne instrument is much more elaborate. Portable sets usually have crystals of 1·5 in. in diameter by 1 in. thick with multiplier tube to match; the airborne versions may have multiple crystals of 9 in. diameter or larger for greater sensitivity and are equipped with a strip recorder or digital readout.

10.3.5 *Gamma-ray spectrometer*

A logical extension of the scintillometer is a spectrometer which would separate characteristic γ-rays of K^{40}, U and Th for identification of the source. Such instruments have been used in airborne surveys and a couple of portable units are also available.

Spectrometers of this type, known as *pulse-height analysers* or 'kick sorters', have been used for γ-ray analysis in nuclear physics laboratories for twenty years. They make use of the fact that the intensity of the light pulse, and hence the amplitude of the voltage pulse from the multiplier, is proportional to the original γ-ray energy. Actually this is only partially true, due to the complex process of γ-ray absorption, and further explanation is necessary.

When the γ-ray loses all of its initial energy at once by photoconversion, the statement above is entirely correct. Even if it is first degraded by scattering and/or pair production, resulting eventually in photoelectrons of lower energy, these will still add up to a pulse of the same amplitude, *provided* the γ-ray does not

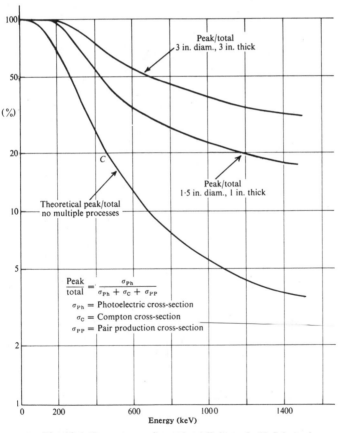

Fig. 10.6 Gamma-ray absorption efficiency in NaI crystals.

escape from the crystal, i.e., it is completely absorbed. This is true because all the processes occur essentially simultaneously (since γ-rays, being electromagnetic radiation, travel with the velocity of light). For example, if the beam of γ-rays entering the crystal were monochromatic, of energy E, and some rays escape with lower energy e, there is a contribution to the pulse-height spectrum of an amplitude corresponding to $E - e$.

Figure 10.6 illustrates the efficiency of NaI crystals in converting the γ-rays into pulses of maximum amplitude by multiple processes. A theoretical curve C shows the ratio of cross-section (effectively absorption) by photoconversion only, to total cross-section, that is, all three conversion processes, for comparison. For energies between 1·5 MeV and 100 keV (below which the photoelectric effect predominates) the larger crystal is on average 35 % more efficient, while the smaller

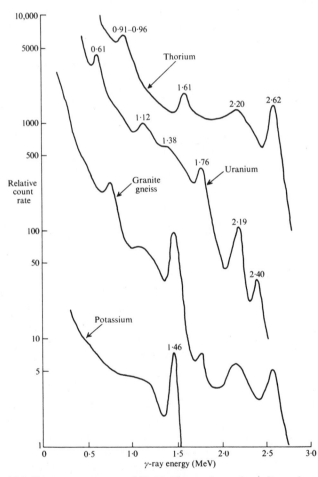

Fig. 10.7 Gamma-ray spectra of K, U, Th samples and granite-gneiss outcrop.

($1\frac{1}{2}$ in. by 1 in.) loses an increasingly larger fraction of γ-rays by scattering out of the crystal.

To obtain 100% efficiency in converting the γ-rays, it would be necessary to mount the radioactive source, as a minute grain, inside the crystal and in fact this is done in laboratory installations. In these circumstances – and provided the crystal is large enough – all the γ-rays would be absorbed in the crystal and the original γ-ray spectrum of the source would be quite faithfully reproduced as a pulse-voltage spectrum in the analyser. In a field measurement, however, the situation is more complicated. Some γ-rays lose energy by scattering in escaping from the source and also during passage through the air to the crystal. This, coupled with the fact that the U and Th series emit numerous γ-rays over a wide energy range, results in a complex pulse-height spectrum, as shown in fig. 10.7. All four curves have characteristic peaks and, in addition, an increasing continuum at low energies, due to Compton scattering. The pure potassium sample produces a relatively simple curve, having only the K^{40} peak at 1·46 MeV. Thorium is characterized by the strong 2·62 MeV peak of Tl^{208}. The uranium spectrum is most complex, although the peak at 1·76 MeV is reasonably distinctive. Potassium and thorium are clearly evident in the granite gneiss, as well as a smaller fraction of uranium.

A prospecting γ-ray spectrograph, then, should be capable of isolating the K, U and Th peaks at 1·46, 1·76 and 2·62 MeV. This is accomplished by replacing the integrator-counter circuit in the scintillation meter with three electronic circuits to select the appropriate pulse heights which correspond to the above γ-ray energies. Considering channel 1 in fig. 10.8, the detail diagram at the upper right of the figure shows that the channel is actually two parallel channels. The discriminator 1A is biased so that it gives an output pulse only for γ-rays with energy greater than 1·3 MeV, while discriminator 1B responds only to γ-rays with energy in excess of 1·6 MeV. Thus neither discriminator registers γ-rays whose energies are less than 1·3 MeV, while for values greater than 1·6 MeV the anticoincidence circuit adds the two outputs out of phase to give zero output as well.

The other two channels operate in the same manner for 1·76 and 2·62 MeV. Generally the channel centres and widths are adjustable. Clearly the window must be wide enough to accommodate the finite width of the peaks in fig. 10.7, but not so wide that the flanks of adjacent peaks may be accepted as well.

The pulses are counted and integrated separately and, in the airborne instrument, applied to a three-channel recorder. Because the radioactive sources may contain both U and Th and even K as well, and since the count level may be considerably higher for one channel than the others (as in fig. 10.7, where the count rate of the thorium sample at 1·46 MeV, even in the absence of a peak, is more than 100 times larger than the potassium peak), some means of subtracting a predetermined fraction of the higher count rate is generally incorporated near the output end of the spectrometer, as shown in fig. 10.8. One model of this type of instrument, designed for airborne work, is illustrated in fig. 10.9. It employs twelve 9 in. × 4 in. crystals; the correct position of the channels is monitored with a Cs^{137} standard γ-source (661 keV).

10.3.6 *Miscellaneous instruments*

Some portable scintillation meters have simple circuit modifications which permit rough discrimination between K, U and Th, as well as measurement of total γ-ray count. A switch provides two bias levels on the pulse amplifier, equivalent to about 2·5 and 1·6 MeV, so that one can, in effect, introduce wide windows, one at a time, for Th and U + Th. Calibration of this type of equipment is discussed at the end of this section.

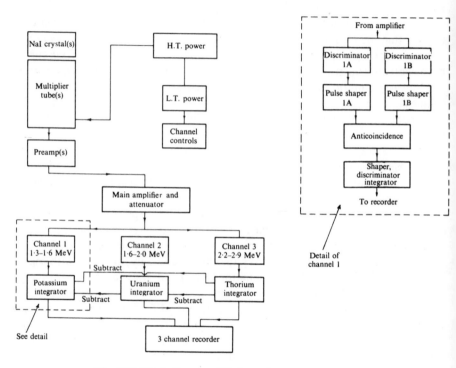

Fig. 10.8 Block diagram of 3-channel gamma-ray spectrometer.

A recently developed instrument known as a the *emanometer*, or *radon sniffer*, has been used to measure the radon content of waters, oils and soils. Since radon is a noble gas, it does not form chemical compounds. It moves freely through pore spaces, joints and faults for distances up to several hundred feet. It will also dissolve in ground water and so move about in the subsurface.

Air samples are obtained from the soil by drilling a short hole (2–3 ft) and pumping air from the hole through filters and a dryer into an ionization chamber or thin ZnS scintillator. Water samples are degassed and the gas-air mixture goes to the detecting chamber. In one model the 4·8 MeV α-particles of Rn^{222} are counted, rather than the 0·51 MeV γ-ray.

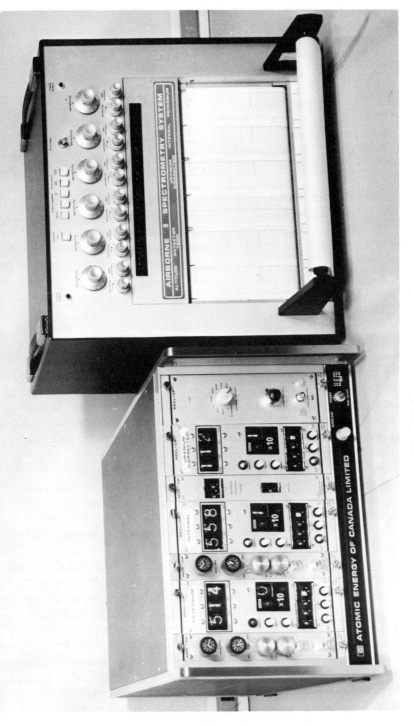

Fig. 10.9 Gamma-ray spectrograph.

If the radon is in equilibrium with the other decay products in the vicinity, the amount of parent product may be determined from eq. (10.5)

$$N_u = N_{rn}\lambda_{rn}/\lambda_u \approx 2 \times 10^{-6}N_{rn}/5 \times 10^{-18} \approx 4 \times 10^{11}N_{rn}$$

assuming the radon originated from U^{238}. The numerical factor would be about 4×10^{15} for radon gas from U^{235} or Th^{232}. These have very short half-lives, however, hence it is much more likely that Rn^{222} is the isotope detected. If the series is not in equilibrium, e.g., some member or members are reduced or missing because of weathering, etc., the count would be reduced, since the above relation would not hold.

A variety of other instruments has been developed for radioactivity measurements in geophysics. Several are adaptations of nuclear physics laboratory equipment for assaying. Others, such as the *beryllium detector*, used for ground prospecting, and the *density logger*, contain their own radioactive sources to initiate artificial radioactive processes in nearby rock. Some of these are discussed in §11.3.3a and §11.3.4.

10.3.7 *Calibration of instruments*

The calibration of instruments for reasonably quantitative measurements in radioactivity is a straightforward procedure in laboratory work, more complicated for field operations. For example, the scintillation meter with threshold bias, described at the beginning of §10.3.6, may be adjusted to determine relative amounts of uranium and thorium by means of a standard Th source. In one model this source, supplied with the instrument, is moved over the instrument barrel, with the bias switch in the 1·6 MeV position, until the meter gives a particular reading. Then the bias is switched to 2·5 MeV (Th only) and the calibration control potentiometer is adjusted to give a reading which is about 30% of the first value. Several successive adjustments may be necessary to obtain this ratio, since the control pot affects both bias settings. When the calibration is correct, the uranium count in the 1·6 MeV bias position is given by: $N_u = N_u' - 3·5N_{th}$, where N_u' is the measured number of counts for the 1·6 MeV bias and N_{th} the measured count for the 2·5 MeV bias.

When an instrument like the 4-channel spectrometer is to be used for ground or airborne surveys, a small standard source is quite unsuitable, since the detection of K, U and Th is carried out under entirely different geometrical conditions from the laboratory. In this case much larger calibration sources are necessary.

The Geological Survey of Canada has constructed five concrete slabs 25 ft by 25 ft by 1·5 ft thick and 50 ft apart, at Uplands Airport near Ottawa for this type of calibration. Embedded in ground adjacent to an aircraft parking area, these slabs contain varying fractions of radioactive material (e.g. 2·20% K, 3·0 ppm eU, 26·10 ppm eTh, + 1 pt. pitchblende). The slabs are effectively of infinite dimensions for detectors located within a few feet of their surfaces. (*Note*: see problem 7 for definitions of eU and eTh.)

10.4 Field operations and interpretation

Ground prospecting is readily carried out with any of the instruments described in §10.3.3 to §10.3.6. As mentioned before, the Geiger counter is suitable only for foot traversing; the other two, particularly the scintillometer, may also be used in vehicles. Ground work is frequently done without cutting line, since the method is very cheap and it often is not integrated with other ground operations. No particular expertise is required for this work. It is sufficient to note the count rate or mr/hour of the instrument and compare it with a background reading. Ratios $\geqslant 3/1$ over background would generally be of interest.

The background itself may vary considerably from place to place, depending on depth of soil cover and potassium content of the local rocks. Some notice must be taken of the geometry of outcropping formations in this regard, since the instrument response is influenced by source-detector separation and the source dimensions. This is particularly true of the Geiger counter. Measurements made in steep cuts and gorges, as well as in areas below drainage slopes, may give abnormally high readings. Background variation in different rocks has already been referred to in table 10.4. In the early days of radioactive prospecting, erratic increases in background occurred occasionally as a result of atomic tests; this is no problem nowadays.

Surface work in radioactivity exploration is relatively a minor effort. Data for 1968 indicate some 10–15,000 line-miles covered and $600,000 expended worldwide. Airborne operations, on the other hand, amounted to 4×10^5 line-miles costing $2·5 million, a sharp increase over the previous year (50,000 line-miles). Radioactivity prospecting from the air is comparatively cheap $\sim$ $6/line-mile) and efficient from the point of view of detecting γ-rays, since the intensity of a 1 MeV beam is reduced by only 50% at 300 ft above surface.

Considerable airborne work was done in the late fifties, using scintillation meters with large crystals. The results were not particularly useful, due to the lack of discrimination in total count measurements. When the demand for uranium fell off, the technique was practically abandoned. The recent revival of interest in uranium, plus the availability of the γ-ray spectrometer, has made the method more attractive.

Spectrometer profiles taken in a helicopter over Elliot Lake and Bancroft areas of Ontario are illustrated in fig. 10.10. Altitude in both cases was 500 ft. The low air speed (25 mph) possible with the helicopter is a decided advantage, both for discrimination of anomalies and amplitude of response. Interpretation, as in ground surveys, is mainly qualitative.

Some attempts have been made to obtain quantitative results in airborne work. By correlation of detailed ground data with airborne surveys made over the same area, it is possible to get a reasonably good fit between the aerial profiles and upward continuation of the ground data, using an empirical expression for the γ-ray attenuation in air. This correlation is valid because the source must outcrop to be detected in either survey. Characteristic curves for elementary shapes are then drawn up for the airborne interpretation in other areas. Three elementary geometries are considered:

(i) the finite or elementary source, assumed to have a circular shape,

(ii) the infinite source, an exposed plane outcrop of great lateral extent in both x- and y-axes,

(iii) the line source, having infinite exposed length along one axis, considered to be the strike direction.

Clearly the finite source is the most usual geometry encountered in the field, since either of the others will generally be covered with overburden at various places. Practically, the elementary source need not be circular, but its largest dimension should not greatly exceed the altitude of the aircraft.

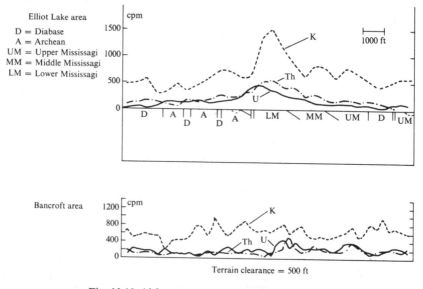

Fig. 10.10 Airborne gamma-ray spectrograph profiles.

The characteristic curves enable us to determine a parameter involving the product of surface area and source intensity; these quantities (as in gravity, EM and other situations) cannot be resolved individually.

Airborne radioactivity surveys have been carried out for minerals other than uranium and thorium, such as titanium and zirconium-bearing heavy minerals, including tantalum, niobium and the rare earths. In this connection carbonatites (such as at Oka and other locations in the province of Quebec) are an intriguing target for γ-ray spectrometer exploration, because they have a very low ratio of uranium to thorium. This distinctive signature ought to apply to kimberlites as well; hence the method may prove useful in prospecting for diamonds. In all these applications, of course, it is the association of the target minerals with small distributions of U and Th (occasionally K as well) which makes the technique possible.

Limited attempts have been made to use the radioactivity method in oil exploration. Surveys in known oil fields sometimes indicated a radioactive low directly over the oil-bearing structure with a halo slightly above background surrounding it. Actually this pattern was reported about forty years ago from crude ground surveys in Texas fields. The source of radioactivity appears to be radon gas, which moves upward through fractures in the perimeter rock to escape at surface; the suggestion has been made that the tight cap rock over the oil pool is relatively impervious to this migration.

The application to oil prospecting, however, does not seem to have been developed to any extent. The same statement holds for the possibility of using airborne radiometrics for surface geological mapping.

10.5 Field examples

1. Figure 10.11 illustrates the application of airborne radioactivity to geologic mapping. The area shown is the Concord quadrangle of North Carolina, where the U.S. Geological Survey carried out an airborne radiometric survey as an aid to detailed mapping of complex geology. Six NaI crystals, 4 in. diameter, 2 in. thick and six photomultiplier tubes connected in parallel were used for the detector. Particular care was taken to correct the data for variations in aircraft altitude.

The compilation of geologic and radioactivity information in fig. 10.11 represents a progressive refinement of data; that is, the original geologic information was changed as a result of additional geologic mapping, which was to a considerable extent guided by the radioactivity results. For example, the radiometric survey outlined a granite stock, northeast of the town of Concord, whose borders could not be well defined by field geology; it also located the smaller granite body in the northwest part of the town.

The granitic zone in the southeast corner of the figure is marked by high radioactivity, while that on the southwest is not. A porphyritic biotite granite in the northern part of the quadrangle shows medium to high radioactivity, but in the vicinity of the gabbro mass in the northwest corner the response is lower. The anomalous high along the east border coincides with a considerable injection of granite into the surrounding gneiss and schist. Finally, the large gabbro-syenite mass in the centre of the quadrangle is fairly well outlined by the radioactivity contours, since the syenite zones on the east and west flanks show higher response than either the gneiss-schist surroundings or the enclosed gabbro.

The variation in radioactivity response over different granitic zones is not definitely explained, although it is suggested that there may be different types of granite within the quadrangle. No mention is made of the depth of overburden; slight variations in the thickness or type of cover could account for the lows in the northwest and southwest corners.

Neither the airborne radiometrics nor the geologic mapping could have produced this interpretation independently. It is the product of a combination of the two, plus some aeromagnetic data. Furthermore, the airborne survey, as men-

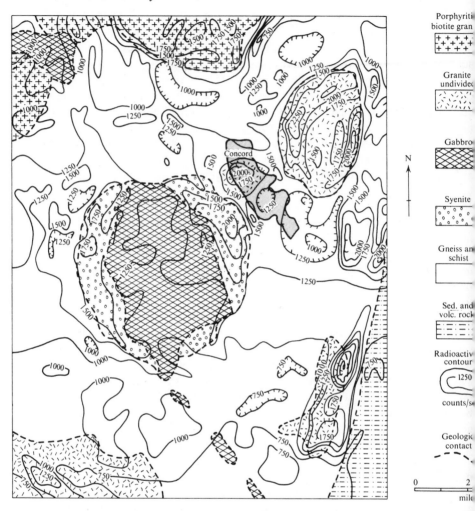

Porphyriti
biotite gran

Granite
undivided

Gabbro

Syenite

Gneiss an
schist

Sed. and
volc. rock

Radioactiv
contour

1250

counts/s

Geologic
contact

0   2

mile

Fig. 10.11 Airborne radiometrics as an aid to geological mapping, Concord Quadrangle
North Carolina. (After Bates, 1966.)

tioned previously, was useful in selecting areas for further detailed mapping, thu
saving time and money.

2. An example of the direct method of radioactivity prospecting is shown in
figs. 10.12 and 10.13 taken from an extensive survey for uranium in Labrador
Following large-scale reconnaissance with airborne radiometrics and magnetics, a
set of targets was selected for detailed ground follow-up. The latter operation
proceeded in two steps. First, the airborne radioactivity anomalies were located
and assessed roughly; then the more promising of these were examined in detail by
scintillometer and magnetometer (in some cases ground EM was also employed

since there were sulphides associated with the radioactive minerals). Stations were 10–25 ft apart along cut lines spaced 100 ft apart.

Figure 10·12 shows the radioactivity and magnetic contours for a small area of the ground survey. Strong anomalies of both types are coincident on L132S at 2E. These were indicated by the airborne survey, the location being almost exactly the same as the ground anomaly. At L112S, 6E, however, there is no abnormal radioactivity associated with the high magnetics.

Magnetic and radiometric profiles on L132S (the latter being obtained by using the uranium channel on the scintillometer), together with a vertical geologic

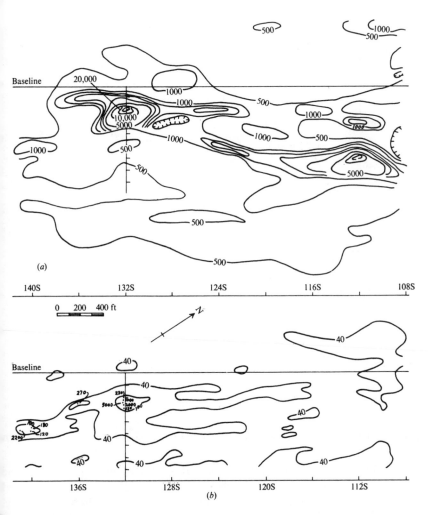

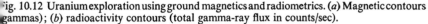

Fig. 10.12 Uranium exploration using ground magnetics and radiometrics. (*a*) Magnetic contours (gammas); (*b*) radioactivity contours (total gamma-ray flux in counts/sec).

section obtained from drilling, are displayed in fig. 10.13. The host rocks are argillites, quartzites and amphibolites. The mineralization, consisting of magnetite and pitchblende with some chalcopyrite, sphalerite and pyrrhotite, occurs in bands of ferruginous quartzites which alternate with diorite dikes. The overburden at the collars of the two drill holes is less than 7 ft thick and the huge magnetic and γ-ray peaks occur directly over exposed mineralization.

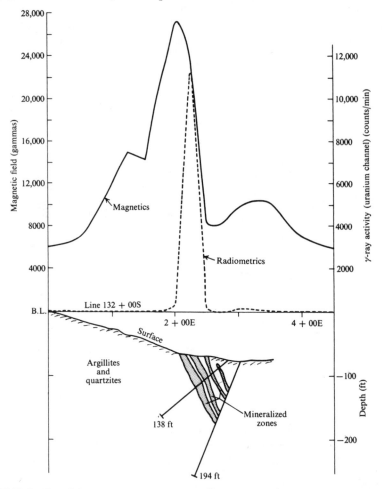

Fig. 10.13 Radioactivity and magnetic profiles plus geologic section, Line 132 + 00S, uranium survey.

Although the U_3O_8 mineralization is probably of economic grade, the volume is small. The maximum depth extent is 130 ft, while additional drilling showed the zone width and strike length to be no greater than 10 ft and 250 ft respectively. Although the contours of figs. 10.12a and 10.12b indicate the small lateral extent of the showing, this evidence is not conclusive in itself for two reasons: firstly, a few feet of overburden would be sufficient to mask the presence of uranium; secondly, the association of Fe_3O_4 and U_3O_8 mineralization does not prevail throughout the area, as proved by drilling in the vicinity of the magnetic anomaly near 6E on L112S.

10.6 Problems

1. Scintillometer readings were taken at station intervals of 25 ft on a set of parallel lines 25 ft apart over an area in Saskatchewan during the course of a uranium exploration programme. The readings, in counts per minute, are tabulated below.

Station	Lines							
	1	2	3	4	5	6	7	8
1	11	11	11	12	11	12	14	15
2	12	12	11	12	12	15	15	17
3	14	14	12	14	14	17	17	17
4	17	15	15	15	15	24	17	17
5	19	18	18	18	17	34	22	16
$5\frac{1}{2}$	—	30	21	24	22	—	—	—
6	16	25	24	23	28	47	27	18
7	17	23	25	28	34	47	30	24
8	16	17	30	36	46	66	47	38
$8\frac{1}{2}$	—	—	—	70	200	—	66	60
9	15	22	36	1200	200	310	260	40
10	14	18	22	50	35	35	46	25
11	13	15	18	22	26	26	33	17
12	12	12	15	18	17	17	14	13

Plot and contour these readings, estimate the strike and width of the anomalous zone. Are there indications of the depth of overburden and variations of the overburden thickness in these values?

2. A 2·5 kg rock sample was tested for radioactivity with a scintillometer, both being enclosed in a large shielded container. The average count rate was 540 cpm. The background count in the container was 35 cpm. Assuming the overall efficiency of the scintillometer to be 30%, and either that the radioactivity is approximately equally distributed between uranium, thorium and potassium, or that there is no potassium in the rock and the U/Th ratio (see problem 7 and fig. 10.17) is about 0·5, determine the content of each element in the sample for the two cases (see table 11.1, §11.3.1b).

3. Scintillometer and vertical-component magnetic contours, taken from a detailed ground survey for uranium in northern Canada are shown in fig. 10.14. An excerpt from the geological report on the region says: 'Geologically these areas are quite featureless and consist of pink quartzites with dark bands of mafic minerals, the latter coinciding with the radioactivity and magnetic anomalies at a number of places.'

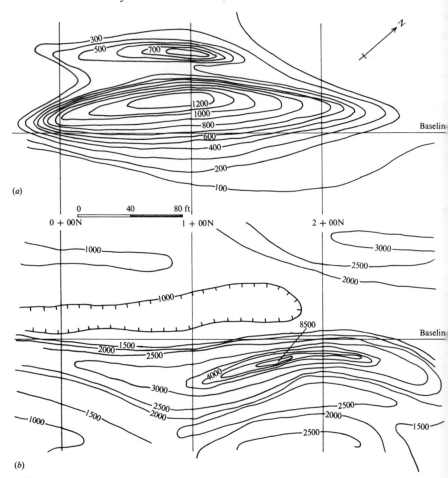

Fig. 10.14 Radioactivity and magnetic contours, Northern Canada. (*a*) Wide-window gamma-ray flux (counts/sec). Background = 60 counts/sec; (*b*) vertical-component of the magnetic field (gammas).

Make an interpretation of this small section using the limited data available and keeping in mind possible coincidences of magnetic and radiometric anomalies, variations in depth of overburden, presence of granitic rock (K^{40} radioactivity), etc. As an aid to interpretation, it is suggested that the two contoured maps be overlaid and also that a few profiles be taken off each map.

4. As mentioned in §10.3.5 the present type of gamma-ray spectrometer has three channels at 1·46, 1·76 and 2·62 MeV to isolate K, U and Th photopeaks respectively. Consequently it is possible to determine the individual amounts of potassium, uranium and thorium contained in a rock sample when the spectrometer has been calibrated by using standard samples of known composition. We obtain the relative content of the three elements in terms of the count rates of the instrument as outlined below.

None of the gamma-rays from uranium or potassium have sufficient energy to be recorded in the thorium (2·62 MeV) channel, hence

$$T = k_1 T_c,$$

where T = thorium content in parts/million, T_c = 2·62 MeV channel count rate less background, k_1 = constant for the thorium channel.

The 1·76 MeV uranium channel records γ-radiation from uranium and thorium but none from potassium. Hence we have

$$U = k_2(U_c - S_1 T_c),$$

where U = uranium content in parts/million, k_2 = constant for the uranium channel, U_c = 1·76 MeV channel count rate less background, S_1 = stripping constant for thorium γ-rays in the uranium channel (see below).

Finally, in the 1·46 MeV potassium channel we may have counts from uranium and thorium as well as potassium. The potassium content, $K_\%$ (expressed as a percentage because it is generally much larger than the amounts of U and Th), is given by

$$K_\% = k_3\{K_c - S_2(U_c - S_1 T_c) - S_3 T_c\},$$

where k_3 = constant for the potassium channel, K_c = 1·45 MeV channel count rate less background, and S_2, S_3 = stripping constants for U and Th γ-rays in the K channel. Since the γ-ray spectrum of an element is invariant, the stripping constant is merely the fixed ratio of the count rates for U or Th at the appropriate energy levels; thus, S_1 is the ratio of the count rates for Th in two channels centred at 1·76 and 2·62 MeV, hence multiplying the observed count rate for the 2·62 MeV channel by S_1 yields the thorium count rate which would be observed at the same instant in the 1·76 MeV channel.

Theoretically only one thorium, two uranium and three potassium standard samples are required to furnish the data to solve the three equations and determine the six constants, k_1, k_2, k_3, S_1, S_2, S_3. In practise it is best to use as many standards as possible because of instrumental and sampling errors.

The following readings were obtained with a γ-ray spectrometer in a traverse perpendicular to foliation across a granite-gneiss outcrop near St Columban, Quebec.

Station	Spectrometer readings (cpm)		
	T_c	U_c	K_c
0 ft	13	28	195
100	8	27	243
170	22	34	265
200	25	36	218
210	18	30	135
230	10	24	223
300	15·	27	193
400	15	30	197
425	12	20	242
500	8	21	233

Given that $k_1 = 0·6$, $k_2 = 0·13$, $k_3 = 0·020$, $S_1 = 1·0$, $S_2 = 1·5$, $S_3 = 1·7$, determine the Th, U and K content at each station and plot profiles for each element as well as a profile of the Th/U ratio.

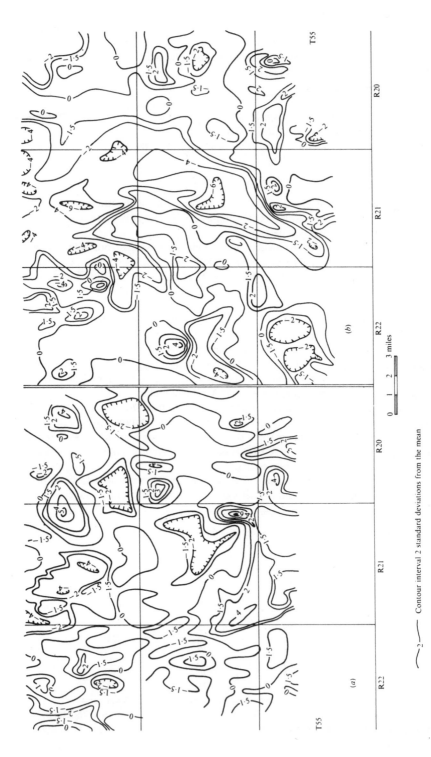

Fig. 10.15 Aeroradiometric survey, Redwater Oilfield, Alberta. (a) Area flown east-west, winter of 1957; (b) area flown east-west, autumn of 1957.

—2— Contour interval 2 standard deviations from the mean

The formation appears to be homogeneous, that is, completely granitized, except for remnants of sedimentary rocks at 35 ft and 215 ft. Do the profiles bear this out over the whole 500 ft traverse or any part of it? On the assumption that high Th/U ratios would be more characteristic of sediments than of the granite-gneiss intrusive, does the Th/U profile provide any additional information? Is the sampling density adequate for a petrogenic study of the rocks?

5. Limited applications of the radioactivity method have been made in oil exploration, as mentioned in §10.4. Figure 10.15 shows radiometric contours from an airborne test survey of the Redwater oil field near Edmonton, Alberta. The area was flown twice, first in 1951 and again in 1957, using an early type of gamma-ray spectrometer which recorded only energies above 1·50 MeV, thus eliminating the effect of potassium.

The rocks of the area are mainly shales and sandstones, with some conglomerate, black quartzite and argillaceous sandstone in the southwest and southeast corners of the survey area. Green shale forms a caprock over the oil pool.

The airborne data have been analysed by Sikka (1959) who made corrections for variations in the radioactivity of the different soils in the area. The arithmetic mean and standard deviation were calculated for the area. The difference between each observed value and the arithemetic mean was divided by the standard deviation and the result plotted as the station value; thus, the contour interval is a multiple of the standard deviation. Sikka also measured the distribution of radon emission on the ground. In the course of this work, he found that the type of soil – for example, sands, sandy loams, loams, etc. – as well as the soil parent material controlled the regional pattern of radioactivity to a considerable extent. Details may be found in the original work.

The airborne radiometric data from 1951 (fig. 10.15b) were corrected only for the soil type. The contours of 1957 (fig. 10.15a) were corrected for soil parent material as well. This was done to show the significance of both factors. Sikka states that the two maps would be similar if the complete corrections were made in both cases.

On the hypothesis that radioactivity lows reflect the relatively impervious caprock over the oil pool, try to outline its boundaries on each map. (Note: the survey covered only about 65 % of the total area of the oilfield.) Can you spot muskeg and swamps from

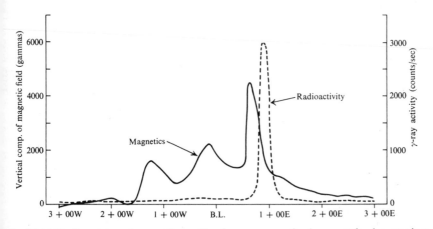

Fig. 10.16 Radiometric and magnetic profiles from a survey for base metals plus uranium.

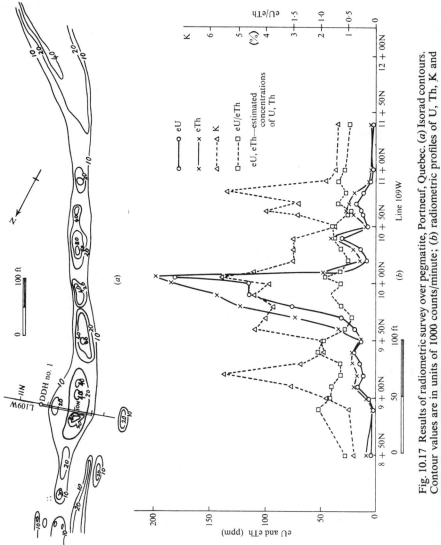

Fig. 10.17 Results of radiometric survey over pegmatite, Portneuf, Quebec. (a) Isorad contours. Contour values are in units of 1000 counts/minute; (b) radiometric profiles of U, Th, K and U/Th ratio on Line 109W

anomalous high radioactivity? Apart from a possible halo of high radioactivity around the oilfield, are there any structures such as faults, indicated by linear high-low contrasts striking in a preferred direction?

6. The profiles in fig. 10.16 are taken from a detailed survey for base metals and uranium. Station readings were made every 10 ft. The magnetic-radiometric correlation is similar to that of fig. 10.13, although the geology is different and this anomaly has a strike length of 1600 ft. Locate the radioactive source, the boundaries of the magnetic anomaly and the dip of the mineralized section. What is the likely source of the magnetic mineralization?

7. Some results from a radiometric survey over a pegmatite in Portneuf County, about 35 miles west of the city of Quebec, are shown in fig. 10.17. This feature outcrops continually for a distance of about 1100 ft. As can be seen in fig. 10.17a the strike is roughly N20°W over most of the length, the main width varying between 30 and 80 ft; there are narrower pegmatitic bands on the flanks. The host rocks are gneisses with some hornblende.

The uranium, thorium and potassium content of the rocks, shown in the profiles of fig. 10.17b, were determined with a multichannel gamma-ray spectrometer by a method similar to that described in Problem 4. Locate the main pegmatite section and any additional pegmatite bands with the aid of this data. Are these sections of shallow or steep dip? Can you determine the direction of dip? Does the uranium mineralization appear to be of economic grade? Compare the U/Th ratios with the actual uranium and thorium concentrations at various places along the profile. Do large ratios correlate with high uranium content?

(Note: the symbols eU and eTh refer to equivalent uranium and equivalent thorium, that is, estimates of uranium and thorium concentration based on γ-ray emission from Bi^{214} and Tl^{208} respectively, assuming radioactive equilibrium.)

References

F – field method; G – general; I – instruments; T – theoretical.

Birks, J. B. (1964). *The theory and practice of scintillation counting*. New York MacMillan. (G)

Bates, R. G. (1966). Airborne radioactivity surveys, an aid to geologic mapping. In *Mining geophysics*, vol. I, pp. 67–76: Tulsa, Soc. Explor. Geophysicists. (F)

Brownell, G. M. (1959). A beryllium detector for field exploration: *Econ. geol.* **54**, 1103–14. (I)

Darnley, A. G. (1970). Airborne gamma-ray spectroscopy. *Can. Inst. Mining Bull.* **63**, no. 694, 145–54. (I)

Darnley, A. G., Grasty, R. L. and Charbonneau, B. W. (1970). Highlights of GSC airborne gamma spectroscopy in 1969. *Can. Min. Jour.* **91**, no. 4, 98–101. (F)

Doig, R. (1967). Current research in the application of natural and induced radioactivity to mineral exploration. In *Mining and Groundwater Geophysics* (ed. L. W. Morley). Econ. Geol. Report no. 26, Geol. Surv. of Canada, 389–400.

Dyck, W. (1968). Radon determination apparatus for geochemical prospecting for uranium. Geol. Sur. Canada, Dept. E.M.R., paper 68–21, no. 10. (I)

Faul, H. (1950). Radioactivity methods. In J. J. Jakosky, *Exploration geophysics*, 2nd ed., pp. 987–1015. Los Angeles, Triza. (G)

Gregory, A. F. (1960). Geological interpretation of aeroradiometric data. *Geol. Sur. Canada, Bull.* **66.** (T)

Gregory, A. F. and Horwood, J. L. (1963). A spectrometric study of the attenuation in air of gamma-rays from mineral sources. Dept. Energy and Mineral Resources (Mines Branch, Ottawa) Research report R110. (F)

Lambert, Roger (1969). Etude spectrometrique de deux anomalies radiometriques. M.Sc. thesis, Dept. Genie geologique, Ecole Polytechnique de Montreal, Montreal. (F)

Lundberg, H., Roulson, K., Pringle, R. W. and Brownell, G. M. (1952). Oil exploration with airborne scintillation counters. *Oil in Canada*, **4**, 40–6. (F)

Moxham, R. M. (1960). Airborne radioactivity surveys in geologic exploration. *Geophysics*, **25**, 408–32. (F)

Nininger, R. D. (1956). *Minerals for atomic energy*, 2nd ed. Princeton, Van Nostrand. (G)

Rankama, K. (1963). *Progress in isotope geology*. New York, Interscience. (G)

Sikka, D. A. (1959). A radiometric survey of Redwater oilfield, Alberta, Canada. Ph.D. thesis, McGill University, Montreal. (F)

Soonawala, N. M. (1968). Correlation of ground and airborne radiometrics. M.Sc. thesis, McGill University, Montreal. (F)

11. Geophysical well logging

11.1 Introduction

11.1.1 *Uses of well logging*

Geophysical methods have been applied to the investigation of drill holes for some forty-five years, using initially the same electrode techniques as in surface exploration. The various instruments and techniques, specifically designed to suit the different environment in drill holes, are used in direct exploration, identification of geologic formations and formation fluids, and correlation between holes.

Since 1928, when the Schlumberger brothers first made electrical well measurements in the Pechelbron oil field in France, geophysical well logging has become a standard operation in petroleum exploration. Correlation and evaluation of the productive capabilities of reservoir formations are usually the principal objectives.

Well logging has not, however, been used extensively in the search for metallic minerals for several reasons. The smaller size holes obtained with the diamond drill, generally less than one-third the diameter of oil wells, impose some limitations on equipment, but this is not a major problem. The complex geologic structure encountered in mineral areas, compared to the relatively uniform sedimentary formations associated with oil, makes identification and correlation more difficult. Finally, it is argued that the complete recovery of core in diamond drilling eliminates the necessity for logging holes, since the information is available laid out in the core box. It is unfortunate that this attitude prevails in mineral exploration. Well logging is cheap compared to drilling. A variety of geophysical logging techniques would be valuable aids to correlation and identification of mineral-associated anomalies, particularly when core is lost or difficult to identify. The resultant control on drilling should more than pay the geophysics bill.

Geophysical methods which have been applied in well logging include resistivity, self-potential, induction, induced-polarization and occasionally other electrical methods, detection of gamma-rays and neutrons in radioactivity methods, acoustic logging, and measurement of magnetic and thermal properties. The emphasis in what follows of necessity will be on logging for petroleum.

11.1.2 *Properties to be evaluated*

Well logging for petroleum usually has the primary objectives of identifying potential reservoir rocks and determining their porosity and permeability and the nature of the fluids present. *Porosity* is the fractional portion of rock volume which is occupied by pore space, often expressed as a per cent. The porosity of reservoir rocks is usually in the range from 30% to 10% although rocks of lesser

771

porosity can also be profitable hydrocarbon reservoirs. The product of porosity, area and thickness of a reservoir gives the volume of fluids which the reservoir contains. Porosity can be determined from resistivity, acoustic velocity (sonic), density, and neutron logs. Each may be subject to distortions and hence better determinations can be made from combinations of logs than from individual logs.

In most reservoirs hydrocarbons fill only part of the pore space, that fraction being the *hydrocarbon saturation*. Where water is the only other fluid present (the usual situation), the water saturation plus the hydrocarbon saturation equals one. The water saturation is calculated from resistivity measurements employing one of Archie's formulae (eq. (11.3)). Calculation of water saturation often provides the distinguishing trait between formations which are, or are not, capable of commercial hydrocarbon production.

Besides porosity, an equally important property is the degree to which the pores are interconnected, that is, the *permeability*. Permeability is usually measured in darcys; a *darcy* is the permeability which will allow a flow of one millilitre per second of fluid of one centipoise viscosity through one square centimetre under a pressure gradient of one atmosphere per centimetre. Commercial reservoirs generally have permeabilities ranging from a darcy to a few millidarcys. Permeability is estimated from logs using empirical rules but only with order-of-magnitude accuracy.

Formation identification and correlation between wells is often as important as the determination of porosity and estimation of permeability. Particular formations may yield log curves of distinctive patterns making it possible to correlate not only major lithologic breaks but many points within the formations themselves. Faults and unconformities often can be located fairly precisely by noting section missing or duplicated in one well compared with others nearby. Stratigraphic details often can be worked out by observing the pattern of systematic variations in log shapes. As logs from more wells in an area become available, the amount of detail which can be extracted increases.

Most routine well logging can be classified, and appropriate logging tools chosen, according to the dominant type of geological section: clastic (sands and shales) or carbonate. Within each of these major groups, fairly standard ensembles of logs are run; occasionally other logs are employed for special purposes. The standard ensemble usually includes a group of electrical response curves (often recorded with the same logging tool or *sonde*) plus porosity logs. Thus, in a clastic section one might expect the following logs:

SP and/or gamma-ray log,
shallow resistivity log (16-inch short-normal or laterolog),
deep resistivity log (induction log or deep laterolog),
possibly a medium-penetration induction log,
porosity log of one or more types (in order of probability: the sonic log, the density log).

In a carbonate section one might expect a somewhat different ensemble:

gamma-ray log,
laterolog or dual laterolog,
two or more porosity logs (in order of probability: the sonic log, the density log, the neutron log).

The dipmeter and microresistivity logs are also run frequently, the latter to delineate permable beds by detecting mud cake (microlog) or residual oil saturation (microlaterolog).

The descriptions which follow cover certain logs which are now used very rarely. They are included because one often has to re-examine old logs and hence needs to understand what they measured. Also, their principles are the same as the modern logs which are refinements of these earlier techniques and they are often simpler to explain.

11.1.3 *Fluid invasion*

The objective of well logging is to measure the properties of the undisturbed rocks and the fluids which they contain. However, the act of drilling a hole produces disturbances. Pressures within rocks are very great and once an avenue of relief, such as a well, is available, the pressure differential tends to expel into the borehole the fluids present in the rock interstices. To prevent such an occurrence, boreholes are filled with drilling *mud*, a complex mixture of solids usually suspended in fresh water. Sufficient solids are added to the mud to make the pressure of the fluid column more-or-less equal to that of the formation fluids. Exact balance is rarely achieved, however, and the usual tendency is for the mud to enter porous formations, pushing the indigenous fluids back from the borehole. This process is called *invasion*. In the invasion process the mud solids tend to become plastered on the borehole wall to form a *mud cake* while the fresh-water fluid portion enters the formation interstices as *mud filtrate*. The mud cake quickly becomes sufficiently thick to prevent further entry of borehole fluid into the formation. However, the mechanical aspects of drilling continually abrade this mud-cake sheath which is repeatedly renewed by additional filtrate invasion. The filtrate itself is usually a chemical solution rarely identical in ionic behaviour to the formation water; this difference in conductivity, resistivity and ionic behaviour is of fundamental importance in understanding the design and interpretation of most electrical logging tools.

Appreciation of the invasion process is essential to interpreting well logs because the rock region which exerts the greatest effect on most log readings is the portion nearest the logging sonde, the very portion altered most by the drilling of the hole and the invasion process. The relative contribution of formation properties at various distances from the logging tool varies with different configurations of sensors. The *effective depth of investigation*, a qualitative term, is the radius, measured from the logging tool, which contains the material whose properties dominate the measurements. Deep penetration implies that the dominant contribution is from formations which have not been disturbed by invasion. At the other extreme, very shallow penetration implies that the properties either of the mud cake or of the borehole mud dominate the measurements, depending on whether the logging tool is pressed against the borehole wall or is centred in the borehole. Intermediate penetration implies domination by the area invaded by mud filtrate in porous formations, for example. Electrical log measurements using various electrode arrangements may give vastly different results because the mud and filtrate are usually fresh and hence highly resistive whereas the indigenous formation water is usually highly saline and hence conductive.

11.2 Electrical methods

11.2.1 *Introduction*

The physical properties of rocks and minerals measured in electrical well logging, as in surface geophysical work, are electrical conductivity and self-potential. The induced-polarization (overvoltage) effect has not yet been developed as a routine logging technique in petroleum applications; this effect may be considered an artificial potential produced by an applied current but the measurement depends on electrical conductivity.

Self-potential and resistivity were the earliest, and are still among the most frequently applied, geophysical techniques in well logging. They were first introduced by the Schlumbergers in 1928 and 1931 as the SP curve and the short-normal curve. Resistivity and SP logs are generally recorded simultaneously as adjacent curves, a combination known as the *electrical survey* (ES).

In most logging in petroleum-exploration boreholes, several logs are recorded simultaneously. Since most measurements can be made only where the hole has not been cased and since most deep holes involve casing portions of the hole at different times, logs are commonly run over different parts of a borehole at different intervals during the drilling and after completion of the hole. Also, since the primary objective of logging usually is to evaluate the productive potential of reservoir sands, these sands are sometimes logged soon after they are drilled, before proceeding to drill the deeper portion of the hole; otherwise the sands may change their log characteristics as a result of standing open to drilling fluid and the consequent penetration of the mud filtrate into the borehole.

11.2.2 *Resistivity logging*

(a) *Rock conductivities*. Resistivities of various rocks and minerals are given in tables 5.2 to 5.6, §5.4.1. Sedimentary formations normally encountered in oil wells are generally poor conductors, having resistivities in the range 1 to 10^6 Ωm. The minerals common in sedimentary rocks – silicates, oxides and carbonates – are practically all non-conductors. However, most sedimentary rocks are porous and contain water in which various salts are dissolved. In solutions these salts disassociate into cations (Na^+, Ca^{++}, Mg^{++}, etc.) and anions (Cl^-, SO_4^{--}, etc.)

which tend to move in an electrical field, thus providing the main vehicle of current-flow in sediments.

Metamorphic and igneous rocks may contain minerals (usually disseminated), such as pyrite, chalcopyrite, graphite, magnetite, galena, etc., which contribute to their conductivity. As in sediments, however, interstitial water is often the controlling factor.

Three significant equations are used in petroleum work to relate the resistivities of rocks, resistivities of fluids in rock pores, the porosity, and the amount of water filling the pore spaces. These are essentially modifications of the empirical formula of Archie (1942), eq. (5.7), §5.2.2c. The first expresses the *bulk water-wet resistivity* of a rock sample, ρ_0, and the resistivity of the water contained in its pores, ρ_w, in terms of a *formation resistivity factor*, F;

$$F = \rho_0/\rho_w. \tag{11.1}$$

(In formation evaluation the symbol R is usually used for resistivity rather than ρ; however, we shall continue to use ρ to be consistent with chapter 5.) Archie (*loc. cit.*) showed that the formation factor is a function of the porosity and, to a lesser degree, of the permeability of the sample. The second relation is

$$F = 1/\phi^m, \tag{11.2a}$$

where ϕ is the porosity of the material and m is a *cementation factor* whose value lies between 1·3 and 2·6. An alternative form of this expression, called the *Humble formula*, applicable to many granular rocks, is

$$F = 0·62\phi^{-2·15}. \tag{11.2b}$$

If the rock pores are not completely filled with water but contain fractions of gas and/or oil, the effective resistivity is larger than ρ_0. The third empirical equation by Archie (*loc. cit.*) accounts for partial *water saturation* of the rock; if S_w is the fraction of the pore volume filled with water,

$$S_w = (\rho_0/\rho_t)^{1/n}, \tag{11.3}$$

where ρ_t is the true resistivity of the sample, derived by applying corrections for tool dimensions and configuration, borehole diameter, mud resistivity, etc., to the measured (or apparent) resistivity, ρ_a, and n the *saturation exponent*, which lies between 1·5 and 3·0; n is usually assumed to be 2 where there is no evidence to the contrary.

(b) *Instrumentation.* The basic methods of resistivity logging are similar to those used in surface-resistivity prospecting. Usually direct current or a low-frequency alternating current is applied between current electrodes and the potential is measured between two or more potential electrodes. The record is then a plot of potential variation (or its equivalent, apparent resistivity) versus depth. Several electrode configurations are shown in fig. 11.1; these are described below.

(c) *Single electrode.* The simple arrangement shown in fig. 11.1*a* is no longer used in oil-well logging. It is suitable for resolving beds of anomalous resistivity in a qualitative manner. The apparent resistivity measured is considerably different from the true value.

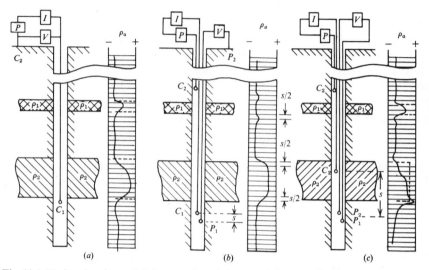

Fig. 11.1 Various logging resistivity spreads and characteristic curves for the case of beds more resistive than the adjacent formations. P = power source (d.c. or a.c.), I = current measuring device, V = device for measuring potential difference, s = electrode spacing. (*a*) Single-electrode spread; (*b*) two-electrode (normal) spread; (*c*) three-electrode (lateral) spread.

(d) *Normal arrangement.* A two-electrode system known as the short and long *normal* was introduced in the early thirties and is still in use. As seen in fig. 11.1*b*, one current and one potential electrode on the logging sonde are closely spaced downhole (16 inches apart for the *short-normal*, 64 inches for the *long-normal*) and the other two are fixed at the top of the hole. From eq. (8.38), §8.5.2, and fig. 8.3, §8.2.4, we get for the apparent resistivity in homogeneous ground,

$$\rho_a = \frac{4\pi \Delta V}{I} \left(\frac{1}{P_1 C_1} - \frac{1}{P_1 C_2} - \frac{1}{P_2 C_1} + \frac{1}{P_2 C_2} \right)^{-1}$$

(The factor is 4π here, not 2π as in eq. (8.38), because this equation holds in the interior of an infinite medium, not at the surface of a semi-infinite medium.) Since the distance $P_1 C_1$ is much smaller than the other three dimensions, this becomes

$$\rho_a \approx \frac{4\pi \Delta V}{I} (P_1 C_1). \tag{11.4}$$

Equation (11.4) is based upon the assumption that the medium is homogeneous. When the medium is not homogeneous but the variations are sufficiently restricted we can obtain an approximate result by replacing ρ_a with appropriately defined average resistivities. We define ρ_{ij} as the average resistivity associated with the interval between the potential electrode P_i and the current electrode C_j such that in the derivation of eq. (8.13), §8.2.4, we get the correct result when we write

$$V_1 = \frac{I}{4\pi}\left(\frac{\rho_{11}}{r_1}\right), \quad V_2 = \frac{I}{4\pi}\left(\frac{\rho_{12}}{r_2}\right), \text{ etc.}$$

Replacing r_1, etc., with P_1C_1, etc., eq. (8.13) becomes

$$\Delta V = \frac{I}{4\pi}\left(\frac{\rho_{11}}{P_1C_1} - \frac{\rho_{12}}{P_1C_2} - \frac{\rho_{21}}{P_2C_1} + \frac{\rho_{22}}{P_2C_2}\right).$$

If we now take P_1C_1 small enough that the other three terms can be ignored, we arrive at eq. (11.4) except that ρ_a has been replaced with ρ_{11}. We conclude from this that the measured ρ_a in eq. (11.4) will depend mainly upon the resistivities of the beds in the vicinity of P_1C_1; this means also that the measurements will be affected by the mud in the borehole nearby and by the penetration of the drilling fluid (filtrate) into porous zones.

The resistivity log of fig. 11.1b shows a curve which is symmetrical with respect to beds whose resistivity differs from those above and below. The interfaces are marked sharply (but not necessarily at their true locations), particularly in the short-normal curve. High-resistivity beds appear thinner than their actual width by a distance equal to, or less than, the electrode spacing, while conductive beds appear thicker. The effective penetration into the formations is about twice the electrode spacing and varies inversely with the hole diameter.

The definition and sharpness of normal logs decrease with an increase in the hole diameter and with a decrease in mud resistivity. The effect of adjacent beds and the invasion of porous zones by drilling-mud fluid are also significant. These effects can be reduced by the use of published correction charts, called *departure curves*.

The short-normal spread is sometimes suitable for measuring the resistivity of porous zones deeply flushed by mud filtrate and hence for determining formation porosity. It is also useful in geological correlation between wells, since the interfaces between beds are usually well defined. The long normal measures an intermediate resistivity which, in theory at least, permits calculation of both the invaded-zone resistivity, ρ_i, and the true formation resistivity, ρ_t.

(e) *Lateral arrangement.* A three-electrode arrangement known as the *lateral curve* has been used since 1936 and is illustrated in fig. 11.1c. The downhole potential electrodes are separated by 32 inches with their centre normally 18 ft 8 in. (called the *spacing*) from the near current electrode. They measure a resistivity of the form

$$\rho_a = \left(\frac{4\pi\Delta V}{I}\right)\frac{(P_1C_1)(P_2C_1)}{(P_1C_1 - P_2C_1)}. \tag{11.5}$$

The most striking feature of lateral curves is their asymmetry; in fig. 11.1c this is particularly apparent at the upper and lower boundaries of the thick bed.

If the current and potential electrodes are interchanged, the asymmetry is reversed. Lateral curves are distorted by borehole effects similar to those described in the two-electrode system, as well as by electrode geometry. The depth of investigation is large and is often taken approximately equal to the spacing; for homogeneous beds of thickness greater than about 40 ft, the lateral curve measures formation resistivity ρ_t unaffected by the invaded zone. A combination of lateral and normal logs permits determination of ρ_i and ρ_t, as well as the extent of fluid invasion.

A double lateral spread, called the *limestone log*, was introduced in 1945 as a means of defining very thin porous sections; it is rarely used today. It employed two short laterals, essentially a pair of potential electrodes above, and another pair below, the current electrodes; the potential pairs are connected in parallel so that the result is a symmetrical curve.

(f) *Microlog.* The *microlog* (*wall-resistivity log*) was developed in 1949 to measure thin beds independently of adjacent zones. However, it proved more efficient as a detector of mud cake and for measuring mud resistivity. The presence of mud cake is a qualitative indication that formations are permeable. Mud cake may not form in a carbonate section with vugular or fracture porosity. Impervious formations (non-shale) are indicated by very high microlog resistivity. The *microlog* is illustrated in fig. 11.2. The button-size electrodes are embedded in an insulating

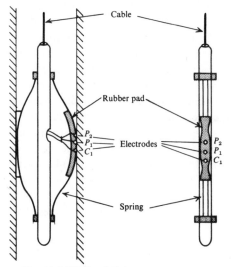

Fig. 11.2 Wall-resistivity spread (microlog).

pad which is pressed against the borehole wall by means of an expansion device, formerly spring-loaded but now power-driven. It is also used as a caliper to measure borehole diameter from 6 to 16 in.

Because the electrodes are against the wall, the effects of hole diameter, mud resistivity and adjacent beds are negligible. Since the electrodes are very closely spaced ($1\frac{1}{2}$ and 2 in. apart), very thin beds can be sharply defined, but the depth of investigation is generally less than 4 in. Differences between resistivities measured with the different electrodes is called *separation* and depends on the thickness of the mud cake. The microlog also measures mud resistivity when the electrode is not pressed against the borehole wall.

(g) *Focused-current logs.* All the resistivity logging devices described so far have serious limitations. The normal and lateral spreads are too large to measure thin beds, while the double lateral and microlog are influenced by mud and mud cake. They are ineffective with saline muds. The possibility of using a sharply focused current, first investigated in 1927, was realized in the *guard log* or *Laterolog-3* in 1949. It is illustrated in fig. 11.3a.

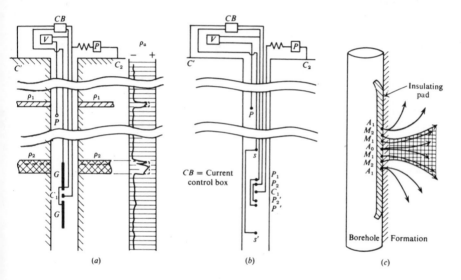

Fig. 11.3 Focused-current logs. (Courtesy Schlumberger.) (*a*) Guard log (Laterolog-3); (*b*) point-electrode focused log (Laterolog-7); (*c*) microlaterolog and schematic current-flow lines.

In order to measure resistivity, ρ_t, with vertical resolution of a few inches and in the vicinity of thin beds and conductive muds, the current is focused into a thin horizontal disc which penetrates the formation laterally instead of flowing up the walls. The focusing is achieved by maintaining P_1 and G at the same potential; this is done by a device which automatically controls the current flow to the guard electrode. The log recorded at detector V is the potential difference between C_1 and P_1. Thus the laterolog and the two-electrode log are similar except for the addition of the guard electrodes.

Depth of investigation, the distance at which the current begins to defocus appreciably, is approximately three times the length of the guards. Thus a very long guard produces greatest penetration but the lower guard prevents logging to the bottom of the hole.

The point-electrode system known as *Laterolog-7* (see fig. 11.3*b*) achieves a focused current sheet about 32 inches thick, somewhat larger than the guard electrode, by maintaining the electrodes P_1 and P_1' at the same potential as the pair P_2 and P_2'. Depth of penetration is about 10 ft if the spacing between E and the nearest guard point is 4 ft. Aside from the thicker current sheet this arrangement gives the same results as the laterolog and measurements can be made closer to the bottom of the hole.

The focusing principle is used with very small electrode spacing in the *microlaterolog*, or *trumpet log*, illustrated in fig. 11.3*c*. The electrodes are mounted like the microlog on a rubber pad which is pressed against the borehole wall; the electrodes are concentric rings, $\frac{9}{16}$ in. apart. Electrodes M_1 and M_2 are maintained at the same potential so that an essentially constant current beam is produced. The depth of penetration is about 3 in. Performance is similar to the microlog with the advantage of current focusing. This device is used to measure resistivity in the flushed zone. It also calipers the hole diameter.

(h) *Induction log.* This instrument energizes the surroundings by induction, as in electromagnetic (EM) prospecting. It was first employed in 1948 for wells drilled with high-resistivity oil-based muds. A schematic diagram is shown in fig. 11.4.

The EM field produced by a transmitting coil induces eddy currents which

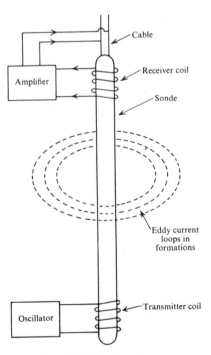

Fig. 11.4 Induction log.

flow in nearby conductive formations in loops centred on the axis of the hole. The eddy currents generate secondary EM fields which induce voltages in the receiver coil; these voltages are then amplified, rectified and transmitted up the cable. Some type of compensation (§7.4.7) must be used to minimize direct coupling of the primary field into the receiver coil. The induction log signal is proportional to the conductivity of the formations (rather than resistivity).

Additional coils are wound on the sonde to obtain a focused field. Depth of investigation for an unfocused system is about 75 % of the spacing between transmitter and receiver coils and twice this amount with focusing. The focusing coils reduce the effect of nearby conducting material greatly and cause the response to peak at a particular distance from the borehole. Thus the influence of mud resistivity, hole diameter and invasion zone can be minimized so that the instrument reads mainly the true conductivity, $1/\rho_t$. The *dual induction log* records separate responses which peak at different depths into the formations. This tool usually includes also a shallow focused log so that the log measures resistivity at three depths. The induction log with the greatest penetration depth is abbreviated ILD (deep induction log), and the one with the intermediate depth emphasis, ILM. Different resistivities at different depths (after correction for borehole effects) indicate permeability which has allowed filtrate to enter the formation; the proportion of filtrate, and hence its effect on resistivity, decreases with distance from the borehole.

Adjacent beds which are less conductive do not affect readings appreciably but those which are more conductive perturb measurements. The focusing arrangement gives generally sharp bed resolution, although resolution is affected by ρ_m and is poorer when highly conductive muds are used. The induction log works satisfactorily in dry holes and in oil-filled holes where other resistivity logging methods fail.

Ground currents generated by atmospheric effects are eliminated in the induction log because of the high frequency (20 kHz) of the transmitter. Although interfaces of low-resistivity beds are sharply marked, the log is insensitive to large contrasts in adjacent beds if both have high resistivity. Anomalous readings may result from a low-resistivity ring called an *annulus* which is sometimes produced in oil-bearing formations by the invasion process; because of their greater mobility, hydrocarbons are displaced farther beyond the invaded zone than conductive formation water, resulting in a high proportion of conductive formation water in a ring around the borehole.

(i) *Resistivity logging in mineral search.* Resistivity logging has not been used in mineral areas except in connection with IP logging and in academic experiments. Probably only a few of the electrode systems, such as the single-electrode or focused two-electrode spread and the induction log, would be useful in mineral search. Since the structure in mineral areas is more complex, the interpretation would probably be only qualitative. It should, however, locate high-conductivity zones and perhaps help in identifying and correlating them.

11.2.3 Self-potential (SP) logging

(a) *Sources of SP*. The self-potential effect has been discussed in §5.2.1 and §6.1.1. All of the principle sources of SP – streaming potential, shale potential, liquid-junction and mineralization potentials – are encountered in SP well logging. In oil well logging these potentials involve principally the boundaries of shale units, especially shale-sand interfaces. The principle effect in oil well logging is due mainly to the electrochemical (shale plus liquid-junction) potential, while in logging in mineral zones the mineralization potential is usually the dominant factor.

Shales are permeable to Na^+ cations but essentially impermeable to Cl^- anions; as a result a shale potential is set up when Na^+ ions pass from saline formation water in sands into adjacent shale beds, then into the fresh water of the mud. In addition, a liquid-junction potential develops at the interface between the fresh-mud filtrate in the invaded zone and saline formation water beyond the invaded zone; as a result of the greater mobility of Cl^- anions over Na^+ cations, there is a net flow of Cl^- into the invaded zone. These effects are illustrated in fig. 11.5.

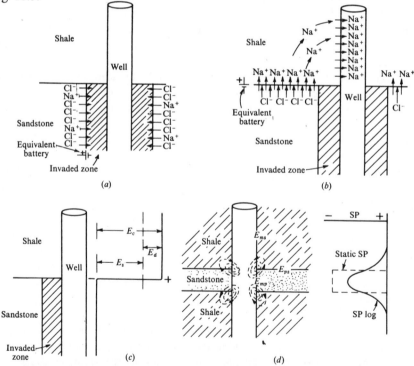

Fig. 11.5 Origin of self-potential effect in a sand-shale section. (Courtesy Schlumberger.) (a) Liquid-junction (diffusion) potential: $E_d = -11 \cdot 6 \log_{10} (\rho_{mf}/\rho_w)$; (b) shale (Nernst) potential: $E_s = -59 \cdot 1 \log_{10} (\rho_{mf}/\rho_w)$; (c) electrochemical potential: $E_c = E_d + E_s = -70 \cdot 7 \log_{10} (\rho_{mf}/\rho_w)$; (d) SP anomaly: E_{ms} = potential at mud-shale interface, E_{mp} = potential at mud-sandstone interface, E_{ps} = potential at shale-sandstone interface.

Equation (5.4) can be modified by replacing the ratio of concentrations with the resistivity ratio to give

$$E_c = -70 \cdot 7 \frac{(273 + T)}{298} \log_{10} (\rho_{mf}/\rho_w), \tag{11.6}$$

where E_c is in millivolts, T is the temperature (°C), ρ_{mf} is the mud-filtrate resistivity, and ρ_w is the resistivity of the original formation water.

The streaming potential arises because of the difference in pressure between fluids in the well and those in the surrounding formations. Equation (5.1) can be expressed in the form

$$E_k = 0 \cdot 039 \Delta P \sqrt{(\rho_{mc} t_{mc} f)}, \tag{11.7}$$

where E_k is in millivolts, ρ_{mc} and t_{mc} are the resistivity and thickness of the mud-cake respectively, f is a filter-loss factor and ΔP is the pressure difference between the borehole mud and the adjacent formation fluid.

The streaming potential usually is much smaller than the electrochemical potential, hence we can take E_c in eq. (11.6) as representing approximately the total SP-anomaly in oil-well logging.

In mineral zones involving sulphides, graphite and/or magnetite, the mineralization potential between the minerals and the surrounding rock is generally much larger than E_c or E_k or their sum, hence E_c and E_k can be ignored. Maximum SP in oil work is normally less than 75 mV whereas sulphide SP may be as much as 700–800 mV.

(b) *Instrumentation.* Equipment for SP-logging is very simple. A recording potentiometer is connected across two electrodes, one downhole and one at the surface (or in the hole near the surface). Sometimes the potential gradient is measured between two downhole electrodes at a fixed small spacing. The electrodes used in oil-well logging are oxidized lead or iron cylinders a few inches long, which should be in equilibrium with the fluids in the well. Since SP and resistivity are logged simultaneously, it is customary to carry both sets of electrodes on the same sonde.

(c) *Characteristic SP curves in oil well logging.* There are three main purposes in SP oil well logging: the location of boundaries between shales and porous beds such as sandstones, correlation between wells, and determination of formation-water resistivity. The shape of the SP curve is characteristic of certain geologic formations and their correlation can be used to indicate convergence, pinching-out, and dip of formations. Changes in formation-water salinity, such as might occur at large-scale unconformities, appear as sudden offsets in the SP baseline. A value for ρ_w is needed for eq. (11.1) as a step in calculating the water and hydrocarbon saturations; it is possible to calculate ρ_w from eq. (11.6) since the potential varies (logarithmically) with the ratio ρ_{mf}/ρ_w. Mud-filtrate resistivity can be measured separately, using samples drawn from the well.

If a thick shale is adjacent to a thick, clean, permeable sand, the maximum potential difference across the sand-shale boundary will develop between two points in the well located some distance from the actual interface. The SP values at these respective points indicate the *shale-base-line value* and the *sand-line value*. The difference is called the *static SP* or *SSP*. The magnitude of the static SP depends on the difference in salinity between the mud and the formation water. If the formation water should change salinity, as might be the case between formations above and below an unconformity, the shale base-line will shift. In fact, if the formation water should be fresher than the mud, the SP phenomena will be inverted and produce a *reverse SP*, sands being positive with respect to the shale baseline rather than negative. When a sand is not clean but contains appreciable clay or disseminated shale, the full static SP may not develop and the SP value is called the *pseudostatic SP* (PSP).

Typical SP curves for thick beds are shown in fig. 11.6*a*. The interface between shales and porous beds can be located by the point of inflection on the SP curve, the porous bed being identified by the negative peak. In thin beds, as illustrated in fig. 11.6*b*, the maximum negative SP measured may be considerably less than the ideal static value.

Figure 11.6*c* shows an SP log through several different formations. Note the appearance of the thin bedding in the laminated shale-sand, the asymmetric curves for colloidal shale-sand and rhyolite, the positive anomaly for a reducing bed and the large anomaly associated with disseminated pyrite.

Although borehole and formation factors influence the SP curve, as with resistivity logs, most of these can be corrected. The effects of hole diameter, adjacent beds and bed thickness can be eliminated by corrections from standard charts. The density and resistivity of the mud affect the curve greatly. Spurious effects due to streaming potentials depend on the mud density and can be removed by use of a correction chart based on eq. (11.7). The ratio of mud resistivity to formation-water resistivity, ρ_{mf}/ρ_w, is the main factor controlling the curve shape. The temperature also affects the curve and is corrected for in eq. (11.6).

The effect of penetration of mud filtrate into porous zones is complicated. Generally the SP deflection decreases with depth of invasion, but occasionally the reverse is true. The potential may change with time during invasion of zones containing water rather than oil; the fact that the curve is not reproducible on successive logging runs may be diagnostic of this situation.

Polarization of electrodes, as in surface SP surveys (§6.1.2), affects the SP measurement greatly. Usually the electrodes can be restored to equilibrium by leaving them in the mud for a time. Junctions of dissimilar metals in the equipment may produce spurious potentials. Abnormal telluric currents, proximity to power lines, local electrical operations, large-scale electrolytic corrosion in the vicinity and cathodic-protection devices affect SP readings. Other disturbing factors are redox potentials due to chemical differences in mud and formation water and formations.

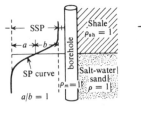
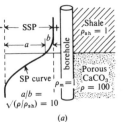
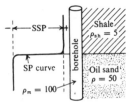

(a)

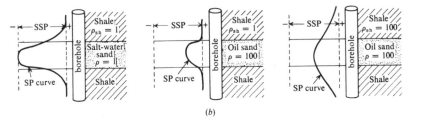

(b)

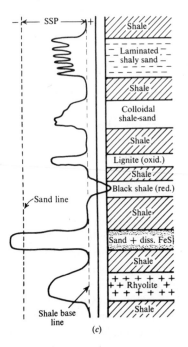

(c)

Fig. 11.6 Characteristic SP curves for a sand-shale section. (After Pirson, 1963.) (a) curves for thick beds; (b) curves for thin beds; (c) characteristic curves for various formations.

(d) *Characteristic SP curves in mineral logging.* Potentials may be considerably larger in the presence of sulphides and graphite than for sedimentary beds; consequently borehole effects are insignificant as long as the hole is filled with water. Three examples of SP logs run in diamond-drill holes (2-in. diameter) are given in fig. 11.7. From the first two curves there appears to be no correlation between the deflection and the mineral content of the anomalous zones. The largest negative potentials seem to occur at the interface between barren rock and disseminated mineralization. However, this is not too significant since 'massive sulphides' are usually inhomogeneous, being composed of many thin sections of high concentration interspersed with disseminated or barren zones.

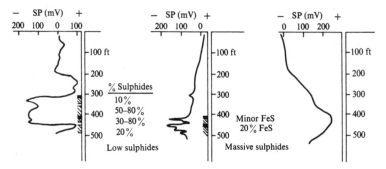

Fig. 11.7 Characteristic SP curves in mineral zones.

Two other effects in these logs are worth noting. In the second curve there is a baseline drift with depth which is not due to temperature. In the third example the broad positive anomaly is caused by a massive pyrite zone near the hole but not intersected by it. Obviously, the current flow in the barren host rock is from depth to surface.

By having both electrodes downhole at small fixed spacing (6 in.) one can measure the potential gradient. The resultant curve shows sharp deflections at the edges of mineral zones. Except for this feature, the regular electrode arrangement provides more information.

11.2.4 *The dipmeter*

In some cases formation dip can be estimated by correlation between holes where no structure intervenes, but often the determination is difficult or impossible. A dipmeter has been in use since 1943. Originally employing EM-response and later SP, it is now based on microresistivity measurements.

The present instrument (fig. 11.8*a*) employs three (or four) pads in the same plane, pressed against the borehole wall at 120° (or 90°) intervals. The upper part of the 16-foot long sonde contains an inclinometer to record the drillhole drift angle and bearing and a magnetometer to determine the azimuth of the reference pad. A hole caliper is also recorded. The log recorded in the field therefore shows the azimuth to the no. 1 electrode, the relative bearing of hole drift with respect to

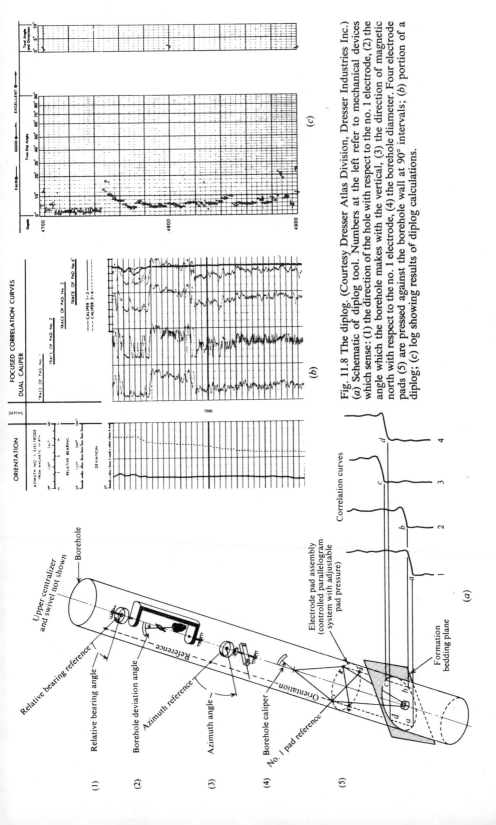

Fig. 11.8 The diplog. (Courtesy Dresser Atlas Division, Dresser Industries Inc.) (a) Schematic of diplog tool. Numbers at the left refer to mechanical devices which sense: (1) the direction of the hole with respect to the no. 1 electrode, (2) the angle which the borehole makes with the vertical, (3) the direction of magnetic north with respect to the no. 1 electrode, (4) the borehole diameter. Four electrode pads (5) are pressed against the borehole wall at 90° intervals; (b) portion of a diplog; (c) log showing results of diplog calculations.

the no. 1 electrode, the deviation of the hole from the vertical, the resistivity curves from the three or four pads, and the caliper log. Such a log is shown in fig. 11.8*b*.

Formation boundaries are sharply defined by the simultaneous recording of the resistivity curves. The dip and strike are determined from the slight differences in depth of the lithological boundaries around the borehole and the data obtained by the orientation devices. Rough interpretation based on major features in the microresistivity curves can be made fairly easily. The full value of the tool, however, is obtained only by correlation of many closely-spaced points and generally requires computational aid. The result of complete processing is usually a 'tadpole plot' such as shown in fig. 11.8*c*, where the result of each correlation is shown as a solid circle indicating the dip angle with a short ray emerging from the circle indicating the dip direction according to the normal compass orientation, i.e., north upward, east to the right, etc.

11.2.5 *Electromagnetic method*

Strictly speaking, the induction log, discussed in §11.2.2*h*, is an electromagnetic method. Modifications of EM surface-prospecting units (see §7.5.4 and §7.5.5) have been employed to a minor extent in mineral hole logging. Usually the aim is to locate mineralized (high-conductivity) zones nearby, rather than to evaluate the formations intersected by the drill hole.

An example of an EM unit is shown schematically in fig. 11.9*a*. The transmitter coil is a conventional vertical loop, set up near the hole collar and pointing

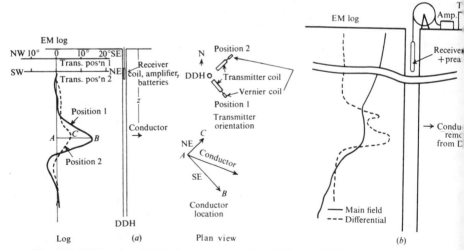

Fig. 11.9 EM logging. (*a*) Vertical transmitter-loop EM unit; (*b*) horizontal transmitter-loop EM unit.

towards it. Connected in series with this loop is a small vertical vernier coil, which can be rotated about its vertical axis. The logging device, which consists of a small receiver coil wound on a high-permeability core, amplifier, and battery power source, is lowered into the hole. By rotating the vernier transmitting coil, it is possible to null the receiver signal; the vernier azimuth angle is plotted against sonde depth. The measurement is then repeated during the uphole passage of the receiver, this time with the transmitter coil and vernier rotated 90° from the original orientation. The maximum azimuth angles of the two curves, when plotted as vector components parallel to the planes of the respective transmitter coils, produce a resultant vector which points towards the conductor, whose depth is established from the curves. The amplitude of azimuth angles is roughly indicative of the horizontal distance and/or the conductivity of the zone and the width of the curve between null points gives some idea of the vertical extent. Range of investigation is said to be several hundred feet.

A somewhat different drill-hole EM log is illustrated in fig. 11.9*b*. The transmitter loop is a large square of one to four turns laid out on the surface. The vertical depth of penetration desired and the type of objective determine the required loop size; loop sides are usually 400 ft or longer. A receiver coil with preamplifier is lowered into the hole, the gain being adjusted to maintain constant meter reading; gain is then plotted against depth. Alternatively, two coils 50 ft apart are connected in opposition to produce a difference signal as with the Turam method. The transmitter square (the hole is located at one corner) is rotated 90° for a successive logging run so that a rough estimate of conductor azimuth may be made. The interpretation is qualitative.

11.2.6 *Induced polarization logging*

Induced polarization has been applied to a minor extent in mineral hole logging. Equipment and electrode spreads are similar to surface IP, with two electrodes downhole and two effectively at infinity. Power requirements are generally smaller than with large spreads on the surface. Resistivity is usually measured at the same time. Various P_1-to-C_1 spacings are used, depending on the depth of penetration desired and the target dimensions. Some azimuthal directivity may be achieved by moving the remote current electrode on the surface. Detection may be either in the time or frequency domain.

11.3 Radioactivity methods

11.3.1 *Radioactive processes*

(a) *Introduction.* As discussed in §10.2.2, §10.3.4b, and §10.3.6, some atomic nuclei emit natural radiations and others can be induced to do so by bombardment. The nuclear radiations are in the form of alpha-, beta- or gamma-rays or neutrons. Both γ-radiation and neutrons possess appreciable penetrating power and are measured in radioactivity logging.

Well logging instruments which measure the radioactivity of nearby formations may be considered under three headings: (i) those which detect γ-radiation resulting from the natural radioactivity of the U, Th and K in the rocks, (ii) those which employ artificial γ-rays, (iii) those which use neutron sources to induce nuclear processes. Instruments employing γ-ray detectors are calibrated by measuring the detector response in the presence of a standard γ-ray source at various distances. The slow-neutron device is calibrated by surrounding it with a standard volume of hydrogen-bearing material.

(b) *Natural radioactivity.* The natural radioactive properties of rocks and minerals are described in §10.2.6. Natural radioactivity results from the presence of small amounts of U, Th and K^{40}; it is, roughly, lowest in basic igneous rocks, intermediate in metamorphic rocks and highest in some sediments. The γ-radiation from either the uranium or thorium series (see table 10.2, §10.2.4) is much larger than that of K^{40}. However, since the potassium isotope is far more common, the total background radiation is roughly equally attributable to the three elements.

Gamma-ray emissions of uranium, thorium and potassium listed in table 11.1 are based on a compilation made by Mott and Ediger (see Pirson (1958)). The energy distribution of γ-rays over a range of 3 MeV is relatively complex for uranium and thorium. The energy spectra of U and Th are broad, although there are also characteristic γ-rays from Th (2.62 MeV) and U (1·7–1·8 MeV). The γ-ray from K^{40} is monoenergetic at 1·46 MeV (see table 10.1). These may be distinguished by using a γ-ray detector sensitive only to the appropriate narrow energy band.

Table 11.1. *Emission of γ-rays by U and Th series and by K*

Energy (MeV)	Uranium series		Thorium series		Potassium	
	(Phot/sec g)	%U	(Phot/sec g)	%Th	(Phot/sec g)	%K
0·2–0·5	$9·4 \times 10^3$	36	$3·9 \times 10^3$	34		
0·5–1·0	$8·0 \times 10^3$	31	$5·5 \times 10^3$	47		
1·0–1·5	$4·3 \times 10^3$	17	$0·4 \times 10^3$	3	3·4	100
1·5–2·0	$3·2 \times 10^3$	12	$0·3 \times 10^3$	3		
2·0–2·5	$1·1 \times 10^3$	4				
2·5–3·0			$1·5 \times 10^3$	13		
Total	$2·6 \times 10^4$	100	$1·2 \times 10^4$	100	3·4	100

(c) *Interaction of γ-rays.* An energetic γ-ray may interact with surrounding material (see also §10.3.4b) by three distinct processes: (i) it may transfer its entire energy to a single atomic electron (photoelectric conversion); (ii) it may lose a fraction at a time to several electrons in a scattering process similar to successive collisions (Compton scattering); (iii) the γ-ray may disappear in the creation of an electron-positron pair. The probability of one process predominating over another depends on the energy of the photon. The photoelectric effect occurs mainly at low energies (< 0.2 MeV), while pair production can take place only if the gamma-ray energy is greater than 1.02 MeV (sufficient to create two particles of 0.51 MeV each) and becomes dominant at very high energies (> 5 MeV). Compton scattering is the most probable process in the intermediate energy range.

The three processes are all related to the density of electrons in the medium, i.e., to the atomic number, Z. Photoconversion is proportional to Z^6, Compton scattering to Z and pair production to Z^2. The attenuation of a γ-ray is thus determined by the type of material through which it passes.

Table 11.2 shows the penetrating power of γ-rays in various media. The attenuation of γ-rays is measured in terms of the material thickness which reduces their

Table 11.2. *Absorption of γ-rays in various materials*

Energy (MeV)	Half-value layer (in.)			
	Water	Sandstone, $CaCO_3$	Iron	Lead
0.2	2.0	0.83	0.26	0.055
1.0	4.0	1.8	0.61	0.34
5.0	9.1	3.9	1.10	0.58

intensity to some definite fraction of the original. The relation is exponential, as for electromagnetic waves generally:

$$I = I_0 \, e^{-\mu x}, \tag{11.8}$$

where μ is the *absorption coefficient* for a particular material.

When $I/I_0 = \frac{1}{2}$, the thickness of the *half-value layer* is $x = (1/\mu) \log 2 = 0.69/\mu$. The average energy of natural γ-rays is about 1 MeV and the range of investigation in sediments is roughly 1 ft. About half the γ-rays detected in a borehole originate within 5 in. of the borehole walls. Casing reduces the intensity by about 30%.

(d) *Interaction of neutrons.* The interaction of neutrons with surrounding matter is also diagnostic of the medium. *Fast neutrons* (KE > 0.1 MeV) are slowed down by elastic and inelastic collisions with nuclei. Elastic collisions result in a partition of energy. In inelastic collisions, the nucleus, in addition to acquiring kinetic energy, is left in an excited state and emits characteristic gamma-rays. The rate at which a neutron loses energy in elastic collisions varies inversely with mass of the target nucleus. When the neutron has been slowed down to a velocity comparable to that of normal atoms (KE < 0.025 eV), it is called a *thermal neutron* and can be captured by a nearby nucleus, which then emits characteristic γ-rays. The probability of neutron capture depends on the capture cross-section, a quantity which is measured in barns (1 *barn* $= 10^{-24}$ cm^2).

Table 11.3 gives neutron capture cross-sections and inelastic scattering cross-sections for a number of elements, as well as some of the characteristic γ-rays emitted on capture.

Table 11.3. *Neutron-capture and inelastic-scattering cross-sections with characteristic γ-rays emitted*

Element	Cross-section (barns)		γ-rays
	Inelast. scatt.*	Capture	(MeV)
Hydrogen		0·33	2·2
Beryllium	0·4	0·01	6·8
Boron		755	
Carbon	0·25	0·003	1·3, 4·9
Oxygen		0·0002	
Sodium	0·5	0·53	3·6, 3·9
Magnesium	0·7	0·27	2·8, 3·9
Aluminum	0·7	0·23	2·8, 7·7
Silicon	0·7	0·16	2·7, 4·9
Sulphur	0·8	0·52	3·0, 5·4
Chlorine	0·8	34	2·0, 6·1, 6·6
Potassium	1·0	2·1	4·4, 7·7
Calcium	1·0	0·43	1·9, 6·4
Manganese		13·3	1·8, 5·3, 7·2
Iron	1·2	2·6	5·5, 6·2
Cobalt	1·4	3·7	
Nickel	1·4	4·6	8·5
Copper	1·5	3·8	7·9
Zinc	1·5	1·1	
Molybdenum	2·0	2·7	
Silver	2·0	6·3	
Cadmium	1·9	2500	1·0–7·0
Tin	1·6	0·63	
Tungsten	2·5	19	4·8
Gold	2·5	99	4·0–7·0
Mercury	2·5	360	3·8
Lead	1·8	0·7	6·7, 7·4
Uranium		7·7	1·0, 3·0

* Average value over the energy range 1–14 MeV

11.3.2 *Gamma-ray logging*

(a) *Instrumentation.* The γ-ray log was first applied in oil wells in 1939. It is used particularly to locate and correlate formations in cased wells where the electric log cannot be employed.

The sonde consists of a detector and amplifier. In the early days of gamma logging the detector was either an ionization chamber or a Geiger counter but these devices have been replaced by the more efficient scintillation counter. All three are described in §10.3.

Several measurement units have been employed in γ-ray logging: $\mu r/hr$ (microröentgens/hour), counts/minute, μg of Ra-eq/ton (micrograms of radium-equivalent/ton) and API (American Petroleum Institute) units (which are different from the API neutron units of §11.3.4b). The radium-equivalent/ton is based on a concentration of uranium in shale (0·002%) and the fixed U/Ra ratio, 3×10^6. Hence common shale has a Ra-equivalent of 6 $\mu g/ton$. Gamma-ray logs now are usually calibrated in API units. The difference between the high- and low-radioactivity sections of cement in the API calibration pit at the University of Houston, Texas, is defined as 200 API units; average shales have values around 100 API units.

(b) *Applications.* In sediments the γ-ray log reflects mainly shale content because radioactive elements tend to concentrate in clays and shales. Hence the log accomplishes essentially the same purposes as the SP log and it is generally correlatable with the SP log. It can be used to replace the SP log in very resistive formations or where there is little difference between the salinity of the mud and formation water. The log can be run in empty holes and in cased holes.

Figure 11.10 shows a typical γ-ray log. Statistical variations, significant at low

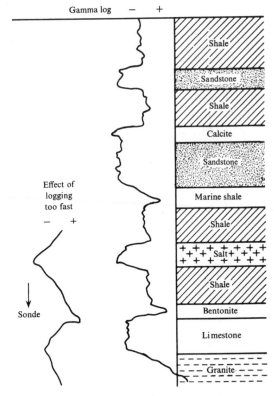

Fig. 11.10 Typical gamma-ray log.

counting rates are smoothed out by integration over a short time interval. If the hole is logged too quickly, however, the smoothing effect leads to erroneous results, as illustrated in fig. 11.10. Barren zones appear shifted in the direction of logging by a foot or so. The length of the detector with respect to the bed thickness also affects the curve shape.

Borehole effects are of minor significance on the γ-ray log and can be corrected. The interface between adjacent barren and radioactive beds can be located fairly accurately at half the maximum deflection when the beds are thick (> 6ft). For thinner zones the centre can be taken as the peak deflection. Best resolution is obtained with a short detector, particularly for thin beds.

11.3.3 *Density log*

(a) *Instrumentation*. When first introduced in 1953, the density log or gamma-gamma log was intended as an aid to gravity prospecting by measuring rock density *in situ*. Its value, however, has been in formation evaluation, because density is closely correlated with porosity.

Figure 11.11 is a schematic diagram of the density logger. The bottom of the sonde contains a concentrated source of monoenergetic γ-rays, usually $_{27}Co^{60}$

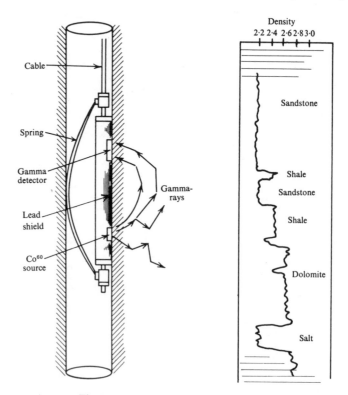

Fig. 11.11 Density logger and typical log.

(1·17 and 1·33 MeV) or Cs^{137} (0·66 MeV). The detector, either a Geiger or scintillation counter, is about 18 in. above the source. The sonde is spring-loaded to bear against the borehole wall and, since both source and detector are surrounded by lead shielding except for a window facing the wall, the only γ-rays which reach the detector are those which have travelled through the adjacent formation. The spring force which presses the sonde against the borehole wall is considerable and the skid has a plough-shaped leading edge so that it cuts through soft mud cakes. The γ-rays from the source interact with the rock elements mainly by Compton scattering, to a lesser extent by photoconversion (and pair production for a Co^{60} source). The detected γ-ray intensity is an exponential function of the rock density. The maximum depth of investigation is about 6 in., with most of the signal coming from the first 3 in. of the wall rock.

The instrument must be calibrated for source intensity, detector sensitivity, mud density, and hole diameter. It may be used in dry holes as well as those filled with mud fluid. A hole-caliper log is an auxiliary output. Since the tool is short, adjacent beds or bed thickness have little distorting effect. The logging speed must be adjusted in relation to the instrument time constant to avoid distortion of the curves and loss of sensitivity.

The newer compensated density-logging sonde employs two detectors at different spacings from the source. The shorter spacing is more affected by mud buildup and the difference between the readings from the two detectors is used to correct for the thickness and density of the mud cake.

(b) *Applications*. The density log responds to the density of electrons in the formation. Most elements have 1/2 or less electrons per atomic-weight unit; hydrogen is an exception and a few other elements also depart slightly from this ratio. The density tool is usually calibrated for fresh-water-saturated limestone. Corrections are needed to give bulk density in sandstone, dolomite and for substances like rock salt, gypsum, anhydrite, coal, and gas, for which the ratio of electron density to atomic weight departs from that assumed.

Porosity ϕ can be derived from bulk density σ_b by the rigorous relation,

$$\phi = \frac{\sigma_{ma} - \sigma_b}{\sigma_{ma} - \sigma_f}, \tag{11.9}$$

where σ_f and σ_{ma} are the densities of the formation fluid and the rock matrix. Errors may result with formations which contain gas (because of erroneous values assumed for σ_f) or disseminated shale. The matrix density of most clay minerals is close enough to that of quartz that no difficulty is encountered in calculating the porosity of shaly sands. The bulk porosity determined with this log includes secondary porosity and porosity which is not interconnected. If interpreted along with other porosity-sensitive logs (see §11.9, example 5), some of these complicating factors can be separated out.

11.3.4 *Neutron logging*

(a) *Neutron sources and detectors.* Neutron logs, which depend on response induced by bombardment of the formations by neutrons from a source in the logging tool, have been in use since 1941. They respond primarily to hydrogen content and are particularly useful in locating porous zones and determining the amount of liquid-filled porosity.

Several neutron sources have been used for the neutron log. These include a combination of beryllium with an α-particle source such as radium, polonium plutonium, or americium. The reaction is

$$_2He^4 + {}_4Be^9 \rightarrow {}_6C^{12} + {}_0n^1. \tag{11.10}$$

The radium-beryllium source has a half-life of 1620 years, but produces a high flux of γ-rays which are a nuisance. The polonium-beryllium source gives few γ-rays, but the neutron yield is lower and the half-life is only 140 days. Plutonium beryllium and americium-beryllium are good neutron sources, being very low in γ-rays and having a long half-life, 24,000 years for Pu-Be, 460 years for Am-Be.

Charged-particle accelerators have also been used to produce neutrons by the reactions

$$_1H^2 + {}_1H^2 \rightarrow {}_2He^3 + {}_0n^1,$$

and

$$_1H^2 + {}_1H^3 \rightarrow {}_2He^4 + {}_0n^1. \tag{11.11}$$

The first reaction produces neutrons of 2·3 MeV and the second, 14 MeV. Compared with the capsule sources, these neutrons are monoenergetic and the source can be shut off.

A neutron-logging instrument is shown schematically in fig. 11.12. Several types of detectors are used. Some are sensitive to both high-energy γ-rays resulting from neutron capture as well as thermal neutrons, others use a proportional counter which is shielded so that only neutrons with energy above some threshold are detected.

(b) *Neutron log (hydrogen-index log).* The neutron log measures porosity by determining the amount of hydrogen, and hence the amount of fluid, filling pore spaces. It was the first nuclear log to be used. Neutrons lose more energy per collision when the nuclei with which they collide have comparable mass. Hence the rate of energy loss (*moderation*) of fast neutrons is proportional to the density of protons (which have nearly the same mass as the neutrons). After the neutrons have been slowed to thermal energies, they may be captured by nuclei which then emit high-energy capture γ-rays. The amount of hydrogen per unit volume is called the *hydrogen index*.

Any one of several types of detectors can be used. Either the capture γ-rays or the neutrons themselves may be counted. In porous formations saturated with water or oil, the neutrons lose energy rapidly so that the counting flux is high most of the response being within 7 in. or so of the borehole. In low-porosity formations the neutrons penetrate farther before losing energy (distances of the order of 24 in.), producing low counting flux. Where thermal neutrons or γ-rays resulting from capture are measured, the nature and abundance of the capturing nucleus has a perturbing effect. Best resolution is obtained when the hole is small so that fewer neutrons are lost in the mud column. Sometimes the source-detector spacing is increased to 30 in. for operation in large diameter holes.

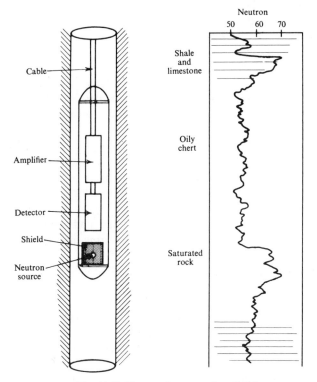

Fig. 11.12 Neutron logger and typical log.

The *sidewall neutron log* measures *epithermal neutrons*, those whose energy is greater than that of thermal neutrons which are involved with capture. Hence, it is not perturbed by neutron absorbers such as chlorine and boron. The *compensated neutron log* employs two detectors spaced at different distances from the source. This makes it possible to correct for mud cake and hole roughness (*rugosity*) effects.

Neutron logs are sometimes scaled in counts per second, more commonly in API neutron units, sometimes in 'limestone porosity' which assumes that the matrix is limestone (porosity for nonlimestone formations can be found from correction charts). *API neutron units* are based on measurements in a standard neutron pit where 19%-porosity water-filled limestone is defined as 1000 API neutron units.

Correction has to be made for the salinity of water. The salinity is usually taken as that of the mud in invaded zones and of the formation water in cased holes.

Neutron logs are affected by all protons, including those in bound water associated with shales or water of crystallization such as contained in gypsum. Consequently, shales and gypsum have distorting effects.

(c) *Pulsed-neutron logging*. Pulsed-neutron logs, which include the thermal-decay-time and neutron-lifetime logs, record the rate of decay of thermal neutrons. Chlorine, which is the most common element with high capture cross-section (see table 11.3), is found in most formations, and is the principal absorber of thermal neutrons. Consequently, this log determines mainly the amount of chlorine, i.e., the amount of saline water present. Thus, it accomplishes essentially the same purpose as resistivity logs and can generally be correlated with them. Its principal advantage over resistivity logs is that it can be used in a cased hole.

A neutron generator emits a burst of high-energy neutrons. These are rapidly slowed to thermal energies and captured, whereupon capture γ-rays are emitted. The γ-rays arriving during a fixed time interval after the burst are detected a short distance away from the source. Measurements are made over two time intervals to determine the *die-away* time; this permits correction for the background effect and the determination of the rate of thermal-neutron capture. Absorption by borehole fluid and casing affects primarily readings made soon after the burst and is largely eliminated by delaying the first measuring interval.

11.4 Elastic-wave propagation methods

11.4.1 *General*

The elastic properties of rocks have been dealt with in chapter 4, §4.2.1 and §4.2.3. Acoustic logs measure elastic or seismic energy in boreholes. In 1954 the continuous-velocity log or sonic log (see §4.5.6b) was introduced to determine seismic velocity as an aid to seismic prospecting. It soon became apparent that seismic velocity correlated with porosity and this is now its principal application. Acoustic logs also have several other applications, including ascertaining the quality of cement bond in cased holes and in questions of fracturing.

11.4.2 *Sonic log*

(a) *Instrumentation*. Early continuous-velocity loggers consisted of a single transmitter a few feet away from a receiver. A burst of sonic energy emitted by the transmitter travelled to the receiver by various paths: through the borehole fluid, as P-waves in the borehole rock, as tube waves, and by other propagation modes. The fastest travel path is for P-waves in the rock surrounding the borehole. The arrival time of the first energy is measured.

Several factors may produce errors in the arrival time. If the sonde is not against the borehole wall, if there is mud build-up or irregularities of the wall, measurements are affected. These errors are eliminated by measuring the difference of arrival time at two receivers at different distances, which eliminates effects opposite the transmitter. Two transmitters are used, one above the receivers and the other below them, to eliminate effects opposite the receivers. The *borehole-compensated sonic log* employs two transmitters (which are pulsed alternately) and four receivers; this sonde is described in §4.5.6b and illustrated in fig. 4.82a.

(b) *Porosity determination.* The sonic log displays the time interval for the sonic wave to travel one foot, expressed in μsec/ft. Porosity usually is determined from the empirical *time-average equation* developed by Wyllie et al. (1958):

$$\Delta t = 1/V_r = (\phi/V_f) + (1 - \phi)/V_m, \tag{11.12a}$$

or,

$$\phi = \frac{\Delta t - \Delta t_m}{\Delta t_f - \Delta t_m}, \tag{11.12b}$$

where $\Delta t = $ *formation transit time*, $V_r = $ measured formation velocity, $V_f = 1/\Delta t_f = $ velocity in the fluid which fills the pore spaces, $V_m = 1/\Delta t_m = $ velocity in the matrix material, and $\phi = $ matrix porosity. Velocities in various matrix materials for use in this calculation are given in table 11.4.

Table 11.4. *Fluid and matrix velocities*

Material	V_f or V_m (ft/sec)	Δt (μsec/ft)
Water (20% salt to pure)	5,250–4,600	190–217
Salt	15,000	66·7
Shale	16,000	62·5
Iron casing	17,500	57·1
Unconsolidated sands	17,000	58·8
Compacted sands	19,000	52·6
Anhydrite	20,000	50·0
Limestone	21,000	47·6
Dolomite	23,000	43·5

The interval time in sands does not depend markedly on the material which fills the pore space (mud filtrate, formation water, oil or silt) for many formations. In shaly sands the indicated porosity is too large. Where the sand is filled with oil or gas, the actual porosity will be 90% to 70% of the measured porosity. In uncompacted or overpressured sands the time-average equation gives porosity values which are too large. In carbonates the sonic log responds mainly to primary porosity rather than to the total porosity in vuggy formations. Empirical correction methods are often used to compensate for inadequacies of the time-average equation.

11.4.3 *Acoustic-amplitude log*

The sonic log measures only the travel time of the first arrival, but the amplitudes of the various arrivals also convey information. This is used in determining the quality of the bond of cement-to-casing and cement-to-formation; it is also used in locating rock fractures. The first-arrival casing signal is strong where the casing is not well cemented to the formation. Fractures cause the sonic amplitude to decrease markedly, often by a factor of 10. However, bedding planes, thin shale streaks, and healed fractures sometimes give much the same response.

11.4.4 *Microseismogram (variable-density) log*

Not only the first arrival but the entire sonic wave train is recorded on the micro-seismogram, or variable-density, log; such a log is shown in fig. 11.13. These logs are often used to determine the quality of the cement bond. Potentially, they also have other applications.

11.5 Magnetic methods

11.5.1 *Introduction*

Magnetic well logging has had limited application. Either the magnetic field or formation susceptibility is measured. Anomalous susceptibility may indicate the presence of magnetic minerals such as magnetite, ilmenite, pyrrhotite, etc. A nuclear-magnetic-resonance log has been used to study the hydrogen in formation fluids and to estimate permeability in reservoir rocks.

11.5.2 *Magnetic field logging*

Both fluxgate and nuclear precession instruments (see §3.4.3 and §3.4.4a) have been adapted for measurements in boreholes. Magnetic field logging has also been used for vertical gradient measurement and to fix the depth of magnetic anomalies.

11.5.3 *Susceptibility log*

Direct measurement of magnetic susceptibility is of some value in the location and correlation of formations, although it is not useful in the identification of rock type.

Instrumentation is similar to the field susceptibility meter (§3.3.8b). The sonde is illustrated in fig. 11.14. The solenoid is wound on a core of low-reluctance material and is connected in one arm of an inductance bridge. If the bridge is balanced in a barren environment, the presence of formations of anomalous susceptibility and conductivity unbalances it, since the susceptibility effect changes the reactance and produces a quadrature voltage while the conductivity produces an in-phase voltage. The two effects are separated by phase detectors and recorded so as to display susceptibility and conductivity independently.

Good correspondence between SP and susceptibility logs indicates that porous

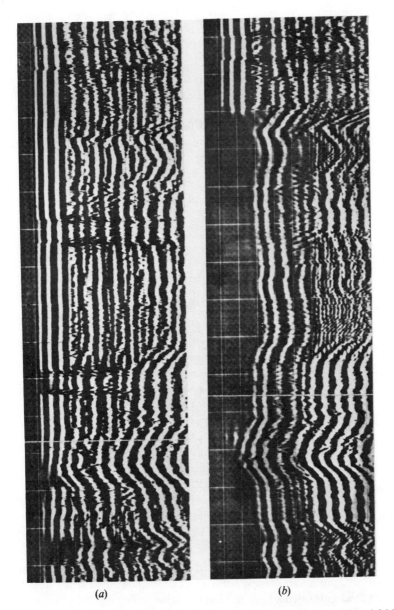

(a)   (b)

Fig. 11.13 Microseismogram log used to indicate cement bonding. (Courtesy Welex.) (a) Before cementing, the first arrival is a strong signal transmitted through the casing; (b) after bonding the casing to the formation with cement, the first arrival is energy which travelled through the formation.

zones have been enriched by ground-water deposition of iron minerals. The susceptibility log is not affected by mud resistivity and can be run in dry holes. The conductivity log has little character in low-resistivity muds, but compares reasonably with a resistivity log for $\rho_m > 2\cdot0$ Ωm. Depth of penetration is about equal to the coil length.

11.5.4 *Nuclear-magnetic-resonance log*

The nuclear precession magnetometer is described in §3.4.4a. By exciting the 'unbound' hydrogen nuclei present in formation fluids (in effect, these fluids replace the bottle of water used in the instrument described in §3.4.4a), it is possible to determine the amount of free fluids. There is no response from such hydrocarbons as tar and asphalt nor from water of hydration in clay lattices.

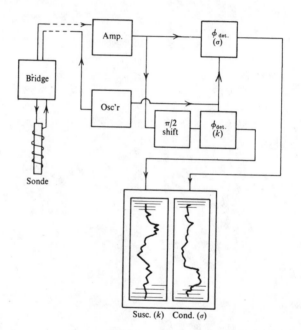

Fig. 11.14 Magnetic susceptibility logger.

Free hydrogen nuclei are first oriented by a strong applied field. When this field is cut off, the nuclei precess for a short time about the earth's field before becoming randomly oriented again. The precession sets up an alternating magnetic field which induces a damped sinusoidal voltage in the receiver coil. Fluid zones may be identified and, potentially, the thermal relaxation time can be measured to distinguish oil from water.

11.6 Thermal measurements

Measurement of temperature in well holes is the oldest logging technique. Although it has been employed mainly to determine heat flow, it has also been used to locate thermal anomalies caused by fluid and gas flow, abnormal radioactivity, and oxidation regions.

Various types of thermometers have been used, including maximum thermometers, platinum-resistance thermometers, and thermistors. The latter two are usually connected in one arm of a Wheatstone bridge. The thermistor has also been used to control the frequency of an oscillator; use of this device reduces measurement time, inevitably rather long with thermal devices. Usually downhole temperature is measured with respect to the surface. Occasionally, two detectors at short spacing are used to measure temperature gradient.

11.7 Gravity logging

Borehole gravimeters are fairly new and have not yet been used extensively. Measuring gravity at stations in a borehole permits calculation of the mean density of the intervening formations, which in turn permits calculation of the porosity. An accuracy of 0·010 mgal and a station spacing of 20 ft permits determining the average density between the stations to 0·01 g/cm^3, which corresponds to about 1 % in average porosity. The depth of investigation is very great (about five times the interval between stations compared with inches for the other porosity-logging tools) so that measurements are little affected by invasion or alteration of the formations by the drilling. Casing does not affect results significantly; this makes it possible to locate porous formations previously not detected, in old, inadequately-logged holes, for example.

Borehole gravimeters also permit more accurate density determination than possible with other logging tools. Borehole gravity also has the potential of locating density changes nearby which were not penetrated by the borehole, such as proximity to a saltdome or vugular porosity in limestone reefs.

The Esso vibrating-string borehole gravimeter utilizes the principle that a change in gravity changes the tension and hence the frequency of a vibrating string. An accuracy of 0·300 mgal can be achieved with this meter.

The LaCoste-Romberg borehole gravimeter is a variation of the shipboard gravimeter mentioned in §2.5.4b. An accuracy of 0·005 to 0·010 mgal can be achieved with it. It is, however, a large tool which cannot be run in a hole smaller than 6 or 6¼ inches. It also has operating limits (110°C and 14,000 psi) which prevent its use in hot or deep boreholes.

Borehole gravimeters have to be at rest during measurements and 10 to 20 minutes are required for each reading. The meters have some drift so that stations have to be repeated to achieve maximum accuracy. The LaCoste-Romberg meter is so expensive that it cannot be risked in open boreholes unless they are in excellent condition.

11.8 Well logging interpretation

Interpretation of well logs for mineral objectives is usually qualitative, i.e., locating and correlating anomalous zones. Interpretation for oil objectives, on the other hand, is highly developed. A great variety of methods is employed and an enormous amount of data is accumulated. The formations to be evaluated are relatively simple and uniform. Oddly enough, geophysicists play a minor role in oil well logging. Routine interpretation is usually carried out by the logging contractor, while the overall assessment of the logs is likely to be made by an oil-company geologist who is specially trained in the techniques and who has all the pertinent data, classified or otherwise, at his disposal. Detailed log interpretation for a complete evaluation of porous and permeable formations for potential production is beyond the scope of this book.

Inspection of the conventional electric logs (SP, normal, lateral, wall-resistivity, induction) often can locate, correlate and identify formations of interest. When this qualitative information is combined with data obtained from additional logs (caliper, acoustic, radioactivity), the interpretation begins to be diagnostic and quantitative. Finally, in favourable situations, identification of fluids and quantitative estimates of porosity, fluid content, water-oil ratio, etc., may be made. The results are controlled by the combination of logs available and borehole and drilling factors.

11.9 Field examples

Although the detailed interpretation of well log data is beyond the scope of this book, a few simple examples may indicate the possibilities. These are taken mainly from Pickett (1970). The variety of minerals encountered in these examples illustrates the versatility of logging techniques, especially combinations of logs.

1. *Analysis of an oil sand.* Figure 11.15 shows SP, resistivity and acoustic-velocity logs for a Miocene sand section containing gas and oil. The SP log has a distinct break of 100 mV from positive to negative at 9270 ft, indicating shale above and sand below this (compare with fig. 11.6). Having found the mud filtrate resistivity by other means, we can use eq. (11.6) to get ρ_w; it is about 0·06 Ωm. Using a combination of the lateral and normal curves with departure charts to correct for borehole, invasion and thin-bed effects, we obtain the formation resistivity, ρ_t. It is about 30 Ωm from 9272 to 9308 ft and 0·6 Ωm between 9308 and 9350 ft. Finally, by means of eq. (11.3), the water saturation S_w is found to be about 15 % between 9272 and 9308 ft and 100 % below 9308 ft.

Significant qualitative information may also be derived from these logs. Separation of the two microlog curves (note that the electrode spacing is different) indicates sections which are more permeable. The resistivity logs suggest which of these contain gas and oil (because of the resultant high resistivity).

In this rather simple example using only four logs, we can estimate a possible 15 ft of gas-bearing sand and 15 ft of oil-bearing sand, both having an average

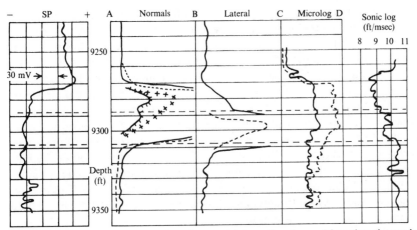

Fig. 11.15 Log suite in Miocene sand section containing oil and gas. Note that the sonic log is given as velocity instead of the usual transit time. The scales for the electric logs are: $A - B = 10\,\Omega m$ for the 16″ normal (solid curve), $A - B = 100\,\Omega m$ for the amplified 16″ normal (solid hachured curve) and for the 64″ normal (dashed curve), $A - B = 1000\,\Omega m$ for the amplified 64″ normal (dashed hachured curve); $B - C = 10\,\Omega m$ for the 18′ 8″ lateral (solid curve) and $100\,\Omega m$ for the amplified 18′ 8″ lateral (dashed). $C - C = 10\,\Omega m$ for the 1″ × 1″ microlog inverse (solid curve) and also for the 2″ micronormal (dashed).

porosity of 30% (calculated from the sonic log and eq. (11.12b)) and a water saturation of about 15%.

2. *Analysis of carbonate section.* The section in fig. 11.16 consists of dolomitic sands, evaporites, carbonates and shaly carbonates. Gamma-ray, sonic, SP and induction logs are shown at the left. The break in the SP curve is less definite than

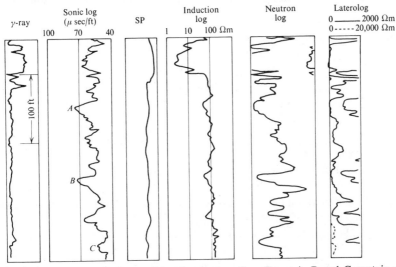

Fig. 11.16 Log suite in Minnelusa oil-bearing sand section. Zones A, B and C contain oil.

in the previous example so that an estimate of ρ_w would be unreliable. The induction log gives a reasonable value for ρ_t in the sands, but not for the higher-resistivity carbonates. The porosity estimate from the acoustic log was questionable because the values of V_f and V_m for eq. (11.12) were not well established. Thus an evaluation of the carbonates was not possible with this log combination.

Neutron and focused resistivity logs were added to aid in determining ϕ and ρ_t in the carbonates. In both these devices the relative response between different zones was reliable but the absolute calibration was not. Sand porosities obtained from the acoustic log were used to calibrate the neutron-log response and the carbonate porosities were then determined. A similar calibration of the Laterolog-7 using the induction-log response in sands permitted estimates of ρ_t in the carbonates. Values of ρ_w were obtained by measurements on cores. This interpretation does not seem as satisfactory as in the previous example; more logs would be required for a more reliable evaluation.

3. *Geological interpretation of SP.* In addition to its use in identifying shales (and especially, distinguishing shales from sands in a clastic sequence) and for correlating corresponding points from well to well, other stratigraphic interpretation can sometimes be inferred from the SP curve. In a somewhat simplistic way, the SP value is read as the degree of shaliness (or the relative abundance of clay minerals) and as the inverse of the 'energy' in the original depositional environment. For example, proximity to a shoreline where there was wave action represents 'high energy', with the consequent removal of clay minerals. Hence an SP curve which gradually increases in shaliness as we approach the surface indicates a receding shoreline and hence a transgressive sea. Conversely, increase in shaliness with depth is interpreted as a regressive sea. This concept is used to develop 'theoretical' SP curve shapes such as in fig. 11.17. This concept can be expanded to many other

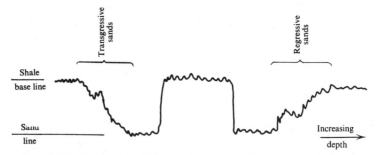

Fig. 11.17 Characteristic SP curves for transgressive and regressive sands.

types of sedimentation patterns. Correct recognition of such patterns in actual SP curves, however, is often not very clear.

Figure 11.18 shows short portions of SP and resistivity logs in four wells located approximately in a north-south line. The correlation of corresponding points on these curves is clear despite minor differences. The resistivity curve in the vicinity of the point *A* is characteristic of a marker bed which can be correlated over a fairly wide region. Obviously the formations dip south about 400 ft in 20 miles; this is a gentle slope of ¼ degree. The SP curve deflection to the left in the region *B* indicates sand in a predominantly shale environment. The shortened distance between marker *A* and the top of this sand in well no. 1 indicates missing section – a normal fault with about 150 ft of throw. The lower portion of sand *B* indicates a regressive pattern as far as well no. 3. Sand *B* is not seen in well no. 4; correlations of points below the sand compared with those above indicate that the sand has not merely been faulted out but rather that shale was being deposited at the location of well no. 4 while sand was being deposited in the other three wells. Hence a sand-shale facies boundary must lie between wells no. 3 and no. 4. Such a pinch-out, of course, represents a potential oil field.

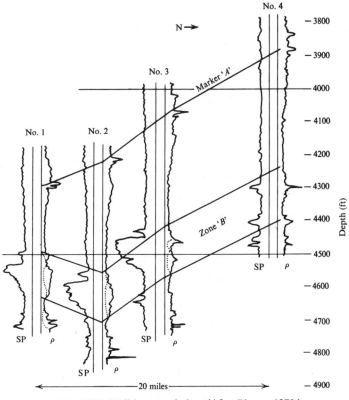

Fig. 11.18 Well-log correlation. (After Pirson, 1970.)

4. *Structural and stratigraphic conclusions from dipmeter logs.* The variation of dip with depth often indicates geological structure, as shown in fig. 11.19. The dip distortions resulting from fault movement may not extend very far from the fault plane. Various types of sedimentation patterns sometimes may be distinguished provided many detailed correlations exist.

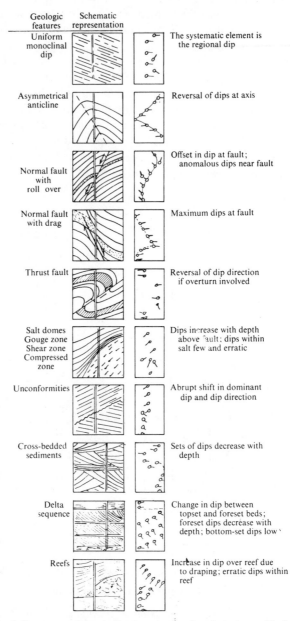

Geologic features	Schematic representation	
Uniform monoclinal dip		The systematic element is the regional dip
Asymmetrical anticline		Reversal of dips at axis
Normal fault with roll over		Offset in dip at fault; anomalous dips near fault
Normal fault with drag		Maximum dips at fault
Thrust fault		Reversal of dip direction if overturn involved
Salt domes Gouge zone Shear zone Compressed zone		Dips increase with depth above fault; dips within salt few and erratic
Unconformities		Abrupt shift in dominant dip and dip direction
Cross-bedded sediments		Sets of dips decrease with depth
Delta sequence		Change in dip between topset and foreset beds; foreset dips decrease with depth; bottom-set dips low
Reefs		Increase in dip over reef due to draping; erratic dips within reef

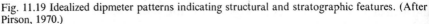

Fig. 11.19 Idealized dipmeter patterns indicating structural and stratigraphic features. (After Pirson, 1970.)

5. *Cross-plotting of porosity logs*. The three major porosity logs which measure different physical quantities yield different porosity values in some situations. This fact is used to determine lithology by means of cross-plots. The porosity calculated from one kind of log is plotted against the porosity calculated from another kind of log; the position of the plotted point indicates the lithology and the actual porosity. Such a crossplot is shown in fig. 11.20*a*. Cross-plots between many types of log measurements are extensively used along with other inter-pretation schemes which employ measurements from combinations of logs. Secondary porosity, for example, affects neutron- and density-log porosity measurements but does not affect sonic-log measurements to the same degree.

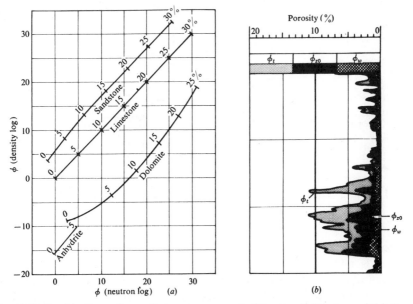

Fig. 11.20 Combinations of porosity-log measurements. (Courtesy Schlumberger.) (*a*) Cross-plot of porosity determined from different logs; (*b*) movable-oil plot. The curve at the left shows porosity calculated from a sonic log, the centre curve shows porosity calculated from a laterolog and the right-hand curve shows porosity calculated from a microlaterolog.

The fact that the porosity determined from different logs is different is used in the '*movable-oil plot*' (fig. 11.20*b*). 'Total porosity' (ϕ_t) is porosity calculated from a sonic log, 'apparent water-filled porosity' (ϕ_w) is calculated from a deep-investigation resistivity log such as a laterolog, and 'flushed-zone porosity' (ϕ_{x0}) is calculated from a shallow resistivity log such as a microlaterolog. The difference between the latter two curves is interpreted as 'movable hydrocarbons', the difference between the first two as 'residual hydrocarbons'.

6. *Coal identification*. Coal may be identified by high resistivity, low density and low acoustic velocity. Electric logs were used as early as 1931 for this purpose.

Figure 11.21 shows a section containing bituminous coal beds from a well in Colorado. The logging program included density, sonic, induction, 16-inch normal, gamma-ray and SP logs. The first three correlate particularly well with coal seams.

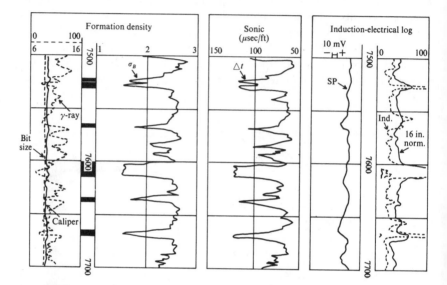

Fig. 11.21 Identification of coal beds by density, sonic and resistivity logs. (From Tixier and Alger, 1970.)

Generally the Δt-values from the sonic log are larger for coal than in the adjacent shale beds, although the contrast depends on the coal grade and depth, both of which affect the compaction. For example, lignite produces a larger Δt-excursion than anthracite, but increasing depth of burial will reduce the variation. The same factors affect the resistivity log, since higher grade coal and deeper beds contain less moisture and consequently have higher resistivity. The density log is probably the most reliable, since coal density is considerably lower than that of the adjacent beds, ranging from a maximum of 1·8 for anthracite to less than 1·0 g/cm³ for lignite. The SP curve occasionally is anomalous opposite a coal seam.

7. *Evaporites.* Caliper, γ-ray and density curves through a section of interbedded shale, halite, and anhydrite are displayed in fig. 11.22. Halite and anhydrite are non-radioactive evaporites. The gamma-ray log would be more useful for potash, sylvite and similar varieties containing potassium. The γ-ray log identifies the shale beds because of their higher radioactivity. The caliper log shows hole enlargement in the salt and shale zones. Anhydrite, with a density of nearly 3 g/cm³, is clearly identified by the density log, while the intercalated shale is indicated by coincident highs and lows in the γ-ray and density curves.

8. *Sulphur*. Sulphur, which occurs mainly in limestone, may be identified by the density or acoustic log because of its low density and low velocity (large Δt). The neutron log is also useful in sulphur detection. Occasionally the neutron log is replaced by a resistivity device for porosity determination.

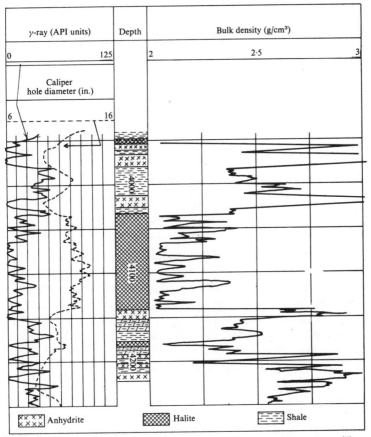

Fig. 11.22 Identification of evaporites by density, gamma-ray and caliper logs. (From Tixier and Alger, 1970.)

In formations containing only limestone, sulphur and water, two of these logs may suffice to provide a quantitative evaluation as well as identification. Where other rock minerals are also present, it may be necessary to employ all three logs.

9. *Slate and chert*. Figure 11.23 shows resistivity, γ-ray and magnetic suscepti-bility logs through a section of slate and chert beds. The γ-ray log clearly shows the slate because of its K-content and the susceptibility curves show the chert because it is enriched with magnetite.

10. *Mineral exploration*. The Lac Dufault orebody northwest of Noranda, Quebec, is a classic example of the use of geophysical well logging in mining

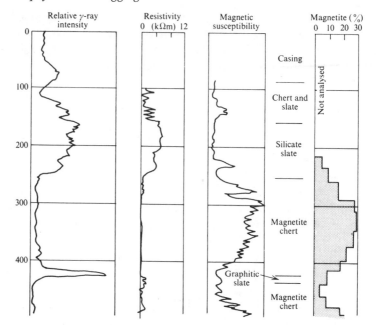

Fig. 11.23 Gamma-ray, susceptibility and resistivity logs in chert and slate beds.

exploration where no other technique is feasible. Massive sulphides, pyrite, pyrrhotite, chalcopyrite and sphalerite, are found in gently dipping contacts between rhyolite and andesite at depths greater than 1000 ft. Although the lateral extent of these ore zones is small, the high grade of chalcopyrite and sphalerite makes an attractive mining operation, provided they can be located. However, a deep-hole diamond drilling program on 200 ft centres is extremely costly.

Salt (1966) described a logging study which was carried out in 1962 using vertical-loop EM, horizontal-loop EM (large Turam-type transmitter loop), induced-polarization and resistivity methods. The problem was to establish the existence of a nearby mineralized zone by logging a hole which missed it. It was found that any one of these methods would detect an orebody roughly 400 × 400 ft in horizontal extent and 150 ft thick using a vertical drill hole within 125 ft of the edge of the orebody. However, it was difficult to determine the direction of the orebody with respect to the hole and to distinguish between massive sulfides and other conductors of unknown character. A plan of the orebody and diamond-drill-hole locations can be seen in fig. 11.24*a*.

Two logs from this study are shown in figs. 11.24*b* and 11.24*c*. Figure 11.24*b* shows metal factor and apparent resistivity measured with a frequency-domain IP unit. Large cable-coupling effects made it necessary to place the current and potential electrodes in separate drill holes (N125 and N135); with one current electrode at a distant point on the surface, a second current electrode was lowered down one hole and the two potential electrodes down the other at the same time. In fig. 11.24*b* the anomaly around 400 ft is not explained, but a definite peak at 1100 ft corresponds to the massive sulphides east of the electrodes. Both the MF and ρ_a peaks decrease slowly to 1300 ft, indicating that the sulphides lie mainly below 1100 ft. The response at 1400 ft is caused by disseminated sulphides below the main ore zone.

The second log, fig. 11.24*c*, is essentially the vertical potential distribution produced by current flow from two orthogonal pairs of current electrodes (connected alternately) at the surface. One potential electrode is fixed near the top of D.D.H. N135 and the other is lowered in the hole, while direct current flows from north to south between surface electrodes 2000 ft apart straddling the hole. Then the movable potential electrode is raised, with current flow east-to-west between similarly-spaced current electrodes. In both curves the potential increases steadily downhole to about 850 ft. The east-west potential curve remains relatively constant between 850 and 1050 ft and then falls off at greater depths. This effect is not apparent in the north-south curve, although the positive gradient is not so pronounced below 850 ft. From the differences between the two curves and the direction of current flow, one concludes that a conductor is located east of drill hole N135 and has a depth extent no greater than 200 ft.

Neither log is conclusive by itself, nor were the other techniques used in the survey. However, the reduction of drilling costs by allowing increased hole spacing would be significant and the possible control of future drilling programs by immediate logging is attractive.

11.10 Problems

1. In field example 10 above, is it possible to conclude from the IP-log that the conductor is definitely located east of drill holes N125 and N135 or merely that it is either east or west?
A direct-current source was used with the surface electrodes for the resistivity log in hole N135. By sketching the current lines and equipotentials, attempt to reproduce the curves in fig. 11.24*c*. How would you change either curve if the electrode polarization were reversed? Sketch the east-west potential curve if the drill holes were east of the orebody. What difference would it make if a.c. were used?

2. The IP log shown in fig. 11.25 was obtained in a base-metal survey in Northwestern Quebec. The mineralization consists of pyrite (up to 20%) and chalcopyrite (maximum 2·6% Cu) in a host rock of tuffs and agglomerates. One current and one potential electrode were lowered in the hole with a fixed separation of 2 ft; the second current and potential electrodes were located on the surface at a considerable distance from the drill collar. Identify the mineral zones and if possible distinguish between chalcopyrite and pyrite sections.

3. Some results from an experimental logging study in base-metal areas are shown in fig. 11.26. The mineralization here occurs in two steeply dipping zones, one containing pyrite and chalcopyrite, the other mainly pyrite. The diamond drill hole from which the logs were obtained was inclined approximately 60°.

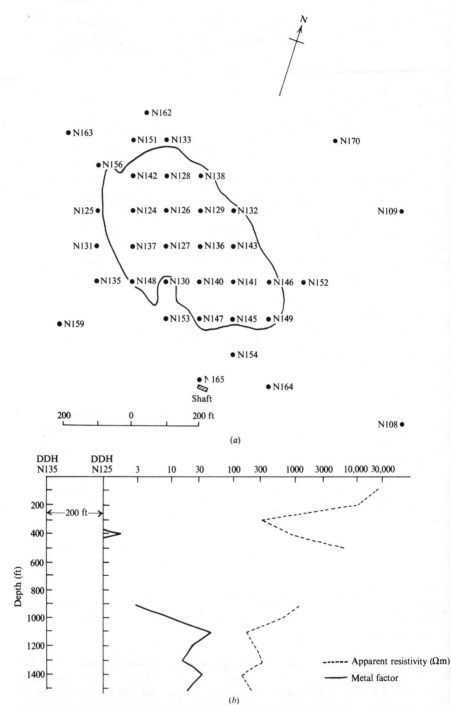

Fig. 11.24 Combined IP-electrical survey to locate ore body at depth, Lac Dufault, Quebec. (From Salt, 1966.) (a) Plan of orebody and diamond-drill holes; (b) IP logs in DDH N125 and N135; (c) log of potential in DDH N135.

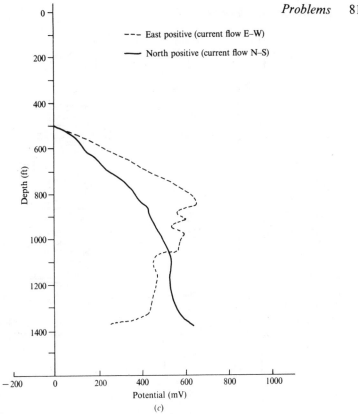

(c)

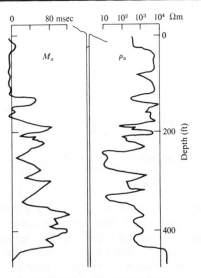

Fig. 11.25 IP log, Northwest Quebec.

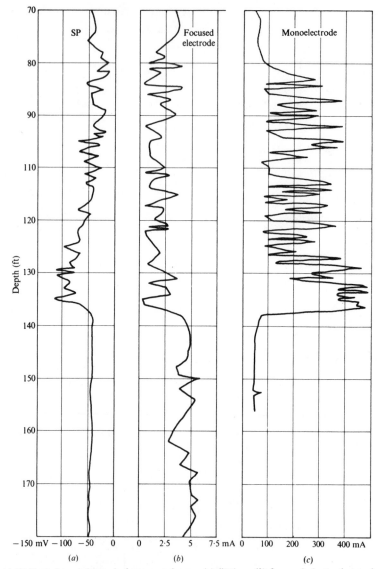

−150 mV −100   −50   0   0   2·5   5   7·5 mA 0   100   200   300   400 mA

(*a*)   (*b*)   (*c*)

Fig. 11.26 Experimental logs in base-metal area. (*a*) SP log; (*b*) focused electrode sonde, using uphole SP electrode fixed for current-return; (*c*) monoelectrode survey using current-return electrode 80–110 ft deep in adjacent hole.

The SP log is conventional, using one fixed electrode in the hole just below the water level (and below the casing). The focused-electrode sonde, similar to the Laterolog illustrated in fig. 11.3*a*, was made from 1-in. diameter lead-antimony pipe with PVC spacers. The guard electrodes were 2 ft long, the measuring electrode 3 in. and the spacers $2\frac{1}{2}$ in. long. Current return was through the uphole fixed electrode used for SP. The current source was a small 60-Hz motor generator.

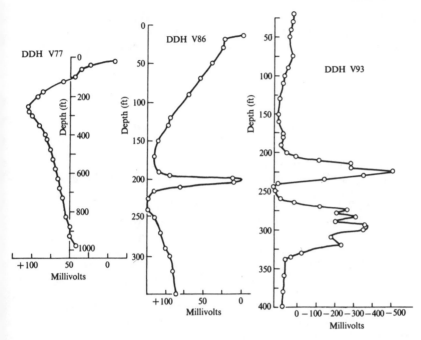

Fig. 11.27 SP logs in base-metal zones.

The third curve, labelled 'monoelectrode', was obtained by measuring current flow between the focused electrode system and an electrode located in an adjacent drill hole, in an attempt to establish continuity of mineralization. The return-current electrode was an aluminum rod, long enough to straddle the main mineral intersections in the second hole. In this example the rod extended from 80–110 ft in the second hole, which had previously been logged by the first two systems and found to be essentially barren. The two holes were about 100 ft apart, roughly east-west.

With this information, identify the mineralized zone or zones. Is there any indication of the lateral extent? Calculate the effective resistivity at a few points on the focused resistivity curve from the formula of Dakhnov (1962),

$$\rho_e = 2\pi \left(\frac{L_m}{L_r}\right)\left(\frac{V}{I}\right) \frac{(L_r^2 - 1)^{\frac{1}{2}}}{\log_{10}(L_r^2 - 1)},$$

where L_m = length of focused electrode, L_r = ratio of length to diameter of the electrode assembly, V = potential of focused electrode (110 V), and I = current in focused electrode.

4. SP logs obtained at a base-metal property in Northwestern Quebec are shown in fig. 11.27. The host rocks are andesites, diorites and rhyolites, the overburden sand and clay. The mineralization consists of pyrite, pyrrhotite, sphalerite, chalcopyrite and in places bands of magnetite. An earlier surface SP survey showed no anomaly.

Based on these three curves, what would be your explanation for the barren SP at the surface? Given the additional information that massive sulphides were found in a fourth hole nearby between 25 and 31 ft and from 65–89 ft, would you change this explanation? Can you account for the pronounced positive excursion in holes V77 and V86 between surface and 200 ft? Make as complete an interpretation as you can.

References

F – field method; G – general; I – instruments; T – theoretical

General references

Borehole geophysics symposium, 1970. *Geophysics*, **35**, 81–152. (G)

Dakhnov, V. N. (1962). Geophysical well logging. *Quarterly of Colorado School of Mines*, **57**, no. 2. (G)

Dresser Atlas (1971). *Log review 1*. Houston, Dresser Atlas.

Evans, H. B. (1970). Status and trends in logging. *Geophysics*, **35**, 93–112. (G)

Haun, J. D. and Leroy, L. W. (ed.) (1958). *Subsurface geology in petroleum exploration*. Golden, Colorado School of Mines.

Lynch, E. J. (1962). *Formation evaluation*. New York, Harper and Row. (G)

Pickett, G. R. (1970). Applications for borehole geophysics in geophysical exploration. *Geophysics*, **35**, 81–92. (G)

Pirson, S. J. (1958). *Oil reservoir engineering*. New York, McGraw-Hill. (G)

Pirson, S. J. (1963). *Handbook of well log analysis*. New York, Prentice-Hall. (T)

Pirson, S. J. (1970). *Geologic well log analysis*. Houston, Gulf. (T)

Salt, D. J. (1966). Tests of drill hole methods of geophysical prospecting on the property of Lake Dufault Mines Ltd, Dufresnoy Twp., Que. In *Mining geophysics*, **1**, 206–16, Tulsa, Society of Exploration Geophysicists. (F)

Schlumberger Well Surveying Corp. *Document no. 2*.

Schlumberger Limited (1972). *Log interpretation*, vol. 1– *Principles*. New York, Schlumberger. (T)

Sheriff, R. E. (1970). Glossary of terms used in well logging. *Geophysics*, **35**, 1116–39. (G)

Tixier, M. P. and Alger, R. P. (1970). Log evaluation of nonmetallic mineral deposits. *Geophysics*, **35**, 124–42. (I)

Tixier, M. P. and Forsythe, R. L. (1951). Application of electrical logging in Canada. *C.I.M.M. Bull.*, **44**, no. 473, 580–91. (G)

Wyllie, M. R. J. (1963). *Fundamentals of well log interpretation*. New York, Academic. (T)

Resistivity, SP

Archie, G. E. (1942). The electrical resistivity log as an aid in determining some reservoir characteristics: *Trans. A.I.M.E.* **146**, 54–62. (T)

Becker, A. and Telford, W. M. (1965). Spontaneous polarization studies. *Geophys. Prosp.* **13**, 173–88. (F)

De Chambrier, P. (1953). The microlog continuous dipmeter. *Geophysics*, **18**, 929–51. (I)

Doll, H. G. (1949). Introduction to induction logging and applications to logging of wells drilled with oil-based mud. *Trans. A.I.M.E.* **186**, 148–62. (F)

Doll, H. G. (1951). The laterolog, a new resistivity logging method with electrodes using an automatic focusing system. *Trans. A.I.M.E.* **192**, 305–16. (F)

Doll, H. G. (1953). The microlaterolog. *Trans. A.I.M.E.* **198**, 17–32. (F)

Lishman, J. R. (1961). Salt bed identification from unfocused resistivity logs. *Geophysics*, **26**, 320–41. (I)

Roy, A. and Dhar, R. L. (1971). Radius of investigation in d.c. well resistivity logging. *Geophysics*, **36**, 754–60. (T)

Salt, D. J. and Clark, A. R. (1951). The investigation of earth resistivities in the vicinity of a diamond drill hole. *Geophysics*, **16**, 659–65. (F)

Wyllie, M. R. J. (1949). A quantitative analysis of the electrochemical component of the SP curve. *Trans. A.I.M.E.* **186**, 17–26. (I)

EM, IP

Bacon, L. O. (1965). Induced-polarization logging in the search for native copper. *Geophysics*, **30**, 246–56.

Dakhnov, V. N. *et al.* (1952). *Investigation of wells by the IP method, SB Promislovaya.* Geofizika.

Drill hole EM exploration for sulphide ores, 1957. 7th Annual Symposium on Exploration Drilling, Univ. Minnesota.

Edwards, J. M. and Stroud, S. G. (1964). Field results of the EM casing inspection log. *Jour. Petr. Tech.* **16**, 377–82.

Wagg, D. M. and Seigel, H. O. (1963). IP in drill holes. *Can. Min. Jour.* **84**, 54–9.

Radioactivity

Kokesh, F. P. (1951). Gamma ray logging. *Oil & Gas Jour.* **50**, 284.

Mero, J. L. (1960). Uses of the gamma-ray spectrometer in mineral exploration. *Geophysics*, **25**, 1054–76.

Pickell, J. J. and Heacock, J. G. (1960). Density logging. *Geophysics*, **25**, 891–904.

Pontecorvo, B. (1941). Neutron well logging. *Oil & Gas Jour.* **40**, 32.

Tittle, C. W., Faul, H. and Goodman, C. (1951). Neutron logging of drill holes; the neutron-neutron method. *Geophysics*, **16**, 626–58.

Elastic wave propagation

Berry, J. E. (1959). Acoustic velocity in porous media. *Trans. A.I.M.E.* **216**, 262–70.

Guyod, H. and Shane, L. E. (1969). *Geophysical well logging*, vol. 1, *Introduction to acoustical logging.* Houston, H. Guyod.

Summers, G. C. and Broding, R. A. (1952). Continuous velocity logging. *Geophysics*, **17**, 598–614.

Tixier, M. P., Alger, R. P. and Doh, C. A. (1959). Sonic logging. *Trans. A.I.M.E.* **216**, 106–14.

Wyllie, M. R. J., Gregory, A. R. and Gardner, G. H. F. (1958). An experimental investigation of factors affecting elastic wave velocities in porous media. *Geophysics*, **23**, 459–93.

Magnetics, gravity

Broding, R. A., Zimmerman, C. W., Somers, E. V., Wilhelm, E. S. and Stripling, A. A. (1952). Magnetic well logging. *Geophysics*, **17**, 1–26.

Brown, R. J. S. and Gamson, B. W. (1960). Nuclear magnetism logging. *Trans. A.I.M.E.* **219**, 199–207.

Howell, L. G., Heintz, K. O. and Barry, A. (1966). The development and use of a high precision downhole gravity meter. *Geophysics*, **31**, 764–72.

12. Integrated geophysical problems

12.1 Introduction

The application of several disciplines – geology, geochemistry, geophysics – con stitutes an integrated exploration program. In a more restricted sense we ma consider the integrated geophysics program as the use of several geophysica techniques in the same area. The fact that this type of operation is so common place is because the exploration geophysicist, by a suitable selection of, say, fou methods, may obtain much more than four times the information he would ge from any one of them alone.

Before elaborating on this topic it is necessary to point out again the paramoun importance of geology in exploration work. Every geologic feature, from tectoni blocks of sub-continental size to the smallest rock fracture, may provide a clue i the search for economic minerals. Thus geologic information exerts a mos significant influence on the whole exploration program, the choice of area geophysical techniques and above all, the interpretation of results. Without thi control the geophysicist figuratively is working in the dark.

The subject of integrated geophysical surveys has received considerabl attention in the technical literature over the past 15 years. In petroleum explora tion the combination of gravity and magnetic reconnaissance, plus seismic fo both reconnaissance and detail (and, of course, various well logging technique during the course of drilling) is well established.

The best combination for an integrated mineral exploration program is not s definite because of the great variety of targets and detection methods available Base-metal search is a case in point. If the area is large enough and the mone available, the program normally would start with a combined airborne magneti and EM survey. On a more modest scale the work might proceed from a study o acquired airborne data or from a reconnaissance geochemical survey. In eithe case the ground follow-up would include magnetics, one or more EM technique and possibly gravity. IP may replace EM or follow it, particularly where th mineralization appears to be diffuse or low grade. There are, of course, additiona possibilities, e.g. SP, tellurics and magnetotellurics. In any event the base-meta program, compared to the standard sequence in oil search, appears eithe pleasantly flexible or somewhat fuzzy, depending on the attitude and experienc of the exploration manager.

There is another significant factor affecting exploration work in general an multiple geophysical surveys in particular which is not sufficiently stressed. This i the time element. In an ideal situation the survey work would be carried ou in a well-ordered sequence, proceeding from reconnaissance to detail and ex racting all possible information from each survey before starting the next one. Practically this orderly, controlled procedure is impossible. Frequently there is pressure from the competition and from the equipment suppliers and contractors several surveys must be done simultaneously, or in the wrong sequence, or one survey cannot be made until it is too late to be of much use). Time is money as well; the financial source may elect to start drilling, or to abandon the operation entirely (for reasons which could be quite valid) before the survey is complete.

Finally it is wise to keep in mind that the exploration program should lead to elimination as well as acquisition of ground. An overabundance of anomalies is, in the end, almost as unattractive as none at all.

The reader who has studied the case histories and problems in the various chapters with anything more than casual interest doubtless will have deduced that many of them are not isolated illustrations of a single geophysical method applied to a particular area. This is indeed the case and in the present chapter we will assemble some of these to make integrated surveys for further consideration. Analyses of the examples and problems given here should be made on the basis of several factors, including the following.

(*a*) How much additional information is provided by the combination of techniques?

(*b*) Is this information positive or negative, definite or indefinite in quality?

(*c*) Is the procedural sequence reasonable (magnetics with EM, followed by gravity, or whatever) and if not why was it done in this way and how would you modify it?

(*d*) Are the number and selection of methods necessary and sufficient to make a decision either to walk away from the prospect or develop it further?

(*e*) Are any of the methods used in a particular example superfluous? Would the money be better spent on drilling?

12.2 Examples and problems

1. Compare the gravity interpretation of problem 14, §2.8, with the analysis of the magnetic feature near St Bruno, Quebec, described in field example 3, §3.7.2. Presumably the same structure produces both anomalies. Are the results satisfactory and if not, is it possible to make reasonable adjustments of certain parameters to obtain better agreement?

2. The large gravity anomaly of field example 1, §2.7, occurs within the area shown in fig. 3.52, problem 13, §3.8. (It is difficult to locate it precisely because the aeromagnetic section is from a much larger map; certain topographic features, such as shorelines in fig. 2.40a, which were not reproduced in fig. 3.52 do not coincide. Furthermore it is surprising that there is no mention in the gravity work report of the 700-ft scarp described in problem 13; such rugged topography would certainly require a terrain correction.) Unfortunately ground magnetics were not carried out during the gravity survey. In spite of these discrepancies, the positive gravity feature appears to correlate with a zone of low magnetics striking SW–NE.

Take off a SE–NW profile through the magnetic low and attempt to match it with the aid of the dolomite section of fig. 2.40c, using eq. (3.55). Alternatively, since the inclination is about 75°, one might approximate F by Z and employ eq. (3.35). The height of the aircraft was 1000 ft and one would expect the susceptibility contrast between the dolomite and adjacent sedimentary rocks to be small and negative.

3. The aeromagnetic contours in fig. 3.53, problem 15, §3.8, are from the same area as the gravity profile of fig. 2.42, field example 3, §2.7. The north end of the gravity profile terminates approximately at the cross on the north direction arrow shown in fig. 3.53.

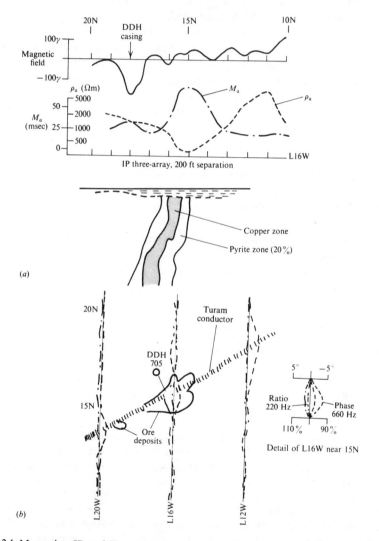

Fig. 12.1 Magnetics, IP and Turam results, Louvicourt area, Quebec. (a) Magnetic and IP profiles; (b) Turam results; loop located at 0 + 00N.

Take off a N–S total-field profile from this point for about 10 miles south and compare it with the gravity profile.

If the geologic section of fig. 2.42 is correct (there is considerable evidence to show that it is), the magnetic profile, like the gravity, should be a reflection of the Precambrian step, since the sedimentary beds have very low and uniform susceptibilities. Hence it should be possible to obtain a fair match to the magnetic profile, using a form of eq. (3.55), §3.6.8g, given that the inclination and declination in the area are about 75° and 22°W respectively and that the aircraft altitude was 1000 ft.

If this interpretation is not satisfactory, can you establish some definite limits on the depth of the magnetic source? (See also §3.6.8f.) Would such a depth be satisfactory for the gravity profile? Are the gravity and magnetic results compatible at all?

4. The magnetic and IP profiles of fig. 12.1a and the Turam profiles of fig. 12.1b are from the Louvicourt copper deposit near Val d'Or, Quebec. Line 16W passes almost directly over the main copper mineralization, shown in vertical section below 15N. This zone has an oval cross-section about 300 ft by 100 ft and is enclosed in a steeply dipping bed of disseminated pyrite (~20%) about 160 ft thick. The pyrite extends for more than a mile along strike, which is parallel to the bedding of acid tuffs and agglomerates. Consider these profiles individually:

(i) magnetic – note the strong negative anomaly caused by the drill hole casing. Can you explain the lack of magnetic signature associated with the ore zone?

(ii) Turam – the original discovery hole was drilled on the strength of the weak Turam anomaly at 15N. Would you consider this a good bet for drilling?

(iii) IP – this profile was obtained after drilling, using the three-electrode array with 200-ft separation. Chargeability and resistivity are both strongly anomalous over the ore zone. Is the former response due to pyrite, chalcopyrite, or both? Would you make a similar interpretation for the resistivity low?

The gravity profile of field example 2, §2.7, along with fig. 2.41, and the EM data given in problem 17, §7.10, are also from line 16W. In addition, the IP and resistivity logs shown in fig. 11.25 (see problem 2, §11.10) were taken in a hole which intersected the ore zone. The geological log gave the following information.

0–25 ft: overburden.
25–175 ft: barren tuff, agglomerate.
175–240 ft: pyrite mineralization, 20% average.
240–300 ft: copper mineralization, 2·6%, pyrite ~5%.
300–330 ft: pyrite ~5%.
330–345 ft: pyrite ~5%, copper 1·3%.
345–435 ft: pyrite 10–15%.
435–480 ft: barren.

Although we have a great variety of geophysical results for this relatively small copper orebody, seven of the eight surveys were performed *after* drilling, to determine which of the methods were best suited for such a target. Assess the geophysical data and consider the results on the basis of the factors listed in §12.1.

5. Figure 12.2 shows a set of total-field ground-magnetic profiles from the Uchi Lake sulphide deposit in northwestern Ontario. Other geophysical results from this zone have

already been displayed in fig. 7.55 (see field example 2 in §7.9), also figs. 7.70, 7.74 and 7.79 of problems 11, 16 and 22, §7.10.

The original reconnaissance work was done by air, including the INPUT and nuclear-precession airborne magnetic profiles shown in fig. 7.79. The line spacing was ¼ mile at an altitude of 380 ft. Although the INPUT anomaly is weak, it stands out clearly against a low background and persists through at least three channels.

Favourable geology, involving siliceous volcanic rocks surrounded by extensive granites, made it attractive to investigate this anomaly in detail with ground geophysics. The first ground follow-up included the vertical-loop broadside method, which located the conductor outlined in fig. 7.55a, and the magnetic profile of fig. 12.2, which indicated a possible weak anomaly associated with it.

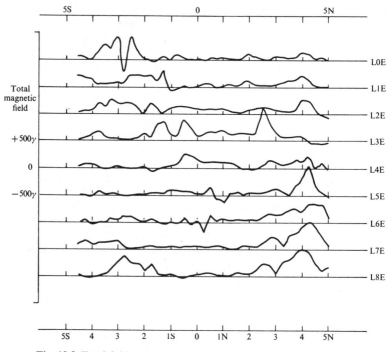

Fig. 12.2 Total-field ground-magnetic profiles, Uchi Lake, NW Ontario.

The sulphide structures are described in field example 2, §7.9. Initial drilling gave good grades of copper (3–8 %), zinc (9–12 %) and silver (2–4 oz/ton). The zone near 1N between 5E and 7E (see fig. 7.55a), is mainly sphalerite with very little iron mineralization and is probably a poor target for EM and magnetics.

The horizontal-loop, Turam and shootback EM profiles shown in figs. 7.70 and 7.74 were obtained in test surveys performed after the preliminary drilling. In this regard the procedure and sequence of events is similar to the previous example at Louvicourt, since the discovery was made on the basis of only part of the geophysics. As in example 4, the problem should be assessed by considering all the results, to establish a necessary and sufficient exploration procedure, if, in fact, one exists.

6. Numerous geophysical techniques have been tested over the Atlantic Nickel base metal property in southern New Brunswick. Examples already included in the text are listed below.

Survey technique	Section	Problem or field example	Figure
Gravity	§2.8	Prob. 13	2.48
Magnetic	§3.8	Prob. 7	3.46
Seismic	§4.9	Prob. 24	4.134
EM16	§7.9	Field ex. 3	7.56*a, b*
Resistivity	§8.8	Prob. 11	8.48
IP	§9.7	Prob. 4	9.19

In addition, the SP log of fig. 6.8, §6.1.4, was made in a diamond drill hole near line 156N. Self-potential, horizontal-loop EM, telluric and further IP results from the same area are shown in figs. 12.3*a* to 12.3*d*.

Although the gravity data plus earlier magnetic and vertical-loop EM surveys were used directly to establish the ore zone on this property, much of the subsequent geophysics was done for test purposes, or to find possible extensions of the mineralization. As a result there is an enormous amount of information here. In general one can see that the gravity and EM anomalies are very strong and relatively simple, the magnetics are complex, while the original IP and the telluric data are poor. However, the IP and resistivity sections in fig. 12.3*d*, obtained by using unusually small electrode spacing, show extremely strong anomalies. What are the reasons for the tremendous difference between the results in the two IP surveys? Make as complete an interpretation as you can.

7. Gravity data of fig. 2.44, problem 7, §2.8, and magnetic contours in fig. 3.43 of problem 2, §3.8, are from the Mobrun sulphide deposit, which is also illustrated in example 1, §7.9, fig. 7.54. The dip angle and gravity profiles in figs. 7.54 and 2.44 are on the line marked N–S in fig. 2.44; the appropriate magnetic profile may be obtained on a line joining the two crosses in fig. 3.43.

A measured telluric profile is shown in fig. 12.4, accompanied by a rough vertical geologic section. This was a N–S traverse about 200 ft west of the dotted line in fig. 3.43. Thus the sections of figs. 7.54 and 12.4 should be much the same.

The mild magnetic low indicates that the sulphides are non-magnetic, but otherwise this survey is of little significance. High background noise from nearby power lines affected the telluric measurements somewhat. The telluric profile is interesting because it reflects a much lower resistivity north of the ore zone than south of it. This large effect partially masks the anomaly directly over the sulphides. It may be due to a considerable resistivity contrast between rhyolite and the volcanics to the north. However, the steep slope near 400N looks very much like an overburden anomaly, as indicated in the theoretical profile obtained by numerical modelling. Would you expect to see this anomaly in the vertical-loop profile?

8. Figures 12.5*a* and 12.5*b* show two-frequency Turam and horizontal-loop EM profiles over zones of pyrite mineralization in Carpentier Township, near Barraute in

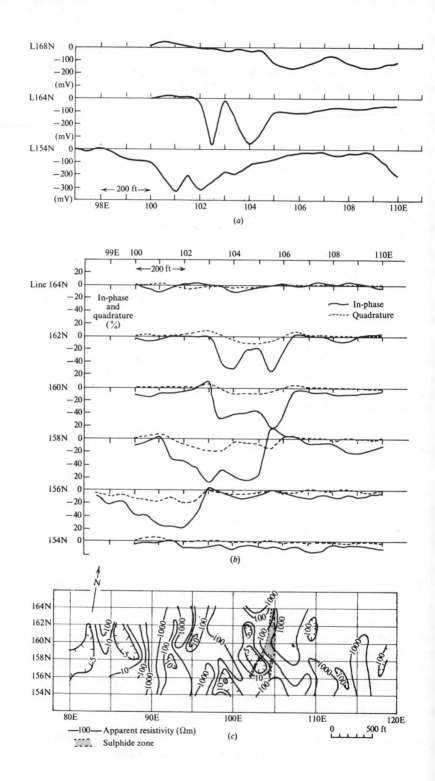

(a)

(b)

(c)

——100—— Apparent resistivity (Ωm)

▨▨▨ Sulphide zone

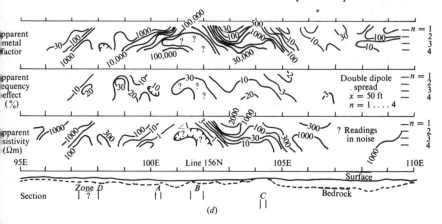

Fig. 12.3 Results of different electrical surveys, Atlantic Nickel Property, Southern New Brunswick. (*a*) Self-potential profiles; (*b*) horizontal-loop EM profiles. Frequency = 876 Hz, coil separation = 200 ft, (*c*) telluric contour map. Frequency 8 Hz, electrode spacing 100 ft; (*d*) IP and resistivity pseudo-depth sections. Double-dipole spread, $x = 50$ ft, $n = 1, 2, 3, 4$.

northwestern Quebec. Magnetic and IP data from this area are found in fig. 3.39 of example 1, §3.7.1, in fig. 8.44 of problem 7, §8.8, and in fig. 9.18 of problem 3, §9.7.

The magnetic results have been discussed in some detail in §3.7.1, where it was concluded that the large magnetic anomalies could only be caused by magnetite or pyrrhotite zones near surface. IP and EM responses, on the other hand, are uniformly negative, which should eliminate the possibility of pyrrhotite.

The geologic section shown in fig. 12.5*c* is based mainly on information from diamond drill hole $T-1$ (see fig. 3.39) and to a lesser degree on the magnetic anomalies. Drilling results establish the depth of overburden to be 82 ft at the collar and the presence of two

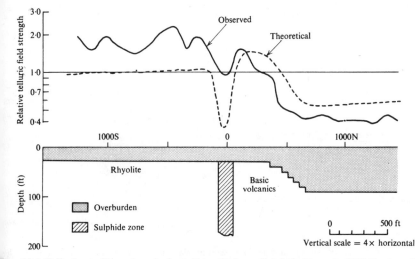

Fig. 12.4 Telluric profile and geologic section, Mobrun Sulphide Deposit, NW Quebec. For theoretical model, $\rho_{sul} = 1$ Ωm, $\rho_{over} = 70$ Ωm, $\rho_{host} = 7000$ Ωm.

zones of 35 % pyrite at about 300 and 450 ft depth. There is some doubt of the bedrock
surface outline shown in fig. 12.5c in view of the seismic results on line 75SE, mentioned
in example 1, §3.7. However, even if this surface is correct, it is difficult to understand
the lack of horizontal-loop EM and Turam response, especially the latter, since the
shallow resistivity values seen in fig. 8.44 would not normally be low enough to mask the

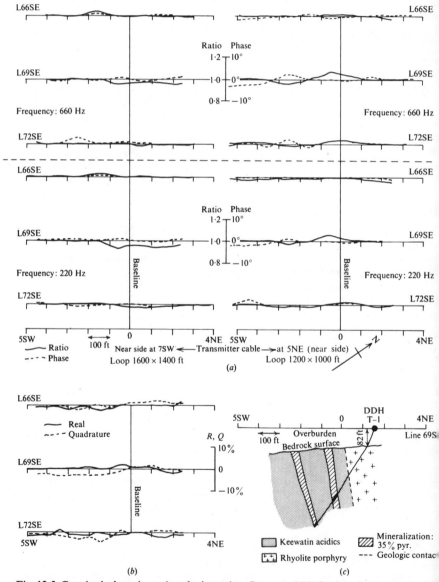

Fig. 12.5 Geophysical results and geologic section, Barraute, NW Quebec. (a) Turam profiles
frequencies 660 Hz, (upper three profiles) and 220 Hz (lower three profiles); (b) horizontal
loop EM profiles. Frequency 1000 Hz, T–R spacing 300 ft; (c) assumed geologic section.

presence of the 50-ft wide pyrite zones at a depth of 80–100 ft. One is forced to conclude either that the pyrite zones do not continue to bedrock surface or that the pyrite is not a good conductor. Try to make an interpretation of these results which is consistent with the magnetic and electrical data.

9. The Murray sulphide deposit in Restigouche County, northern New Brunswick, was used as an illustration of the Turam method in example 7, §7.9, fig. 7.60. Additional geophysical results are displayed in fig. 12.6 for magnetics, self-potential and horizontal-loop EM.

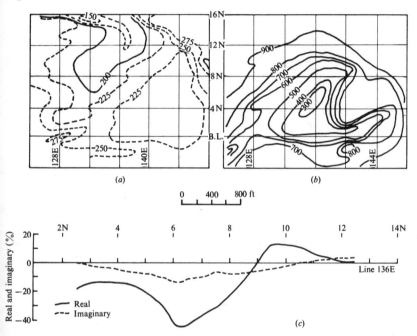

Fig. 12.6 Magnetic and electrical data, Murray sulphide deposit, New Brunswick. (*a*) Magnetic contours. Contour interval 25γ; (*b*) self-potential contours. Contour interval 100 mV; (*c*) horizontal-loop EM profile. Frequency 440 Hz, T–R spacing 300 ft.

Although there is no magnetic reflection of the sulphide mineralization (for a description, see §7.9), the SP and HLEM anomalies are very strong, possibly as a result of the gossan cover. In particular the horizontal-loop profile indicates a wide conductor dipping north, with the negative maxima falling between the Turam peaks. Make an estimate of the geometry and conductivity of the zone from horizontal-loop characteristic curves. Take off a SP profile from line 136E and interpret the SP results.

10. The VLF dip-angle and telluric profiles of fig. 12.7 were obtained from the same road traverse which supplied the data listed below.

Self-potential, example 1(SP), §6.3.1, fig. 6.21
Magnetotellurics, problem 9, §6.4, fig. 6.31
Magnetics, problem 9, §3.8, fig. 3.48.

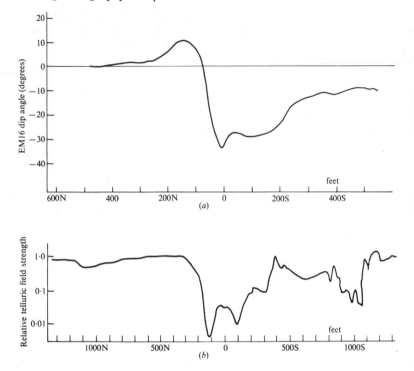

Fig. 12.7 EM and telluric results, Bartouille, NW Quebec. (*a*) VLF-EM16 dip-angle profile. (*b*) telluric profile, 25 ft electrode spacing.

This massive sulphide showing, in Bartouille Township, near Senneterre in NW Quebec, is described in the SP example.

There is good correlation here between the various geophysical methods, although it is by no means exact. Both SP and EM16 profiles indicate the strongest anomaly to be a thin conductor dipping north, located at 40N (SP) and 70N (EM16). (The latter is measured from the zero crossover, which may be slightly downdip.) Both show a second conductor in the swampy area at 180S, although it cannot be located with any precision in the EM16 profile. There is a third smaller SP anomaly at 110N.

Magnetic anomalies are similarly sharp. The smaller positive is at 25N, while the main peak is at 125N. A definite magnetic low at 150S correlates with the swamp zone. Minor peaks at 200S and 350S are not associated with any SP or EM16 anomalies.

The telluric and *MT* profiles of fig. 6.31 reflect a wide conductive zone extending at least 200 ft north and south of station 0, while the *MT* profile also has a slight dip near 200S. When fig. 6.31 is compared with fig. 12.7*b*, there is clearly a lack of discrimination in the former. This is caused by the larger electrode spacing, 100 ft, rather than 25 ft as in fig. 12.7*b*. With the 25 ft spacing the wide conductor is resolved into two peaks, the main one being at 110N and the second at 100S. There is also a minor low at 350S, coinciding with the magnetic anomaly. The telluric low at 1000S, which was equally strong with 100 ft electrode spacing, has no correlation with the magnetics.

11. Induced polarization, resistivity and magnetic data, shown in fig. 12.8, are taken from the same area discussed in: (i) field example 6, for horizontal loop EM, §7.9, fig. 7.59; (ii) the telluric profiles of problem 6, §6.4, fig. 6.28; (iii) drill hole logs in problem 3, §11.10, fig. 11.26. The zones of mineralization, which are mainly pyrite, are shown in fig. 7.59. There is, in addition, a section of chlorite schist containing traces of pyrite between $7 + 00E$ and $7 + 50E$. All zones dip about 70° west.

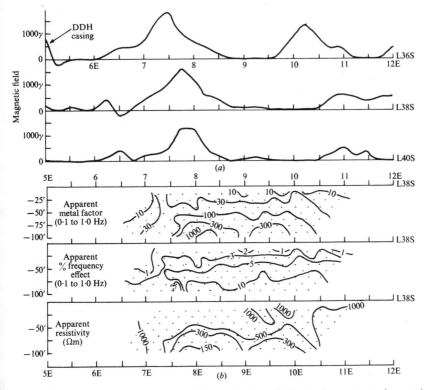

Fig. 12.8 Magnetic and electrical results, Woburn, Quebec. (*a*) Vertical component (magnetic field) profiles; (*b*) IP and apparent-resistivity pseudo-depth plots.

Considering line 38S, the three-frequency telluric profiles in fig. 6.28 have two peaks, at $8 + 30E$ and $9 + 30E$, which correlate very well with the two sulphide zones. The high frequency response is most pronounced, while the 8 Hz profile indicates that the better conductor is at $8 + 30E$. The horizontal-loop EM profile, on the other hand, reflects only the massive sulphides. Neither sulphide zone appears to have any magnetic contrast on line 38S, although there is a small tail on the west flank of the magnetic peak at $7 + 75E$, which falls between the chlorite bed and the massive sulphide.

The IP traverse was made with a double-dipole array, with $x = 25$ ft and $n = 1,...,7$. The small spacing was used purposely in an attempt to discriminate between the conductors. Clearly this was only partly successful, since the frequency-effect contours show a single wide zone from $7 + 75E$ to $10 + 50E$. Apparent-resistivity and metal-factor results are better in this regard, with the stronger response over the better conductor, but only for $n = 6$ and 7, the largest dipole separations.

On lines 36S and 40S, the HLEM peaks again occur over the massive sulphide zon
while the telluric profiles (which, it should be noted, were made with 50 ft electro
separation) locate both zones. In addition, the strongest response in all three lines is
145 Hz. However, this is not particularly significant, since the average resistivity
sufficiently high to make the skin depth at 145 Hz considerably greater than the dept
extent of the conductive zones.

This is a relatively straightforward example of non-magnetic sulphide conducto
lying at shallow depth. The magnetic data are, at best, merely an indicator of mild
anomalous susceptibility contrast on the flanks of the sulphides. All the electric
techniques responded to the anomalous zones in some fashion. The horizontal-loo
survey located only the better conductor, indicated its dip and provided an estimate o
depth and conductivity-thickness product; these profiles also show that the zone

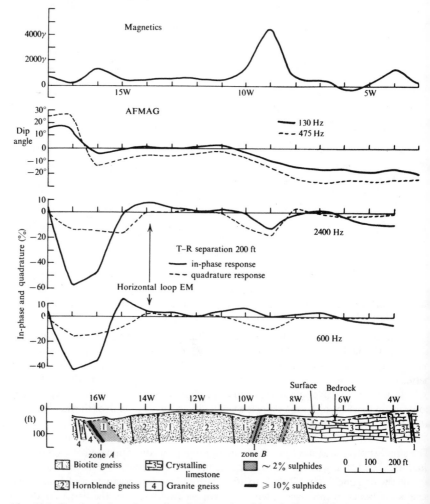

Fig. 12.9 Geophysical results and geologic section, Line *C*, Cavendish Township, Ontario

thin. As mentioned in the discussion of field example 6 in chapter 7, it might have been possible to detect the two zones by using 100 ft spacing between the transmitter and receiver. The telluric measurement resolved the two conductors with 50 ft electrode spreads, but failed to do so when the separation was 100 ft; there is little evidence of dip on lines 38S and 40S, while the profile on 36S suggests a dip to the east. Finally, the IP survey successfully established the presence of two conductors by employing extremely small dipole spreads. Had the conventional 100 ft separation been used, it is practically certain that this resolution would have been lost.

12. Vertical geologic sections obtained by drilling the Cavendish Township area in central Ontario are displayed in figs. 12.9 and 12.10. Lines C and D are 400 ft apart.

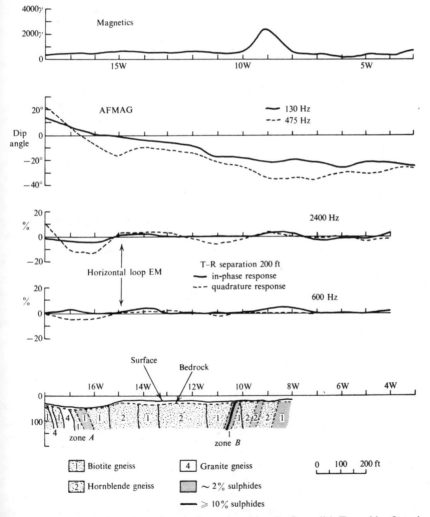

Fig. 12.10 Geophysical results and geologic section, Line *D*, Cavendish Township, Ontario.

This has been a favourite site for testing geophysical equipment since 1967, when a case history was prepared by McPhar Geophysics for the Canadian Centennial Conference on Mining and Groundwater Geophysics. Practically all the electrical techniques, plus magnetics, have been used here.

Magnetic, AFMAG and horizontal loop EM profiles are also shown in the same diagrams. Other examples from Cavendish already incorporated in this book are listed below.

Section	Problem	Method	Figure	Correlation of lines
6·4	4	SP	6·27	Line 12N is line C, 8N is D
7·10	4	VL Broadside	—	Line 12N is line C, 8N is D
7·10	18	AEM Quadrature	7.75	
9.7	7	IP Freq. domain	9.21	Data obtained on Line C

A shallow drilling program, recently carried out by the Geological Survey of Canada, produced the sections of figs. 12.9 and 12.10. The rocks are Grenville mafic gneisses, crystalline limestones and granite gneiss, which generally dip steeply to the east. Pyrrhotite is associated with calcite and pyrite with quartz in calcareous-silicate sulphide alteration zones and there is a general background content of 0·1% sulphides in the rocks of the area. Both zones A and B strike approximately N30° E for more than 1600 ft, the B zone apparently splitting into two or three zones, over a greater width, to the south. The sulphides, almost entirely pyrrhotite and pyrite, occur in widths up to 80 ft at concentrations of 1 or 2%. Within these zones there are narrower sections of 10% and higher grade; for example, in zone A, the 10 ft wide section on line C extends 100 ft north and south, while within zone B there appears to be a similar, but much narrower, section which is about 1000 ft long. The mineralization occurs as stringers which are mainly parallel to the foliation, although zone B appears to dip west.

Neither the depth nor the depth extent of these zones has been established by the drilling to date. However, they seem to be no more than 25 ft deep and probably outcrop at bedrock surface in many places. The depth extent is at least 125 ft.

With such a wealth of geophysical data available here, it should be possible to make a fairly complete interpretation of these zones. Considering line C, the magnetic profile shows a good peak at 16W and a much stronger one at 9W, correlating well with zone A, centred at 16W and reasonably well with the main section of B at about 9 + 25W. There is not much indication of the dips of these beds, although the peak at 16W has a slightly steeper slope on the west flank and the larger anomaly at 9W decreases more rapidly to the east, suggesting east and west dips respectively. The two peaks, of course, are a reflection of the pyrrhotite which apparently is in greater concentration and/or closer to the surface at zone B. Depth estimates from the width at half-maximum indicate about 100 ft in both cases, but these are obviously unreliable because both zones are probably wider than the depth to the top.

If we take a SP profile from line 12N in fig. 6.27, §6.4, we find good correlation between the negative peak of 373 mV and zone A. The broader minimum from 8W to 11W, however, indicates a second SP source at 11W, which is not apparent from the geologic section. The asymmetry of the SP anomalies suggests a general steep dip to the west. Clearly no valid depth estimates can be made from this profile.

On plotting the broadside vertical-loop profiles for 12N (line C) from the data in problem 4, §7.10, we find that zone A produces a good anomaly at all four frequencies. The crossovers lie between 16 + 50 and 16 + 60W and the asymmetry of the 1000 and 5000 Hz profiles indicates a westward dip. Employing the characteristic curve of fig. 7.49a, §7.8.7b (admittedly this is a very crude approach, because the marked asymmetry of the profiles does not imply a steeply dipping sheet), we find that $z/l \approx 0\cdot3$, or $z \approx 120$ ft. This value appears to be much too large. Further evidence that it is unreliable is provided when we attempt to estimate the response parameter from fig. 7.50b, §7.8.7d. The peak-to-peak response is much too large, at both frequencies, for a conductor 120 ft deep.

Using the same curves for zone B, we find that $z \approx$ 100, 120 and 340 ft, for frequencies of 5000, 2400 and 1000 Hz respectively. (There is no response at 600 Hz.) From fig. 7.50b, we then obtain values of $\sigma s \approx 5$ at 5000 and 2400 Hz; the depth z corresponding to 1000 Hz is again too large to fit the curves.

The AFMAG profiles in fig. 12.9 show a good crossover approximately at 16 + 50W, corresponding to zone A, while the asymmetry indicates an eastward dip. There is little more than a suggestion of zone B on these profiles. It should be noted that in both the vertical-loop and AFMAG methods (and in the horizontal-loop results as well), the high-frequency response is the larger, which tells us that the conductors are shallow.

Figure 12.9 also shows a very strong response for the horizontal-loop EM, centred at 16 + 50W over zone A and a relatively weak anomaly at 9W for zone B. This type of EM equipment is not particularly sensitive to dip; one might guess that the profiles indicate a general dip to the west. Employing the characteristic curves of fig. 7.51b, §7.8.7e, we can estimate the parameters tabulated below.

Conductor	Frequency Hz	z/l	z (ft)	$\mu\omega\sigma sl$	σs (mhos)	s (ft)	σ (mhos/m)
Zone A	600	0·1	40	45	140	70	7
	2400	0·05	20	60	55	100	2
Zone B	600	0·2	80	3	10	—	—
	2400	0·1	40	4	4	<10	1

The values for s are obtained from the profiles in fig. 12.9, being the excess of the horizontal distance between zero crossovers over the transmitter-receiver separation.

Finally, the IP data for line C (fig. 9.21, §9.7) presents a somewhat different picture. The top of Zone A is located at least 100 ft east of its actual position. Both frequency-effect and metal factor contours suggest it to be a conductor of rather limited depth extent dipping east. Zone B, on the other hand, appears to be located in about the right place; it also appears to dip west and to increase in conductivity at depth. In the pseudo-depth plots for $x = 200$ ft there is an indication that the zones come together at $n = 3$ (supposedly about 400 ft). Although the frequency-effect values continue to $n = 4$, mainly below zone B, the metal-factor anomaly appears to pinch out at this depth.

Judging from the evidence presented by the shallow drilling program, this site is not a particularly good target for IP. However, a limited amount of deep drilling (say, 750 ft holes) would be very interesting, both to determine the depth extent of the two zones and to check the IP indication that they are joined at depth.

The information obtained from the various surveys on line *C* is summarized in the table opposite.

We may conclude that all these methods locate the zones reasonably well, with the exception of the IP and possibly SP. The dip of zone *B* is established to be steep and to the west, while there is considerable doubt about the attitude of zone *A*. Only the horizontal-loop technique gives reasonable depth estimates; the SP and vertical loop, qualitatively at least, suggest that both zones are shallow and that *B* may lie somewhat deeper than *A*, although the magnetic profile would not necessarily bear this out. Depth extent cannot properly be determined by any of these methods, while the width is roughly indicated only by the horizontal-loop. Finally, conductivity or conductivity-width product, as estimated by horizontal-loop and partially by vertical-loop, tells us that zone *A* is a better conductor than zone *B* (probably because of the width of the 10%-sulphide section), but that neither zone is a very good conductor.

It is suggested that the interested reader carry out a similar interpretation of Line *D*.

In the following examples no specific subsurface geological and geophysical informa-tion has been supplied, although in many instances further geophysical survey results have been added. Practically speaking, these cases remain unsolved. The combined geophysical data are to be reinterpreted (assuming, of course, that the original problems were done first, a procedure which is strongly recommended). The value in this exercise lies in the experience gained through assessment and comparison of the individual methods and correlation between them.

13. Problem 3, §3.8, and problem 1, §6.4, give magnetic and SP readings over the same traverse line.

14. Gravity readings in problem 5, §2.8, were taken from the same line surveyed with horizontal-loop EM in problem 8, §7.10, fig. 7.68.

15. The magnetic profiles in problem 4, §3.8, fig. 3.44, the horizontal-loop EM profiles in problem 10, §7.10, fig. 7.69 and the vertical-loop EM data in problem 2, §7.10 are all from the Margie Lake area of northwest Quebec.

16. The problems listed below are all from the same survey area.

Survey	Section	Problem	Figure	Lines surveyed	Stations
Gravity	2·8	1, 9	—	0, 2S	53W–24W
Tellurics	6·4	7	6·29	0, 2S	55W–23W
VLEM	7·10	1	—	0, 2S	42W–34W, 43W–37W
AFMAG	7·10	7	7·67	6(N), 4, 2, 0	52W–0
Resistivity	8·8	8	8·45	0	42W–35W
IP	9·7	6	—	0, 2S	44W–36W, 40W–35W

Method	Location		Dip		Depth (ft)		Depth ext. (ft)		Width (ft)		σ_S (mhos)	
	Zone A	Zone B	A	B	A	B	A	B	A	B	A	B
Magnetic	16W	9W	E	W	100?	100?	—	—	—	—	—	—
SP	16W	9–11W	W	W	—	—	—	—	—	—	—	5
VL Broad.	16 + 55	9 – 9 + 65	W	W	120	>100	—	—	—	—	?	—
AFMAG	16 + 50	?	E	?	—	—	—	—	—	—	—	—
HL	16 + 55	9	W	W	20–40	40–80	400?	—	70–100	⩽10	55–140	4–10
IP	15 – 15 + 50	9–11	E	W	<100	<100	>125	—	150?	100?	—	—
	16 + 30 (conc).	9 + 30							10	0·3		
Drilling	15 + 60	9 + 15	E	W	~20	~20		>125	80	30	—	—
	– 16 + 40 (disseminated)	– 9 + 45										

As noted in problem 9, §2.8, the base line for this survey, which is at station 0, strikes 20° east, making an angle of 110° with the traverse lines, which are due east-west. Thus it is necessary to shift the AFMAG lines in fig. 7.67; that is to say, lines 2 4 and 6 (lines are 200 metres apart measured on the base line) should be moved 68, 137 and 205 metres respectively east of line 0.

17. The gravity and SP data, problem 8, §2.8, and problem 2, §6.4, fig. 6.25, are on the same grid. To these have been added the Slingram (horizontal-loop EM) and spot IP results shown in fig. 12.11.

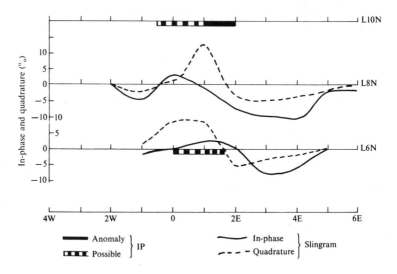

Fig. 12.11 Slingram and IP results, Northern New Brunswick. EM frequency 876 Hz, T–R separation 200 ft.

18. Figure 12.12 displays horizontal-loop EM profiles on two lines taken from a survey in northwestern Quebec. Further geophysical results, all from line 4 + 00W, are given in the following problems:

Type of survey	Section	Problem	Figure
Magnetic	3·8	5	—
Turam	7·10	14	—
Resistivity	8·8	9	8·46
IP	9·7	8	9·22

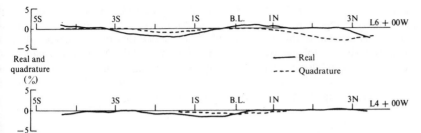

Fig. 12.12 Horizontal-loop EM profiles, Abitibi West, Quebec. Frequency 1000 Hz, T–R separation 300 ft.

19. The profiles in fig. 12.13 are a compilation for a single traverse line from a survey carried out in eastern Nova Scotia. Other problems dealing with this area are listed below:

Type of Survey	Section	Problem	Figure	Type of display
Topographic	2·8	4	2·43	Terrain contours
Gravity	2·8	12	2·47	Bouguer gravity contours
Magnetic	3·8	11	3.50a	Terrain and magnetic contours
EM16	7·10	6	—	Profiles

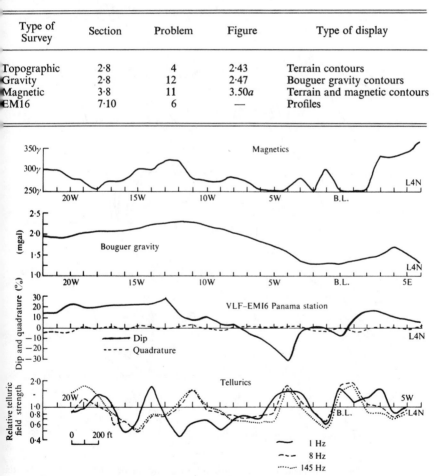

Fig. 12.13 Magnetic, gravity, VLF and telluric profiles, Eastern Nova Scotia.

The contour map of fig. 3.50a may be located with respect to the other survey data by noting that the upper left hand corner of the margin is station 22W, line 12N. As an aid to the assessment of this area, it should be explained that the original geophysical reconnaissance was a follow-up of a geochemical anomaly with EM16, SP and magnetics. The first two methods located a mild anomaly just west of the baseline between lines 0 and 8N. Following the layout of a small grid, horizontal-loop EM and limited gravity surveys produced promising results. The survey area then began to grow.

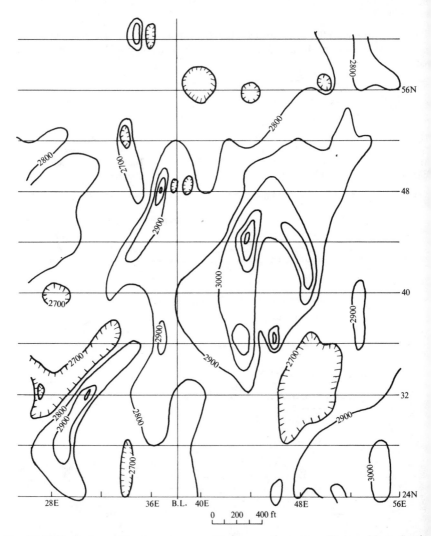

Fig. 12.14 Vertical-component contours, ground magnetic survey, Eastern Nova Scotia. Contour interval 100 gammas.

20. The data contained in the following problems originated in the same area.

Type of survey	Section	Problem	Figure	Type of display
Magnetic	3·8	11	3.50b	Terrain and magnetic contours
Telluric	6.4	5	—	Single profile
Resistivity	8·8	6	—	3 profiles
IP	9·7	9	9·23	3 profiles (lines 10N, 0, 10S)

21. Figure 12.14 shows magnetic contours from a ground survey in eastern Nova Scotia, as additional data to be integrated with the problems listed below:

Type of survey	Section	Problem	Figure	Display
SP	6·4	3	6·26	Profiles
Telluric	6·4	8	6·30	Contours
EM16	7·10	5	7·66	Profiles
Resistivity	8·8	10	8·47	Contours
IP	9·7	5	9·20	Contours

The IP contours of fig. 9.20 may be tied in with the remaining data by locating station 28E, line 60N, at the upper left-hand corner of the margin.

Author index

Subject index